Lecture Notes in Computer Science 1953
Edited by G. Goos, J. Hartmanis and J. van Leeuwen

Springer
Berlin
Heidelberg
New York
Barcelona
Hong Kong
London
Milan
Paris
Singapore
Tokyo

Gunilla Borgefors   Ingela Nyström
Gabriella Sanniti di Baja (Eds.)

Discrete Geometry for Computer Imagery

9th International Conference, DGCI 2000
Uppsala, Sweden, December 13-15, 2000
Proceedings

Series Editors

Gerhard Goos, Karlsruhe University, Germany
Juris Hartmanis, Cornell University, NY, USA
Jan van Leeuwen, Utrecht University, The Netherlands

Volume Editors

Gunilla Borgefors
Swedish University of Agricultural Sciences, Centre for Image Analysis
Lägerhyddvägen 17, 752 37 Uppsala, Sweden
E-mail: gunilla@cb.uu.se

Ingela Nyström
Uppsala University, Centre for Image Analysis
Lägerhyddvägen 17, 752 37 Uppsala, Sweden
E-mail: ingela@cb.uu.se

Gabriella Sanniti di Baja
National Research Council of Italy, Institute of Cybernetics
Via Toiano 6, 80072 Arco Felice (Naples), Italy
E-mail: gsdb@imagm.cib.na.cnr.it

Cataloging-in-Publication Data applied for

Die Deutsche Bibliothek - CIP-Einheitsaufnahme

Discrete geometry for computer imagery : 9th international conference ; proceedings / DGCI 2000, Uppsala, Sweden, December 13 - 15, 2000. Gunilla Borgefors ... (ed.). - Berlin ; Heidelberg ; New York ; Barcelona ; Hong Kong ; London ; Milan ; Paris ; Singapore ; Tokyo : Springer, 2000
(Lecture notes in computer science ; Vol. 1953)
ISBN 3-540-41396-0

CR Subject Classification (1998):I.3.5, I.4, G.2, I.6.8, F.2.1

ISSN 0302-9743
ISBN 3-540-41396-0 Springer-Verlag Berlin Heidelberg New York

This work is subject to copyright. All rights are reserved, whether the whole or part of the material is concerned, specifically the rights of translation, reprinting, re-use of illustrations, recitation, broadcasting, reproduction on microfilms or in any other way, and storage in data banks. Duplication of this publication or parts thereof is permitted only under the provisions of the German Copyright Law of September 9, 1965, in its current version, and permission for use must always be obtained from Springer-Verlag. Violations are liable for prosecution under the German Copyright Law.

Springer-Verlag Berlin Heidelberg New York
a member of BertelsmannSpringer Science+Business Media GmbH
© Springer-Verlag Berlin Heidelberg 2000
Printed in Germany

Typesetting: Camera-ready by author, data conversion by DA-TeX Gerd Blumenstein
Printed on acid-free paper SPIN: 10780979 06/3142 5 4 3 2 1 0

Preface

This proceedings volume includes papers for presentation at DGCI 2000 in Uppsala, Sweden, 13–15 December 2000. DGCI 2000 is the 9th in a series of international conferences on Discrete Geometry for Computer Imagery. DGCI 2000 is organized by the Centre for Image Analysis (Swedish University of Agricultural Sciences and Uppsala University). It is sponsored by the International Association for Pattern Recognition (IAPR) and by the Swedish Society for Automated Image Analysis (SSAB). This is the first DGCI conference to take place outside France.

The number of researchers active in the field of discrete geometry and computer imagery is increasing. In comparison with the previous editions, DGCI 2000 has attracted a significantly larger number of research contributions from academic and research institutions in different countries. In fact, 62 papers were submitted. The contributions focus on topology, discrete images, surfaces and volumes, shape representation, and shape understanding. After careful reviewing by an international board of reviewers, 40 papers from 16 countries were accepted, leading to a high-quality conference. Of the 40 accepted papers, 28 were selected for oral presentation and 12 for poster presentation. All accepted contributions have been scheduled in plenary sessions. The programme also has been enriched by three invited lectures, presented by three internationally known speakers: Pieter Jonker (Delft University of Technology, The Netherlands), Pierre Soille (ISPRA, Italy), and Jayaram K. Udupa (University of Pennsylvania, USA).

We hope that we have made DGCI 2000 an unforgettable event, especially for researchers from outside Sweden. In fact, the uniquely Swedish Lucia celebration of returning light in the middle of winter has been scheduled in the early morning of 13 December, just before starting the DGCI technical programme. Also, the conference dates were chosen to overlap with the week when the Nobel Prizes are awarded in Stockholm and the Nobel Laureates are giving lectures in Uppsala. In our opinion, these events should be very stimulating for all researchers.

Hereby, we would like to thank the contributors who responded to the call for papers, the invited speakers, all reviewers and members of the Steering, Programme, and Local Committees, as well as DGCI participants. We are also grateful to the sponsoring organizations that provided financial help, indispensable to a successful conference.

September 2000

Gunilla Borgefors
Ingela Nyström
Gabriella Sanniti di Baja

Organization

DGCI 2000 is organized by the Centre for Image Analysis, Swedish University of Agricultural Sciences, and Uppsala University in Uppsala. The conference venue is the Mathematics and Information Technology campus (MIC) of Uppsala University.

Conference Chairs

General Chair:	Gunilla Borgefors Centre for Image Analysis, Swedish University of Agricultural Sciences, Uppsala, Sweden
Programme Chair:	Gabriella Sanniti di Baja Istituto di Cibernetica, National Research Council of Italy, Arco Felice (Napoli), Italy
Local Arrangements Chair:	Ingela Nyström Centre for Image Analysis, Uppsala University, Uppsala, Sweden

Steering Committee

Ehoud Ahronovitz	LIRMM, Montpellier	France
Gilles Bertrand	ESIEE, Marne-la-Vallée (Paris)	France
Jean-Marc Chassery	TIMC-IMAG, Grenoble	France
Annick Montanvert	ENSIEG-LIS, Grenoble	France
Maurice Nivat	LIAFA-Paris VI	France
Denis Richard	LLAIC1, Clermont-Ferrand	France

Programme Committee

Yves Bertrand	IRCOM-SIC, Poitiers	France
Alberto Del Lungo	Università di Siena	Italy
Ulrich Eckhardt	Universität Hamburg	Germany
Christophe Fiorio	LIRMM, Montpellier	France
Richard W. Hall	University of Pittsburgh	USA
T. Yung Kong	City University of New York	USA
Vladimir Kovalevsky	Universität Rostock	Germany
Walter Kropatsch	Technische Universität Wien (Vienna)	Austria
Attila Kuba	József Attila University, Szeged	Hungary
Rémy Malgouyres	LLAIC1, Clermont-Ferrand	France
Serge Miguet	ERIC, Université Lyon 2	France
Ingemar Ragnemalm	Linköping University	Sweden
Gabor Szekely	ETH, Zürich	Switzerland
Mohamed Tajine	ULP, Strasbourg	France

Local Organizing Committee

Stina Svensson
Lena Wadelius

Referees

Ehoud Ahronovitz
Eric Andres
Dominique Attali
Jean-Pierre Aubert
Emese Balogh
Gilles Bertrand
Yves Bertrand
Gunilla Borgefors
Sara Brunetti
Jean-Marc Chassery
David Coeurjolly
Michel Couprie
Guillaume Damiand
Alain Daurat
Alberto Del Lungo
Marianna Dudásné
Jean-François Dufourd
Ulrich Eckhardt
Christophe Fiorio
Sébastien Fourey
Yan Gerard
Michel Habib
Richard W. Hall
Christer Kiselman
T. Yung Kong
Vladimir Kovalevsky
Walter Kropatsch
Attila Kuba
Christophe Lohou
Rémy Malgouyres
Serge Miguet
Annick Montanvert
Antal Nagy
Phillippe Nehlig
Maurice Nivat
Ingela Nyström
Kálmán Palágyi
Ingemar Ragnemalm
Denis Richard
Christian Ronsé
Gabriella Sanniti di Baja
Pierre Soille
Gabor Szekely
Mohamed Tajine
Larue Tougne
Jayaram K. Udupa

Sponsors

City of Uppsala
Swedish Foundation for Strategic Research (SSF)
through the Visual Information Technology (VISIT) programme
Swedish Research Council for Engineering Sciences, TFR
Faculty of Forestry, Swedish University of Agricultural Sciences
Faculty of Information Technology, Uppsala University
Faculty of Science and Technology, Uppsala University

Swedish Society for Automated Image Analysis, SSAB
International Association for Pattern Recognition, IAPR

Table of Contents

Topology

Discrete Images

Surfaces and Volumes

Shape Representation

Shape Understanding

[illegible]

Topology

Homotopy in Digital Spaces*

Rafael Ayala[1], Eladio Domínguez[2], Angel R. Francés[2], and Antonio Quintero[1]

[1] Dpt. de Geometría y Topología, Facultad de Matemáticas. Universidad de Sevilla
Apto. 1160. E-41080, Sevilla, Spain
quintero@cica.es

[2] Dpt. de Informática e Ingeniería de Sistemas, Facultad de Ciencias
Universidad de Zaragoza. E-50009, Zaragoza, Spain
ccia@posta.unizar.es

Abstract. The main contribution of this paper is a new "extrinsic" digital fundamental group that can be readily generalized to define higher homotopy groups for arbitrary digital spaces. We show that the digital fundamental group of a digital object is naturally isomorphic to the fundamental group of its continuous analogue. In addition, we state a digital version of the Seifert–Van Kampen theorem.

Keywords: Digital homotopy, digital fundamental group, lighting functions, Seifert–Van Kampen theorem.

1 Introduction

Thinning is an important pre-processing operation in pattern recognition whose goal is to reduce a digital image into a "topologically equivalent skeleton". In particular, thinning algorithms must preserve "tunnels" when processing three–dimensional digital images. As it was pointed out in [4], this requirement can be correctly established by means of an appropriate digital counterpart of the classical fundamental group in algebraic topology; see [11].

The first notion of a digital fundamental group (and even of higher homotopy groups) is dued to Khalimsky ([3]). He gave an "extrinsic" definition of this notion for a special class of digital spaces based on a topology on the set $\mathbb{Z}^n$, for every positive integer n. However, this approach is not suitable for other kinds of digital spaces often used in image processing, as the (α, β)-connected spaces, where $(\alpha, \beta) \in \{(4,8),(8,4)\}$ if $n = 2$ and $(\alpha, \beta) \in \{(6,26),(26,6),(6,18),(18,6)\}$ if $n = 3$. Within the graph-theoretical approach to Digital Topology, Kong solved partially this problem in [4] by defining an "intrinsic" digital fundamental group for the class of *strongly normal digital picture spaces* (SN-DPS), which include as particular cases both the (α, β)-connected spaces and the 2- and 3-dimensional Khalimsky's spaces. Nevertheless, Kong's definition seems not be general enough to give higher homotopy groups.

* This work has been partially supported by the projects DGICYT PB96-1374 and DGES PB96-0098C04-01.

G. Borgefors, I. Nyström, and G. Sanniti di Baja (Eds.): DGCI 2000, LNCS 1953, pp. 3–14, 2000.
© Springer-Verlag Berlin Heidelberg 2000

The goal of this paper is to introduce, via the framework of the multilevel architecture for Digital Topology in [2], a new notion of digital fundamental group (denoted by π_1^d) that, at least from a theoretical point of view, presents certain advantages over the notions of Khalimsky and Kong. Firstly, the group π_1^d is defined by using an "extrinsic" setting that can be readily generalized to define higher digital homotopy groups (see Section 3). Secondly, this group is available on a larger class of digital spaces than Khalimsky's and Kong's digital fundamental groups, since the digital spaces described in the multilevel architecture quoted above include as examples all Khalimsky's spaces and the (α, β)-connected spaces (see [1]), and even most of the SN-DPS. And finally, our digital fundamental group of a digital object O turns out to be naturally isomorphic to the fundamental group of its continuous analogue; that is, of the continuous object perceived when one looks at O (see Section 4). This isomorphism shows that the group π_1^d is an appropriate counterpart of the ordinary fundamental group in continuous topology. In particular, this fact leads us to obtain a (restricted) digital version of the Seifert–Van Kampen Theorem (see Section 5). Although this theorem provides a powerful theoretical tool to obtain the group π_1^d for certain digital objects, it remains as an open question to find an algorithm that computes this group for arbitrary objects; that is, to resemble in our framework the well–known algorithm for the fundamental group of polyhedra ([8]). This problem could be tackled by adapting to our multilevel architecture the algorithm recently developed by Malgouyres in [7], which computes a presentation of the digital fundamental group of an object embedded in an arbitrary graph.

2 The Multilevel Architecture

In this section we briefly summarize the basic notions of the multilevel architecture for digital topology developed in [2] as well as the notation that will be used through all the paper.

In that architecture, the spatial layout of pixels in a digital image is represented by a *device model*, which is a homogeneously n-dimensional locally finite polyhedral complex K. Each n-cell in K is representing a pixel, and so the digital object displayed in a digital image is a subset of the set $\mathrm{cell}_n(K)$ of n-cells in K. A *digital space* is a pair (K, f), where K is a device model and f is *weak lighting function* defined on K. The function f is used to provide a continuous interpretation, called *continuous analogue*, for each digital object $O \subseteq \mathrm{cell}_n(K)$. Next we recall the notion of weak lighting function. For this we need the following notation.

Let K be a device model and γ, σ cells in K. We shall write $\gamma \leq \sigma$ if γ is a face of σ, and $\gamma < \sigma$ if in addition $\gamma \neq \sigma$. If $|K|$ denotes the underlying polyhedron of K, a centroid-map is a map $c : K \to |K|$ such that $c(\sigma)$ belongs to the interior of σ; that is, $c(\sigma) \in \sigma - \partial\sigma$, where $\partial\sigma = \cup\{\gamma; \gamma < \sigma\}$ stands for the boundary of σ. Given a cell $\alpha \in K$ and a digital object $O \subseteq \mathrm{cell}_n(K)$, the *star of α in O* and *the extended star of α in O* are respectively the digital objects

$\mathrm{st}_n(\alpha; O) = \{\sigma \in O; \alpha \leq \sigma\}$ and $\mathrm{st}_n^*(\alpha; O) = \{\sigma \in O; \alpha \cap \sigma \neq \emptyset\}$. The *support of* O, $\mathrm{supp}(O)$, is the set of all cells $\alpha \in K$ such that $\alpha = \cap\{\sigma; \sigma \in \mathrm{st}_n(\alpha; O)\}$. To ease the writing, we shall use the following notation: $\mathrm{supp}(K) = \mathrm{supp}(\mathrm{cell}_n(K))$, $\mathrm{st}_n(\alpha; K) = \mathrm{st}_n(\alpha; \mathrm{cell}_n(K))$ and $\mathrm{st}_n^*(\alpha; K) = \mathrm{st}_n^*(\alpha; \mathrm{cell}_n(K))$. Finally, we shall write $\mathcal{P}(A)$ for the family of all subsets of a given set A.

Given a device model K, a *weak lighting function* (w.l.f.) on K is a map $f : \mathcal{P}(\mathrm{cell}_n(K)) \times K \to \{0, 1\}$ satisfying the following five properties for all $O \in \mathcal{P}(\mathrm{cell}_n(K))$ and $\alpha \in K$:

1. if $\alpha \in O$ then $f(O, \alpha) = 1$;
2. if $\alpha \notin \mathrm{supp}(O)$ then $f(O, \alpha) = 0$;
3. $f(O, \alpha) \leq f(\mathrm{cell}_n(K), \alpha)$;
4. $f(O, \alpha) = f(\mathrm{st}_n^*(\alpha; O), \alpha)$; and,
5. if $O' \subseteq O \subseteq \mathrm{cell}_n(K)$ and $\alpha \in K$ are such that $\mathrm{st}_n(\alpha; O) = \mathrm{st}_n(\alpha; O')$, $f(O', \alpha) = 0$ and $f(O, \alpha) = 1$, then the set of cells $\alpha(O'; O) = \{\beta < \alpha; f(O', \beta) = 0,\ f(O, \beta) = 1\}$ is not empty and connected in $\partial\alpha$; moreover, if $O \subseteq \overline{O} \subseteq \mathrm{cell}_n(K)$, then $f(\overline{O}, \beta) = 1$ for every $\beta \in \alpha(O'; O)$.

A w.l.f. f is said to be *strongly local* if $f(O, \alpha) = f(\mathrm{st}_n(\alpha; O), \alpha)$ for all $\alpha \in K$ and $O \subseteq \mathrm{cell}_n(K)$. Notice that this strong local condition implies both properties 4 and 5 in the definition above.

To define the continuous analogue of a given digital object O in a digital space (K, f), we need to introduce several other intermediate models (the levels of this multilevel architecture) as follows.

The *device level* of O is the subcomplex $K(O) = \{\alpha \in K; \alpha \leq \sigma,\ \sigma \in O\}$ of K induced by the cells in O. Notice that the map f_O given by $f_O(O', \alpha) = f(O, \alpha) f(O', \alpha)$, for all $O' \subseteq O$ and $\alpha \in K(O)$, is a w.l.f. on $K(O)$, and we call the pair $(K(O), f_O)$ the *digital subspace* of (K, f) induced by O.

The *logical level* of O is an undirected graph, $\mathcal{L}_O^f$, whose vertices are the centroids of n-cells in O and two of them $c(\sigma)$, $c(\tau)$ are adjacent if there exists a common face $\alpha \leq \sigma \cap \tau$ such that $f(O, \alpha) = 1$.

The *conceptual level of* O is the directed graph $\mathcal{C}_O^f$ whose vertices are the centroids $c(\alpha)$ of all cells $\alpha \in K$ with $f(O, \alpha) = 1$, and its directed edges are $(c(\alpha), c(\beta))$ with $\alpha < \beta$.

The *simplicial analogue* of O is the order complex $\mathcal{A}_O^f$ associated to the digraph $\mathcal{C}_O^f$. That is, $\langle x_0, x_1, \ldots, x_m \rangle$ is an m-simplex of $\mathcal{A}_O^f$ if $x_0, x_1, \ldots, x_m$ is a directed path in $\mathcal{C}_O^f$. This simplicial complex defines the simplicial level for the object O in the architecture and, finally, the *continuous analogue* of O is the underlying polyhedron $|\mathcal{A}_O^f|$ of $\mathcal{A}_O^f$.

For the sake of simplicity, we will usually drop "f" from the notation of the levels of an object. Moreover, for the whole object $\mathrm{cell}_n(K)$ we will simply write $\mathcal{L}_K$, $\mathcal{C}_K$ and $\mathcal{A}_K$ for its levels.

Example 1. In this paper it will be essential the role played by the archetypical device model R^n, termed the *standard cubical decomposition* of the Euclidean n-space $\mathbb{R}^n$. Recall that the device model R^n is the complex determined by the collection of unit n-cubes in $\mathbb{R}^n$ whose edges are parallel to the coordinate axes and whose centers are in the set $\mathbb{Z}^n$. The centroid-map we will consider in R^n associates to each cube σ its barycenter $c(\sigma)$. In particular, if $\dim \sigma = n$ then $c(\sigma) \in \mathbb{Z}^n$, where $\dim \sigma$ stands for the dimension of σ. So that, every digital object O in R^n can be identified with a subset of points in $\mathbb{Z}^n$. Henceforth we shall use this identification without further comment. In particular, we shall consider the family of digital spaces (R^n, g), for every positive integer n, where g is the w.l.f. given by $g(O, \alpha) = 1$ if and only if $\mathrm{st}_n(\alpha; R^n) \subseteq O$, for any digital object $O \subseteq \mathrm{cell}_n(R^n)$ and any cell $\alpha \in R^n$. Notice that the w.l.f. g induces in R^n the $(2n, 3^n - 1)$–connectedness (see [1, Def. 11]); that is, the generalization to arbitrary dimension of the $(4, 8)$–connectedness on $\mathbb{Z}^2$.

3 A Digital Fundamental Group

We next introduce an "extrinsic" digital fundamental group that readily generalizes to higher digital homotopy groups of arbitrary digital spaces. For this purpose, we will first define a digital-map (Def. 2), and then we will focus our interest in a special class of such digital-maps termed digital homotopies (Defs. 5 and 6).

Definition 1. *Let $S \subseteq cell_n(K)$ be a digital object in a digital space (K, f). The* light body of K shaded with S *is the set of cells*

$$Lb(K/S) = \{\alpha \in K; f(cell_n(K), \alpha) = 1, f(S, \alpha) = 0\};$$

that is, $Lb(K/S) = \{\alpha \in K; c(\alpha) \in |\mathcal{A}_K| - |\mathcal{A}_S|\}$. Notice that if $S = \emptyset$ is the empty object then $Lb(K/\emptyset) = Lb(K) = \{\alpha \in K; f(cell_n(K), \alpha) = 1\}$. Moreover, $Lb(K/cell_n(K)) = \emptyset$.

Definition 2. *Let $(K_1, f_1), (K_2, f_2)$ be two digital spaces, with $\dim K_i = n_i$ $(i = 1, 2)$, and let $S_1 \subset cell_{n_1}(K_1)$ and $S_2 \subset cell_{n_2}(K_2)$ be two digital objects. A map $\phi : Lb(K_1/S_1) \to Lb(K_2/S_2)$ is said to be a* (digital) (S_1, S_2)-map from (K_1, f_1) into (K_2, f_2) *(or, simply, a d-map denoted $\Phi_{S_1,S_2} : (K_1, f_1) \to (K_2, f_2)$) provided*

1. *$\phi(cell_{n_1}(K_1) - S_1) \subseteq cell_{n_2}(K_2) - S_2$; and,*
2. *for $\alpha, \beta \in Lb(K_1/S_1)$ with $\alpha < \beta$ then $\phi(\alpha) \leq \phi(\beta)$.*

Example 2. (1) Let $S' \subset S \subseteq \mathrm{cell}_n(K)$ be two digital objects and $(K(S), f_S)$ the digital subspace of (K, f) induced by S. Then, the inclusion $\mathrm{Lb}(K(S)/S') \subseteq \mathrm{Lb}(K/S')$ is a (S', S')-map from $(K(S), f_S)$ into (K, f). And, similarly, the inclusion $\mathrm{Lb}(K/S') \subseteq \mathrm{Lb}(K/\emptyset)$ defines a $(S', \emptyset)$-map from (K, f) into itself.

(2) Let $S_1 \subset \mathrm{cell}_{n_1}(K_1)$ and $\sigma \in \mathrm{cell}_{n_2}(K_2)$. For any digital object $S_2 \subseteq \mathrm{cell}_{n_2}(K_2) - \{\sigma\}$, the constant map $\phi^\sigma : \mathrm{Lb}(K_1/S_1) \to \mathrm{Lb}(K_2/S_2)$, given by $\phi^\sigma(\alpha) = \sigma$, for all $\alpha \in \mathrm{Lb}(K_1/S_1)$, defines a (S_1, S_2)-map from (K_1, f_1) into (K_2, f_2) .

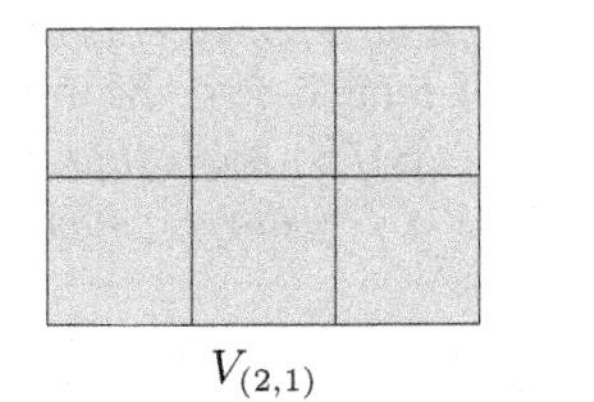

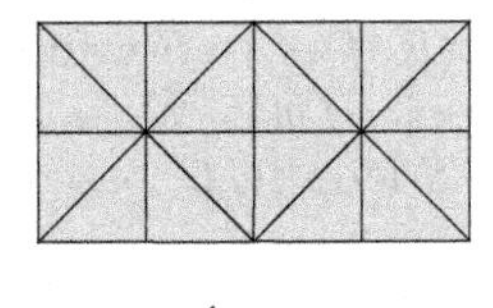

$V_{(2,1)}$ $\mathcal{A}_{V_{(2,1)}}$

Fig. 1. A $(2,1)$-window in R^2 and its simplicial analogue

(3) The composition of digital maps is a digital map. Namely, if

$$\Phi_{S_1,S_2} : (K_1, f_1) \to (K_2, f_2) \text{ and } \Phi_{S_2,S_3} : (K_2, f_2) \to (K_3, f_3)$$

are d-maps, then their composite $\Phi_{S_1,S_2} \circ \Phi_{S_2,S_3}$ is also a d-map from (K_1, f_1) into (K_3, f_3).

A d-map from (K_1, f_1) into (K_2, f_2) naturally induces a simplicial map between the simplicial analogues of K_1 and K_2. More precisely, if $L_1, L_2 \subseteq L$ are simplicial complexes and $L_1 \setminus L_2 = \{\alpha \in L_1; \alpha \cap |L_2| = \emptyset\}$ denotes the simplicial complement of L_2 in L_1, then it is straightforward to show

Proposition 1. *Any d-map $\Phi_{S_1,S_2} : (K_1, f_1) \to (K_2, f_2)$ induces a simplicial map $\mathcal{A}(\Phi_{S_1,S_2}) : \mathcal{A}_{K_1} \setminus \mathcal{A}_{S_1} \to \mathcal{A}_{K_2} \setminus \mathcal{A}_{S_2}$.*

Given two points $x = (x_1, \dots, x_m), y = (y_1, \dots, y_m) \in \mathbb{R}^m$, we write $x \preceq y$ if $x_i \leq y_i$, for all $1 \leq i \leq m$, while $x+y$ will stand for the point $(x_1+y_1, \dots, x_m+y_m) \in \mathbb{R}^m$.

Definition 3. *Given two points $r, x \in \mathbb{Z}^m$, with $r_i \geq 0$ for $1 \leq i \leq m$, we call a* window of size r *(or* r-window*) of R^m based at x to the digital subspace V_r^x of (R^m, g) induced by the digital object $O_r^x = \{\sigma \in cell_m(R^m); x \preceq c(\sigma) \preceq x+r\}$, where (R^m, g) is the digital space defined in Example 1.*

Notice that the simplicial analogue of an r-window V_r^x of R^m is (simplicially isomorphic to) a triangulation of a unit n-cube, where n is the number of non-zero coordinates in r (see Figure 1). Moreover, the set $\{y \in \mathcal{Z}^m; x \preceq y \preceq x+r\}$ are the centroids of the cells in $\mathrm{Lb}(V_r^x/\emptyset)$ which actually span the simplicial analogue of V_r^x. Here $\mathcal{Z} = \frac{1}{2}\mathbb{Z}$ stands for the set of points $\{z \in \mathbb{R}; z = y/2, y \in \mathbb{Z}\}$.

For the sake of simplicity, we shall write V_r to denote the r-window of R^m based at the point $x = (0, \dots, 0) \in \mathbb{Z}^m$. Moreover, if V_r is an r-window of R^1, then $\mathrm{Lb}(V_r/\emptyset) = \{\sigma_0, \sigma_1, \dots, \sigma_{2r-1}, \sigma_{2r}\}$ consists of $2r+1$ cells such that $c(\sigma_i) = i/2$.

With this notation we are now ready to give "extrinsic" notions of walks and loops, in a digital object, which will lead us to the definition of a digital fundamental group.

Definition 4. *Let (K,f) be a digital space and $S,O \subseteq cell_n(K)$ two disjoint digital objects in (K,f). A* S-walk in O of length $r \in \mathbb{Z}$ from σ to τ *is a digital $(\emptyset,S)$-map $\phi_r : Lb(V_r/\emptyset) \to Lb(K(O\cup S)/S)$ such that $\phi_r(\sigma_0)=\sigma$ and $\phi_r(\sigma_{2r})=\tau$. A* S-loop in O based at σ *is a S-walk ϕ_r such that $\phi_r(\sigma_0)=\phi_r(\sigma_{2r})=\sigma$.*

The juxtaposition *of two given S-walks ϕ_r,ϕ_s in O, with $\phi_r(\sigma_{2r})=\phi_s(\sigma_0)$, is the S-walk $\phi_r * \phi_s : Lb(V_{r+s}/\emptyset) \to Lb(K(O\cup S)/S)$ of length $r+s$ given by*

$$\phi_r * \phi_s(\sigma_i) = \begin{cases} \phi_r(\sigma_i) & \text{if } 0 \le i \le 2r \\ \phi_s(\sigma_{i-2r}) & \text{if } 2r \le i \le 2(r+s) \end{cases}$$

Notice that the notion of a S-walk is compatible with the definition of S-path given in [1, Def. 5]. Actually, each S-walk ϕ_r defines a S-path given by the sequence $\varphi(\phi_r) = (\phi_r(\sigma_{2i}))_{i=0}^r$. And, conversely, a S-path $(\tau_i)_{i=0}^r$ in O can be associated with a family Φ_r of S-walks such that $\phi_r \in \Phi_r$ if and only if $\phi_r(\sigma_{2i}) = \tau_i$ $(0 \le i \le r)$ and $\phi_r(\sigma_{2i-1}) \in \{\alpha \le \tau_{2i-2} \cap \tau_{2i}; f(O\cup S,\alpha)=1, f(S,\alpha)=0\}$ $(1 \le i \le r)$. However, this "extrinsic" notion of S-walk will be more suitable to define the digital fundamental group of an object since, together with the notion of r-window, it allows us to introduce the following definition of digital homotopy.

Definition 5. *Let ϕ_r^1,ϕ_r^2 two S-walks in O of the same length $r \in \mathbb{Z}$ from σ to τ. We say that ϕ_r^1,ϕ_r^2 are* digitally homotopic *(or, simply,* d-homotopic*)* relative $\{\sigma,\tau\}$, *and we write $\phi_r^1 \simeq_d \phi_r^2$ rel. $\{\sigma,\tau\}$, if there exists an (r,s)-window $V_{(r,s)}$ in R^2 and a $(\emptyset,S)$-map $H : Lb(V_{(r,s)}/\emptyset) \to Lb(K(O\cup S)/S)$, called a* d-homotopy, *such that $H(i/2,0)=\phi_r^1(\sigma_i)$ and $H(i/2,s)=\phi_r^2(\sigma_i)$, for $0 \le i \le 2r$, and moreover $H(0,j/2)=\sigma$ and $H(r,j/2)=\tau$, for $0 \le j \le 2s$. Here we use the identification $H(x,y)=H(\alpha)$, where $c(\alpha)=(x,y) \in \mathbb{Z}^2$ is the centroid of a cell $\alpha \in Lb(V_{(r,s)}/\emptyset)$.*

Clearly, the previous definition of d-homotopy induces an equivalence relation between the S-walks in O from σ to τ of the same length. Moreover, it is easy to show that d-homotopies are compatible with the juxtaposition of S-walks.

The definition of d-homotopy between S-walks of the same length can be extended to arbitrary S-walks as follows.

Definition 6. *Let ϕ_r,ϕ_s two S-walks in O from σ to τ of lengths $r \ne s$. We say that ϕ_r is d-homotopic to ϕ_s relative $\{\sigma,\tau\}$, $\phi_r \simeq_d \phi_s$ rel. $\{\sigma,\tau\}$, if there exist constant S-loops $\phi_{r'}^\tau$ and $\phi_{s'}^\tau$ such that $r+r'=s+s'$ and $\phi_r * \phi_{r'}^\tau \simeq_d \phi_s * \phi_{s'}^\tau$ rel. $\{\sigma,\tau\}$.*

Proposition 2. *Let ϕ_r be a S-walk in O from σ to τ, and ϕ_s^σ, ϕ_s^τ two constant S-loops of the same length s. Then, $\phi_s^\sigma * \phi_r \simeq_d \phi_r * \phi_s^\tau$ rel. $\{\sigma,\tau\}$.*

The proof of this proposition, although it is not immediate, can be directly obtained from definitions by means of an inductive argument. Moreover, from Proposition 2 and the remarks above, it can be easily derived that d-homotopy defines an equivalence relation in the set of S-walks in O of arbitrary length. So, we next introduce the digital fundamental group as follows.

Definition 7. *Let O be a digital object in a digital space (K, f). The* digital fundamental group of O at σ *is the set $\pi_1^d(O, \sigma)$ of d-homotopy classes of $\emptyset$-loops in O based at σ provided with the product operation $[\phi_r] \cdot [\psi_s] = [\phi_r * \psi_s]$.*

Remark 1. Definition 7 can be easily extended to the definition of a digital fundamental group for the complement of an object S in a digital space using the notion of S-loop. Moreover, this last notion readily generalizes to give higher digital homotopy groups by replicating the same steps as above but starting with a suitable notion of m-dimensional S-loop. More explicitly, let $r \in \mathbb{Z}^m$ be a point with positive coordinates, and call boundary of an r-window V_r to the set of cells $\partial V_r = \{\alpha \in \mathrm{Lb}(V_r/\emptyset); c(\alpha) \in \partial \mathcal{A}_{V_r}\}$. Notice that the boundary $\partial \mathcal{A}_{V_r}$ is well-defined since $\mathcal{A}_{V_r}$ triangulates the unit m-cube. Then define an *m-dimensional S-loop* in O of size r at σ as any $(\emptyset, S)$-map $\phi_r : \mathrm{Lb}(V_r/\emptyset) \to \mathrm{Lb}(K(O \cup S)/S)$ such that $\phi_r|_{\partial V_r} = \sigma$.

4 Isomorphism with the Continuous Fundamental Group

As usual the fundamental group of a topological space X, $\pi_1(X, x_0)$, is defined to be the set of homotopy classes of paths $\xi : I = [0,1] \to X$ that send 0 and 1 to some fixed point x_0 (*loops* at x_0). The set $\pi_1(X, x_0)$ is given the structure of a group by the operation $[\alpha] \cdot [\beta] = [\alpha * \beta]$, where $\alpha * \beta$ denotes the juxtaposition of paths. However, for a polyhedron $|K|$ there is an alternative definition of the fundamental group $\pi_1(|K|, x_0)$ that is more convenient for our purposes, so we next explain it briefly. Recall that an *edge–path* in $|K|$ from a vertex v_0 to a vertex v_n is a sequence α of vertices $v_0, v_1, \ldots, v_n$, such that for each $k = 1, 2, \ldots, n$ the vertices v_{i-1}, v_i span a simplex in K (possibly $v_{i-1} = v_i$). If $v_0 = v_n$, α is called an *edge–loop*.

Given another edge–path $\beta = (v_j)_{j=n}^{m+n}$ whose first vertex is the same as the last vertex of α, the *juxtaposition* $\alpha * \beta = (v_i)_{i=0}^{m+n}$ is defined in the obvious way. The *inverse* of α is $\alpha^{-1} = (v_n, v_{n-1}, \ldots, v_0)$.

Two edge–paths α and β are said to be *equivalent* if one can be obtained from the other by a finite sequence of operations of the form:

(a) if $v_{k-1} = v_k$, replace $\ldots, v_{k-1}, v_k, \ldots$ by $\ldots, v_k, \ldots$, or conversely replace $\ldots, v_k, \ldots$ by $\ldots, v_{k-1}, v_k, \ldots$; or
(b) if v_{k-1}, v_k, v_{k+1} span a simplex of K (not necessarily 2-dimensional), replace $\ldots, v_{k-1}, v_k, v_{k+1}, \ldots$ by $\ldots, v_{k-1}, v_{k+1}, \ldots$, or conversely.

This clearly sets up an equivalence relation between edge–paths, and the set of equivalence classes $[\alpha]$ of edge–loops α in K, based at a vertex v_0, forms a group $\pi_1(K, v_0)$ with respect to the juxtaposition of edge–loops. This group will be called the *edge–group* of K. Moreover it can be proved

Theorem 1. (Maunder; 3.3.9) *There exists an isomorphism $\pi_1(|K|, v_0) \to \pi_1(K, v_0)$ which carries the class $[f]$ to the class $[\alpha_f]$, where α_f is an edge–loop defined by a simplicial approximation of f.*

Corollary 1. *Let O, S be two disjoint digital objects of a digital space (K, f). Then $\pi_1(\mathcal{A}_{O \cup S} \setminus \mathcal{A}_S, c(\sigma)) \cong \pi_1(|\mathcal{A}_{O \cup S}| - |\mathcal{A}_S|, c(\sigma))$ for any $\sigma \in O$.*

The proof of this corollary is a consequence of Theorem 1 and next lemma.

Lemma 1. *Let $K, L \subseteq J$ be two full subcomplexes. Then $|K \backslash L| = |K \backslash K \cap L|$ is a strong deformation retract of $|K| - |L| = |K| - |K \cap L|$.*

This lemma is actually Lemma 72.2 in [9] applied to the full subcomplex $K \cap L \subseteq K$ (this fact is shown using that L is full in J).

We are now ready to prove the main result of this paper. Namely,

Theorem 2. *Let O be a digital object in the digital space (K, f). Then there exists an isomorphism $h : \pi_1^d(O, \sigma) \to \pi_1(|\mathcal{A}_O|, c(\sigma))$.*

The function h is defined as follows. Let ϕ_r be any $\emptyset$-loop in O based at σ. Then the sequence $c(\phi_r) = (c(\phi_r(\sigma_i)))_{i=0}^{2r}$ defines and edge–loop in $\mathcal{A}_O$ based at $c(\sigma)$, and so we set $h([\phi_r]) = [c(\phi_r)]$. Lemma 2 below, and the two immediate properties (1), (2), show that the function h is a well defined homomorphism of groups.

(1) $c(\phi_r * \phi'_s) = c(\phi_r) * c(\phi'_s)$
(2) if ϕ_r is a constant $\emptyset$-loop then $c(\phi_r)$ is also a constant edge–loop.

Lemma 2. *If $\phi_r \simeq_d \phi'_s$ are equivalent $\emptyset$-loops then $c(\phi_r)$ and $c(\phi'_s)$ define both the same element in $\pi_1(|\mathcal{A}_O|, c(\sigma))$.*

Proof. According to (2) above and the definition of equivalence between two $\emptyset$-loops it will be enough to show that any d-homotopy $H : \mathrm{Lb}(V_{(r,s)}/\emptyset) \to \mathrm{Lb}(K(O)/\emptyset)$ between two $\emptyset$-loops ϕ_r, ϕ'_r of the same length r induces a continuous homotopy $\tilde{H} : [0,1] \times [0,1] \to |\mathcal{A}_O|$ between $c(\phi_r)$ and $c(\phi'_r)$. This fact is readily checked since Proposition 1 yields a simplicial map $\mathcal{A}(H) : \mathcal{A}_{V_{(r,s)}} \to \mathcal{A}_O$, and $\mathcal{A}_{V_{(r,s)}}$ is (simplicially isomorphic to) a triangulation of the unit square (see Figure 1). Moreover $\mathcal{A}(H)$ restricted to the top and the bottom of that unit square define $c(\phi_r)$ and $c(\phi'_r)$ respectively, and the result follows.

Now, let γ be any loop in $|\mathcal{A}_O|$ based at $c(\sigma)$. By Corollary 1 we can assume that $\gamma = (c(\gamma_i))_{i=0}^{k}$ is an edge–loop based at $c(\sigma)$ in $\mathcal{A}_O$. After applying equivalence operations (a) and (b) above we can reduce the edge–loop γ to a new edge–loop $\overline{\gamma} = (c(\overline{\gamma}_i))_{i=0}^{2r}$ equivalent to γ and such that $\overline{\gamma}_{2i-1}$ are k-cells in K with $k < n$ and $\overline{\gamma}_{2i-1} \leq \overline{\gamma}_{2i-2} \cap \overline{\gamma}_{2i}$ $(1 \leq i \leq r)$. By the use of $\overline{\gamma}$ we define the following set $F(\gamma)$ of $\emptyset$-loops at σ of length r.

The set $F(\gamma)$ consists of all $\emptyset$-loops ϕ_r for which $\phi_r(\sigma_0) = \phi(\sigma_{2r}) = \sigma$, $\phi_r(\sigma_{2i-1}) = \overline{\gamma}_{2i-1}$ $(1 \leq i \leq r)$ and $\phi_r(\sigma_{2i}) \in \mathrm{st}_n(\overline{\gamma}_{2i}; O)$ $(0 \leq i \leq r)$. Notice that $\mathrm{st}_n(\overline{\gamma}_{2i}; O) = \{\overline{\gamma}_{2i}\}$ if and only if $\overline{\gamma}_{2i} \in O$, while $\mathrm{st}_n(\overline{\gamma}_{2i}; O)$ contains at least two elements otherwise. This is clear since $c(\overline{\gamma}_{2i}) \in \mathcal{A}_O$ yields $\overline{\gamma}_{2i} \in \mathrm{supp}(O)$ by Axiom 2 of w.l.f.'s. Notice that $F(\gamma) = \{\overline{\gamma}\}$ if and only if $\overline{\gamma}_{2i} \in O$ for all $0 \leq i \leq r$; and moreover, in any case, $F(\gamma) \neq \emptyset$ is a non-empty set.

Lemma 3. *For each $\phi_r \in F(\gamma)$, $c(\phi_r)$ is homotopic to the loop γ. Hence h is onto.*

Proof. Let us consider the set $\overline{F}(\gamma)$ of edge–loops $\alpha = (\alpha_i)_{i=0}^{2r}$ at $c(\sigma)$ such that $\alpha_0 = \alpha_{2r} = c(\sigma)$, $\alpha_{2i-1} = \overline{\gamma}_{2i-1}$ $(1 \leq i \leq r)$ and $\alpha_{2i} \in \mathrm{st}_n(\overline{\gamma}_{2i}; O) \cup \{\overline{\gamma}_{2i}\}$. Notice that $\{c(\phi_r); \phi_r \in F(\gamma)\} \cup \{\overline{\gamma}\} \subseteq \overline{F}(\gamma)$. Since $\overline{\gamma}$ was obtained from γ by transformations of types (a) and (b), they are equivalent edge–loops by Theorem 1. So, it will suffice to show that α is homotopic to $\overline{\gamma}$, for any $\alpha \in \overline{F}(\gamma)$. This will be done by induction on the number $t(\alpha)$ of vertices $\alpha_{2i} \neq \overline{\gamma}_{2i}$ in α.

For $t(\alpha) = 0$ we get $\alpha = \overline{\gamma}$. Assume that $\alpha \in \overline{F}(\gamma)$ is equivalent to $\overline{\gamma}$ if $t(\alpha) \leq t-1$. Then, for an edge–loop $\alpha \in \overline{F}(\gamma)$ with $t(\alpha) = t$ let α_{2i} any vertex in α such that $\alpha_{2i} \neq \overline{\gamma}_{2i}$. Then we have $\overline{\gamma}_{2i+1}, \overline{\gamma}_{2i-1} < \overline{\gamma}_{2i} < \alpha_{2i}$, and we obtain a new edge–loop $\tilde{\alpha} \in \overline{F}(\gamma)$, replacing α_{2i} by $\overline{\gamma}_{2i}$ in α, with $t(\tilde{\alpha}) = t-1$ and two equivalence transformations of type (b) relating α and $\tilde{\alpha}$. Hence α is an edge–loop equivalent to $\tilde{\alpha}$ and, by induction hypothesis, to $\overline{\gamma}$.

Lemma 4. *Any two $\emptyset$-loops ϕ_r^1, ϕ_r^2 in $F(\gamma)$ are d-homotopic.*

Proof. It is enough to observe that the map $H : \mathrm{Lb}(V_{(r,1)}/\emptyset) \to \mathrm{Lb}(K(O)/\emptyset)$ given by $H(i/2, 0) = \phi_r^1(\sigma_i)$, $H(i/2, 1) = \phi_r^2(\sigma_i)$ and $H(i/2, 1/2) = \overline{\gamma}_i$, for $0 \leq i \leq 2r$, and $H(i-1/2, k) = \overline{\gamma}_{2i-1}$, for $1 \leq i \leq 2r$ and $k \in \{0, 1/2, 1\}$, is a d-homotopy relating ϕ_r^1 and ϕ_r^2. Here we use again the identification $H(x,y) = H(\alpha)$, where $c(\alpha) = (x,y) \in \mathcal{Z}^2$ is the centroid of a cell $\alpha \in \mathrm{Lb}(V_{(r,1)}/\emptyset)$.

Lemma 5. *Let γ^1 and γ^2 be two edge-loops at $c(\sigma)$ in $\mathcal{A}_O$ such that they are related by an equivalence transformation of type (a) or (b). Then there exist $\emptyset$-loops $\phi_{r_i}^i \in F(\gamma^i)$ $(i = 1, 2)$ and a d-homotopy such that $\phi_{r_1}^1 \simeq_d \phi_{r_2}^2$ rel. σ. Hence h is injective.*

Proof. In case γ^1 is related to γ^2 by a transformation of type (a), it is readily checked that $\overline{\gamma}^1 = \overline{\gamma}^2$. Hence $F(\gamma^1) = F(\gamma^2)$ and the result follows. And it suffices to check the essentially distinct twelve ways for deriving γ^2 from γ^1 by transformations of type (b) to complete the proof.

5 A Digital Seifert–Van Kampen Theorem

The Seifert–Van Kampen Theorem is the basic tool for computing the fundamental group of a space which is built of pieces whose groups are known. The statement of the theorem involves the notion of push–out of groups, so we begin by explaining this bit of algebra. A group G is said to be the push–out of the solid arrow commutative diagram

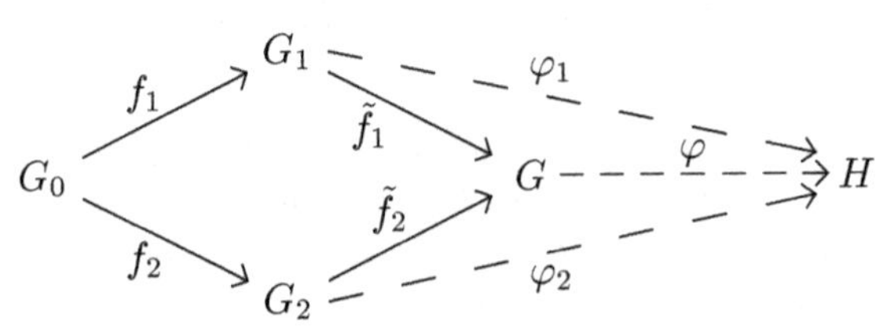

if for any group H and homomorphisms φ_1, φ_2 with $\varphi_1 f_1 = \varphi_2 f_2$ there exists a unique homomorphism φ such that $\varphi \tilde{f_i} = \varphi_i$ $(i = 1, 2)$. Then, the Seifert–Van Kampen Theorem is the following

Theorem 3. (Th. 7.40 in [10]) *Let K be a simplicial complex having connected subcomplexes K_1 and K_2 such that $K = K_1 \cup K_2$ and $K_0 = K_1 \cap K_2$ is connected. If $v_0 \in K_0$ is a vertex then $\pi_1(K, v_0)$ is the push–out of the diagram*

$$\begin{array}{ccc} \pi_1(K_0, v_0) & \xrightarrow{i_{1*}} & \pi_1(K_1, v_0) \\ {\scriptstyle i_{2*}}\downarrow & & \downarrow{\scriptstyle j_{1*}} \\ \pi_1(K_2, v_0) & \xrightarrow{j_{2*}} & \pi_1(K, v_0) \end{array}$$

where i_{k*} and j_{k*} are the homomorphisms of groups induced the obvious inclusions.

By using explicit presentations of the groups $\pi_1(K_i, v_0)$ $(i = 0, 1, 2)$ the Seifert–Van Kampen Theorem can be restated as follows. Suppose there are presentations $\pi_1(K_i, v_0) \cong (x_1^i, x_2^i, \ldots ; r_1^i, r_2^i, \ldots)$ $(i = 0, 1, 2)$ Then the fundamental group of K has the presentation

$$\begin{aligned} \pi_1(K, v_0) \cong (&x_1^1, x_2^1, \ldots, x_1^2, x_2^2, \ldots ; \\ &r_1^1, r_2^1, \ldots, r_1^2, r_2^2, \ldots, i_{1*}(x_1^0) = i_{2*}(x_1^0), i_{1*}(x_2^0) = i_{2*}(x_2^0), \ldots). \end{aligned}$$

In other words, one puts together the generators and relations from $\pi_1(K_1, v_0)$ and $\pi_1(K_2, v_0)$, plus one relation for each generator x_i^0 of $\pi_1(K_0, v_0)$ which says that its images in $\pi_1(K_1, v_0)$ and $\pi_1(K_2, v_0)$ are equal.

The digital analogue of the Seifert–Van Kampen Theorem is not always true as the following example shows.

Example 3. Let O_1, O_2 be the two digital objets in the digital space (R^2, g) shown in Figure 2. It is readily checked that both $\pi_1^d(O_1, \sigma)$ and $\pi_1^d(O_2, \sigma)$ are trivial groups, but $\pi_1^d(O_1 \cup O_2, \sigma) = \mathbb{Z}$ despite of O_1, O_2 and $O_1 \cap O_2$ are connected digital objects.

However we can easily derive a Digital Seifert–Van Kampen Theorem for certain objects in a quite large class of digital spaces. Namely, the locally strong digital spaces; that is, the digital spaces (K, f) for which the lighting function f satisfies $f(O, \alpha) = f(\mathrm{st}_n(\alpha; O), \alpha)$. We point out that all the (α, β)-connected digital spaces on $\mathbb{Z}^3$ defined within the graph-theoretical approach to Digital Topology, for $\alpha, \beta \in \{6, 18, 26\}$, are examples of locally strong digital spaces; see [1, Example 2].

Theorem 4. (Digital Seifert–Van Kampen Theorem) *Let (K, f) be a locally strong digital space, and let $O \subseteq cell_n(K)$ be a digital object in (K, f) such that $O = O_1 \cup O_2$, where O_1, O_2 and $O_1 \cap O_2$ are connected digital objects. Assume in addition that $\mathcal{A}_{O_1 \cap O_2} \subseteq \mathcal{A}_{O_1} \cap \mathcal{A}_{O_2}$ and $\mathcal{A}_{O_i} \subseteq \mathcal{A}_O$ $(i = 1, 2)$. Moreover assume* *that for each cell $\sigma \in O_1 - O_2$ any cell $\tau \in O$ which is adjacent to σ in O lies in O_1. Then, for $\sigma \in O_1 \cap O_2$, $\pi_1^d(O,\sigma)$ is the push–out of the diagram*

$$\begin{array}{ccc} \pi_1^d(O_1 \cap O_2, \sigma) & \longrightarrow & \pi_1^d(O_1, \sigma) \\ \downarrow & & \downarrow \\ \pi_1^d(O_2, \sigma) & \longrightarrow & \pi_1^d(O, \sigma) \end{array}$$

where the homomorphisms are induced by the obvious inclusions.

The proof of this theorem is immediate consequence of Theorem 2 and Theorem 3 if we have at hand the equalities $|\mathcal{A}_{O_1 \cap O_2}| = |\mathcal{A}_{O_1}| \cap |\mathcal{A}_{O_2}|$ and $|\mathcal{A}_O| = |\mathcal{A}_{O_1}| \cup |\mathcal{A}_{O_2}|$. We devote the rest of this section to check these equalities.

Lemma 6. *If $f(O, \alpha) = 1$ then one of the following statements holds:*
(1) $st_n(\alpha; O) = st_n(\alpha; O_1 \cap O_2) = st_n(\alpha; O_1) = st_n(\alpha; O_2)$; or
(2) $st_n(\alpha; O) = st_n(\alpha; O_i)$ and $st_n(\alpha; O_j) = st_n(\alpha; O_1 \cap O_2)$, $\{i, j\} = \{1, 2\}$.

Proof. In case $\mathrm{st}_n(\alpha; O) = \mathrm{st}_n(\alpha; O_1 \cap O_2)$, we obtain (1) from the inclusions

$$\mathrm{st}_n(\alpha; O_1 \cap O_2) \subseteq \mathrm{st}_n(\alpha; O_i) \subseteq \mathrm{st}_n(\alpha; O),\ (i = 1, 2).$$

Otherwise, there exists $\sigma \in O - (O_1 \cap O_2)$ with $\alpha \leq \sigma$. Assume $\sigma \in O_1 - O_2$, then for all $\tau \in \mathrm{st}_n(\alpha; O)$ we have $\tau \in O_1$ by hypothesis and hence $\mathrm{st}_n(\alpha; O) = \mathrm{st}_n(\alpha; O_1)$. Moreover $\mathrm{st}_n(\alpha; O_2) \subseteq \mathrm{st}_n(\alpha; O) = \mathrm{st}_n(\alpha; O_1)$ yields $\mathrm{st}_n(\alpha; O_1 \cap O_2) = \mathrm{st}_n(\alpha; O_2) \subset \mathrm{st}_n(\alpha; O)$.

The case $\sigma \in O_2 - O_1$ is similar since then $\tau \in O_2$ ($\tau \notin O_2$ yields $\tau \in O_1 - O_2$ and hence $\sigma \in O_1$ by hypothesis).

Lemma 7. $\mathcal{A}_{O_1} \cap \mathcal{A}_{O_2} \subseteq \mathcal{A}_{O_1 \cap O_2}$ *and* $\mathcal{A}_O \subseteq \mathcal{A}_{O_1} \cup \mathcal{A}_{O_2}$. *And so the equalities follow by hypothesis.*

Proof. Let $c(\alpha) \in \mathcal{A}_{O_1} \cap \mathcal{A}_{O_2}$, then $f(O_i, \alpha) = 1$ for $i = 1, 2$ and hence $\mathrm{st}_n(\alpha; O_1 \cap O_2) = \mathrm{st}_n(\alpha; O_i)$ for some i by Lemma 6. Thus $f(O_1 \cap O_2, \alpha) = 1$ by the strong local condition of f, and so $c(\alpha) \in \mathcal{A}_{O_1 \cap O_2}$. Finally $\mathcal{A}_{O_1} \cap \mathcal{A}_{O_2} \subseteq \mathcal{A}_{O_1 \cap O_2}$ since $\mathcal{A}_{O_1 \cap O_2}$ is a full subcomplex.

Now let $\gamma = \langle c(\gamma_0), \ldots, c(\gamma_k) \rangle \in \mathcal{A}_O$. Then $\mathrm{st}_n(\gamma_k; O) \subseteq \mathrm{st}_n(\gamma_{k-1}; O) \subseteq \cdots \subseteq \mathrm{st}_n(\gamma_0; O)$. By Lemma 6 and the strong local condition we easily obtain $\gamma \in \mathcal{A}_{O_i}$ whenever $\mathrm{st}_n(\gamma_0; O) = \mathrm{st}_n(\gamma_0; O_i)$ $(i = 1, 2)$.

6 Future Work

The Digital Seifert–Van Kampen Theorem provides us a theoretical tool that, under certain conditions, computes the digital fundamental group of an object. Nevertheless, the effective computation of the digital fundamental group requires an algorithm to compute a presentation of this group directly at the object's logical level. In a near future we will intend to develop such an algorithm for general digital spaces, as well as to compare the digital fundamental group in Def. 7 with those already introduced by Khalimsky [3] and Kong [4].

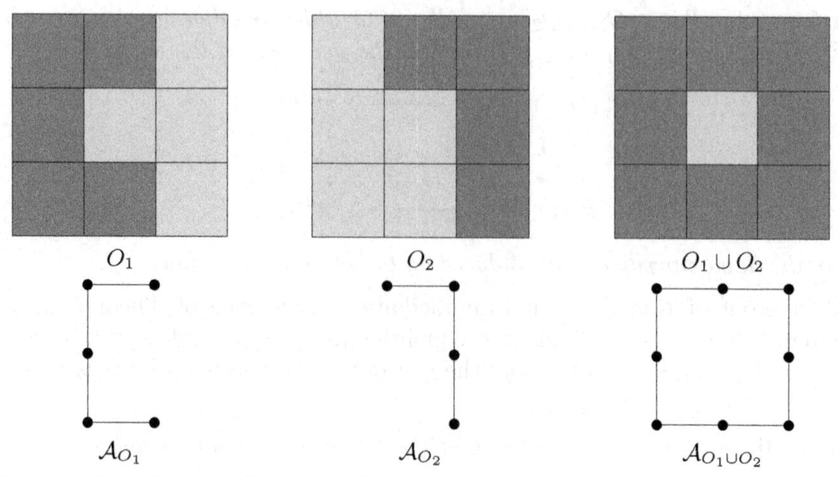

Fig. 2. A digital object for which the Digital Seifert–Van Kampen Theorem does not hold

References

1. R. Ayala, E. Domínguez, A. R. Francés, A. Quintero. Digital Lighting Functions. *Lecture Notes in Computer Science.* 1347(1997) 139-150. 4, 6, 8, 12
2. R. Ayala, E. Domínguez, A. R. Francés, A. Quintero. Weak lighting functions and strong 26-surfaces. To appear in *Theoretical Computer Science.* 4
3. E. Khalimsky. Motion, deformation and homotopy in finite spaces. *Proc. of the 1987 IEEE Int. Conf. on Systems., Man and Cybernetics, 87CH2503-1.* (1987) 227-234. 3, 13
4. T. Y. Kong. A Digital Fundamental Group. *Comput. & Graphics, 13(2).* (1989) 159-166. 3, 13
5. R. Malgouyres. Homotopy in 2-dimensional digital images. *Lecture Notes in Computer Science.* 1347(1997) 213-222.
6. R. Malgouyres. Presentation of the fundamental group in digital surfaces. *Lecture Notes in Computer Science.* 1568(1999) 136-150.
7. R. Malgouyres. Computing the fundamental group in digital spaces. *Proc. of the 7th Int. Workshop on Combinatorial Image Analysis IWCIA'00.* (2000) 103-115. 4
8. C. R. F. Maunder. *Algebraic Topology.* Cambridge University Press. 1980. 4
9. J. R. Munkres. *Elements of algebraic topology.* Addison-Wesley. 1984. 10
10. J. J. Rotman. *An introuduction to algebraic topology.* GTM, 119. Springer. 1988. 12
11. J. Stillwell. *Classical topology and combinatorial group theory.* Springer. 1995. 3

Tesselations by Connection in Orders

Michel Couprie and Gilles Bertrand

Laboratoire A[2]SI, ESIEE Cité Descartes
B.P. 99, 93162 Noisy-Le-Grand Cedex France
{coupriem,bertrand}@esiee.fr

Abstract. The watershed transformation is a powerful tool for segmenting images, but its precise definition in discrete spaces raises difficult problems. We propose a new approach in the framework of orders. We introduce the tesselation by connection, which is a transformation that preserves the connectivity, and can be implemented by a parallel algorithm. We prove that this transformation possesses good geometrical properties. The extension of this transformation to weighted orders may be seen as a generalization of the watershed transformation.

Keywords: discrete topology, order, discrete distance, influence zones, watershed

1 Introduction

The watershed transformation [1] is a powerful tool for segmenting images. Initially introduced in the field of topography, the notion of watershed is often described in terms of steepest slope paths, watercourses and catchment basins. Here we prefer the following presentation, which is precise enough to be implemented by a computer program.

Consider a grayscale image as a topographic relief: the gray level of each point corresponds to its altitude. Watersheds may be obtained by piercing a hole at each minimum of this relief, and immersing the relief into a lake. The water will progressively fill up the different basins around each minimum. When the waters coming from different minima are going to merge, a dam is built to prevent the merging. At the end of this process, the set of points immersed by the water coming from one minimum m_i is called the catchment basin associated to m_i, and the points that have not been immersed, and where dams have been built, constitute the watersheds.

This transformation may be seen as a set transformation, guided by the gray levels, which transforms the set of the points belonging to minima into the set of the points belonging to catchment basins. This set transformation has an important property: it preserves the connected components of the set (Fig. 1). On the other hand, the connected components of the complementary set are not preserved, as shown in Fig. 1(b_2, c_2). It is important to note that a topology-preserving transformation, such as an homotopic kernel [2] (which may be seen as an "ultimate skeleton"), should preserve both the connected components of

G. Borgefors, I. Nyström, and G. Sanniti di Baja (Eds.): DGCI 2000, LNCS 1953, pp. 15–26, 2000.
© Springer-Verlag Berlin Heidelberg 2000

the set and those of the complementary set (see Fig. 1(b_1, d_1 and b_2, d_2)), and in 3D, it should also preserve the tunnels of both the set and the complementary set.

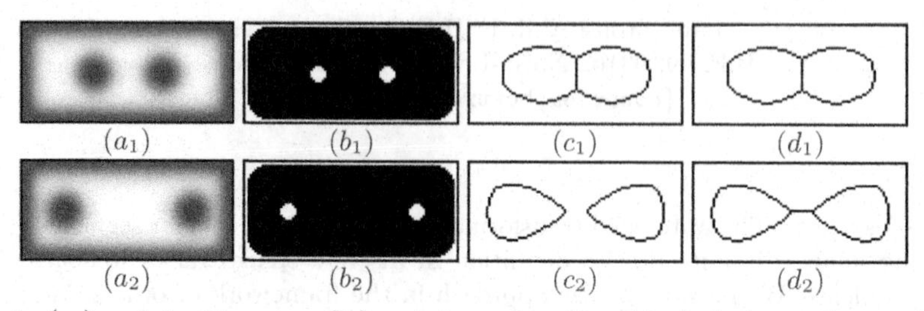

Fig. 1. (a_i): original images; (b_i): minima of a_i (in white); (c_i): catchment basins of a_i (in white); d_i: an upper homotopic kernel of b_i (in white)

Many definitions and algorithms for the watershed transformation have been proposed [3, 4, 5, 6, 7]. Particularly difficult is the problem of correctly specifying the watersheds that are located in plateaus, especially in the discrete spaces. Several authors have used a notion of distance in order to impose a "good centering" of the watersheds in plateaus, but in the usual discrete grids this centering is not always perfect. Their approach is based on the notion of influence zones: to a subset X of $\mathcal{Z}^n$, which is composed of k connected components $X_1, \ldots, X_k$, we can associate the influence zones $V_1, \ldots, V_k$ such that a point x belongs to V_i if x is nearer from X_i than from any other component X_j of X. This constitutes a set transformation, which does not preserve any topological characteristic (as defined in the framework of digital topology [8]), not even the number of connected components: see Fig. 2.

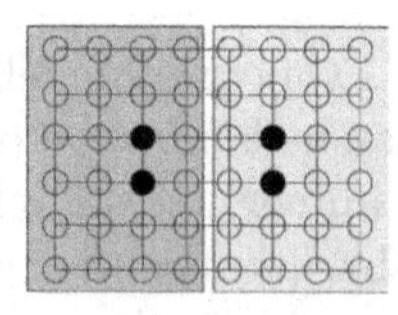

Fig. 2. Two regions (in black) and their respective influence zones (shaded) for the usual 4-distance. We see that the union of the two influence zones is 4-connected, while the original set (black points) is not 4-connected

This paper extends a work presented in [9], in the framework of graphs. We propose a new approach based on the notion of order [10]. An order is equivalent to a discrete topological space (in the sense of Alexandroff [11]). In such a space, we prove that the influence zones transformation, defined thanks to the "natural" distance and applied to a closed set, preserves the connected components. We introduce the notion of uniconnected point, which allows to define a set transformation that preserves the connected components of the set: the tesselation by connection. From this set transformation, we derive a transformation on grayscale images, and more generally on weighted orders, that can be seen as a generalization of a watershed transformation. We propose some parallel algorithms to implement such transformations in orders and weighted orders. We also show that the link between these transformations and the influence zones transformation ensures that our algorithms produce a "well centered" result. In this paper, we give some examples based on orders that "modelize" the discrete grid $\mathcal{Z}^n$. Nevertheless, we emphasize the fact that all the presented properties and algorithms are valid for any order.

2 Basic Notions

In this section, we introduce some basic notions relative to orders (see also [10]).

If X is a set, $\mathcal{P}(X)$ denotes the set composed of all subsets of X, if S is a subset of X, $\overline{S}$ denotes the complement of S in X. If S is a subset of T, we write $S \subseteq T$, the notation $S \subset T$ means $S \subseteq T$ and $S \neq T$. If γ is a map from $\mathcal{P}(X)$ to $\mathcal{P}(X)$, the *dual of* γ is the map $*\gamma$ from $\mathcal{P}(X)$ to $\mathcal{P}(X)$ such that, for each $S \subseteq X$, $*\gamma(S) = \overline{\gamma(\overline{S})}$. Let δ be a binary relation on X, *i.e.*, a subset of $X \times X$. We also denote by δ the map from X to $\mathcal{P}(X)$ such that, for each x of X, $\delta(x) = \{y \in X,\ (x,y) \in \delta\}$. We define $\delta^\square$ as the binary relation $\delta^\square = \delta \setminus \{(x,x); x \in X\}$.

An *order* is a pair $|X| = (X, \alpha)$ where X is a set and α is a reflexive, antisymmetric, and transitive binary relation on X. An element of X is also called a *point.* The set $\alpha(x)$ is called the α*-adherence* of x, if $y \in \alpha(x)$ we say that y is α*-adherent to* x.

Let (X, α) be an order. We denote by α the map from $\mathcal{P}(X)$ to $\mathcal{P}(X)$ such that, for each subset S of X, $\alpha(S) = \cup\{\alpha(x);\ x \in S\}$, $\alpha(S)$ is called the α*-closure of* S, $*\alpha(S)$ is called the α*-interior of* S. A subset S of X is α*-closed* if $S = \alpha(S)$, S is α*-open* if $S = *\alpha(S)$.

Let (X, α) be an order. We denote by β the relation $\beta = \{(x,y); (y,x) \in \alpha\}$, β is the *inverse* of the relation α. We denote by θ the relation $\theta = \alpha \cup \beta$. The *dual of the order* (X, α) is the order (X, β).
Note that $*\alpha(S) = \{x \in S;\ \beta(x) \subseteq S\}$, and $*\beta(S) = \{x \in S;\ \alpha(x) \subseteq S\}$.

The set $\mathcal{O}_\alpha$ composed of all α-open subsets of X satisfies the conditions for the family of open subsets of a topology, the same result holds for the set $\mathcal{O}_\beta$ composed of all β-open subsets of X; we denote respectively by $\mathcal{T}_\alpha = (X, \mathcal{O}_\alpha)$ and by $\mathcal{T}_\beta = (X, \mathcal{O}_\beta)$ these two topologies. These topologies are Alexandroff topologies, *i.e.*, topologies such that every intersection of open sets is open [11].

An order (X, α) is *countable* if X is countable, it is *locally finite* if, for each $x \in X$, $\theta(x)$ is a finite set. A *CF-order* is a countable locally finite order.

If (X, α) is an order and S is a subset of X, the *order relative to* S is the order $|S| = (S, \alpha \cap (S \times S))$.

Let (X, α) be a CF-order. Let x_0 and x_k be two points of X. A *path* from x_0 to x_k is a sequence $x_0, x_1, ..., x_k$ of elements of X such that $x_i \in \theta(x_{i-1})$, with $i = 1, ..., k$. The number k is called the *length* of the path. We consider the relation $\{(x,y);$ there is a path from x to $y\}$. It is an equivalence relation, its equivalence classes are called the *connected components of* X. We say that (X, α) is *connected* if it has exactly one connected component.

Let $|X| = (X, \alpha)$ be a CF-order. The α*-rank* of a point x is the length of a longest α-path having x as origin. It is denoted by $r(x)$.

Let (X, α) be an order. An element x such that $\alpha^{\square}(x) = \emptyset$ (*i.e.*, such that $r(x) = 0$) is said to be α*-terminal (for* X*)*.

3 Orders Associated to $\mathcal{Z}^n$

We give now a presentation of some orders which may be associated to $\mathcal{Z}^n$ [12,10].

Let $\mathcal{Z}$ be the set of integers. We consider the families of sets H_0^1, H_1^1, H^1 such that, $H_0^1 = \{\{a\};\ a \in \mathcal{Z}\}$, $H_1^1 = \{\{a, a+1\};\ a \in \mathcal{Z}\}$, $H^1 = H_0^1 \cup H_1^1$.
A subset S of $\mathcal{Z}^n$ which is the Cartesian product of exactly m elements of H_1^1 and $(n-m)$ elements of H_0^1 is called a *m-cube* of $\mathcal{Z}^n$. We denote H^n the set composed of all m-cubes of $\mathcal{Z}^n$, $m = 0, ..., n$.
An *m-cube* of $\mathcal{Z}^n$ is called a *singleton* if $m = 0$, a *unit interval* if $m = 1$, a *unit square* if $m = 2$, a *unit cube* if $m = 3$.

In this paper, the basic order associated to $\mathcal{Z}^n$ is the order (H^n, α), where $\alpha = \supseteq$, thus $y \in \alpha(x)$ if $x \supseteq y$. In Fig. 3(a), an example of a subset S of H^2 is given. The object S is made of two connected components S_1 (to the left) and S_2 (to the right). It may be seen that S_1 contains one singleton (α-terminal, α-rank 0), two unit intervals (α-rank 1), and two unit squares (β-terminals, α-rank 2). In Fig. 3(b), an alternative representation of the same object is presented, we call it the *array representation.* We use the following conventions: a singleton is depicted by a circle ($\circ$), a unit interval by a rectangle ($\square$), and a unit square by a square ($\square$).

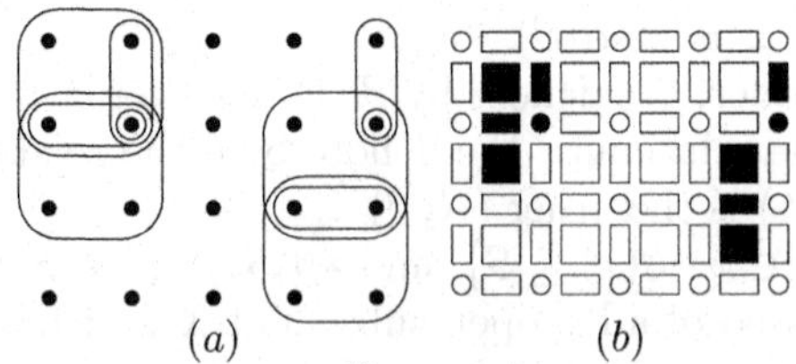

Fig. 3. (a): a subset S of H^2, and (b): its array representation

In order to build consistent topological notions for a subset S of $\mathcal{Z}^n$, we associate to S a subset $\Psi(S)$ of H^n; thus we recover the structure of a (discrete) topological space by considering the order $(H^n, \supseteq)$. In this paper, the transformation Ψ is chosen in such a way that the induced topological notions may be seen as "compatible" with the notions derived from the digital topology framework. A natural idea for defining Ψ is to consider "hit or miss" transformations [13]. Thus we consider the set S^h composed of all elements of H^n which have a non-empty intersection with S. In a dual way, we consider the set S^m composed of all elements of H^n which are included in S:
$\Psi^h(S) = S^h = \{x \in H^n, x \cap S \neq \emptyset\}$; $\Psi^m(S) = S^m = \{x \in H^n, x \subseteq S\}$

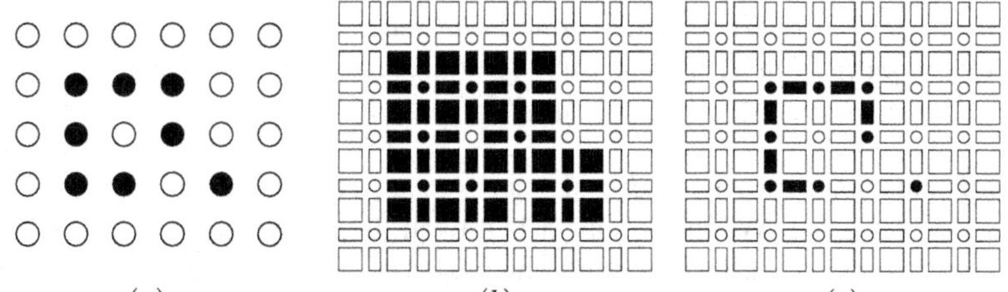

(a) (b) (c)

Fig. 4. (a): a subset S of $\mathcal{Z}^2$ (black disks), (b): $S^h = \Psi^h(S)$, (c): $S^m = \Psi^m(S)$

Fig. 4(a) presents an object S of $\mathcal{Z}^2$ (in black). The adjacency used for the object is denoted by a, the adjacency used for the background is denoted by $\overline{a}$ (see [8]). Note that if $(a, \overline{a}) = (4, 8)$, then S is composed of two connected components and $\overline{S}$ is connected. On the other hand if $(a, \overline{a}) = (8, 4)$, then S is connected and $\overline{S}$ is composed of two connected components. The images S^h and S^m are shown in Fig. 4(b) and (c) respectively. We note that in S^h we retrieve the connectivity that corresponds to $(8, 4)$, and in S^m we retrieve the connectivity that corresponds to $(4, 8)$. For a more complete discussion about a model of the digital topology notions in the framework of orders, see [14].

Note also that S^h is always an α-open set (*i.e.*, a β-closed set), and that S^m is always an α-closed set.

4 Distance and Influence Zones

Let (X, α, β) be a CF-order, let x and y be two points of X. Let $\pi = (x_0, \ldots, x_k)$, with $x = x_0, y = x_k$, be a path from x to y. The length of π is denoted by $l(\pi)$ and is equal to k. We denote by $\Pi_{x,y}$ the set of all the paths from x to y, and we define $d(x, y) = \min\{l(\pi), \pi \in \Pi_{x,y}\}$, the *distance between x and y*. A path σ from x to y such that $l(\sigma) = d(x, y)$ is a *shortest path from x to y*.

It may be easily checked that d is a (discrete) distance on X, that is, a map from $X \times X$ to $\mathcal{N}$ that verifies the properties of symmetry, separation ($d(x, y) = 0$ implies $x = y$) and triangular inequality ($d(x, y) \leq d(x, z) + d(z, y)$).

If Y is a subset of X, we define $d(x, Y) = \min\{d(x, y), y \in Y\}$, and call this value the *distance between x and Y*. A path σ from x to a point $y \in Y$ such that $l(\sigma) = d(x, Y)$ is called a *shortest path from x to Y*.

Let R be a subset of X, and let $\{R_1, \ldots, R_m\}$ be the connected components of R. For each R_i, $i = 1, \ldots, m$, we define the *influence zone V_i associated to R_i*: $V_i = \{x \in X / \forall j \in [1, \ldots, m], j \neq i, d(x, R_i) < d(x, R_j)\}$.

Ideally, the influence zones should possess the following, desirable properties: first, they should be connected sets, and second, they should be mutually disconnected, i.e. any union of two different influence zones should be a non-connected set. If we consider the discrete plane $\mathcal{Z}^2$, and the classical 4- or 8-distance, the property of mutual disconnection is not verified (see Fig. 2).

Here, we prove that the influence zones associated to a family of α-closed or β-closed subsets of a CF-order possess these two fundamental properties. We have to prove first these two intermediate lemmas:

Lemma 1 *Let (X, α, β) be a CF-order, and let $\pi = (x_0, \ldots, x_k)$ be a path. If π is a shortest path from x_0 to x_k, then $\forall i = 0, \ldots, k-2$ we have:*
$x_{i+1} \in \alpha^\square(x_i) \Rightarrow x_{i+2} \in \beta^\square(x_{i+1})$, *and* $x_{i+1} \in \beta^\square(x_i) \Rightarrow x_{i+2} \in \alpha^\square(x_{i+1})$.

Proof: clearly, $\forall i = 0, \ldots, k-1;\ x_i \neq x_{i+1}$. Suppose that $x_{i+1} \in \alpha^\square(x_i)$ and $x_{i+2} \notin \beta^\square(x_{i+1})$. We know that π is a path, hence we have $x_{i+2} \in \alpha^\square(x_{i+1})$. Thus, $x_{i+2} \in \alpha^\square(x_i)$, and we see that the path $(x_0, \ldots, x_i, x_{i+2}, \ldots, x_k)$ is shorter than π, a contradiction. □

Lemma 2 *Let (X, α, β) be a CF-order, let R be an α-closed subset of X, let x be a point of $X \setminus R$. Then, there exists a shortest path π from x to R such that the last point y of π is an α-terminal.*

Proof: let π be a shortest path from x to R, let w, y be the two last points of π ($y \in R$). We have $y \in \alpha(w)$, because if we suppose $w \in \alpha(y)$, then w belongs to R (R is α-closed), which contradicts the hypothesis "π is a shortest path from x to R".

Suppose that y is not an α-terminal, i.e. $\alpha(y) \neq \emptyset$. As R is α-closed, we have $\alpha(y) \subseteq R$ and $\alpha(y)$ contains α-terminals, let y' be one of them. We have $y \in \alpha(w)$ and $y' \in \alpha(y)$, thus $y' \in \alpha(w)$. Let us consider the path π' identical to π except that y is replaced by y': it is also a shortest path from x to R, and its last point is an α-terminal. □

We are now ready to prove the aforementioned properties:

Property 3 *Let (X, α, β) be a CF-order, let R be a subset of X, let $R_1, \ldots, R_m$ be the connected components of R, and let $V_1, \ldots, V_m$ be the influence zones associated to $R_1, \ldots, R_m$, respectively. For each $i = 1, \ldots, m$, the set V_i is connected.*

Proof: we shall prove that for any x in V_i, a shortest path between x and R_i is entirely included in V_i. The property follows immediately from this result and from the connectedness of R_i.

Let $x \in V_i$. Following the definition of influence zones, there exists a shortest path π from x to R_i such that $l(\pi) = d(x, R_i)$ and $l(\pi) < d(x, R_j), \forall j \in [1, \ldots, m], j \neq i$. Let y be any point in π, and suppose that $y \notin V_i$, which means that there is a $k \in [1, \ldots, m]$, $k \neq i$, such that $d(y, R_k) \leq d(y, R_i)$.

As π is a shortest path from x to R_i, the subpath π' of π from y to R_i is a shortest path from y to R_i, hence $l(\pi') = d(y, R_i)$. Also, the subpath π'' of π from x to y is a shortest path from x to y, hence $l(\pi'') = d(x, y)$.

Let σ be a shortest path from y to R_k, we have $l(\sigma) = d(y, R_k)$ and hence $l(\sigma) \leq l(\pi')$. Then using the triangular inequality: $d(x, R_k) \leq l(\pi'') + l(\sigma) \leq l(\pi'') + l(\pi')$, that is $d(x, R_k) \leq d(x, R_i)$, a contradiction. □

Property 4 *Let (X, α, β) be a CF-order, let R be an α-closed subset of X, let $R_1, \ldots, R_m$ be the connected components of R, and let $V_1, \ldots, V_m$ be the influence zones associated to $R_1, \ldots, R_m$, respectively. Then, the V_i's are mutually disconnected, i.e. $\forall i, j \in [1, \ldots, m], i \neq j, V_i \cup V_j$ is not connected.*

Proof: suppose that there is an x in V_i and a y in V_j such that $x \in \theta(y)$. We shall: **a.** prove that $d(x, R_i) = d(y, R_j)$, and **b.** raise a contradiction using lemmas 1 and 2.

a. Suppose that $d(x, R_i) < d(y, R_j)$. Let π be a shortest path from R_i to x, and let σ be a shortest path from R_j to y, we have: $l(\pi) = d(x, R_i), l(\sigma) = d(y, R_j)$. The sequence πy is a path from R_i to y, and we have $l(\pi y) = l(\pi) + 1$, hence $l(\pi y) \leq l(\sigma)$. But $d(y, R_i) \leq l(\pi y)$ (by definition of d) , hence $d(y, R_i) \leq d(y, R_j)$, a contradiction with the fact that $y \in V_j$. A symmetric argument shows that $d(x, R_i) > d(y, R_j)$ is also false, hence $d(x, R_i) = d(y, R_j)$.

b. Let π be a shortest path from R_i to x, begining by an α-terminal (see lemma 2) and ending by the sequence (x', x). Let σ be a shortest path from R_j to y, begining by an α-terminal and ending by the sequence (y', y). These two paths having the same length (see **a.**), and beginning by α-terminals, we deduce from lemma 1: $x \in \alpha(x') \Leftrightarrow y \in \alpha(y')$. Our hypothesis $x \in \theta(y)$ implies either $x \in \alpha(y)$ or $y \in \alpha(x)$. For example, if $x \in \alpha(y)$ and $y \in \alpha(y')$, then we have $x \in \alpha(y')$, hence $l(\sigma x) = l(\sigma) = l(\pi)$. This implies that $d(x, R_j) \leq d(x, R_i)$, a contradiction with the fact that $x \in V_i$. There are three other possibilities, that lead to a contradiction in similar ways. □

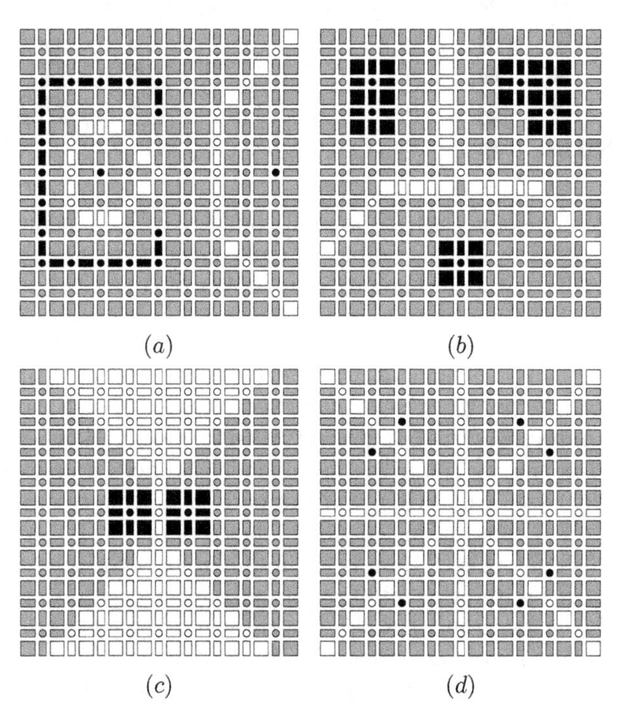

Fig. 5. Examples of families of α-closed (a,d) or β-closed (b,c) sets (in black), and their influence zones (in gray)

5 Tesselation by Connection

In Fig. 5, we show some examples of families of α-closed (Fig. 5(a), in black) and β-closed (Fig. 5(b), in black) subsets of H^2, and their influence zones (in gray). In these cases, the complementary of the influence zones is "thin"; nevertheless, we cannot guarantee such a property in the general case. Fig. 5(c,d) shows counter-examples. In many image analysis applications, we need "thin" frontiers between the influence zones, this motivates the introduction of the following notions. Intuitively, a point is *uniconnected* if its addition preserves the connected components, in other terms, it preserves a connection in the sense of [15].

Let (X, α, β) be a finite CF-order, let $R \subseteq X$, and let $x \in \overline{R}$. The point x is *uniconnected (for R)* if the number of connected components of R equals the number of connected components of $R \cup \{x\}$. The point x is *α-uniconnected (for R)* if x is uniconnected for R and if $\alpha(x) \cap R \neq \emptyset$.

We can easily see that a point $x \in \overline{R}$ is uniconnected if and only if $\theta(x)$ intersects exactly one connected component of R. A point $x \in \overline{R}$ is said to be *multiconnected (for R)* if $\theta(x)$ intersects at least two different connected components of R. It is *isolated (for R)* if $\theta(x) \cap R = \emptyset$.

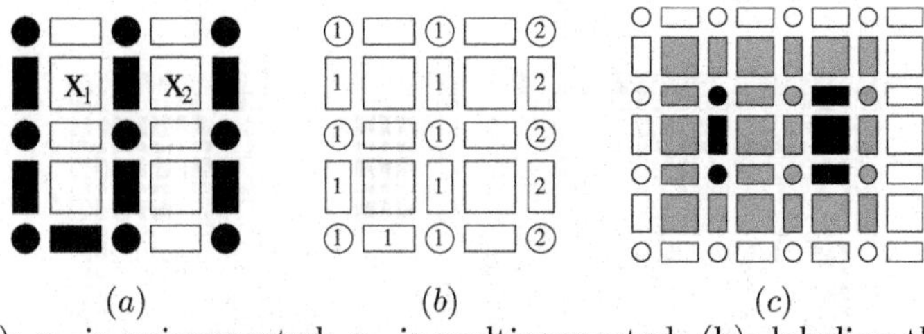

Fig. 6. (a): x_1 is uniconnected, x_2 is multiconnected; (b): labeling the components makes the local checking possible; (c): uniconnected points (in gray) cannot be added in parallel without changing the number of connected components

In Fig. 6, we can see that the information contained in $\theta(x)$ is not sufficient to check whether a point x is uniconnected or multiconnected. On the other hand, if we assume that the points of each connected component of R are labeled with an index which represents this component, then we can check whether a point x is uniconnected or not by counting the number of different indexes carried by the points in $\theta(x)$ (see Fig. 6(b)). Furthermore, a uniconnected point x is α-uniconnected if $\alpha(x)$ contains at least one labeled point.

Let $R \subseteq X$, we say that $T \subseteq X$ is a *thickening by connection* of R if T may be derived from R by iterative addition of uniconnected points. We say that T is a *tesselation by connection* of R if T is a thickening by connection of R and if all the points of $X \setminus T$ are not uniconnected.

In general, there are several tesselations by connection for a set R. This is due to the iterative nature of the definition: depending on the order of selection of the uniconnected points, one can get different results. Nevertheless, in many applications we want to obtain a "well centered" result, that could be uniquely defined. This is why we introduce, by the way of a parallel algorithm, a particular tesselation by connection that possesses good geometrical properties.

We can see in Fig. 6(c) that, in general, uniconnected points cannot be added in parallel to a set R without changing the number of connected components of R. In fact, α-uniconnected points can indeed be added in parallel to R without changing the number of connected components of R. This can be proved by induction thanks to the following property.

Property 5 *Let (X, α, β) be a CF-order, let $R \subseteq X$, and let $x, y \in \overline{R}$ be two α-uniconnected points for R. Then, x is α-uniconnected for $R \cup \{y\}$.*

Proof: if $y \notin \theta(x)$, then the property is obvious. If $y \in \alpha(x)$ or $x \in \alpha(y)$, then the component of R which is α-adherent to y is clearly the same as the component of R which is α-adherent to x. Thus in both cases, adding the point y to R does not change the fact that x is α-uniconnected.□

The following algorithm computes a tesselation by connection of a closed subset of X.

Algorithm 1
Input data:
    (X, α, β), *a finite CF-order,*
    $R_1, \ldots, R_m$, *the m connected components of an α-closed subset R of X*
Output: a tesselation by connection T associated to R
Initialization:
    $T^0 := R$; $B^0 := \emptyset$; $n := 1$; *label the points of each R_i with the index i*
Repeat until stability:
    Compute the set S^n of the points that are uniconnected for T^{n-1},
        and the set U^n of the points that are multiconnected for T^{n-1}
        (this can easily be done using the indexes)
    Label the points x of S^n with the index of the component found in $\theta(x)$
    Label the points y of U^n with the index 0
    $T^n := T^{n-1} \cup S^n$; $B^n := B^{n-1} \cup U^n$; $n := n + 1$
End Repeat
$T := T^n$

It can be easily seen that for any even (resp. odd) value of n, the set T^n is α-closed (resp. β-closed) in $|X \setminus B^n|$. From this, it follows that S^{n+1} is only composed of α-uniconnected (resp. β-uniconnected) points if n is even (resp. odd). Thus (Prop. 5), the result T of this algorithm is a thickening by connection of the original α-closed set R. In addition, at the end of this algorithm no uniconnected point remains. Thus, the result T is a tesselation by connection of the original set R.

The following property makes a link between the tesselation by connection computed by algorithm 1 an the influence zones, ensuring that this tesselation by connection is "well centered".

The following lemma and the following property derive from the fact that the labeling process is guided by a breadth-first strategy.

Lemma 6 *Every point x that receives an index during the step n of the algorithm, is such that $d(x, R) = n$.*
Every point x that receives an index $i > 0$ during the step n of the algorithm, is such that $d(x, R_i) = n$.

Property 7 *Let (X, α, β) be a CF-order, let $R_1, \ldots, R_m$ be the m connected components of an α-closed subset R of X, let $V_1, \ldots, V_m$ be the influence zones associated to $R_1, \ldots, R_m$, respectively, and let $T_1, \ldots, T_m$ be the connected components of the tesselation by connection of R computed by algorithm 1. Then we have : $\forall i = 1, \ldots, m$, $V_i \subseteq T_i$.*

This property establishes that the influence zones are included in the components of the tesselation by connection computed by algorithm 1. The converse is true only in some cases, like in Fig. 5(a,b,d); but it is not true in general, as shown by the counter-examples of Fig. 7.

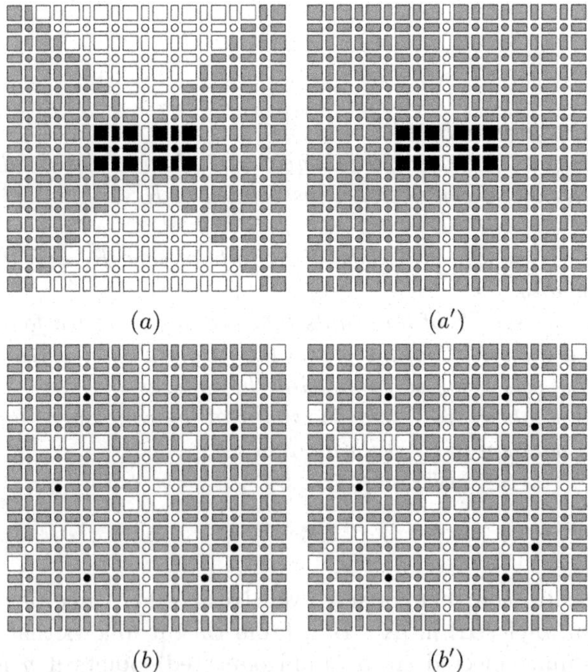

Fig. 7. (a,b): some objects (in black) and their influence zones (in gray). (a',b'): the same objects (in black) and their tesselation by connection (in gray)

6 Tesselation by Connection for Weighted Orders

In this section, we extend the notions of uniconnected point and tesselation by connection to weighted orders. A tesselation by connection for weighted orders may be considered as a generalization of the watershed transformation[1]: the result of the proposed transformation is a function, whereas the result of a watershed transformation is a set. We recover a set which corresponds to the complementary of the watersheds, by extracting the regional maxima of the tesselation by connection.

Let $|X| = (X, \alpha, \beta)$ be a finite CF-order, and let W be a mapping from X to $\mathcal{Z}$. The couple $(|X|, W)$ is called a *weighted order on* $|X|$. For applications to digital image processing, $W(x)$ typically represents the graylevel of the point x. We denote by W_k, and call the *cross-section of* W *at level* k, the set $W_k = \{x \in X, W(x) \geq k\}$ with $k \in \mathcal{Z}$. The weighted order $(|X|, W)$ is *α-closed* if, $\forall k \in \mathcal{Z}$, the order $|W_k|$ is α-closed.

Let $(|X|, W)$ be a weighted order, and let $x \in X$. The point x is *uniconnected for* W if x is uniconnected for W_{k+1}, with $k = W(x)$.

Let $(|X|, W)$ and $(|X|, M)$ be weighted orders, we say that M is a *thickening by connection* of W if M may be derived from W by iteratively selecting a uniconnected point x and raising its value $W(x)$ by one. We say that M is a *tesselation by connection* of W if M is a thickening by connection of W and if all the points of X are not uniconnected for M.

Algorithm 2
Input data: $(|X|, W)$, *a finite* α*-closed weighted order,*
Output: a tesselation by connection M *associated to* W.
Initialization: Let $L = (k_0, \ldots, k_s)$ *the list of values taken by* $W(x)$,
sorted in increasing order.
For all i *from* 1 *to* s
Compute the tesselation by connection Z *of* W_{k_i}
in the order induced by $W_{k_{i-1}}$, *using algorithm 1*
For all $z \in Z \setminus W_{k_{i-1}}$ *do* $W[z] := k_i$; *End For*
End For
$M := W$

The following transformations (introduced in [16]) may be used to construct a closed weighted order from a grayscale image.

Let $\mathcal{F}$ be the set of functions from $\mathcal{Z}^n$ to $\mathcal{N}$. Let $\mathcal{W}$ be the set of functions from $\mathcal{H}^n$ to $\mathcal{N}$. We define the transformations Ψ^h and Ψ^m from $\mathcal{F}$ to $\mathcal{W}$ which associate to each F in $\mathcal{F}$ the functions W^h and W^m respectively, defined by:

$$\forall x \in \mathcal{H}^n, W^m(x) = \min\{F(y), y \in \mathcal{Z}^n, \{y\} \in \alpha(x)\}$$
$$\forall x \in \mathcal{H}^n, W^h(x) = \max\{F(y), y \in \mathcal{Z}^n, \{y\} \in \alpha(x)\}$$

It can be easily seen that $(|X|, W^m)$ is an α-closed weighted order, and that $(|X|, W^h)$ is a β-closed weighted order. In Fig. 8, we show the weighted order obtained by applying Ψ^m to a "real" image, and the tesselation by connection of this weighted order computed by algorithm 2.

References

1. S. Beucher, Ch. Lantuejoul, "Use of Watersheds in Contour Detection", *Proc. Int. Workshop on Image Processing, Real-Time Edge and Motion Detection/Estimation*, Rennes, France, 1979. 15, 24

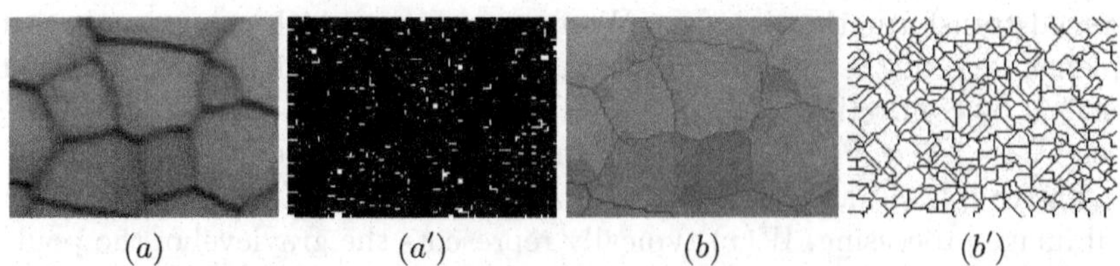

Fig. 8. (a): a weighted order obtained thanks to the transformation Ψ^m; (b): tesselation by connection of (a); (a',b'): regional maxima of (a,b) respectively

2. C. Lohou, G. Bertrand, "Poset approach to 3D parallel thinning", *SPIE Vision Geometry VIII.*, Vol. 3811, pp. 45-56, 1999. 15
3. S. Beucher, "Segmentation d'images et morphologie mathématique", *PhD Thesis*, École des Mines, Paris, 1990. 16
4. L. Vincent, P. Soille, "Watersheds in Digital Spaces: An Efficient Algorithm Based on Immersion Simulations", *IEEE Trans. on Pattern Analysis and Machine Intelligence*, Vol. 13, No. 6, pp. 583-598, 1991. 16
5. F. Prêteux, "Watershed and Skeleton by Influence Zones: A Distance-Based Approach", *Journal of Mathematical Imaging and Vision*, No. 1, pp. 239-255, 1992. 16
6. S. Beucher and F. Meyer, "The morphological approach to segmentation: the watershed transformation", *Mathematical Morphology in Image Processing*, Chap. 12, pp. 433-481, Dougherty Ed., Marcel Dekker, 1993. 16
7. F. Meyer, "Topographic distance and watershed lines", *Signal Processing*, No. 38, pp. 113-125, 1994. 16
8. T. Y. Kong and A. Rosenfeld, "Digital topology: introduction and survey", *Comp. Vision, Graphics and Image Proc.*, 48, pp. 357-393, 1989. 16, 19
9. M. Couprie and G. Bertrand: "Topological Grayscale Watershed Transformation" *SPIE Vision Geometry V Proceedings*, Vol. 3168, pp. 136-146, 1997. 16
10. G. Bertrand, "New notions for discrete topology", *8th Conf. on Discrete Geom. for Comp. Imag.*, Vol. 1568, Lect. Notes in Comp. Science, Springer Verlag, pp. 218-228, 1999. 16, 17, 18
11. P. Alexandroff, "Diskrete Räume", *Mat. Sbornik*, 2, pp. 501-518, 1937. 16, 17
12. E. Khalimsky, R. Kopperman, P. R. Meyer, "Computer Graphics and Connected Topologies on Finite Ordered Sets", *Topology and its Applications*, 36, pp. 1-17, 1990. 18
13. J. Serra, *Image Analysis and Mathematical Morphology*, Academic Press, 1982. 18
14. G. Bertrand, M. Couprie, "A model for digital topology", *8th Conf. on Discrete Geom. for Comp. Imag.*, Vol. 1568, Lect. Notes in Comp. Science, Springer Verlag, pp. 229-241, 1999. 19
15. G. Matheron, J. Serra, *Strong filters and connectivity*, in *Image Analysis and Mathematical Morphology, Vol. II: Theoretical Advances*, Chap. 7, pp. 141-157, Serra J. ed., Academic Press, 1988. 22
16. V. A. Kovalevsky, "Finite Topology as Applied to Image Analysis", *Computer Vision, Graphics, and Image Processing*, 46, pp. 141-161, 1989. 25

A Concise Characterization of 3D Simple Points

Sébastien Fourey and Rémy Malgouyres

GREYC, ISMRA, 6 bd Maréchal Juin 14000 Caen - France
{Fourey,Malgouyres}@greyc.ismra.fr

Abstract. We recall the definition of simple points which uses the digital fundamental group introduced by T. Y. Kong in [Kon89]. Then, we prove that a not less restrictive definition can be given. Indeed, we prove that there is no need of considering the fundamental group of the complement of an object in order to characterize its simple points. In order to prove this result, we do not use the fact that "the number of holes of X is equal to the number of holes in $\overline{X}$" which is not sufficient for our purpose but we use the linking number defined in [FM00]. In so doing, we formalize the proofs of several results stated without proof in the literature (Bertrand, Kong, Morgenthaler).

Keywords: digital topology, linking number, simple points

1 Introduction

The definition of a simple point is the key notion in the context of thinning algorithms. Indeed, this definition leads to the most commonly admitted definition of the fact that a given thinning algorithm does preserve the topology of a digital image. Usually, one says that an image $\mathcal{I}_1$ is *topologically equivalent* to an image $\mathcal{I}_2$ if $\mathcal{I}_1$ can be obtained from $\mathcal{I}_2$ by sequential addition or deletion of simple points. In this case, a simple point is defined as a point the deletion of which does not change the topology of the image. The problem with topology preservation in 3D is that taking care not to change the number of connected components in the image as well as in its bakground is not sufficient as in 2D. In 3D, one must also take care not to change the number and the location of the *tunnels* as donuts have. Now, different characterizations have been proposed by several authors which all lead to local characterizations which are equivalent. A first set of characterizations mainly use the Euler characteristic in order to count the number of tunnels, but even if this kind of characterization leads to a *good* local characterization it is limited by the fact that no localization of the tunnels is provided by the use the Euler characteristic (see Figure 3). Another definition has been proposed by Kong & al. in [KR89] which is based on remarks made by Morgenthaler in [Mor81], with a new formalism which inlvolves the digital fundamental group introduced by Kong in [Kon89]. In this latter definition, topology preservation is expressed in term of existence of a canonical isomorphism between the fundamental group of the object and the fundamental group of the object without the considered point to be removed, and a similar

G. Borgefors, I. Nyström, and G. Sanniti di Baja (Eds.): DGCI 2000, LNCS 1953, pp. 27–36, 2000.
© Springer-Verlag Berlin Heidelberg 2000

isomorphism must exist in the background of the image (see Definition 3). In this paper, we prove that this second condition is in fact implied by the first one. In other words, we show that preserving the tunnels of the objects will imply the preservation of tunnels in the background. In order to prove that a such more concise characterization can be given, we use the linking number between paths of voxels as defined in [FM00] which provides an efficient way to prove that a given path cannot be homotopic to a degenerated path. Furtermore, the proof of Theorem 2 given in Section 3 uses propositions an lemmas, some of which are the direct answers to some open questions left by Morgenthaler in [Mor81] such as: do any two paths which can be be continuously deformed one into each other in an object X keep this property after removal of a simple point of X ?

2 Definitions

2.1 Digital Image, Paths, Connectivity

In this paper, we consider objects as subsets of the 3 dimensional space $\mathbb{Z}^3$. Elements of $\mathbb{Z}^3$ are called *voxels* (short for "volume elements"). The set of voxels which do not belong to an object $O \subset \mathbb{Z}^3$ constitute the complement of the object and is denoted by $\overline{O}$. Any voxel is identified with a unit cube centered at a point with integer coordinates $v = (i, j, k) \in \mathbb{Z}^3$. Now, we can define some binary symmetric and anti-reflexive relations between voxels. Two voxels are said $6-$*adjacent* if they share a face, $18-$*adjacent* if they share an edge and $26-$*adjacent* if they share a vertex. By transitive closure of these adjacency relations, we can define another one: connectivity between voxels. We first define an *n-path* π with a length l from a voxel a to a voxel b in $O \subset \mathbb{Z}^3$ as a sequence of voxels $(y_i)_{i=0\dots l}$ such that for $0 \leq i < l$ the voxel y_i is n-adjacent or equal to y_{i+1}, with $y_0 = a$ and $y_l = b$. The path π is a *closed path* if $y_0 = y_l$ and is called a *simple path* if $y_i \neq y_j$ when $i \neq j$ (except for y_0 and y_l if the path is closed). The voxels y_0 and y_l are called the *extremities* of π even in the case when the path is closed, and we denote by π^* the set of voxels of π. A closed path (x, x) with a length 1 for $x \in \mathbb{Z}^3$ is called a *trivial path.* If x is a voxel of $\mathbb{Z}^3$ and $n \in \{6, 18, 26\}$ then we denote by $N_n(x)$ the set of voxels of $\mathbb{Z}^3$ which are $n-$adjacent to x. We call $N_n(x)$ the $n-$neighborhood of x. A subset C of $\mathbb{Z}^3$ is called a *simple closed $n-$curve* if it is $n-$connected and any voxel of C is $n-$adjacent to exactly two other voxels of C; then, if c is a simple closed $n-$path such that $c^* = C$, c is called a *parameterized simple closed $n-$curve.*

Given a path $\pi = (y_k)_{k=0,\dots,l}$, we denote by π^{-1} the sequence $(y'_k)_{k=0,\dots,l}$ such that $y_k = y'_{l-k}$ for $k \in \{0, \dots, l\}$.

Now we can define the connectivity relation: two voxels a and b are called *n-connected* in an object O if there exists an n-path π from a to b in O. This is an equivalence relation between voxels of O, and the *$n-$connected components* of an object O are equivalence classes of voxels according to this relation.

In order to avoid topological paradoxes, we always study the topology of an object using an $n-$adjacency for the object and a complementary adjacency $\overline{n}$ for its complement. We sum up this by the use of a couple $(n, \overline{n})$ in $\{(6, 26), (6+, 18), (18, 6+), (26, 6)\}$. The notation 6+ is used in order to distinguish the relation of 6−connectivity associated to the 26−connectivity from the (6+)−connectivity associated to the 18−connectivity.

If $\pi = (y_i)_{i=0,\dots,p}$ and $\pi' = (y'_k)_{k=0,\dots,p'}$ are two n−paths such that $y_p = y'_0$ then we denote by $\pi.\pi'$ the path $(y_0, \dots, y_{p-1}, y'_0, \dots, y'_{p'})$ which is the concatenation of the two paths π and π'.

2.2 Geodesic Neighborhoods and Topological Numbers

The geodesic neighborhoods have been introduced by Bertrand ([Ber94]) in order to locally characterize in an efficient way the simple points of an object.

Definition 1 (geodesic neighborhood [Ber94]). *Let $x \in X$, we define the* geodesic neighborhood of x in X *denoted by $G_n(x, X)$ as follows:*
- $G_6(x, X) = (N_6(x) \cap X) \cup \{y \in N_{18}(x) |$ *y is 6−adjacent to a voxel of* $N_6(x) \cap X\}$.
- $G_{26}(x, X) = N_{26}(x) \cap X$.

Definition 2 (topological numbers [Ber94]). *Let $X \subset \mathbb{Z}^3$ and $x \in \mathbb{Z}^3$. We define the* topological number *associated to x and X, and we denote by $T_n(x, X)$ for $(n, \overline{n}) \in \{(6, 26), (26, 6)\}$, the number of connected components of $G_n(x, X)$.*

2.3 Digital Fundamental Group

In this section, we define the digital fundamental group of a subset X of $\mathbb{Z}^3$ following the definition of Kong in [Kon89] and [Kon88].

First, we need to introduce the *n−homotopy relation* between n−paths in X. Intuitively, a path c is homotopic to a path c' if c can be "continuously deformed" into c'. Let us consider $X \subset \mathbb{Z}^3$. First, we introduce the notion of an *elementary n−deformation.* Two closed n−paths c and c' in X having the same extremities are *the same up to an elementary n−deformation (with fixed extremities) in X*, and we denote $c \sim_n c'$, if they are of the form $c = \pi_1.\gamma.\pi_2$ and $c' = \pi_1.\gamma'.\pi_2$, the n−paths γ and γ' having the same extremities and being both included in a $2 \times 2 \times 2$ cube if $(n, \overline{n}) = (26, 6)$, in a 2×2 square if $(n, \overline{n}) = (6, 26)$. Now, the two n−paths c and c' are said to be *n−homotopic (with fixed extremities) in X* if there exists a finite sequence of n−paths $c = c_0, \dots, c_m = c'$ such that for $i = 0, \dots, m-1$ the n−paths c_i and c_{i+1} are the same up to an elementary n−deformation (with fixed extremities). In this case, we denote $c \simeq_n c'$. A closed n−path $c = (x_0, \dots, x_q = x_0)$ in X is said to be n−reducible in X if $c \simeq_n (x_0, x_0)$ in X.

Let $B \in X$ be a fixed voxel of X called the *base surfel.* We denote by $A^n_B(X)$ the set of all closed n−paths $c = (x_0, \dots, x_p)$ which are included in X and such that $x_0 = x_p = B$. The n−homotopy relation is an equivalence relation on $A^n_B(X)$, and we denote by $\Pi^n_1(X, B)$ the set of the equivalence classes of this equivalence relation. If $c \in A^n_B(X)$, we denote by $[c]_{\Pi^n_1(X,B)}$ the equivalence class of c under this relation.

The concatenation of closed $n-$paths is compatible with the $n-$homotopy relation, hence it defines an operation on $\Pi_1^n(X,B)$, which to the class of c_1 and the class of c_2 associates the class of $c_1.c_2$. This operation provides $\Pi_1^n(X,B)$ with a group structure. We call this group the *$n-$fundamental group of X*. The $n-$fundamental group defined using a voxel B' as base surfel is isomorphic to the $n-$fundamental group defined using a voxel B as base surfel if X is $n-$connected.

Now, we consider $Y \subset X \subset \mathbb{Z}^3$ and $B \in Y$ a base voxel. A closed $n-$path in Y is a particular case of a closed $n-$path in X. Furthermore, if two closed $n-$paths of Y are $n-$homotopic (with fixed extremities) in Y, they are $n-$homotopic (with fixed extremities) in X. These two properties enable us to define a canonical morphism $i_* : \Pi_1^n(Y,B) \longrightarrow \Pi_1^n(X,B)$, which we call the morphism induced by the inclusion map $i : Y \longrightarrow X$. To the class of a closed $n-$path $\alpha \in A_B^n(Y)$ in $\Pi_1^n(Y,B)$ the morphism i_* associates the class of the same $n-$path in $\Pi_1^n(X,B)$.

2.4 The Digital Linking Number

The digital linking number, denoted by $L_{\pi,c}$, has been defined in [FM00] for a couple (π, c) of closed paths of $\mathbb{Z}^3$ which do not intersect each other. It is the digital analogue of the linking number defined in knot theory (see [Rol]) and it is immediately computable (see [FM00]) for any couple (π, c) such that π is an $n-$path and c is an $\overline{n}-$path with $(n, \overline{n}) \in \{(6, 26), (26, 6), (6+, 18), (18, 6+)\}$. This number counts the number of times two digital closed paths are interlaced one with each other, as illustrated in Figure 1. Since the precise definition of the linking number is too long to be recalled here, we simply give the idea of the way it can be computed in Figure 2. In this figure, we have depicted several configurations which illustrate the contributions of superposed voxels, one in each paths, in a 2D projection of the two paths. The digital linking number is nothing but the sum of such contributions for all couples of overlapping voxels.

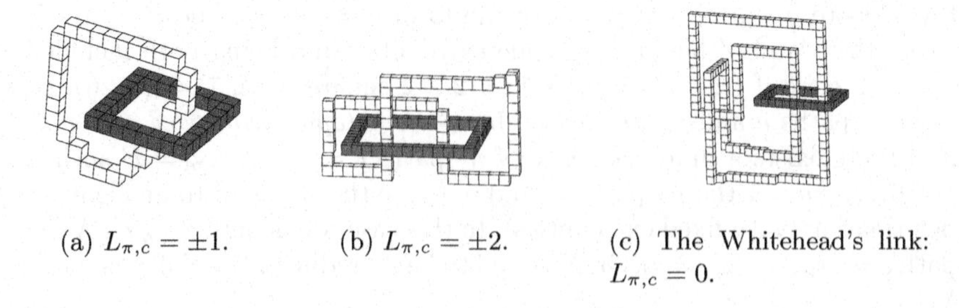

(a) $L_{\pi,c} = \pm 1$. (b) $L_{\pi,c} = \pm 2$. (c) The Whitehead's link: $L_{\pi,c} = 0$.

Fig. 1. Three kinds of links: a $6-$path π in black and a $18-$path c in white

The two following theorems have been proved in [FM00] and allows to say that the linking number is a new topological invariant in the field of digital topology.

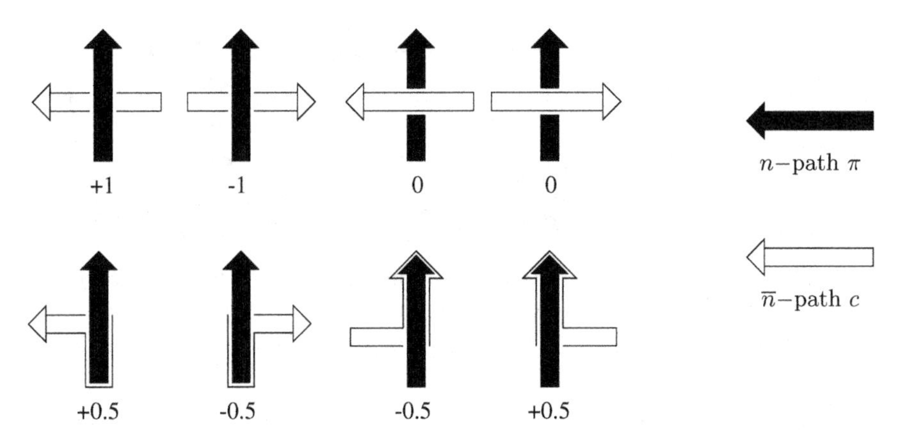

Fig. 2. Contributions associated with voxels where the two paths of a link overlap in a 2D projection of a link

Theorem 1. *Let π and π' be two closed $n-$path and c be a closed $\overline{n}-$path of $\mathbb{Z}^3$ such that $\pi^* \cap c^* = \emptyset$ and $\pi'^* \cap c^* = \emptyset$. If $\pi \simeq_n \pi'$ in $\mathbb{Z}^3 \setminus c^*$ then $L_{\pi,c} = L_{\pi',c}$ and $L_{c,\pi} = L_{c,\pi'}$.*

Remark 1. If c is a trivial path, then $L_{c,\pi} = 0$ for any closed $n-$path such that $c^* \cap \pi^* = \emptyset$. It follows that if a closed $n-$path c in $X \subset \mathbb{Z}^3$ is $n-$reducible in X, then $L_{c,\pi} = 0$ for all closed $\overline{n}-$path π in $\overline{c^*}$.

2.5 Characterization of Simple Points

A *simple point* for $X \subset \mathbb{Z}^3$ is a point the deletion of which does not change the topology of X. Now, topology preservation in 3D is not as simple to express as in the 2D case because of the existence of *tunnels*. A few authors have used two main tools to study topology preservation: the Euler characteristic which only allows to count the number of tunnels of an object ([TF82]), and the digital fundamental group ([Kon89]) which allows to "localize" the tunnels. Indeed, as depicted in Figure 3, counting the number of tunnels is not sufficient to characterize the fact that the topology is preserved. In this paper, we are interested by a definition of simple points which uses the digital fundamental group and which avoids the problem previously mentioned. The following definition appears as the most convenient for the property "the deletion of x preserves topology of X". It comes from the criterion given in [Kon89] for a thinning algorithm to preserve topology.

Definition 3. *Let $X \subset \mathbb{Z}^3$ and $x \in X$. The point x is said to be $n-$simple if:*

i) *X and $X \setminus \{x\}$ have the same number of $n-$connected components.*
ii) *$\overline{X}$ and $\overline{X} \cup \{x\}$ have the same number of $\overline{n}-$connected components.*
iii) *For each voxel B in $X \setminus \{x\}$, the group morphism $i_* : \Pi_1^n(X \setminus \{x\}, B) \longrightarrow \Pi_1^n(X, B)$ induced by the inclusion map $i : X \setminus \{x\} \longrightarrow X$ is an isomorphism.*

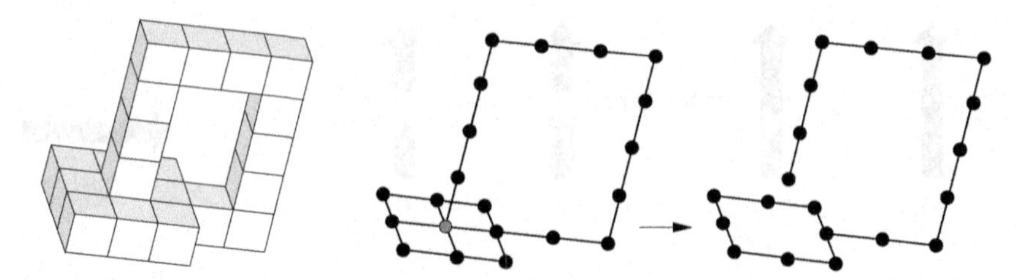

Fig. 3. The gray point can be removed without changing the Euler characteristic of the object which is equal to zero. However, this point is obviously not $n-$simple (if $(n, \overline{n}) \in \{(6, 18), (6, 26)\}$)

iv) *For each voxel* B' *in* $\overline{X}$, *the group morphism* $i'_* : \Pi_1^{\overline{n}}(X, B')\overline{X} \longrightarrow \Pi_1^{\overline{n}}(\overline{X} \cup \{x\}, B')$ *induced by the inclusion map* $i' : \overline{X} \longrightarrow \overline{X} \cup \{x\}$ *is an isomorphism.*

Bertrand, in [BM94], gave a local characterization for 3D simple points in term of number of connected components in geodesic neighborhoods. However, the definition of simple point given in [BM94] differs from the definition used here since it does not consider any morphism between digital fundamental groups but only require the preservation of cavities and "tunnels". One purpose of this paper is to well formalize the fact that the local characterization given by Bertrand is a consequence of the three first conditions of Definition 3 and conversly that the four conditions are themselves consequences of the local characterization using the topological numbers. We recall here the characterization given by Bertrand and Malandain in [BM94]. Note that the definition of *simple points* used in this proposition slightly differs from Definition 3.

Proposition 1 ([BM94]). *Let* $x \in X$ *and* $(n, \overline{n}) \in \{(6, 26), (26, 6)\}$. *The point* x *is a* $n-$*simple point if and only if* $T_n(x, X) = 1$ *and* $T_{\overline{n}}(x, \overline{X}) = 1$.

3 A New Characterization of 3D Simple Points

In the sequel of this paper $(n, \overline{n}) \in \{(6, 26), (26, 6)\}$.

In this section, we state the main result of this paper which is that a not less restrictive criterion for topology preservation is obtained using the only conditions *i*), *ii*) and *iii*) of Definition 3. In other words, we prove the following theorem:

Theorem 2. *Let* $X \subset \mathbb{Z}^3$ *and* $x \in X$. *The point* x *is* $n-$*simple for* X *if and only if:*

i) X *and* $X \setminus \{x\}$ *have the same number of connected components.*
ii) $\overline{X}$ *and* $\overline{X} \cup \{x\}$ *have the same number of connected components.*
iii) *For each voxel* B *in* $X \setminus \{x\}$, *the group morphism* $i_* : \Pi_1^n(X \setminus \{x\}, B) \longrightarrow \Pi_1^n(X, B)$ *induced by the inclusion map* $i : X \setminus \{x\} \longrightarrow X$ *is an isomorphism.*

In order to prove this theorem, we first prove (Subsection 3.1) that a voxel which satisfies the three conditions of Theorem 2 also satisfies the the local characterization given by Proposition 1 and then, we show (Subsection 3.2) that this characterization itself implies that the four conditions of Definition 3 are satisfied.

In the sequel of the paper, we may suppose without loss of generality that X is an $n-$connected subset of $\mathbb{Z}^3$; and that x and B are two distinct voxels of X whereas B' is a voxel of $\overline{X}$. Furthermore, $i_* : \Pi_1^n(X\backslash\{x\}, B) \longrightarrow \Pi_1^n(X, B)$ is the group morphism induced by the inclusion of $X\backslash\{x\}$ in X; and $i'_* : \Pi_1^{\overline{n}}(\overline{X}, B') \longrightarrow \Pi_1^{\overline{n}}(\overline{X}\cup\{x\}, B')$ is the group morphism induced by the inclusion of $\overline{X}$ in $\overline{X}\cup\{x\}$.

Remark 2. In the sequel, we shall admit the basic property that, if $Y \subset X$ are $n-$connected subsets of $\mathbb{Z}^3$, the group morphism from $\Pi_1^n(Y, B)$ to $\Pi_1^n(X, B)$ induced by the inclusion of Y in X for a base surfel $B \in Y$ is an isomorphism if and only if the group morphism between $\Pi_1^n(Y, B')$ and $\Pi_1^n(X, B')$ is an isomorphism for any base surfel $B' \in Y$.

3.1 First Step of the Proof of Theorem 2

The purpose of this section is to prove the following proposition.

Proposition 2. *If the conditions i), ii) and iii) of Definition 3 are satisfied, then $T_n(x, X) = 1$ and $T_{\overline{n}}(x, \overline{X}) = 1$.*

In order to prove this proposition, we introduce several other propositions and lemmas.

Proposition 3. *If $T_n(x, X) \geq 2$, then either an $n-$connected component of X is created by deletion of x, or the morphism i_* is not onto.*

Sketch of proof of Proposition 3: The proof of a similar proposition given in [BM94] may be adapted to our formalism. It involves the definition of a number $\nu(x, \alpha, C)$ which counts the number of time a given $n-$path α goes from a component C of $G_n(x, X)$ to x minus the number of times α goes from x to C (see Figure 4(a)). This number is shown to be invariant under $n-$homotopic deformations of the $n-$path α inside X. Then, it allows to prove that, when connected components of $G_n(x, X)$ are $n-$connected in $X \setminus \{x\}$, an homotopy class of paths distinct from the class of the trivial path, cannot be reached by the morphism i_* (class of the path α of Figure 4(a)). □

Here, we state the following new proposition.

Proposition 4. *If $T_n(x, X) = 1$ and $T_{\overline{n}}(x, \overline{X}) \geq 2$ then two $\overline{n}-$connected component of $\overline{X}$ are merged by deletion of x or the morphism i_* is not one to one.*

The main idea of this paper is to use the linking number in order to prove Proposition 4. Indeed, until this paper and the possible use of the linking number, one could prove that when $T_n(x, X) = 1$ and $T_{\overline{n}}(x, \overline{X}) \geq 2$ and no $\overline{n}-$connected

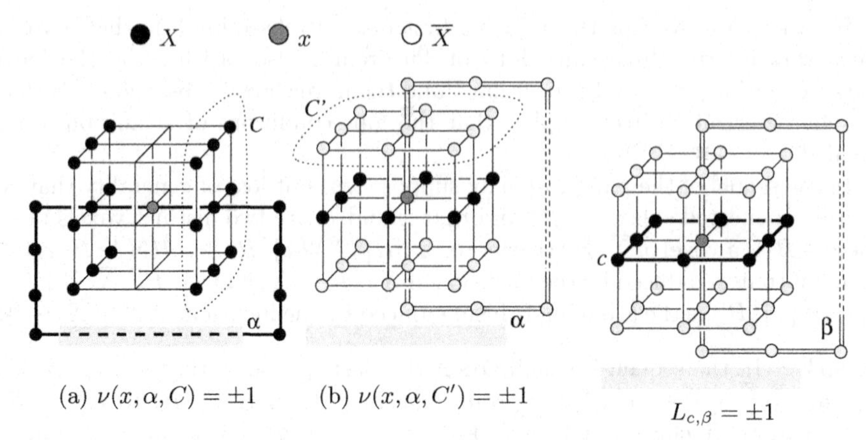

(a) $\nu(x, \alpha, C) = \pm 1$ (b) $\nu(x, \alpha, C') = \pm 1$

Fig. 4. Parts of the proofs of Propositions 3 and 5

$L_{c,\beta} = \pm 1$

Fig. 5. Idea of the proof of Proposition 4

component of $\overline{X}$ are merged by deletion of x, then, the morphism i_*' induced by the inclusion of $\overline{X}$ in $\overline{X} \cup \{x\}$ is not onto. In other words, "a hole is created in $\overline{X} \cup \{x\}$". Indeed, a similar proof to Proposition 3 would lead to the following proposition (see Figure 4(b)).

Proposition 5. *If $T_n(x, X) = 1$ and $T_{\overline{n}}(x, \overline{X}) \geq 2$ then two $\overline{n}-$connected component of $\overline{X}$ are merged by deletion of x or the morphism i'_* is not onto.*

In this paper, we show that in this case "a hole is created in $X \setminus \{x\}$" or more formally, i_* is not one to one. This is proved using the linking number as illustrated in Figure 5 where the closed path c is reducible in X (Lemma 2 below) whereas it is not reducible in $X \setminus \{x\}$ since $L_{c,\beta} = \pm 1$ (Remark 1). This shows that a condition on the preservation of tunnels in the object (Condition *iii* of Definition 3) is sufficient to ensure that tunnels of the complement are also left unchanged. And the proof of this result is obtained with the only formalism provided by the use of the digital fundamental group.

Before proving Proposition 4, we must state the two following lemmas.

Lemma 1. *Let $x \in X$ such that $T_n(x, X) = 1$ and $T_{\overline{n}}(x, \overline{X}) \geq 2$ and let $A = G_n(x, X)$. Then there exists a parameterized simple closed $n-$curve c in A (see Figure 5) and a closed $\overline{n}-$path β in $\overline{N_{26}(x) \cap X}$ such that $L_{c,\beta} = \pm 1$.*

The latter lemma has been proved by investigating, using a computer, all the possible local configurations in the neighborhood of a voxel x such that $T_n(x, X) = 1$ and $T_{\overline{n}}(x, \overline{X}) \geq 2$; showing for each one the existence of a simple closed curve in $G_n(x, X)$. The existence of the $\overline{n}-$path β of Lemma 1 is then proved using local considerations.

Lemma 2. *Let x be a point of X such that $T_n(x, X) = 1$. Then, any closed $n-$path c in $G_n(x, X)$ is $n-$reducible in X.*

Sketch of proof of Proposition 4: Let x be a point of X such that $T_n(x, X) = 1$ and $T_{\overline{n}}(x, \overline{X}) \geq 2$. Let c and β be the paths of Lemma 1; and let a and b be the two voxels of β which are $\overline{n}$−adjacent to x. If a and b are not $\overline{n}$−connected in $\overline{X}$ then it is clear that two $\overline{n}$−connected components of $\overline{X}$ are merged by deletion of x from X.

If a and b are connected by an $\overline{n}$−path α' in $\overline{X}$. Then, it follows that β is $\overline{n}$−homotopic to the path $\alpha = (a).\alpha'.(b, x, a)$ in $\overline{(N_{26}(x) \cap X)}$. Now, from Theorem 1 and since $\overline{(N_{26}(x) \cap X)} \subset \overline{c^*}$ we have $L_{c,\beta} = L_{c,\alpha} = \pm 1$.

From Theorem 1 (and Remark 1), it follows that the path c is not n−reducible in $\overline{\alpha'^*}$ and since $\alpha^* \subset \overline{X} \cup \{x\}$ then $X \setminus \{x\} \subset \overline{\alpha^*}$ so that *a fortiori* α cannot be n−reducible in $X \setminus \{x\}$. Formally, if c is a closed path from a voxel A to A in $X \setminus \{x\}$, we have $[c]_{\Pi_1^n(X\setminus\{x\},A)} \neq [1]_{\Pi_1^n(X\setminus\{x\},A)}$. Now, from Lemma 2, $c \simeq_n (A, A)$ in X so that $i_*([c]_{\Pi_1^n(X\setminus\{x\},A)}) = [c]_{\Pi_1^n(X,A)} = [1]_{\Pi_1^n(X,A)} = i_*([1]_{\Pi_1^n(X\setminus\{x\},A)})$, i.e. i_* is not one to one. □

Proof of Proposition 2: Suppose that properties *i*), *ii*) and *iii*) of Definition 3 are satisfied. From Proposition 3 we deduce that $T_n(x, X) < 2$. Furthermore, if no connected component of X is removed then $T_n(x, X) \neq 0$. Finally, $T_n(x, X) = 1$.

From Proposition 4 we deduce that $T_{\overline{n}}(x, \overline{X}) < 2$. Then, if no connected component of $\overline{X}$ is created then $T_{\overline{n}}(x, \overline{X}) \neq 0$. Finally, $T_{\overline{n}}(x, \overline{X}) = 1$. □

3.2 Second Step of the Proof of Theorem 2

We prove the following proposition.

Proposition 6. *If* $T_n(x, X) = 1$ *and* $T_{\overline{n}}(x, \overline{X}) = 1$, *then conditions i), ii), iii) and iv) of Definition 3 are satified.*

The main difficult part of the proof of Proposition 6 relies on the following proposition.

Proposition 7. *If* $T_n(x, X) = 1$ *and* $T_{\overline{n}}(x, \overline{X}) = 1$ *then* i_* *is an isomorphism.*

Corollary 1. *If* $T_n(x, X) = 1$ *and* $T_{\overline{n}}(x, \overline{X}) = 1$ *then* i'_* *is an isomorphism.*

In order to prove Proposition 7 we first state Lemma 3 and Lemma 4.

Lemma 3. *If* $T_n(x, X) = 1$ *and* $T_{\overline{n}}(x, \overline{X}) = 1$ *then for all* $B \in X \setminus \{x\}$ *and all* n*−path* c *of* $A_n^B(X)$, *there exists a path* c' *in* $A_n^B(X \setminus \{x\})$ *such that* $c \simeq_n c'$ *in* X.

Lemma 4. *If* $T_n(x, X) = 1$ *and* $T_{\overline{n}}(x, \overline{X}) = 1$ *then, for any voxel* $B \in X \setminus \{x\}$, *two closed* n*−paths* c *and* c' *of* $A_n^B(X \setminus \{x\})$ *which are* n*−homotopic in* X *are* n*−homotopic in* $X \setminus \{x\}$.

Sketch of proof of Proposition 7: Lemma 3 allows to prove that i_* is onto, and Lemma 4 allows to prove that i_* is one to one. □

Proof of Proposition 6: It is readily seen that $T_n(x, X) = T_{\overline{n}}(x, \overline{X}) = 1$ implies conditions *i*) and *ii*) of Definition 3. Then, from Proposition 7 and Corollary 1, we have $T_n(x, X) = 1$ and $T_{\overline{n}}(x, \overline{X}) = 1 \Rightarrow$ *iii*) and *iv*). □

3.3 End of the Proof of Theorem 2

Proof of Theorem 2: Following Definition 3, a simple voxel obviously satisfies the three conditions of Theorem 2. Now, from Proposition 2, a voxel which satisfies the three conditions of Theorem 2 is such that $T_n(x, X) = T_{\overline{n}}(x, \overline{X}) = 1$. Finally, from Proposition 6, if $T_n(x, X) = T_{\overline{n}}(x, \overline{X}) = 1$ then x satisfies the four conditions of Definition 3. □

4 Conclusion

The digital linking number allows us to formalize in a comprehensive way the characterization of 3D simple points for the complementary adjacency couples $(6, 26)$ and $(26, 6)$. The new theorem which is proved here shows the usefulness of the linking number in order to prove new theorems which involve the digital fundamental group in $\mathbb{Z}^3$. Now, even if the linking number is well defined for $(n, \overline{n}) \in \{(6+, 18), (18, 6+)\}$, it has not been used yet to provide a characterization of 3D simple points, similar to Theorem 2, for the latter couples of adjacency relations. This, because an open question remains about the existence of a simple closed curve, analogue to the curves c of Lemma 1, in this case.

References

Ber94. G. Bertrand. Simple points, topological numbers and geodesic neighborhoods in cubics grids. *Patterns Recognition Letters*, 15:1003–1011, 1994. 29

BM94. G. Bertrand and G. Malandain. A new characterization of three-dimensional simple points. *Pattern Recognition Letters*, 15:169–175, February 1994. 32, 33

FM00. S. Fourey and R. Malgouyres. A digital linking number for discrete curves. In *Proceedings of the 7th International Workshop on Combinatorial Image Analysis (IWCIA'00)*, pages 59–77. University of Caen, July 2000. 27, 28, 30

Kon88. T. Y. Kong. Polyhedral analogs of locally finite topological spaces. In R. M. Shortt, editor, *General Topology and Applications: Proceedings of the 188 Northeast Conference, Middletown, CT (USA)*, volume 123 of *Lecture Notes in Pure and Applied Mathematics*, pages 153–164, 1988. 29

Kon89. T. Y. Kong. A digital fundamental group. *Computer Graphics*, 13:159–166, 1989. 27, 29, 31

KR89. T. Y. Kong and Azriel Rosenfeld. Digital topology : introduction and survey. *Computer Vision, Graphics and Image Processing*, 48:357–393, 1989. 27

Mor81. D. G. Morgenthaler. Three-dimensional simple points : serial erosion, parallel thinning,and skeletonization. Technical Report TR-1005, Computer vision laboratory, Computer science center, University of Maryland, February 1981. 27, 28

Rol. Dale Rolfsen. *Knots and Links.* Mathematics Lecture Series. University of British Columbia. 30

TF82. Y. F. Tsao and K. S. Fu. A 3d parallel skeletonwise thinning algorithm. In *Proceeddings, IEEE PRIP Conference*, pages 678–683, 1982. 31

Digital n-Pseudomanifold and n-Weakmanifold in a Binary $(n+1)$-Digital Image

Mohammed Khachan[1], Patrick Chenin[2], and Hafsa Deddi[3]

[1] University of Poitiers, Department of Computer Science (IRCOM-SIC). Bd 3, Teleport 2, Bp 179, 86960 Futuroscope cedex, France
khachan@sic.sp2mi.univ-poitiers.fr

[2] University Joseph Fourier, LMC-IMAG, 51 Rue des Mathematiques, BP 53, 38420 Grenoble cedex 9, France
pchenin@imag.fr

[3] University of Quebec at Montreal, LACIM, C.P 8888, Succ. Centre-ville. Montreal (Quebec) H3C 3P8 Canada
deddi@math.uqam.ca

Abstract. We introduce the notion of digital n-pseudomanifold and digital n-weakmanifold in $(n+1)$-digital image, in the context of $(2n, 3^n-1)$-adjacency, and prove the digital version of the Jordan-Brouwer separation theorem for those classes. To accomplish this objective, we construct a polyhedral representation of the n-digital image, based on cubical complex decomposition. This enables us to translate some results from polyhedral topology into the digital space. Our main result extends the class of "thin" objects that are defined locally and verifies the Jordan-Brouwer separation theorem.

Keywords: digital topology, combinatorial topology, discrete spaces, combinatorial manifolds

1 Introduction

A binary $(n+1)$-digital image is an ordered uplet $\mathcal{P} = (\mathbb{Z}^{n+1}, \mathcal{R}, H)$, where H is a finite subset of $\mathbb{Z}^{n+1}$ and $\mathcal{R}$ represents the adjacency relation in the whole lattice.

Morgenthaler and Rosenfeld in [16] introduced for a binary 3-digital image the concept of a **simple surface point**, where $\mathcal{R} \in \{(6,26),(26,6)\}$, to characterize a class of objects H that verifies the digital version of the Jordan-Brouwer separation theorem. They define a simple surface point axiomatically by three local conditions, these conditions reflect basic properties of triangulated simple and closed surface's vertices in $\mathbb{R}^3$. They show that any finite and connected subset of $\mathbb{Z}^3$ consisting entirely of simple surfaces points verifies the digital version of the Jordan-Brouwer separation theorem. Thus, simple surface points criterion determines a class of thin object called **digital simple surface** of $\mathbb{Z}^3$.

Kong and Roscoe in [11] reveal precisely the geometric meanning of the surface simple points and extend its properties to other classes of $\mathcal{R}$-adjacency, i.e,

G. Borgefors, I. Nyström, and G. Sanniti di Baja (Eds.): DGCI 2000, LNCS 1953, pp. 37–45, 2000.
© Springer-Verlag Berlin Heidelberg 2000

$\mathcal{R} \in \{(18,6),(6,18),(26,18),(18,26)\}$. They introduce the notion of continuous analog for transfering statements from continuous topology to digital space, and show that a connected finite subset H of $\mathbb{Z}^3$ is a digital simple surface if and only if the continuous analogs of H is a simple and closed polyhedral surface of $\mathbb{R}^3$. Note that [14] extends the concept of continuous analogs to the class of strongly normal 3D digital image.
In [6] Bertrand and Malgouyres defined strong surfaces and in [7] Couprie and Bertrand introduced simplicity surfaces. Those classes of surfaces are more richer than Morgenthaler-Rosenfeld surfaces because they contain more local configurations and discrete analogs for these classes of surfaces can be built (see [5] for strong surfaces).
In [4] Ayala, Dominguez, Frances and Quintero prove that any digital n-manifold in $\mathbb{Z}^{n+1}$ verifies the Jordan-Brouwer separation theorem (the digital n-manifold is the generalization of the notion of digital simple surface for $n \geq 2$).
Malgouyres in [15] investigated the reverse problem in $\mathbb{Z}^3$, in the context of the (26,6)-adjacency relation. He shows that there is not a local characterization of subset of $\mathbb{Z}^3$ which separates $\mathbb{Z}^3$ in exactly two 6-connected components.
It is not difficult to realize that there is still a large margin between the local characterization of digital simple surface and the global characterization represented by the digital version of the Jordan-Brouwer separation theorem. Intuitively, the problem is that there exists polyhedra with singularity which verify the Jordan-Brouwer separation theorem. In this context, the notion of n-pseudomanifold is very interesting, it is a formulation of weak singularities corresponding to degenarate n-manifold (see property 4 in remark 1).
This work is a first attempt to bring closer local and global properties that define the class of object verifying the digital version of the Jordan-Brouwer separation theorem. Our approach is based on combinatorial topology and uses topological properties in addition to local neighborhood structures. Definitions and statements of n-manifold and n-pseudomanifold are given in section 2. Section 3 is devoted to review basic of digital topology in $\mathbb{Z}^{n+1}$. In section 4, we give a simple method to construct the continuous analogs of binary $(n+1)$-digital image based on cubical complex decomposition. This enables us to translate results from polyhedral topology into digital topology. Section 5 is aimed to state our main results.

2 n-Manifold and n-Pseudomanifold

In this section we review some definitions of algebraic topology.

Definition 1. *(see [1], pp. 42)*
A triangulation of a space X is a pair (K, ϕ), where K is a simplicial complex and $\phi : |K| \to X$ is a homeomorphism. The complex K is said to triangulate X. A polyhedron is any space which admits a triangulation.

Definition 2. *(see [8], pp.20)*
Let K be an homogeneous n-dimensional simplicial complex. A combinatorial n-

sphere is a polyhedron $|K|$ such that certain subdivision of K is simplicially homeomorphic to a subdivision of the boundary of the $(n+1)$-simplex.

Let K be a simplicial complex and $e \in K$, the link of e in K, denoted by $\mathrm{Lk}(e;K)$, is a subcomplex of K such that $\sigma \in Lk(e;K)$ if and only if

1. e is not a face of σ, and
2. $\exists\ \sigma' \in K$ such that
   2.1. σ is a face of σ', and
   2.2. e is a face of σ'.

Definition 3. *(see [8], pp. 26)*
A connected polyhedron X is said to be an n-manifold without boundary if it admits a triangulation (K,ϕ) satisfying:
1. *K is homogeneously n-dimensional.*
2. *for all $\sigma \in K$, $Lk(\sigma;K)$ is a combinatorial $(n - dim(\sigma) - 1)$-sphere.*
   K is called a combinatorial n-manifold without boundary.

Definition 4. *(see [1], pp. 195)*
A connected polyhedron X is said to be an n-pseudomanifold without boundary if it admits a triangulation (K,ϕ) such that:
1. *K is homogeneously n-dimensional.*
2. *Every $(n-1)$-simplex of K is a face of precisely two n-simplices of K.*
3. *If σ and σ' are two distincts n-simplices of K, then there exists a sequence $\sigma_1, \ldots, \sigma_u$ of n-simplices in K such that $\sigma_1 = \sigma$, $\sigma_u = \sigma'$ and σ_i meets σ_{i+1} in an $(n-1)$-dimensional face for $1 \le i < u$.*
   K is called a combinatorial n-pseudomanifold without boundary.

Remark 1. (see [1])
1. An n-sphere, denoted by $\mathbb{S}^n$, is an n-manifold without boundary.
2. Every n-manifold without boundary is an n-pseudomanifold without boundary.
3. Let X be an n-pseudomanifold and let (L,ψ) be any triangulation of X, then L is a combinatorial n-pseudomanifold without boundary.
4. An octahedron is a 2-manifold without boundary. If we identify two opposite vertices, we obtain a 2-dimensional curved polyhedron, whose triangulations are examples of 2-pseudomanifold without boundary which are not closed 2-manifolds without boundary.

Theorem 1. *(see [3], pp. 94)*
Every n-pseudomanifold without boundary in $\mathbb{S}^{n+1}$ divides $\mathbb{S}^{n+1}$ into two domains [1] and is the common boundary of both domains.

This theorem obviously remains true if $\mathbb{R}^{n+1}$ is substitued for $\mathbb{S}^{n+1}$: it is sufficient to map $\mathbb{S}^{n+1}$ onto $\mathbb{R}^{n+1}$ by means of a stereographic projection with center of projection lying outside the given pseudomanifold, see [3] page 94.

Theorem 2. *Every n-pseudomanifold without boundary in $\mathbb{R}^{n+1}$ divides $\mathbb{R}^{n+1}$ into two domains and is the common boundary of both domains.*

[1] A domain of $\mathbb{S}^{n+1}$ is an open and connected subset of $\mathbb{S}^{n+1}$

3 General Definitions

In digital image, the study of neighborhood is a very important and significant concept. For a given lattice point, a neighborhood of a point is defined typically using a metric distance. Let p, $q \in \mathbb{Z}^{n+1}$ with the coordinates $\left(x_i(p)\right)_{i=1}^{n+1}$ and $\left(x_i(q)\right)_{i=1}^{n+1}$ respectively. We will consider two types of distance between elements in $\mathbb{Z}^{n+1}$:

$$d_1(p,q) = \sum_{i=1}^{n+1} |x_i(p) - x_i(q)| \text{ and } d_\infty(p,q) = \underset{i=1\ldots n+1}{Max} |x_i(p) - x_i(q)|.$$

Let $\beta \in \{1, \infty\}$; two points $p, q \in \mathbb{Z}^{n+1}$ are said to be d_β-adjacent if $d_\beta(p,q) = 1$; we denote by $\mathbb{V}_\beta(p)$ the set of all d_β-neighbors of p. Consequently we have:

i) $\mathbb{V}_1(p) = \{q \in \mathbb{Z}^{n+1} \ / \ d_1(p,q) = 1\}$ and $\mathrm{Card}(\mathbb{V}_1(p)) = 2(n+1)$.
ii) $\mathbb{V}_\infty(p) = \{q \in \mathbb{Z}^{n+1} \ / \ d_\infty(p,q) = 1\}$ and $\mathrm{Card}(\mathbb{V}_\infty(p)) = 3^{n+1} - 1$.

In the literature ([13], [14] and [17]) the d_1-neighbor is referred as $(2(n+1))$-neighbor and the d_∞-neighbor as $(3^{n+1} - 1)$-neighbor.

Let $T \subset \mathbb{Z}^{n+1}$, we denote by T^c the **complement** of T in $\mathbb{Z}^{n+1}$; $T^c = \mathbb{Z}^{n+1} - T$.

A binary (n+1)-digital image is an ordered uplet $\mathcal{P} = (\mathbb{Z}^{n+1}, \mathcal{R}, H)$, where H is a finite subset of $\mathbb{Z}^{n+1}$ and $\mathcal{R}$ represents the adjacency relation in the whole lattice. In this paper the adjacency relation $\mathcal{R}$ will be taken as follows :

two elements of H are said to be $\mathcal{R}$-adjacent **if** they are d_1-adjacent, two elements of H^c are said to be $\mathcal{R}$-adjacent **if** they are d_∞-adjacent, and an element in H is $\mathcal{R}$-adjacent to an element in H^c **if** they are d_∞-adjacent.

Let $T \subset \mathbb{Z}^{n+1}$ and $p \in \mathbb{Z}^{n+1}$, p is said to be $\mathcal{R}$-adjacent to T if p is $\mathcal{R}$-adjacent to some point in T. T is said to be **strongly thin** if and only if any element of T is $\mathcal{R}$-adjacent to all $\mathcal{R}$-components of T^c. It is said to be **separating** if and only if T^c has exactly two $\mathcal{R}$-components.

If $T \subset \mathbb{Z}^{n+1}$ and $p, q \in T$ then an $\mathcal{R}$-**path** from p to q in T is a sequence of distincts points $(p_1, \ldots, p_m)$ in T such that $p_1 = p$ and $p_m = q$ and p_i is $\mathcal{R}$-adjacent to p_{i+1}, $1 \leq i < m$. T is $\mathcal{R}$-**connected** if given any two elements p and q in T there is an $\mathcal{R}$-path in T from p to q; an $\mathcal{R}$-**component** of T is a maximal $\mathcal{R}$-connected subset of T.

4 Polyhedral Representation of a Binary (n+1)-Digital Image

Let I_i denote the open unit interval $]r_i, r_{i+1}[$ or the single point r_i for some integer r_i. A k-**cube** e_k is defined as $e_k = \prod_{i=1}^{n+1} I_i$ where k of the I_i's are intervals and $n+1-k$ are single points. Thus e_k is an open k-cube embedded in $\mathbb{R}^n$. The closure of e_k, will be denoted by $\overline{e_k}$. Note that $\overline{e_k} = \prod_{i=1}^{n+1} \overline{I_i}$. We denote by $Som(e_k)$ the set of all vertices of $\overline{e_k}$. Obviously we have $Som(e_k) \subset \mathbb{Z}^{n+1}$. The subscript of a k-cube will be omitted when irrelevant to the argument at hand. A face of a k-cube e is a k'-cube e' such that $Som(e') \subseteq Som(e)$. We will write $e' \preceq e$.

A **cubical complex** K, is a finite collection of k-cubes in some $\mathbb{R}^{n+1}$, $0 \leq k \leq n+1$, such that:
 ⋆ if $e \in K$, then all faces of e belong to K, and
 ⋆ if $e, e' \in K$, then $e \cap e' = \emptyset$.

Now, we will construct the continuous analog of a binary $(n+1)$-digital image $\mathcal{P} = (\mathbb{Z}^{n+1}, \mathcal{R}, H)$. Intuitively, this construction consists of *'filling in the gaps'* between black points of $\mathcal{P}$, and must be consistent with the $\mathcal{R}$-adjacency relation of $\mathcal{P}$. Precisely, let $\mathbb{C}(H)$ be a collection of k-cubes in $\mathbb{R}^{n+1}$, $0 \leq k \leq n+1$, defined as follows:

$$\mathbb{C}(H) = \{e : k - cube,\ 0 \leq k \leq n+1 \ /\ Som(e) \subset H\}$$

$\mathbb{C}(H)$ is a cubical complex, the underlying polyhedron $|\mathbb{C}(H)|$ will be called the **polyhedral representation** of $\mathcal{P}$.

For each $x \in \mathbb{R}^{n+1}$, $e(x)$ will denote the k-cube, $0 \leq k \leq n+1$, that contains x. From the construction of $|\mathbb{C}(H)|$, we can deduce some natural properties:

Remark 2. Let $\mathcal{P} = (\mathbb{Z}^{n+1}, \mathcal{R}, H)$. We have :
1. if $x \in |\mathbb{C}(H)|$, then $e(x) \in \mathbb{C}(H)$.
2. if $x \in \mathbb{R}^{n+1} - |\mathbb{C}(H)|$, then at least one vertex of $e(x)$ belongs to H^c.
3. each component of $|\mathbb{C}(H)|$ or of $\mathbb{R}^{n+1} - |\mathbb{C}(H)|$ meets $\mathbb{Z}^{n+1}$.

The following theorem expresses the fundamental properties that permit us to relate digital topology to Euclidean space topology.

Theorem 3. *Let $\mathcal{P} = (\mathbb{Z}^{n+1}, \mathcal{R}, H)$.*
1. $|\mathbb{C}(H)| \cap \mathbb{Z}^{n+1} = H$ *and* $(\mathbb{R}^{n+1} - |\mathbb{C}(H)|) \cap \mathbb{Z}^{n+1} = H^c$.
2. *Two points in H are in the same $\mathcal{R}$-component of H* **iff** *they are in the same component of $|\mathbb{C}(H)|$.*
3. *Two points in H^c are in the same $\mathcal{R}$-component of H^c* **iff** *they are in the same component of $\mathbb{R}^{n+1} - |\mathbb{C}(H)|$.*
   The boundary of a component A of $|\mathbb{C}(H)|$ meets the boundary of a component B of $\mathbb{R}^{n+1} - |\mathbb{C}(H)|$ if and only if there is a point in $A \cap \mathbb{Z}^{n+1}$ that is $\mathcal{R}$-adjacent to a point in $B \cap \mathbb{Z}^{n+1}$.

The proof of the above theorem and other properties related to the concept of continuous analogs in a binary $(n+1)$-digital image are given in [10].

5 Digital n-Weakmanifold and Digital n-Pseudomanifold

We denote by K a cubical complex such that $|K|$ is a connected polyhedron.

Definition 5. *n-weakmanifold*
K will be called a cubical n-weakmanifold without boundary if and only if
1. *K is homogeneously n-dimensional.*
2. *for each vertex $p \in K$, $Lk(p; K)$ is a combinatorial $(n-1)$-sphere.*

The underlying polyhedron $|K|$ is called an n-weakmanifold without boundary.

The definition of the n-weakmanifold is a weak formulation of n-manifold given in definition 3, it uses uniquely the property of link in vertices of K. For $1 \leq n \leq 2$ those two notions are equivalent.

Definition 6. *n-pseudomanifold*
K will be called a cubical n-pseudomanifold without boundary if and only if
1. *K is homogeneously n-dimensional.*
2. *Every $(n-1)$-cube of K is a face of exactly two n-cubes of K.*
3. *If e and e′ are two distincts n-cubes of K, then there exists a sequence $e_1, \ldots, e_u$ of n-cubes in K such that $e_1 = e$, $e_u = e'$ and e_i meets e_{i+1} in an $(n-1)$-cube for $1 \leq i < u$.*

The underlying polyhedron $|K|$ is called an n-pseudomanifold without boundary.

Remark 3. It is not difficult to see that
1. If K is a cubical n-pseudomanifold without boundary then the barycentric subdivision of K is a combinatorial n-pseudomanifold without boundary.
2. Let p be a vertex of K. If $\mathrm{Lk}(p; K)$ is a combinatorial $(n-1)$-sphere, then $\mathrm{Lk}(p; K)$ is a cubical $(n-1)$-pseudomanifold without boundary.

Let σ_1 be a k-cube of K and p be a vertex of K such that $p \notin Som(\sigma_1)$. We denote by $p.\sigma_1$ the $(k+1)$-cube, if it exists, such that $\left(\{p\} \cup Som(\sigma_1)\right) \subset Som(p.\sigma_1)$.
Let σ and σ_1 be respectively a $(k+1)$-cube and a k-cube of K such that $\sigma_1 \prec \sigma$. If $p \in Som(\sigma) - Som(\sigma_1)$ then $p.\sigma_1 = \sigma$.

Proposition 1. *Any cubical n-weakmanifold without boundary K is a cubical n-pseudomanifold without boundary.*

Proof. We have to prove properties 2 and 3 in the definition 6.

$\star$ Let σ be an $(n-1)$-cube of K, we will show that σ is a face of exactly two n-cubes of K.
Let $p \in Som(\sigma)$, and σ' an $(n-2)$-cube of $\mathrm{Lk}(p, K)$ such that $\sigma' \prec \sigma$. It is easy to see that $p.\sigma' = \sigma$. $\mathrm{Lk}(p, K)$ is a combinatorial $(n-1)$-sphere, by using part 2 of remark 2, we deduce that σ' is a face of exactly two $(n-1)$-cubes (σ_1, σ_2) of $\mathrm{Lk}(p, K)$. This implies that $p.\sigma' = \sigma$ is a face of exactly two n-cubes $(p.\sigma_1, p.\sigma_2)$ of K, otherwise σ would be a face for more than two $(n-1)$-cubes of $\mathrm{Lk}(p, K)$.

$\star$ Let σ and σ' be two n-cubes of K, we will show that there exists a sequence $e_1, \ldots, e_u$ of n-cubes in K such that $e_1 = \sigma$, $e_u = \sigma'$ and e_i meets e_{i+1} in an $(n-1)$-cube for $1 \leq i < u$.
Let $p \in Som(\sigma)$ and $p' \in Som(\sigma')$. Since $|K|$ is a connected polyhedron, there

is a sequence $(p_0, \dots, p_k)$ of vertices of K that joins $p = p_0$ to $p' = p_k$, i.e, $\bigcup_{i=0}^{k-1} [p_i, p_{i+1}] \subset |K|$. For i, $0 < i < k$, p_{i-1} and p_{i+1} belong to $\mathrm{Lk}(p_i, K)$.
Moreover, any two $(n-1)$-cubes e_i and e'_i of $\mathrm{Lk}(p_i, K)$ can be joined by a sequence of $(n-1)$-cubes $e_i^0, \dots, e_i^{k_i}$ in $\mathrm{Lk}(p_i, K)$ such that consecutive $(n-1)$-cubes of this sequence share a common $(n-2)$-cube. So, any two n-cubes of K having p_i as a vertex can be joined by a sequence of n-cubes in K such that consecutive n-cubes of this sequence share a common $(n-1)$-cube.
Let σ_{n,p_i} denotes an n-cube of K such that p_i and p_{i+1} belong to $Som(\sigma_{n,p_i})$. Thus, σ_{n,p_i} and $\sigma_{n,p_{i+1}}$ can be joined by a sequence of n-cubes in K such that consecutive n-cubes of this sequence share a common $(n-1)$-cube, for $0 \leq i < k$. It is the same thing for the pair (σ, σ_{n,p_0}) and $(\sigma_{n,p_k}, \sigma')$. So there exists a sequence of n-cubes in K that joins σ to σ' such that any consecutive n-cubes of this sequence share a common $(n-1)$-cube.
This completes the proof. □

Definition 7. *Let $\mathcal{P} = (\mathbb{Z}^{n+1}, \mathcal{R}, H)$, and H be $\mathcal{R}$-connected.*
1. *H will be called* **digital n-weakmanifold** *if $\mathbb{C}(H)$ is a cubical n-weakmanifold without boundary.*
2. *H will be called* **digital n-pseudomanifold** *if $\mathbb{C}(H)$ is a cubical n-pseudomanifold without boundary.*

Proposition 2. *Let $\mathcal{P} = (\mathbb{Z}^{n+1}, \mathcal{R}, H)$.*
1. *If H is a digital n-weakmanifold, then H is a digital n-pseudomanifold.*
2. *If H is a digital n-pseudomanifold, then H is a separating set (see section 3 for the definition of separating set).*
3. *If H is a digital n-pseudomanifold, then H is strongly thin.*

Proof.
1. Let H be a digital n-weakmanifold.
$\mathbb{C}(H)$ is a cubical n-weakmanifold without boundary. By using proposition 1, we deduce that $\mathbb{C}(H)$ is a cubical n-pseudomanifold without boundary. This implies that H is a digital n-pseudomanifold.
2. Let H be a digital n-pseudomanifold.
$\mathbb{C}(H)$ is a cubical n-pseudomanifold without boundary. By using the part 1 of the remark 3 and part 3 of the remark 1, we deduce that $|\mathbb{C}(H)|$ is an n-pseudomanifold without boundary. So, the theorem 2 enables us to assert that $|\mathbb{C}(H)|$ divides $\mathbb{R}^{n+1}$ in two domains (Int, Ext) and is the common boundary of both domains. By using Theorem 3, we deduce that H^c has exactly two $\mathcal{R}$-components ($Int \cap \mathbb{Z}^{n+1}$ denoted by $Int(H)$, and $Ext \cap \mathbb{Z}^{n+1}$ denoted by $Ext(H)$). Furthermore $\partial Int = |\mathbb{C}(H)| = \partial Ext$.
3. Let $p \in H$. Since $\partial Int = |\mathbb{C}(H)|$ and $p \in |\mathbb{C}(H)|$, we can assert that :

$$\forall \epsilon > 0, \mathbb{B}(p, \epsilon) \neq \emptyset, \text{ with } \mathbb{B}(p, \epsilon) = \{x \in \mathbb{R}^{n+1} \ / \ d_\infty(p, x) \leq \epsilon\}.$$

Let $\epsilon = \frac{1}{3}$. $\exists x \in Int \ / \ d_\infty(p, x) \leq \frac{1}{3}$.
Let e be an $(n+1)$-cube such that $p \in Som(e)$ and $x \in \overline{e}$.
$e(x) \notin \mathbb{C}(H)$, otherwise $x \in |\mathbb{C}(H)|$ (for the definition of $e(x)$ see section 4). So, $\exists p_1 \in Som(e(x)) \ / \ p_1 \notin H$. Note that $e(x) \preceq e$.

$[x, p_1] \subset e(x)$. Since $p_1 \in H^c$ and $x \in Int$, then $p_1 \in Int$ (Int is a connected component of $\mathbb{R}^{n+1} - |\mathbb{C}(H)|$). $Int \cap \mathbb{Z}^{n+1} = Int(H)$, so $p_1 \in Int(H)$.
Since $e(x) \preceq e$, then p and p' are vertices of e. This implies that p is d_∞-adjacent to p', i.e, p is $\mathcal{R}$-adjacent to $Int(H)$.
In the same way, we prove that p is $\mathcal{R}$-adjacent to $Ext(H)$.
This completes the proof. □

Theorem 4. *Let* $\mathcal{P} = (\mathbb{Z}^{n+1}, \mathcal{R}, H)$.

1. *If H is a digital n-pseudomanifold, then H verifies the digital version of the Jordan-Brouwer separation theorem.*
2. *If H is a digital n-weakmanifold, then H verifies the digital version of the Jordan-Brouwer separation theorem.*

Proof. It is easy to deduce this theorem from proposition 2.

6 Conclusion

We have defined in a binary $(n + 1)$-digital image $\mathcal{P} = (\mathbb{Z}^{n+1}, \mathcal{R}, H)$, where $\mathcal{R} = (2(n + 1), 3^{n+1} - 1)$, a new class of objects H (digital n-weakmanifold and digital n-pseudomanifold) that verifies the digital version of the Jordan-Brouwer separation theorem, for $n \geq 2$.

References

1. Agoston, M. K.: Algebraic topology. Marcel Dekker, New York, 1976. 38, 39
2. Aleksandrov, P. S.: Combinatorial topology. Vol 2. Rochester, New York, 1957.
3. Aleksandrov, P. S.: Combinatorial topology. Vol 3. Rochester, New York, 1957. 39
4. Ayala, R., E. Dominguez, A. R. Francés, A. Quintero: Determining the components of the complement of a digital $(n-1)$-manifold in $\mathbb{Z}^n$. 6th International Workshop on Discrete Geometry for Computer Imagery, DGCI'96, Lectures Notes in Computer Science.1176 (1996), 163–176. 38
5. Ayala, R., E. Dominguez, A. R. Francés, A. Quintero: A Digital Lighting Function for strong 26-Surfaces. Discrete Geometry for Computer Imagery, DGCI'96, Lectures Notes in Computer Science, Springer Verlag, 1568 (1999), 91–103. 38
6. Bertrand, G., R. Malgouyres: Local Property of Strong Surfaces. SPIE Vision Geometry, Vol 3168, (1997), 318–327. 38
7. Couprie, M., G. Bertrand : Simplicity surfaces: a new definition of surfaces in $\mathbb{Z}^3$ SPIE Vision Geometry, Vol 3454, (1998), 40–51. 38
8. Hudson, J. F. P.: Piecewise linear topology. W.A Benjamin, New York, 1969. 38, 39
9. Kenmochi, Y., A. Imiya, N. Ezquerra: Polyhedra generation from lattice points. 6th International Workshop on Discrete Geometry for Computer Imagery, DGCI'96, Lectures Notes in Computer Science.1176 (1996), 127–138.
10. Khachan, M., P. Chenin, H. Deddi: Polyhedral representation and adjacency graph in n-digital image, **CVIU**, Computer Vision and Image Understanding (accepted, to appear). 41

11. Kong, T. Y., A. W. Roscoe: Continuous analogs of axiomatized digital surfaces. CVGIP **29** (Computer Vision, Graphics, and Image Processing) (1985), 60–86. 37
12. Kong, T. Y., A. W. Roscoe: A Theory of Binary Digital Pictures. CVGIP **32** (1985), 221–243.
13. Kong, T. Y., A. Rosenfeld: Survey Digital Topology: Introduction and Survey. CVGIP **48** (1989), 357–393. 40
14. Kong, T. Y., A. W. Roscoe, A. Rosenfeld: Concepts of digital topology. Topology and its Application **46** (1992), 219–262. 38, 40
15. Malgouyres, R.: About surfaces in $\mathbb{Z}^3$. Fifth International Workshop on Discrete Geometry for Computer Imagery, DGCI'95, pp.243-248, 1995. 38
16. Morgenthaler, D. G., A. Rosenfeld: Surfaces in three-dimensional digital images. Information and Control **51** (1981), 227–247. 37
17. Rosenfeld, A., T. Y. Kong, A.Y Wu: Digital surfaces. CVGIP **53**, 1991, 305–312. 40

Digital Jordan Curve Theorems

Christer O. Kiselman

Uppsala University, Department of Mathematics
P. O. Box 480, SE-751 06 Uppsala, Sweden
kiselman@math.uu.se
http://www.math.uu.se/~kiselman

Abstract. Efim Khalimsky's digital Jordan curve theorem states that the complement of a Jordan curve in the digital plane equipped with the Khalimsky topology has exactly two connectivity components. We present a new, short proof of this theorem using induction on the Euclidean length of the curve. We also prove that the theorem holds with another topology on the digital plane but then only for a restricted class of Jordan curves.

1 Introduction

The classical Jordan curve theorem says that the complement of a Jordan curve in the Euclidean plane $\mathbf{R}^2$ consists of exactly two connectivity components. Efim Khalimsky's digital Jordan curve theorem states the same thing for the digital plane $\mathbf{Z}^2$. Of course we must use a suitable definition of the concept of digital Jordan curve, as well as a suitable topology on $\mathbf{Z}^2$. In this case $\mathbf{Z}^2$ is given the Cartesian product topology of two copies of the digital line $\mathbf{Z}$ equipped with the Khalimsky topology.

A proof of Khalimsky's theorem was published in 1990 by Khalimsky, Kopperman and Meyer [1990]. They refer to earlier proofs by Khalimsky (E. D. Halimskiĭ) [1970, 1977]. Our first purpose in this note is to present a new, short proof.

The idea of the proof is simple. For the smallest Jordan curves (having four or eight points) the conclusion of the theorem can be proved by inspection. Given any other Jordan curve J, we construct a Jordan curve J' which has shorter Euclidean length and is such that its complement has as many components as the complement of J. Since the possible Euclidean lengths form a discrete set, this procedure will lead to one of the smallest Jordan curves, for which the theorem is already established. The construction of J' can intuitively be described as follows: attack J where its curvature is maximal and shorten it there; it cannot offer resistance from within.

We then consider a topology on $\mathbf{Z}^2$ which is not a product topology. In contrast to the Khalimsky topology it has the property that every point is either open or closed—there are no mixed points as in the Khalimsky plane. We prove that the Jordan curve theorem holds for this topology for a restricted class of Jordan curves.

G. Borgefors, I. Nyström, and G. Sanniti di Baja (Eds.): DGCI 2000, LNCS 1953, pp. 46–56, 2000.
© Springer-Verlag Berlin Heidelberg 2000

2 Connectedness and Adjacency

A topological space is said to be *connected* if it is nonempty and the only sets which are both open and closed are the empty set and the whole space. A subset of a topological space is called *connected* if it is connected as a topological space with the induced topology.[1] A *connectivity component* of a topological space is a connected subset which is maximal with respect to inclusion. A component is always closed. A connected subset which is both open and closed is a component (but not necessarily conversely).

If X and Y are topological spaces and $f\colon X \to Y$ a continuous mapping, then the image $f(A)$ of a connected subset A of X is connected. We may apply this result to a situation where we define a topology on a set Y using a mapping $f\colon X \to Y$ from a topological space X:

Proposition 1. *Let $f\colon X \to Y$ be a surjective mapping from a connected topological space X onto a set Y. Equip Y with the strongest topology such that f is continuous. Then Y is connected.*

In particular we shall use this result with $X = \mathbf{R}$ and $Y = \mathbf{Z}$ to define connected topologies on the digital line $\mathbf{Z}$.

In any topological space we shall denote by $N(x)$ the intersection of all neighborhoods of a point x. Spaces such that $N(x)$ is always a neighborhood of x were introduced and studied by Aleksandrov [1937]. Equivalently, they are the spaces where the intersection of any family of open sets is open. In such a space all components are open.

The closure of a subset A of a topological space will be denoted by $\overline{A}$. We note that $x \in \overline{\{y\}}$ if and only if $y \in N(x)$.

A *Kolmogorov space* (Bourbaki [1961:I:§1:Exerc. 2]; also called a T_0-space) is a topological space such that $x \in N(y)$ and $y \in N(x)$ only if $x = y$. It is quite reasonable to impose this axiom; if x belongs to the closure of $\{y\}$ and vice versa, then x and y are indistinguishable from the point of view of topology. (We should therefore identify them and consider a quotient space.)

The separation axiom T_1 states that $N(x) = \{x\}$. It is too strong to be of interest for the spaces considered here.

Two points x and y in a topological space Y are said to be *adjacent* if $x \neq y$ and $\{x, y\}$ is connected. We note that $\{x, y\}$ is connected if and only if either $x \in N(y)$ or $y \in N(x)$. We shall say that two points x, z are *second adjacent* if $x \neq z$; x and z are not adjacent; and there exists a third point $y \in Y$ such that x and y are adjacent and y and z are adjacent.

3 Topologies on the Digital Line

It is natural to think of $\mathbf{Z}$ as an approximation of the real line $\mathbf{R}$ and to consider mappings $f\colon \mathbf{R} \to \mathbf{Z}$ expressing this idea. We may define $f(x)$ to be the integer

[1] According to Bourbaki [1961:I:§11:1] the empty space is connected. Here I follow the advice of Adrien Douady (personal communication, June 26, 2000). In the present paper it will not matter whether the empty set is said to be connected or not.

closest to x; this is well-defined unless x is a half-integer. So when $x = n + \frac{1}{2}$ we have a choice for each n: shall we define $f(n + \frac{1}{2}) = n$ or $f(n + \frac{1}{2}) = n + 1$? If we choose the first alternative for every n, thus putting $f^{-1}(n) = \,]n - \frac{1}{2}, n + \frac{1}{2}]$, the topology defined in Proposition 1 is called the *right topology* on $\mathbf{Z}$; if we choose the second, we obtain the *left topology* on $\mathbf{Z}$; cf. Bourbaki [1961:I:§1: Exerc. 2]. *Khalimsky's topology* arises if we always choose an even integer as the best approximant of a half-integer. Then the closed interval $[-\frac{1}{2}, \frac{1}{2}]$ is mapped to 0, so $\{0\}$ is closed for the Khalimsky topology, whereas the inverse image of 1 is the open interval $]\frac{1}{2}, \frac{3}{2}[$, so that $\{1\}$ is open.

A *Khalimsky interval* is an interval $[a, b] \cap \mathbf{Z}$ equipped with the topology induced by the Khalimsky topology on $\mathbf{Z}$. A *Khalimsky circle* is a quotient space $\mathbf{Z}_m = \mathbf{Z}/m\mathbf{Z}$ of the Khalimsky line for some even integer $m \geqslant 4$. (If m is odd, the quotient space receives the chaotic topology, which is not interesting.)

4 Khalimsky Jordan Curves

Khalimsky, Kopperman and Meyer [1990: 3.1] used the following definitions of path and arc in the Khalimsky plane. We just extend them here to any topological space. We modify slightly their definition of a Jordan curve [1990: 5.1]. A Jordan curve in the Euclidean plane $\mathbf{R}^2$ is a homeomorphic image of the circle $\mathbf{R}/\mathbf{Z}$, and similarly a Khalimsky Jordan curve is a homeomorphic image of a Khalimsky circle.

Definition 1. *Let Y be any topological space. A* **Khalimsky path** *in Y is a continuous image of a Khalimsky interval. A* **Khalimsky arc** *is a homeomorphic image of a Khalimsky interval. A* **Khalimsky Jordan curve** *in Y is a homeomorphic image of a Khalimsky circle.*

Sometimes Khalimsky Jordan curves are too narrow. We impose a condition on them to make their interior fatter:

Definition 2. *Let J be a Khalimsky Jordan curve in a topological space Y. We shall say that J is* **strict** *if every point in J is second adjacent to exactly two points in J.*

We note that if $x, z \in J$ are second adjacent, then the intermediary y required by the definition need not belong to J. Thus the concept of strict Jordan curve is not intrinsic.

A three-set $\{x, y, z\}$ such that all three points are adjacent to each other can be a Khalimsky path but never a Khalimsky arc. This follows from the fact that in a Khalimsky interval $[a, b]$, the endpoints a and b are not adjacent unless $b = a + 1$. Let us say that a three-set $\{x, y, z\}$ in a topological space is a *forbidden triangle* if all points are adjacent to each other. The absence of forbidden triangles is therefore a necessary condition for Khalimsky arcs and consequently for Khalimsky Jordan curves, and it is often easy to check.

Different topologies may induce the same adjacency structure. However, when the adjacency structure is that of a Khalimsky circle, the topology of the space must also be that of a Khalimsky circle. More precisely we have the following result.

Theorem 1. *Given a subset J of a topological space Y, the following conditions are equivalent.*
(A) J is a Khalimsky Jordan curve.
(B) J has at least four points, and for every $a \in J$, $J \setminus \{a\}$ is homeomorphic to a Khalimsky interval.
(C) J is finite, connected, with cardinality at least 4, and each of its elements has exactly two adjacent points.
(D) J has the adjacency structure of a Khalimsky circle, i.e., $J = \{x_1, x_2, ..., x_m\}$ for some even integer $m \geqslant 4$ and for each $j = 1, ..., m$, x_{j-1} and x_{j+1} and no other points are adjacent to x_j. (Here we count indices modulo m.)

Proof. If (A) holds, then for every $a \in J$, $J \setminus \{a\}$ is homeomorphic to a Khalimsky circle minus one point, thus to a Khalimsky interval. Conversely, suppose that (B) holds and consider $J \setminus \{a\}$ and $J \setminus \{b\}$ for two points $a, b \in J$ which are not adjacent. Then we have homeomorphisms of $J \setminus \{a\}$ and $J \setminus \{b\}$ into a Khalimsky circle $\mathbf{Z}_m$. We can modify them by rotating the circle so that the two mappings agree on $J \setminus \{a, b\}$. Then they define a local homeomorphism of J onto $\mathbf{Z}_m$, thus a homeomorphism; we have proved (A).

It is clear that (A) implies (C) and (D).

Suppose that (C) holds. Then call an arbitrary point x_1 and one of its adjacent points x_2 and then go on, always choosing x_{j+1} after x_j so that x_{j+1} is adjacent to x_j but not equal to any of the already chosen $x_1, ..., x_{j-1}$. After a while we must arrive at a situation where there are no new points left, i.e., we arrive at x_m and the two points adjacent to x_m are x_{m-1} and a point which has already been chosen, say x_k. A priori k may be any of $1, 2, ..., m-2$, but in fact the only possibility is $k = 1$—any other choice would mean that x_k had three adjacent points contrary to the assumption. It remains to be seen that m is even. That x_j and x_{j+1} are adjacent means that we have either $x_j \in N(x_{j+1})$ or $x_{j+1} \in N(x_j)$. If $x_j \in N(x_{j+1})$, then we cannot have $x_{j+1} \in N(x_{j+2})$, for that would imply that x_j belonged to $N(x_{j+2})$, so that x_{j+2} would have three adjacent elements, viz. x_j, x_{j+1} and x_{j+3}. So the statement $x_j \in N(x_{j+1})$ holds only for j of a certain parity. Since this is true modulo m, that number must be even. Thus we have proved (D). Conversely, (D) obviously implies (C) since (D) is just a more detailed version of (C).

It remains to be seen that (D) implies (A). First of all it is clear that, assuming (D), $N(x)$ can never have more than three elements—a fourth element would mean that x had at least three adjacent points. So $N(x_j) \subset \{x_{j-1}, x_j, x_{j+1}\}$. Considering the three points x_{j-1}, x_j, x_{j+1}, we note that either $x_{j-1} \in N(x_j)$ or $x_j \in N(x_{j-1})$, and that $x_j \in N(x_{j+1})$ or $x_{j+1} \in N(x_j)$. However, these alternatives cannot be chosen at will, for as we have seen in the previous paragraph $x_{j-1} \in N(x_j)$ implies $x_j \notin N(x_{j+1})$. Consider now the case $x_{j-1} \in N(x_j)$. Then $x_{j+1} \in N(x_j)$, so that $N(x_j) \supset \{x_{j-1}, x_j, x_{j+1}\}$. On the other hand we know already that $N(x_j)$ has at most three elements; we conclude that $N(x_j) =$

$\{x_{j-1}, x_j, x_{j+1}\}$. By the same argument, $N(x_{j+2}) = \{x_{j+1}, x_{j+2}, x_{j+3}\}$. Therefore $N(x_{j+1}) = \{x_{j+1}\}$, and we have proved that Y is a Khalimsky circle where points with indices of the same parity as j have three-neighborhoods and points with indices of the other parity are open. The other possibility, viz. that $x_j \in N(x_{j-1})$, can be reduced to the former by just shifting the indices one step.

It follows from property (C) that two Khalimsky Jordan curves can never be contained in each other. More precisely, if J and K are Khalimsky Jordan curves and $J \subset K$, then $J = K$.

A point on a Khalimsky Jordan curve J consisting of at least six points has at least two second adjacent points; with the order introduced in property (D), x_{j-2} and x_{j+2} are second adjacent to x_j and $x_{j-2} \neq x_{j+2}$ when $m > 4$. Then $x_{j\pm1}$ serve as intermediaries, but there may also exist other intermediaries. When a Jordan curve is not strict and $m > 4$, then some point, say x_j, has at least one second adjacent point in addition to x_{j-2} and x_{j+2}, say x_k. Then an intermediary b such that x_j and b are adjacent and b and x_k are adjacent cannot belong to J.

Suppose now that Y is a metric space with metric d. Since every Khalimsky arc Γ is homeomorphic either to $[0, m-1] \cap \mathbf{Z}$ or to $[1, m] \cap \mathbf{Z}$ for some m, it can be indexed as $\{x_1, ..., x_m\}$, where the indices are uniquely determined except for inversion. We may define its length as

$$\operatorname{length}(\Gamma) = \sum_{1}^{m-1} d(x_{j+1}, x_j).$$

Similarly, a Khalimsky Jordan curve can be indexed as $\{x_1, ..., x_m\}$, where the indices are uniquely determined up to inversion and circular permutations, and its length can be defined as

$$\operatorname{length}(J) = \sum_{1}^{m} d(x_{j+1}, x_j),$$

where we count the indices modulo m.

We shall use the following norms in $\mathbf{R}^2$ to measure distances in $\mathbf{Z}^2$:

$$\|x\|_p = \|(x_1, x_2)\|_p = \begin{cases} \left(|x_1|^p + |x_2|^p\right)^{1/p}, & x \in \mathbf{R}^2, \quad 1 \leqslant p < +\infty; \\ \max\left(|x_1|, |x_2|\right), & x \in \mathbf{R}^2, \quad p = \infty. \end{cases}$$

5 Khalimsky's Digital Jordan Curve Theorem

The Khalimsky topology of the digital plane is the Cartesian product topology of two copies of the Khalimsky line $\mathbf{Z}$. A point $x = (x_1, x_2)$ in the product $\mathbf{Z}^2 = \mathbf{Z} \times \mathbf{Z}$ is closed if and only if both x_1 and x_2 are closed, thus if and only if both x_1 and x_2 are even; similarly x is open if and only if both coordinates are odd. These points are called *pure*; the other points, which are neither open nor closed, are called *mixed*.

Perhaps the quickest way to describe Khalimsky's topology τ_∞ on $\mathbf{Z}^2$ is this: We first declare the nine-set

$$U_\infty = \{x \in \mathbf{Z}^2;\ \|x\|_\infty \leqslant 1\} = \{(0,0), \pm(1,0), \pm(1,1), \pm(0,1), \pm(-1,1)\} \quad (1)$$

to be open, as well as all translates $U_\infty + c$ with $c_1, c_2 \in 2\mathbf{Z}$. Then all intersections of such translates are open, as well as all unions of the sets so obtained. As a consequence, $\{(1,-1),(1,0),(1,1)\}$, the intersection of U_∞ and $U_\infty + (2,0)$, and $\{(1,1)\}$, the intersection of U_∞ and $U_\infty + (2,2)$, are open sets, and $\{(0,0)\}$ is a closed set. The sets $\{(1,0)\}$ and $\{(0,1)\}$ are neither open nor closed.

Theorem 2. *Given a subset J of $\mathbf{Z}^2$ equipped with the Khalimsky topology, the conditions A, B, C and D of Theorem 1 are all equivalent to the following.*
(E) $J = \{x^{(1)}, x^{(2)}, ..., x^{(m)}\}$ for some even integer $m \geqslant 4$ and for all j, $x^{(j-1)}$ and $x^{(j+1)}$ and no other points are adjacent to $x^{(j)}$; moreover each path consisting of three consecutive points $\{x^{(j-1)}, x^{(j)}, x^{(j+1)}\}$ turns at $x^{(j)}$ by 45° or 90° or not at all if $x^{(j)}$ is a pure point, and goes straight ahead if $x^{(j)}$ is mixed.

Here we use the informal expression "turn by 45°" etc. with reference to angles in the Euclidean plane of which we consider the Khalimsky plane to be a subset.

Proof. If (D) holds, we see that J cannot turn at a mixed point and cannot turn 135° at a pure point—otherwise we would have a forbidden triangle. So (D) implies (E). Conversely, (E) is just a more precise version of (D), so (E) implies (D).

In this section we shall measure the lengths of Khalimsky Jordan curves using the Euclidean metric, $d(x,y) = \|x - y\|_2$. It is not possible to use $\|\cdot\|_1$ or $\|\cdot\|_\infty$ in the proof of the Jordan curve theorem.

The smallest possible Jordan curve in $\mathbf{Z}^2$ is the four-set

$$J_4 = \{x \in \mathbf{Z}^2;\ \|x - (1,0)\|_1 = 1\} = \{(0,0), (1,-1), (2,0), (1,1)\}. \quad (2)$$

We add all translates of J_4 by a vector $c \in \mathbf{Z}^2$ with $c_1 + c_2$ even and call these the *Jordan curves of type* J_4.

There is also a Jordan curve having eight points,

$$J_8 = \{x \in \mathbf{Z}^2;\ \|x\|_\infty = 1\} = U_\infty \smallsetminus \{(0,0)\}. \quad (3)$$

This curve and all its translates by a vector $c \in \mathbf{Z}^2$ with $c_1 + c_2$ even we call the *Jordan curves of type* J_8.

Let us agree to call the three-set

$$T = \{(1,1), (0,0), (1,-1)\} \quad (4)$$

and rotations of T by 90°, 180° and 270°, as well as all translates of these sets by vectors $c \in \mathbf{Z}^2$ with $c_1 + c_2$ even, a *removable triangle*. It turns out that elimination of removable triangles is a convenient way to reduce Jordan curves, as shown by the following lemma.

Lemma 1. *Let J be a Jordan curve in the Khalimsky plane and assume that J contains the three-set T defined by (4). Define*

$$J' = (J \setminus \{(0,0)\}) \cup \{(1,0)\}.$$

Then either $J = J_4$ or else J' is a Jordan curve such that $\complement J'$ and $\complement J$ have the same number of components, and $\operatorname{length}(J') = \operatorname{length}(J) - 2\sqrt{2} + 2$.

Proof. Assume first that $(2,0) \in J$, thus that $J \supset J_4$. Then necessarily $J = J_4$.

Next we suppose that $(2,0) \notin J$. Then J' is a Jordan curve: J' is a set where the new point $(1,0)$ plays exactly the same role topologically as the old point $(0,0)$ in J. Thus J' is also homeomorphic to a Khalimsky circle.

Finally we must check that the number of components in $\complement J'$ is the same as that of $\complement J'$. Indeed, $(1,0)$ and $(2,0)$ belong to the same component of $\complement J$, and $(0,0)$ and $(-1,0)$ belong to the same component of $\complement J'$.

Theorem 3 (Khalimsky's Jordan curve theorem). *Let us equip the digital plane $\mathbf{Z}^2$ with the Khalimsky topology τ_∞ (see (1)). Then for any Khalimsky Jordan curve J in $\mathbf{Z}^2$, the complement $\complement J = \mathbf{Z}^2 \setminus J$ has exactly two connectivity components.*

Proof. The complement of J_4 consists of $A = \{(1,0)\}$ and the set B of all points x with $|x_1 - 1| + |x_2| > 1$. It is obvious that these two sets are connected. Moreover, A is closed and open in $\complement J_4$, so it is a component. Therefore, also B is closed and open in $\complement J_4$ and also a component. The proof for J_8 is similar.

Thus we know that the conclusion of the theorem holds for Jordan curves of types J_4 and J_8.

Next we shall prove that if J is not of the kind already treated, then there exists a Jordan curve J' of strictly smaller Euclidean length such that $\complement J$ and $\complement J'$ have the same number of components. After a finite number of steps we must arrive at a situation where the hypothesis is no longer satisfied, which means that we have a Jordan curve of type J_4 or J_8, for which the complement has two components as we already proved.

The construction of J' is as follows. First we may assume, in view of Lemma 1, that J contains no removable triangles. Define

$$a_2 = \inf(x_2; x \in J).$$

Thus $x_2 \geqslant a_2$ for all points $x \in J$ with equality for at least one x. Consider a horizontal interval

$$H = \{(x_1, a_2)\} + \{(0,0), (1,0), ..., (p,0)\}$$

which is maximal with respect to inclusion and consists of points in J with ordinate equal to a_2. The maximality implies that the two points $(x_1 - 1, a_2)$ and $(x_1 + p + 1, a_2)$ do not belong to J. Then we see that p must be an even number, but we cannot have $p = 0$, since that would imply that J contained a removable triangle, contrary to the assumption. Thus H contains at least three points. Moreover, at the endpoints of H, J must turn upwards. Indeed, since $(x_1 - 1, a_2)$ does not belong to J, exactly one of the points $(x_1 - 1, a_2 + 1)$, $(x_1, a_2 + 1)$ belongs to J; when we go left from (x_1, a_2), the curve must turn upwards by either 45° or 90°; it cannot turn downwards. Similarly, the curve turns upwards by 45° or 90° when we go right from the last point in H, viz. from $(x_1 + p, a_2)$.

We now consider the set $\mathcal{I}$ of all maximal horizontal intervals I in J such that J turns upwards at the endpoints of I. The previous argument served just to prove that there exists such an interval. Now there exists an interval $K \in \mathcal{I}$ of smallest length,

$$K = \{y\} + \{(0,0), (1,0), ..., (q,0)\},$$

containing $q+1$ points for some even number $q \geqslant 2$. We shall assume that K is of smallest length also among all intervals that can be obtained from the intervals in $\mathcal{I}$ by rotating them 90°, 180° or 270°.

To simplify the notation we may assume (after a translation if necessary) that $y = (0,0)$, so that

$$K = \{(0,0), (1,0), ..., (q,0)\} = [(0,0), (q,0)] \cap \mathbf{Z}^2.$$

Case 1. J turns upwards by 45° at both ends of K. This means that $(-1,1)$ and $(q+1,1)$ both belong to J. In this case, we define

$$J' = (J \smallsetminus K) \cup (K + (0,1)).$$

This operation shortens the Euclidean length by $2\sqrt{2}-2$ (but it does not shorten the l^∞ length). We note that the interval $K + (0,1)$ is disjoint from J; otherwise some point in K would have three adjacent points. Moreover $K + (0,2)$ must be disjoint from J. Indeed, if $(K + (0,2)) \cap J$ were nonempty, then either J would contain a removable triangle (contrary to our assumption) or there would exist a subinterval K' of $K + (0,2)$ contained in J and such that J turns upwards at its endpoints; thus $K' \in \mathcal{I}$. This subinterval must have fewer than $q+1$ points, since $(0,2)$ and $(q,2)$ cannot belong to J—otherwise there would be a removable triangle in J. Now a shorter interval is impossible, since K is by assumption an interval in $\mathcal{I}$ of shortest length. One checks that J' is a Jordan curve. Indeed, the points of $K + (0,1)$ play the same role topologically in J' as do the points of K in J. The number of components in the complement of J' is the same as for J.

Case 2. J turns upwards by 90° at one end of K. Assume that $(0,1) \in J$, the case $(q,1) \in J$ being symmetric. Then also $(0,2) \in J$. We consider the subcases 2.1 and 2.2.

Case 2.1. $(2,2) \notin J$. We cut off a corner, i.e., we remove $(0,1)$, $(0,0)$, $(1,0)$, and add $(1,1)$. This operation shortens the Euclidean length by $4 - 2\sqrt{2}$ (but J' has the same l^1-length as J). Since $(1,1)$ and $(2,2)$ belong to the same component of $\complement J$, and $(0,1)$, $(0,0)$, $(1,0)$, and $(-1,0)$ belong to the same component of $\complement J'$, the number of components in the respective complements are the same.

Case 2.2. $(2,2) \in J$. We consider four subcases, 2.2.1.1, 2.2.1.2, 2.2.2.1 and 2.2.2.2.

Case 2.2.1.1. $(2,1) \in J$, $(1,2) \in J$. Then J contains a Jordan curve of type J_8, more precisely $J \supset (1,1) + J_8$. So J must be equal to that curve.

Case 2.2.1.2. $(2,1) \in J$, $(1,2) \notin J$. Remove the five points $(0,1)$, $(0,0)$, $(1,0)$, $(2,0)$, $(2,1)$, and add $(1,2)$. Thus J' is shorter by 4. We can check that J' has all desired properties.

Case 2.2.2.1. $(2,1) \notin J$, $(1,2) \in J$. Turn 90° to reduce to case 2.2.1.2.

Case 2.2.2.2. $(2,1) \notin J$, $(1,2) \notin J$. This case cannot occur since q is smallest possible. To see this, define I' as the set of all points $(2,2), (3,2), ..., (q',2) \in J$ with q' as large as possible. If J turns upwards at $(q',2)$, then I' belongs to $\mathcal{I}$ with $q' < q$, which contradicts the definition of K and q. If on the other hand J turns downwards at $(q',2)$, then there exists a vertical interval consisting of three points, which becomes an interval in $\mathcal{I}$ if we turn it 90°, thus again contradicting the definition of $\mathcal{I}$.

6 The Jordan Curve Theorem for Another Topology

We define a topology τ_1 on $\mathbf{Z}^2$ by first declaring the five-set

$$U_1 = \{x \in \mathbf{Z}^2;\ \|x\|_1 \leqslant 1\} = \{(0,0), \pm(1,0), \pm(0,1)\} \tag{5}$$

to be open, then all translates $U + c$ with $c \in \mathbf{Z}^2$, $c_1 + c_2 \in 2\mathbf{Z}$ to be open, as well as all intersections of such translates (cf. (1)). This implies that $\{(1,0)\}$ is open, and that the origin is closed. In fact, all points $x \in \mathbf{Z}^2$ with $x_1 + x_2 \in 2\mathbf{Z}$ are closed, and all points with $x_1 + x_2 \notin 2\mathbf{Z}$ are open; there are no mixed points. This topology was described by Wyse et al. [1970] and Rosenfeld [1979: 624].

The four-set

$$J_4' = \{(0,0), (1,0), (1,1), (0,1)\} \tag{6}$$

is a Jordan curve for τ_1. However, it is not strict, for a point in J_4' has only one second adjacent point. Its complement is connected, so the Jordan curve theorem does not hold. The set J_8 defined by (3) is a Khalimsky Jordan curve and its complement has exactly two components. Also J_8 is not strict, because the point $(1,0)$ has three second adjacent points, viz. $(0,1)$, $(0,-1)$ and $(-1,0)$.

Another example is the twelve-set

$$J_{12} \\ = \{(0,0), (1,0), (2,0), (2,1), (3,1), (3,2), (3,3), (2,3), (1,3), (1,2), (0,2), (0,1)\}.$$

It is a Jordan curve, not strict, and its complement has three connectivity components, viz. an infinite component and the two singleton sets $\{(1,1)\}$ and $\{(2,2)\}$.

Theorem 4. *Let $\mathbf{Z}^2$ be equipped with the topology τ_1 just defined (see (5)). Then the complement of every strict Jordan curve has exactly two components.*

Proof. For the proof we shall use the fact that $\mathbf{Z}^2$ equipped with the topology τ_1 is homeomorphic to the subspace of all pure points in the Khalimsky plane. This fact was used also by Kong, Kopperman and Meyer [1991: 915].

Let X be the digital plane $\mathbf{Z}^2$ with the topology τ_1, and Y the Khalimsky plane ($\mathbf{Z}^2$ with the topology τ_∞). Consider the mapping $\varphi\colon X \to Y$ defined by $\varphi(x) = (x_1 - x_2, x_1 + x_2)$. Its image $\varphi(X)$ is the set of all pure points in Y, and if we equip it with the topology induced by Y it is homeomorphic to X. Moreover, the image of any Khalimsky Jordan curve J in X is a Khalimsky Jordan curve in Y. Therefore $Y \smallsetminus \varphi(J)$ has exactly two components by Theorem 3. We claim that $\varphi(X) \smallsetminus \varphi(J)$ has exactly two components. It is clear that this set has at least two components, so the problem is to prove that a component A of $Y \smallsetminus \varphi(J)$ gives rise to a connected set $A \cap \varphi(X)$, i.e., that the pure points in A form a connected set.

To this end, assume that $a, a' \in A \cap \varphi(X)$, and consider a Khalimsky arc

$$\{a = a^{(0)}, a^{(1)}, ..., a^{(s)} = a'\}$$

contained in $Y \smallsetminus \varphi(J)$. (Connectedness in Y is the same as arcwise connectedness; cf. Khalimsky et al. [1990: Theorem 3.2].) We shall prove that this arc can be replaced by another consisting only of pure points. So assume that $a^{(j)}$ is a mixed point. Then its predecessor $a^{(j-1)}$ and its successor $a^{(j+1)}$ are both pure points. Without loss of generality, we may assume that $a^{(j-1)} = (0,0)$, $a^{(j)} = (0,1)$, and $a^{(j+1)} = (0,2)$. We may then replace $a^{(j)}$ by one of the pure points $(-1,1)$, $(1,1)$, because both of them cannot belong to $\varphi(J)$. To see this, suppose that $(1,1), (-1,1) \in \varphi(J)$. Then $(-1,1)$ would be a second adjacent point to $(1,1)$, and this point has, by hypothesis, exactly two second adjacent points in $\varphi(J)$ (considering everything in the space $\varphi(X)$). However, none of them can be equal to $(-1,1)$, for the only possible intermediaries would then be $(0,0)$ and $(0,2)$, none of which belongs to $\varphi(J)$. (In a strict Jordan curve, one of the possible intermediaries to a second adjacent point must belong to the curve.) This contradiction shows that not both of $(1,1)$ and $(-1,1)$ can belong to $\varphi(J)$. Thus we may define $b = (-1,1)$ or $b = (1,1)$ so that $b \notin \varphi(J)$ and observe that

$$\{a^{(0)}, ..., a^{(j-1)}, b, a^{(j+1)}, ..., a^{(s)}\}$$

is a Khalimsky arc with a mixed point replaced by a pure point. After finitely many such replacements we obtain an arc connecting a and a' and consisting only of pure points. This shows that $\varphi(X) \smallsetminus \varphi(J)$ has at most as many components as $Y \smallsetminus \varphi(J)$; therefore exactly two components, and then the same is true of $X \smallsetminus J$.

7 Conclusion

We have showed that digital Jordan curves can be subject to a simple induction process: their lengths in a suitable metric form a discrete set, and if we show that a curve has a certain property when a shorter curve has the same property, the induction works, starting from some very short curves. Using this idea, we have presented a short proof of Khalimsky's digital Jordan curve theorem, which is valid for the Khalimsky topology on the digital plane, where each point has eight adjacent points. We have also considered a topology for the digital plane where each points has four adjacent points. For this topology the Jordan curve theorem does not hold in general, but it does for a restricted class of curves.

Acknowledgment

I am grateful to Ingela Nyström and Erik Palmgren for helpful comments on an early version of this paper.

References

1. Aleksandrov, P. S. 1937, Diskrete Räume. *Mat. Sb.* **2** (44), 501–519.
2. Bourbaki, Nicolas 1961, *Topologie générale.* Éléments de mathématique, première partie, livre III, chapitres 1 & 2. Third edition. Paris: Hermann.
3. Halimskiĭ, E. D. 1970, Applications of connected ordered topological spaces in topology. Conference of Math. Departments of Povolsia.
4. 1977, *Uporyadochennnye topologicheskie prostranstva.* Kiev: Naukova Dumka. pp. 92.
5. Khalimsky, Efim; Kopperman, Ralph; Meyer, Paul R. 1990, Computer graphics and connected topologies on finite ordered sets. *Topology Appl.* **36**, 1–17.
6. Kong, Yung; Kopperman, Ralph; Meyer, Paul R. 1991, A topological approach to digital topology. *Amer. Math. Monthly* **98**, 901–917.
7. Rosenfeld, Azriel 1979, Digital topology. *Amer. Math. Monthly* **86**, 621–630.
8. Wyse, Frank, et al. 1970, Solution to problem 5712. *Amer. Math. Monthly* **77**, 1119.

A New Means for Investigating 3-Manifolds

Vladimir Kovalevsky

Institute of Computer Graphics, University of Rostock
Albert-Einstein-Str. 21, 18051 Rostock, Germany
kovalev@tfh-berlin.de

Abstract. The paper presents a new method of investigating topological properties of three-dimensional manifolds by means of computers. Manifolds are represented as finite cell complexes. The paper contains definitions and a theorem necessary to transfer some basic knowledge of the classical topology to finite topological spaces. The method is based on subdividing the given set into blocks of simple cells in such a way, that a *k*-dimensional block be homeomorphic to a *k*-dimensional ball. The block structure is described by the data structure known as "cell list" which is generalized here for the three-dimensional case. Some experimental results are presented.

1 Introduction

Topological knowledge plays an important role in computer graphics and image analysis. Images may be represented in computers only as finite sets. Therefore it is usual to perform topological investigations in a Hausdorff space and then to transfer the results to finite sets. One of the aims of the present investigation is to demonstrate that topological investigations may be performed *directly in finite sets* on which a T_0-topology is defined. Such a topological space can be represented in computers. We demonstrate here a new tool for investigating 3-manifolds by means of computers: the three-dimensional cell list. The same tool may be implemented for economically encoding and analyzing three-dimensional images, e.g. in computer tomography.

2 State of the Art

It is known from the topological literature that the problem of the complete classification of 3-manifolds is still unsolved while the classification of 2-manifolds is known since about hundred years [3]. In recent time some efforts have been made to use computers for investigating 3-manifolds.

In [9] the following method based on the notion of a spine was suggested. A spine [2] is some kind of a two-dimensional skeleton of the 3-manifold: if K is a polyhedron, if K collapses to L [12, p. 123], and if there is no elementary collapse

G. Borgefors, I. Nyström, and G. Sanniti di Baja (Eds.): DGCI 2000, LNCS 1953, pp. 57-68, 2000.
© Springer-Verlag Berlin Heidelberg 2000

of L, then L is a *spine* of K. A standard [2] or special [9] spine of a manifold has the same fundamental group as the manifold. In [9] the notion of complexity k of a 3-manifold was introduced. It is the number of vertices (0-cells) in the so called *almost special spine* of the manifold under consideration. It has been shown that the singular graph of a special spine is a regular graph of degree 4. There are only finitely many different spines corresponding to a given regular graph of degree 4. Thus it is possible to enumerate all spines with a given number of vertices.

Matveev has also introduced the so called T-transformation, which transforms a spine of a given manifold to another spine of the same manifold, which may be simpler. He also uses topological invariants introduced in [14]. More than 1000 3-manifolds of complexity k up to nine have been analyzed by means of this method [9].

We suggest here another method of using computers for the investigation of 3-manifolds. According to our method a 3-manifold is represented as an abstract cell complex (ACC) [6] with a minimum number of cells. It is encoded by the cell list as described in Section 5. It is easy to see that homeomorphic cell lists correspond to combinatorially homeomorphic manifolds. The question whether the minimum cell list of a 3-manifold is unique is recently open. There is the hope that in the case that it is not unique, the number of different cell lists of a 3-manifold (of a limited complexity) with a minimum number of cells is not too large, so that all such lists may be exhaustively tested by a computer whether they are combinatorially homeomorphic to a cell list of some already known manifold. In this presentation we describe our method of computing the cell list with a minimum number of cells for a given 3-manifold.

3 Basic Notions

It is known [10] that any 3-manifold may be triangulated and that homeomorphic 3-manifolds are combinatorially homeomorphic. Two complexes are called *combinatorially homeomorphic* if their simplicial schemata become isomorphic after finite sequences of elementary subdivisions [12, p. 24]. However, simplicial complexes contain too many elements and are therefore difficult to process. Simplices may be united to greater cells by an operation inverse to the subdivision: a subcomplex combinatorially homeomorphic to a k-simplex (or equivalently to a k-ball) may be declared to be a k-dimensional cell or a k-cell. In what follows we shall write "homeomorphic" for "combinatorially homeomorphic".

While simplices are mostly considered as subsets of a Euclidean space we prefer to work with ACC's [6]. An ACC is a set of *abstract cells*. A non-negative integer is assigned to each cell. It is called the *dimension* of the cell. The set is provided with an antisymmetric, irreflexive and transitive binary relation called *bounding relation*. If the cell c_1 bounds the cell c_2 it is usual to write $c_1 < c_2$. A cell can only bound another cell of higher dimension.

ACC's differ both from simplicial and Euclidean complexes in so far that a cell is never a part of another cell. This property makes it possible to easily introduce the notion of *open subsets* of an ACC and thus to define a T_0-topology on it in accordance with classical axioms [6]. Although an ACC is a quotient of some Hausdorff space we do not consider the cells as subsets of a Hausdorff space, which subsets are infinite ones and therefore not representable in computers. We rather consider cells as elements of an abstract finite set. This is another advantage of the ACC's since the topological space of a finite ACC may be *directly and completely represented in a computer*. Thus there is no necessity to consider theoretical problems in a Hausdorff space (which is not representable in computers) and then to transfer the results to a different set represented in the computer. This advantage of the ACC's is widely used in the present investigation.

One of our methods of representing 3-manifolds in computers consists in constructing a four-dimensional ACC in the computer, in defining a strongly connected subset of the ACC and in calculating the boundary of the subset.

To make the number of cells as little as possible we subdivide the ACC A representing a 3-manifold into subsets each of which is homeomorphic to an open k-ball. We call such a subset a *k-dimensional block cell* or a *k-block* of A. A block b_1 of A is said to bound another block b_2 of A if b_1 contains a cell of A which bounds another cell of A contained in b_2. In this way a *bounding relation* is defined on the set of blocks of A and the set becomes an ACC B called the *block complex* of A. The blocks are cells of B. The topology of the block complex B is a quotient topology of that of the underlying ACC A, however, there is no necessity to consider it as a quotient topology of an Euclidean space.

4 Incidence Structures

4.1 The Main Idea

In topological literature manifolds are often represented as cell complexes. Thus e.g. the surface of a torus may be represented as a complex consisting of a 0-cell, two 1-cells and one 2-cell (Fig. 1a). This representation has the advantage of being very simple.

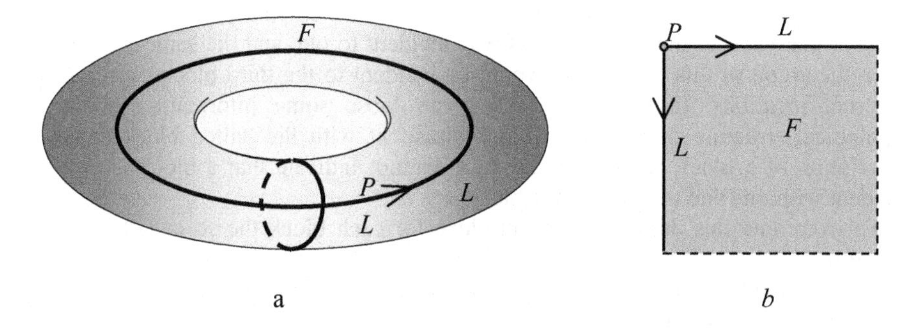

Fig. 1. Representations of the surface of a torus (a) and of a simple complex (b)

However, if one would try to interpret this representation as an ACC, difficulties would occur since e.g. the ACC's corresponding to Fig. 1a and Fig. 1b are the same: the same sets of four cells, the same bounding relation and the same dimensions of the cells. The difference between these two complexes is that each of the 1-cells in Fig. 1a bounds the 2-cell *two times*, on both sides. This may be seen, if one considers the embedding of the complex in an Euclidean space: a neighborhood of a point on the 1-cell contains two half-disks each of which lies in one and the same 2-cell. However, there is no possibility to describe this relation in the language of the ACC's.

Since one of our aims is to consider a purely combinatorial approach with no relation to a Euclidean space we consider the possibility to overcome this difficulty by introducing the notion of an incidence structure.

Definition PB: A k-block is called *proper* if its closure is homeomorphic to a closed k-ball. A block complex is called proper if all its blocks are proper.

Thus when considering Fig. 1a as a representation of a block complex then it is not a proper one: though each k-block is homeomorphic to an open k-ball the closures of the blocks are not homeomorphic to closed k-balls.

An example of a proper block complex for the surface of a torus (n=2) is shown in Fig. 2. The only drawback of this representation is that it has too many blocks as compared to Fig. 1a.

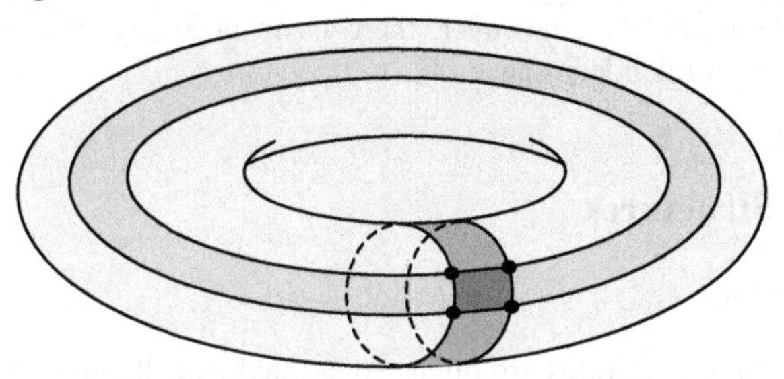

Fig. 2. A proper block complex of the surface of a torus

It is possible to reduce the number of blocks of a proper block complex while uniting two adjacent k-blocks which are not mutually simple (see Section 6 for the definition of "simple") by an operation inverse to the elementary subdivision. It may happen, that each of the united blocks was incident to one and the same third block. Then the union of this two blocks would be incident to the third block twice, at two different locations. In such a case we can loose some information about the topological structure of the set of blocks incident with the united block since the description of a block complex as an ACC cannot indicate that a block is multiply incident with one and the same other block.

To overcome this drawback we introduce for each block the so called incidence structure.

Definition IN: Two cells of an ACC are called *incident* with each other if either they are identical, or one of them bounds another one.

Definition IS: The *incidence structure* of a block BC of a proper block complex K is a subcomplex of K containing all blocks incident with BC except BC itself.

This subcomplex may be described as an ACC: by the set of its blocks, each block represented by its label, and the bounding relation.

To preserve the topological information about the set of blocks incident with a given block *BC* the incidence structure of *BC* must be stored *before the uniting* of blocks. During the uniting of two blocks the label of one of them in the incidence structure must be replaced by the label of another one. In this way it becomes possible that the label of one and the same block multiply occurs in the incidence structure of another block.

The incidence structures of all blocks of a block complex must be stored in a data structure which is a generalization of the cell list [6]. The former cell list was designed to describe two-dimensional Cartesian ACC's [7, 8] where a point may be incident with at most four lines. In the generalized three-dimensional cell list the number of blocks incident with a point or with a curve in 3D is not limited. This property is important for transformations of block complexes during topological investigations.

4.2 Incidence Structures in Multidimensional Spaces

We will show in what follows that the incidence structure of any block of a multidimensional block complex representing a closed manifold is similar to the union of two topological spheres. This fact is the basis of the development of data structures enabling an economical representation of multidimensional block complexes in computers. To prove the necessary theorem we remain the reader some definitions.

An ACC A is an *Alexandroff space* [1, 5] and hence there exists in A the *smallest open neighborhood* of each cell $c \in A$. It is the set containing c and all cells of A, bounded by c. We denote it by SON(c, A). For the incidence structure we need the subcomplex SON(c, A) without the cell c itself: we denote it by $\mathrm{SON}^*(c, A) = \mathrm{SON}(c, A) - \{c\}$.

The *closure* Cl(c, A) is a notion dual to SON(c, A). It is the set containing c and all cells of A, bounding c. Again, we need the set without c itself: $\mathrm{Cl}^*(c, A) = \mathrm{Cl}(c, A) - \{c\}$. The incidence structure of a proper block *BC* is the union:

$$IS(BC, A) = \mathrm{SON}^*(BC, A) \cup \mathrm{Cl}^*(BC, A). \quad (1)$$

The incidence structure of a non-proper block containing fewer cells must be computed while starting with that of the original proper block complex and uniting some blocks which are not mutually simple.

Definition BI: An isomorphism between two complexes, which retains the bounding relation, is called *B-isomorphism*. BI: $A \rightarrow B$ is a B-isomorphism iff for any $a_1, a_2 \in A$, $a_1 < a_2$ implies $BI(a_1) < BI(a_2)$.

Theorem SN: The set $\mathrm{SON}^*(c^k, M^n)$ of any k-cell c^k of an n-manifold M^n is B-isomorphic to an $(n-k-1)$-dimensional sphere if c^k does not belong to the boundary ∂M^n and if $0 \le k \le n-1$. The set $\mathrm{Cl}^*(c^k, M^n)$ is then B-isomorphic to an $(k-1)$-dimensional sphere.

To prove the Theorem we prove at first the particular case of $k=0$, which is the contents of the following

Lemma: The set $\mathrm{SON}^*(c^0, M^n)$ of a 0-cell $c^0 \in M^n$ is B-isomorphic to an $(n-1)$-dimensional sphere.

Proof of Lemma: According to the definition of an n-manifold M^n the SON of a point (i.e. of a 0-cell) $c^0 \in M^n$ is an open n-ball B^n. The frontier of B^n is an $(n-1)$-sphere $S^{(n-1)}$. Consider the set S=SON*(c^0, M^n). Each cell $c^k \in S$ has some cells in $S^{(n-1)}$ which bound c^k. Let us join them to a $(k-1)$-dimensional block $b^{(k-1)}$: a block $b^{(k-1)}$ corresponding to c^k is a subset of $S^{(n-1)}$ homeomorphic to an open $(k-1)$-dimensional ball, containing all $(k-1)$-dimensional cells $c^{(k-1)} \in S^{(n-1)}$ which bound c^k and containing also the closures of all cells of dimension $k-2$ each of which bounds at least two of the $c^{(k-1)}$. With other words:

$$b^{(k-1)}(c^k)=U(c^k)-\partial U(c^k), \qquad \text{where } U(c^k)=\mathrm{Cl}^*(c^k)\cap S^{(n-1)}. \tag{2}$$

All such blocks compose an $(n-1)$-dimensional block complex $SB^{(n-1)}$ of $S^{(n-1)}$, homeomorphic to $S^{(n-1)}$ and thus being a topological $(n-1)$-sphere. The map I: $S \rightarrow SB^{(n-1)}$ takes each k-dimensional cell $c^k \in S$ to a $(k-1)$-dimensional block of $SB^{(n-1)}$ corresponding to c^k. Under rather general suppositions about M^n the map I retains the bounding relation: for any two cells a, $b \in S$, $a<b$ implies $I(a)<I(b)$. Thus I is a B-isomorphism.

Proof of the Theorem: Consider the SON of a 0-cell c^0 and a k-cell $c^k \in$ SON*(c^0, M^n), $1 \le k \le n-1$. According to Lemma c^k will be mapped (as an element of SON*(c^0, M^n)) by I onto a $(k-1)$-dimensional cell $a^{(k-1)}$ of an $(n-1)$-dimensional sphere $S^{(n-1)}$. Suppose, the Theorem is true for a $(k-1)$-dimensional cell of a manifold. Since $S^{(n-1)}$ is a manifold, SON*$(a^{(k-1)}, S^{(n-1)})$ must be B-isomorphic to a sphere of the dimension:

$$(n-1)-(k-1)-1=n-1-k+1-1=n-k-1. \tag{3}$$

However, I maps SON*(c^k, M^n) onto SON*$(a^{(k-1)}, S^{(n-1)})$ and the letter onto $S^{(n-k-1)}$. Thus, if the Theorem is true for a $(k-1)$-dimensional cell it is also true for a k-dimensional one. According to Lemma the Theorem is true for k=1. Therefore, it is true for any $1 \le k \le (n-1)$.

To prove the assertion concerning Cl* it is sufficient to consider a set dual to M^n, where each k-cell is replaced by an $(n-k)$-cell, the bounding relation is reversed and the SON of a cell c is replaced by its closure.

Fig. 3 shows the SON* of a point in a 3D Cartesian ACC and its B-isomorphic map onto the surface of an octahedron, which surface is a S^2. The SON* of a point contains 8 cubes V_1 to V_8 (V_2 is removed), 12 faces and 6 edges. The B-isomorphic surface of an octahedron contains 8 faces, 12 edges and 6 points.

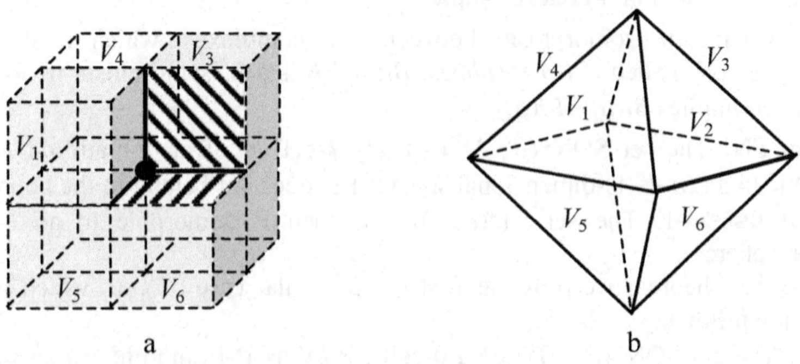

Fig. 3. The SON of a point (a) and the B-isomorphic surface of an octahedron (b)

The above results are illustrated in Table 1 showing the incidence structures of interior cells (or blocks) of a three-dimensional Cartesian ACC A^3. In cases of spaces of dimension 2 and 3 the union of SON* with Cl* happens to be B-isomorphic to a two-dimensional sphere for cells of any dimension. It should be noted that this fact is of no importance for applications since the implementation of a data structure isomorphic to the union $S^1 \cup S^0$ is simpler then that of S^2. Table 1 shows the incidence structures of cells c^k of all dimensions k and the 2-spheres B-isomorphic to them.

We use the incidence structures to describe non-proper block complexes. Such a description is the list of incidence structures of all blocks of a complex, called the *cell list* [6, 7]. The cell list for 3D complexes is described below in Section 5. According to Theorem SN the incidence structures of a k-block in an n-dimensional manifold M^n consists of two complexes one of which is B-isomorphic to $S^{(n-k-1)}$ and the other to $S^{(k-1)}$. Thus the topological structure of M^n may be described as a list of descriptions isomorphic to spheres of lower dimensions. Therefore it may be recursively composed of structures isomorphic to S^0 and S^1 which are a pair of points and a cyclically closed sequence respectively.

Table 1. Incidence structures in a 3D space

	Cl*(c^k, A^3)		SON*(c^k, A^3)		Cl* ∪ SON*
dimension k	complex	min. sphere	complex	min. sphere	min. sphere
1	*2*	*3*	*4*	*5*	*6*
0	∅	∅	dim=3	dim=2	dim=2
1	dim=0	dim=0	dim=3	dim=1	dim=2
2	dim=1	dim=1	dim=3	dim=0	dim=2
3	dim=2	dim=2	∅	∅	dim=2

5 The Three-Dimensional Cell List

On the base of Theorem SN it becomes possible to construct the three-dimensional cell list as a set of tables while each line of a table describes the incidence structure of a block of the block complex of a 3-manifold. Each incidence structure is described as one or two ACC's each of which is B-isomorphic to a k-sphere with $k \leq 2$.

Let us demonstrate an example. The tables below are constituents of the 3-dimesnional cell list of the 3-manifold with boundary shown in Fig. 4.

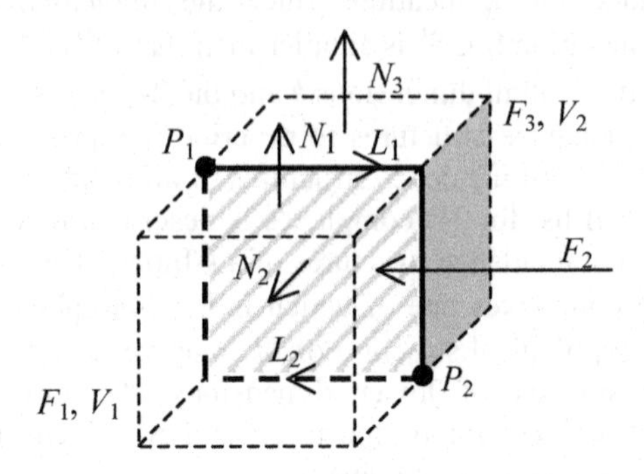

Fig. 4. Example of a simple 3-manifold with boundary

Table 2. List of the branch points (0-blocks)

Label	N_{lin}	Line	N_{SON}	Pointer	Chained list
P_1	2	$-L_1$	5	Z_1→	$-F_1$→V_1→$-F_2$→V_2→$+F_3$→0
		$+L_2$	5	Z_2→	$-F_3$→V_2→$+F_2$→V_1→$+F_1$→0
P_2	2	$+L_1$	5	Z_3→	$-F_1$→V_1→$-F_2$→V_2→$+F_3$→0
		$-L_2$	5	Z_4→	$-F_3$→V_2→$+F_2$→V_1→$+F_1$→0

The value N_{lin} denotes the number of lines (1-blocks) incident with P_j, j=1,2, while the value N_{SON} denotes the number of blocks in the SON* of the corresponding line L_i, i=1,2. These blocks compose the chained list where they are represented in the order of the rotation around L_i corresponding to the right screw. The sign "–" before the label of an oriented face F_k shows that the normal to F_k points against the rotation. The pointer Z_m points to the first element of the chained list. A zero symbol at the end of the chained list denotes that the list is not cyclically closed which may be the case for manifolds with boundary.

The list of the lines (1-blocks) contains two special blocks: the starting and the end point of the line L_i and besides that the cyclic sequence of faces and volumes incident with L_i as described above. The value N_{SON} denotes the number of blocks in the SON* of L_i.

Table 3. List of the lines (1-blocks)

Label	Start	End	N_{SON}	Pointer	Chained list
L_1	P_1	P_2	5	Z_5→	$-F_1$→V_1→$-F_2$→V_2→$+F_3$→0
L_2	P_2	P_1	5	Z_6→	$-F_3$→V_2→$+F_2$→V_1→$+F_1$→0

The list of faces (2-blocks) has a similar structure: the incidence structure of a face F_i contains two special blocks (these are now the two incident 3-blocks) and a cyclic sequence of N_{Cl} blocks in Cl*(F_i). The first symbol is repeated at the end of the chained list to show that the sequence is closed.

Table 4. List of faces (2-blocks)

Label	+Vol	–Vol	N_{Cl}	Pointer	Chained list
F_1	–	V_1	4	Z_7→	P_1→$-L_2$→P_2→$-L_1$→P_1
F_2	V_1	V_2	4	Z_8→	P_1→$-L_2$→P_2→$-L_1$→P_1
F_3	–	V_2	4	Z_9→	P_1→$+L_1$→P_2→$+L_2$→P_1

Table 5. List of volumes (3-blocks)

Label	N_f	Face	N_{Cl}	Pointer	Chained list
V_1	2	$+F_1$	4	Z_{10}→	P_1→$-L_2$→P_2→$-L_1$→P_1
		$-F_2$	4	Z_{11}→	P_1→$+L_1$→P_2→$+L_2$→P_1
V_2	2	$+F_2$	4	Z_{12}→	P_1→$-L_2$→P_2→$-L_1$→P_1
		$+F_3$	4	Z_{13}→	P_1→$+L_1$→P_2→$+L_2$→P_1

The value N_f denotes the number of faces F_j incident with V_i, i=1,2, while the value N_{Cl} denotes as before the number of blocks in Cl*(F_j).

A presentation of the fundamental group of a given complex may be computed from its cell list by the method suggested by Poincaré [11] and proved by Tietze [13]. According to this method it is necessary to find the spanning tree of the 1-dimesnional skeleton of the complex and ignore all 1-cells in the tree. Each of the remaining 1-cells is a generator, the concatenation of the generators in the perimeter of each 2-cell, being equated to identity, is a relation of the presentation of the fundamental group.

In a similar way cell lists for manifolds (with and without boundary) of greater dimension may be constructed. The list consists of incidence structures each of which is B-isomorphic to a sphere of some lower dimension and thus may be described by a cell list of lower dimension. Thus e.g. in the cell list of a 5-manifold the incidence structure of a 0- and of a 5-block is a cell list of dimension 4. The incidence structures of other blocks are lists of lower dimensions.

The gained understanding shows that cell lists for block complexes of manifolds of any dimension may be constructed by means of a recursion: the cell list of an n-manifold consists of lists of lower dimensions.

6 Computer Experiments

6.1 Generating Block Complexes and Cell Lists of 3-Manifolds

We use two methods of producing block complexes and their cell lists in the computer. The first method implements the classical idea of gluing (or identifying) the faces of a polyhedron. The description of a polyhedron may be input into the computer manually, just in the form of a three-dimensional cell list containing a single 3-block and as many 2-blocks as the number N_f of faces. Also a list of desired identifications of the faces and their closures (specifying the homeomorphism of the gluing) must be input. The corresponding computer program replaces the labels of some blocks and calculates the new incidence structures which are unions of the initial ones. The result is a cell list of a 3-manifold.

The second method consists in defining a strongly connected subset of a four-dimensional Cartesian ACC represented as a four-dimensional array in the computer. The boundary of the subset is then the desired 3-manifold. It is a three-dimensional ACC.

The block complex of a given ACC may be computed as follows. Consider two closed n-balls whose boundary intersection is an $(n-1)$-ball. Then the union of the n-balls is again an n-ball since uniting is a procedure inverse to the elementary subdivision of an n-cell. We call such two n-balls *mutually simple* or *simple relative to each other*. The union of the closures of two mutually simple n-cells or n-blocks is a closed subcomplex homeomorphic to a closed n-ball B^n.

The program selects an arbitrary n-cell of the given n-dimensional complex as the seed of B^n. Then all n-cells, which are simple relative to the growing ball B^n, are *sequentially* united with it, one cell at each step. The closures of the united n-cells are labeled as belonging to the closure of the n-block. When there are no more simple cells, the rest consisting of n-cells which are not simple relative to B^n can be subdivided into handles of indices 0 to 2 [4, p. 28]. All n-cells of a handle and the cells of their closures which are not jet labeled get a label of the handle. This is accomplished in the order of decreasing indices: all handles of index 2 first, etc.

Each handle of index k is then contracted to a k-block, i.e. a corresponding k-block is recorded in the cell list. The incidence structure of a block may be directly read from the closure of the n-cells c^n of the corresponding handle H, which cells are incident with the base of H: the cells of $Cl^*(c^n)$ contain the labels of other handles having a common boundary with H, due to the labeling procedure described above.

We have developed a computer program which automatically calculates according to the described algorithm the cell list of a two- or tree-dimensional orientable manifold without boundary. The manifold must be defined as the frontier of a strongly connected subsets of a four-dimensional Cartesian ACC. The program also minimizes the number of blocks while uniting pairs of 0-blocks (points) incident with a line (1-block) until a block complex with a single 0-block, a single 3-block, m 1-blocks and m 2-blocks is obtained (the Euler number $N^0-N^1+N^2-N^3=1-m+m-1$ must be 0).

Several examples of manifolds were successfully tested. As an example we show the results of investigating a well-known 3-manifold $S^1\times S^1\times S^1$ which may be obtained by identifying opposite faces of a cube. The minimized cell list of this manifold contains a single point, three closed curves, three faces each spanned by two curves and a single volume. Since the cell list is redundant (its redundancy is necessary for the purpose of a fast search) its contents may be represented by that of the incidence structure of the single volume. As demonstrated above, the incidence structure of any block of a 3-manifold is B-isomorphic to a 2-sphere and hence may be projected onto the plane. Fig. 5 shows a planar projection of the incidence structure of the 3-block.

In Fig. 5 elements of the same hatching represent identified blocks. The 2-blocks are denoted by *a*, *b* and *c*. Primed symbols correspond to opposite orientations. The 0- and 1-dimensional blocks are denoted by combinations of the symbols of the bounded 2-blocks.

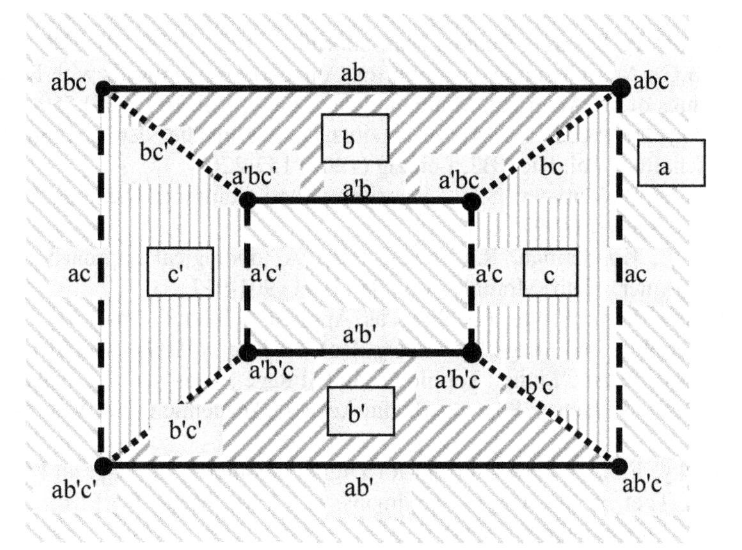

Fig. 5. Incidence structure of the 3-block of the block complex of the glued cube

A presentation of the fundamental group of this manifold may be found as follows: there is a single vertex which is the spanning tree. Therefore all three 1-blocks *ab*, *ac* and *bc* are generators. Let us denote them by $x=ab$, $y=ac$ and $z=bc$. Then the perimeter of the face *a* (the exterior area) contains the sequence $xyx'y'=1$, the perimeter of the face *b* the sequence $xzx'z'=1$ and the perimeter of the face *c* the sequence $yzy'z'=1$. Thus the fundamental group is the free abelian group of rank 3.

7 Conclusion

The described method gives the possibility to compute automatically a representation of a two- or three-dimensional manifold as a cell complex with the (almost) minimum number of cells. This evokes the hope to test by means of a fast computer the combinatorial homeomorphism of 3-manifolds as the isomorphism of cell complexes.

The method also makes it possible to compute automatically a presentation of the fundamental group of the given manifold. The method may be useful for further investigations of 3-manifolds and may be a contribution for the solution of the still unsolved problem of classifying 3-manifolds.

The tree-dimensional cell list developed here may be also used for economically encoding and analyzing tree-dimensional images, e.g. in computer tomography, or time sequences of two-dimensional images in digital television.

References

1. Alexandroff, P.: Diskrete Räume, Mat. Sbornik, Vol. 2 (1937) 501-518
2. Casler, B.G.: An Imbedding Theorem for Connected 3-Manifolds with Boundary, Proceedings of the American Mathematical Society, Vol. 16 (1965) 559-566
3. Dehn, M., Heegaard, P.: Analysis situs, Encyklopädie der mathematischen Wissenschaften, Vol. III, AB3, Leipzig (1907) 153-220
4. Fomenko, A.T., Matveev, S.V.: Algorithmic and Computer Methods for Three-Manifolds, Kluwer (1997)
5. Kong, T.Y., Kopperman, R., Meyer, P.R.: A topological approach to digital topology, Amer. Math. Monthly, Vol. 98 (1991) 901-917
6. Kovalevsky, V.A.: Finite Topology as Applied to Image Analysis, Computer Vision, Graphics and Image Processing, Vol. 45, No. 2 (1989) 141-161
7. Kovalevsky, V.A.: Finite Topology and Image Analysis, in Advances in Electronics and Electron Physics, P. Hawkes ed., Academic Press, Vol. 84 (1992) 197-259
8. Kovalevsky, V.A.: A New Concept for Digital Geometry, in Shape in Picture, O, Ying-Lie, Toet, A., Foster, D., Heijmans, H.J.A.M., Meer, P. (Eds.), Springer-Verlag Berlin Heidelberg (1994) 37-51
9. Matveev, S.V.: Computer classification of 3-manifolds, in Russian, TR, University of Cheliabinsk, Russia, (1999)
10. Moise, E.E.: Affine structures in 3-manifolds. V. The triangulation theorem and Hauptvermutung, Ann. Math. Vol. 56 (1952) 865-902
11. Poincaré, H.: Analysis situs, J. de l'Ècole Polyt. (2) Vol. 1 (1895) 1-123
12. Stillwell, J.: Classical Topology and Combinatorial Group Theory, Springer (1995)
13. Tietze, H.: Über die topologischen Invarianten mehrdimensionaler Mannigfaltigkeiten, Monatsh. Math. Phys. Vol. 19, (1908) 1-118
14. Turaev, V.G., Viro, O.Y.: State sum invariants of 3-manifolds and quantum 6j-symbol, Topology, Vol. 31, N 4 (1992) 865-902

Nearness in Digital Images and Proximity Spaces*

Pavel Pták[1] and Walter G. Kropatsch[2]

[1] Faculty of Electrical Engineering, CMP – Czech Technical University
Center for Machine Perception
Karlovo nám. 13, 121 35 Praha 2, Czech Republic
ptak@math.feld.cvut.cz

[2] Institute of Computer Aided Automation 183/2, Vienna University of Technology
Pattern Recognition and Image Processing Group
Favoritenstr.9, A-1040 Wien, Austria
krw@prip.tuwien.ac.at

Abstract. The concept of "nearness", which has been dealt with as soon as one started studying digital images, finds one of its rigorous forms in the notion of proximity space. It is this notion, together with "nearness preserving mappings", that we investigate in this paper. We first review basic examples as they naturally occur in digital topologies, making also brief comparison studies with other concepts in digital geometry. After this we characterize proximally continuous mappings in metric spaces. Finally, we show by example that the "proximite complexity" of a finite covering in a digital picture may be too high to be adequately depicted in a finite topological space. This combinatorial result may indicate another conceptual advantage of proximities over topologies.

1 Introduction

Recently there have been quite an intense investigation of topological structures in image processing, mostly in connection with the analysis of connectivity and the operation of thinning (see e.g. [2,4,9,13], etc.). An interesting attempt to introduce richer structures than those of topology, and replacing thus "local" continuity properties by a global notion of nearness, has been done in [12] where the authors contemplated the so called semi–proximity spaces as a theoretical tool in the image processing studies. In this paper we want to go on in a similar vain by shedding light on certain questions which implicitly announced themselves in the paper [12], and by complementing the results of [12] with some new findings.(Our investigation here is relatively technical. The reader is supposed to be acquainted with the motivation for investigating digital topology and thus for the selection of the problems we pursue. Reading the papers [8,12,18] would

* This work was supported by the Austrian Science Foundation under grant S 7002-MAT, by the project of Czech Ministry of Education No. VS96049, and by the Grant Agency of the Czech Republic under the grant GACR 102/00/1679.

G. Borgefors, I. Nyström, and G. Sanniti di Baja (Eds.): DGCI 2000, LNCS 1953, pp. 69–77, 2000.
© Springer-Verlag Berlin Heidelberg 2000

certainly be instrumental as it was for the authors. We hope, however, that our exposition is reasonably self-contained.)

The paper is organized as follows: In the section 2 the proximity spaces are introduced and related to the topological spaces, in particular to the topology introduced by Marcus and Wyse. Section 3 investigates the proximities in metric spaces. In the section 4 we discuss the problem whether nearness of finite partitions allows for a suitable finite topology on the index set. The conclusion summarizes the paper.

2 Proximity Spaces – In Relation to Topologies and Metrics

We shall deal with proximity spaces as implicitly defined by Riesz [16] and intensely pursued by several authors ([3,5,6,14]). Our definition is a modified version of the definition by E. Čech [3].

Definition 1 (Proximity Space). *A pair (X, π) is called a proximity space if X is a set and π is a binary relation on the power set of X, $\exp X$, which is subject to the following four requirements ($A, B, C, D, E \in \exp X$, the symbol $A \; non\pi \; B$ means that $A \; \pi \; B$ is false):*

(i) $(A \cup B) \; \pi \; C \Leftrightarrow A \; \pi \; C$ *or* $B \; \pi \; C$,
(ii) $A \; \pi \; B \Rightarrow A \neq \emptyset$ *and* $B \neq \emptyset$,
(iii) $A \cap B \neq \emptyset \Rightarrow A \; \pi \; B$,
(iv) $A \; \pi \; B \Rightarrow B \; \pi \; A$.

If (X, π) fulfils the following additional requirement (v), it is called a T-proximity:

(v) if $\{a\} non\pi B$, then there is a set A such that $\{a\} non\pi \; (X - A)$ and $\{d\} non\pi \; B$ for each $d \in A$.

The axioms of a proximity space reflect the properties we observe when we consider the common–sense nearness. It should be noted that the less plausible axiom of T-proximity – the axiom *(v)* – guarantees that the proximity in question induces a topology (the proximities induce only the closure spaces – see [3] and [12]). There is an important link of proximity spaces and topological spaces.

Theorem 1 (Proximity Space, Topological Space).

(i) Let (X, π) be a proximity (resp. T-proximity) space. Let $\bar{A} = \{x \in X | \{x\} \pi A\}$. Then $(X, ^-)$ is a closure (resp. topological) space (the closure $^-$ is said to be induced by the proximity π).
(ii) Let $(X, ^-)$ be a closure (resp. topological) space and let $(X, ^-)$ fulfil the following condition (the Čech R_0 condition): If $a, b \in X$, then $a \in \overline{\{b\}}$ if and only if $b \in \overline{\{a\}}$. Let $A \; \pi \; B \Leftrightarrow \bar{A} \cap \bar{B} \neq \emptyset$. Then (X, π) is a proximity (resp. T-proximity) space and, moreover, (X, π) induces $(X, ^-)$. k

The proof of the above result is slightly technical (though essentially easy – one would use Th.23.B3 and Th.25.B9 of Čech [3]) and we omit it.

By the previous result, we can associate closures (resp. topologies) to proximities (resp. T-proximities) although not all topologies are proximizable. On the other hand, observe that many proximities on X may induce the same topology on X (and, vice versa, a topology may give rise to several proximities which induce it). For the reader's intuition, let us note that e. g. the following two proximities p_1, p_2 on an infinite set X define the same (discrete) topology: $A\, p_1\, B \Leftrightarrow A \cap B \neq \emptyset$, $A\, p_2\, B \Leftrightarrow$ either $A \cap B \neq \emptyset$ or both the sets A and B are infinite.

It is possible to adopt the notion of proximity as primary and view the notion of topology as secondary. One of the reasons for doing it is that the topology describes only *the local character of points* in a picture (or, if we want, the proximity of points and sets). But if we are to treat geometrical qualities of a picture – a situation which typically arises in image processing – it is *the proximity of sets* which matters most. E.g. in the case of scanned text documents the characters are the connected components of the black pixels of the thresholded image. The pixel sets of adjacent characters of words are closer to each other than between the words.

Thus, besides the concrete reason obvious from the study of thinning and shape deformation [12], the general reason for investigating the proximity relation lies in its fundamental role in all kinds of geometrically oriented considerations.

In view of Theorem 1 and the investigation of connectedness in digital images [20], the following proximity in Z^2 is worth recording. Recall that if $[r,s] \in Z^2$, then by the 4-neighbourhood of $[r,s]$ we mean the set $\{[r,s], [r \pm 1, s], [r, s \pm 1]\}$.

Definition 2 (Marcus–Wyse proximity). *Let Z^2 denote the subset of R^2 consisting of all points with the integer coordinates. Let $A \subset Z^2, B \subset Z^2$. Let us write $A\ \pi\ B$ if either $A \cap B \neq \emptyset$ or there exists a point $p = [r,s] \in A \cup B$ such that the following assertion holds true:*

(i) $r + s$ is an odd number,
(ii) if $p \in A$, then the 4-neighbourhood of p intersects B,
(iii) if $p \in B$, then the 4-neighbourhood of p intersects A.

The result of [20] (see also [11]) can be now reformulated proximity-wise. Following [12], let us agree to say that a set S in a proximity space (P, π) is connected if S cannot be written in the way $S = A \cup B$, where A, B are two non-proximal sets (i.e. $A\ non\,\pi\ B$).

Theorem 2. *The graph–theoretic connectedness in Z^2 induced by the 4-neighbourhood adjacency relation coincides with the proximite connectedness induced by the Marcus-Wyse proximity.*

Let us take up proximities on metric spaces (here all proximities will automatically be T-proximities). Let us consider a metric space, (M, ρ).Then (M, ρ)

can be viewed as a topological space with the closure $\bar{A} = \{x \in M \mid \rho(x, A) = 0\}$. In this way the metric space (M, ρ) induces a "topological" proximity, π_t, defined as follows: $A\ \pi_t\ B \Leftrightarrow \bar{A} \cap \bar{B} \neq \emptyset$. But (M, ρ) also induces a metric proximity, π_m. Write, for two subsets A and B, $\rho(A, B) = \inf_{a \in A, b \in B} \rho(a, b)$ and set $A\ \pi_m\ B \Leftrightarrow \rho(A, B) = 0$. It is easily seen that if $A\ \pi_t\ B$, then $A\ \pi_m\ B$ but not necessarily the other way round. If, for instance, A is the graph in R^2 of the function $f(x) = \frac{1}{x}$ and B is the x–axis in R^2, then $A\ \pi_m\ B$ with respect to the Euclidean metric, but not $A\ \pi_t\ B$.

For a reader not trained in topology, let us explicitly clarify why (and when) the latter phenomenon (of the difference of π_t and π_m) may occur. Recall that a metric space M is called *compact* ([3,5], etc.) if each sequence in M allows for a convergent subsequence in M. As known, a subset P of R^n endowed with the Euclidean metric (taken form R^n) is compact if and only if it is closed and bounded.

Theorem 3. *Let (M, ρ) be a compact metric space. Then $\pi_t = \pi_m$, i.e. in compact metric spaces the topological proximity agrees with the metric proximity.*

Proof. Suppose that $A, B \subset M$. It only remains to show that if $A\ \pi_m\ B$, then $A\ \pi_t\ B$, the other implication is always valid. Let $A\ \pi_m\ B$. We are to show that $\bar{A} \cap \bar{B} \neq \emptyset$. The relation $A\ \pi_m\ B$ means $\rho(A, B) = 0$. It follows that for each $n \in N$ we can find points $a_n \in A$, $b_n \in B$ such that $\rho(a_n, b_n) \leq \frac{1}{n}$. By compactness, there is a subsequence, a_{n_k}, of $\{a_n\}$ which converges to some element $a \in M$, and there is a subsequence, b_n, of $\{b_{n_k}\}$ which converges to some $b \in M$. Obviously, $a = b$ and therefore the sequences $\{a_{n_\ell}\}$ and $\{b_{n_\ell}\}$ converge to a common element $a (= b)$. Since $a \in \bar{A}$ and $b \in \bar{B}$ and since $a = b$, we see that $\bar{A} \cap \bar{B} \neq \emptyset$. This completes the proof.

The morphisms in the category of proximity spaces are the proximally continuous mappings.

Definition 3 (proximally continuous mapping). *Let (X_1, π_1), (X_2, π_2) be proximity spaces. A mapping $f\colon X_1 \to X_2$ is called proximally continuous if the following implication is true:*

If $A, B \subset X_1$ and $A\ \pi_1\ B$, then $f(A)\ \pi_2\ f(B)$.

Thus, proximally continuous mappings are those mappings which preserve proximity. It is easily seen that a proximally continuous mapping is automatically continuous when understood as a mapping between the respective topological spaces induced by proximities. A continuous mapping does not have to be proximally continuous even if we consider it with respect to the metric proximity.

Example: Consider the function $f(x) = x^2\colon (0, +\infty) \to (0, +\infty)$. This mapping is obviously continuous but not proximally continuous. Indeed, let p_n be such a sequence that $(p_n + \frac{1}{2})^2 - p_n^2 \geq 1$. Let $A = \{p_n \mid n \in N\}$ and $B = \{p_n + \frac{1}{n} \mid n \in N\}$. Then $A\ \pi_m\ B$ but $f(A) non\, \pi_m\ f(B)$.

3 Metric Proximities

In this section we are going to prove that the metric proximity of a space determines, up to a metric equivalence, the metric of the space. This relatively deep result has found applications in a number of geometric problems (see e. g. [5] for relevant comments). This result can be expressed in terms of "small" (=countable) subsets of the metric space in question and therefore it may have bearing on digital geometry. Also, we see that metric considerations of digital pictures ([17,18,19]) have a proximity character (i.e. can be expressed in proximity terms).

Let us take up the proof of the result. We provide a simplified transparent proof based on elementary reasonings only. We also point out other features of metric spaces relevant to proximity. Recall first two standard definitions.

Definition 4. *Let (M, ρ) be a metric space and let $\{a_n\}$ be a sequence in M. We say that $\{a_n\}$ is a* **Cauchy subsequence** *if for each $\varepsilon > 0$ there is $n_0 \in N$ such that $\rho(x_m, x_n) \leq \varepsilon$ provided $n \geq n_0, m \geq n_0$. We say that a sequence$(x_n)_{n \in N}$ in M is* **metrically discrete** *(of order ε) if for any $n, m \in N$ we have $\rho(x_m, x_n) \geq \varepsilon$.*

The following proposition is essential in our argument. It may be interesting in its own right. Before we formulate it, let us agree to call the set $B_r(a) = \{b \in M \mid \rho(a, b) \leq r\}$ the r-ball around a.

Proposition 1 (Sequence principle in metric spaces). *Each sequence in a metric space contains either a Cauchy sequence or a metrically discrete subsequence.*

Proof. Take a sequence in a metric space and form the collection of all 1 -balls centered at each of its points. If each of these balls contains only finitely many points of the sequence, we can easily construct a subsequence of the given sequence which is metrically discrete of order 1. It not, there is a 1-ball around a point of the sequence which contains infinitely many points of the sequence. Take these points and form the collection of all $\frac{1}{2}$-balls centered at these points. If each of these $\frac{1}{2}$-balls contains only finitely many points, we can easily construct a metrically discrete subsequence of order $\frac{1}{2}$. If not, there is a point such that the $\frac{1}{2}$-ball around it contains infinitely many points. Going on this way inductively, either the procedure stops at the n-th step and we have an $\frac{1}{n}$-discrete subsequence, or we obviously obtain a Cauchy subsequence of the given sequence.

We shall need one more metric notion. Let us recall it together with proximal continuity in metric space.

Definition 5. *Let (M_1, ρ_1), (M_2, ρ_2) be metric spaces. Let $f\colon M_1 \to M_2$ be a mapping. In accord with our general definition, we say that f is* **proximally continuous** *if the following property is fulfilled:*

If P and Q are subsets of M_1 such that $\rho_1(P, Q) = 0$, then $\rho_2(f(P), f(Q)) = 0$.

We say that f is **metrically continuous** *if for any $\varepsilon > 0$ there exists $\delta > 0$ such that whenever $\rho_1(x, y) < \delta$, then $\rho_2(f(x), f(y)) < \varepsilon$ (such a mapping f is sometimes called uniformly continuous).*

Let us now formulate and prove the main result of this section. In the effort to make the proof accessible for nonspecialists in topology, we use only elementary reasonings. The novelty seems to be the utilization of the sequence principle as established in Prop. 1.

Theorem 4. *A mapping between metric spaces is proximally continuous if and only if it is metrically continuous.*

A consequence: If two metrics on a set induce the same proximity, they have to be (metrically) equivalent.

Proof. A metrically continuous mapping between metric spaces is obviously proximally continuous. Let us take up the nontrivial implication of the theorem.

Let $f\colon M_1 \to M_2$ be proximally continuous. Suppose f is not metrically continuous. It means that for some $\varepsilon > 0$ there exist sequences $(a_n)_{n\in N}$ and $(b_n)_{n\in N}$ in M_1 such that $\rho_1(a_n, b_n) \to 0$ whereas $\rho_2(f(a_n), f(b_n)) \geq \varepsilon$. Let us look at the sequence $f(a_n)_{n\in N}$. If it contains a Cauchy subsequence, then there is an infinite subset of this sequence which is all contained in an $\frac{\varepsilon}{4}$-ball. Let us denote this subsequence by $f(a_{n_k})_{k\in N}$. Then the sets $\{f(a_{n_k})\colon k \in N\}$ and $\{f(b_{n_k})\colon k \in N\}$ are obviously $\frac{\varepsilon}{2}$-apart (the triangular inequality), which is absurd. If there is a Cauchy subsequence of $f(b_n)_{n\in N}$, we can apply an analogous reasoning. Suppose that neither of the former two cases applies. By the sequence principle, we can easily construct metrically discrete subsequences $f(a_{n_k})_{k\in N}$ and $f(b_{n_k})_{k\in N}$ of $f(a_n)_{n\in N}$ and $f(b_n)_{n\in N}$, respectively. For simplicity, let us denote them again by $f(a_n)_{n\in N}$ and $f(b_n)_{n\in N}$.

Let us assume that these sequences are metrically discrete of order α. Define $r = \frac{1}{2}\min\{\varepsilon, \alpha\}$ and form the collection of all the r-balls centered at $f(a_n)_{n\in N}$. Note that each of them contains at most one of the elements $f(b_n)_{n\in N}$. Let us now proceed inductively. Put $n_1 = 1$. Take $n_2 \in N$, $n_2 > 1$ such that

$n_2 > h$ if $f(b_h)$ belongs to the r-ball centered at $f(a_1)$, and

$n_2 > k$ if the r-ball centered at $f(a_k)$ contains $f(b_1)$,

(if neither h or k exists, we simply take $n_2 = n_1 + 1$). By induction, given $n_1 < n_2 < \cdots < n_i$, take $n_{i+1} \in N$, $n_{i+1} > n_i$ such that

$n_{i+1} > h$ if $f(b_h)$ belongs to the r-ball centered at $f(a_{n_i})$, and

$n_{i+1} > k$ if the r-ball centered at $f(a_k)$ contains $f(b_{n_i})$.

Note that, by our construction, if $j \in \{n_k | k \in N\}$, then $f(b_j)$ does not belong to any of the r-balls centered in $f(a_{n_k})$. The sets $f(a_{n_k})_{k\in N}$ and $f(b_{n_k})_{k\in N}$ are therefore apart of the order r. Write $A = \{a_{n_k}\}_{k\in N}$, $B = \{b_{n_k}\}_{k\in N}$. The $\rho_1(A, B) = 0$ but $\rho_1(f(A), f(B)) \geq r$. This means that f is not proximally continuous - a contradiction. This completes the proof.

The main consequence of the latter theorem (and its proof) as far as the potential application in image processing goes is contained in the next theorem. It does not seem to be explicitly formulated in the literature but it is this formulation which may have relevant bearing on the digital images studies – it reduces proximities of general sets to proximities of small (=countable, digitally accessible) sets. (Recall that a set is said to be countable if it has the smallest possible infinite cardinality, i.e. if it has the cardinality of natural numbers.)

Theorem 5. *Let ρ_1, ρ_2 be two metrics on a set M. If $\rho_1(A, B) = 0 \Leftrightarrow \rho_2(A, B) = 0$ for all* **countable** *subsets of M, then the metrics ρ_1, ρ_2 are metrically equivalent. In other words, if the proximities given by ρ_1 and ρ_2 agree when restricted to countable subsets of M, then ρ_1 and ρ_2 are metrically indistinguishable.*

4 Near and Far Sets in a Finite Partition – Could "nearness" Be Controlled by a Finite Topology?

Each picture can be viewed as a partition of the underlying set into a finite family of sets. The main point in understanding the partition is specifying which sets are "near" (proximal) and which are "far" (non-proximal). In this section we exhibit an example which shows that the proximal relation of sets in a finite partition may be too complex to be described with the help of the notion of finite topology. This relates in a natural way our investigation here with topological studies in image processing (see [13,8,11,18,10,20], etc.). (Recall that by a finite partition of a set S we mean a mutually disjoint collection $\{S_k \ (k = 1, 2, \ldots, n)\}$ of non-empty subsets of S such that $\bigcup_{k=1}^{n} S_k = S$.)

Let us now introduce an auxiliary notion.

Definition 6. *Let (X, π) be the metric proximity space of the metric space (M, ρ). Let $P = \{S_k \mid k = 1, 2, \ldots, n\}$ be a partition of X. Let us say that P is* **controlled by topology** *if there is a topology t on the set $T = \{1, 2, \ldots, n\}$ such that $S_k \ \pi \ S_\ell$ if and only if the set $\{k, \ell\}$ is topologically connected as a topological subspace of T.*

If each partition could be topologically controlled, which seems conceivable at first sight, we would find ourselves in an advantages situation in view of the understanding of finite topologies (see e.g. [18] and [10]). However, it is not necessarily the case. We will illustrate it by a simple example. Before, observe that we can restrict ourselves to open sets of the partition if we allow for a small degree of overlapping (in practice, we have to allow for it anyhow in view of the imperfection of our measurement).

Proposition 2. *Let (M, ρ) be a metric space. Let $\{S_k \mid k = 1, 2, \ldots, n\}$ be a partition of M. Then there is an ε, $\varepsilon > 0$ such that the following statement holds true: If $T_k = \{x \in M \mid \rho(x, S_k) < \varepsilon\}, k = 1, 2, \ldots, n$, then each T_k is an open subset of M, and $T_k \cap T_\ell \neq \emptyset$ if and only if $S_k \ \pi_\rho \ S_\ell$.*

Proof. All sets T_k, $k = 1, \ldots, n$ are obviously open. Further, consider all couples S_k, S_ℓ such that S_k *non* π_ρ S_ℓ. Since there are only finitely many couples with this property, there is a sufficiently small ε, $\varepsilon > 0$ such that $T_k \cap T_\ell = \emptyset$

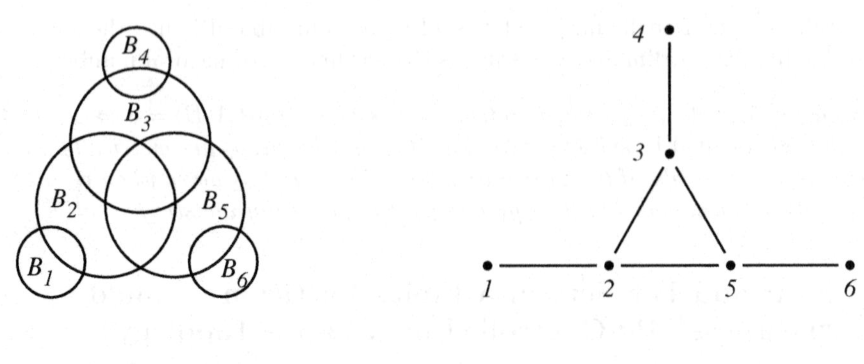

Fig. 1. **Fig. 2.**

for all couples of T_k, T_ℓ corresponding to the couples S_k, S_ℓ considered. Since we have $S_k \ \pi_m \ S_\ell$ for the remaining couples, we infer from the definition of metric proximity that the neighbourhood sets T_k, T_ℓ, corresponding to these S_k, S_ℓ fulfil $T_k \cap T_\ell \neq \emptyset$. The proof is complete.

We see from the above proposition that we can simply construct the counterexample on the disjointness – intersection basis of an open covering.

Example: Consider the Fig. 1 above. Let $\{B_i | i \leq 6\}$ be the family of balls in R^2 as drawn in Fig. 1. Then the proximity of the family $\{B_i | i \leq 6\}$ cannot be topologically controlled.

Indeed, considering the couples of balls which are disjoint (resp. which overlap), we see that the existence of the required topology would amount to having a topology on the set $\{1, 2, 3, 4, 5, 6\}$ such that precisely the following pairs would constitute connected subspaces: $\{1, 2\}, \{2, 3\}, \{2, 5\}, \{3, 4\}, \{3, 5\}$,and $\{5, 6\}$. This would mean the existence of topology on the Fig. 2 above which would produce the 8-adjacency connectedness but this is impossible as shown in [13] and [11].

5 Conclusions

We suggest that the notion of proximity space might be a useful tool for theoretical studies in image processing. We exhibit basic examples and we link them with previous topological investigations (for instance, with the Marcus–Wyse topologies). Then we analyze more thoroughly the metric proximities. As a main result we show that the metric proximity of small sets determines in a way the metric of the underlying space. The interpretation of this result reads, roughly, that if we can verify which sequences of points in a metric space are proximal and which are not, we can in a certain sense recover the metric. In the end we show that a topological result known from the investigation of 8-adjacency relation in Z^2 disproves a conjecture about "topologizing" proximities in a partition.

References

1. P. Alexandroff: *Elementary Concepts of Topology*, A translation of P. Alexandroff, 1932, by A. E. Farley, Dover, New York, 1961
2. L. Boxer: *Digitally continuous functions*, Pattern Recognition Lett. 15 (1994), 833–839 69
3. E. Čech: *Topological Spaces*, J. Wiley-Interscience Publ., New York, (1966) 70, 71, 72
4. J.–M. Chassery: *Connectivity and consecutivity in digital pictures*, Computer Graphics and Image Processing 9 (1979), 294–300 69
5. J. Isbell: *Uniform Spaces*, Publ. AMS, Providence, (1968) 70, 72, 73
6. H. Herrlich: *A concept of nearness*, General Topology Appl. 5 (1974), 191–212 70
7. E. R. Khalinsky, E. R. Kopperman and P. R. Meyer: *Computer graphics and connected topologies on finite ordered sets*, Topology Appl. 36, 117, (1990)
8. T. Y. Kong and A. Rosenfeld: *Digital topology: introduction and survey*, Computer Vision, Graphics, and Image Processing 48 (1989), 367–393 69, 75
9. T. Y. Kong: *On the problem of determining whether a parallel reduction operator of n-dimensional binary images always preserves topology*, Proc. SPIE's Conf. on Vision Geometry, (1993), 69–77 69
10. V. A. Kovalevski: *Finite topology as applied to image analysis*, Computer Vision, Graphics and Image Processing 46, (1989), 141–161 75
11. W. Kropatsch, P. Pták: *The path–connectedness in Z^2 and Z^3 and classical topologies.* Advances in Pattern Recognition, Joint IAPR International Workshops SSPR'98 and SPR'98, Sydney 1998, Springer Verlag – Lecture Notes in Computer Science Vol. 1451, 181–189 71, 75, 76
12. L. Latecki, F. Prokop: *Semi–proximity continuous functions in digital images*, Pattern Recognition Letters 16 (1995), 1175–1187 69, 70, 71
13. L. Latecki: *Topological connectedness and 8-connectedness in digital pictures*, Computer Vision, Graphics and Image Processing: Image Understanding 57, (1993), 261–262 69, 75, 76
14. S. A. Naimpally and B. D. Warrack: *Proximity Spaces*, Cambridge Univ. Press, Cambridge 1970 70
15. J. Pelant, P. Pták: *The complexity of σ-discretely decomposable families in uniform spaces*, Comm. Math. Univ. Carolinae 22 (1981), 317–326
16. F. Riesz: *Stetigkeitsbegriff und abstrakte Mengenlehre*, Atti IV Congresso Internazionale dei Matematici, Roma, 1908, Vol. II, 18–24 70
17. A. Rosenfeld: *Digital topology*, Am. Math. Monthly 86, (1979), 621–630 73
18. A. Rosenfeld: *"Continuous" functions on digital pictures*, Pattern Recognition Lett. 4, (1986), 177–184 69, 73, 75
19. A. Rosenfeld and J. L. Pfaltz: *Sequential operations in digital picture processing*, J. Assoc. Comput. Mach. 13, (1966), 471–494 73
20. F. Wyse and D. Marcus et al.: *Solution to Problem 5712*, Am. Math. Monthly 77, 1119, 1970 71, 75

Morphological Operators with Discrete Line Segments

Pierre Soille

EC Joint Research Centre, Space Applications Institute
TP 441, I-21038 Ispra, Italy
Pierre.Soille@jrc.it
http://ams.egeo.sai.jrc.it/soille

Abstract. The morphological approach to image processing consists in probing the image structures with a pattern of known shape called *structuring element*. In this paper, we concentrate on structuring elements in the form of discrete line segments, including periodic lines. We investigate fast algorithms, decomposition/cascade schemes, and translation invariance issues. Several application examples are provided.

1 Introduction

In mathematical morphology [15,20,21,9,23], image structures are extracted or filtered out by letting them interact with a pattern of known shape called structuring element. While discrete approximations of disk structuring elements are desirable in many applications, discrete line segments of a given length and orientation are best suited for processing thin structures such as roads in satellite images or ridges and valleys in fingerprints. In addition, cascades of morphological operators with line segments can be used for producing discrete approximations of disks. The goal of this paper is to study morphological operators based on discrete line segments.

The paper is organised as follows. Issues related to the definition of discrete line segments are detailed in Sec. 2. Recursive van Herk's algorithm for erosions and dilations is recalled in Sec. 3 together with its extension to arbitrary directions, grey scale structuring elements, and 3-dimensional images. Section 4 concentrates on cascades of morphological operators with line segments. Parallel combinations of a bank of openings and closings with line segments for varying orientations are investigated in Sec. 5. Before concluding, we show in Sec. 6 that binary and grey scale convex hulls can be obtained by performing max computations along a series of periodic lines.

2 Discrete Line Segments

2.1 Connected Line Segments

Rosenfeld [19] showed that a digital arc L is a digitization of a straight line segment if and only if it satisfies the so-called *chord property*, i.e., the line segment

G. Borgefors, I. Nyström, and G. Sanniti di Baja (Eds.): DGCI 2000, LNCS 1953, pp. 78–98, 2000.
© Springer-Verlag Berlin Heidelberg 2000

joining any two points of L lies everywhere within a distance 1 of S. Note that the Kim's area property [13] and chord property are equivalent as pointed out later by the same author [14].

Bresenham [2] has proposed an attractive algorithm for generating discrete lines satisfying the chord property (see [4] for further improvements). This algorithm is especially attractive on systems not equipped for fast floating point operations because it requires only integer arithmetic operations. The pixel selection criterion is based on a distance criterion. For instance, Fig. 1 shows a continuous line segment of slope 3/5 over a square grid of pixels. The pixels belonging to the discrete line approximating this continuous line segment are determined using distance measurements along each vertical line linking the centres of the pixels of the grid. These vertical lines are represented by vertical dotted lines in Fig. 1. For each vertical line, the pixel whose centre is closest to the continuous line belongs to the corresponding discrete line. When the slope is larger than 1, the horizontal dotted lines linking the centres of the pixels must be considered instead of the vertical ones.

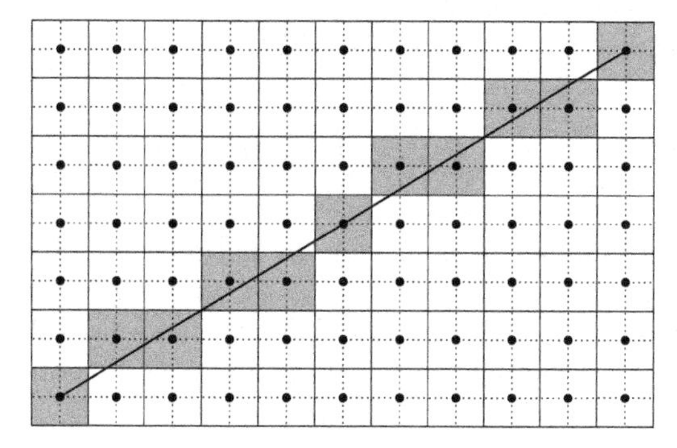

Fig. 1. Euclidean line of slope 3/5 and corresponding Bresenham discrete line: each time the Euclidean line cuts the vertical lines linking the centres of the pixels (i.e., the vertical dotted lines), the centre of the pixel which is closest to the Euclidean line defines a pixel of the discrete line

Note that for a Bresenham line of slope in the form of an irreducible fraction y/x, the number of distinct line segments occurring along the discrete line equals $\max(|x|, |y|)$. It is convenient to include the forms 0/1 and 1/0 for referring to horizontal and vertical lines respectively. The number of distinct line segments corresponds to the periodicity k of the elementary pattern occurring along the Bresenham line, i.e., $k = \max(|x|, |y|)$. In the sequel, we denote by $L_{\lambda_i,(x,y)}$ the connected line segment obtained by considering λ successive pixels of a Bresenham line of slope y/x, starting from the ith pixel of the line ($i \in \{1, \dots, k\}$). For example, Fig. 2 shows the five line segments of length equal to 11 pixels occurring for a slope of 3/5: $L_{11_1,(5,3)}$, $L_{11_2,(5,3)}$, $L_{11_3,(5,3)}$, $L_{11_4,(5,3)}$, and $L_{11_5,(5,3)}$.

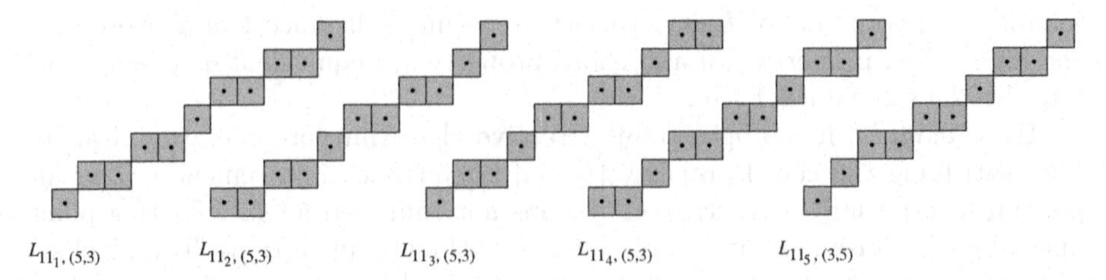

Fig. 2. The five possible line segments occurring along a Bresenham line of slope 3/5 (here for a length of 11 pixels). Note that all these line segments satisfy the chord property

2.2 Periodic Lines

Digital connected line segments at arbitrary orientation are broad approximations of Euclidean line segments. This led Jones and Soille [11,12] to introduce the concept of *periodic* lines by considering only those points of the Euclidean line that fall exactly on grid points.

In mathematical terms, a periodic line $P_{\lambda,\boldsymbol{v}}$ is defined as follows:

$$P_{\lambda,\boldsymbol{v}} = \bigcup_{i=0}^{i=\lambda-1} i\boldsymbol{v}, \tag{1}$$

where $\lambda > 1$ is the number of points in the periodic line and $\boldsymbol{v}$ is a constant vector[1]. The vector $\boldsymbol{v}$ is in the form $\boldsymbol{v} = (x, y)$ where $x, y \in \mathbf{Z}$. Similarly to Bresenham lines, we define the periodicity k of the periodic line as follows: $k = \max(|x|, |y|)$. For example, Figs. 3a, b, and c correspond to the periodic lines $P_{3,(1,0)}$, $P_{3,(1,1)}$ and $P_{3,(2,1)}$ respectively.

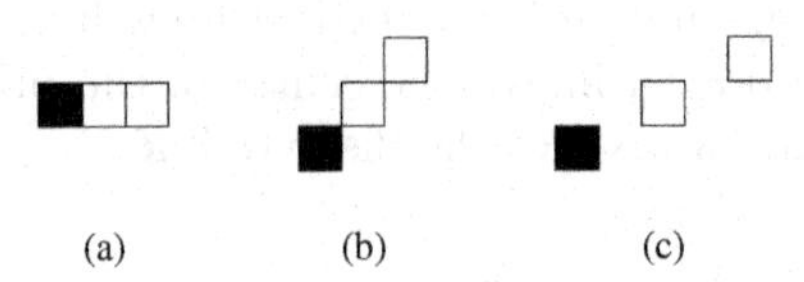

Fig. 3. Examples of periodic lines. (a) $P_{3,(1,0)}$. (b) $P_{3,(1,1)}$. (c) $P_{3,(2,1)}$. Points belonging to a line are indicated by squares. The position of the origin of each line is indicated by a black square

An interesting property of a Bresenham line of slope in the form of an irreducible fraction y/x is that it is exactly covered by $\max(|x|, |y|)$ Euclidean lines

[1] Note that periodic lines were originally defined [12] as follows: $\cup_{i=0}^{i=\lambda} i\boldsymbol{v}$. Here, we have adapted the definition of periodic lines so that λ equals to the number of pixels of the periodic line (rather than $\lambda + 1$ in the original definition), in accordance with the parameter λ used for Bresenham connected line segments L.

of slope y/x. This property is illustrated in Fig. 4, for a line whose slope equals 2/3. Indeed, due to the construction of a Bresenham line, the periodicity of the pixels in a line of slope y/x equals $\max(|x|, |y|)$ and all these pixels belong to a Euclidean line of slope y/x. In other words, any Bresenham lines of slope y/x is covered by $\max(|x|, |y|)$ periodic lines of the same slope. This property is at the basis of the translation invariant implementation of half-plane closings described in Sec. 6.

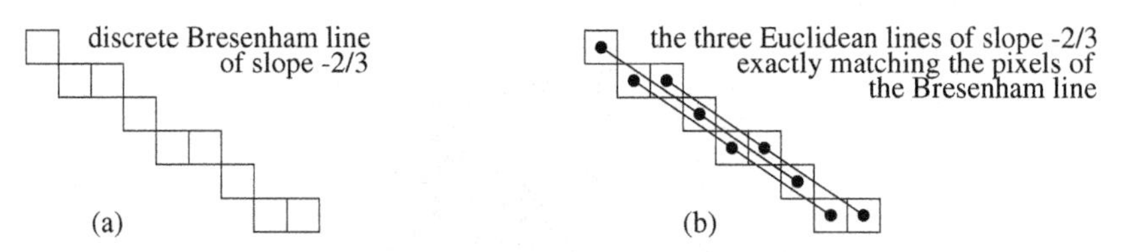

Fig. 4. (a) A Bresenham line segment of slope $-2/3$. (b) The three Euclidean line segments matching all points of *(a)*. Note that the intersection between each Euclidean line and the Bresenham line defines a periodic line $P_{3,(3,-2)}$

2.3 Angular Resolution

In a discrete grid, the angular resolution of a discrete line segment depends on its length. In a square grid, only $2n - 2$ directions can be defined with a connected line segment of odd length equal to n pixels, and whose middle and extreme pixels are matched by the Euclidean line of the same orientation. By relaxing the condition regarding the middle pixel, there are $2(2n-2)$ possible orientations for any $n \geq 2$. For example, Fig. 5 illustrates the 8 (reps. 16) possible orientations for a connected line segments containing 5 pixels.

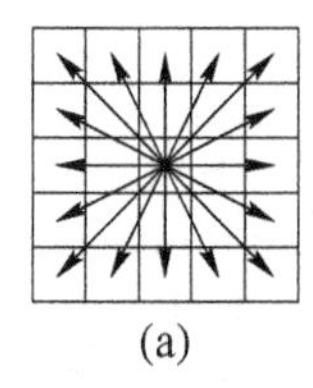

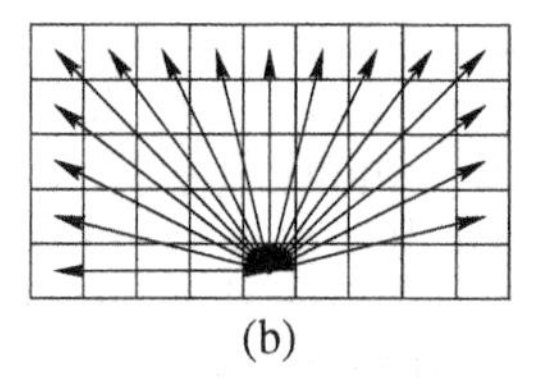

Fig. 5. (a) The 8 possible orientations for a line segment of 5 pixels and whose middle and extreme pixels are matched by the Euclidean line of the same orientation. (b) The 16 possible orientations by relaxing the condition concerning the middle pixel

If we merely impose to the line segments to contain at least two pixels exactly matching the corresponding Euclidean line, the number of distinct orientations equals $4\,\text{card}(F_i) - 4$, where F_i is the Farey sequence [3] of order i, and card($\cdot$) returns the number of elements (i.e., cardinal number). The Farey sequence F_i of order $i \geq 1$ is the ascending sequence of all fractions p/q for which $0 \leq p/q \leq 1$, $q \leq i$, and p and q are nonnegative integers with no common divisors other than 1 (note that the form $0/1$ is included in the sequence). A graphical representation of all possible discrete slopes in the range $[0, 1]$ for a line segment whose length is less than or equal to 32 pixels is shown in Fig. 6.

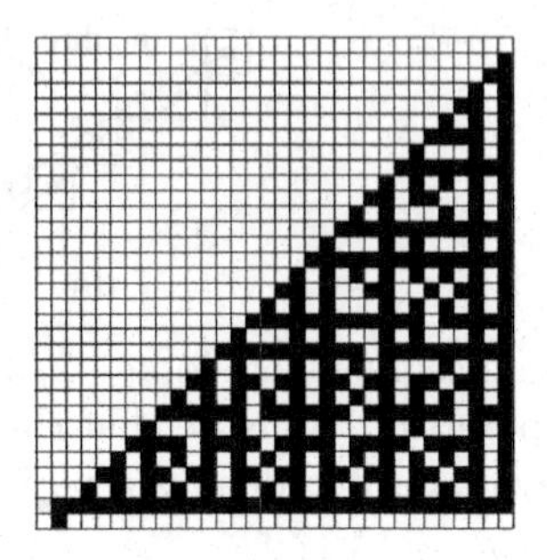

Fig. 6. Graphical representation of all possible discrete slopes in the range $[0, 1]$ and for a length less than or equal to 32 pixels. Each slope is obtained by linking the origin (lower left pixel) to a given dark pixel

3 Recursive Algorithm for Erosions and Dilations

3.1 Principle [34,8]

The recursive algorithm is due to van Herk [34] and, almost simultaneously, Gil & Werman [8] (see also [6]). We follow here the notations and description proposed in [23, pp. 77-78]. A 1-dimensional input image f of length nx is divided into blocks of size λ, where λ is the length of the line segment in number of pixels. The elements of f are indexed by indices running from 0 to $nx - 1$. It is also assumed that nx is a multiple of λ. Two temporary buffers g and h of length nx are also required. In the case of dilation, the maximum is taken recursively inside the blocks in both the right and left directions (right for g and left for h). When both g and h have been constructed, the result for the dilation r at any coordinate x is given by considering the maximum value between g at position $x + \lambda - o - 1$ and h at position $x - o$, o denoting the coordinate of the origin of the structuring element (e.g., 0 for the first pixel of the line segment). This recursive dilation algorithm can be written as follows:

$$g(x) = \begin{cases} f(x), & \text{if } x = 0, \lambda, \ldots, (m-1)\lambda, \\ \max[g(x-1), f(x)], & \text{otherwise.} \end{cases}$$

$$h(x) = \begin{cases} f(x), & \text{if } x = m\lambda - 1, (m-1)\lambda - 1, \ldots, \lambda - 1, \\ \max[h(x+1), f(x)], & \text{otherwise.} \end{cases}$$

$$r(x) = \max[g(x + \lambda - o - 1), h(x + o)].$$

In [7], minor improvements to the recursive procedure are proposed. This is achieved by reducing the number of computations in the generation the forward and backward buffers and the merge procedure. Other enhancements for the simultaneous computation of min and max filters and openings/closings are also developed. When dealing with binary images, a recursive algorithm based on directional distance transforms also running in constant time is described in [17].

3.2 Extension to Arbitrary Directions and Periodic Lines

Principle [26,27] The recursive procedure is directly applied to the image pixels falling along a line at given angle. The line is then translated and the whole procedure is repeated until all image pixels have been processed, i.e., until the translations of the line have swept the whole image plane. The direction of the translation depends on the slope of the line (see Fig. 7). By doing so, each

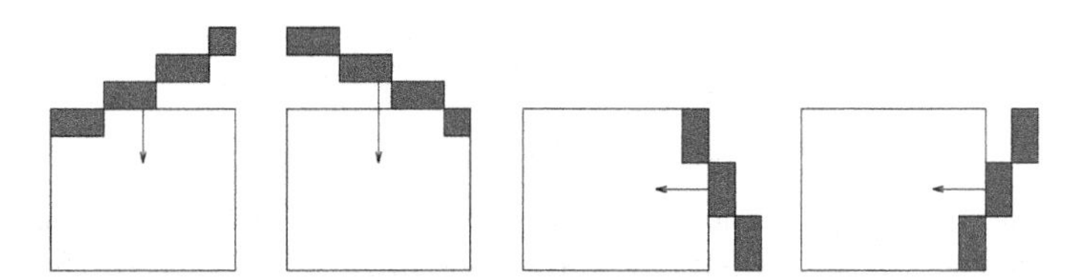

Fig. 7. Depending on its slope, the discrete line is drawn from an appropriate image corner. It is then translated in a unique direction in order to sweep the whole image plane while avoiding overlapping pixels (the arrows indicate the translation direction)

pixel is processed only once. When translating the line from its original position, the number of pixels falling within the image plane first increases, then remains constant, and finally decreases. If the image size along the translation direction is shorter than the distance between the extremity of the line falling off the image plane and the image plane, the constant zone is replaced by a zone where the line is increasing at one end while decreasing at the other. The number of pixels falling within the image after each translation can be efficiently updated using a run length coding of the discrete line. The algorithm is suited to line erosions and dilations in 3-dimensional images but requires a careful analysis of border effects.

The procedure extends directly to periodic lines. Once the periodic structuring element has been defined, the discrete Bresenham line having the same orientation is traced from one of the image corners. Since the connected periodic line has the same orientation as the Bresenham line, the latter can be translated along the line so as to process all pixels of the line falling within the image plane. As described in the previous paragraph, the line is then translated in an appropriate direction and the procedure is repeated until the whole image plane has been swept. Figure 8 shows an example with the periodic line $P_{3,(2,1)}$.

Translation Invariance Issues As already noticed in [27], the shape of the connected line segment varies slightly from one pixel to another for all orientations not matching one of the principal directions of the digitisation grid. These variations are studied in detail in [31] while proposing solutions for achieving translation invariance wherever necessary. In general, there are k possible outputs for a neighbourhood image operator Ψ by a line segment of length λ pixels applied along a Bresenham line of slope y/x and sweeping the whole image definition domain. We denote them by $\Psi_{BL_{\lambda_i},(x,y)}$ where $i \in \{1,\ldots,k\}$ and $k = \max(|x|,|y|)$.

The recursive translation invariant strategy relies on the following structuring element decomposition:

$$L_{nk_i,(x,y)} = L_{k_i,(x,y)} \oplus P_{n,(x,y)}, \tag{2}$$

where $\oplus$ denotes the Minkowski addition [16] and $i \in \{1,\ldots,k\}$. Therefore, the following relationships hold

$$\delta_{L_{nk_i,(x,y)}} = \delta_{L_{k_i,(x,y)}} \delta_{P_{n,(x,y)}}, \tag{3}$$

$$\varepsilon_{L_{nk_i,(x,y)}} = \varepsilon_{L_{k_i,(x,y)}} \varepsilon_{P_{n,(x,y)}}. \tag{4}$$

We also show in [31] that the union (i.e., point-wise maximum $\vee$ for discrete grey scale images) of all k possible non-TI openings is identical to the union of the openings by the k possible line segments (the same result holds for closings ϕ by replacing the union with the point-wise minimum $\wedge$):

$$\bigvee_{i=1}^{i=k} \gamma_{BL_{\lambda_i},(x,y)} = \bigvee_{i=1}^{i=k} \gamma_{L_{\lambda_i},(x,y)}, \tag{5}$$

$$\bigwedge_{i=1}^{i=k} \phi_{BL_{\lambda_i},(x,y)} = \bigwedge_{i=1}^{i=k} \phi_{L_{\lambda_i},(x,y)}. \tag{6}$$

3.3 Extension to Grey Scale Structuring Elements

The dilation of an image f with a grey scale (also referred to as volumic or non-flat) structuring element B_v is denoted by $\delta_{B_v}(f)$ and is defined as follows for each point x:

$$[\delta_{B_v}(f)](x) = \max_{b \in B_v}\{f(x+b) + B_v(b)\}. \tag{7}$$

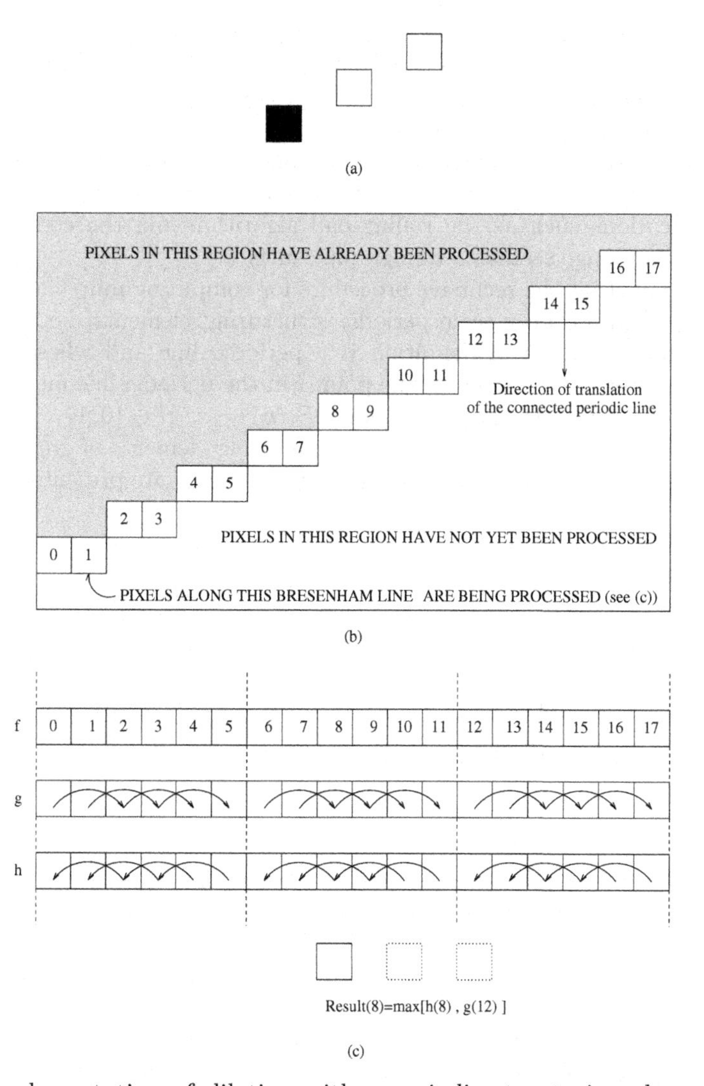

Fig. 8. Implementation of dilation with a periodic structuring element using a modified van Herk algorithm. (a) A periodic structuring element $P_{3,(2,1)}$, its origin being the black pixel. (b) The image plane after 9 translations of the corresponding Bresenham line. (c) Schematic of the algorithm: the value in g or h at the starting point of a cycle of arrowed arcs equals the original value in the input image f (e.g., $g(6) = f(6)$ or $h(11) = f(11)$). The value at the pixel pointed by an arrowed arc equals the maximum between the value of f at this position and the value at the beginning of the arrowed arc (e.g., $h(2) = \max[f(2), h(4)]$). This algorithm requires 3 max comparisons, whatever the number of pixels of the periodic line

The grey scale weights of a grey scale structuring element should be set according to the image intensity values. The erosion is defined by duality with respect to set complementation: $[\varepsilon_{B_v}(f)](x) = \min_{b \in B_v}\{f(x+b) - B_v(b)\}$.

Grey scale structuring elements should be used with care because the corresponding erosions and dilations do not commute with scalings of the pixel intensity values [18,29]. Nevertheless, these structuring elements are useful for some applications such as the rolling-ball algorithm and the computation of shadows of an image seen as a topographic surface [32].

Let us show that the recursive procedure for computing min/max filters (see Sec. 3.2) extends to grey scale periodic structuring elements, i.e., structuring elements whose domain of definition is a periodic line and whose grey scale values are defined as the index of the point i in the periodic line multiplied by a real number s defining the grey scale slope: $P(i\boldsymbol{v}) = is$, $\forall i \in \{0, 1, \ldots, \lambda-1\}$ [28]. We denote by $P_{\lambda,\boldsymbol{v},s}$ grey scale periodic structuring elements of grey scale slope s. Examples of grey scale periodic structuring elements are presented in Fig. 9.

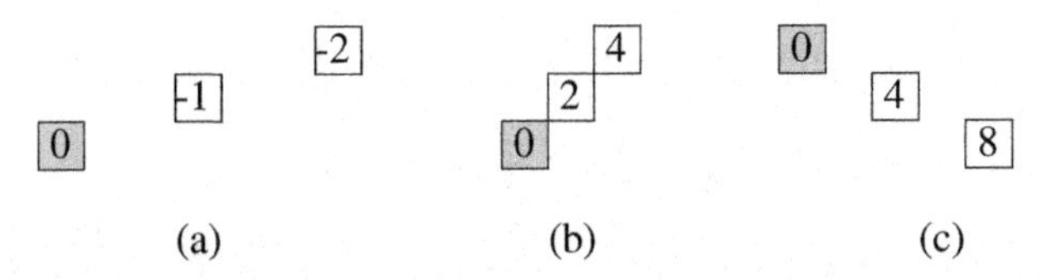

Fig. 9. Grey scale periodic structuring elements. (a) $P_{3,(3,1),-1}$. (b) $P_{3,(1,1),2}$. (c) $P_{3,(2,-1),4}$

The algorithm requires an additional buffer f'. The values of the input image are copied in this buffer which is partitioned into blocks as described in Section 3.2. The periodic structuring element is then positioned at the first pixel of each block and the weights are added to each pixel of the block whose intersection with the structuring element is non-empty. The structuring element is then translated by one pixel to the right and the procedure is repeated until a weight has been added to all pixels of f' (there are $k-1$ translations per block). Once the values of the buffer f' have been calculated, the buffers g and h are computed from the buffer f' using the recursive procedure detailed in Section 3.2. Finally, the resulting value at each position equals the maximum value between the value in the buffer h at the current position and the value in the buffer g at the current position plus $x + k(\lambda - 1)$. However, in order to ensure that the same weights are used for all positions, appropriate multiples of the slope s must be added to g and subtracted from h beforehand. More precisely, $x \bmod \lambda$ times s must be removed from h and $(\lambda-1) - (x + \lambda - 1) \bmod \lambda$ times s must be added to g. For clarity and conciseness, the case of a connected periodic line of λ pixels (i.e., the periodicity k is equal to 1) with the origin matching the first pixel of the periodic line (i.e., $o = 0$) is presented hereafter:

$$
\begin{aligned}
f'(x) &= f(x) + (x \bmod \lambda)s \\
g(x) &= \begin{cases} f'(x) & \text{if } x \bmod \lambda = 0, \\ \max[g(x-1), f'(x)] & \text{otherwise.} \end{cases} \\
h(x) &= \begin{cases} f'(x) & \text{if } (x+1) \bmod \lambda = 0, \\ \max[h(x+1), f'(x)] & \text{otherwise.} \end{cases} \\
r(x) &= \max[g(x+\lambda-1) + ((\lambda-1) - (x+\lambda-1) \bmod \lambda)s,\ h(x) - (x \bmod \lambda)s].
\end{aligned}
$$

Note that all divisions are integer divisions. This algorithm adapts directly to non-unitary periodicities (i.e., $k > 1$). An example is given in Fig. 10 for the periodic line $P_{3,(2,0),s}$. For conciseness, we use a structuring element of 3 pixels only. Since our algorithm requires 3 max comparisons per pixel whatever the number of pixels in the structuring element, speed gains are obtained for larger structuring elements.

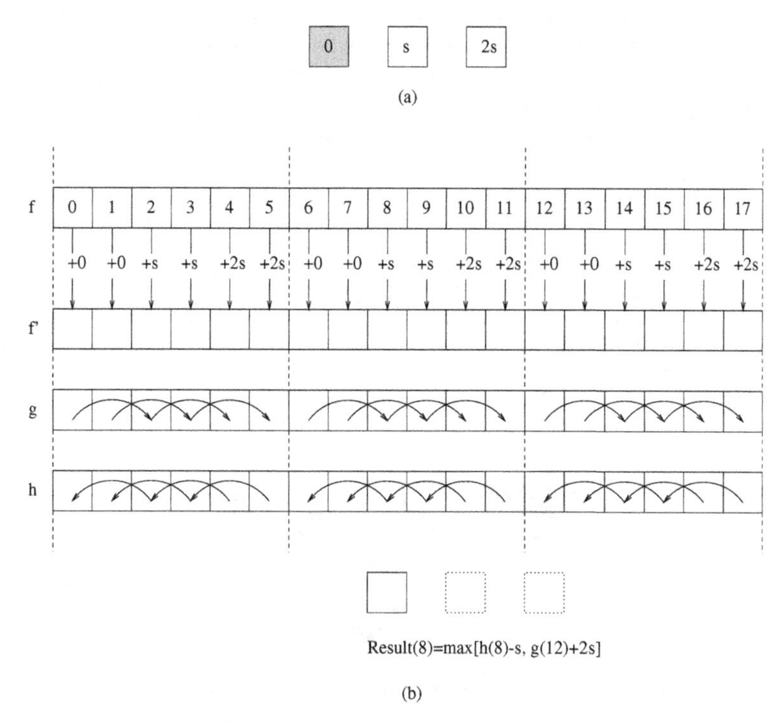

Fig. 10. Recursive min/max filter with a grey scale periodic structuring element $P_{3,(2,0),s}$ (shaded origin). The structuring element is shown in (a) and the schematic of the algorithm in (b)

3.4 Extension to 3-Dimensional Images

This extension is trivial in the sense that the very nature of the algorithm is 1-dimensional. In practice however, the implementation of a sweeping procedure allowing for the processing of the image volume along discrete lines while avoiding processing a voxel more than once requires a very careful analysis of the translation directions and handling of the image borders. The original position (image corner) of the line and the two translations to consider depend on the orientation of the line (there are 16 cases).

4 Line Segment Cascades

By cascading two erosions (resp. dilations) with vertical and horizontal line segments, one achieves erosions (resp. dilations) with square structuring elements:

$$\square_n = L_{n,(1,0)} \oplus L_{n,(0,1)}, \tag{8}$$

where $\square_n$ is a square a width n pixels. It has long been know that discrete diamond-shaped structuring elements cannot be generated by cascading erosions (resp. dilations) with structuring elements at 45 and -45 degrees. Although logarithmic decompositions have been proposed [33] for speeding up operations with diamond-shaped sets, the following simple and efficient decomposition can be used instead:

$$\diamond_n = L_{n-1,(1,1)} \oplus L_{n-1,(1,-1)} \oplus \diamond_2, \tag{9}$$

where $\diamond_n$ is the diamond-shaped structuring element with a side of n pixels ($n \geq 2$), i.e., $\diamond_2$ is the 4-connected neighbourhood plus its central pixel. By definition, $\diamond_1$ is a single pixel. An example is shown in Fig. 11. All these decompositions

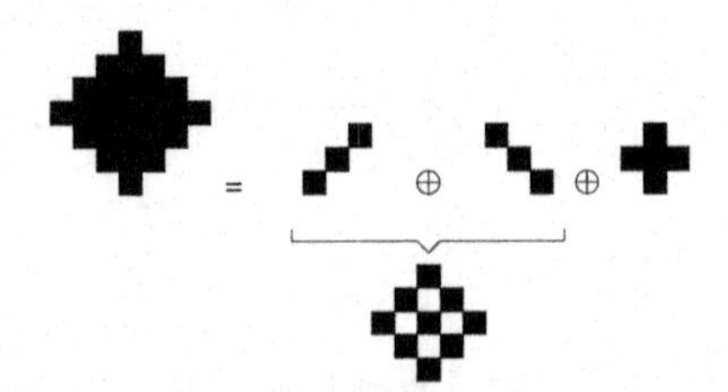

Fig. 11. Decomposition of a diamond-shaped structuring element of a width of 4 pixels using Eq. 9

substantially speed up operations for n large enough (6 (resp. 10) min/max comparisons per pixel whatever the width n of the square (resp. diamond) when using the recursive algorithm instead of a number in $O(n^2)$ for the brute force algorithm). Note also that cascades of erosions/dilations by $\diamond_2$ currently used for generating operations with diamond-shaped structuring elements of larger size are not only less efficient than Eq. 9 but also only lead to diamonds of odd width. In the sequel of this section, we concentrate on cascades starting from periodic lines and leading to connected lines and discrete disks.

4.1 Leading to Connected Lines

Cascades of periodic lines can be used for generating connected line segments and granulometric families of disks. This idea has originally been developed in [12]. We summarise here the main results. It is apparent from the example shown in Fig. 3c that in general a periodic line is not connected. In fact, in the two-dimensional case using 8-connectivity, the only examples of connected periodic lines are when both the horizontal and vertical components of the periodicity are either -1, 0, or 1. However, periodic lines can be cascaded with other periodic lines or carefully chosen structuring elements to form a connected line (this principle it at the very basis of Eq. 2). Indeed, if $P_{\lambda,\boldsymbol{v}}$ is a periodic line, where $\lambda > 1$, and A is any connected line with end points given by $\mathbf{0}$ and $\boldsymbol{v}$, then $L_\lambda = A \oplus P_{\lambda,\boldsymbol{v}}$ is a connected line with the same end points as $P_{\lambda,\boldsymbol{v}}$. An example is shown in Fig. 12 using the periodic line $P_{3,(2,1)}$ from Fig. 3c. In Fig. 12a are

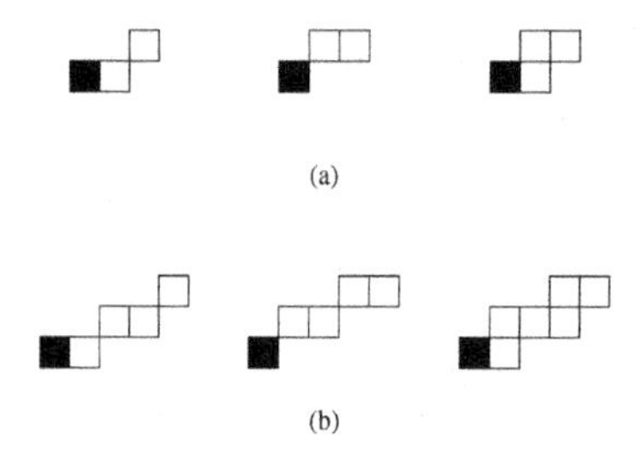

Fig. 12. Connected periodic lines. (a) Examples of the connected line A for the periodicity $(2,1)$. (b) Corresponding connected periodic lines $L_2 = A \oplus P_{2,(2,1)}$

examples of the line A that have the two end points $(0,0)$ and $(2,1)$. Note that the line A may be of any type so long as it is connected; the first two we show here are 8-connected Bresenham lines and the third is 4-connected. Although it would be advantageous if the connected line A was also periodic, as this would admit a fast and translation-invariant implementation of A, this is only possible for a restricted class of periodicities. The corresponding connected periodic lines, given by the cascade $A \oplus P_{2,(2,1)}$, are shown below each line A in Fig. 12b.

The granulometric properties of periodic shapes are summarised by the following theorem [12]: Given any set A and periodic shape $S_{\boldsymbol{\lambda},\mathbf{v}}$, the opening $\Gamma_{\boldsymbol{\lambda}}(f) = \gamma_{A\oplus S_{\boldsymbol{\lambda},\mathbf{v}}}(f)$ admits $\Gamma_{\boldsymbol{\lambda}+\boldsymbol{\mu}} \leq \Gamma_{\boldsymbol{\lambda}}$, where $\boldsymbol{\lambda}$, $\boldsymbol{\mu} \geq \mathbf{0}$. It is therefore a granulometric function with size vector $\boldsymbol{\lambda}$.

A particular application is to linear granulometries. Since the line $P_{\lambda,\boldsymbol{v}}$ is an example of a periodic shape $S_{\boldsymbol{\lambda},\mathbf{v}}$, we may state the following: If $\Gamma_\lambda(f) = \gamma_{L_\lambda}(f)$, where L_λ is a connected periodic line, then Γ_λ is a granulometric function with size parameter $\lambda > 1$. Figure 13 shows an example using the periodic line $P_{3,(2,1)}$ from Figs. 3 and 12. Figure 13a is a connected line A with end points $(0,0)$ and $(2,1)$. Figures 13a, b and c respectively are the connected periodic lines $A \oplus P_{1,(2,1)}$, $A \oplus P_{2,(2,1)}$ and $A \oplus P_{3,(2,1)}$ which are used to generate the first three members of the family of the granulometric function $\Gamma_\lambda(f) = \gamma_{L_\lambda}(f)$. Note

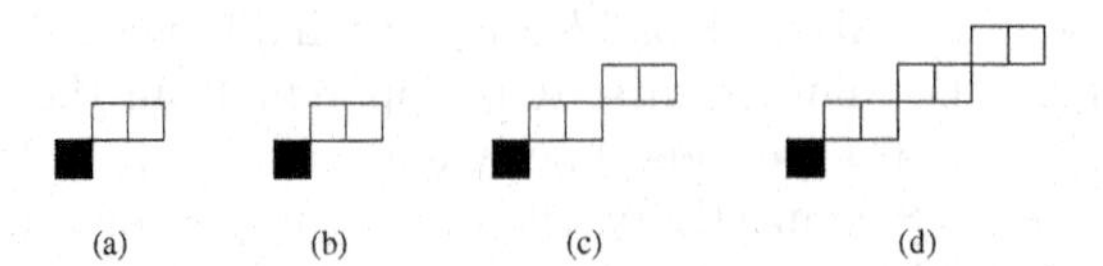

Fig. 13. Granulometric functions can be generated from a connected periodic line. (a) Connected line A. (b), (c) and (d) Growth of the connected periodic line

that the increase in length of L_λ equals k pixels, the periodicity of the underlying periodic line. By considering the union of openings described in Sec. 3.2, one can however increase the length of the structuring element one pixel at a time while satisfying all axioms of a granulometry. The resulting directional granulometry Γ for a given slope y/x and size parameter λ is then defined as follows:

$$\Gamma_{\lambda,(x,y)} = \bigvee_{i=1}^{i=k} \gamma_{L_{\lambda_i,(x,y)}} . \tag{10}$$

4.2 Leading to Discrete Disks

In Euclidean morphology, Matheron [15, p. 94] has shown that for $\theta_1, \theta_2, \ldots \theta_n$ distinct in $[0, \pi)$, $k_1, k_2, \ldots, k_n > 0$, then $k_1 L_{\theta_1} \oplus k_2 L_{\theta_2} \oplus \cdots \oplus k_n L_{\theta_n}$ is a convex polygon of $2n$ sides whose opposite edges are of length $2k_i$ and have orientation given by θ_i. Adams [1] used this principle for generating disks (and spheres) of increasing size from cascades of dilations by discrete Bresenham line segments. In [12], it has been show that cascades of periodic lines lead to better results in the sense that the resulting disks are symmetric. The approximation of a Euclidean disk (ball) B can be written as follows: $B \approx S_{\boldsymbol{\lambda},\boldsymbol{v}}$. However, no clue was given on how to select the vector of sizes and the corresponding vector of periodic lines. This can be achieved as follows. Suppose we look for the best discrete approximation of an Euclidean disk using cascades of line segments whose Euclidean length equals l. We then consider Fig. 6 and look for all slopes defined for this length and use the corresponding periodic lines with the maximal number of pixels so that the corresponding Euclidean line segment does not exceed l. We denote by B_l the corresponding approximation. For example we have the following cascades for the five first approximations:

$$\begin{aligned}
B_1 &= P_{2,(1,0)} \oplus P_{2,(0,1)}, \\
B_{\sqrt{2}} &= P_{2,(1,0)} \oplus P_{2,(0,1)} \oplus P_{2,(1,1)} \oplus P_{2,(1,-1)}, \\
B_2 &= P_{3,(1,0)} \oplus P_{3,(0,1)} \oplus P_{2,(1,1)} \oplus P_{2,(1,-1)}, \\
B_{\sqrt{5}} &= P_{3,(1,0)} \oplus P_{3,(0,1)} \oplus P_{2,(1,1)} \oplus P_{2,(1,-1)} \\
&\quad \oplus P_{2,(2,1)} \oplus P_{2,(2,-1)} \oplus P_{2,(1,2)} \oplus P_{2,(1,-2)}, \\
B_{2\sqrt{2}} &= P_{3,(1,0)} \oplus P_{3,(0,1)} \oplus P_{3,(1,1)} \oplus P_{3,(1,-1)} \\
&\quad \oplus P_{2,(2,1)} \oplus P_{2,(2,-1)} \oplus P_{2,(1,2)} \oplus P_{2,(1,-2)}.
\end{aligned}$$

Figure 14 displays the eight first disks. Notice that, by construction, these disks

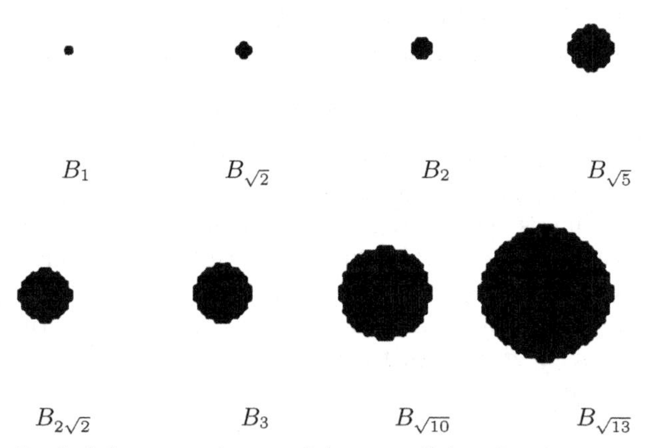

Fig. 14. Radial decompositions of discrete disks of increasing size using cascades of dilations with periodic lines. Each disk is symmetric, convex (see Sec. 6), and is opened for all disks of smaller size: $\gamma_{B_i}(B_j) = \emptyset$ for all $i \leq j$ (see text for the definition of each B_i)

form a granulometry with size parameter given by l. In contrast, radial decompositions using Bresenham lines [1] cannot be used to generate a granulometric function, neither the disks obtained by thresholding the Euclidean distance computed from a center pixel (the first element would be $\diamond_2$ and the second $\square_3$ but it is not opened by the first, i.e., the absorption property is not satisfied). The generation of a granulometric family of discrete spheres S from cascades of periodic lines can be achieved using a similar approach. For example, we have the following two first decompositions:

$$\begin{aligned}
S_1 &= P_{2,(1,0,0)} \oplus P_{2,(0,1,0)} \oplus P_{2,(0,0,1)}, \\
S_{\sqrt{2}} &= S_1 \oplus P_{2,(1,1,0)} \oplus P_{2,(1,-1,0)} \\
&\quad \oplus P_{2,(0,1,1)} \oplus P_{2,(0,1,-1)} \oplus P_{2,(1,0,1)} \oplus P_{2,(1,0,-1)}.
\end{aligned}$$

5 Opening/Closings by Line Segments

5.1 Basics

Directional openings and closings are useful for a wide range of practical applications such as fingerprint and fibre analysis, document interpretation (where thin lines have to be tracked), industrial inspection (where cracks, scratches, and other elongated defects have to be detected), and satellite imagery for the detection of road networks. A bank of directional openings and/or closings can also be used for extracting long thin objects of an image. This approach is illustrated in Fig. 15 for the extraction of bus-like structures in a 1 metre resolution satellite images. Other applications of directional morphological filters are detailed in [25] for the processing of veins appearing on plant leaves and [35] for the filtering of laid lines appearing in paper watermarks. When the elongated structures contains gaps, better results are obtained by considering rank-max directional openings [31]. The recursive algorithm described in Sec. 3 is then replaced by the moving histogram algorithm [5,10] suited for rank filters.

5.2 Orientation Field

The orientation at a given pixel x in an image is defined as the orientation of the line segment that minimises the difference between the grey-level value in the original image at x and the grey-level value at the same location in the image filtered by the considered line segment [30,31]. Openings should be used for image structures that are brighter than their background (i.e., 'positive' image structures) and closings for image structures darker than their background (i.e., 'negative' image structures).

In mathematical terms, we define the positive orientation at a given image pixel x and for a given scale λ as the orientation of the directional morphological opening of length λ which modifies the least the original image value at position x. We denote the the positive orientation by Dir^+, the negative orientation Dir^- being defined by duality:

$$Dir^+_\lambda(f)(x) = \{\theta_i \mid \gamma_{L_{\lambda,\theta_i}}(f)(x) \geq \gamma_{L_{\lambda,\theta_j}}(f)(x),\ \forall\,\theta_i \neq \theta_j\}, \tag{11}$$

$$Dir^-_\lambda(f)(x) = \{\theta_i \mid \phi_{L_{\lambda,\theta_i}}(f)(x) \leq \phi_{L_{\lambda,\theta_j}}(f)(x),\ \forall\,\theta_i \neq \theta_j\}. \tag{12}$$

The positive (resp. negative) directional signature at a given pixel can be obtained by plotting the normalised opened (resp. closed) values versus the orientation of the line segment. This signature can then be used to detect crossing lines, flat zones, etc. We also define the following quantities for each point x of the input image f:

$$Max^+_\lambda(f)(x) = \{\gamma_{L_{\lambda,\theta_i}}(f)(x) \mid \gamma_{L_{\lambda,\theta_i}}(f)(x) \geq \gamma_{L_{\lambda,\theta_j}}(f)(x),\ \forall\,\theta_i \neq \theta_j\}, \tag{13}$$

$$Min^+_\lambda(f)(x) = \{\gamma_{L_{\lambda,\theta_i}}(f)(x) \mid \gamma_{L_{\lambda,\theta_i}}(f)(x) \leq \gamma_{L_{\lambda,\theta_j},\lambda}(f)(x),\ \forall\,\theta_i \neq \theta_j\}, \tag{14}$$

$$Gdir^+_\lambda(f)(x) = Max^+_\lambda(f)(x) - Min^+_\lambda(f)(x). \tag{15}$$

(a) Input satellite image (IKONOS) showing Piazza Venetia in Roma

(b) Union of openings by line segments slightly shorter than buses (followed by reconstruction of original image)

(c) Union of openings by line segments slightly longer than buses (followed by reconstruction of original image)

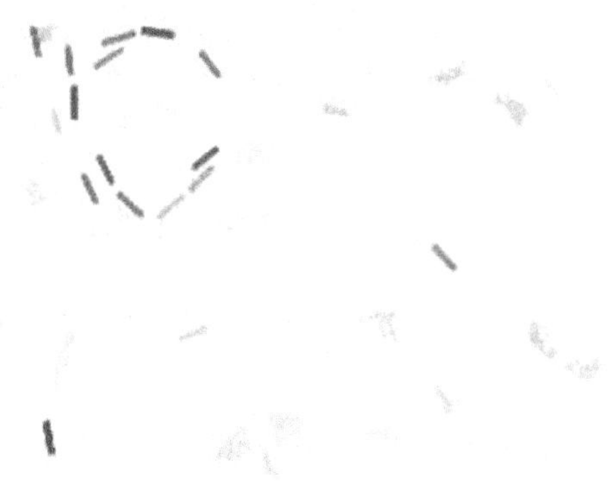

(d) Difference between image *(b)* and *(c)*

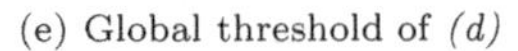

(e) Global threshold of *(d)*

Fig. 15. Extraction of bus-like structures appearing in a 1 metre resolution satellite image (IKONOS) using union of openings by line segments

$Gdir_\lambda^+$ can be interpreted as the strength of the positive orientation: it will output a small value if there is no predominant orientation for a structuring element length of λ pixels. Max_λ^-, Min_λ^-, and $Gdir_\lambda^-$ are defined by replacing the opening with the closing in Eqs. 13–14.

When comparing the values of $Gdir^+$ and $Gdir^-$ of a given pixel, it is possible to detect whether it belongs to a positive or negative image structure: positive, if $Gdir^+ > Gdir^-$, negative, otherwise. We denote by $Gdir$ the point-wise maximum between the images $Gdir^+$ and $Gdir^-$: $Gdir = Gdir^+ \vee Gdir^-$. The image of directions Dir is then defined as follows:

$$Dir_\lambda(f)(x) = \begin{cases} Dir_\lambda^+(f)(x), \text{ if } Gdir_\lambda(f)(x) = Gdir_\lambda^+(f)(x), \\ Dir_\lambda^-(f)(x), \text{ otherwise.} \end{cases} \tag{16}$$

A colour representation of the local orientation information is then simply achieved by equating the orientation information (i.e., either Dir^+, Dir^-, or Dir) to the hue and strength of the orientation information (i.e., either $Gdir^+$, $Gdir^-$, or $Gdir$) to the lightness component of the image, the colours being fully saturated in all cases. For example, the colour representation of the local orientation by opening is illustrated in Fig. 16 on IRS satellite image of Athens.

6 Closing by a Half-Plane and Convex Hulls

In [22], we have been shown that the convex hull transformation can be defined in terms of an intersection (point-wise minimum $\wedge$) of half-plane closings. Hence, denoting by ϕ the closing transformation, π_θ a closed half-plane having a given slope $\theta = \arctan(y/x)$, and $\breve{\pi}_\theta$ the reflected half-plane, the convex hull transformation CH of a grey scale image f is defined as follows:

$$CH(f) = \bigwedge_\theta [\phi_{\pi_\theta}(f) \wedge \phi_{\breve{\pi}_\theta}(f)]. \tag{17}$$

In the discrete case and for a bounded image, only a finite number of directions need to be considered. More precisely [24], let us first define the convex hull CH_0 of order 0 as the intersection of the horizontal and vertical half-planes (i.e., slopes in the form 0/1 and 1/0). The convex hull of order 0 is nothing but the strong convex hull in the 4-connected graph. The convex hull CH_i of order $i \in \mathbf{N}$ is then defined as the intersection of all half-planes whose slopes are in the form of y/x where x and y are integers in the range $[-i, i]$ with no common divisors other than 1. The corresponding number of slopes is given by $4\,\text{card}(F_i) - 4$ (see also Sec. 2.3). Since the orientations considered for any order $i \geq 0$ is a subset of the orientations considered for the order $i+1$, the following ordering relationship is satisfied: $CH_{i+1} \leq CH_i$.

For a $n \times n$ image, convergence is reached at the latest for the order $n - 1$ (upper bound). The actual order number depends on the shape of the image objects. In practice, the number of half-plane closings corresponding to the upper bound may be too large for applications where speed is an issue. In this case,

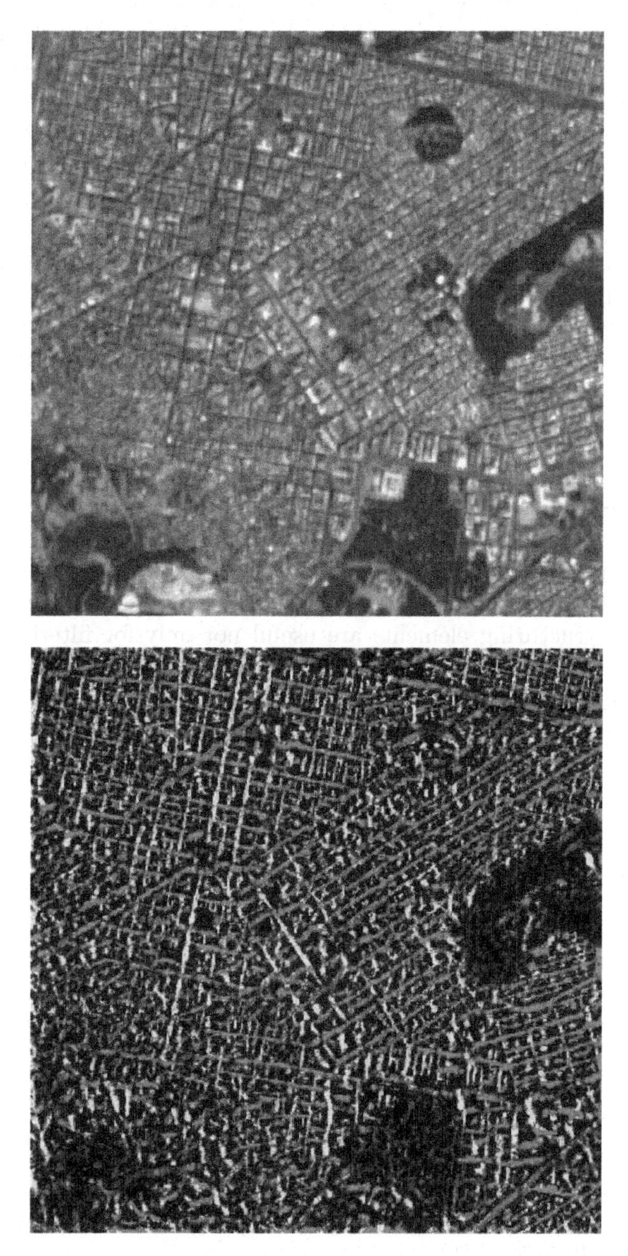

Fig. 16. Image of the local orientation enhancing the road network of a city. Top: input IRS image of Athens. Bottom: colour representation of Dir^{-}. The structuring element used in this experiment is a line segment of 11 pixels using the moving histogram TI implementation and a rank of 4 for the corresponding rank-max openings

a trade-off between accuracy and computation time must be considered. For example, a convex approximation of the discrete convex hull of order $n \geq 1$ is achieved by performing half-plane closings whose slopes are in the form $\pm i/n$ and $\pm n/i$, where $i \in \{0, \dots, n\}$ (there are only $4n - 4$ such slopes). We denote by $\widetilde{CH}_n$ the corresponding approximation. Notice that the following ordering relation holds: $CH_n \leq \widetilde{CH}_n$. In [24], we show that if CH_w outputs the exact discrete convex hull (i.e., $CH = CH_w$), then $\widetilde{CH}_{o>w} \setminus CH$ is at most one-pixel thick.

Finally, note that the translation-invariant implementation of half-plane closings can be achieved by processing the pixels in the order they are reached when progressively translating each Euclidean half-plane π_θ so as to sweep the whole image definition domain. By doing so, the pixels reached at any given step correspond to those falling along the periodic line having the same slope as the half-plane.

7 Conclusion

The study of discrete line segments for morphological operators leads to a wide variety of developments of interest to both discrete geometry and image analysis. The resulting structuring elements are useful not only for filtering thin image structures but also for generating sound approximations of discrete disks and grey scale convex hulls.

References

1. R. Adams. Radial decomposition of discs and spheres. *Computer Vision, Graphics, and Image Processing: Graphical Models and Image Processing*, 55(5):325–332, September 1993. 90, 91
2. J. Bresenham. Algorithm for computer control of digital plotter. *IBM System Journal*, 4:25–30, 1965. 79
3. J. Farey. On a curious property of vulgar fractions. *Philosophical Magazine*, 1816. See http://cut-the-knot.com/blue/FareyHistory.html for a historical note and a copy of Farey's original letter. 82
4. J. Foley, A. van Dam, S. Feiner, and J. Hughes. *Computer graphics —Principles and Practice—*. Addison-Wesley, 2nd edition, 1990. 79
5. G. Garibotto and L. Lambarelli. Fast on-line implementation of two dimensional median filtering. *Electronic Letters*, 15(1):24–25, January 1979. 92
6. D. Gevorkian, J. Astola, and S. Atourian. Improving Gil-Werman algorithm for running Min and Max filters. *IEEE Transactions on Pattern Analysis and Machine Intelligence*, 19(5):526–529, May 1997. 82
7. J. Gil and R. Kimmel. Efficient dilation, erosion, opening, and closing algorithms. In J. Goutsias, L. Vincent, and D. Bloomberg, editors, *Mathematical Morphology and Its Applications to Image and Signal Processing*, volume 18 of *Computational Imaging and Vision*, pages 301–310, Boston, 2000. Kluwer Academic Publishers. Proc. of ISMM'2000, Palo Alto, June 26–29. 83

8. J. Gil and M. Werman. Computing 2-D min, median, and max filters. *IEEE Transactions on Pattern Analysis and Machine Intelligence*, 15(5):504–507, May 1993. 82
9. H. Heijmans. *Morphological image operators.* Advances in Electronics and Electron Physics. Academic Press, Boston, 1994. 78
10. T. Huang, G. Yang, and G. Tang. A fast two-dimensional median filtering algorithm. *IEEE Transactions on Acoustics, Speech, and Signal Processing*, 27(1):13–18, February 1979. 92
11. R. Jones and P. Soille. Periodic lines and their applications to granulometries. In P. Maragos, W. Schafer, and M. Butt, editors, *Mathematical Morphology and its Applications to Image and Signal Processing*, pages 264–272. Kluwer Academic Publishers, 1996. Proc. of ISMM'96. 80
12. R. Jones and P. Soille. Periodic lines: Definition, cascades, and application to granulometries. *Pattern Recognition Letters*, 17(10):1057–1063, September 1996. 80, 89, 90
13. C. Kim. On the cellular convexity of complexes. *IEEE Transactions on Pattern Analysis and Machine Intelligence*, 3:617–625, 1981. 79
14. C. Kim. On cellular straight line segments. *Computer Graphics and Image Processing*, 18(4):369–381, April 1982. 79
15. G. Matheron. *Random sets and integral geometry.* Wiley, New York, 1975. 78, 90
16. H. Minkowski. Über die Begriffe Länge, Oberfläche und Volumen. *Jahresbericht der Deutschen Mathematiker Vereinigung*, 9:115–121, 1901. 84
17. D. Nadadur and R. Haralick. Recursive binary dilation and erosion using digital line structuring elements in arbitrary orientations. *IEEE Transactions on Image Processing*, 9(5):749–759, May 2000. 83
18. J.-F. Rivest, J. Serra, and P. Soille. Dimensionality in image analysis. *Journal of Visual Communication and Image Representation*, 3(2):137–146, 1992. 86
19. A. Rosenfeld. Digital straight line segments. *IEEE Transactions on Computers*, 23:1264–1269, 1974. 78
20. J. Serra. *Image analysis and mathematical morphology.* Academic Press, London, 1982. 78
21. J. Serra, editor. *Image analysis and mathematical morphology. Volume 2: Theoretical advances.* Academic Press, London, 1988. 78
22. P. Soille. Grey scale convex hulls: definition, implementation, and application. In H. Heijmans and J. Roerdink, editors, *Mathematical Morphology and its Applications to Image and Signal Processing*, volume 12 of *Computational Imaging and Vision*, pages 83–90. Kluwer Academic Publishers, Dordrecht, 1998. Proc. of ISMM'98. 94
23. P. Soille. *Morphological Image Analysis.* Springer-Verlag, Berlin, Heidelberg, New York, 1999. 78, 82
24. P. Soille. From binary to grey scale convex hulls. *Fundamenta Informaticae*, 41(1–2):131–146, January 2000. 94, 96
25. P. Soille. Morphological image analysis applied to crop field mapping. *Image and Vision Computing*, 18(13), October 2000. 92
26. P. Soille, E. Breen, and R. Jones. A fast algorithm for min/max filters along lines of arbitrary orientation. In I. Pitas, editor, *Proc. of 1995 IEEE Workshop on Nonlinear Signal and Image Processing*, pages 987–990, Neos Marmaras, June 1995. 83
27. P. Soille, E. Breen, and R. Jones. Recursive implementation of erosions and dilations along discrete lines at arbitrary angles. *IEEE Transactions on Pattern Analysis and Machine Intelligence*, 18(5):562–567, May 1996. 83, 84

28. P. Soille and R. Jones. Periodic lines: Fast implementation and extensions to greyscale structuring elements and 3-D images. Technical report, Fraunhofer IPK (Berlin)/CSIRO DMS (Sydney), 1995. 86
29. P. Soille and J.-F. Rivest. Dimensionality of morphological operators and cluster analysis. In E. Dougherty, P. Gader, and J. Serra, editors, *Image algebra and morphological image processing IV*, volume 2030, pages 43–53. Society of Photo-Instrumentation Engineers, July 1993. 86
30. P. Soille and H. Talbot. Image structure orientation using mathematical morphology. In A. Jain, S. Venkatesh, and B. Lovell, editors, *14th International Conference on Pattern Recognition*, volume 2, pages 1467–1469, Brisbane, August 1998. IAPR, IEEE Computer Society. 92
31. P. Soille and H. Talbot. Directional morphological filtering. *IEEE Transactions on Pattern Analysis and Machine Intelligence*, Submitted. 84, 92
32. S. Sternberg. Grayscale morphology. *Computer Graphics and Image Processing*, 35:333–355, 1986. 86
33. R. van den Boomgaard and R. van Balen. Methods for fast morphological image transforms using bitmapped binary images. *Computer Vision, Graphics, and Image Processing: Graphical Models and Image Processing*, 54(3):252–258, May 1992. 88
34. M. van Herk. A fast algorithm for local minimum and maximum filters on rectangular and octogonal kernels. *Pattern Recognition Letters*, 13:517–521, 1992. 82
35. P. Whelan, P. Soille, and A. Drimbarean. Real-time registration of paper watermarks. *Real-Time Imaging*, page In Press, 2000. 92

Hausdorff Discretizations of Algebraic Sets and Diophantine Sets

Mohamed Tajine and Christian Ronse

Laboratoire des Sciences de l'Image, de l'Informatique et de la Télédétection
(LSIIT, UPRES-A CNRS 7005)
Boulevard Sébastien Brant, F-67400 Illkirch, France
{tajine,ronse}@dpt-info.u-strasbg.fr

Abstract. This paper is a continuation of our works [12,13,15,16,17,18] [21,22] in which we study the properties of a new framework for discretization of closed sets based on Hausdorff metric. Let F be a non-empty closed subset of $\mathbb{R}^n$; $S \subseteq \mathbb{Z}^n$ is a Hausdorff discretization of F if it minimizes the Hausdorff distance to F.
We study the properties of Hausdorff discretizations of algebraic sets. Actually we give some decidable and undecidable properties concerning Hausdorff discretizations of algebraic sets and we prove that some Hausdorff discretizations of algebraic sets are diophantine sets. We refine the last results for algebraic curves and more precisely for straight lines.

Keywords: Algebraic set, diophantine set, Hausdorff discretization, homogeneous metric.

1 Introduction

We have introduced a new framework for discretization of closed sets based on the Hausdorff metric : Hausdorff discretization. The basic idea is to select as possible discretizations of a Euclidean set F all discrete sets S such that the Hausdorff distance between F and S is minimal. This leads to several possible choices for such a discretization.

Our framework is as follows: given a metric d on $\mathbb{R}^n$, let F be a non-empty closed subset of $\mathbb{R}^n$; then $S \subseteq \mathbb{Z}^n$ is a Hausdorff discretization of F if it minimizes the Hausdorff distance $H_d(S, F)$ to F. Some geometrical and topological properties of Hausdorff discretization and its comparison with other discretization schemes are studied in the serie of papers [12,13,15,16,17,18,21,22].
In this paper, we study the Hausdorff discretizations of particular sets : real algebraic sets. After recaling the fundametal properties of Hausdorff discretizations and the refinement of their properties for the class of homogeneous metrics, we give some decidable and undecidable properties concerning Hausdorff discretizations of algebraic sets and we prove that some Hausdorff discretizations of algebraic sets are diophantine sets. Some of these results are proven by using the Tarski's elimination algorithm concerning the decidability of first order logic for elementary algebra [9,14,19,20]. The time complexity of Tarski's elimination

G. Borgefors, I. Nyström, and G. Sanniti di Baja (Eds.): DGCI 2000, LNCS 1953, pp. 99–110, 2000.
© Springer-Verlag Berlin Heidelberg 2000

algorithm, is in the general case, very high [5]. We refine also the last results for algebraic curves and more precisely for straight lines. Actually we prove that the Hausdorff discretizations of straight lines can be discribed by using two periodic diophantine sets. This last result generalize the results of [1] concerning the supercover discretization of straight lines.

In this paper, for simplifying the notation in the notions which depend on a metric d, we do not refer explicitely to d, except when there is an ambiguity.

This paper is divided into eight sections. In the second section we briefly recall classical notions of metric space and Hausdorff space. In the third section, we introduce the Hausdorff discretizations. Section 4 deals with homogeneous metrics, while Section 5 introduc real algebraic sets, real semi-algebraic sets and diophantine sets. In the sixth section we characterize some Hausdorff discretizations of real algebraic sets. In the seventh section, we refine the last results for algebraic curves and more precisely for straight lines. The last section is a conclusion.

The proofs are not given here, the proofs the the results of sections 2, 3 and 4 can be found in [12,15].

2 Some Metric Notions and Hausdorff Metric

We assume that the reader is familiar with classical notions of topological space, metric space and normed space, see for example [3,6,7]. We introduce here our notations, most of them are recalled in [12,15,21,22].

Definition 1. *Let $(\mathcal{E}, d)$ be a metric space, and let $p \in \mathcal{E}$ and $r \in I\!R^+$,*

$$\mathcal{B}_r^d(p) = \{x \in \mathcal{E} \mid d(x,p) \leq r\}.$$

$\mathcal{B}_r^d(p)$ is called the ball of center p and of radius r.
Let $E \subseteq \mathcal{E}$. $int(E) = \{x \in E \mid \exists r > 0,\ \mathcal{B}_r^d(x) \subseteq E\}$, $int(E)$ is called the interior *of E (i.e. $int(E)$ is the largest open set contained in E).*

In all the following, all topological notions in a metric space $(\mathcal{E}, d)$ are considered relatively to the topology induced by d. All metrics used in this paper are induced by norm. So if N is a norm over $\mathcal{E}$, then the function d_N such that: $\forall x, y \in \mathcal{E},\ d_N(x,y) = N(x-y)$ is called the metric over $\mathcal{E}$ induced by N.

Example:
Let $x = (x_1, x_2, ..., x_n) \in I\!R^n$. Then $\forall p \geq 1,\ |x|_p = \sqrt[p]{|x_1|^p + ... + |x_n|^p}$ and $|x|_\infty = max_{1 \leq i \leq n}|x_i|$ are a norms over $\mathcal{E} = I\!R^n$. The metrics induced by these norms are denoted d_p and d_∞ respectively.

2.1 Hausdorff Metric

The definitions and results presented in this subsection can be found in [3,6].

Definition 2. *Let $(\mathcal{E}, d)$ be a metric space; $\mathcal{H}(\mathcal{E})$ is the set of non-empty compact subsets of $\mathcal{E}$, $\mathcal{F}(\mathcal{E})$ the set of closed subsets of $\mathcal{E}$, and $\mathcal{F}'(\mathcal{E})$ the set of non-empty closed subsets of $\mathcal{E}$ (i.e. $\mathcal{F}'(\mathcal{E}) = \mathcal{F}(\mathcal{E}) \setminus \{\emptyset\}$).*

On $\mathcal{H}(\mathcal{E})$, we will define a metric H_d, such that if $(\mathcal{E}, d)$ is a complete metric space then $(\mathcal{H}(\mathcal{E}), H_d)$ is a complete metric space.

Definition 3. *Let $(\mathcal{E}, d)$ be a metric space and let $A, B \in \mathcal{H}(\mathcal{E})$. We define* the oriented Hausdorff metric *from A to B by $h_d(A, B) = max_{a \in A}(d(a, B))$ where $d(a, B) = min_{b \in B}(d(a, b))$.*

Definition 4. *Let $(\mathcal{E}, d)$ be a metric space. The* Hausdorff distance *between two compact sets A, $B \in \mathcal{H}(\mathcal{E})$ is defined by $H_d(A, B) = max(h_d(A, B), h_d(B, A))$.*

Remark:
Let $\mathcal{F}'(\mathcal{E})$ be the set of non empty closed set of $\mathcal{E}$. Then, the functions h_d and H_d can be extended in natural way as function from $\mathcal{F}'(\mathcal{E}) \times \mathcal{F}'(\mathcal{E})$ to $\mathbb{R}^+ \cup \{+\infty\}$ (note: in Definition 3, the "max" (resp. "min") becomes a "sup" (resp. "inf"))
.

H_d is a "generalized metric" on $\mathcal{F}'(\mathcal{E})$ in the sense that it satisfies the axioms of metric, but can take infinite values.

3 Hausdorff Discretization

In this section, we recal our framework of discretization based on Hausdorff metric. Our results are proved in [15]. In the rest of this paper we assume that we have as metric space $(\mathbb{R}^n, d)$, where d is a metric induced by a norm on $\mathbb{R}^n$, and as a discrete space $\mathcal{D}_\rho = \rho\mathbb{Z}^n$, for $\rho > 0$. For a such distance, $M \subseteq \mathcal{D}_\rho$ implies that $M \in \mathcal{F}(\mathbb{R}^n)$.
In all the following, we assume that $\rho > 0$.

Definition 5. *Let d be a metric on $\mathbb{R}^n$ and $\rho > 0$. The* covering radius *of the metric d in $\mathcal{D}_\rho$ is*

$$r_c(\rho) = sup_{x \in \mathbb{R}^n}(d(x, \mathcal{D}_\rho)).$$

Notation:
Let d be a metric on $\mathbb{R}^n$ and $r \geq 0$;

$$\forall F \in \mathcal{F}'(\mathbb{R}^n),\ \Delta_r^d(F, \rho) = \{p \in \mathcal{D}_\rho \mid \mathcal{B}_r^d(p) \cap F \neq \emptyset\}.$$

Let F be a non-empty closed subset of $\mathbb{R}^n$; $M \subseteq \mathcal{D}_\rho$ is a Hausdorff discretization of F in $\mathcal{D}_\rho$ if it minimizes the Hausdorff distance to F. In this section, we study the properties of Hausdorff discretizations. In [12,13,21,22], we have studied the Hausdorff discretizations when F is a compact set.

Definition 6. *Let $F \in \mathcal{F}'(\mathbb{R}^n)$.*

- *A set $M \subseteq \mathcal{D}_\rho$ is a* Hausdorff discretization *of F in $\mathcal{D}_\rho$ if $H_d(F,M) = inf(\{H_d(F,S) \mid S \subseteq \mathcal{D}_\rho\})$.*
- *$\mathcal{M}_H(F,\rho) = \{M \subseteq \mathcal{D}_\rho \mid H_d(F,M) = inf_{S \subseteq \mathcal{D}_\rho}(H_d(F,S))\}$ is the set of Hausdorff discretizations of F.*
- *$\Delta_H(F,\rho) = (\bigcup_{M \in \mathcal{M}_H(F,\rho)} M)$ is called* the maximal Hausdorff discretization *of F in $\mathcal{D}_\rho$ (i.e. The justification of this definition is given in the following Theorem).*
- *The value $r_H(F,\rho) = sup_{x \in F}(d(x,\mathcal{D}_\rho))$ is called* the Hausdorff radius *of F in $\mathcal{D}_\rho$ for the metric d.*

We characterize now the Hausdorff discretizations.

Theorem 1. *Let $F \in \mathcal{F}'(\mathbb{R}^n)$; then:*

- *$\mathcal{M}_H(F,\rho)$ is nonvoid and for $M \in \mathcal{M}_H(F,\rho)$, $H_d(F,M) = r_H(F,\rho)$;*
- *for a family $(M_i)_{i \in I}$ of members of $\mathcal{M}_H(F,\rho)$, $\bigcup_{i \in I} M_i \in \mathcal{M}_H(F,\rho)$ and so $\Delta_H(F,\rho) \in \mathcal{M}_H(F,\rho)$;*
- *if $(M_n)_{n \in \mathbb{N}}$ is a decreasing sequence in $\mathcal{M}_H(F,\rho)$ (relatively to the set inclusion) then $\bigcap_{n \in \mathbb{N}} M_n \neq \emptyset$ and $\bigcap_{n \in \mathbb{N}} M_n \in \mathcal{M}_H(F,\rho)$;*
- *$\Delta_H(F,\rho) = \{p \in \mathcal{D}_\rho \mid d(p,F) \leq r_H(F,\rho)\}$;*
- *$\mathcal{M}_H(F,\rho)$ is the set of all $M \in \mathcal{D}_\rho$ such that $M \subseteq \Delta_H(F,\rho)$ and F is included in the union of balls of radius $r_H(F,\rho)$ centered about points of M;*
- *$r_H(F,\rho) \leq r_c(\rho)$;*
- *$\Delta_H(F,\rho) = \Delta^d_{r_H(F,\rho)}(F,\rho)$.*

Definition 7. *Let $p \in \mathcal{D}_\rho$, we define the* the cell *of center p as*

$$\mathcal{C}(p,\rho) = \{x \in \mathbb{R}^n \mid d_\infty(x,p) \leq \frac{\rho}{2}\}$$

Note that cells are closed and overlap only at their boundaries.

In Figure 1 we illustrate the construction of Hausdorff discretizations for a closed set F: computing the Hausdorff radius (maximal distance from points of F to the discrete space), one takes for $\Delta_H(F,\rho)$ all discrete points p such that the ball of center p and Hausdorff radius intersects F; any subset M of $\Delta_H(F,\rho)$ such that the corresponding balls for $p \in M$ cover F, will be a Hausdorff discretization.

Definition 8. *Let d be a metric on $\mathbb{R}^n$ and $F \in \mathcal{F}'(\mathbb{R}^n)$,* the skeleton *of $\Delta_H(F,\rho)$ is the set*

$$\mathcal{S}k(F,\rho) = \bigcap_{M \in \mathcal{M}_H(F,\rho)} M.$$

Note that $\mathcal{S}k(F,\rho)$ is not generally a Hausdorff discretization of F.

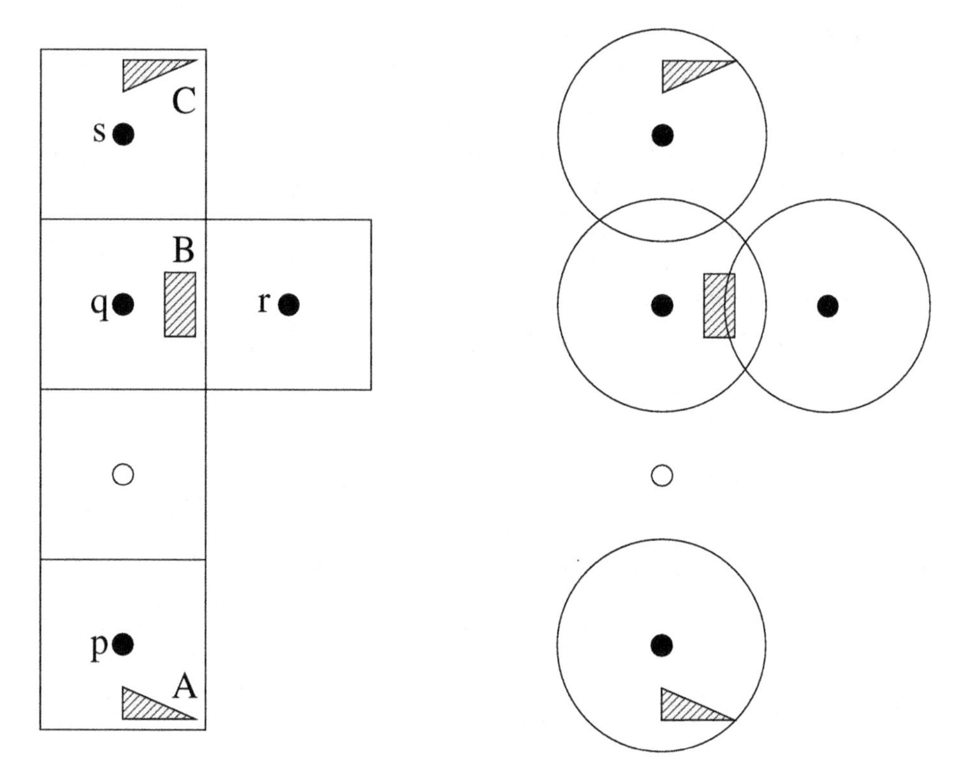

Fig. 1. Left: The set $F = A \cup B \cup C$ overlayed with discrete points p, q, r, s and their square cells $C(p,1), C(q,1), C(r,1), C(s,1)$. Right: For $d = d_2$ (the Euclidean distance), the maximal Hausdorff discretization of F is $\{p, q, r, s\}$; indeed, we show the circles of radius $r_H(F,1)$ centered about these points. The unique other Hausdorff discretizing set of F is $\{p, q, s\}$

4 Homogeneous Metric and Hausdorff Discretization

We present some properties of a homogeneous metric and we refine the characterization of Hausdorff discretizations for such metrics. Actually, we study the relationship between Hausdorff discretizations and covering discretizations. Again, results of this section are proved in [15].

Definition 9. *A norm N on $\mathbb{R}^n$ is* homogeneous *if for every $(x_1, ..., x_n) \in \mathbb{R}^n$, every $(\varepsilon_1, ..., \varepsilon_n) \in \{-1, 1\}^n$, and for every permutation σ of $\{1, ..., n\}$, we have $N(\varepsilon_1 x_{\sigma(1)}, ..., \varepsilon_n x_{\sigma(n)}) = N(x_1, ..., x_n)$. A metric induced by a homogeneous norm is called a* homogeneous *metric.*

Theorem 2. *Let d be a homogeneous metric induced by a norm N, then:*

- $r_c(1) = \frac{1}{2} N(1, ..., 1)$ *and*
- $\mathcal{B}^{d_\infty}_{\frac{1}{2}}(O) \subseteq \mathcal{B}^{d}_{r_c(\rho)}(O) \subseteq \mathcal{B}^{d_1}_{\frac{n}{2}}(O)$, *for all $O \in \mathbb{R}^n$.*

Examples:
In $\mathbb{R}^n$, $\forall p \geq 1$, d_p is a homogeneous metric and thus, $r_c(1) = (\frac{n^{\frac{1}{p}}}{2})$, and d_∞ is also a homogeneous metric and thus, $r_c(1) = \frac{1}{2}$.

Definition 10. *Let $E \subseteq \mathbb{R}^n$, a subset $S \subseteq \mathcal{D}_\rho$ is called a* covering discretization *of E in $\mathcal{D}_\rho$, if $\forall p \in S$, $E \cap \mathcal{C}(p,\rho) \neq \emptyset$ and $E \subseteq \bigcup_{p \in S} \mathcal{C}(p,\rho)$*

An example of covering discretization is the *supercover discretization* Δ_{SC} which associates to every $X \in \mathcal{P}(\mathbb{R}^n)$ the set of all $p \in \mathcal{D}_\rho$ such that $\mathcal{C}(p,\rho)$ intersects X (see Figure 2):

$$\forall X \subseteq \mathbb{R}^n, \qquad \Delta_{SC}(X,\rho) = \{p \in \mathcal{D}_\rho \mid C(p,\rho) \cap X \neq \emptyset\}.$$

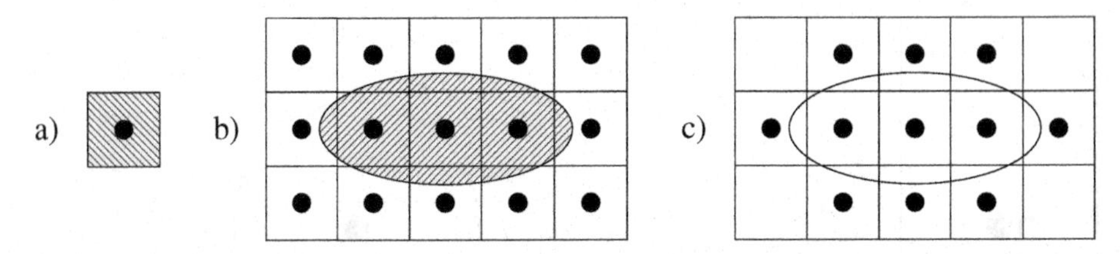

Fig. 2. (a) The cell $C(o,1)$. **(b)** A Euclidean set X overlayed with the discrete points p and their cells $C(p,1)$. **(c)** $\Delta_{SC}(X,1)$

Theorem 3. *Let d be an homogeneous metric and let $F \in \mathcal{F}'(\mathbb{R}^n)$. If S is a covering discretization of F in $\mathcal{D}_\rho$, then $S \in \mathcal{M}_H(F,\rho)$.*

Definition 11. *A metric d on $\mathbb{R}^2$ is called* strictly homogeneous *if d is homogeneous and $\mathcal{B}^d_{r_c(1)}(0,0) \cap \mathcal{B}^d_{r_c(1)}(1,1) = \{(\frac{1}{2},\frac{1}{2})\}$.*

In other words, the balls of covering radius centered about diagonally adjacent discrete points intersect only at their corners. For example d_p is strictly homogeneous for all $p > 1$ and for $p = \infty$.

5 Algebraic Sets and Diophantine Sets

In this section, we give the definitions of real algebraic sets, real semi-algebraic sets and diophantine sets. We recall also the decidability of first order logic for elementary algebra corresponding to Tarski's elimination algorithm.

Definition 12. *A subset $\mathcal{A}$ of $\mathbb{R}^n$ is an* algebraic set, *if there exists a polynomial $P \in \mathbb{R}[X_1,...,X_n]$ such that $\mathcal{A}$ is the set of zeros of P in $\mathbb{R}^n$ (i.e. $\mathcal{A} = \{x \in \mathbb{R}^n \mid P(x) = 0\}$).*
If the coefficients of P are integers (i.e. $P \in \mathbb{Z}[X_1,...,X_n]$) then $\mathcal{A}$ is called a $\mathbb{Z}$-algebraic set.

Definition 13. *A subset $\mathcal{S}$ of $\mathbb{R}^n$ is a* semi-algebraic set*, if it can be written as a finite union of set of the form:*

$$\{x \in \mathbb{R}^n \mid P_1(x) = ... = P_l(x) = 0, Q_1(x) > 0, ..., Q_m(x) > 0\},$$

where $P_1, ..., P_l, Q_1, ..., Q_m$ are in $\mathbb{R}[X_1, ..., X_n]$. If the coefficients of polynomials involved in a definition of the semi-algebraic set $\mathcal{S}$ are integers (i.e. $P_i, Q_i \in \mathbb{Z}[X_1, ..., X_n]$) $\mathcal{S}$ is called a $\mathbb{Z}$-semi-algebraic set.

Actually every subset of $\mathbb{R}^n$ defined by a boolean combination (obtained by disjunction, conjunction and negation) of atomic formulas of the form $P(x) > 0$, $P(x) < 0$ or $P(x) = 0$, where $P \in \mathbb{R}[X_1, ..., X_n]$, is a semi-algebraic set of $\mathbb{R}^n$.

Definition 14. *A subset D of $\mathbb{Z}^n$ is a* diophantine set*, if there exists a polynomial $P \in \mathbb{Z}[X_1, ..., X_n]$ such that D is the set of zeros of P in $\mathbb{Z}^n$ (i.e. $D = \{x \in \mathbb{Z}^n \mid P(x) = 0\}$).*

Remarks:
• A finite intersection of algebraic (resp. diophantine) sets is an algebraic (resp. diophantine) set. Actually if $A_1, ..., A_k$ are in $\mathbb{R}$ (resp. in $\mathbb{Z}$) then
$A_1 = 0$ and $A_2 = 0$ and ... and $A_k = 0$ if and only if $A_1^2 + A_2^2 + ... + A_k^2 = 0$.
• A finite union of algebraic (resp. diophantine) sets is an algebraic (resp. diophantine) set. Actually if $A_1, ..., A_k$ are in $\mathbb{R}$ (resp. in $\mathbb{Z}$) then
$A_1 = 0$ or $A_2 = 0$ or ... or $A_k = 0$ if and only if $A_1.A_2...A_k = 0$.

In the following theorem, we present the result corresonding to Tarski's elimination algorithm which proves the decidability of first order logic corresponding to elementary algebra.

Theorem 4. *[2,14,19,20]*
Let $P_1(X_1, ..., X_n, Y), ..., P_k(X_1, ..., X_n, Y)$ be a sequence of polynomials in $n+1$ variables with coefficients in $\mathbb{Z}$. Then there exists an effectively computable boolean combination $\mathcal{B}(X_1, ..., X_n)$ of polynomial equations and inequalities in the variables $X_1, ..., X_n$ with coefficients in $\mathbb{Z}$, such that for every $(x_1, ..., x_n) \in \mathbb{R}^n$, the system of polynomial equations and inequalities

$$\begin{cases} P_1(x_1, ..., x_n, Y) \succ_1 0 \\ ... \\ P_k(x_1, ..., x_n, Y) \succ_k 0 \end{cases}$$

where $\succ_i \in \{=, \geq, >\}$ for $i = 1, ..., k$,
has a solution y in $\mathbb{R}$ if and only if $\mathcal{B}(x_1, ..., x_n)$ holds true in $\mathbb{R}$.

Corollary 1. • *The boolean combination $\mathcal{B}$ in the last theorem define a semi-algebraic set (i.e. $\{x \in \mathbb{R}^n \mid \mathcal{B}(x)$ holds true in $\mathbb{R}\}$ is a semi-algebraic set of $\mathbb{R}^n$).*

- *Let F be a field which is a finite extention of $\mathbb{Q}$ and let $\mathcal{B}'(X_1, ..., X_n, Y)$ be a boolean combination of polynomial equations and inequalities in the variables $X_1, ..., X_n, Y$ with coefficients in F; then there exists an effectively computable boolean combination $\mathcal{B}''(X_1, ..., X_n)$ of polynomial equations and inequalities in the variables $X_1, ..., X_n$ with coefficients in F, such that for every $(x_1, ..., x_n) \in \mathbb{R}^n$, there exists $y \in \mathbb{R}$ such that $\mathcal{B}'(x_1, ..., x_n, y)$ holds true in $\mathbb{R}$ if and only if $\mathcal{B}''(x_1, ..., x_n)$ holds true in $\mathbb{R}$. $\mathcal{B}''$ is obtained from $\mathcal{B}'$ by using Tarski's elimination algorithm corresponding to transformation rules permitting the elimination of the variable Y.*

6 Hausdorff Discretizations of Algebraic Sets

In this section we study the Hausdorff discretizations of real algebraic sets. We prove that some Hausdorff discretizations of real algebraic sets are diophantine sets and we give some decidable and undecidable properties concerning Hausdorff discretizations of real $\mathbb{Z}$-algebraic sets.

Definition 15. *A metric d is called* algebraic *(resp.* $\mathbb{Z}$-algebraic*) if the unit ball of d is a semi-algebraic (resp. $\mathbb{Z}$-semi-algebraic) set.*

For example d_p is $\mathbb{Z}$-algebraic for all $p \in \mathbb{Q}$ such that $p \geq 1$ and for $p = \infty$.

The following property is a consequence of the undecidability of diophantine equations (*Hilbert's Tenth Problem*) [10].

Property 1. Let $n \geq 3$. The following problem is undecidable:
Is $r_H(\mathcal{A}, \rho) = r_c(\rho)$ for a given $\mathbb{Z}$-algebraic set $\mathcal{A}$ of $\mathbb{R}^n$, a $\mathbb{Z}$-algebraic metric d on $\mathbb{R}^n$ and $\rho \in \mathbb{Q}^+$?

The following property is obtained by successive applications of Tarski's elimination algorithm, to a polynomially constrained formula F_n whith n unknows corresponding to the problem, until obtaining a formula F_0 without unknowns.

Property 2. The following problem is decidable:
Is $p \in \Delta_r^d(\mathcal{A}, \rho)$ for a given $p \in \rho\mathbb{Z}^n$, a $\mathbb{Z}$-algebraic set $\mathcal{A}$ of $\mathbb{R}^n$, a $\mathbb{Z}$-algebraic metric d on $\mathbb{R}^n$ and $r, \rho \in \mathbb{Q}^+$?

Theorem 5. *Let $\mathcal{A}$ be a $\mathbb{Z}$-algebraic set of $\mathbb{R}^n$ and d be a $\mathbb{Z}$-algebraic metric on $\mathbb{R}^n$, then*

- $\frac{1}{\rho}\Delta_r^d(\mathcal{A}, \rho)$ *is a diophantine set for all $r, \rho \in \mathbb{Q}^+$.*
- $\frac{1}{\rho}Sk(\mathcal{A}, \rho)$ *is a diophantine set for all $\rho \in \mathbb{Q}^+$.*

Corollary 2. *Let $\mathcal{A}$ be a $\mathbb{Z}$-algebraic set of $\mathbb{R}^n$ and d be $\mathbb{Z}$-algebraic metric on $\mathbb{R}^n$, then*

- *The maximal Hausdorff discretization $\Delta_H(\mathcal{A}, 1)$ of $\mathcal{A}$ is a diophantine set and $\frac{1}{\rho}\Delta_H(\mathcal{A}, \rho)$ is a diophantine set for all $\rho \in \mathbb{Q}^+$.*
- *The supercover discretization $\Delta_{SC}(\mathcal{A}, 1)$ of $\mathcal{A}$ is a diophantine set and $\frac{1}{\rho}\Delta_{SC}(\mathcal{A}, \rho)$ is a diophantine set for all $\rho \in \mathbb{Q}^+$.*

7 Hausdorff Discretization of Algebraic Curves

In this section we study the Hausdorff discretizations of $\mathbb{Z}$-algebraic curves *(i.e. a $\mathbb{Z}$-algebraic set of $\mathbb{R}^2$)*. We refine this study for straight lines by using Fourier's elimination algorithm because some of the constraits corresponding to this problem are linear. So Fourier's elimination algorithm is a special case of Tarski's elimination algorithm. Our result concerning Hausdorff discretizations of straight lines generalize the results [1] concerning the supercover discretization of straight lines. Several properties of Hausdorff discretizations of straight lines are decidable because these discretizations are defined by linear constraints, which is not the case for Hausdorff discretizations of non linear algebraic sets of $\mathbb{R}^n$ for $n \geq 3$ because, it corresponds to diophantine equations on n unknows [10]. Notice that the decidability of a diophantine equation in two unknowns is still an open problem.

Theorem 6. *Let $\rho \in \mathbb{R}^+$ and let $\mathcal{C}$ be a $\mathbb{Z}$-algebraic curve such that the radius of curvature of $\mathcal{C}$ is strictly greater than ρ and let d be an homogeneous and $\mathbb{Z}$-algebraic metric, then*

- *If $r_H(\mathcal{C},\rho) < r_c(\rho)$, then there exists two diophantine sets D_1 and D_2 such that*

$$M \in \mathcal{M}_H(\mathcal{C},\rho) \text{ if and only if } M = \rho(D_1 \cup S) \text{ with } S \subseteq D_2.$$

- *If $r_H(\mathcal{C},\rho) = r_c(\rho)$ then there exists tree diophantine sets D_0, D_1 and D_2 such that*
  $M \in \mathcal{M}_H(\mathcal{C},\rho)$ if and only if $M = \rho(D_0 \cup S_1 \cup S_2)$ with $S_1 \subseteq D_1$, $S_2 \subseteq D_2$ and if $p \in (D_1 \backslash S_1)$, then there exists $S \subseteq \mathcal{V}_4(p)$, $S \subseteq S_2$, $\mathcal{B}^d_{r_c(\rho)}(q) \cap \mathcal{C} \neq \emptyset$ for all $q \in S$ and $\mathcal{C} \cap \mathcal{C}(p,\rho) \subseteq \bigcup_{q \in S} \mathcal{B}^d_{r_c(\rho)}(q)$, where $\mathcal{V}_4(p)$ is the 4-neighbourhood of p in $\rho\mathbb{Z}^2$.

Corollary 3. *Let $\rho \in \mathbb{R}^+$ and let $\mathcal{C}$ be a $\mathbb{Z}$-algebraic curve such that the radius of curvature of $\mathcal{C}$ is a strictry greater than ρ and assume that d is strictly homogeneous and $\mathbb{Z}$-algebraic metric, then there exists two diophantine sets D_1 and D_2 such that,*

$$M \in \mathcal{M}_H(\mathcal{C},\rho) \text{ if and only if } M = \rho(D_1 \cup S) \text{ with } S \subseteq D_2.$$

7.1 Hausdorff Discretizations of Straight Lines

In this subsection we study the Hausdorff discretizations of straight lines for homogeneous metrics. Some of the results of this subsection are proved by using Fourier's elimination algorithm [4,8,11] which transforms linear constraints. Fourier's elimination algorithmis is a special case of Tarski's elimination algorithm. The time complexity of Tarski's elimination algorithm, is in the general case, very high [5].

Before giving the results corresponding to Hausdorff discretizations of straight lines, we treat two special cases of horizontal and vertical straight lines.

Let L be a straight line such that $L = \{(x,y) \in \mathbb{R}^2 \mid x + \rho(m+\frac{1}{2}) = 0\}$ *(resp.* $L = \{(x,y) \in \mathbb{R}^2 \mid y + \rho(m+\frac{1}{2}) = 0\}$ *)* for $m \in \mathbb{Z}$, then $r_H(L,\rho) = r_c(\rho)$ and $\forall M \in \mathcal{M}_{H_d}(F,\rho)$ $\forall i \in \rho\mathbb{Z}$, if $(\rho m, i) \notin M$ then $(\rho(m+1), i) \in M$ *(resp.* $\forall i \in \rho\mathbb{Z}$, *if* $(i, \rho m) \notin M$ *then* $(i, \rho(m+1)) \in M$ *).*

In all the following, we assume that the considered straight lines are not of the last forms.

Definition 16. *$S \subset \mathbb{Z}^2$ is a* periodic set *if there exists $p \in \mathbb{Z}^2$ and $p \neq (0,0)$ such that $q + p \in S$ for all $q \in S$.*

Theorem 7. *Let $\mathcal{L} = \{(x,y) \in \mathbb{R}^2 \mid ax + by + c = 0\}$ such that $a, b, c \in \rho\mathbb{Z}$*

1- *If d is an homogeneous metric (the unit ball of d is not necessarily a semi-algebraic set) and $r_H(\mathcal{L},\rho) < r_c(\rho)$, then there exist two periodic diophantine sets D_1 and D_2 such that*

$$M \in \mathcal{M}_H(\mathcal{L},\rho) \text{ if and only if } M = \rho(D_1 \cup S) \text{ with } S \subseteq D_2.$$

2- *If d is a strictly homogeneous metric (the unit ball of d is not necessarily a semi-algebraic set), then there exist two periodic diophantine sets D_1 and D_2 such that*

$$M \in \mathcal{M}_H(\mathcal{L},\rho) \text{ if and only if } M = \rho(D_1 \cup S) \text{ with } S \subseteq D_2.$$

Sketch of proof:
This theorem is a consequence of the following two facts:

$\mathcal{L} = \{(x,y) \in \mathbb{R}^2 \mid ax + by + c = 0\}$,
• Let $r > 0$ and $D_r(\mathcal{L},\rho) = \{(m,n) \in \rho\mathbb{Z}^2 \mid \mathcal{L} \cap int(\mathcal{C}((m,n),r)) \neq \emptyset\}$. So, $D_r(\mathcal{L},\rho)$ is the set of $(m,n) \in \rho\mathbb{Z}^2$ such that there exists $t, t' \in \mathbb{R}$ and

$$\begin{cases} a(m+t) + a(n+t') + c = 0 \\ t > -\frac{r}{2} \\ -t > -\frac{r}{2} \\ t' > -\frac{r}{2} \\ -t' > -\frac{r}{2} \end{cases}$$

So, $D_r(\mathcal{L},\rho) = \{(m,n) \in \rho\mathbb{Z}^2 \mid -\frac{r}{2}(|a|+|b|) < am+bn+c < \frac{r}{2}(|a|+|b|)\}$. Thus, $\frac{1}{\rho}D_r(\mathcal{L},\rho)$ is a diophantine set and if $(m,n) \in D_r(\mathcal{L},\rho)$ then $(m,n) + z(b,-a) \in D_r(\mathcal{L},\rho)$ for all $z \in \mathbb{Z}$.

• Let $r \geq 0$ and $(m,n) \in \rho\mathbb{Z}^2$ such that $\mathcal{B}^d_r(m,n) \cap \mathcal{L} \neq \emptyset$, then $\mathcal{B}^d_r((m,n) + z(b,-a)) \cap \mathcal{L} \neq \emptyset$ for all $z \in \mathbb{Z}$. Moreover, if for some r', $\mathcal{L} \cap int(\mathcal{C}((m,n),r')) = \emptyset$, then $\mathcal{L} \cap int(\mathcal{C}((m,n) + z(b,-a), r')) = \emptyset$.

In Figure 3, we illustrate the sets D_1 and D_2 of the last theorem.

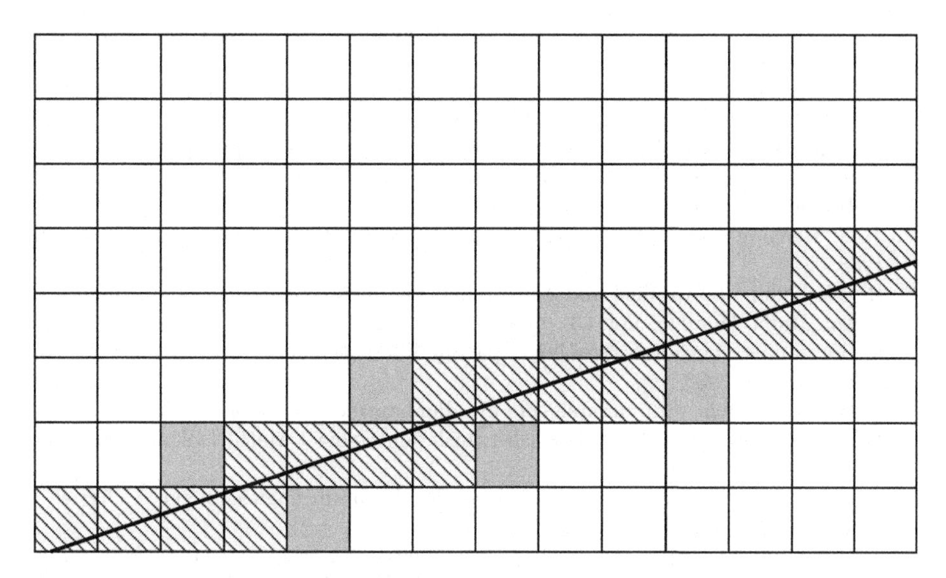

Fig. 3. The hatched squares correspond to points of D_1 and the grey squares correspond to points of D_2

8 Conclusion

Throughout several papers we have introduced a new framework for the discretization of a non-empty closed set, based on the Hausdorff distance. In this paper we have studed Hausdorff discretizations of real algebraic sets. Actually we have given some decidable and undecidable properties concerning Hausdorff discretizations of algebraic sets and we have proved that some Hausdorff discretizations of algebraic sets are diophantine sets. We have refined the last results for algebraic curves and more precisely for straight lines. Some of these results are proven by using the Tarski's elimination algorithm concerning the decidability of first order logic for elementary algebra.
We intend to do further investigations on:

- effective computation of Hausdorff discretizations of non linear algebraic sets;
- topological properties of Hausdorff discretizations of algebraic sets;
- complexity of representation of some Hausdorff discretizations of algebraic sets.

References

1. E. Andres. *Standard Cover: a new class of discrete primitives. Internal Report*, IRCOM, Université de Poitiers. 100, 107
2. J. Bochnak, M. Coste, M.-F. Roy. Real Algebraic Geometry. *Serie of Modern Surveys in Mathematics*, Vol. 36, Springer, 1998. 105

3. H. Busemann *The geometry of geodesics.* Academec Press, New York, 1955. 100, 101
4. J. B. J. Fourier. Solution d'une question particulière du calcul des inégalités. *Oeuvre II*, Paris, pp. 317–328, 1826. 107
5. D. Grigor'ev, N. Vorobjov. Solving systems of polynomial inequalities in subexponential time. *J. Symbolic Comput.*, Vol. 5, pp. 37–64, 1988. 100, 107
6. F. Haudorff. *Set Theory.* Chelsea, New York, 1962. 100, 101
7. J. G. Hocking and G. S. Young *Topology.* Dover Publications Inc., New York, 1988. 100
8. H. W. Kuhn. Solvability and consistency for linear equations and inequalities. *Amer. Math. Monthly*, Vol. 63, pp 217–232, 1956. 107
9. L. Hörmander. The analysis of linear partial differential operators. *Springer-Verlag Berlin*, Vol. 2, 1983. 99
10. Y. Matiiassevitch. Enumerable sets are diophantine. *Doklady Akad.Nauk SSSR*, Vol. 191, pp 279–282, 1970 (English translation : Soviet Math. Doklady, pp 354–357, 1970). 106, 107
11. T. S. Motzkin. Beitrage zur theorie der linearen ungleichungen. *Azriel :* Jerusalem, 1936. 107
12. C. Ronse and M. Tajine. Discretization in Hausdorff Space. *Journal of Mathematical Imaging & Vision*, Vol. 12, no 3, pp. 219-242, 2000. 99, 100, 101
13. C. Ronse and M. Tajine. Hausdorff discretization for cellular distances, and its relation to cover and supercover discretization. *To be revised*, 2000. 99, 101
14. A. Seidenberg. A new decision method for elementary algebra. *Ann. of Math.*, Vol. 60, pp 365–374, 1954. 99, 105
15. M. Tajine and C. Ronse. Preservation of topology by Hausdorff discretization and comparison to other discretization schemes. *Submitted*, 1999. 99, 100, 101, 103
16. M. Tajine and C. Ronse. Hausdorff sampling of closed sets in a boundedly compact space. *In preparation*, 2000. 99
17. M. Tajine and C. Ronse. Topological Properties of Hausdorff discretizations. *International Symposium on Mathematical Morphology 2000 (ISMM'2000)*, Palo Alto CA, USA. Kluwer Academic Publishers, pp. 41-50, 2000. 99
18. M. Tajine, D. Wagner and C. Ronse. Hausdorff discretization and its comparison with other discretization schemes. *DGCI'99*, Paris, LNCS Springer-Verlag, Vol. 1568, pp. 399–410, 1999. 99
19. A. Tarski. Sur les ensembles définissables de nombres réels. *Fund. Math.*, Vol. 17, pp. 210–239, 1931. 99, 105
20. A. Tarski. A decision method for elementary algebra and geometry. *Tech. Rep.*, University of California Press, Berkeley and Los Angeles, 1951. 99, 105
21. D. Wagner. Distance de Hausdorff et problème discret-continu. Mémoire de D. E. A. (M.Sc. Dissertation), Université Louis Pasteur, Strasbourg (France). 1997. 99, 100, 101
22. D. Wagner, M. Tajine and C. Ronse. An approach to discretization based on the Hausdorff metric.*ISMM'1998*. Kluwer Academic Publishers. pp. 67–74, 1998. 99, 100, 101

Discrete Images

An Algorithm for Reconstructing Special Lattice Sets from Their Approximate X-Rays*

Sara Brunetti[1], Alain Daurat[2], and Alberto Del Lungo[3]

[1] Dipartimento di Sistemi e Informatica
Via Lombroso 6/17, Firenze, 50134, Italy
brunetti@dsi.unifi.it

[2] Laboratoire de Logique et d'Informatique de Clermont-1 (LLAIC1)
Ensemble Universitaire des Cézeaux
B.P. n° 86, 63172 Aubière Cedex, France
daurat@llaic.u-clermont1.fr

[3] Dipartimento di Matematica, Via del Capitano 15, 53100, Siena, Italy
dellungo@unisi.it

Abstract. We study the problem of reconstructing finite subsets of the integer lattice $\mathbb{Z}^2$ from their approximate X-rays in a finite number of prescribed lattice directions. We provide a polynomial-time algorithm for reconstructing Q-convex sets from their "approximate" X-rays. A Q-convex set is a special subset of $\mathbb{Z}^2$ having some convexity properties. This algorithm can be used for reconstructing convex subsets of $\mathbb{Z}^2$ from their exact X-rays in some sets of four prescribed lattice directions, or in any set of seven prescribed mutually nonparallel lattice directions.

1 Introduction

The problem of reconstructing two-dimensional lattice sets from their X-rays has been studied in discrete mathematics and applied in several areas. In this context, a *two-dimensional lattice set* is a finite subsets of the integer lattice $\mathbb{Z}^2$, and an *X-ray* of a set in a direction u is a function giving the number of its points on each line parallel to u. This problem is the basic reconstruction problem in discrete tomography [10] and it has various interesting applications in image processing [15], statistical data security [12], biplane angiography [14] and reconstructing crystalline structures from X-rays taken by an electron microscope [13]. The computational complexity of this reconstruction problem is studied in [9]. It is shown that the question of reconstructing two-dimensional lattice sets from their X-rays in a set of $m \geq 3$ pairwise nonparallel directions is $\mathbb{NP}$-hard. In most practical applications we have some a priori properties about the sets to be reconstructed. The algorithms can take advantage of this information to reconstruct the set. Mathematically, these properties can be described

* This work is partially supported by MURST project: *Modelli di calcolo innovativi: metodi sintattici e combinatori.* and by the University Siena project: *Problemi Inversi Discreti: Tomografia Discreta*

G. Borgefors, I. Nyström, and G. Sanniti di Baja (Eds.): DGCI 2000, LNCS 1953, pp. 113–125, 2000.
© Springer-Verlag Berlin Heidelberg 2000

in terms of a subclass of subsets of $\mathbb{Z}^2$ to which the solution must belong. For instance, there are polynomial-time algorithms to reconstruct hv-convex polyominoes (i.e., two-dimensional lattice subsets which are 4-connected and convex in the horizontal and vertical directions) from their X-rays in horizontal and vertical directions [2,5]. The class of convex lattice subsets (i.e., finite subsets F with $F = \mathbb{Z}^2 \cap \text{conv}F$) is another well-known and studied class in discrete tomography. Gardner and Gritzmann [8] proved that the X-rays in four suitable or any seven prescribed mutually nonparallel lattice directions uniquely determine all the convex lattice subsets. The complexity of the reconstruction problem on this class is an open problem raised by Gritzmann during the workshop: *Discrete Tomography: Algorithms and Complexity* (1997).

In this paper, we study the problem of reconstructing Q-convex sets from their X-rays in a finite number of prescribed directions. The Q-convexity is a weak convexity property linked to a finite number of directions. The class of Q-convex sets contains all the convex lattice subsets. We provide a polynomial-time algorithm for reconstructing Q-convex sets from their "approximate" X-rays. This means that, the algorithm decides whether or not there is a Q-convex set whose X-rays all lie within prescribed bounds. If there is at least one Q-convex set having X-rays lying within these bounds, the algorithm reconstructs one of them in polynomial time. Boufkhad et al. [3] studied the problem of reconstructing h, v-convex polyominoes from their "approximate" horizontal and vertical X-rays. We show that our algorithm solves this problem in polynomial time. The algorithm can be used for reconstructing Q-convex sets from their "exact" X-rays. A greedy algorithm for solving this problem has been defined in [4], and our new approach is faster than this algorithm for a number of directions equal to two and three. We point out that recently, it is proved in [6] that the uniqueness Gardner's and Gritzmann's results can be extended to the class of Q-convex sets. From this uniqueness result for Q-convex sets it follows that our algorithm can be used for reconstructing convex lattice subsets from their exact X-rays in some sets of four prescribed lattice directions, or in any set of seven prescribed mutually nonparallel lattice directions. This means that, our algorithm and the one defined in [4] solve Gritzmann's problem for these special sets of directions.

2 A Reconstruction Algorithm for Two X-Rays

In this section, we are going to define an algorithm for reconstructing Q-convex sets from their approximate X-rays in two directions. The basic idea of the algorithm is to determine a polynomial transformation of our problem to the 2-Satisfiability problem which can be solved in linear time [1,2,5].

2.1 Definitions and Preliminaries

Let p and q be two independent linear forms on $\mathbb{Q}^2$. We can assume that: $p(x, y) = ax + by$ and $q(x, y) = cx + dy$ with $a, b, c, d \in \mathbb{Z}$, $ad - bc \neq 0$, $gcd(a, b) = 1$, $gcd(c, d) = 1$. Assuming that $M = (x_M, y_M)$, we denote $p(x_M, y_M)$

by $p(M)$. The *X-ray* of a lattice set F along direction $p(M) = const$ is the function $X_pF : \mathbb{Z} \to \mathbb{N}_0$ defined by:

$$X_pF(i) = \text{card}(\{N \in F \mid p(N) = i\}).$$

This, in turn, means that an X-ray of a lattice set F in a direction p is a function giving the number of points of F on each line parallel to this direction (see Fig. 2(a)). We define four zones around a point M of $\mathbb{Z}^2$ (see Fig. 1) as follows:

$$Z_0(M) = \{N \in \mathbb{Z}^2 \mid p(N) \leq p(M) \text{ and } q(N) \leq q(M)\},$$
$$Z_1(M) = \{N \in \mathbb{Z}^2 \mid p(N) \geq p(M) \text{ and } q(N) \leq q(M)\},$$
$$Z_2(M) = \{N \in \mathbb{Z}^2 \mid p(N) \geq p(M) \text{ and } q(N) \geq q(M)\},$$
$$Z_3(M) = \{N \in \mathbb{Z}^2 \mid p(N) \leq p(M) \text{ and } q(N) \geq q(M)\}.$$

We can now introduce the definition of Q-convex set around two directions.

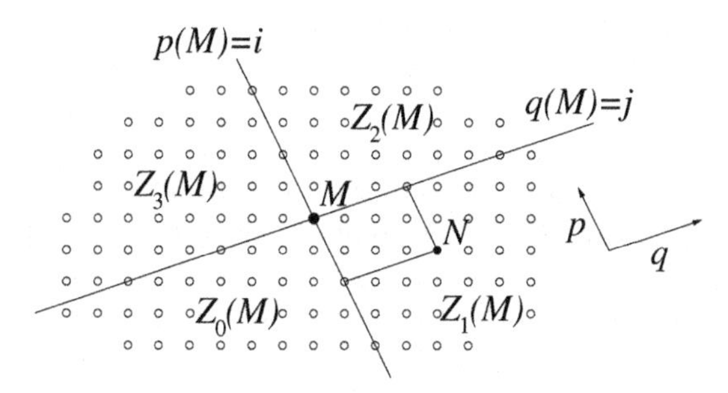

Fig. 1. The four zones around a point M of $\mathbb{Z}^2$

Definition 1. *A lattice set F is* Q-convex *around p and q if and only if for each $M \notin F$ there exists k such that $Z_k(M) \cap F = \emptyset$ for $k \in \{0, 1, 2, 3\}$.*

By the definition, if there is at least one point of F in every zone $Z_0(M)$, $Z_1(M)$, $Z_2(M)$ and $Z_3(M)$, the point M has to belong to F. Fig. 2(a) shows some examples of Q-convex sets. We point out that, from the definition it follows that a Q-convex set around p and q is a discrete set which is convex along p and q. A discrete set F is convex along p if for each pair of points (M, N) of F such that $p(M) = p(N)$, the discrete segment $[MN] \cap \mathbb{Z}^2$ is contained in F. Let us now take the following problem into consideration.

Problem 1. **Approximate Consistency with two directions**
Instance: four vectors $P = (p_1, \dots, p_n)$, $P' = (p'_1, \dots, p'_n)$, $Q = (q_1, \dots, q_m)$, $Q' = (q'_1, \dots, q'_m)$ whose elements are no-negative integer numbers and p_1, p_n, p'_1, p'_n, q_1, q_m, q'_1, q'_m are positive integer numbers.
Question: is there a Q-convex set F around p and q such that $p_i \leq X_pF(i) \leq p'_i$ for $i = 1, \dots, n$ and $q_j \leq X_qF(j) \leq q'_j$ for $j = 1, \dots, m$?

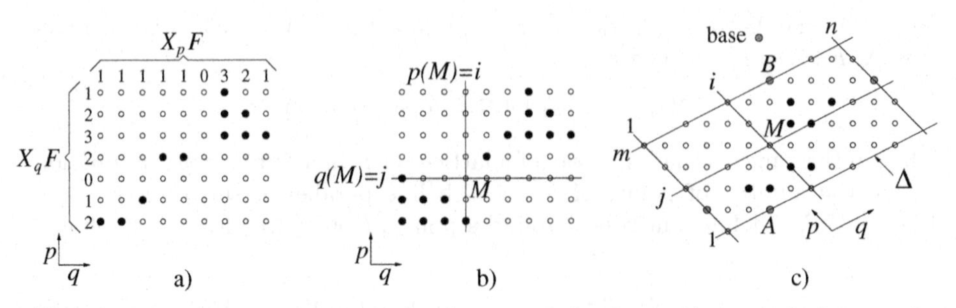

Fig. 2. a) A Q-convex set around $p(x, y) = x$ and $q(x, y) = y$. b) A set which is not Q-convex set around $p(x, y) = x$ and $q(x, y) = y$. c) A Q-convex set around $p(x, y) = x+y$ and $q(x, y) = x - 2y$

The problem is to decide whether or not there is a Q-convex set around p and q whose X-rays in these two directions lie within prescribed bounds. If $P = P'$ and $Q = Q'$, we have the *Exact Consistency* problem with two directions.
In the following subsections, we determine a polynomial transformation of Problem 1 to the 2-Satisfiability problem (2SAT).

2.2 Q-Convexity

The intersection of the p-line $p(M) = i$ with the q-line $q(M) = j$ is not always in $\mathbb{Z}^2$. It is easy to prove that the point M intersection of $p(M) = i$ with $q(M) = j$ belongs to $\mathbb{Z}^2$ if and only if $j \equiv \kappa i \pmod{\delta}$, where: $\delta = |ad - bc|$, $\kappa = (cu + dv)\text{sign}(ad - bc) \pmod{\delta}$ and $au + bv = 1$ (see Fig. 3(a) and [7]).
Without any loss of generality, we can assume that a Q-convex set F around p and q whose X-rays are such that: $p_i \leq X_pF(i) \leq p'_i$ and $q_j \leq X_qF(j) \leq q'_j$ for all i, j, is contained in the lattice parallelogram:

$$\Delta = \{N \in \mathbb{Z}^2 \mid 1 \leq p(N) \leq n \text{ and } 1 \leq q(N) \leq m\}.$$

We denote the point $M \in \Delta$ intersection of $p(M) = i$ with $q(M) = j$ by $M = (i, j)$ (see Fig. 2(c)). Let $K = \{0, 1, 2, 3\}$. We associate four boolean variables $V_k(M)$, with $k \in K$, at every point $M \in \Delta$ (i.e., one variable for each zone $Z_k(M)$). The idea of the algorithm is to build a 2SAT formula APPROX on the variables $(V_k(M))_{k\in K, M\in\Delta}$ so that there is a solution F of Problem 1 if and only if APPROX is satisfiable. If there is an evaluation V of the variables $(V_k(M))_{k\in K, M\in\Delta}$ satisfying APPROX, the corresponding lattice set F solving Problem 1 is defined by function Φ as follows:

$$F = \Phi(V) \text{ iff } F = \{M \in \Delta \mid \neg V_k(M) \text{ is true}, \forall k \in K\},$$

where $\neg V_k(M)$ is true if and only if $V_k(M)$ is false. Conversely, if F is a subset of Δ solving Problem 1, the corresponding evaluation V of the variables $(V_k(M))_{k\in K, M\in\Delta}$ satisfying APPROX is defined by function Ψ as follows:

$$V = \Psi(F) \text{ iff } V_k(M) = \text{“}Z_k(M) \cap F = \varnothing\text{”}, \text{ with } k \in K,\ M \in \Delta.$$

We assume that all literals outside Δ are true. The boolean formula APPROX is made up of three sets of clauses expressing: the *Q-convexity* (QCONV), a *lower bound* (LB) and an *upper bound* (UB) on the X-rays. The Q-convexity can be expressed with the boolean variables by the formulas:

$$\mathcal{Z}_0 = \bigwedge_{M=(i,j)\in\Delta} \left(\begin{array}{l} (V_0(i,j) \Rightarrow V_0(i-\delta,j)) \wedge (V_0(i,j) \Rightarrow V_0(i,j-\delta)) \\ \wedge \bigwedge_{\substack{0<u<\delta \\ 0<v<\delta \\ v\equiv\kappa u(\mathrm{mod}\delta)}} \Big(V_0(i,j) \Rightarrow V_0(i-u,j-v)\Big) \end{array} \right)$$

$$\mathcal{Z}_1 = \bigwedge_{M=(i,j)\in\Delta} \left(\begin{array}{l} (V_1(i,j) \Rightarrow V_1(i+\delta,j)) \wedge (V_1(i,j) \Rightarrow V_1(i,j-\delta)) \\ \wedge \bigwedge_{\substack{0<u<\delta \\ 0<v<\delta \\ v\equiv\kappa(\mathrm{mod}\delta)}} \Big(V_1(i,j) \Rightarrow V_1(i+u,j-v)\Big) \end{array} \right)$$

$$\mathcal{Z}_2 = \bigwedge_{M=(i,j)\in\Delta} \left(\begin{array}{l} \Big(V_2(i,j) \Rightarrow V_2(i+\delta,j)\Big) \wedge (V_2(i,j) \Rightarrow V_2(i,j+\delta)) \\ \wedge \bigwedge_{\substack{0<u<\delta \\ 0<v<\delta \\ v\equiv\kappa u(\mathrm{mod}\delta)}} \Big(V_2(i,j) \Rightarrow V_2(i+u,j+v)\Big) \end{array} \right)$$

$$\mathcal{Z}_3 = \bigwedge_{M=(i,j)\in\Delta} \left(\begin{array}{l} (V_3(i,j) \Rightarrow V_3(i-\delta,j)) \wedge (V_3(i,j) \Rightarrow V_3(i,j+\delta)) \\ \wedge \bigwedge_{\substack{0<u<\delta \\ 0<v<\delta \\ v\equiv\kappa u(\mathrm{mod}\delta)}} \Big(V_3(i,j) \Rightarrow V_3(i-u,j+v)\Big) \end{array} \right)$$

The points in the grey zone around $M = (i,j)$ in Fig. 3(a) are the points of Δ used in $\mathcal{Z}_0, \mathcal{Z}_1, \mathcal{Z}_2$ and $\mathcal{Z}_3$ (i.e., the points $(i \pm \delta, j \pm \delta)$, $(i \pm u, j \pm v)$ with $0 < u < \delta$, $0 < v < \delta$). Let us denote $\mathcal{Z}_0 \wedge \mathcal{Z}_1 \wedge \mathcal{Z}_2 \wedge \mathcal{Z}_3$ by QCONV.

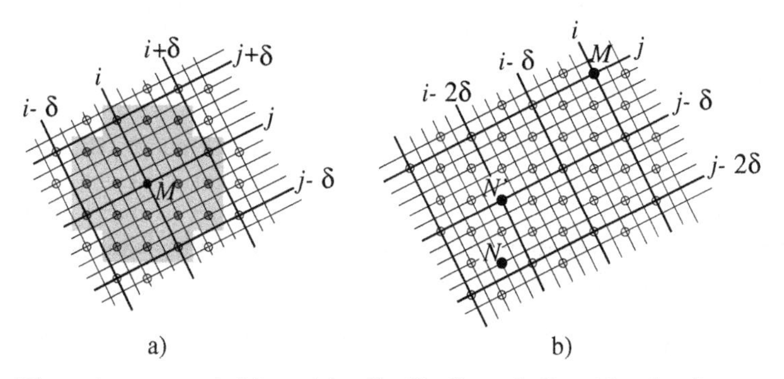

Fig. 3. The points around M used in $\mathcal{Z}_0, \mathcal{Z}_1, \mathcal{Z}_2$ and $\mathcal{Z}_3$ with $p(x,y) = 2x + y$ and $q(x,y) = x - 2y$

Lemma 1. *Let V be an evaluation of the variables $(V_k(M))_{k\in K, M\in\Delta}$ satisfying QCONV. If $M \in \Delta$ and $V_k(M)$ is true, then $V_k(N)$ is true for all $N \in Z_k(M)$.*

Proof. Assume that $k = 0$ and let $M = (i_M, j_M)$. We have: $V_0(i,j) \Rightarrow V_0(i-\delta, j)$ and $V_0(i,j) \Rightarrow V_0(i, j-\delta)$ are satisfied, for all i,j. Therefore, by induction we can prove that $V_0(i_M - k\delta, j_M - l\delta)$ is true, for all $k, l \in \mathbb{N}$. Let N be a point of $Z_0(M)$. Let N' be the point of Δ such that:

$$\frac{p(M)-p(N')}{\delta} = \left\lfloor \frac{p(M)-p(N)}{\delta} \right\rfloor, \qquad \frac{q(M)-q(N')}{\delta} = \left\lfloor \frac{q(M)-q(N)}{\delta} \right\rfloor$$

By the previous statement we have $V_0(N')$ is true (see Fig. 3(b)) and, since the formula $V_0(N') \Rightarrow V_0(N)$ is in $\mathcal{Z}_0$, we finally obtain $V_0(N)$ is true. We proceed in the same way for k equal to 1,2 and 3.

Thus, we can characterize the Q-convexity by means of the formula QCONV.

Lemma 2.

- *For any set $F \subset \Delta$ the evaluation $V = \Psi(F)$ of the boolean variables $(V_k(M))_{k \in K, M \in \Delta}$ satisfies the formula QCONV.*
- *If an evaluation V of the boolean variables $(V_k(M))_{k \in K, M \in \Delta}$ satisfies the formula QCONV, then $F = \Phi(V)$ is Q-convex around p and q.*

Proof.

- Assume that $V = \Psi(F)$ does not satisfy QCONV. By this assumption, there exists k such that at least a clause of $\mathcal{Z}_k$ is not satisfied. Then $V_k(M)$ and $\neg V_k(N)$ are true, where $N \in Z_k(M)$. As a consequence, $Z_k(M) \cap F = \varnothing$, $Z_k(N) \cap F \neq \varnothing$ and $Z_k(N) \subset Z_k(M)$. We got a contradiction and so $V = \Psi(F)$ satisfies QCONV.
- The second statement is just a consequence of Lemma 1. If $M \notin F$, by the definition of Φ there exists k such that $V_k(M)$ is true. Therefore, by Lemma 1, we have $V_k(N)$ for each $N \in Z_k(M)$. Consequently, $Z_k(M) \cap F = \varnothing$.

2.3 A Lower Bound

Now we want to express that X-ray values of a lattice set in the direction p are greater than some prescribed integers. Let us take the line $p(M) = i$ into consideration. Let $qmin_i = \min\{j|(i,j) \in \Delta\}$ and $qmax_i = \max\{j|(i,j) \in \Delta\}$. Notice that, if $\delta \neq 1$ these numbers are not always equal to 1 and m. We define the formula LB(p,i,l) in the following way:

$$\mathrm{LB}(p,i,l) = \mathit{TRUE} \qquad \textbf{if } l = 0$$

$$\mathrm{LB}(p,i,l) = \left(\begin{array}{l} \bigwedge_{\substack{1 \leq j \leq m-\delta l \\ j \equiv \kappa i \pmod{\delta}}} (L_1(j) \wedge L_2(j) \wedge L_3(j) \wedge L_4(j)) \\ \wedge L'_1 \wedge L'_2 \wedge L'_3 \wedge L'_4 \end{array} \right) \qquad \textbf{otherwise}$$

where

$$
\begin{aligned}
L_1(j) &= V_0(i,j) \Rightarrow \neg V_2(i,j+\delta l) \\
L_2(j) &= V_0(i,j) \Rightarrow \neg V_3(i,j+\delta l) \\
L_3(j) &= V_1(i,j) \Rightarrow \neg V_2(i,j+\delta l) \\
L_4(j) &= V_1(i,j) \Rightarrow \neg V_3(i,j+\delta l) \\
L'_1 &= \neg V_2(i, qmin_i + \delta(l-1)) \\
L'_2 &= \neg V_3(i, qmin_i + \delta(l-1)) \\
L'_3 &= \neg V_0(i, qmax_i - \delta(l-1)) \\
L'_4 &= \neg V_1(i, qmax_i - \delta(l-1))
\end{aligned}
$$

Lemma 3.

- *If a lattice set F is Q-convex around p and q and its X-ray along p is such that $X_pF(i) \geq l$, then the evaluation $V = \Psi(F)$ of the variables $(V_k(M))_{k\in K, M\in\Delta}$ satisfies $QCONV \wedge \mathrm{LB}(p,i,l)$.*
- *If an evaluation V of the variables $(V_k(M))_{k\in K, M\in\Delta}$ satisfies $QCONV \wedge \mathrm{LB}(p,i,l)$, then the X-ray of $F = \Phi(V)$ along p is such that $X_pF(i) \geq l$.*

We do not show the proof of this Lemma for brevity's sake. We define the formula $\mathrm{LB}(q,j,l)$ for the lines in the direction q in a similar way.

2.4 An Upper Bound

Now we want to express that X-ray values of a lattice set in the direction q are smaller than some prescribed integers. For this upper bound, we need to fix two points A and B such that $p(A) = 1$ and $p(B) = n$. We call *bases* these two points (see Fig. 2(c)). Let us take the line $q(M) = j$ into consideration. We introduce the formula:

$$
\mathrm{UB}(q,j,l,A,B) = \mathrm{IN}(A) \wedge \mathrm{IN}(B) \wedge \bigwedge_{\substack{1\leq i\leq n-\delta l \\ j\equiv \kappa i \pmod{\delta}}} U(i)
$$

where : $\mathrm{IN}(M) = \neg V_0(M) \wedge \neg V_1(M) \wedge \neg V_2(M) \wedge \neg V_3(M)$ and

a) If $j \leq \min\{q(A), q(B)\}$, $U(i) = \neg V_0(i,j) \Rightarrow V_1(i+\delta l, j)$
b) If $q(A) \leq j \leq q(B)$ $U(i) = \neg V_3(i,j) \Rightarrow V_1(i+\delta l, j)$
c) If $q(B) \leq j \leq q(A)$ $U(i) = \neg V_0(i,j) \Rightarrow V_2(i+\delta l, j)$
b) If $j \geq \max\{q(A), q(B)\}$ $U(i) = \neg V_3(i,j) \Rightarrow V_2(i+\delta l, j)$

Lemma 4.

- *If a lattice set F containing the bases A, B is Q-convex around p and q, and its X-ray along q is such that $X_q F(j) \leq l$, then the evaluation $V = \Psi(F)$ of the variables $(V_k(M))_{k \in K, M \in \Delta}$ satisfies $QCONV \wedge \mathrm{UB}(q, j, l, A, B)$.*
- *If an evaluation V of the boolean variables $(V_k(M))_{k \in K, M \in \Delta}$ satisfies the formula $QCONV \wedge \mathrm{UB}(q, j, l, A, B)$, then $F = \Phi(V)$ contains the bases A, B, and its X-ray along q is such that $X_q F(j) \leq l$.*

We do not show the proof of this Lemma for brevity's sake. We define the formula $\mathrm{UB}(p, i, l, A, B)$ for the lines in the direction p in a similar way .

2.5 The Reconstruction Algorithm

Let (P, P', Q, Q') be an instance of Problem 1. We fix four bases A, B, C, D such that $p(A) = 1$, $p(B) = n$, $q(C) = 1$, $q(D) = m$ and then we build the formula:

$$\mathrm{APPROX}(P, P', Q, Q', A, B, C, D) = QCONV \wedge$$

$$\bigwedge_{1 \leq i \leq n} (\mathrm{LB}(p, i, p_i) \wedge \mathrm{UB}(p, i, p'_i, C, D)) \wedge \bigwedge_{1 \leq j \leq m} (\mathrm{LB}(q, j, q_j) \wedge \mathrm{UB}(q, j, q'_j, A, B))$$

As a consequence of Lemmas 2,3 and 4, we get:

Theorem 1. $\mathrm{APPROX}(P, P', Q, Q', A, B, C, D)$ *is satisfiable if and only if there is a Q-convex set F around p and q containing the bases A, B, C, D and having X-rays along p and q such that $p_i \leq X_p F(i) \leq p'_i$, for $i = 1, \ldots, n$, and $q_j \leq X_q F(j) \leq q'_j$, for $j = 1, \ldots, m$.*

Since $\mathrm{APPROX}(P, P', Q, Q', A, B, C, D)$ is a boolean formula in conjunctive normal form with at most two literals in each clause, from Theorem 1 we have a transformation of Problem 1 to 2SAT problem. The algorithm chooses four bases A, B, C, D, and builds $\mathrm{APPROX}(P, P', Q, Q', A, B, C, D)$. Each formula $\mathrm{APPROX}(P, P', Q, Q', A, B, C, D)$ has size $O(mn)$ and can be constructed in $O(mn)$ time. This is a 2SAT formula and so it can be solved in $\mathrm{O}(mn)$ time (see [1]). If the formula is satisfiable and V is the evaluation of the boolean variables, then $F = \Phi(V)$ is solution of Problem 1. On the contrary, the reconstruction attempt fails and the algorithm chooses a different position of the four bases A, B, C, D, and repeats the procedure. The number of reconstruction attempts is bounded by the number of different positions of the four bases A, B, C, D, and this is at most $m^2 n^2$. Consequently:

Corollary 1. *Problem 1 can be solved in $O(m^3 n^3)$ time.*

Remark 1. An 8-connected hv-convex set is a Q-convex set around $p(x, y) = x$ and $q(x, y) = y$ with at least one point in each row and column. This class of lattice sets is a well-know generalization of the class of hv-convex polyominoes [2,5] which are 4-connected and convex in horizontal and vertical directions. Boufkhad et al. [3] studied the "approximate consistency with two directions" problem on the class of hv-convex polyominoes. In detail, given a pair of vectors $V = (v_1, \dots, v_n)$ and $H = (h_1, \dots, h_m)$, they want to reconstruct an hv-convex polyomino whose X-rays along vertical and horizontal directions are such that: $|X_pF(i) - v_i| \leq 1$ for $i = 1, \dots, n$ and $|X_qF(j) - h_j| \leq 1$ for $j = 1, \dots, m$ (i.e., $P = V - 1, P' = V + 1, Q = H - 1$ and, $Q' = H + 1$). Our algorithm solves this problem in polynomial time on the classes of 8-connected hv-convex sets and hv-convex polyominoes (with an extra condition on the boolean variables). We point out that the goal of Boufkhad et al. is to solve the corresponding optimization problem. They want to reconstruct hv-convex polyominoes from these "approximate" horizontal and vertical X-rays and such that the sum of the absolute differences is minimum. By means of a SAT solver, the authors defined a heuristic algorithm for solving this problem. We do not know at this time if our algorithm can be used for solving this optimization problem in polynomial time.

Remark 2. Mirsky [11] proved that "approximate consistency with two directions" problem on the class of all lattice sets can be solved in polynomial time.

The Exact Reconstruction. If $P = P'$ and $Q = Q'$, Problem 1 becomes the *Exact Consistency* problem with two directions on the Q-convex sets around p and q. Notice that, $\sum_{i=1}^{n} p_i = \sum_{j=1}^{n} q_j$ is a necessary condition. This problem has been studied in [4] and the authors propose a greedy algorithm whose computational cost is $O(m^2n^2(m+n)\min\{m^2, n^2\})$. We are going to show an algorithm which is faster than the algorithm defined in [4]. Let A, B be two bases such that $p(A) = 1$ and $p(B) = n$. We build the following formula:

$$\mathrm{EXACT}(P, Q, A, B) = QCONV \bigwedge_{1 \leq i \leq n} \mathrm{LB}(p, i, p_i) \bigwedge_{1 \leq j \leq m} \mathrm{UB}(q, j, q_j, A, B).$$

Proposition 1. EXACT(P, Q, A, B) *is satisfiable if and only if there is a Q-convex set F around p and q containing A, B and having X-rays along p and q such that $X_pF(i) = p_i$, for $i = 1, \dots, n$, and $X_qF(j) = q_j$, for $j = 1, \dots, m$.*

Proof.

- Assume that EXACT(P, Q, A, B) is satisfied by an evaluation V of $(V_k(M))_{k \in K, M \in \Delta}$. By Lemmas 2,3 and 4, the set $F = \Phi(V)$ satisfies the conditions: $A, B \in F$, $X_pF(i) \geq p_i$ and $X_qF(j) \leq q_j$ for all i, j. Then,

$$\sum_{j=1}^{m} q_j \geq \sum_{j=1}^{m} X_qF(j) = \sum_{i=1}^{n} X_pF(i) \geq \sum_{i=1}^{m} p_i$$

  and, since $\sum_{j=1}^{m} q_j = \sum_{i=1}^{n} p_i$, $X_pF(i) = p_i$ and $X_qF(j) = q_j$ for all i, j.
- If F is Q-convex around p and q and satisfies $A, B \in F$, $X_pF(i) = p_i$, and $X_qF(j) = q_j$, by Lemmas 2,3 and 4, the evaluation $V = \Psi(F)$ satisfies EXACT(P, Q, A, B).

The number of reconstruction attempts for the exact consistency problem is bounded by the number of different positions of the two bases A, B, and this is at most $\min\{m^2, n^2\}$. Consequently:

Corollary 2. *Exact Consistency problem is solved in* $O(mn\min\{m^2, n^2\})$ *time.*

3 More than Two Directions

We now outline an algorithm for reconstructing Q-convex sets from their X-rays in more than two directions. Let us introduce a definition of Q-convex set around more than two directions. Let U be a set of d directions $\{\boldsymbol{u}_h = (a_h, b_h)_{h=1}^d\}$ (i.e., pairs of coprime integers, with $b_h \geq 0$). The linear form corresponding to vector $\boldsymbol{u}_h = (a_h, b_h)$ is $u_h(x, y) = b_h x - a_h y$. Given two directions $u_i, u_j \in U$, we define four zones $Z_k^{(i,j)}(M)$ around every $M \in \mathbb{Z}^2$ as in the previous section. Therefore, there are $2d(d-1)$ zones for each $M \in \mathbb{Z}^2$ and we are going to select $2d$ of these zones. A point M of a line in direction u_h splits it into the following two semi-lines having origin in M:

$$s_h^+(M) = \{N \in \mathbb{Z}^2 \mid u_h(N) = u_h(M) \text{ and } \boldsymbol{u}_h \cdot \overrightarrow{ON} \geq \boldsymbol{u}_h \cdot \overrightarrow{OM}\}$$
$$s_h^-(M) = \{N \in \mathbb{Z}^2 \mid u_h(N) = u_h(M) \text{ and } \boldsymbol{u}_h \cdot \overrightarrow{ON} \leq \boldsymbol{u}_h \cdot \overrightarrow{OM}\}$$

where "$\cdot$" denotes the scalar product of two vectors and O is any origin point.

Definition 2. *An almost-semi-plane (or ASP) along U is a zone $Z_k^{(i,j)}(M)$ such that for each direction u_h of U only one of the two semi-lines $s_h^+(M), s_h^-(M)$ is contained in $Z_k^{(i,j)}(M)$.*

Let $\Pi_0(M)$ be the ASP containing $s_h^+(M)$ for each $h = 1, \ldots, d$. We denote the other almost-semi-planes encountered clockwise around M from $\Pi_0(M)$ by $\Pi_1(M), \ldots, \Pi_{2d-1}(M)$. For example, let $U = \{u_1, u_2, u_3\}$, with $u_1 = y$, $u_2 = x$, $u_3 = x+y$. The six ASP around a point M are: $\Pi_0(M) = Z_2^{(1,3)}(M), \Pi_1(M) = Z_2^{(2,3)}(M), \Pi_2(M) = Z_1^{(1,2)}(M), \Pi_3(M) = Z_0^{(1,3)}(M), \Pi_4(M) = Z_0^{(2,3)}(M)$ and $\Pi_5(M) = Z_3^{(1,2)}(M)$ (see Fig. 4). Now we can generalize the Q-convexity to any set of directions:

Definition 3. *A lattice set F is* strongly Q-convex *around U if and only if for each $M \notin F$ there exists an ASP $\Pi_k(M)$ around M such that $F \cap \Pi_k(M) = \varnothing$.*

Let us consider the approximate consistency problem on this class.

Problem 2. **Approximate Consistency with more than two directions**
Instance: $2d$ vectors $P_1 = (p_{1,1}, \ldots, p_{1,n_1}), P_1' = (p_{1,1}', \ldots, p_{1,n_1}'), \ldots, P_d = (p_{d,1}, \ldots, p_{d,n_d})$ and $P_d' = (p_{d,1}', \ldots, p_{d,n_d}')$ whose elements are no-negative integer numbers and $p_{1,1}, p_{1,n_1}, p_{1,1}', p_{1,n_1}', \ldots, p_{d,1}, p_{d,n_d}, p_{d,1}', p_{d,n_d}'$ are positive integer numbers.
Question: is there a strongly Q-convex set F around U such that:
$p_{h,i} \leq X_{u_h}F(i) \leq p_{h,i}'$ for $i = 1, \ldots, n_h$ and $h = 1, \ldots, d$?

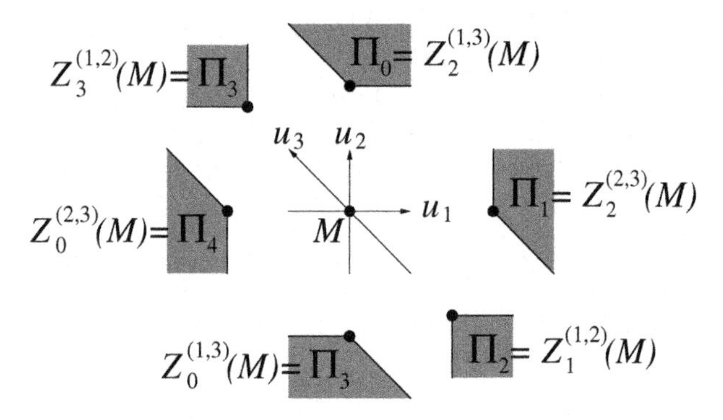

Fig. 4. The six ASP around M, with $U = \{u_1, u_2, u_3\}$, $u_1 = y$, $u_2 = x$, $u_3 = x + y$

A Q-convex set F around U whose X-rays are such that: $p_{h,i} \leq X_{u_h}F(i) \leq p'_{h,i}$ for all h, i, is contained in the lattice polygon:

$$\Delta = \{N \in \mathbb{Z}^2 \mid 1 \leq u_h(N) \leq n_h \; \forall 1 \leq h \leq d\}.$$

We use the strategy of the previous section, by replacing the zone $Z_k(M)$ with $\Pi_k(M)$. Assuming that $K = \{0, 1, \ldots, 2d-1\}$, we associate $2d$ boolean variables $V_k(M)$, with $k \in K$, at every point M in Δ (i.e., one variable for each $\Pi_k(M)$ around M). We build a 2SAT formula APPROX on the variables $(V_k(M))_{k\in K, M\in\Delta}$, so that there is a solution F of Problem 2 if and only if APPROX is satisfiable. If there is an evaluation V of $(V_k(M))_{k\in K, M\in\Delta}$ satisfying APPROX, the corresponding set F solving Problem 2 is $F = \Phi(V)$, with Φ defined as in the previous section. Conversely, if F is a subset of Δ solving Problem 1, the corresponding evaluation V of $(V_k(M))_{k\in K, M\in\Delta}$ satisfying APPROX is $V = \Psi(F)$, with Ψ defined as in the previous section. Since every ASP $\Pi_k(M)$ is equal to a zone $Z_h^{(i,j)}(M)$ defined by directions u_i, u_j, it is easy to generalize the formulas to contexts having n directions. The generalization SQCONV of QCONV is such that SQCONV$=\mathcal{Z}_0 \wedge \ldots \wedge \mathcal{Z}_{2d-1}$, where $\mathcal{Z}_k$ corresponds to $\Pi_k(M)$ (i.e., $Z_h^{(i,j)}(M)$). Thus, we have the following extensions of Lemma 2:

Lemma 5.

- *For any set $F \subset \Delta$ the evaluation $V = \Psi(F)$ of the variables $(V_k(M))_{k\in K, M\in\Delta}$ satisfies SQCONV.*
- *If an evaluation V of the variables $(V_k(M))_{k\in K, M\in\Delta}$ satisfies SQCONV, then $F = \Phi(V)$ is strongly Q-convex around U.*

A lower bound of l on the X-ray of F along the line $u_h(M) = i$ is expressed by a 2SAT formula. This formula can be constructed by proceeding as in the two-directions case. We do not describe the construction of the formula LB(u_h, i, l) for brevity's sake. Finally, we fix $d-1$ pairs of bases for each directions, in order to give an upper bound of l on the X-ray of F in the direction u_h. The chosen bases correspond to the other $d-1$ directions. The upper bound is expressed by the formula $\text{UB}(u_h, i, l, A_1, B_1, \ldots, A_{h-1}, B_{h-1}, A_{h+1}, B_{h+1}, \ldots, A_d, B_d)$, where A_j, B_j are the bases of the direction u_j. This formula is a simple generalization of the upper bound formula for the two-directions case. Let $(P_1, P_1', \ldots, P_d, P_d')$ be an instance of Problem 2. We fix $2d$ bases $A_1, B_1, \ldots, A_d, B_d$ and then we build the following boolean formula:

$$\text{APPROX}(P_1, P_1', \ldots, P_d, P_d', A_1, B_1, \ldots, A_d, B_d) = SQCONV \wedge$$
$$\bigwedge_{\substack{1\leq h\leq d \\ 1\leq i\leq n_h}} \text{LB}(u_h, i, p_i) \wedge \text{UB}(p, i, p_i', A_1, B_1, \ldots, A_{h-1}, B_{h-1}, A_{h+1}, B_{h+1}, \ldots, A_d, B_d)$$

We deduce that:

Theorem 2. $\text{APPROX}(P_1, P_1' \ldots, A_d, B_d)$ *is satisfiable if and only if there is a strongly Q-convex set* F *around* U *containing the bases* $A_1, B_1, \ldots, A_d, B_d$ *and having X-rays along* u_h *such that* $p_{h,i} \leq X_{u_h} F(i) \leq p_{h,i}'$, *for* $i = 1, \ldots, n_h$ *and* $h = 1, \ldots, d$.

The algorithm chooses d pair of bases, and builds the 2SAT expression APPROX. Assuming that $n = \max\{n_1, \ldots, n_d\}$, we have that each formula APPROX has size $O(n^2)$ and can be constructed in $O(n^2)$ time. The number of reconstruction attempts is bounded by the number of different positions of the $2d$ bases, and this is at most n^{2d}. Consequently:

Corollary 3. *Problem 2 can be solved in* $O(n^{2d+2})$ *time.*

If $P_h = P_h'$ for each $1 \leq h \leq d$, Problem 2 become the exact consistency problem with more than two directions. In this case, we have to choose $d-1$ pair of bases for the upper bound and we complexity of algorithm for solving this problem is $O(n^{2d})$. The convex lattice sets are special Q-convex sets, and so by uniqueness results for Q-convex sets proved in [6], the algorithm can be used for reconstructing convex lattice subsets from their exact X-rays in some sets of four suitable lattice directions, or in any set of seven prescribed mutually nonparallel lattice directions. This means that the two algorithms solve Gritzmann's problem for these special sets of directions.

References

1. B. Aspvall, M. F. Plass and R. E. Tarjan, A linear-time algorithm for testing the truth of certain quantified Boolean formulas, *Information Processing Letters*, **8** 3 (1979) 121-123. 114, 120
2. E. Barcucci, A. Del Lungo, M. Nivat and R. Pinzani, Reconstructing convex polyominoes from their horizontal and vertical projections, *Theoretical Computer Science*, **155** (1996) 321-347. 114, 120
3. Y. Boufkhad, O. Dubois and M. Nivat, Reconstructing (h,v)-convex bidimensional patterns of objects from approximate horizontal and vertical projections, preprint. 114, 120

4. S. Brunetti, A. Daurat, Reconstruction of Discrete Sets From Two or More Projections in Any Direction, *Proc. of the Seventh International Workshop on Combinatorial Image Analysis* (IWCIA 2000), Caen, (2000) 241-258. 114, 121
5. M. Chrobak, C. Dürr, Reconstructing hv-Convex Polyominoes from Orthogonal Projections, *Information Processing Letters*, **69** 6 (1999) 283-289. 114, 120
6. A. Daurat, Uniqueness of the reconstruction of Q-convex from their projections, preprint. 114, 124
7. A. Daurat, A. Del Lungo and M. Nivat, The medians of discrete sets according to some linear metrics, *Discrete & Computational Geometry*, **23** (2000) 465-483. 116
8. R. J. Gardner and P. Gritzmann, Discrete tomography: determination of finite sets by X-rays, *Trans. Amer. Math. Soc.*, **349** (1997) 2271-2295. 114
9. R. J. Gardner, P. Gritzmann and D. Prangenberg, On the computational complexity of reconstructing lattice sets from their X-rays, *Discrete Mathematics*, **202** (1999) 45-71. 113
10. A. Kuba and G. T. Herman, Discrete Tomography: A Historical Overview, in *Discrete Tomography: Foundations, Algorithms and Applications*, editors G. T. Herman and A. Kuba, Birkhauser, Boston, MA, USA, (1999) 3-34. 113
11. L. Mirski, Combinatorial theorems and integral matrices, *Journal of Combinatorial Theory*, **5** (1968) 30-44. 121
12. R. W. Irving and M. R. Jerrum, Three-dimensional statistical data security problems, *SIAM Journal of Computing*, **23** (1994) 170-184. 113
13. C. Kisielowski, P. Schwander, F. H. Baumann, M. Seibt, Y. Kim and A. Ourmazd, An approach to quantitative high-resolution transmission electron microscopy of crystalline materials, *Ultramicroscopy*, **58** (1995) 131-155. 113
14. G. P. M. Prause and D. G. W. Onnasch, Binary reconstruction of the heart chambers from biplane angiographic image sequence, *IEEE Transactions Medical Imaging*, **15** (1996) 532-559. 113
15. A. R. Shliferstein and Y. T. Chien, Switching components and the ambiguity problem in the reconstruction of pictures from their projections, *Pattern Recognition*, **10** (1978) 327-340. 113

A Question of Digital Linear Algebra

Yan Gérard

Laboratoire de Logique et d'Informatique de Clermont1 (LLAIC1) IUT
Département d'Informatique, Ensemble Universitaire des Cézeaux
B.P. 86, 63172 Aubière Cedex, France
gerard@llaic.u-clermont1.fr

Abstract. In classical linear algebra, the question to know if a vector $v \in \mathbb{R}^n$ belongs to the linear space $Vect\{v^1, v^2, \cdots, v^k\}$ generated by a familly of vectors, is solved by the GAUSS pivot. The problem investigated in this paper is very close to this classical question: we denote $\lfloor . \rfloor_n$ the function of $\mathbb{R}^n$ defined by $\lfloor (x_i)_{1 \leq i \leq n} \rfloor_n = (\lfloor x_i \rfloor)_{1 \leq i \leq n}$ and the question is now to determine if a given vector $v \in \mathbb{Z}^n$ belongs to $\lfloor Vect\{v^1, v^2, \cdots, v^k\} \rfloor_n$. This problem can be easily seen as a sytem of inequalities and solved by using linear programming but in some special cases, it can also be seen as a particular geometrical problem and solved by using tools of convex geometry. We will see in this framework that the question $v \in \lfloor Vect\{v^1, v^2, \cdots, v^k\} \rfloor_n$? generalizes the problem of recognition of the finite parts of digital hyperplanes and we will give equivalent formulations which allow to solve it efficiently.

Keywords: chord, convex hull, digital hyperplane, simplex, slice.

1 Introduction

The problem that we are going to consider in this paper belongs to a set of questions of geometry of numbers, which have been deeply investigated since the beginning of the century. The problem which can be considered as the origin of the others, is the one which consists in computing if a given vector $v \in \mathbb{R}^n$ belongs to the linear subspace generated by a familly v^1, v^2,....v^k of $\mathbb{R}^n$:

$$v \in Vect\{v_i / 1 \leq i \leq k\}?$$

The method of computation used for solving this question is the very classical GAUSS pivot. Many interessant problems are close to this easy and basic question: if we consider no more the linear space $Vect\{v_i/1 \leq i \leq k\}$ but only the additive group $Group\{v_i/1 \leq i \leq k\} = \{\sum_{i=1}^{k} a_i v^i / \forall 1 \leq i \leq k, a_i \in \mathbb{Z}\}$ generated by the vectors v^i, the problem is a system of diophantine equations. We could also be interested in the distance between v and $Vect\{v_i/1 \leq i \leq k\}$ or v and $Group\{v_i/1 \leq i \leq k\}$ instead of the single membership of v to these sets. If the choosen distance comes from a defined quadratic form, it leads to classical computations but it is not the case if one makes the choice of the distance d_1

G. Borgefors, I. Nyström, and G. Sanniti di Baja (Eds.): DGCI 2000, LNCS 1953, pp. 126–136, 2000.
© Springer-Verlag Berlin Heidelberg 2000

associated with the norm $N_1(x_i)_{1\leq i\leq n} = \sum_{i=1}^{n} |x_i|$ or of the distance d_∞ associated with the classical norm $N_\infty(x_i)_{1\leq i\leq n} = max\{|x_i|/1 \leq i \leq n\}$. We can mention KRONECKER's approximation theorem which provides a characterization of the vectors v whose distance $d_\infty(v, Group\{v_i/1 \leq i \leq k\})$ to the group $Group\{v_i/1 \leq i \leq k\}$ is stricly less than a given ϵ [S88].

The question investigated in this paper is of the same kind: we introduce the function $\lfloor . \rfloor_n : \mathbb{R}^n \longrightarrow \mathbb{Z}^n$ defined by $\lfloor (x_i)_{1\leq i\leq n} \rfloor_n = (\lfloor x_i \rfloor)_{1\leq i\leq n}$ where $\lfloor r \rfloor$ denotes the integer part of a real number r (in other words, the largest integer i verifying $i \leq r$) and the question is to determine whether a vector $v \in \mathbb{Z}^n$ belongs to the image of $Vect(v^1, v^2, \ldots, v^k)$ by the map $\lfloor . \rfloor_n$:

$$v \in \lfloor Vect(v^1, v^2, \ldots, v^k) \rfloor_n \ ?$$

The problem of determining if $v \in \lfloor Vect(v^1, v^2, \ldots, v^k) \rfloor_n$ does not appear in the literature of geometry of numbers and we can give two reasons to explain this absence:

- the first one is that the function integer part has never been considered as a serious source of problems in classical mathematics and if it is now the case, it is especially a preoccupation of computer scientists for which the truncature is a usual practical tool.
- the second reason comes from the fact that the analytical form of the question is quite ordinary. It is simply a system of inequalities: by denoting v_j the coordinates of $v \in \mathbb{Z}^n$ and v_j^i the coordinates of $v^i \in \mathbb{R}^n$, the problem is to determine whether there exist k real numbers a_1, a_2, ..., a_k verifying the system of n double inequalities

$$v_j \leq \sum_{i=1}^{k} a_i v_j^i < v_j + 1$$

for $1 j \leq n$. The systems of linear inequalities is of the domain of linear programming and there exists nowadays many different methods to solve them. The famoust one is the simplex method [Ga60] and we will come back later on its geometrical meaning in order to provide geometrical efficient algorithms. A second solution is the classical method of FOURIER-MOTZKIN which consists in eliminating the variables. A third one is an algorithmic technique eliminating the redundant linear inequalities and described in [E87]. There exists many others but the aim of this paper is not to present them. Our ambition is mainly to bring out the geometry which can be hidden in our special kind of systems. It is the reason why we are going to choose a geometrical point of view. We are however unable to do it in the general case and then we have to restict ourself to two special cases:
- when the rank of the familly v^1, v^2,.... v^k is n or $n-1$, technics of linear algebra allow to conclude,
- when the constant vector $(1)_{1\leq i\leq n}$ belongs to the linear subspace

$Vect(v^1, v^2, \dots, v^k)$, the question to know whether $v \in \lfloor Vect(v^1, v^2, \dots, v^k) \rfloor_n$ generalizes a problem which has been discussed many times in the previous DGCI and which consists in recognizing the finite parts contained by naive digital hyperplanes ([DR94,D95,FST96,R95]). We will give a geometrical characterization of the solutions by introducing a new tool, the chords' set of a part of $\mathbb{R}^n$, and a classical notion, the convex hull. This geometrical characterization provides two natural algorithms and as the question generalizes the recognition of the digital hyperplanes, they can naturally be used to determine if a finite part belongs to a digital hyperplane of double inequality $h \leq \sum_{i=1}^{n} a_i x_i < h + |a_n|$.

As conclusion of this introduction, we have to say a word about the applications and we can begin with the recognition of the digital hyperplanes. A good algorithm of recognition can be the origin of many applications in digital imagery and it can lead to a whole set of geometrical methods of treatment of images. We have however to stress the applications of the main problem that we have considered because it has an inexhaustible number of numeric applications: one of them is the use of what we could call the *polynomial approximations* of the digital functions. We call polynomial approximation of a digital function $f : \mathbb{Z}^d \longrightarrow \mathbb{Z}$ under a domain $D \subset \mathbb{Z}^d$ a polynomial $P(X_i)_{1 \leq i \leq n}$ verifying for all $x \in D$,

$$f(x) = \lfloor P(x) \rfloor.$$

These objects provides values $(d^k P/dx_{i_1}.dx_{i_2}......dx_{i_k})$ which can be seen as the differentials of the function f. The computation of a polynomial approximation of degree less than r for a function f under a domain D supposed ordered and of cardinality n consists in determining a linear combination of the vectors $(1)_{i \in D}$, $(i)_{i \in D}$, $(i^2)_{i \in D}$, ..., $(i^r)_{i \in D}$ whose image by $\lfloor . \rfloor_n$ is $(f(i))_{i \in D}$. Then it is a direct application of the second case that we have planned to investigate. We can also approach the function f by other things as polynomials and use it to create digital filters or many other things that we have not place enough to describe.

2 Easy Cases

The general question in which we are interested, consists in determining if a given vector $v \in \mathbb{Z}^n$ belongs to the image of the linear subspace $Vect\{v^i / 1 \leq i \leq k\}$ by the map $\lfloor . \rfloor_n$.

The first easy case is trivial: when the rank of the familly v^1, v^2, ..., v^k is n, the linear subspace $Vect\{v^i / 1 \leq i \leq k\}$ is equal to the whole space $\mathbb{R}^n$ and as $\lfloor . \rfloor_n$ is sujective, it is clear that the intersection of the reciprocal image of any $v \in \mathbb{Z}^n$ by $\lfloor . \rfloor_n$ with $Vect\{v^i / 1 \leq i \leq k\}$ can not be empty.

The second easy case is not trivial but it is not necesary to use the double inequalities to understand what happens : when the rank of the familly v^1, v^2,..., v^k is $n-1$, the linear subspace $Vect\{v^i / 1 \leq i \leq k\}$ generated by these elements can be characterized by a linear form φ (then we have $Vect\{v^i / 1 \leq i \leq k\} = \{x \in \mathbb{R}^n / \varphi(x) = 0\}$) . We denote a a point of the set $\{0, 1\}^n$ where φ is minimum and b a point of $\{0,1\}^n$ where φ is maximum. By denoting $\lfloor\rfloor_n^{-1}(v)$ the reciprocal image of v by $\lfloor.\rfloor_n$, we can see that

$$]\varphi(v+a);\varphi(v+b)[\subset\varphi(\lfloor\rfloor_n^{-1}(v))\subset[\varphi(v+a);\varphi(v+b)].$$

There are two trivial consequences :

- if 0 belongs to $]\varphi(v+a);\varphi(v+b)[$, then v belongs to $\lfloor Vect\{v^i/1\leq i\leq k\}\rfloor_n$.
- if 0 does not belong to $[\varphi(v+a);\varphi(v+b)]$, then v does not belong to the set $\lfloor Vect\{v^i/1\leq i\leq k\}\rfloor_n$.

and there remains only one case: the one where one of the two values $\varphi(v+a)$ or $\varphi(v+b)$ is null. In that case, we just have to compute $\varphi(v)$:

- if $\varphi(v)=0$ then v belongs to $Vect\{v^i/1\leq i\leq k\}$ and then also to its image $\lfloor Vect\{v^i/1\leq i\leq k\}\rfloor_n$.
- if $\varphi(v)\neq 0$ then the boundary 0 is not reached by $\varphi(\lfloor\rfloor_n^{-1}(v))$ and it follows that v does not belong to $\lfloor Vect\{v^i/1\leq i\leq k\}\rfloor_n$.

As conclusion, it proves that we just have to compute $\varphi(v)$. If it is null then v belongs to $\lfloor Vect\{v^i/1\leq i\leq k\}\rfloor_n$ and otherwise v belongs to $\lfloor Vect\{v^i/1\leq i\leq k\}\rfloor_n$ if and only if $\varphi(v)\in]-\varphi(b);-\varphi(a)[$. The complexity of this computation is very low. It is the sum of $O(n^2)$ for the computation of φ with $O(2^n)$ for the computation of a and b, and this global time has nothing to do with the complexity of the investigation of linear inequalities.

3 Geometrical Case

We are still interested in the question: does a given vector $v\in\mathbb{Z}^n$ belong to the image of $Vect\{v^i/1\leq i\leq k\}$ by the map $\lfloor.\rfloor_n$?

We suppose now that the constant vector $(1)_{1\leq j\leq n}$ belongs to $Vect\{v^i/1\leq i\leq k\}$. In order to simplify the notations and the computations, we are even going to assume that the vectors v^1, v^2,.... , v^k are free and that $v^k=(1)_{1\leq j\leq n}$.

3.1 A Geometrical Equivalent Problem

Before explaining the consequences of the previous hypothesis, let us see what happens for the question $v\in Vect\{v^i/1\leq i\leq k\}$? (for the moment, we forget the integer part $\lfloor.\rfloor_n$). The system $\sum_{i=1}^k a_i v^i=v$ can be written $\sum_{i=1}^{k-1}a_i v_j^i-v_j=-a_k$ for $1\leq j\leq n$ and it follows that if there exists a solution, the n vectors $(v_j^1,v_j^2,....,v_j^{k-1},v_j)$ containing the j coordinates of v^1, v^2,.... v^{k-1} and v all belong to an affine hyperplane of $\mathbb{R}^k$. Conversely, if the n vectors of coordinates $(v_j^1,v_j^2,....,v_j^{k-1},v_j)$ belong to an affine hyperplane, there exist real coefficients a_1, a_2, ..., a_k and a such that $\sum_{i=1}^{k-1}a_i v^i+av=-a_k(1)_{1\leq j\leq n}$. The familly v^1, v^2, ..., $v^k=(1)_{1\leq j\leq n}$ was supposed free and it follows that a can not be null (excepting the particular case where $v = 0$). It proves that v can be written $\sum_{i=1}^{k} a_i v^i$ if and only if there exists an affine hyperplane of $\mathbb{R}^k$ containing the vectors of j coordinates $(v_j^1, ..., v_j^{k-1}, v_j)$. It means that the linear question "$v \in Vect\{v^i/1 \leq i \leq k\}$?" has been transformed by using duallity into a problem of affin geometry. We can do the same with the question "$v \in \lfloor Vect\{v^i/1 \leq i \leq k\} \rfloor_n$?" with the difference that the part taken by the affine hyperplanes is played by the subsets of $\mathbb{R}^n$ characterized by double inequations of the form $h \leq \sum_{i=1}^{n} \alpha_i x_i < h + |\alpha_n|$. In order to lighten the notations, these sets are simply called the *slices* of $\mathbb{R}^n$:

Definition 3.1 .– *A* slice *of* $\mathbb{R}^n$ *is a subset of* $\mathbb{R}^n$ *which can be characterized by a double inequation of form* $h \leq \sum_{i=1}^{n} \alpha_i x_i < h + |\alpha_n|$ *where* h, α_1, α_2,.... α_n *are real numbers and where* α_n *is not null.*

We can already notice that the traces of the slices under $\mathbb{Z}^n$ are digital hyperplanes [R91]. The affine transformation of the problem in which we are interested provides the following proposition.

Proposition 3.1 .– *Let* v^1, v^2,.... , v^{k-1}, $v^k = (1)_{1 \leq j \leq n}$ *be a free familly of vectors of* $\mathbb{R}^n$ *and let* v *be an element of* $\mathbb{Z}^n$. *We have* $v \in \lfloor \mathrm{Vect}\{v^i/1 \leq i \leq k\} \rfloor_n$ *if and only if the set of the* n *vectors* $(v_j^1, v_j^2, \ldots, v_j^{k-1}, v_j)$ *of the* j *coordinates is contained by a slice of* $\mathbb{R}^k$.

Proof. If $v \in \lfloor Vect\{v^i/1 \leq i \leq k\} \rfloor_n$, it means that the system of n inequalities

$$v_j \leq \sum_{i=1}^{k} a_i v_j^i < v_j + 1$$

for $1 \leq j \leq n$, has some solutions. We can rewrite it

$$-a_k \leq \sum_{i=1}^{k-1} a_i v_j^i - v_j < -a_k + 1.$$

and it means that the subset of $\mathbb{R}^k$ characterized by the double inequation of variable $x = (x_i)_{1 \leq i \leq n} \in \mathbb{R}^n$ and of expression

$$-a_k \leq \sum_{i=1}^{k-1} a_i x_i - x_k < -a_k + 1$$

contains the n vectors $(v_j^1, v_j^2, \cdots, v_j^{k-1}, v_j)$ of the j coordinates for $1 \leq j \leq n$. By setting $h = -a_k$, $\alpha_k = -1$ and $\alpha_i = a_i$ for i between 1 and $k-1$, we obtain a first part of the proposition.

Now, let us prove the converse. We suppose that there exists a slice of $\mathbb{R}^k$ characterized by a double inequation $h \leq \sum_{i=1}^{k} \alpha_i x_i < h + |\alpha_k|$ containing the n vectors $(v_j^1, v_j^2, \cdots, v_j^{k-1}, v_j)$. Before anything, we shall begin by showing that

α_k can be supposed negative. The set of the values taken by the linear form $\sum_{i=1}^{k} \alpha_i x_i$ for the n vectors $(v_j^1, v_j^2, \cdots, v_j^{k-1}, v_j)$ is finite. Then there exists a real value $\epsilon > 0$ such that the points $(v_j^1, v_j^2, \cdots, v_j^{k-1}, v_j)$ belong to the set defined by

$$h \leq \sum_{i=1}^{k} \alpha_i x_i \leq h + |\alpha_k| - \epsilon.$$

By multiplying by -1, it follows that they also belong to the slice defined by

$$-h - |\alpha_k| + \epsilon \leq \sum_{i=1}^{k} -\alpha_i x_i < -h + \epsilon.$$

By setting $h' = -h - |\alpha_k| + \epsilon$ and $\alpha_i' = \alpha_i$, we see that α_k can be supposed negative and from now, we assume it.

We have for any j between 1 and n

$$h \leq \sum_{i=1}^{k-1} \alpha_i v_j^i + \alpha_k v_j < h + |\alpha_k|.$$

It follows that

$$h/|\alpha_k| \leq \sum_{i=1}^{k-1} (\alpha_i/\alpha_k).v_j^i - v_j < h/|\alpha_k| + 1.$$

By setting $a_k = -h/|\alpha_k|$ and $a_i = \alpha_i/|\alpha_k|$ for all i between 1 and $k-1$, we have

$$-a_k \leq \sum_{i=1}^{k-1} a_i.v_j^i - v_j < -a_k + 1$$

and it ends the proof. □

The proposition 3.1 transforms the question to know if $v \in \lfloor Vect\{v^i/1 \leq i \leq k\} \rfloor_n$ into a problem which consists in recognizing the finite parts of slices. In the particular case of finite parts of $\mathbb{Z}^n$, this problem has arroused the curiosity of several authors in the previous DGCI ([FST96,DR94]) and it is known as the problem of recognition of digital hyperplanes.

3.2 Recognition of the Finite Parts of Slices

The recognition of the finite parts of slices is a problem of convex geometry, which leads to introduce the chords' notion.

Definition 3.2 .– *Given a part* $\mathrm{A} \subset \mathbb{R}^n$, *the elements* $\mathrm{a}' - \mathrm{a}$ *where* $\mathrm{a} \in \mathrm{A}$ *and* $\mathrm{a}' \in \mathrm{A}$ *are called the* chords *of* A. *The chords'set of* A *is the subset of* $\mathbb{R}^n$ *denoted* chord(A) *and defined by*

$$\mathrm{chord}(\mathrm{A}) = \{\mathrm{a}' - \mathrm{a} / \mathrm{a} \in \mathrm{A},\ \mathrm{a}' \in \mathrm{A}\}.$$

We also introduce the notion of convex hull: the smallest convex part of $\mathbb{R}^n$ containing a set A is called the convex hull of A and is denoted $conv_{R^n}(A)$.

These two operators have several properties as the one to commute: we have $chord \circ conv_{R^n} = conv_{R^n} \circ chord$.

Theorem 3.1 .– *There exists a slice of* $\mathbb{R}^n$ *containing a given finite subset* $A \subset \mathbb{R}^n$ *if and only if the point* u *of coordinates all null except the last one equal to* 1, *does not belong to* $\text{conv}_{\mathbb{R}^n}(\text{chord}(A))$.

It means that we just have to compute if $u = (0, 0, \ldots, 0, 1)$ belongs to the set $conv_{R^n}(chord(A))$ in order to know if there exists a slice containing the finite part A. Theorem 3.1 can be proved directly but we are going to use an other way based on the elimination of variables according to the method of FOURIER-MOTZKIN.

Proof. According to proposition 3.1, there exists a slice containing the finite set $A = \{a^1, a^2, \ldots, a^m\}$ if and only if there exist real numbers h, α_1, α_2,..., α_{n-1} satisfying the system (S) of the n double inequalities

$$a_n^j \leq \sum_{i=1}^{n-1} \alpha_i a_i^j - h < a_n^j + 1$$

for $1 \leq j \leq n$. By eliminating the variable h, we obtain the system (S')

$$1 < \sum_{i=1}^{n-1} \alpha_i (a_i^j - a_i^{j'}) + (a_n^{j'} - a_n^j) < 1$$

which according to FOURIER-MOTZKIN, is satisfiable if and only if (S) also is. It means that the existence of a slice containing A is equivalent to the existence of a linear form $\varphi(x_i)_{1 \leq i \leq n} = \sum_{i=1}^{n-1} \alpha_i x_i - x_n$ sending the set $chord(A)$ in $]-1, 1[$. As $chord(A)$ is a symmetric set, it is also equivalent to the existence of a linear form $\varphi(x_i)_{1 \leq i \leq n} = \sum_{i=1}^{n-1} \alpha_i x_i - x_n$ sending the set $chord(A)$ in $]\varphi(u), +\infty[$. It implies that if there exists a slice containing A, then the hyperplane of equation $\varphi(x) = h$ where $\varphi(u) < h < min(\varphi(chord(A)))$ stricly separates $chord(A)$ and the point u. The converse is also true: if there exists an hyperplane stricly separating the set $chord(A)$ and the point u, we can write its equation $\varphi(x_i)_{1 \leq i \leq n} = h$ or again $\sum_{i=1}^{n-1} \alpha_i x_i + \alpha_n x_n = h$. The coefficient α_n can be supposed equal to -1 because it can not be null ($0 \in chord(A)$ would be on the same hyperplane as u). It follows from the separation and from $\varphi(0) > \varphi(u)$ (the origin belongs to $chord(A)$) that $\varphi(chord(A)) \subset]\varphi(u), +\infty[$. Then we obtain a new equivalence : there exists a slice containing A if and only if there exists an hyperplane stricly separating $chord(A)$ and u. At this step, the theorem of separation of compact sets [V64] ends the proof. □

Theorem 3.1 gives a geometrical characterization of the finite subsets of $\mathbb{Z}^n$ contained by slices or in other words, by naive digital hyperplanes and at first sight, one could regret that the tools used in the theorem, namely the chords and the convex hull, are continues operators. This critic is howewer not founded because the chords and the discrete convex hull can also be defined in $\mathbb{Z}^n$. By denoting them *chord* and $conv_{Z^n}$, the property of commutation is preserved: for any finite set $A \subset \mathbb{Z}^n$, we have $chord(conv_{Z^n})(A) = conv_{Z^n}(chord(A))$ (it is not a trivial result) and it is exactly the trace of the continues set $conv_{R^n}(chord(A))$ on $\mathbb{Z}^n$. It proves the discrete corollary of theorem 3.1 :

Corollary 3.1 .– *There exists a digital hyperplane of double inequality* $\mathrm{h} \leq \sum_{\mathrm{i}=1}^{\mathrm{n}} \alpha_{\mathrm{i}} \mathrm{x}_{\mathrm{i}} < \mathrm{h} + |\alpha_{\mathrm{n}}|$ *containing a finite set* $\mathrm{A} \subset \mathbb{Z}^{\mathrm{n}}$ *if and only if the vector* $\mathrm{u} = (0, \cdots, 0, 1)$ *does not belong to* $\mathrm{conv}_{\mathbb{Z}^{\mathrm{n}}}(\mathrm{chord}(\mathrm{A}))$.

In the framework of digital geometry, corollary 3.1 shows that the recognition of finite subsets of slices can be solved by computing the digital convex hull of the chords of the set.

3.3 Algorithms

Theorem 3.1 provides a constructive characterization of the finite sets of $\mathbb{R}^n$ contained by slices and it provides different geometrical algorithms to recognize them.

The first one is natural. In order to determine if there exists a slice containing the finite part $A \subset \mathbb{R}^n$, we begin by computing its chords' set. We obtain a set and we compute its convex hull, by a given method of algorithmic geometry. It only remains to know if the point u belongs to the polytope $conv_{R^n}(chord(A))$ by determining if there exists a face strictly separating u from the set. If there exists one, then its direction is the direction of a slice containing A and otherwise, there does not exist any solution.

This algorithm can be made incrementaly and several technics improve its effectiveness. The central point of the algorithm is the computation of the convex hull of $chord(A)$. When we add a point x to the set A (we denote $A' = A \cup \{x\}$), the chords' set becomes $chord(A') = chord(A) \cup translation_x(A) \cup translation_{-x}(A)$ and its convex hull can be obtained from $conv_{R^n}(chord(A))$ and $conv_{R^n}(A)$. We can even notice that if $x \in conv_{R^n}(A)$, then $chord(A') = chord(A)$. Then an incremental algorithm can be obtained by repeating the following process :

We assume that we have computed $conv_{R^n}(A)$, $conv_{R^n}(chord(A))$ and that u does not belong to this last set. We add the point x to A and obtain $A' = A \cup \{x\}$. If $x \in conv_{R^n}(A)$, then $conv_{R^n}(A') = conv_{R^n}(A)$ and $conv_{R^n}(chord(A')) = conv_{R^n}(chord(A))$. If $x \notin conv_{R^n}(A)$, then we can use the central step of the incremental Beneath-Beyond method [E87] to compute $conv_{R^n}(A')$. We also have to compute $conv_{R^n}(chord(A'))$ which is equal to the convex hull of the three convex sets $conv_{R^n}(conv_{R^n}(chord(A)) \cup translation_{-x}(conv_{R^n}(A)) \cup translation_x(conv_{R^n}(A)))$. An efficient method should be a "wrapping" process using the structures of $conv_{R^n}(A)$ and $conv_{R^n}(chord(A))$ but as we are in dimension n, it is not easy. We are then reduced to use the Beneath-Beyond method to obtain $conv_{R^n}(chord(A'))$ without expoiting the convex structure of $conv_{R^n}(A)$.

This algorithm and its incremental version are natural but they are not optimal. We can wait much better because the structure of the convex polytope $conv_{R^n}(chord(A))$ (equal to $chord(conv_{R^n}(A))$) seems able to be directly constructed from the structure of the convex polytope $conv_{R^n}(A)$. It is however not yet completly clear.

After this really constructive algorithm, let us present an algorithm coming from the simplex method. There are many different ways of thinking this famous method, and we are going to choose the one which involves the simplexes. Given a finite subset S of $\mathbb{R}^n$, the question to determine if a point $y \in \mathbb{R}^n$ belongs to the convex hull of S can be solved in searching a simplex whose vertices belong to S and which contains y but instead of going all over the set of all the simplexes, the simplex method is based on the following strategy:

Let us assume that the smallest affine space containing S is $\mathbb{R}^n$ because we can also come back to this case, and let us take a n-simplex whose vertices x^1, x^2,..., x^{n+1} belong to S. If it contains y, then y is in the convex hull of S and it ends the computation, otherwise almost one of the vertices x^k in on the opposite side of y towards the face which does not contain it. It provides a linear form φ which is negative in x^k, null in the other vertices and positive in y. Then we compute the maximum of φ in S. If it is null, y does not belong to the convex hull of S and otherwise, the vertices $x^1, \ldots, x^{k-1}, x'^k, x^{k+1}, \ldots, x^{n+1}$ where x'^k maximizes φ in S define our new simplex. We continue in this way until finding a null maximum or a simplex containing y. Problems of cycles can sometimes happen but a simple labelling of all the simplexes already met allow to avoid it.

In our problem, we have to determine if the point u belongs to the convex hull of *chord*(A) and an easy solution is to use the simplexe method. The points used in the algorithm are, except at the beginning, the one which maximize linear forms and we can notice that given a linear form φ and a finite set A, a point maximizing φ in *chord*(A) is the difference $x - x'$ between a point x maximizing φ in A and x' minimizing φ in A. It means that we do not need to compute the whole set *chord*(A) to obtain its maximum by a linear form: with the simplex method, we do not need to compute the chords' set *chord*(A). It makes the algorithm very simple and easy to make incremental.

The complexities of the two previous algorithms are not easy to evaluate and we are only going to increase them by rough boundaries. By denoting a the cardinality of the set A and n the dimension, the complexity of the first algorithm is $O(a^{2\lfloor (n+1)/2 \rfloor})$ and the fact to make it incremental does not modify it. The complexity of the simplex method is the product between $O(a^{2(n+1)})$ and the sum of $O(a)$ with the complexity of the integer computation of the inverse of a matrix of size n. This boundary is however excessive because it corresponds to the completly impracticable case where all the n-simplexes have to be investigated. We have to compare these values with the complexities of the classical methods solving systems of linear inequalities. The one of the FOURIER-MOTZKIN algorithm is near from $O(a^{2^n})$ and the one of the method described by EDELSBRUNNER in [E87] is linear in a but still doubling exponential in n.

The complexities of the two geometrical algorithms that we have suggested, are better in n but not in a. It remains however probable that in practice, the best of all should be the simplest one, namely the algorithm based on the simplex method.

4 Conclusion

The aim of this paper was to bring out the algebra and the geometry hidden in the question "does v belong to the image of a given linear subspace H of $\mathbb{R}^n$ by the map $\lfloor . \rfloor_n$?" but we were not able to do it in the general case. Then we have placed us in particular frameworks and we have obtained the following results:

- If the dimension of H is $n-1$ or n, then classical tools of linear algebra allow to conclude.

- If $(1)_{1 \leq i \leq n} \in H$, then the problem is equivalent to the one which consists in determining if there exists a slice containing a given finite part of $\mathbb{R}^{rank(H)}$ and it can be solved by using the chords and the convex hull. This geometrical translation of the problem can not be generalized to any case. It is specific to this one because the vector $(1)_{1 \leq i \leq n}$ plays a particular role being a kind of step of the function $\lfloor . \rfloor_n$. Then the hypothesis is important but in practical, it is not too restrictive. We can as example notice that the problem which consists in computing polynomial approximations belongs to this framework, and then can be solved by the algorithms described at the end.

Generally speaking, the problem of determining if v belongs to $\lfloor H \rfloor_n$ has an inexhaustible number of applications and its practical resolution would be a step in the development of what we could call a real digital technology.

References

DR94. I. Debled-Renesson & J. P. Reveillès, *An incremental algorithm for digital plane recognition.* 4th DGCI, Grenoble, France, 1994. 128, 131

D95. I. Debled-Renesson, *Étude et reconnaissance des droites et plans discrets.* Thèse de doctorat soutenue à l'Université Louis Pasteur de Strasbourg, 1995. 128

D86. A. Douady, *Algorithms for computing angles in the Mandelbrot set.* Chaotic dynamics and fractals (Atlanta, GA., 1985) pp155-168, Notes Rep. Math. Sci. Engrg., 2, Academic Press, Orlanda, FL, 1986.

E87. H. Edelsbrunner, *Algorithms in Combinatorial Geometry.* EATCS, Monographs on Theoritical Computer Science, Springer-Verlag, 1987. 127, 133, 134

FST96. J. Françon, J. M. Schramm & M. Tajine, *Recognizing arithmetic straight lines and planes.* 6th Conference on Discrete Geometry in Computer Imagery, Lyon, 1996. 128, 131

GA60. D. Gale, *The Theory of Linear Economic Models.* McGraw-Hill, 1960.

GE99. Y. Gerard, Contribution à la géométrie discrète, thèse de doctorat soutenue ê l'Université d'Auvergne, 1999.

R91. J. P. REVEILLÈS, *Géométrie discrète, calcul en nombre entiers et algorithmique.* Thèse de doctorat soutenue à l'Université Louis Pasteur de Strasbourg, 1991. 130

R95. J. P. REVEILLÈS, *Combinatorial pieces in digital lines and planes.* Vision Geometry IV, vol 2573 SPIE 1995. 128

S88. C. L. SIEGEL, *Lectures on the Geometry of Numbers.* Springer-Verlag, 1988. 127

V64. F. VALENTINE, *Convex sets.* McGraw-Hill, series in Higher Mathematics, 1964. 132

Reconstruction of Discrete Sets with Absorption

Attila Kuba[1] and Maurice Nivat[2]

[1] Department of Applied Informatics, University of Szeged,
Árpád tér 2, H-6720 Szeged, Hungary
kuba@inf.u-szeged.hu

[2] Laboratoire d'Informatique Algorithmique: Fondements et Applications,
Université Paris 7 Denis-Diderot, Paris, France
tcsmn@club-internet.fr

Abstract. A generalization of a classical discrete tomography problem is considered: Reconstruct binary matrices from their absorbed row and columns sums, i.e., when some known absorption is supposed. It is mathematically interesting when the absorbed projection of a matrix element is the same as the absorbed projection of the next two consecutive elements together. We show that, in this special case, the non-uniquely determined matrices contain a certain configuration of 0s and 1s, called alternatively corner-connected components. Furthermore, such matrices can be transformed into each other by switchings the 0s and 1s of these components.

1 Introduction

Consider a ray (such as light or X-ray) transmitting through homogeneous material. It is a well-known physical phenomenon that a part of the ray will be absorbed in the material. Quantitatively, let I_0 and I denote the initial and the detected intensities of the ray. Then

$$I = I_0 \cdot e^{-\mu x} , \tag{1}$$

where $\mu \geq 0$ denotes the absorption coefficient of the material and x is the length of the path of the ray in the material.

Consider now the 2-dimensional integer lattice $\mathbb{Z}^2$. Let m and n be positive integers. A *discrete rectangle* with size $m \times n$ is a special discrete set of $\mathbb{Z}^2$ determined as the intersection of m consecutive horizontal lattice lines with n consecutive vertical lattice lines. If F is a discrete set in $\mathbb{Z}^2$, then there is an $m \times n$ discrete rectangle containing F, it is called *containing* (*discrete*) *rectangle.* (For the sake of simplicity we take the smallest containing rectangle in the following.)

Let us suppose that the horizontal and vertical projections of F are measured by detectors placed in the next column to left and in the next row to above, respectively, of the containing rectangle (see Fig. 1.a).
Then the projections can be computed according to (1). For example, in the case of the discrete set F given in Fig. 1a, the projections along the horizontal

G. Borgefors, I. Nyström, and G. Sanniti di Baja (Eds.): DGCI 2000, LNCS 1953, pp. 137–148, 2000.
© Springer-Verlag Berlin Heidelberg 2000

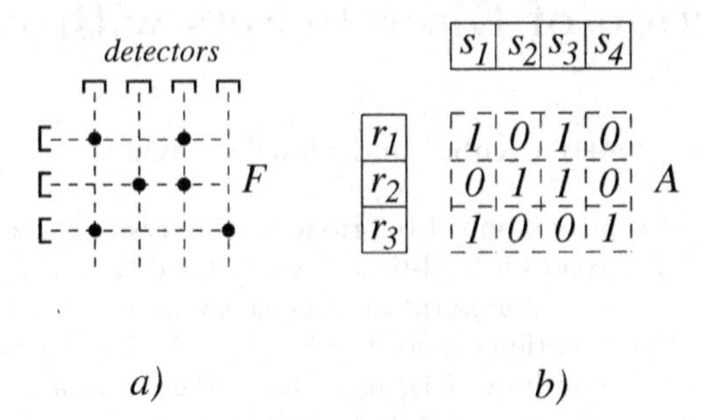

Fig. 1. A 2D discrete set F, the corresponding binary matrix A, and the horizontal and vertical projections. (a) A 2D discrete set F (its elements are indicated by points) and the detectors measuring its horizontal and vertical projections. (b) The binary matrix A and its row and column sums

lattice lines of the containing rectangle are $r_1 = e^{-\mu\cdot 1}+e^{-\mu\cdot 3}$, $r_2 = e^{-\mu\cdot 2}+e^{-\mu\cdot 3}$, and $r_3 = e^{-\mu\cdot 1}+e^{-\mu\cdot 4}$ (from top to down).

Let us use an equivalent representation of the 2D discrete sets and their horizontal and vertical projections. The containing rectangle including F can be represented by a binary matrix $A = (a_{ij})_{m\times n}$ as follows (see Fig. 1.b): $a_{ij} = 1$ if the lattice point corresponding to (i, j) is an element of F, $a_{ij} = 0$ otherwise. In order to use the generally accepted notation of numeration systems [5] let us introduce

$$\beta = e^{\mu} . \tag{2}$$

Clearly, $\beta \geq 1$, because $\mu > 0$. Then we can define the *absorbed row* and *column sums* of A, $R_\beta(A)$ and $S_\beta(A)$, respectively, as

$$R_\beta(A) = (r_1, \ldots, r_m) , \qquad \text{where} \qquad r_i = \sum_{j=1}^{n} a_{ij}\beta^{-j} , \quad i = 1, \ldots, m, \tag{3}$$

and

$$S_\beta(A) = (s_1, \ldots, s_n) , \qquad \text{where} \qquad s_j = \sum_{i=1}^{m} a_{ij}\beta^{-i} , \quad j = 1, \ldots, n. \tag{4}$$

For example, in the case of matrix A given in Fig. 1.b) $R_\beta(A) = (\beta^{-1} + \beta^{-3},\ \beta^{-2} + \beta^{-3},\ \beta^{-1} + \beta^{-4})$ and $S_\beta(A) = (\beta^{-1} + \beta^{-3},\ \beta^{-2},\ \beta^{-1} + \beta^{-2},\ \beta^{-3})$.

Then the *reconstruction problem of binary matrices with absorption* knowing the projections along horizontal and vertical lines can be posed as

RECONSTRUCTION $DA2D(\beta)$.

Given: $\beta \geq 1$, m, n, $R \in \mathbb{N}^m$, *and* $S \in \mathbb{N}^n$.

Task: *Construct a binary matrix* A *with size* $m \times n$ *such that*

$$R_\beta(A) = R \qquad \text{and} \qquad S_\beta(A) = S . \tag{5}$$

We say that a binary matrix A is a *solution of the* $DA2D(\beta)$ *reconstruction problem* if (5) is satisfied.

If $\beta = 1$ then we have the classical reconstruction problem of binary matrices without absorption (as summaries see e.g. [1,3]). Select, now, a mathematically interesting case when $\beta = \beta_0$ where

$$\beta_0^{-1} = \beta_0^{-2} + \beta_0^{-3} \tag{6}$$

giving

$$\beta_0 = \frac{1+\sqrt{5}}{2} . \tag{7}$$

(The other solution of (6), namely $(1-\sqrt{5})/2$, is not applicable in this case, see (2).)

In this paper we discuss the problem of reconstruction of binary matrices from their row and column sums in the case of absorption corresponding to β_0. Because of the page limit, the proofs of theorems and lemmas are omitted here (but they are given in [4]).

2 β_0-Representation

Consider the absorbed row and column sums of the binary matrix A in the case of $\beta = \beta_0$:

$$r_i = \sum_{j=1}^{n} a_{ij}\beta_0^{-j} , \ i = 1, \ldots, m, \qquad \text{and} \qquad s_j = \sum_{i=1}^{m} a_{ij}\beta_0^{-i} , \ j = 1, \ldots, n. \tag{8}$$

Using the terminology of numeration systems we can say that the finite (binary) word $a_{i1} \cdots a_{in}$ is a *(finite) representation in base* β_0 (or a *finite* β_0*-representation*) of r_i for each $i = 1, \ldots, m$, and, similarly, $a_{1j} \cdots a_{mj}$ is a β_0-representation of s_j for each $j = 1, \ldots, n$. The equations (8) mean also that the row and column sums of A are nonnegative real numbers having a finite β_0-representation with n and m binary digits, respectively (including the eventually ending zeros).

Let B_k denote the set of nonnegative real numbers having a β_0-representation with k binary digits ($k > 1$), formally,

$$B_k = \left\{ \sum_{i=1}^{k} a_i \beta_0^{-i} \mid a_i \in \{0,1\} \right\} . \tag{9}$$

Then

$$r_i \in B_n, \; i = 1, \ldots, m, \qquad \text{and} \qquad s_j \in B_m, \; j = 1, \ldots, n, \tag{10}$$

are necessary conditions for the existence of a matrix A with

$$R_{\beta_0}(A) = (r_1, \ldots, r_m) \qquad \text{and} \qquad S_{\beta_0}(A) = (s_1, \ldots, s_n) . \tag{11}$$

2.1 Switching in β_0-Representations

The β_0-representation is generally non-unique, because there are binary words with the same length representing the same number. For example, on the base of (6) it is easy to check the following equality between the 3-digit-length β_0-representations

$$100 = 011 . \tag{12}$$

Furthermore, Equality (12) may allow us to generate newer β_0-representations: If in a finite length β_0-representation there is one of the sub-words 100 or 011, then it can be replaced by the other one. This transformation of binary words is called *(1D) elementary switching.*

It is clear that one or more 1D elementary switchings give(s) new β_0-representation(s) of the same number. As an example, see the 5-digit-length β_0-representations of $1/\beta_0$:

$$10000 = 01100 = 01011 , \tag{13}$$

where the second β_0-representation is created from the first representation by switching on the first three positions and the third representation is created from the second one by switching on the last three positions.

As direct consequences of (12), it is easy to see that

$$\begin{aligned} 10x00 &= 01x11 \\ 10x0x00 &= 01x1x11 \\ 10x0x0x00 &= 01x1x1x11 \\ &\cdots \end{aligned} \tag{14}$$

where x denotes the positions where both β_0-representations have the same binary digit. (That is, such kind of transformation $1(0x)^k00 \leftrightarrow 0(1x)^k11$ ($k \geq 0$) between the sub-words of the β_0-representations can be performed without changing the represented value and without changing the values in the positions indicated by x.) The transformations described by (12) and (14) are called *(1D) switchings.*

Now, we are going to prove that any finite β_0-representation of a number can be get from its any other β_0-representation by switchings.

Lemma 1. *Let $a_1 \cdots a_k$ and $b_1 \cdots b_k$ be different, k-digit-length β_0-representations of the same number. Then $b_1 \cdots b_k$ can be get from $a_1 \cdots a_k$ by a finite number of 1D switchings.*

Proof. A procedure by which the suitable switchings can be found is described in [4].

Consequence. If $a_1 \cdots a_k$ and $b_1 \cdots b_k$ are different, k-digit-length β_0-representations of the same number, then there are positions $i, i+1, i+2$ $(1 \leq i \leq k-2)$ such that there is a switching between $a_1 \cdots a_k$ and $b_1 \cdots b_k$ on these positions.

3 2D Switchings

In this section we determine the 2D switchings, i.e., the transformations of binary matrices by which some of their 0s and 1s can be switched to each other such that the absorbed row and column sums remain the same.

3.1 Connectedness

Consider the class of $m \times n$ $(m, n \geq 3)$ binary matrices with given row and column sum vectors in the case of absorption corresponding to β_0. Let $S_{(i,j)}$ $(1 < i < m, 1 < j < n)$ denote the 3×3 discrete square

$$S_{(i,j)} = \{i-1, i, i+1\} \times \{j-1, j, j+1\} .$$

Let Σ be a set of 3×3 discrete squares of $\{1, \ldots, m\} \times \{1, \ldots, n\}$ and let $S_{(i,j)}, S_{(i',j')} \in \Sigma$. Two kinds of connectedness will be defined on the set of 3×3 discrete squares: side-connectedness and corner-connectedness. There is *side-connection* between $S_{(i,j)}$ and $S_{(i',j')}$ if $(i', j') \in \{(i-2, j), (i, j-2), (i, j+2), (i+2, j)\}$. The squares being side-connected to $S_{(i,j)}$ are called the *side-neighbors* of $S_{(i,j)}$. As an illustration see Fig. 2.a. The squares $S_{(i,j)}$ and $S_{(i',j')}$ are *corner-connected* if $(i', j') \in \{(i-2, j-2), (i-2, j+2), (i+2, j-2), (i+2, j+2)\}$. The squares being corner-connected to $S_{(i,j)}$ are called the *corner-neighbors* of $S_{(i,j)}$ (see Fig. 2.b).

There is a *side-chain between $S_{(i,j)}$ and $S_{(i',j')}$ in Σ* if a sequence of elements of Σ can be selected such that the first element is $S_{(i,j)}$ and the last element is $S_{(i',j')}$ and any two consecutive elements of the sequence are side-connected. (A sequence consisting of only one square is a side-chain by definition.)

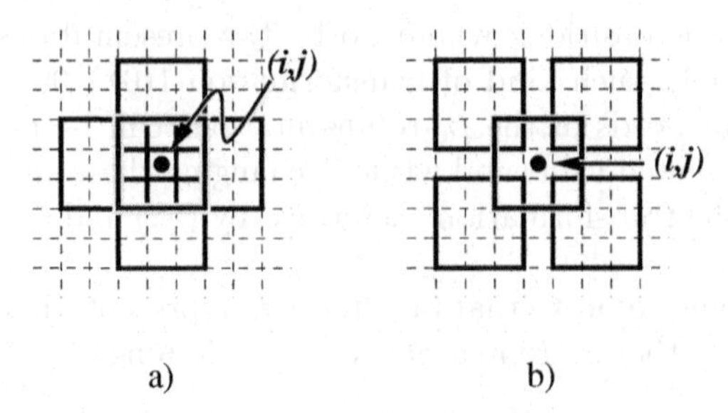

Fig. 2. The discrete square $S_{(i,j)}$ and a) its side-neighbors and b) its corner-neighbors

The set Σ is *side-connected* if there is a side-chain in Σ between its any two different elements (see Fig. 3.a). A side-connected set Σ is *strongly side-connected* if whenever $S_{(i,j)}$, $S_{(i',j')} \in \Sigma$ and $S_{(i,j)} \cap S_{(i',j')} \neq \emptyset$ then they are side-connected or they have a common side-connected neighbor (see Fig. 3.b). A maximal strongly side-connected subset of Σ is called a *strongly side-connected component* of Σ.

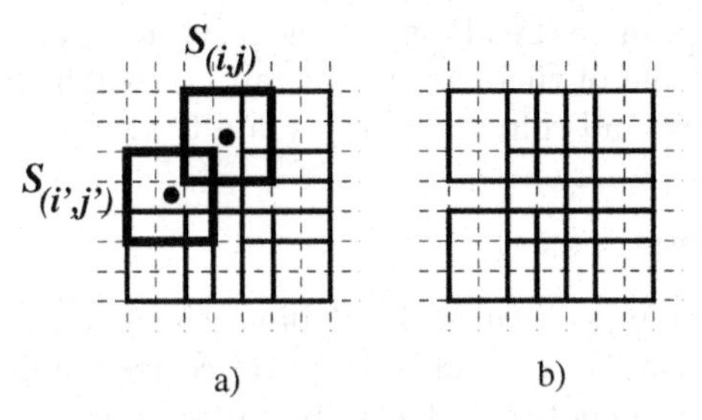

Fig. 3. Side-connected sets of discrete squares. The set a) is not strongly side connected because $S_{(i,j)} \cap S_{(i',j')} \neq \emptyset$ but they are not side-connected and they have no common side-neighbor. b) A strongly side-connected set

Let σ be a set of strongly side-connected components $\Sigma^{(1)}, \ldots, \Sigma^{(k)}$ $(k \geq 1)$. Let $\Sigma^{(l)}, \Sigma^{(l')} \in \sigma$ $(1 \leq l, l' \leq k)$. $\Sigma^{(l)}$ and $\Sigma^{(l')}$ are *corner-connected* if whenever $S \in \Sigma^{(l)}$ and $S' \in \Sigma^{(l')}$ have a common position then S and S' are corner-connected squares. (Since $\Sigma^{(l)}$ and $\Sigma^{(l')}$ are maximal, S and S' cannot be side-connected squares.) There is a *corner-chain between* $\Sigma^{(l)}$ *and* $\Sigma^{(l')}$ *in* σ if a sequence of elements of σ can be selected such that the first element is $\Sigma^{(l)}$ and the last element is $\Sigma^{(l')}$ and any two consecutive elements of the sequence are corner-connected. (A sequence consisting of only one component is a corner-chain by definition.) The set σ is *corner-connected* if there is a corner-chain in σ between its any two elements (see Fig. 4.a).

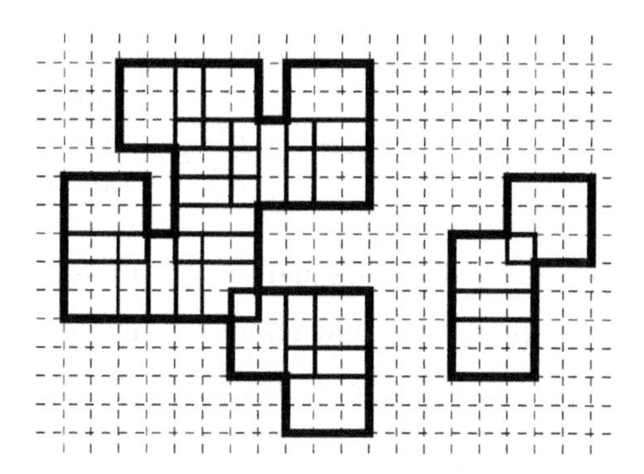

Fig. 4. Two corner-connected sets, both of them have two strongly side-connected components

3.2 Switching Patterns

In order to identify not necessarily rectangular parts of binary matrices, we introduce the concept of *binary patterns* (or shortly, *patterns*) as binary valued functions defined on an arbitrary non-empty subset of $\{1,\ldots,m\} \times \{1,\ldots,n\}$. (In this terminology binary matrices are binary patterns on discrete rectangles.)

Let P be a binary pattern, its domain will be denoted by $\mathrm{dom}(P)$. The absorbed row and column sums of P are denoted by $R_{\beta_0}(P)$ and $S_{\beta_0}(P)$, respectively, where the ith component of $R_{\beta_0}(P)$ is

$$\sum_{(i,j)\in\mathrm{dom}(P)} P(i,j)\cdot\beta_0^{-j}$$

for $1 \le i \le m$ and the jth component of $S_{\beta_0}(P)$ is

$$\sum_{(i,j)\in\mathrm{dom}(P)} P(i,j)\cdot\beta_0^{-i}$$

for $1 \le j \le n$.

Let us define the switching pair of P, P', as

$$P'(i,j) = 1 - P(i,j)$$

on $\mathrm{dom}(P)$. We say that P is a *switching pattern* if

$$R_{\beta_0}(P) = R_{\beta_0}(P') \qquad \text{and} \qquad S_{\beta_0}(P) = S_{\beta_0}(P') ,$$

and P and P' are a *switching pair*. That is, the patterns of a switching pair can be get from each other by switching their 0s and 1s and still they have the same absorbed row and column sums.

Consider the following binary patterns:

$$E^{(0)}_{(i,j)} = \begin{matrix} 0\,1\,1 \\ 1\,0\,0 \\ 1\,0\,0 \end{matrix} \quad \text{and} \quad E^{(1)}_{(i,j)} = \begin{matrix} 1\,0\,0 \\ 0\,1\,1 \\ 0\,1\,1 \end{matrix} ,$$

both of them are defined on the discrete square $S_{(i,j)}$. It is easy to check that $E^{(0)}_{(i,j)}$ and $E^{(1)}_{(i,j)}$ are a switching pair. They play an important role in the generation of switching patterns, $E^{(0)}_{(i,j)}$ and $E^{(1)}_{(i,j)}$ are called the *0-type* and *1-type elementary (2D) switching pattern*, respectively.

3.3 Composition of Patterns

The *composition* of two patterns P and P' is the function

$$P * P' : \ \mathrm{dom(P)} \triangle \mathrm{dom(P')} \longrightarrow \{0, 1\}$$

($\triangle$ denotes the symmetric difference) such that

$$[P * P'](i,j) = \begin{cases} P(i,j), & \text{if}(i,j) \in \mathrm{dom}(P) \setminus \mathrm{dom}(P'), \\ P'(i,j), & \text{if}(i,j) \in \mathrm{dom}(P') \setminus \mathrm{dom}(P) . \end{cases}$$

(That is, $P * P'$ is undefined on $\mathrm{dom}(P) \cap \mathrm{dom}(P')$.)

For example,

$$E^{(0)}_{(i,j)} * E^{(0)}_{(i+2,j)} = \begin{matrix} 0\,1\,1 \\ 1\,0\,0 \\ x\,x\,x \\ 1\,0\,0 \\ 1\,0\,0 \end{matrix} , \tag{15}$$

defined on $S_{(i,j)} \triangle S_{(i+2,j)} = \{i-1, i, i+2, i+3\} \times \{j-1, j, j+1\}$. (Just for the sake of simple presentation, on the right side of (15) the whole sub-matrix on the rectangle $\{i-1, \ldots, i+3\} \times \{j-1, j, j+1\}$ is indicated and x denotes the positions in the sub-matrix where the composition is undefined.) Similarly,

$$E^{(0)}_{(i,j)} * E^{(0)}_{(i,j+2)} = \begin{matrix} 0\,1\,x\,1\,1 \\ 1\,0\,x\,0\,0 \\ 1\,0\,x\,0\,0 \end{matrix} . \tag{16}$$

defined on $S_{(i,j)} \triangle S_{(i,j+2)} = \{i-1, i, i+1\} \times \{j-1, j, j+2, j+3\}$. It is easy to see that $E^{(0)}_{(i,j)} * E^{(0)}_{(i+2,j)}$ and $E^{(0)}_{(i,j)} * E^{(0)}_{(i,j+2)}$ are switching patterns, their switching pairs are $E^{(1)}_{(i,j)} * E^{(1)}_{(i+2,j)}$ and $E^{(1)}_{(i,j)} * E^{(1)}_{(i,j+2)}$, respectively, where

$$E^{(1)}_{(i,j)} * E^{(1)}_{(i+2,j)} = \begin{matrix} 1\,0\,0 \\ 0\,1\,1 \\ x\,x\,x \\ 0\,1\,1 \\ 0\,1\,1 \end{matrix} \quad \text{and} \quad E^{(1)}_{(i,j)} * E^{(1)}_{(i,j+2)} = \begin{matrix} 1\,0\,x\,0\,0 \\ 0\,1\,x\,1\,1 \\ 0\,1\,x\,1\,1 \end{matrix} . \tag{17}$$

These examples show how new switching patterns can be created from elementary switching patterns by composition. Now, we show a general way how two (not only elementary) switching patterns can be used to generate another switching pattern.

3.4 Composition of Elementary Switching Patterns

Let $\mathcal{E} = \{E_1, \ldots, E_k\}$ $(k \geq 1)$ be a set of elementary switching patterns of the same type on a strongly side-connected set $\Sigma = \{S_1, \ldots, S_k\}$. The composition of $\mathcal{E}$ on Σ, denoted by $\mathcal{C}(\mathcal{E})$, is defined as follows. If $k = 1$ then $\mathcal{C}(\mathcal{E}) = E_1$. If $k > 1$ then let us suppose that $S_1, \ldots, S_k$ are indexed such that for each $l(< k)$ $\{S_1, \ldots, S_l\}$ is a strongly side-connected set and one of its squares is side-connected with S_{l+1}. (It is easy to see that such an indexing exists.) Then let

$$C = \mathcal{C}(\mathcal{E}) = ((E_1 * E_2) * \cdots) * E_k .$$

Now, we are going to show that this definition is independent from the indexing of $\Sigma = \{S_1, \ldots, S_k\}$. There are four cases depending on how many times the position (i, j) is in the sets $\{S_1, \ldots, S_k\}$:

(i) If there is exactly one l such that $(i, j) \in S_l$, then

$$C(i,j) = e^{(l)}_{ij} ,$$

where $e^{(l)}_{ij}$ denotes the value of E_l in the position (i, j).

(ii) If there are exactly two different l_1 and l_2 such that $(i, j) \in S_{l_1}$ and $(i, j) \in S_{l_2}$, then $C(i, j)$ is undefined.

(iii) If there are exactly three different l_1, l_2, and l_3 such that $(i, j) \in S_{l_1} \cap S_{l_2} \cap S_{l_3}$, then two of the elementary switching patterns (say, E_{l_1} and E_{l_3}) has the same value in the position (i, j) and the other one (E_{l_2}) has a different value here, i.e., $e^{(l_1)}_{ij} = e^{(l_3)}_{ij} = 1 - e^{(l_2)}_{ij}$. In this case

$$C(i,j) = e^{(l_1)}_{ij} = e^{(l_3)}_{ij} .$$

(iv) If there are exactly four different l_1, l_2, l_3, and l_4 such that $(i, j) \in S_{l_1} \cap S_{l_2} \cap S_{l_3} \cap S_{l_4}$, then $C(i, j)$ is undefined.

That is, the value of $C(i, j)$ can be decided simply on the base of the parity of the number of discrete squares of Σ covering (i, j) (independently from the indexing of Σ). Accordingly, if $\mathcal{E}$ is the set of 2D elementary switching patterns of the same type on a strongly side-connected set, then we can simply write $\mathcal{C}(\mathcal{E})$ to denote the compositions of the elements of $\mathcal{E}$. As an example, see Fig. 5.

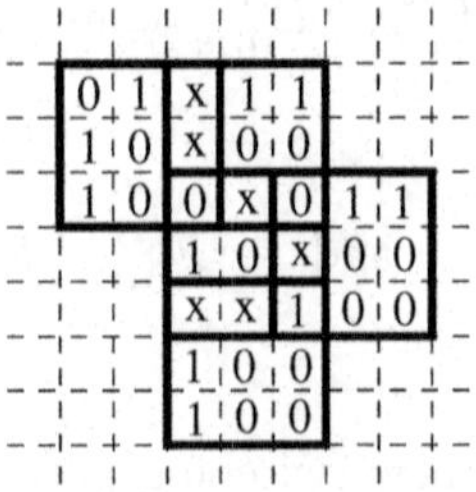

Fig. 5. The composition of the elementary switching patterns $E^{(0)}_{(2,2)}$, $E^{(0)}_{(2,4)}$, $E^{(0)}_{(4,4)}$, $E^{(0)}_{(4,6)}$, and $E^{(0)}_{(6,4)}$ ('x' denotes 'don't care' position)

Lemma 2. *Let $\mathcal{E}$ be a set of elementary switching patterns of the same type on a strongly side-connected set of 3×3 squares. Then $\mathcal{C}(\mathcal{E})$ is a switching pattern.*

Proof. The proof is given in [4].

Henceforth, the switching patterns constructed from elementary switching patterns of the same type by composition are called *composite switching patterns.* We say that the composite switching pattern has 0-type/1-type if it is the composition of 0-type/1-type elementary switching patterns. For example, (15)-(17) and Fig. 5 show composite switching patterns of types 0, 0, 1, 1, and 0.

3.5 Composition of Corner-Connected Components

Lemma 3. *Let $\Sigma^{(0)}$ and $\Sigma^{(1)}$ be corner-connected components. Let $C^{(0)}$ and $C^{(1)}$ be 0-type and 1-type composite switching patterns on $\Sigma^{(0)}$ and on $\Sigma^{(1)}$, respectively. Then $C^{(0)} * C^{(1)}$ is a switching pattern.*

Proof. The proof is given in [4].

Let σ be a corner-connected set of strongly side-connected components $\Sigma^{(1)}$, $\ldots, \Sigma^{(k)}$ $(k \geq 1)$. Let $C^{(1)}, \ldots, C^{(k)}$ be 0- or 1-type composite switching patterns on $\Sigma^{(l)}, \ldots, \Sigma^{(k)}$, respectively. Let us suppose also that if $\Sigma^{(l)}$ and $\Sigma^{(l')}$ are corner-connected then $C^{(l)}$ and $C^{(l')}$ have different type. In this case we say that $\gamma = \{C^{(1)}, \ldots, C^{(k)}\}$ is a set of *alternatively corner-connected components* on σ.

The composition of γ on σ, denoted by $\mathcal{C}(\gamma)$, is defined as follows. If $k = 1$ then $\mathcal{C}(\gamma) = C^{(1)}$. If $k > 1$ then let us suppose that $\Sigma^{(1)}, \ldots, \Sigma^{(k)}$ are indexed such that for each l $(l < k)$ $\{\Sigma^{(1)}, \ldots, \Sigma^{(k)}\}$ is a corner-connected set and one of its elements is corner-connected with $\Sigma^{(l+1)}$. (It is easy to see that such an indexing exists.) Then let

$$C = \mathcal{C}(\gamma) = ((C^{(1)} * C^{(2)} * \cdots) * C^{(k)} .$$

It is easy to show that this definition is independent from the indexing of σ, because $\mathrm{dom}(C) = \cup_l \Sigma^{(l)} \setminus \cup_{l,l'}(\Sigma^{(l)} \cap \Sigma^{(l')})$ and $C(i,j) = C^{(l)}(i,j)$, where $l \in \{1,\ldots,k\}$ is the only index such that $(i,j) \in \Sigma^{(l)}$. Henceforth, we may simply write $\mathcal{C}(\gamma)$ to denote the composition of the elements of γ. As an example, see Fig. 6.

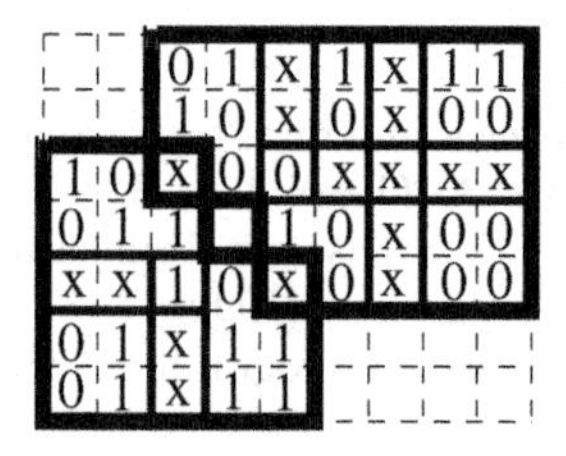

Fig. 6. Composition of two alternatively corner-connected components (the two components are indicated with thick lines)

Theorem 1. *Let* $\gamma = \{C^{(1)},\ldots,C^{(k)}\}$ $(k \geq 1)$ *be a set of alternatively corner-connected components. Then* $\mathcal{C}(\gamma)$ *is a switching pattern.*

Proof. It follows from Lemma 3 directly. □

Theorem 2. *If a binary matrix has a switching pattern* C *then there is a finite number of sets* $\gamma_1,\ldots,\gamma_l$ $(l > 1)$ *of alternatively corner-connected components such that* C *is the composition of* $\mathcal{C}(\gamma_1),\ldots,\mathcal{C}(\gamma_l)$.

Proof. The proof is given in [4].

Therefore, on the base of Theorems 1 and 2, it is proven that the 2D switching patterns are just the patterns created as the composition of alternatively corner-connected components.

Finally, two consequences are mentioned (c.f. [6]):

(i) A binary matrix is uniquely determined by its absorbed row and column sums if and only if it has no sub-pattern created by composition from a set of alternatively corner-connected components.

(ii) If A and B are different binary matrices with the same absorbed row and column sums then A is transformable into B by a finte number of switchings of switching patterns created by compositions from sets of alternatively corner-connected components.

Acknowledgements

The authors thank to the reviewer to point out that one of the lemmas in the previous form of the paper was not correct. A part of this research was done while the first author visited the Laboratoire d'Informatique Algorithmique: Fondements et Applications, Université Paris 7 Denis-Diderot, Paris, France. Special thanks to Prof. Gabor T. Herman (Temple University, Philadelphia) and Laurent Vuillon (LIAFA, Université Paris 7) for their valuable comments and discussions. This work was supported by the grants FKFP 0908/1997 and OTKA T 032241, and MTA-NSF 123 joint grant supplement to NSF grant DMS 9612077 (Aspects of Discrete Tomography).

References

1. Brualdi, R. A.: Matrices of zeros and ones with fixed row and column sums. Linear Algebra and Its Applications **33** (1980) 159-231 139
2. Herman, G. T., Kuba, A. (Eds.): Discrete Tomography: Foundations, Algorithms and Applications. Birkhäuser, Boston (1999) 148
3. Kuba, A., Herman, G. T.: Discrete tomography: A historical overview. In [2] (1999) 3-34 139
4. Kuba, A., Nivat, M.: Reconstruction of discrete sets from projections in the case of absorption. Technical Reports, Institute of Informatics, University of Szeged (2000) 139, 141, 146, 147
5. Lothaire, M.: Combinatorics on Words. Cambridge University Press, Cambridge (1997) 138
6. Ryser, H. J.: Combinatorial properties of matrices of zeros and ones. Canad. J. Math. **9** (1957) 371-377 147

Some Properties of Hyperbolic Networks

Christophe Papazian[1] and Eric Rémila[1,2]

[1] Laboratoire de l'Informatique du Parallélisme CNRS UMR 5668
Ecole Normale Supérieure de Lyon
46 Allée d'Italie, 69364 Lyon Cedex 07, France
Christophe.Papazian@ens-lyon.fr
[2] GRIMA, IUT Roanne, Université J. Monnet
20 Avenue de Paris, 42334 Roanne Cedex, France
Eric.Remila@ens-lyon.fr

Abstract. Many mathematical results exist about continuous topological surfaces of negative curvature. We give here some properties of discrete regular tessellations on such objects and explain a characterization of discrete geodesics and areas that shows how such hyperbolic networks can be seen as intermediary structures between Euclidean infinite tessellations (like square grid) and regular infinite trees.
We do not use some possible group structures of this networks (Cayley graphs) but only geometrical arguments in our constructive proofs. Hence we can see that there are few geodesics in hyperbolic networks and that large areas have very unsmooth borders.

1 Introduction

A lot of interesting and deep results (see for example [1], [4] and the proceedings of the previous DGCI conferences) have been found about the topology of the three classical regulars tessellations of the Euclidean plane, (with squares, equilateral triangles or hexagons). These tessellations are seen as discrete versions of the plane, they are often called Euclidean discrete planes. In computer imagery, each cell of such a tessellation is viewed as a pixel of a flat screen.

At the opposite, there are very few results about regular tessellations of hyperbolic planes, even though these tessellations also are very interesting : to be more explicit, if a screen were a part of an hyperboloid, then it can canonically be divided in "hyperbolic pixels", which are cells of a tessellation of a hyperbolic plane.

In this paper, we are focused on geodesic lines of these tessellations. We have two main results, which show large structural differences between hyperbolic discrete planes and the Euclidean discrete plane. The proofs of these results are based on precise isoperimetric inequalities in hyperbolic planes (we also give a general isoperimetric inequality).

Our first result describes the set of geodesic lines between two fixed points. We prove that this set is "thin": precisely, there exists a positive constant c such

G. Borgefors, I. Nyström, and G. Sanniti di Baja (Eds.): DGCI 2000, LNCS 1953, pp. 149–158, 2000.
© Springer-Verlag Berlin Heidelberg 2000

that if a_1, a_2, a' and a" are points such that $d(a_1, a') + d(a', a_2) = d(a_1, a'') + d(a'', a_2) = d(a_1, a_2)$ and $d(a_1, a') = d(a_1, a'')$, then $d(a', a'') \leq c$.

Our second result proves that for each 3-tuple (a_1, a_2, a_3) of points of the hyperbolic plane, the triangle (a_1, a_2, a_3) is (nearly) empty. This result can be seen as a discrete version of the classical result about triangles of a continuous hyperbolic plane [3].

2 Definitions

A graph is a set $G = (V, E)$ with V a set of vertices, and E a symmetric subset of V^2, whose elements are called "edges".

$a, b \in V$ are called neighbors if $(a, b) \in E$.

$a \in V$ has a degree $n \in \mathbb{N}$ if a has n differents neighbors.

A path c of a graph is a sequence of vertices, $(v_1, v_2, ..., v_n)$, such that for all i, v_i and v_{i+1} are neighbors.

A planar graph is a graph with a planar representation (no crossing edges).

A cycle is a path such that $v_1 = v_n$. Given a planar representation of the graph, a cycle c cuts the set of vertices into three subsets : $Int(c)$, the finite set of vertices inside the cycle, c, the vertices of the cycle, and $Ext(c)$, the set of vertices outside the cycle.

A cell is a cycle with interior void and no edge inside.

Definition 1. *A network $\Gamma(k, d)$ is the unique planar graph, with every cell of size k and with every vertex of degree d whose planar representation does not contain an accumulation point.*

So, $\Gamma(4, 4)$ is the infinite grid isomorphic to $\mathbb{Z}^2$.

Definition 2. *A geodesic xy is a path from the vertex x to the vertex y such that there is no strictly shorter path.*

For all $\Gamma(k, d)$, we note $\gamma = dk - 2d - 2k = (k-2)(d-2) - 4$ the *curvature* of the network. If $\gamma \geq 0$ then the network is infinite. If $\gamma = 0$, then the network is called Euclidean, if $\gamma \geq 1$ then the network is hyperbolic. Now, we will only consider hyperbolic networks $\Gamma(k, d)$ (see [2] for details).

For a path, or a set of vertices c, we note $|c|$ the number of vertices it contains.

3 Fundamental Properties

First, some basic geometric properties, from Euler Formula :

Lemma 1. *A completely triangulated planar graph with n vertices on the border and m interior vertices is composed of $n + 2m - 2$ triangles (cells of size 3), and $2n + 3m - 3$ edges.*

Then, from the previous Lemma, we can deduce the more general lemma :

Lemma 2. *A planar graph whose every cell is of size k (with every vertex into some cell, no vertex of degree 1), with n vertices on the border and m interior vertices is composed of* $\frac{(n+2m-2)}{(k-2)}$ *cells and* $\frac{(n+2m-2)}{(k-2)} \times \frac{k}{2} + \frac{n}{2} = n+m-1+\frac{n+2m-2}{k-2}$ *edges.*

Proof. We just triangulate cells : there are $k-2$ triangles in a cell of size k, and the triangulation needs $k-3$ additional edges for each cell.

Theorem 1 (Isoperimetric inequality). *For a cycle c of a $\Gamma(k,d)$, we have :*

$$|Int(c)| \leq \frac{k}{(k-1)\gamma}.(|c| - 2(k-1))$$

Proof. We have to consider the planar subgraph $G=(V,E)$ of a network $\Gamma(k,d)$ such that $V = c \cup Int(c)$ and E is the set of all edges inside c or in c. So G is composed of cells of size k.

G has at least 2 leaving arc for each border vertex plus one every $k-1$ border vertices (if we consider the border larger than a cell) to close the cells and d leaving edges for each interior vertex. So we can deduce the following inequality (using lemma 2, and $m = |Int(c)|$ et $n = |c|$) :

$$n+m-1+\frac{(n+2m-2)}{(k-2)} \geq \frac{md}{2} + n + \frac{1}{2}.\left\lceil \frac{n}{k-1} \right\rceil$$

As the border is a cycle, we have to obviously round up the fractional part (if we note p the number of leaving edges, it verifies $p(k-1) \geq n$ so $p \geq \frac{n}{k-1}$).

$$\frac{(n+2m-2)}{(k-2)} - 1 \geq \frac{m(d-2)}{2} + \frac{n}{2.(k-1)}$$

$$(n+2m-2)-(k-2) \geq \frac{m(\gamma+4)}{2} + \frac{n}{2} - \frac{n}{2(k-1)}$$

$$n - 2k + \frac{n}{k-1} \geq m\gamma$$

$$k(n-2(k-1)) \geq m\gamma(k-1)$$

□

Hence, we obtain a linear upper bound of area by perimeter, that is a specific property of hyperbolic networks, whereas area is quadratically majored by perimeter in Euclidean networks. We remark that the curvature of the plane naturally comes into the inequality. Areas are as little as curvature is great: d alone doesn't make any differences, and k is of little importance in the factor $\frac{k-1}{k}$ and in the constant $2(k-1)$.

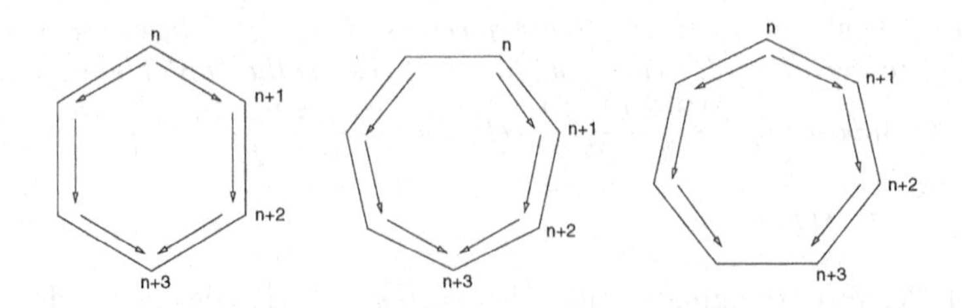

Fig. 1. Orientation of cells for $k = 6$ and $k = 7$

Lemma 3. *Every cell is correctly oriented according to the distance from a given vertex. If k is even, there is one closest point and one farvest point, if k is odd, there is one closest point or one farvest point and "a transversal edge" : one edge that links two edges at the same distance (see figure 1).*

Proof. By induction on the building of the network from the given vertex (figure 6).

So, for any vertex and any cell (with k even), there is only one vertex on the cell that minimize the distance with the given vertex. Informally, this property could be seen as a topological uniformity of the metric space. We must note that this property is still true on Euclidean networks, but is false on finite network (when $\gamma < 0$).

4 Characterization of Geodesics

Given two points x and y of a network $\Gamma(k, d)$, with distance λ from each other, we consider the subgraph formed by all vertices on any geodesic xy and the edges of this geodesic (There are no transversal edges).

We must note that, for any point z of a geodesic xy, the path composed by two arbitrary geodesics xz and zy is a geodesic xy.

Lemma 4. *The graph of geodesics xy is a planar subgraph delimited by two exterior geodesics.*

Proof. (By contradiction) We consider both half-borders of the subgraph that links x to y (it is possible because x and y are compulsory on the border, due to the fact that we consider only *infinite* surfaces). If one of this path is not a geodesic, then there exists an oriented arc (a, b) of this path that is a part of a geodesic yx and not a xy one. But, as a and b are on the border, we note that, due to planarity (see figure 2), this geodesic yx must cross itself : it must pass trough the same point twice. It is obviously not possible for a geodesic. □

The graph of geodesics is planar (as a subgraph of a planar graph), and due to lemma 3, the graph of geodesics is only composed of cell of size k if k is even, and $2k - 2$ if k is odd.

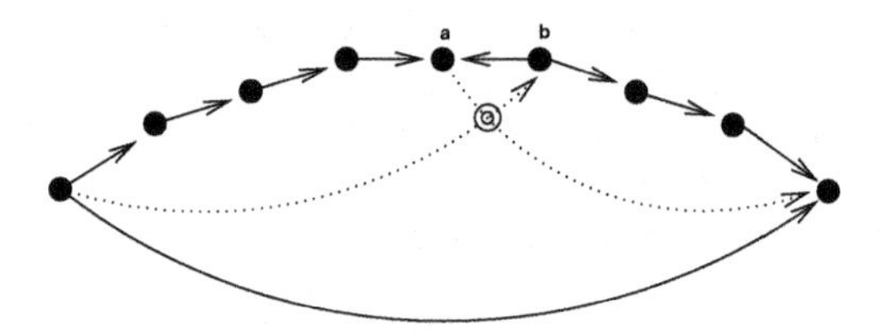

Fig. 2. The impossible geodesic

Theorem 2 (Theorem of neighbor geodesics). *Given two vertices x and y of a network $\Gamma(k,d)$, and two geodesics xy (that we note u and v). Then, $Int(uv^{-1})$ is empty.*

Proof. We first consider k as an even number. We use the same method as for Theorem 1 of isoperimetric inequality. We consider that the cycle (border of the area) is $c = uv^{-1}$, and we define the subgraph G of border c. We suppose (with any loss of generality) that u and v are disjoint (they share no vertex).

G has at least 2 leaving edges for each border vertex and d leaving edges for each interior vertex. Additionally, if we consider $k-1$ consecutive vertices on u (or v), we remark the necessary presence of at least 2 supplementary leaving edges, because we have to respect the fact that u (and v) is a geodesic and that cells are of size k, as we can see from figure 3 : u can follow the boundary of a cell during at most the half of the perimeter.

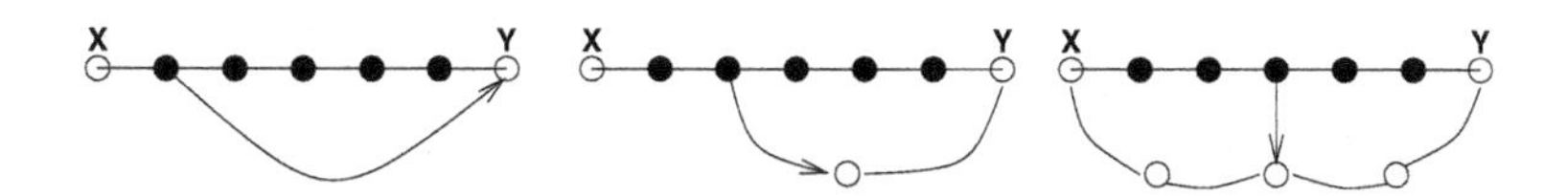

Fig. 3. Impossible to add only one leaving arc on $k-1$ consecutive vertices with the necessity of having a geodesic (between X and Y for $k = 6$ here)

So, we can deduce the following formula (using Lemma 2 and $m = Int(c)$ and $n = |c|$). We note that here, the fractional part has to be rounded down, as the proposition "at least 2 leaving arc for k-1 vertices" is true on two paths u and v (but not on the entire cycle). (We remark that $\lfloor \nu/\omega \rfloor \geq \frac{\nu+1}{\omega} - 1$) :

$$\frac{(n+2m-2)}{(k-2)} \times \frac{k}{2} + \frac{n}{2} \geq \frac{md}{2} + n + \frac{n+1}{k-1} - 1$$

that we can simplify (as for Theorem 1) :

$$2n - 6k + 6 \geq m\gamma(k-1)$$

Now, we will prove that m is not large compare to $n/2$. Using that fact, we will split our problem P_m into two smaller ones $P'_{m'}$ and $P''_{m''}$ with the property $m = m' + m''$. To solve the entire problem, we will split every subproblem until they are trivial (the graph of geodesics is reduced to a single cell where $m = 0$) which implies that $m = 0$ for the initial problem.

The trivial cases are $n = k$ and $n = 2k - 2$ where obviously $m = 0$.
If we consider $m > n/2 - k$, we can deduce by using the last inequality :

$$2n - 6k + 6 \geq m\gamma(k-1) > \left(\frac{n}{2} - k\right)\gamma(k-1)$$

that we can simplify into :

$$(n - 2k)(\gamma(k-1) - 4) < 12 - 4k$$

As $k \geq 4$, we remark that $(\gamma(k-1) - 4) > 0$, so we have $n < 2k$ for $m > 0$! So $m \leq n/2 - k$.

In every graph of geodesics of border $c = uv^{-1}$, with u and v disjoint, $m \leq n/2 - k$ means that for some δ such that $k/2 \leq \delta \leq d(x, y) - k/2$, there is no interior vertex μ such that $d(x, \mu) = \delta$ (because there are $n/2 - k + 1$ such different δ, and not as many interior points). So there are only two vertices a and b with distance δ from x, one on u and the other on v. Moreover, there is no edge ab (it would be a transversal edge !) and no edge between a and b because such an edge would link a vertex with distance $d < \delta$ to another vertex with distance $d > \delta$. Then, a et b are on a cell of the graph of geodesics : There is a cell F which contains a and b (see figure 4. As $k/2 \leq \delta$, F does not contain x, and as $\delta \leq d(x, y) - k/2$, F does not contain y.

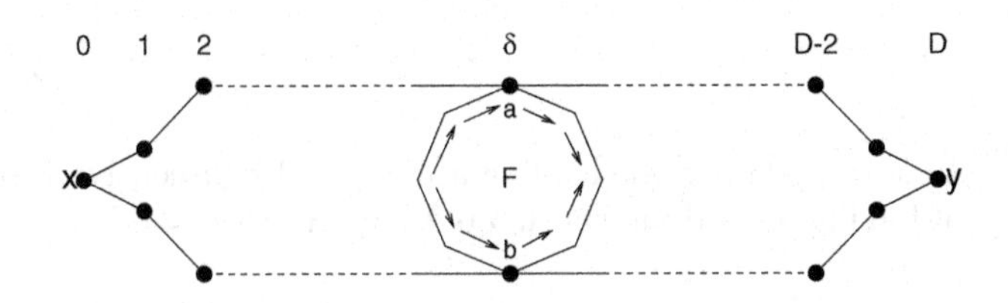

Fig. 4. A cell in the middle

We can now cut this problem into two strictly smaller ones, by using lemma 3 : one that consider the graph of geodesics xF_x where F_x is the vertex of F farvest from x and the graph of geodesics yF_y.

We must remark that any geodesic xy can be obtained by concatenation of two geodesics xa and ay, or xb and by, due to the fact that a and b are the only vertices v that verify $\{d(x, v) = \delta; d(x, y) = d(x, v) + d(v, y)\}$. This is still true for any geodesic xF_x or yF_y. So the graph of geodesics xy is actually “split” into two smaller ones xF_x and yF_y.

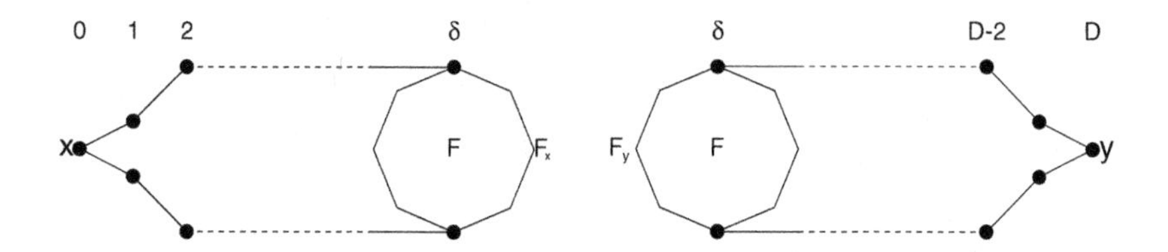

Fig. 5. Two smaller problems

The interior vertices of the initial problem will be interior vertices of exactly one of the two smaller problems, that we can cut until they are reduce to a very small trivial problem (c is one cell or two) where $m = 0$.

So m of the initial problem is a sum of interior of subproblems that all are equal to zero. This proves the theorem for k even.

If k is an odd number, we just remark that graph of geodesics have no transversal edges, by definition. As every cell in such a $\Gamma(2i+1, d)$ has a transversal edge, we remark that every cell in a graph of geodesic has size $2k-2$. Removing transversal edges reduces the degree of vertices, but we note that for $k = 3$ there are at most two transversal leaving edges for a vertex, and for $k > 4$ there is at most one.

So the proof is the same for $\Gamma(3, d)$ as for $\Gamma(4, d-2)$, and the same for $\Gamma(2j+5, d)$ as for $\Gamma(4j+8, d-1)$. For $d = 3$, we have to rearrange the initial proof but the result is the same. □

Corollary 1. *Let x, y be two points of a $\Gamma(k, d)$ and $0 < \delta < d(x, y)$. There are at most two points with distance δ from x and distance $d(x, y) - \delta$ from y.*

Hence, the set of geodesics between two points is a "narrow" graph (of a width of at most one cell). We must remark that, in every hyperbolic network $\Gamma(k, d)$ we can find two points as far from each other as we want such that there is only one unique geodesic between them and we can find, in the same way, two points with two disjoint geodesics.

5 Triangle Inequality

In this section, we not only show that the metric space of an hyperbolic network is hyperbolic (it is a known result), but give computable results about geometrical properties due to hyperbolic metric.

Theorem 3 (Theorem of void triangles). *Let x, y and z be three points of an hyperbolic $\Gamma(k, d)$. Let the border (the triangle) be the cycle composed by three arbitrary geodesics xy, yz and zx. Then, if the triangle has a minimal area (it means that it doesn't contain another xyz triangle in its interior), it does not contain any interior point for $d > 3$ and at most one point for $d = 3$.*

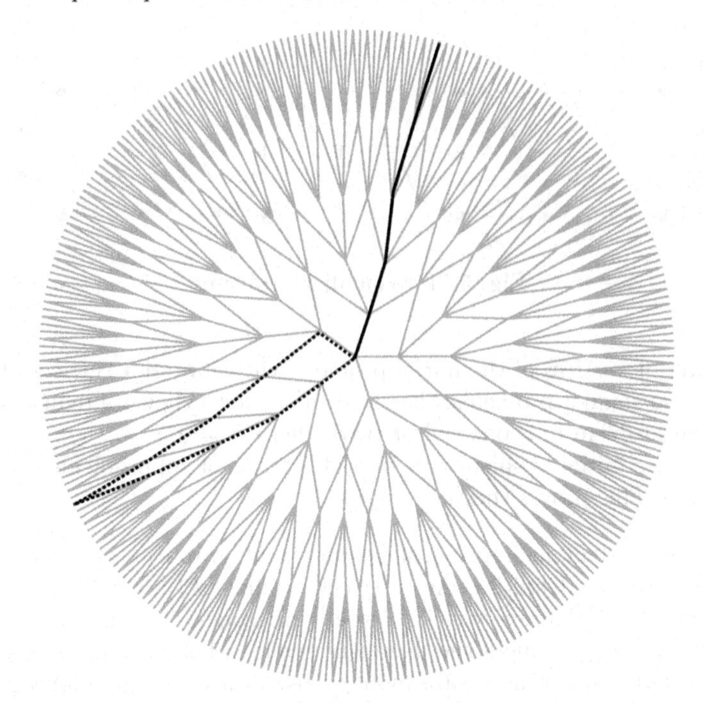

Fig. 6. Two types of geodesics

Proof. As the previous theorem, we suppose $m > 0$ and we use an upper bound of edges on the graph of border $c = xy\ yz\ zx$. On each third of the border, we can remark the necessary presence of two leaving arcs every $k-2$ consecutive vertices, due to the fact that the considered path is a geodesic, and that we do not have another geodesic "inside".

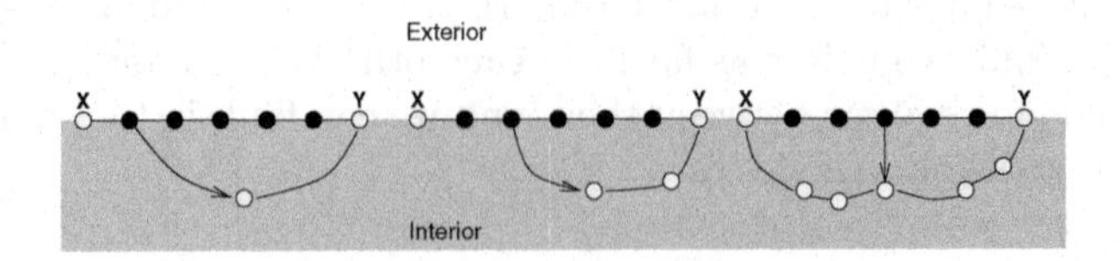

Fig. 7. Impossible to have only one leaving arc on $k-2$ vertices with the necessity of having only one geodesic between X and Y (for $k = 7$ here)

So we have the following inequality :

$$n + m - 1 + \frac{(n + 2m - 2)}{(k-2)} \geq \frac{md}{2} + n + \frac{n}{k-2} - \frac{3}{2}$$

that we can simplify :

$$(k-6) \geq m\gamma$$

Hence if $d \geq 4$ we can verify that we have $m \leq 0$ for any hyperbolic $\Gamma(k, d)$.
If $d = 3$, we can verify that the case $m = 1$ can be obtained in some little triangle (even minimal for k odd) as we can see from figure 5. □

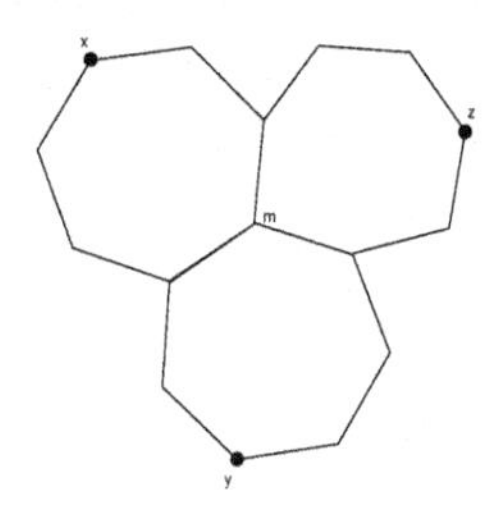

Fig. 8. A xyz triangle with an interior point

Corollary 2. *We note $D(x|y, z)$ the maximum distance from x to a point in the intersection of the subgraphs of geodesics xy and xz (it is always defined as x must be in this intersection). We consider x', such a point with distance $D(x|y, z)$ from x. If we look at the triangle $x'yz$, we can deduce :*

$$D(x, z) + D(x, y) \geq D(y, z) \geq D(x, z) + D(x, y) - 2.D(x|y, z) - 2k$$

We must note that $D(x'|y, z) = 0$ in that example. Hence, distance on a hyperbolic discrete network is working in a similar way as in trees.

Corollary 3. *Due to the characterization of geodesics, we can remark that, if $d > 3$ and $k > 4$, every triangle (even non minimal) can not have any interior points.*

This Theorem could be extended from triangles (3-gones) to g-gones. The result would be the same for $d \geq g$. It means that it is difficult to have a border with a large area. When d is great, so is the curvature γ, and areas are delimited by very unsmooth large borders.

6 Conclusion

Now, we know the properties of a hyperbolic network : areas are small, perimeters very unsmooth and discrete geodesics are thin. So we can see such networks like a compromise between common Euclidean network, with quadratic areas, simple perimeter and a lots of geodesics; and infinite trees, with infinite areas and unicity of geodesics.

For network algorithms, hyperbolic networks have the qualities of Euclidean grids, like good connectivity but as we can see from this article, they have qualities of tree networks, like efficient broadcast (exponential number of neighbors), and specific routing methods (thinness of geodesics).

For graphical purpose, the main problem is still in the representation of such networks. They "grow" very fast and are very difficult to show in an efficient way. Mathematicians use some finite nice way of representation that is completely useless for practical uses. The solution seems to implement some three dimensional representation using the underlying hyperbolic structure of the network. This could lead to a better understanding of every hyperbolic three dimensional surface, allowing best polygonalization of special shapes.

References

1. J. M. Chassery, A. Montanvert, *Géométrie discrète en analyse d'images*, Hermès, Paris (1991). 149
2. H. S. M. Coxeter, W. O. J. Moser, Generators and relations for discrete groups, Springer-Verlag, Berlin . Göttingen 1957. 150
3. E. Ghys, C. de la Harpe, *Sur les groupes hyperboliques d'après Michaël Gromov*, Birkhäuser Ed., (1990). 150
4. J. P. Reveillès, *Géométrie discrète, calcul en nombre entiers et algorithmique*, Thèse d'Etat, Université L. Pasteur, Strasbourg (1991). 149

The Reconstruction of the Digital Hyperbola Segment from Its Code

Nataša Sladoje

Faculty of Engineering
Trg D. Obradovića 6, 21000 NoviSad, Yugoslavia
sladoje@uns.ns.ac.yu

Abstract. It is known that a set consisting of the digital curves whose "original curves" are graphs of continuous functions, having at most two intersection points, pairwise, on a given interval, can be uniquely coded by five parameters. This result is applied to the set of digital hyperbola segments, corresponding to the hyperbolas of the form $y = \frac{\alpha}{x-\beta} + \gamma$ inscribed into the $(m \times m)$-integer grid. An $\mathcal{O}(m \cdot (\log(m + |\beta|))^2)$ algorithm for recovering the digital hyperbola segment from its proposed code, is presented.

Keywords: image processing, pattern analysis, reconstruction, digital hyperbola.

1 Introduction

The most often appearing digital curves in practice are those obtained by digitizing straight lines and conic sections. For the most of them the basic problems considered in computer vision and image processing, such as the efficient representation and reconstruction, as well as the parameter estimation, are already solved, by using different approaches [2]-[7],[10]. The coding scheme developed for the sets of the digital straight line and parabola segments, based on the least squares technique, is successfully applied to the set of the digital hyperbola segments of the form $y = \frac{\alpha}{x} + \beta$ [9]. Unfortunately, that concept, which has the advantage of saving information of the approximate form and position of the original object, could not be applied to the segment of the general hyperbola because, in that case, the determination of the coefficients of the least squares hyperbola fit leads to a nonlinear problem of great complexity. The solution of a coding problem for the general digital hyperbola segment is suggested in [8]. It is shown that the set consisting of the digital curves whose "original curves" are graphs of continuous functions, having at most two intersection points, pairwise, on a given interval, can be uniquely coded by five parameters. This result, specified to the set of the digital hyperbola segments, is briefed in Section 2 and the optimality of the proposed code is considered. One-to-one correspondence established between the digital hyperbola segments and their representations enables

G. Borgefors, I. Nyström, and G. Sanniti di Baja (Eds.): DGCI 2000, LNCS 1953, pp. 159–170, 2000.
© Springer-Verlag Berlin Heidelberg 2000

reconstruction of the digital hyperbola segment from its proposed code. The recovering algorithm is described in Section 3. As an illustration, some numerical results are presented. In Section 4 some concluding remarks are given.

As it is usual in the digital picture analysis, we assume through the paper that all the appearing points have positive coordinates. In other words, the origin is placed in the left-lower corner of the given picture.

2 Preliminaries

Consider a hyperbola h, in the Euclidean plane, given by $y = \frac{\alpha}{x-\beta} + \gamma$. It is digitized by digitizing method in which the first digital points (points with integer coordinates) below the curve are taken.

We will be dealing with digital hyperbola segment obtained by digitizing part of the hyperbola lying between lines $x = x_1$ and $x = x_2$ for some x_1 and x_2. Without loss of generality we can assume that x_1 and x_2 are integers. We also require that $\beta \notin [x_1, x_2]$; otherwise, the problem can be divided onto two independent digitization processes: the first one on the interval $[x_1, \lfloor\beta\rfloor]$ and the second one on the interval $[\lceil\beta\rceil, x_2]$. If $x_1 = k$ and $x_2 = k + n - 1$, then the associated set of digital points for the segment of the hyperbola h is the digital hyperbola segment $H_k(h, n)$, defined by

$$H_k(h,n) = \left\{ \left(i, \left\lfloor \frac{\alpha}{i-\beta} + \gamma \right\rfloor \right), \; i = k, k+1, \ldots, k+n-1 \right\} ,$$

where $\lfloor t \rfloor$ is the largest integer no bigger than t. Then n is the length of the digital hyperbola segment $H_k(h, n)$, i.e. the number of digital points of $H_k(h, n)$.

So, if we denote $y_i = \lfloor \frac{\alpha}{i-\beta} + \gamma \rfloor$, for $i = k, k+1, \ldots, k+n-1$ and $\beta \notin [k, k+n-1]$, then the digital hyperbola segment $H_k(h, n)$ of an arbitrary length, is coded by five parameters $(k, n, a(h), b(h), c(h))$, where:

- k is abscissa of the left endpoint of the digital hyperbola segment $H_k(h, n)$;
- n is the length of the digital hyperbola segment $H_k(h, n)$;
- $a(h) = \sum\limits_{i=k}^{k+n-1} y_i$;
- $b(h) = \sum\limits_{i=k}^{k+n-1} i \cdot y_i$;
- $c(h) = \sum\limits_{i=k}^{k+n-1} g(i) \cdot y_i$, where $g(x)$ is an arbitrary strictly convex or strictly concave function having a second derivative on $[k, k+n-1]$.

From the general result proved in [8], it follows that there do not exist two different digital hyperbola segments with the same proposed code. The proof of the statement, specified to the digital segments corresponding to the hyperbolas in the general form, is briefly reported here, since some of the details will be needed in the recovering procedure analysis. As it is already mentioned, the size of the observed grid is denoted by m. An auxiliary statement is used:

Lemma 1 *Let a sequence of nonnegative numbers, satisfying*

$$0 \le a_1 \le a_2 \le \ldots \le a_t < b_1 \le b_2 \le \ldots \le b_n < a_{t+1} \le \ldots \le a_n$$

and $a_1 + a_2 + \ldots + a_n = b_1 + b_2 + \ldots + b_n$, *be given, and let $g(x)$ be either a strictly convex or a strictly concave function, which has a second derivative on the interval $[a_1, a_n]$. Then*

$$g(a_1) + g(a_2) + \ldots + g(a_n) \ne g(b_1) + g(b_2) + \ldots + g(b_n) \ .$$

The proof of this statement is also given in [8].

Theorem 1 *Let $g(x)$ be a strictly convex (concave) function having a second derivative on the interval $[0, m-1]$ and let $H_r(l,p)$ and $H_s(t,q)$ be digital hyperbola segments, with $(r, p, a(l), b(l), c(l))$ and $(s, q, a(t), b(t), c(t))$ - codes, respectively. Then*

$$H_r(l,p) = H_s(t,q) \qquad \text{is equivalent to}$$
$$(r = s \ \wedge \ p = q \ \wedge \ a(l) = a(t) \ \wedge \ b(l) = b(t) \ \wedge \ c(l) = c(t)).$$

Proof. From the definition of the proposed code follows that

$$H_r(l,p) = H_s(t,q) \qquad \text{implies}$$
$$(r = s \ \wedge \ p = q \ \wedge \ a(l) = a(t) \ \wedge \ b(l) = b(t) \ \wedge \ c(l) = c(t)) \ .$$

The opposite direction will be proved by a contradiction.

For the given digital hyperbola segment $H_k(h,n)$, the set $D_k(h,n)$ containing all the digital points lying above the x-axis and below the hyperbola h, with abscissas belong to the interval $[k, k+n-1]$ is observed . Then the interpretation of the parameters $a(h)$, $b(h)$, $c(h)$ can be given as follows:

- $a(h) = \sum\limits_{i=k}^{k+n-1} y_i = \sum\limits_{(x,y)\in D_k(h,n)} 1$ is the number of the points of $D_k(h,n)$;
- $b(h) = \sum\limits_{i=k}^{k+n-1} i \cdot y_i = \sum\limits_{(x,y)\in D_k(h,n)} x$
  is the sum of the x-coordinates of the points of $D_k(h,n)$;
- $c(h) = \sum\limits_{i=k}^{k+n-1} g(i) \cdot y_i = \sum\limits_{(x,y)\in D_n(h,k)} g(x)$,
  is the sum of the values of the function $g(x)$ in all the points of $D_k(h,n)$.

From the assumption that two different digital hyperbola segments $H_r(l,p)$ and $H_s(t,q)$ have the same representation, we conclude that they are observed on the same interval $[r, r+p-1] = [s, s+q-1]$, and that the relations

$$D_r(l,p) \ne D_s(t,q) \tag{1}$$

$$a(l) = a(t) \quad , \quad b(l) = b(t) \quad , \quad c(l) = c(t) \tag{2}$$

are satisfied.

From (1) and the first relation in (2) follows that

$$\#(D_r(l,p) \setminus D_s(t,q)) = \#(D_s(t,q) \setminus D_r(l,p)) \neq 0 \tag{3}$$

which means that hyperbolas l and t must have an intersection point $x_0 \in [r, r+p-1]$. Assuming $b(l) = b(t)$, we have

$$\sum_{(x,y)\in D_r(l,p)\setminus D_s(t,q)} x = \sum_{(x,y)\in D_s(t,q)\setminus D_r(l,p)} x \ . \tag{4}$$

But the sets $D_r(l,p) \setminus D_s(t,q)$ and $D_s(t,q) \setminus D_r(l,p)$ can be separated by the straight line $x = x_0$ in such way that, let's say, $x < x_0$ for $x \in D_r(l,p) \setminus D_s(t,q)$ and $x > x_0$ for $x \in D_s(t,q) \setminus D_r(l,p)$ (2). Contradiction with (3) follows.

From (3) and the assumption $c(l) = c(t)$ follows that

$$\sum_{(x,y)\in D_r(l,p)\setminus D_s(t,q)} g(x) = \sum_{(x,y)\in D_s(t,q)\setminus D_r(l,p)} g(x) \ . \tag{5}$$

If we suppose that l and t have two intersection points $x_1, x_2 \in [r, r+p-1]$ $(x_1 < x_2)$, the sets $D_r(l,p) \setminus D_s(t,q)$ and $D_s(t,q) \setminus D_r(l,p)$ are separated by straight lines, $x = x_1$ and $x = x_2$ in such way that, let's say, $x < x_1$ or $x > x_2$ for $x \in D_r(l,p) \setminus D_s(t,q)$ and $x_1 < x < x_2$ for $x \in D_s(t,q) \setminus D_r(l,p)$ (2).

In other words, taking into account (3) and (4), we have that the abscissas of the points of $D_r(l,p) \setminus D_s(t,q)$ and $D_s(t,q) \setminus D_r(l,p)$, are divided into two sets in such way that the conditions of the Lemma 1 are satisfied. But the result following from Lemma 1 is in contradiction to (5).

Since two hyperbola arcs can have at most two intersection points on the observed interval, the proof is finished. □

Since for the appropriate choice of $g(x)$ all of its parameters are integers, this code is optimal, which is proved by the next lemma.

Lemma 2 *The $(k, n, a(h), b(h), c(h))$-code requires an asymptotically optimal number of bits.*

Proof. For coding numbers k, n, $a(h)$, $b(h)$ and $c(h)$ (for $g(x)$ chosen to be, for example, any polynomial function) $\mathcal{O}(\log m)$ bits are required. Since there are at least m different digital hyperbola segments which can be inscribed into the $m \times m$ integer grid, at least $\log m$ bits are required for representing them. □

Remark 1.: If $l : y = \frac{\alpha_l}{x-\beta_l} + \gamma_l$ and $t : y = \frac{\alpha_t}{x-\beta_t} + \gamma_t$ are two hyperbolas corresponding to different digital segments on the interval $[r, r+p-1]$, from the proofs of Lemma 1 (conditions assumed) and Theorem 1 it follows:

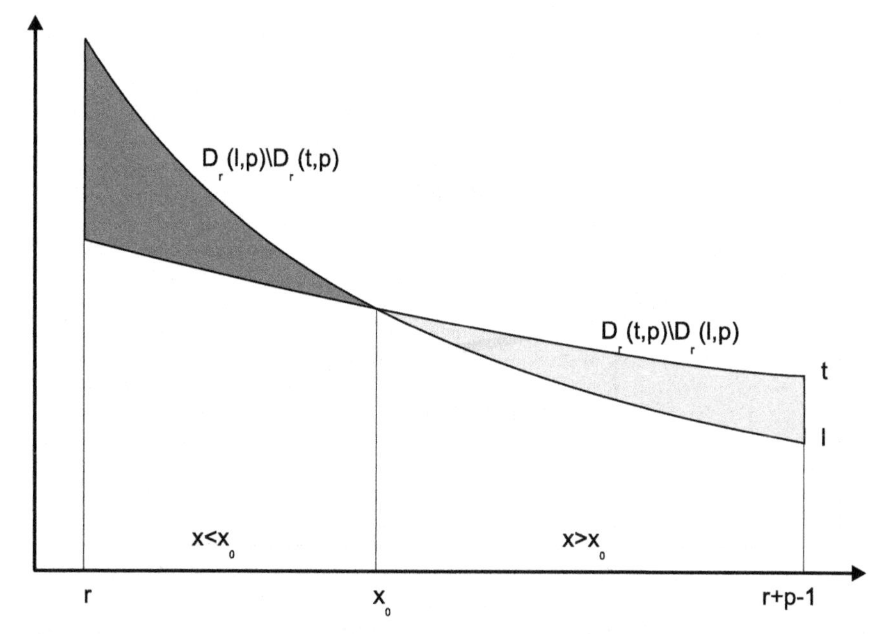

Fig. 1. l and t have one intersection point $x_0 \in [r, r+p-1]$, so $D_r(l,p) \setminus D_r(t,p)$ and $D_r(t,p) \setminus D_r(l,p)$ are separated by $x = x_0$. Assuming $a(t) = a(l)$, $b(t) > b(l)$

• If l and t have one intersection point on $[r, r+p-1]$ and $a(l) = a(t)$, then

$$\frac{\alpha_l}{r-\beta_l} + \gamma_l > \frac{\alpha_t}{r-\beta_t} + \gamma_t \quad \Leftrightarrow \quad b(l) < b(t)$$

• If l and t have two intersection points on $[r, r+p-1]$ and $a(l) = a(t)$ and $b(l) = b(t)$ hold, then

$$\frac{\alpha_l}{r-\beta_l} + \gamma_l > \frac{\alpha_t}{r-\beta_t} + \gamma_t \quad \Leftrightarrow \quad c(l) > c(t) \quad \text{for} \quad g''(x) > 0$$

$$\frac{\alpha_l}{r-\beta_l} + \gamma_l > \frac{\alpha_t}{r-\beta_t} + \gamma_t \quad \Leftrightarrow \quad c(l) < c(t) \quad \text{for} \quad g''(x) < 0.$$

3 Recovering Algorithm

The one-to-one correspondence between the set of the digital hyperbola segments and the set of their proposed representations enables reconstruction of the segment from its code. In this section we describe a recovering algorithm.

We assume that $\alpha > 0$ and $\beta < k$. Other cases are treated in similar way.

First, we describe a procedure $Set\_a(k, n, A, B, a, H)$ which, for a given interval $[k, k+n-1]$, parameters A, B and a, finds a hyperbola $H : y = \frac{A}{x-B} + C$, such that $a = a(H)$

Procedure $Set\_a(k, n, A, B, a, H)$

Input: The numbers k, n, A, B, a

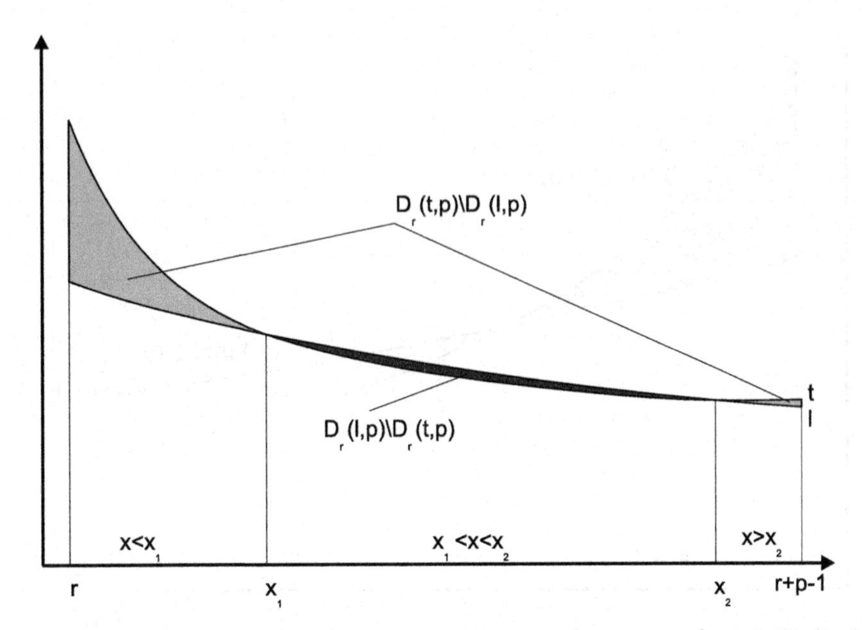

Fig. 2. l and t have two intersection points $x_1, x_2 \in [r, r+p-1]$, and $D_r(l,p) \setminus D_r(t,p)$ and $D_r(t,p) \setminus D_r(l,p)$ are separated by $x = x_1$ and $x = x_2$. Assuming $a(l) = a(t)$ and $b(l) = b(t)$, $c(t) > c(l)$ for $g''(x) > 0$ and $c(t) < c(l)$ for $g''(x) < 0$

Output: Hyperbola H of the form $y = \frac{A}{x-B} + C$, satisfying $a(H) = a$.

Step 1: Draw a hyperbola segment $h_{temp} : y = \frac{A}{x-B}$ and compute $a(h_{temp})$.

Step 2: Translate h_{temp} by $\lfloor \frac{a-a(h_{temp})}{n} \rfloor$ in the vertical direction (upward if $a - a(h_{temp}) > 0$ and downward otherwise) and denote obtained hyperbola by h'. Compute $a(h')$. (In this way, $|a - a(h')| < n$ is preserved).

Step 3: Let $(k+i-1, y_i)$, $i = 1, 2, \ldots, n$ be the digital points obtained by digitizing the hyperbola h'. Find the numbers d_i, $i = 1, 2, \ldots, n$, where d_i is the vertical distance between the point $(k+i-1, y_i+1)$, $i = 1, 2, \ldots n$ and the hyperbola h'.

Step 4: Find the $(a - a(h'))$-th smallest number among the numbers d_i. (Note: Any algorithm that finds the r-th median of the set can be used - for example see [1] which is linear time algorithm.)

Step 5: Translate h' to its final position H by $(a - a(h'))$-th smallest distance, in the vertical direction.

Lemma 3 *The time complexity of the procedure Set_a(k, n, A, B, a, H) is $O(m)$.*

We continue with the lemma which, for the given coefficient β, determines the interval $I(\alpha)$ containing coefficient α:

Lemma 4 *If $H_k(h,n) = \{(i, y_i),\ i = k, \dots k+n-1\}$ is the digital hyperbola segment obtained by digitizing the hyperbola $h : y = \frac{\alpha}{x-\beta} + \gamma$, then*

$$\alpha \in \left(\frac{2b(h)-a(h)\cdot(2k+n-1)+n\cdot(2k+n-1)}{\sum\limits_{i=1}^{n} \frac{2i-n-1}{k+i-1-\beta}},\ \frac{2b(h)-a(h)\cdot(2k+n-1)-n\cdot(2k+n-1)}{\sum\limits_{i=1}^{n} \frac{2i-n-1}{k+i-1-\beta}} \right).$$

Proof. According to the digitizing method, for $i = 1, 2, \dots, n$ it follows:

$$\frac{\alpha}{k+i-1-\beta} + \gamma \in [y_{k+i-1}, y_{k+i-1}+1)$$

and

$$\frac{\alpha\cdot(k+i-1)}{k+i-1-\beta} + (k+i-1)\cdot\gamma \in [(k+i-1)\cdot y_{k+i-1}, (k+i-1)\cdot (y_{k+i-1}+1)).$$

This implies the following system of inequalities:

$$\alpha \sum_{i=1}^{n} \frac{1}{k+i-1-\beta} + n\gamma \in [a(h),\ a(h)+n)$$

$$\alpha \sum_{i=1}^{n} \frac{k+i-1}{k+i-1-\beta} + \gamma \sum_{i=1}^{n} (k+i-1) \in \left[b(h),\ b(h)+nk+\frac{n(n-1)}{2}\right).$$

By solving it, the statement follows. (Note: $\sum\limits_{i=1}^{n} \frac{2i-n-1}{k+i-1-\beta} < 0.$) □

Remark 2. From above, it is not difficult to conclude that, for the different values of α, while keeping β and $a(h)$ fixed, the parameter $b(h)$ can take all the integer values between 1 and $n(2k+n-1)$.

The following lemma permits a binary searching over the coefficient α, while β is fixed:

Lemma 5 *Let H' and H'' be hyperbolas corresponding to different digital segments, obtained by calling Set_a(k,n,A',B,a,H') and Set_a(k,n,A'',B,a,H''). Then $A' < A''$ is equivalent to $b(H') > b(H'')$.*

Proof. Let H' and H'' be given, respectively, by the equations

$y = \frac{A'}{x-B} + C'$ and $y = \frac{A'+\triangle A'}{x-B} + C''$, where $A' + \triangle A' = A''$.

From $H_k(H',n) \neq H_k(H'',n)$ and $a(H') = a(H'') = a$ we conclude that the (only) intersection point x_0 of H' and H'' is in $x_0 \in (k, k+n-1)$, and that the relation $(C'-C'')(x_0-B) = \triangle A'$ holds. Since $B < k$ and $x_0 > k$, we have $(x_0-B) > 0$. According to the assumption that $A' > 0$ and $A'' > 0$, the condition $b(H') > b(H'')$ is equivalent to $C' > C''$ (see *Remark 1.* and Fig.1).

So, we have $b(H') > b(H'') \Leftrightarrow \triangle A' > 0 \Leftrightarrow A' < A''$. □

There is a possibility that the hyperbola H of the form $y = \frac{A}{x-B} + C$, satisfying $a = a(H)$ and $b = b(H)$, for the chosen B is not found, which means that $|b(H)-b|$ is unreachable for any A. It could happen only if the parameter β, locating the vertical asymptote of the original (digitized) curve is very close to (far from) k (the beginning of the segment). So, in that case, in order to increase (decrease) $b(H)$, the chosen B should be increased (decreased).

Now, we give the procedure which, for a given parameter B, interval $[k, k+n-1]$ and numbers a and b, finds a hyperbola H of the form $y = \frac{A}{x-B} + C$, such that $a(H) = a$ and $b(H) = b$. (We mention again that m is the grid size).

Procedure *Set_ab*(k, n, B, a, b, H)

Input: The numbers k, n, B, a, b.
Output: Hyperbola H of the form $y = \frac{A}{x-B} + C$, with $a(H) = a$ and $b(H) = b$.
Step 1: Set $I(\alpha) = \left(\frac{2b - a\cdot(2k+n-1) + n\cdot(2k+n-1)}{\sum_{i=1}^{n} \frac{2i-n-1}{k+i-1-B}}, \frac{2b - a\cdot(2k+n-1) - n\cdot(2k+n-1)}{\sum_{i=1}^{n} \frac{2i-n-1}{k+i-1-B}} \right)$;
Choose a coefficient $A \in I(\alpha)$;
Step 2: Set_a(k, n, A, B, a, H);
Step 2: WHILE $(b(H)) \neq b$ and the weight of $I(\alpha) > \frac{1}{16m^8})$ DO
{chose the next $A \in I(\alpha)$ by using binary searching; (* Lemma 5 *)
Set_a(k, n, A, B, a, H) } ;
Step 3: IF $(b(H) \neq b)$
(* required accuracy is reached without finding appropriate A *)
{increase B if $b > b(H)$ or decrease B if $b < b(H)$;
Set_ab(k, n, B, a, b, H) }.

As a result of executing the Procedure *Set_ab*(k, n, B, a, b, H), we have the segment $\{(i, y_i),\ i = k, \ldots, k+n-1\}$ of the hyperbola $H : y = \frac{A}{x-B} + C$ for which $a(H) = a$ and $b(H) = b$. If $c(H) = c$, the recovering process is finished. If not, we need the following result which permits binary searching over both of the coefficients α and β:

Lemma 6 *Let H' and H'' be defined by $y = \frac{A'}{x-B'} + C'$ and $y = \frac{A'+\triangle A'}{x-(B'+\triangle B')} + C''$, respectively and correspond to different digital segments, obtained by calling Set_ab(k, n, B', a, b, H') and Set_ab(k, n, B'', a, b, H'').*

Then $c(H') > c(H'') \Leftrightarrow (A' < A''$ *and* $B' > B'')$ *if* $g''(x) > 0$,
and $c(H') > c(H'') \Leftrightarrow (A' > A''$ *and* $B' < B'')$ *if* $g''(x) < 0$.

Proof. From
$H_k(H', n) \neq H_k(H'', n), \qquad a(H') = a(H'') = a, \qquad b(H') = b(H'') = b,$
we conclude that H' and H'' have two intersection points, $x_1, x_2 \in (k, k+n-1)$, satisfying $(C' - C'')(x - B')^2 - (\triangle B'(C' - C'') + \triangle A')(x - B') - A'\triangle B' = 0$. According to the *Remark 1*, (see Fig.2) we have $c(H') > c(H'') \Leftrightarrow C' > C''$, if $g''(x) > 0$ on $[k, k+n-1]$, and
$c(H') > c(H'') \Leftrightarrow C' < C''$, if $g''(x) < 0$ on $[k, k+n-1]$.
Both of the intersection points of H' and H'' belong to $(k, k+n-1)$, so

$$(x_1 - B') \cdot (x_2 - B') > 0 \Leftrightarrow \quad -\frac{A'\triangle B'}{C'-C''} > 0,$$
$$(x_1 - B') + (x_2 - B') > 0 \Leftrightarrow \triangle B' + \frac{\triangle A'}{C'-C''} > 0,$$

which implies $\triangle B' < 0$ and $\triangle A' > 0$, if $g''(x) > 0$
and $\triangle B' > 0$ and $\triangle A' < 0$, if $g''(x) < 0$. □

Finally, here is the recovering algorithm for the digital hyperbola segment from its proposed (k, n, a, b, c)-code:

Algorithm *HYPERBOLA*(k, n, a, b, c, H)

Input: Numbers k (the abscissa of the left endpoint of the segment), n (length of the segment), a, b and c.
Output: Hyperbola $H : y = \frac{A}{x-B} + C$, which satisfies $a(H) = a$, $b(H) = b$ and $c(H) = c$, i.e. which is corresponded to the coded digital segment.
Step 1: Choose the coefficient $B < k$;
Set_ab(k, n, B, a, b, H) ;
Step 2: IF $(c(H) \neq c)$
{ choose the next B by binary searching; (* Lemma 6 *);
Set_ab(k, n, B, a, b, H) } .

The procedure will be terminated when the solution (one of the hyperbolas corresponding to the given code) is found. If the initial lower bound for B is given, the algorithm can be used for solving the recognition problem as well, since it will be shown that the solution, if exists, should be reached before the width of the interval containing β becomes less then $\frac{1}{4m^4}$.

In the end, complexity of the algorithm *HYPERBOLA* will be discussed.

Lemma 7 *Every digital hyperbola segment can be obtained as the digitization of the hyperbola of the form* $y = \frac{\alpha-\varepsilon_\alpha}{x-(\beta-\varepsilon_\beta)} + \gamma_1$, *where* α *and* β *are the coefficients of the hyperbola* $y = \frac{\alpha}{x-\beta} + \gamma$ *passing through three digital points, and* ε_α *and* ε_β *are positive, small enough numbers.*

Proof. Consider a hyperbola $h : y = \frac{a}{x-b} + c$.

By translating h downward until it reaches the nearest digital point, (x_γ, y_γ), we get hyperbola $h_1 : y = \frac{a}{x-b} + \gamma$. By changing parameter b till the next nearest point, (x_β, y_β), is reached, we get hyperbola $h_2 : y = \frac{a}{x-\beta} + \gamma$. Finally, changing the parameter a and reaching the third digital point (x_α, y_α) leads to the hyperbola $h_3 : y = \frac{\alpha}{x-\beta} + \gamma$.

So, h_3 passes through three digital points and $y = \frac{\alpha-\varepsilon_\alpha}{x-(\beta-\varepsilon_\beta)} + \gamma_1$ is determined by the points $(x_\alpha \pm \varepsilon, y_\alpha \pm \varepsilon)$, $(x_\beta \pm \varepsilon, y_\beta \pm \varepsilon)$ and $(x_\gamma \pm \varepsilon, y_\gamma \pm \varepsilon)$, where ε is positive, small enough number, and sign "+" is taken for those of the points (x_α, y_α), (x_β, y_β), (x_γ, y_γ) which are elements of the digital segment, else "-" is taken. Obviously, $y = \frac{\alpha-\varepsilon_\alpha}{x-(\beta-\varepsilon_\beta)} + \gamma_1$ and h have the same digitization. □

Theorem 2 *Any digital hyperbola segment, corresponding to the hyperbola of the form* $y = \frac{\alpha}{x-\beta} + \gamma$ *and presented on the* $m \times m$ *integer grid, represented by its* (k, n, a, b, c)*-code, can be recovered in* $\mathcal{O}(m \cdot (\log(m + |\beta|))^2)$ *time.*

Proof. The coefficients α and β of the hyperbola $y = \frac{\alpha}{x-\beta} + \gamma$ passing through three given points (x_1, y_1), (x_2, y_2), (x_3, y_3) are determined as follows:

$$\alpha = \frac{(x_1 - x_2)(x_1 - x_3)(x_3 - x_2)(y_2 - y_1)(y_2 - y_3)(y_3 - y_1)}{((y_2 - y_1)(x_1 - x_3) - (y_3 - y_1)(x_1 - x_2))^2},$$

$$\beta = \frac{x_2(y_2 - y_1)(x_1 - x_3) - x_3(y_3 - y_1)(x_1 - x_2)}{(y_2 - y_1)(x_1 - x_3) - (y_3 - y_1)(x_1 - x_2)}.$$

So, for the difference between different coefficients α_1 and α_2, and analogously, β_1 and β_2, of the hyperbolas $y = \frac{\alpha_1}{x-\beta_1} + \gamma_1$ and $y = \frac{\alpha_2}{x-\beta_2} + \gamma_2$, each passing through three points of a grid of size $m \times m$, we have

$$|\alpha_1 - \alpha_2| > \frac{1}{16m^8}, \quad \text{and} \quad |\beta_1 - \beta_2| > \frac{1}{4m^4}.$$

The width of $I(\alpha)$ is (see Lemma 4)

$$\frac{2n(2k+n-1)}{\sum_{i=1}^{n} \frac{n+1-2i}{k+i-1-\beta}} < 4(2k+n-1)(k+n-1-\beta)^2,$$

since
$$\sum_{i=1}^{n} \frac{n+1-2i}{k+i-1-\beta} > \frac{n}{2(k+n-1-\beta)^2}.$$

Taking into account that k and n are at most equal to m, we have $I(\alpha) < 12m(2m+|\beta|)^2$, which implies that $\mathcal{O}(\log(m+|\beta|))$ iterations will be needed to obtain $I(\alpha)$ less then $\frac{1}{16m^8}$.

For locating β, $\mathcal{O}(\log(m+|\beta|))$ iterations will be needed, too; $I(\beta)$ will be less then $\frac{1}{4m^4}$ after that. With Lemma 3, the statement follows. □

Table 1. contains some numerical examples.

Table 1. The digital segment $H_k(h,n)$, obtained by digitizing $h : y = \frac{\alpha}{x-\beta} + \gamma$ on the interval $[k, k+n-1]$, is coded by (k,n,a,b,c)-code, where g is chosen to be x^2. After applying the recovering procedure, the parameters A, B, C of the hyperbola $h_{rec} : y = \frac{A}{x-B} + C$, having the same corresponding digital segment on the observed interval, are obtained

α	β	γ	(k,n,a,b,c)	A	B	C
5.12	2.88	4.31	(3,7,80,350,1842)	5.278417477	2.875	4.223145
			(4,42,179,4182,125968)	4.529032519	3	4.393433
			(45,27,108,6264,369864)	0.890407522	44	4
17.19	-0.53	2.77	(1,4,36,77,209)	10.758620689	0	4
			(10,22,70,1399,30927)	0.744398733	9	3.83251
			(25,67,184,10247,633817)	0.702846342	24	2.98608

4 Comments and Conclusion

Our research has been motivated by the fact that the conic sections are the most often appearing digital curves in practice of computer vision and image processing. For all of them the different solutions of the coding problem are known, and for most of them either a recovering procedure or the parameter estimation of the coded digital curve can be found in the literature. But as far as the digital segment corresponding to the general hyperbola is considered, only the problem of representation can be solved, by applying the general result developed in [8]. The proposed code consists of five parameters. Four of them are strictly determined, while the fifth can be chosen according to the practical reasons, since it depends on an arbitrary strictly convex (concave) function $g(x)$, having the second derivative on the observed interval. In the previous sections the recovering procedure for the digital hyperbola segments, represented by that code, is described. If the particular function $g(x)$ is noted, the recovering algorithm can be applied for any choice of $g(x)$. In practice, if $y = \frac{\alpha}{x-\beta} + \gamma$ is coded on the interval $[k, k+n-1]$, the information about the sign of α and $(k-\beta)$ is supposed to be included in the code, as well; the conclusion analogous to those of Lemma 5 and Lemma 6, permitting binary searching, can be derived for any of the cases. The result is obtained in $\mathcal{O}(m \cdot (\log(m + |\beta|))^2)$ time.

The main contribution of the paper is that it gives a complete and efficient solution to the representation and the reconstruction problem connected with one class of the digital images. The proposed code is simple, uniquely determined, recoverable and optimal, while the recovering algorithm enables the exact and fast reconstruction of the digital segment. For comparing, it might be noted that the algorithm for the exact reconstruction of the digital ellipse (circle) does not exist, even though the efficient coding scheme (based on the similar concept as proposed here) enables very efficient estimation of the parameters of the original object [10]. So the idea for the further research could be to find a constant time approximate reconstruction for the relevant parameters of the hyperbola from the code of its digital segment.

References

1. Blum, M., Floyd, R. W., Pratt, V., Rivest, R. L., Tarjan, R. E.: Time bounds for selection. J. Comput. System Sci. 7(4). (1973) 448-461 164
2. Dorst, L., Smeulders, A. W. M.: Discrete representation of straight lines. IEEE Trans. Pattern Analysis and Machine Intelligence, Vol. 6. (1984) 450-463 159
3. Kim, C. E.: Digital disks. IEEE Trans. Pattern Analysis and Machine Intelligence, Vol. 6. (1984) 372-374
4. Lindenbaum, M., Koplowitz, J.: A new parametrization of digital straight lines. IEEE Trans. Pattern Analysis and Machine Intelligence, Vol. 13. No.4. (1991) 847-852
5. Melter, R. A., Stojmenović, I., Žunić, J.: A new characterization of digital lines by least square fits. Pattern Recognition Letters, Vol. 14. (1993) 83-88

6. Woring, M., Smeulders, A. W. M.: Digitized Circular Arcs: Characterization and Parameter Estimation. IEEE Trans. Pattern Analysis and Machine Intelligence, Vol. 17. (1995) 587-597
7. Žunić, J., Koplowitz, J.: A representation of digital parabolas by least square fits. SPIE Proc., Vol. 2356. (1994) 71-78 159
8. Žunić, J.: A coding scheme for certain sets of digital curves. Pattern Recognition Letters, Vol. 16. (1995) 97-104 159, 160, 161, 169
9. Žunić, J.: A Representation of Digital Hyperbolas $y = \frac{1}{x}\alpha + \beta$. Pattern Recognition Letters, Vol. 17. (1996) 975-983 159
10. Žunić J., Sladoje, N.: Efficiency of Characterizing Ellipses and Ellipsoids by Discrete Moments. IEEE Trans. Pattern Analysis and Machine Intelligence, Vol. 22. (2000) 407-414 159, 169

Determining Visible Points in a Three-Dimensional Discrete Space

Grit Thürmer[1,2*], Arnault Pousset[2], and Achille J.-P. Braquelaire[2]

[1] CoGVis/MMC – Computer Graphics, Visualization, Man-Machine Communication Group, Faculty of Media, Bauhaus-University Weimar,
99421 Weimar, Germany
thuermer@medien.uni-weimar.de

[2] LaBRI – Laboratoire Bordelais de Recherche en Informatique – UMR 5800, University Bordeaux 1
351, cours de la Libération, 33405 Talence, France
{pousset,braquelaire}@labri.u-bordeaux.fr

Abstract. A method is proposed which computes the visible points of surfaces in a 3-dimensional discrete space. The occlusion of surface points of an object by other object points is determined by shooting a discrete ray from each surface point towards the center of projection considering the intersection of the ray with other object points. Since the projection of points onto the viewing plane is done by a continuous mapping, additionally to the discrete ray, the location of the continuous projection ray is examined regarding its location to the surface points that are intersected by the discrete ray.

1 Introduction

The growing interest of computer graphics in the three-dimensional discrete space $\mathbb{Z}^3$ has led to new application fields of volume data: e.g. virtual reality in medicine [19], volume-based interactive design and sculpturing [16,11]. For such application fields, synthetic objects, i.e. geometrically defined objects in $\mathbb{Z}^3$ or data sets obtained by rastering [1,15] geometric descriptions of objects given in $\mathbb{R}^3$, are generated and the boundaries of these objects are required to be visualised.

Three principal approaches for rendering discrete objects can be distinguished: *backward* or *image-order projection*, *forward* or *object-order projection* techniques, and *hybrid techniques* [14,10,9] that combine advantages of both the backward and forward projection methods. Backward projection algorithms traverse the pixels and solve the visibility problem for each pixel by casting a ray from the viewing point through each pixel of the image into the data space. This group includes *ray-casting* [5,6,3] and *ray-tracing* [18] techniques, which have

* This work has been supported by a postdoctoral grant of the DFG (Deutsche Forschungsgemeinschaft).

G. Borgefors, I. Nyström, and G. Sanniti di Baja (Eds.): DGCI 2000, LNCS 1953, pp. 171–182, 2000.
© Springer-Verlag Berlin Heidelberg 2000

been specifically developed for the rendering of volumes. Forward projection algorithms, e.g. *splatting* [17,4], traverse the three-dimensional discrete scene and project its components onto the viewing plane.

In contrast to splatting techniques, where a point of $\mathbb{Z}^3$ is represented by a kernel function in the viewing plane, we assume that a point of $\mathbb{Z}^3$ is mapped to the viewing plane such that its image is a point in $\mathbb{R}^2$. Consequently, generally the image of a discrete surface in the viewing plane is a set of scattered of points. Moreover, assuming a continuous viewing plane requires a method for the determination of visible points, which provides results independently from the resolution of the final image. Therefore, rendering methods applying the z-buffer algorithm are not suitable to solve the visibility problem. Instead, a technique is needed which determines the visible points of a scene in object space, i.e. in $\mathbb{Z}^3$. A similar problem has to be solved in discrete ray-tracing [2] when it has to be determined if a point is in shadow, i.e. if the point cannot be seen from the light source. After the visible points are mapped onto the viewing plane, the final image could be obtained by applying a technique, e.g., as proposed in [12].

In this paper, a method is proposed which determines the visible points of closed surfaces in $\mathbb{Z}^3$. Each surface forms the boundary of a discrete object. In a first step, front facing points are computed in the common way, comparing the normal vectors at the surface points with the viewing direction. The computation of occluded front facing points is performed utilising the idea of forward ray-casting: a discrete ray is shot from each surface point towards the centre of projection considering the intersection with other non-empty points, i.e. points which belong to objects of the scene. Since the actual projection of surface points onto the viewing plane is done by a continuous mapping, additionally to the discrete ray, the location of the continuous projection ray is examined regarding its location to the surface points that are intersected by the discrete ray.

The paper is organised as follows: Section 2 states the definitions used throughout the paper. In Sect. 3, at first the problem is examined in detail. Afterwards, a solution is proposed in 2D which is then extended to 3D. Experimental results are presented and discussed in Sect. 4. Finally, Sect. 5 summarises the paper.

2 Definitions

The three-dimensional discrete space is constituted by $\mathbb{Z}^3$, that is the 3D array of points with integer coordinates in the Cartesian coordinate system. Two points $p(x_p, y_p, z_p)$ and $q(x_q, y_q, z_q)$ of $\mathbb{Z}^3$ where $(\mid x_p - x_q \mid \leq 1) \wedge (\mid y_p - y_q \mid \leq 1) \wedge (\mid z_p - z_q \mid \leq 1)$, are said to be *6-adjacent* if $\mid x_p - x_q \mid + \mid y_p - y_q \mid + \mid z_p - z_q \mid = 1$, *18-adjacent* if $0 < \mid x_p - x_q \mid + \mid y_p - y_q \mid + \mid z_p - z_q \mid \leq 2$, and *26-adjacent* if $0 < \mid x_p - x_q \mid + \mid y_p - y_q \mid + \mid z_p - z_q \mid \leq 3$. Points k-adjacent to p, where $k \in \{6, 18, 26\}$, are called *k-neighbours* of p. A *k-component* of a set A is a maximal subset of A in which between each pair of points p, q exists a sequence of distinct points $P = \{p = p_0, p_1, ..., p_n = q\}$ of A whereby two consecutive points p_i and p_{i+1} with $0 \leq i < n$ along the sequence are k-adjacent.

The surface $S \subset \mathbb{Z}^3$ to be rendered is assumed as a closed surface that forms the boundary of a finite object $O \subset \mathbb{Z}^3$, whereby O is a 6-connected component. The points of O are called *object points*. The surface $S \subset O$ is constituted of all points of O that are 6-adjacent to some point of $\overline{O} = \mathbb{Z}^3 - O$. Each point of S is also required to be 26-adjacent to some point of $O - S$. Note that S may consist of more than one component. The points of S are denoted as *surface points*. In contrast, the points in $O - S$ are called *inside points* and the points of $\overline{O}$ are named *empty points*.

3 Determination of Visible Surface Points

In general, a point on a closed surface is *visible* only if it is front facing, i.e. the surface in this point is oriented towards the observer, and the point is not occluded by another point.

The computation of front facing points is rather trivial and can be done as described in Sect. 3.1. In contrast, the determination of occluded points in discrete space needs further investigations. Section 3.2 describes the basic difficulties of the solution of the occlusion problem, which arise while employing the idea of discrete forward ray casting. In Sect. 3.3, an approach is introduced to determine occluded points in 2D space. Then, in Sect. 3.4 this approach is extended to 3D.

3.1 Front Facing Surface Points

The culling of back facing surface points in discrete space is done in the same way as for polygons: the cosine of the angle δ between the normal vector at a surface point and the vector representing the viewing direction is determined. If $cos(\delta) \leq 0$ the point is back facing, otherwise it is front facing.

Clearly, a normal vector associated with each point of the discrete surface is essential to cull back facing points. The normals are required to represent the local surface geometry. The determination of normal vectors at discrete surfaces is an active field of research. For example, the method introduced in [13] provides results that are suitable for the purpose of this work. However, to deal with the problem of visibility independently of the computation of the normal vectors, one could also associate the discrete surface points with the true normals of the underlying continuous surface.

3.2 Forward Ray-Casting

To determine the surface points that are not occluded by other points, the idea of forward ray-casting is utilised: shoot a discrete ray, denoted as *discrete viewing ray*, from each front facing surface point along the viewing direction and check if the ray intersects any other object point. A discrete viewing ray is the rasterization of the straight line segment between a surface point and the centre of projection.

The basic problems, which arise for the discrete viewing rays and are matter of this section, address the features of a ray in order to define a point as occluded and the connectivity of the rays. There are two basic approaches possible to solve the first problem: a point is not occluded only if its discrete viewing ray does not intersect any other object point, or the ray is allowed to be tangent to the surface, i.e. the ray may intersect other surface points but no inside point. These two approaches and their related problems are discussed below.

Viewing Rays do not Intersect any Object Point. The following assumption would be the simplest approach when solving the visibility problem in $\mathbb{Z}^3$ by forward ray-casting: a point is not occluded if its discrete viewing ray does not intersect any object point. However, this assumption is too strict. Frequently, the viewing ray of a surface point close to the contour of a surface is almost tangent to the surface, i.e. the ray intersects surface points but no inside point. Examples for this fact are shown in Fig. 1(a) and (b) for orthographic and perspective projection, respectively. For simplification, these examples are given in 2D. Clearly, the viewing rays shown in the Fig. 1 intersect other surface points. The condition of non-occlusion, as stated above, would lead to a missing of such points. A similar problem has been reported by Delfosse et al. [2] for shadow rays in discrete ray-tracing. They solve the problem by considering the real boundary of the underlying continuous object. This approach is not suitable for our work since we consider only the discrete objects, without any assumption on the underlying continuous objects. Consequently, we need a further specification of the condition for non-occluded points to allow viewing rays to be tangent to the surface, so that the discrete viewing ray of a visible point may intersect other surface points.

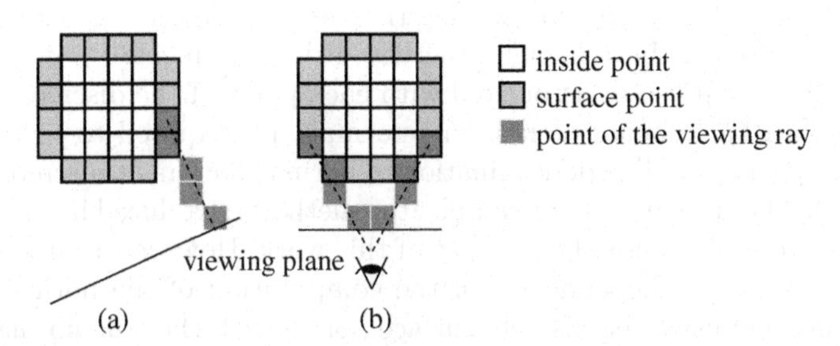

Fig. 1. Discrete viewing rays intersecting a surface point for (a) orthographic projection and (b) perspective projections

Connectivity of Viewing Rays. If there is no further restriction on the topology of the surface, i.e. S is defined as in Sect. 2, the discrete viewing rays have to be 6-connected to avoid a traversal of the object by the rays while intersecting only surface points. In such cases, the viewing rays are not tangent to the surface. An example in 2D is shown in Fig. 2(a). For the points p and q, a 26-connected viewing ray would pass through the object, hitting only surface points.

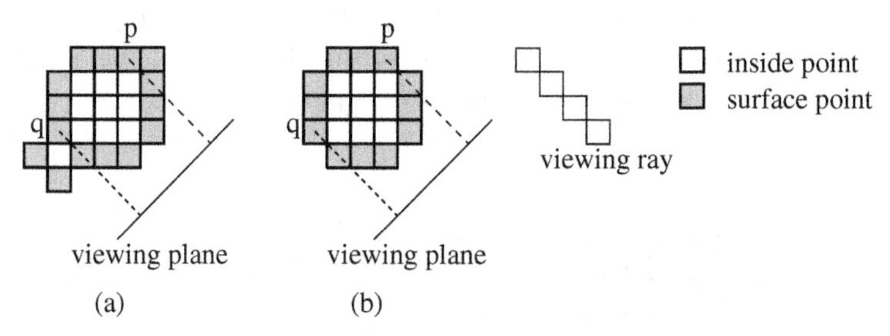

Fig. 2. 26-connected viewing rays and (a) a surface with no restriction and (b) a surface satisfying the definition from discrete topology

The generation of 26-connected rays is twice as fast than that of 6-connected rays since a 6-connected ray consists of about twice as many points as the 26-connected ray. For optimisation purposes, a technique proposed in [18] can be applied to speed up the usage of 6-connected rays: to traverse the empty space between the objects of a scene quickly, the rays are 26-connected until they come close to an object. Then the connectivity of the rays changes to 6-connectivity. Nevertheless, 26-connected rays would be preferred for practical applications because of performance reasons.

Assume S satisfies a surface definition as known from discrete topology [8,7]. Then S can be viewed as the discrete analog of a closed two-manifold surface, and S is the minimal set which separates $\mathbb{Z}^3 - S$ into two non-empty 6-connected sets. Thus, each point of S is 6-connected to some point of $O - S$ and to some point of $\overline{O}$. Experiments have shown, that 26-connected viewing rays are sufficient if S has these properties. Fig. 2(b) shows an example for this case in 2D. It is beyond the scope of this paper to proof this fact. However, this will be a matter of future work. In Sect. 4, results are shown for both the cases 6-connected and 26-connected viewing rays.

Viewing Rays may Intersect other Surface Points. Allowing the discrete viewing rays to intersect surface points but no inside point may cause problems. It cannot be ensured that the projection of the visible points leads to a correct image. The problem is illustrated in Fig. 3. Consider a convex discrete object O, i.e. O is the discrete representation of a convex continuous object $\tilde{O}$, located in the viewing space with coordinate axes x and y. The image of O, denoted O', is the projection of O onto the x-axis. If viewing rays of visible points may hit any surface point, the point $p \in O$ would be visible. Since p bounds the set of visible points of O shown in light grey in Fig. 3(a), one would expect that the image p' of p bounds O', i.e. it should be the point of O' with the smallest x-value in Fig. 3(a). This case is illustrated for the continuous object $\tilde{O}$ in Fig. 3(b). However for the discrete object in Fig. 3(a), after the projection of the visible points the image q' of the visible point q forms the boundary of O' instead. This problem arises because discrete rays are applied to determine the visible points in $\mathbb{Z}^3$ and, in contrast, a continuous mapping is used for the projection of the visible points onto the viewing plane. Therefore, a criterion is needed which ensures that points like p are excluded from the set of visible points while allowing viewing rays to intersect other surface points.

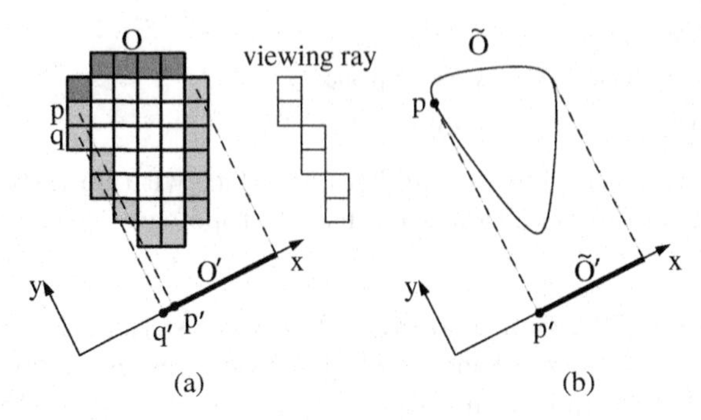

Fig. 3. Problem arising after mapping visible points onto the viewing plane: (a) for the discrete case, the projection of the point p does not form an outline of the image as it does in (b) the corresponding continuous case

3.3 Visibility in 2D

To solve the problem described in the previous section, a condition is introduced to exclude points from the set of visible surface points whose projection would lead to artefacts in the rendered image. This problem may arise only for surface points whose discrete viewing rays intersect other surface points.

At first, the problem is examined in 2D. Consider a front facing surface point p associated with a discrete viewing ray r_p, which intersects only empty points or other surface points. In case r_p intersects any inside point, p is not visible. The point p is not occluded if for each surface point q along r_p the continuous projection ray $\tilde{r}_p$ of p does not intersect O. Note that $\tilde{r}_p$ is the continuous representation of r_p. If $\tilde{r}_p$ is referred to as a vector subsequently then it is considered as the normalised vector representing the projection ray of p pointing towards the centre of projection.

Assume a continuous surface patch S_q of differential size with a normal vector N_q located at a surface point $q \in r_p$ and check the location of $\tilde{r}_p$ with respect to S_q. The vector N_q is assumed to be oriented towards $\overline{O}$. Thus, it gives us some notion on which side of S_q the inside of the object O is located. Consequently, it can be determined if $\tilde{r}_p$ crosses the inside of O in the neighbourhood of q.

The formal realization of this test is described in the following and is illustrated in Fig. 4. The straight line defined by $\tilde{r}_p$ separates the 2D space into two half-planes. This is illustrated in Fig. 4(a). By translating the normal vector N_q into p, it can be checked if $\tilde{r}_p$ intersects O in the neighbourhood of q depending on which half-plane $p + N_q$ belongs to and where q is located. If $p + N_q$ and q are in different half-planes, p is not occluded by q. This case is shown Fig. 4(b). The line through q represents the surface patch S_q in q. The projection ray of p goes along the outside of the surface in the neighbourhood of q. This is additionally illustrated by the projections p', q' and N'_q of p, q and N_q, respectively, onto the viewing plane. If $p + N_q$ belongs to the same half-plane containing q or $q \in \tilde{r}_p$, p is not visible. In the first case, the projection ray of p goes along the inside of the object in the neighbourhood of q as illustrated in Fig. 4(c). In the latter case, the projections of p and q are identical, but q is closer to the observer and, therefore, it occludes p.

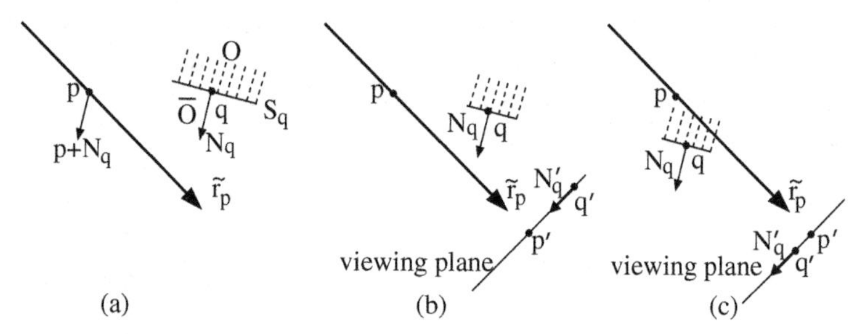

Fig. 4. (a) Subdivision of the space by $\tilde{r}_p$ into half-planes, (b) and (c) location of q in the half-planes with respect to the normal vector N_q of q

The two cases discussed above do not cover a normal vector N_q having the same direction like $\tilde{r}_p$, i.e. $\tilde{r}_p = N_q$ or $\tilde{r}_p = -N_q$. If the normal vectors associated with the discrete surface points reflect the local surface configurations correctly, one can expect for $\tilde{r}_p = -N_q$, that r_p hits also some inside point of the object since it crosses S with respect to N_q from the outside to the inside. If $\tilde{r}_p = N_q$, r_p would pass through S in q from the inside to the outside and should also hit some inside point. In fact, these cases could not appear for a discrete viewing ray which is tangent to a closed surface. If they arise nevertheless in a practical application, we assume for these cases that q occludes p as presented above.

3.4 Visibility in 3D

For the extension of the approach described in the previous section to 3D, an additional dimension for the location of the local surface in q with respect to $\tilde{r}_p$ has to be taken into account.

Consider the surface patch S_q in q and its projection onto the viewing plane. Without loss of generality, assume the viewing plane is located in q. (More generally, the viewing frustum is assumed to be bounded by a front clipping plane located in q and parallel to the projection plane.) Unless the direction of N_q or $-N_q$ is equal to the projection ray $\tilde{r}_q$ of q, the intersection of the projection plane and S_q leads to a curve γ_q with a normal vector N'_q that is the projection of N_q. This is illustrated in Fig. 5(a). Note, that for $\tilde{r}_p = N_q$

and $\tilde{r}_p = -N_q$, we make the same assumption as in 2D, i.e. p is occluded by q. For orthographic projection $\tilde{r}_p = \tilde{r}_q$. For perspective projection, the difference between $\tilde{r}_p$ and $\tilde{r}_q$ are very small since q is intersected by r_p. Thus, we neglect this difference and assume in the following additionally if $\tilde{r}_q = N_q$ or $\tilde{r}_q = -N_q$ that q occludes p.

To determine if $\tilde{r}_p$ intersects the surface in q, we translate the tangent line at γ_q to p'. This line separates the viewing plane into half-planes and enables a similar approach as described for 2D to determine if q occludes p: if $p' + N_q'$ and q' are in the same half-plane, q occludes p. If they are in different half-planes, q does not occlude p. The distance d shown in Fig. 5(a) between p' and q' depends on the location of p and q in 3D and is limited since p and q belong to the same discrete viewing ray, which in turn is the rasterization of the projection ray. Consider the straight line that contains p and is perpendicular to the tangent line at γ_q. If d is neglected, we obtain on this line the same scenario as illustrated on the viewing plane of Fig. 4(c) and (d).

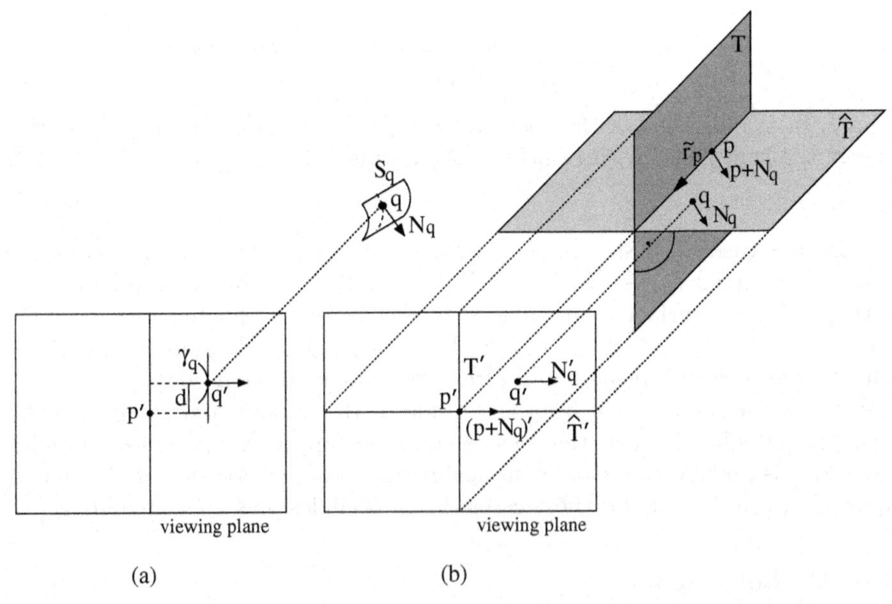

Fig. 5. Computation of the plane T

For practical applications, we perform the test described above directly in 3D. For this, a plane is needed in 3D that separates the space into half-spaces. Clearly the plane, denoted with T in the following, must contain $\tilde{r}_p$ and the projection T' of T must be perpendicular to N_q'. Then T can be computed as follows: determine a plane $\hat{T}$ containing the vectors $\tilde{r}_p$ and N_q, i.e. $\hat{T} : \tilde{r}_p \times N_q$. Then T is the plane perpendicular to $\hat{T}$ containing $\tilde{r}_p$. More formally: $(T \perp \hat{T}) \wedge (\tilde{r}_p \in T)$. This is illustrated in Fig. 5(b). The point q occludes p, if q and $p + N_q$ are in the same half-space or $q \in T$. If they are in different half-spaces, p is not occluded by q.

4 Experimental Results

Experimental results are shown below for the orthographic as well as the perspective projection. The normal vectors for the examples were computed employing the method introduced in [13]. The images in Fig. 6 show the results for two spheres using 26-connected viewing rays. The spheres were rastered in an array of 75^3 points. The occlusion was checked only for front facing points. The results shown in Fig. 6 were obtained by applying our method (upper row), by not allowing the discrete viewing rays to hit any object point (middle row) and, finally by allowing the discrete viewing rays to hit any surface point without any further restriction (lower row). Clearly for the second case, many points particularly along the outline of the objects are missed. This leads to a gap between the two spheres. In the last case, consider the part of the outline of the sphere in the foreground which occludes the other sphere: especially for perspective projection in the example, points of the occluded sphere were projected "into" the image of the occluding sphere. In contrast, this problem does not arise for our method. The lines near the outlines of the images in the upper and lower row of Fig. 6 are visible points which are very close to each other in the viewing plane such that these points are represented in the final image by neighbouring pixels. If more than one visible point would be represented by a pixel in the image, the point closest to the observer is viewed.

Figure 7 shows the necessity of the application of 6-connected discrete viewing rays, depending on the properties of the surface. The two cuboids were rastered in an array of 50^3 points. The problem arising for 26-connected rays along the left edge of the cuboid in the foreground corresponds with the situation illustrated in Fig. 2(a) by the point p: viewing rays of points of the occluded cuboid pass through the surface so that they are projected "into" the image of the occluding cuboid. This can be avoided by using 6-connected viewing rays instead.

5 Summary

The determination of the visibility of points is a major task of rendering. In this paper we have proposed a technique which determines the visibility of points on a surface in $\mathbb{Z}^3$, which is defined as the boundary of an object. The method introduced works in object space and employs the basic idea of discrete forward ray-casting. The visibility of each point is computed in $\mathbb{Z}^3$ considering a continuous mapping for the projection of the points onto the viewing plane. This assumption makes the results independent of the resolution of the final image and enables an application of the method, e.g., for discrete ray-tracing to solve the shadow problem. We have shown by examples, that our method is suitable to solve the visibility problem and thus can be employed for rendering techniques such as those proposed in [12].

A remaining problem of this work, that is matter of future work, is a proof that 26-connected rays are sufficient if the surfaces to be rendered satisfy certain conditions. This would lead to a performance increase by using 26-connected rays instead of 6-connected rays for our method.

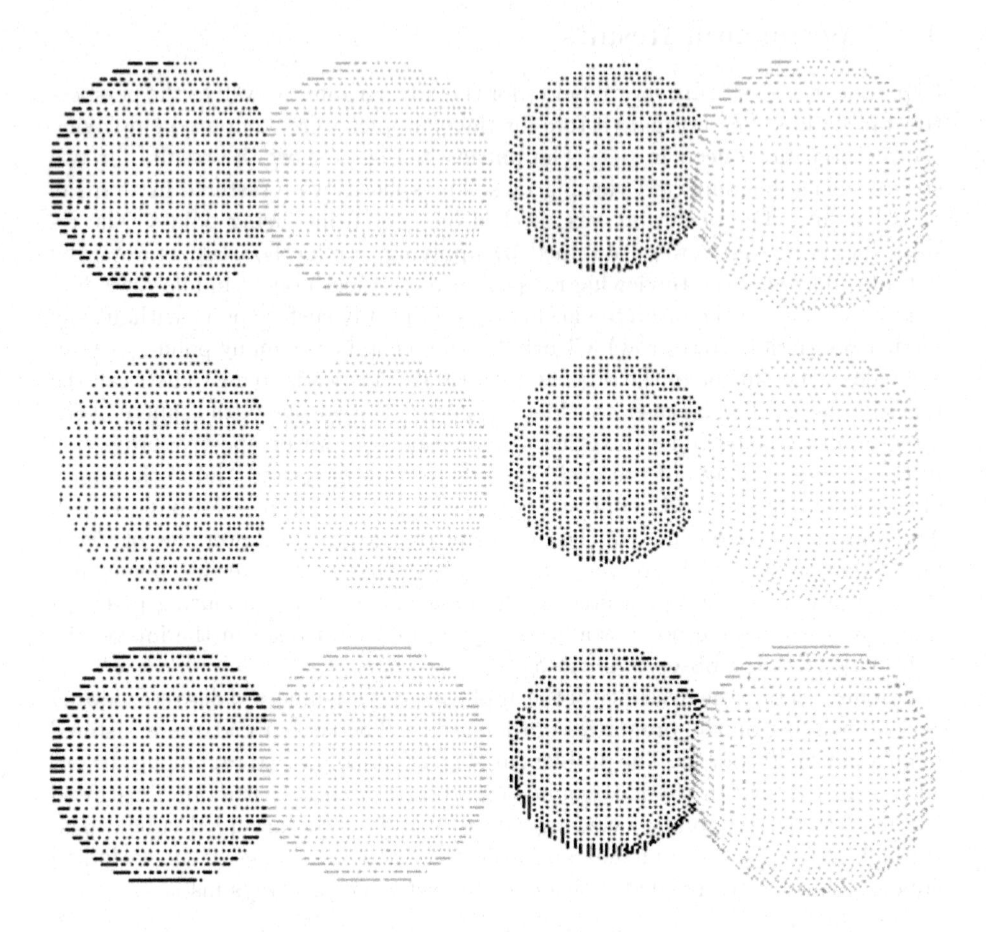

Fig. 6. Visible points of two spheres determined for orthographic (left column) and perspective projection (right column) by allowing discrete viewing rays to hit other surface points only if the projection rays do not intersect the object (upper row), not allowing intersections of any object point (middle row), and allowing intersections of any surface point (lower row) by the discrete viewing ray

References

1. Cohen, D., and Kaufman, A. Scan-conversion algorithms for linear and quadratic objects. In *Volume visualization*, A. Kaufman, Ed. IEEE Computer Society Press, 1991, pp. 280–301. 171
2. Delfosse, J., Hewitt, W. T., and Mériaux, M. An investigation of discrete ray-tracing. In *Proc. of the 4th Colloquium on Discrete Geometry for Computer Imagery (DGCI'94)* (Grenoble (France), 1994), pp. 65–74. 172, 174
3. Ke, H. R., and Chang, R. C. Sample buffer: A progessive refinement ray-casting algorithm for volume rendering. *Computers and Graphics 17*, 3 (1993), 277–283.

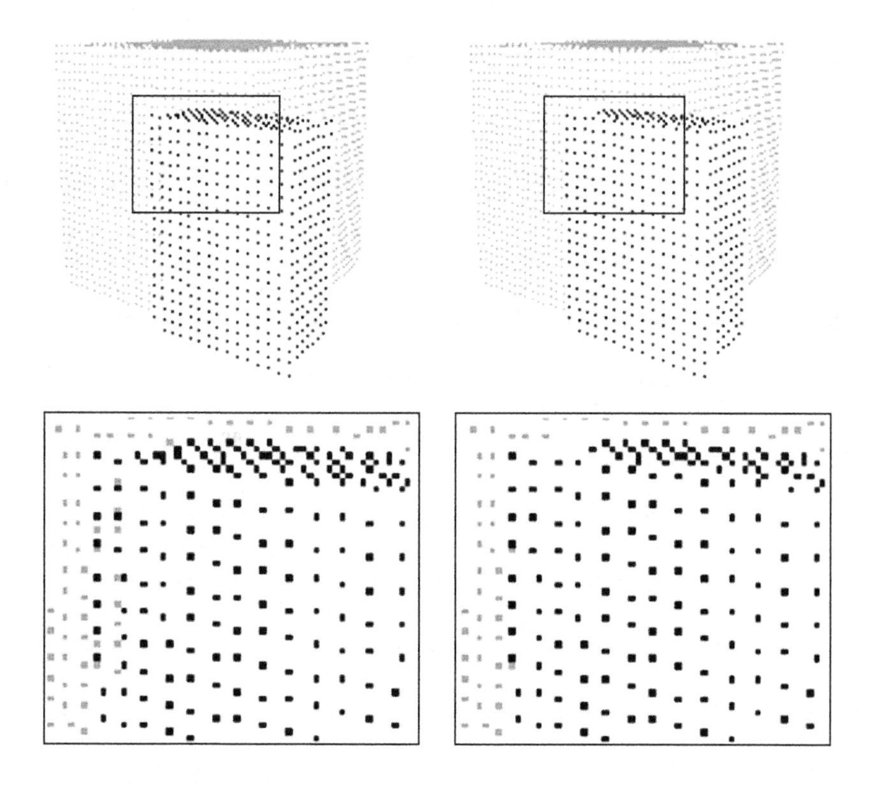

Fig. 7. Visible points obtained from 26-connected discrete viewing rays (left) and 6-connected viewing rays (right), and an enlargement of a detail (lower row) showing the upper left corner of the cuboid in the foreground

171
4. LAUR, D., AND HANRAHAN, P. Hierarchical splatting: A progressive refinement algorithm for volume rendering. *Computer Graphics 25*, 4 (1991), 285–288. 172
5. LEVOY, M. Display of surfaces from volume data. *IEEE Computer Graphics and Applications 8*, 5 (1988), 29–37. 171
6. LEVOY, M. Efficient ray tracing of volume data. *ACM Transactions on Graphics 9*, 3 (1990), 245–261. 171
7. MALGOUYRES, R. A definition of surfaces of Z^3: A new 3D discrete Jordan theorem. *Theoretical Computer Science 186* (1997), 1–41. 175
8. MORGENTHALER, D. G., AND ROSENFELD, A. Surfaces in three-dimensional digital images. *Information and Control 51* (1981), 227–247. 175
9. MUELLER, K., AND YAGEL, R. Fast perspective volume rendering with splatting by utilizing a ray-driven approach. In *Proc. of Visualization '96* (San Francisco, CA, 1996), pp. 65–72. 171
10. NEHLIG, P., AND MONTANI, C. A discrete template based plane casting algorithm for volume viewing. In *Proc. of the 5th Colloquium on Discrete Geometry for*

Computer Imagery (DGCI'95) (Clermont-Ferrand (France), 1995), pp. 71–81. 171
11. Shareef, N., and Yagel, R. Rapid previewing via volume-based solid modeling. In *Proc. of Solid Modeling '95* (1995), pp. 281–292. 171
12. Thürmer, G. Rendering surfaces of synthetic solid objects. *submitted for publication.* 172, 179
13. Thürmer, G., and Wüthrich, C. A. Normal computation for discrete surfaces in 3D space. *Computer Graphics Forum (Proc. Eurographics'97) 16*, 3 (1997), C15–C26. 173, 179
14. Upson, C., and Keeler, M. V-buffer: Visible volume rendering. *Computer Graphics (SIGGRAPH'88) 22*, 4 (1988), 59–64. 171
15. Wang, S. W., and Kaufman, A. E. Volume sampled voxelisation of geometric primitives. In *Proc. of Visualization '93* (1993), IEEE Computer Society Press, pp. 78–84. 171
16. Wang, S. W., and Kaufman, A. E. Volume-sampled 3D modeling. *IEEE Computer Graphics and Applications 14*, 5 (1994), 26–32. 171
17. Westover, L. Footprint evaluation for volume rendering. *Computer Graphics (SIGGRAPH'90) 24*, 4 (1990), 144–153. 172
18. Yagel, R., Cohen, D., and Kaufman, A. Discrete ray tracing. *IEEE Computer Graphics and Applications 12*, Sep. (1992), 19–28. 171, 175
19. Yagel, R., Stredney, D., Wiet, G. J., Schmalbrock, P., Rosenberg, L., Sessanna, D. J., and Kurzion, Y. Building a virtual environment for endoscopic sinus surgery simulation. *Computers and Graphics 20*, 6 (1996), 813–823. 171

Surfaces and Volumes

Extended Reeb Graphs for Surface Understanding and Description

Silvia Biasotti, Bianca Falcidieno, and Michela Spagnuolo

Istituto per la Matematica Applicata,
Consiglio Nazionale delle Ricerche
{silvia,falcidieno,spagnuolo}@ima.ge.cnr.it
http://www.ima.ge.cnr.it/

Abstract. The aim of this paper is to describe a conceptual model for surface representation based on topological coding, which defines a sketch of a surface usable for classification or compression purposes. Theoretical approaches based on differential topology and geometry have been used for surface coding, for example Morse theory and Reeb graphs. To use these approaches in discrete geometry, it is necessary to adapt concepts developed for smooth manifolds to discrete surface models, as for example piecewise linear approximations. A typical problem is represented by degenerate critical points, that is non-isolated critical points such as plateaux and flat areas of the surface. Methods proposed in literature either do not consider the problem or propose local adjustments of the surface, which solve the theoretical problem but may lead to a wrong interpretation of the shape by introducing artefacts, which do not correspond to any shape feature. In this paper, an Extended Reeb Graph representation (ERG) is proposed, which can handle degenerate critical points, and an algorithm is presented for its construction.

1 Introduction

Reasoning about shape is a common way of describing real objects in engineering, architecture, medicine, biology, physics and in daily life. So far, research in modelling shapes has mainly focused on geometry, with the aim of defining effective representations and accurate approximations of objects [1]. To describe *geometric* objects, however, different levels of mental models can be used. It is possible to use natural-language terms to qualitatively describe the external shape of an object, or to draw a sketch of it, or to describe it by listing its differences with respect to some other similar objects, or also to define it according to what it is used for, and so on.

Shape recognition and classification are therefore basic steps for constructing descriptions of objects. These abstraction processes have been effectively described for example for biological taxonomy, where shape descriptions have been constructed by finding a good representation and then discarding irrelevant details [2,3]. The important point here is that reasoning about shape is an efficient approach not only for building high-level modelling environments, but

G. Borgefors, I. Nyström, and G. Sanniti di Baja (Eds.): DGCI 2000, LNCS 1953, pp. 185–197, 2000.
© Springer-Verlag Berlin Heidelberg 2000

also for devising top-down simplification methods based on the comprehension of the object shape [1].

Differential topology and geometry have been often used for surface coding, [4,5,6], and building shape descriptions which nicely correspond to intuitive mechanisms of shape cognition and recognition.

For example an interesting method for coding the critical points of smooth Morse functions of two variables has been studied by Nackman [4]. Pfaltz proposed a similar approach for discrete surfaces defining the so-called surface networks [6]. Bajaj et al. [7] propose two different approaches, one bottom-up the other top-down, to preserve the topology during the simplification of discrete data. Their method, however, does not guarantee an exhaustive description of characteristics of the surface. Moreover, it does not allow to extract global properties of the surface and to detect degenerate configurations.

Major research on the use of topology for surface description has been proposed by Kunii, Shinagawa et al. in [5,8] where a surface coding based on Morse theory and Reeb graphs has been defined (see next section). Takahashi et al. [9], use an approach based on surface-networks to reconstruct the Reeb graph of a $2.5D$ surface. As the previous references also this approach does not consider degenerate configurations which are typical of discrete surfaces, such as plateaux or flat areas.

In this paper, the Extended Reeb Graph (ERG) representation is proposed, which can handle degenerate critical points, and an algorithm is presented for constructing the ERG of bi-variate surfaces (scalar fields). The proposed extension does not distort the semantic meaning of the Reeb graph and faithfully represent the surface morphology.

The reminder of this paper is organised as follows: first, basic results of Morse and Reeb graph theory are given to introduce the proposed method; then our extension from the critical points concept to the critical areas is presented for discrete surfaces; finally, the proposed ERG representation and an algorithm for its construction from a set of contour levels are described. Some examples of its application to real surfaces are shown.

2 Theoretical Background

Morse theory originates from the calculus of variations and it allows describing differentiable manifolds using a limited number of information, for example by coding the topological relationships between critical points [10,11]. Using Morse theory, it is possible to construct topological spaces equivalent to a given smooth surface, which describe the surface as a decomposition into primitive topological cells [12]. Some basic results of Morse theory are given.

Definition 1. *(Morse function) Let f, $f : M \to \mathbb{R}$, be a real smooth function defined on the smooth manifold M; f is called Morse function if all of its critical points are non-degenerate. A critical point is non-degenerate if the Hessian matrix H of f is non-singular at that point.*

Non-degenerate critical points are isolated, therefore scalar fields having plateaux or volcano rims do not comply with the definition of Morse function.

An example of a simple Morse function is given in figure 1(a). The figure 1(b) depicts an example of a complex Morse function, i.e. a function having more than one critical point at the same level, while figure 1(c) does not comply with the Morse's function definition.

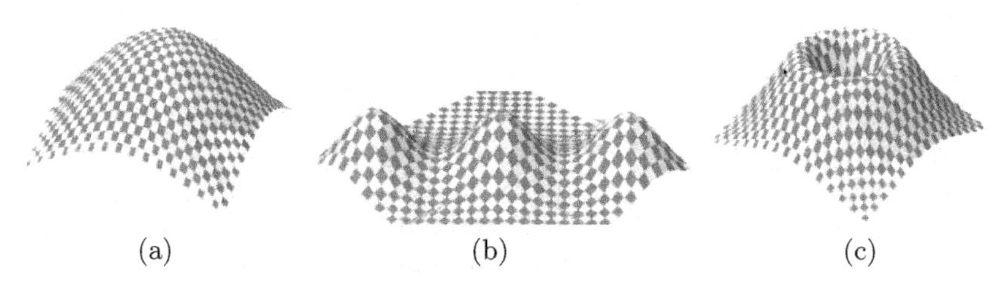

Fig. 1. An example of height function with an isolated maximum, (a), a non simple height function, (b), and a circle of non isolated maxima, (c). In (b) the saddle points as well as the maximum points have indeed the same elevation

The height function can be effectively used to study the surface shape. Intuitively, the height function of a smooth manifold M, embedded into the usual three-dimensional Euclidean space, is the real function which associates to each point on the surface its elevation.

Therefore, the level sets of a height function associated to a surface are the intersections of the surface with planes orthogonal to a fixed directions, which may be shaped in arbitrarily complex ways. If the height function is Morse, then the contour configuration can be quite simply classified. Moreover, smooth surfaces satisfy the Euler formula which states that the number of non-degenerate maximum (M), saddle (p) and minimum (m) points satisfies the relation $M - p + m = 2(1 - g)$ where g represents the genus of the surface.

Reeb defined a graph to code the evolution and the arrangement of level curves [5,13]. The Reeb graph of a function is defined as follows:

Definition 2. *(Reeb graph) Let $f : M \rightarrow \mathbb{R}$ be a real valued function on a compact manifold M. The Reeb graph of M wrt f is the quotient space of $M x \mathbb{R}$ defined by the equivalence relation "$\sim$", given by*

$$(X1, f(X1)) \sim (X2, f(X2)) \Leftrightarrow f(X1) = f(X2)$$

and $X1$ and $X2$ are in the same connected component of $f^{-1}(f(X1))$

A Reeb graph of a compact manifold collapses into one element all points having the same value under a real function and being in the same connected component. Moreover, theorem 1 allows us to define a further equivalence relation among elements of the Reeb's quotient space.

Theorem 1. *Let f be a real smooth function on a compact smooth surface M, and assume that the segment $[a, b]$ (where $a < b$) contains no critical value of f. Then:*

- *the level set $f^{-1}(a)$ and $f^{-1}(b)$ are diffeomorfic and they consist of the same number of smooth circles diffeomorfic to the standard circle;*
- *let M_a be the subset of M defined $M_a = \{x : f(x) \leq a\}$, and M_b defined accordingly. Then M_a and M_b are diffeomorfic as two-manifolds with boundary.*

In figure 2(a) the points drawn on the manifold represent the equivalence classes: here the manifold considered is a bi-torus and the function defined over it is the height function. In figure 2(b) the Reeb's quotient space is represented as a "traditional" graph where the equivalence classes are grouped into arcs if they are representative of diffeomorfic contours, as stated in the theorem 1. Obviously, since the choice of the mapping function is not unique, a manifold admits different Reeb graphs.

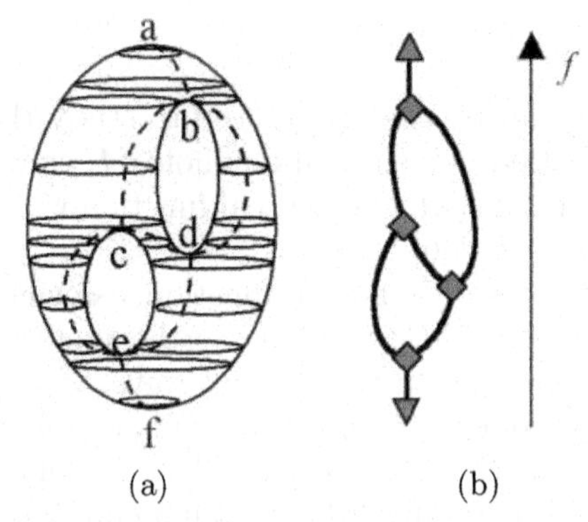

Fig. 2. A bi-torus, its contour lines and the equivalence classes defined by Reeb's quotient space (a), a possible representation of the Reeb graph wrt the function f, (b)

A Reeb graph describes a manifold surface by considering the evolution of the surface sections under a given real valued function. If we have to describe the shape of a manifold surface M embedded into the Euclidean space, moreover, the Reeb graph of M under its "natural" height function codes the shape description of M in terms of critical points of h, corresponding to meaningful semantic labels, such as peaks, pits or passes. Moreover, under the assumption that the height function is Morse, the structure of the Reeb graph is rather simple. A Reeb graph of M under h can be defined as $RG_h(M) = (P_h(M), A_h(M))$ where the node set is defined by $P_h(M) = \{P_i \in M, P_i \textit{ is a critical point of } h(M)\}$ and the arc set is given by $A_h(M) = \{(P_i, P_j), P_i \textit{ and } P_j \in P_h(M) \textit{ and the topological type of } M \textit{ does not change between } P_i \textit{ and } P_j\}$ (see Theorem 1).

If $h(M)$ is Morse, moreover, the arcs of $RG_h(M)$ can be oriented and the nodes have almost degree three.

3 Extended Reeb Graph Representation

Reeb graph theory provides a powerful abstraction mechanism to describe the shape of smooth surfaces. Even if there are no restrictions in the Reeb graph definition on the type of f, several authors have in practice limited the use of the Reeb graph to Morse mapping functions (see section 1). When dealing with discrete surface representations, such as triangular meshes, problems may arise due to the loss of properties of the resulting surface which can be assumed to be at most continuous. The straightforward application of the Reeb graph definition to a generic polyhedral surface (e.g. a 3D triangulated manifold) requires at first the definition and extraction of the critical points. The possible definitions of discrete critical points, however, generally suffer of instability since small perturbations of the vertex coordinates may result in rather different configurations [9,7].

This problem is particularly crucial for surfaces defined by measurements of some natural or physical objects, as in digital terrain modelling or reverse engineering, where the position of the mesh vertices may have some uncertainty associated. Therefore, we have adopted a different approach based on the computation of a "sufficiently" dense number of contour lines and the definition of the Reeb graph from the contour set. The evolution of contour lines is less sensible to local perturbation of vertex position and, globally, provides a faithful description of the surface shape. The same approach has been adopted by Kunii, Shinagawa et al., [5,8], for smooth surfaces for which the height function is Morse, while here generic continuous surfaces are considered for which a set of contour levels is given (or computed).

An extension of the Reeb graph representation (*ERG*) is introduced, which can handle degenerate and non simple critical points, and an algorithm is presented for constructing the ERG automatically from the set of contour levels of a discrete surface. The proposed extension does not distort the semantic meaning of the Reeb graph and faithfully represents the surface morphology.

The innovation of this method is both the way of constructing the graph and the efficiency in dealing with degenerate situations. The proposed approach is actually not an extension of the Reeb graph itself, but rather a full application of its definition in the discrete domain, which does not require the height function to be Morse. The quotient space defined by the Reeb equivalence relation can be represented in terms of a graph representation, which formally represents the contour containment and the adjacency relationships. In this way, with an extended definition of critical areas, the application domain can be enlarged to generic continuous scalar fields, without any artifacts [14].

3.1 Definition of Critical Areas and Influence Zones

First of all, to get a more general and unambiguous shape description, we extend the idea of critical points to critical *areas* as well. With respect to other definitions, the idea of critical areas provides a more general approach as it takes into consideration the behaviour of a surface in a neighbourhood rather than locally around a point.

Formally, let M be a continuous surface defined by a scalar field $f : D \subset \mathbb{R}^2 \rightarrow \mathbb{R}$, that is, $M = \{P = (x, y, z) | z = f(x, y)\}$ and let h be the height function naturally defined over M, $h(P) : M \rightarrow \mathbb{R}$ such that $h(P) = h((x, y, f(x, y)))$ $= f(x, y)$ and let P_i be the critical points of $h(M)$. Under the assumption that M is Morse, $f^{-1}(P_i)$ is an isolated critical point of the surface, while if we do not require M is Morse, $f^{-1}(P_i)$ generally represents a critical *area* of the surface, a level region of M. Let $C(M)$ be the set of contour levels of M; we assume that $C(M)$ is a set of unorganised polygonal contours, which are either simple closed polygons or polygonal lines with the end points on the boundary of the surface.

Let $C_T(M)$ be the Delaunay triangulation of $C(M)$ constrained to all contours lines. The critical points of $h(M)$ can be efficiently detected by classifying the flat regions of $C_T(M)$, that is, the connected regions of $C_T(M)$ composed by triangles having all the three vertices at the same elevation [15,16,17].

The flat regions of $C_T(M)$ correspond either to simply or to multiply-connected areas. Let $B_R(M)$ be the boundary of a critical area R of $C_T(M)$, in general $B_R(M) = B_1 \cup B_2 \cup \ldots \cup B_n$ where B_i represents the $i - th$ connected component of the boundary. According to the definition of critical area, each B_i may be either a contour in $C(M)$, having elevation $h(B_i)$ or may be a sequence of edges of $C_T(M)$, as for saddle areas. Simply-connected critical areas, for which $B_R(M) = B_1$, correspond either to isolated or degenerate critical points of M. A multiply- connected critical area divides the surface into two parts: an "*outer*" part which is defined by the portion of the surface outside the boundary of the multiply-connected area, and as many "*inner*" parts as the multiplicity of the critical area boundary. Let B_1 be the outer boundary component and B_i, with $i > 1$, the inner ones. The following classification scheme is adopted:

- $B_R(M)$ is a simple maximum area (resp. simple minimum area) iff $n = 1$ and the outgoing direction from B_1 is descendent (resp. ascending), see figure 3(a);
- $B_R(M)$ is a complex maximum area (resp. complex minimum area) iff $n > 1$ and the outgoing directions from each B_i are descendent (resp. ascending), (see figure 3(b));
- $B_R(M)$ is a simple saddle area iff $n = 1$ and there are at least four outgoing directions alternatively descendent and ascendent;
- $B_R(M)$ is a complex saddle area iff $n > 2$ and the outgoing directions are descendent for some B_i and ascending for some other B_j or the outgoing direction for B_1 are either descendent and ascending.

When considering surfaces defined by scalar fields or more generally manifold surfaces with boundary it is necessary to give a unique interpretation to critical points occurring along the surface boundary by considering a global virtual minimum point as a virtual closing of the surface with descending directions from the boundary to a global minimum.

Obviously, the flat regions of $C_T(M)$ do not always correspond to critical points of the height function: for example, ridges and ravines of M will cause level regions to appear as well, with a step-like effect which is a very well-known problem in digital terrain modelling. These areas, however, can be easily detected considering the number of the ascending or descending directions of the region boundary, as explained in details in [14,17]. Therefore, let us assume from now that the critical areas only correspond to critical points of $h(M)$, either isolated or degenerate. Generally for polyedral surfaces the Eulero formula given in section 2 is not verified. Nevertheless considering the above defined critical areas for $C_T(M)$ the Eulero formula is still satisfied if they correspond to non-degenerate critical points .

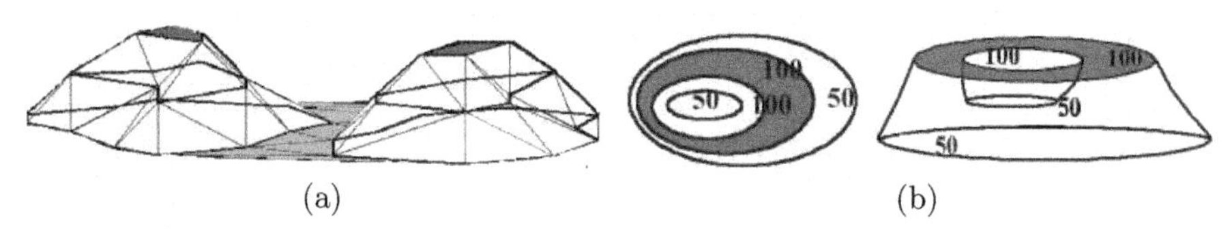

(a) (b)

Fig. 3. Critical areas of M correspond to flat regions in $C_T(M)$: areas of maximum are depicted in dark and saddle areas in light grey

It is important to show the link between critical areas and Reeb graph nodes. By applying the definition of Reeb graph (cf. definition 2) all points belonging to a simply-connected critical area are Reeb-equivalent and may therefore collapse into the same node. If the isolated critical points of M were known, moreover, a simple labelling of the graph's node set would be sufficient to distinguish them from degenerate critical points. Similarly, the behaviour of arcs incident to simple critical areas is equivalent to the behaviour of arcs incident to isolated critical points. Therefore, simple critical areas can be represented in the ERG by simple nodes as in the normal Reeb graph representation.

Multiply-connected critical areas correspond to macro-nodes: that is particular nodes having at least one arc connected to an inner node.

With respect to the Reeb graph arcs we have introduced the concepts of *influence zones* which are defined using adjacency among contours: two contours are adjacent if they are edge-adjacent in the constrained triangulation. To connect critical areas in the Reeb graph, an influence zone is associated to critical areas, which intuitively spans the surface between adjacent Reeb graph nodes.

First of all, if R is a simple saddle critical area, its influence zone is defined as the surface region delimited by the contours which are directly edge-adjacent to R, in other words contours which are at a distance 1 from the saddle area.

Then, if R is a maximum or minimum critical area, the influence zone of R is the portion of the surface identified by growing B_1, the outmost component of the region boundary, until the boundary of another critical area is reached. Therefore, if the critical area is a simple maximum its influence area is the maximal region in $C_T(M)$ containing R and non containing other critical areas.

Similarly, the influence zone of a multiply-connected area does not contain any other critical areas except those located in its internal boundary components. In practice the influence zone definition for minimum and maximum areas strictly relates with the theorem 1: in fact the expansion process allows to connect level sets which are diffeomorfic. For saddle areas, indeed, the influence zone represents the portion of the surface where the topological change arises. In figure 4, some critical areas are shown with the corresponding influence zones. The dark colour identifies maximum (a) and saddle areas (c), their corresponding influence zones are depicted in medium grey colour, (b) and (d).

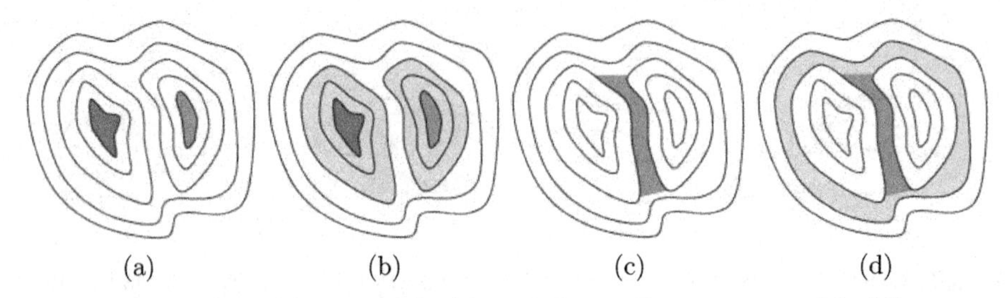

Fig. 4. Critical areas, (a), (c) and their influence zones (b), (d)

3.2 Construction of the ERG Representation

In this paragraph a short description of the algorithm to extract the critical areas is given. First of all it is important to notice that each contour $C_T(M)$ is edge-adjacent to the previous and the next one, in the height sequence of $C(M)$; therefore the ascending/descending directions between contours can be simply checked. Then, critical points of $h(M)$ can be efficiently detected by classifying the level regions of $C_T(M)$, that is, the connected regions of $C_T(M)$ composed by triangles having all the vertices at the same elevation [15,16,17]. The classification of flat areas as critical areas, simple or complex, is done by checking the number of non-constrained edges in the boundary:

- simple saddle areas are detected in $C_T(M)$ as flat areas with more than one edge in the boundary, which does not belong to the constraints, that is to the contours;
- maximum or minimum areas, either simple or complex, and complex saddles are detected in $C_T(M)$ as flat areas whose boundary is fully composed by constrained edges of $C_T(M)$.

The distinction among the different types of critical areas is done by analyzing the coordinates of the vertices which are edge-adjacent to the region boundary, that is, by checking the ascending/descending directions, as in section 3.1.

According to the graph representation of the Reeb's quotient space, each node of the graph corresponds to a critical point, or area, and each arc corresponds to a connected component of the manifold between two critical levels of the height function. This suggests to follow a similar approach for automatically extracting the Reeb graph, based on tracking the evolution of contour lines. First, the critical areas are recognised, and the ERG is initialized by creating the node corresponding to the virtual minimum, VM. The VM is connected to the saddle (complex or simple) having the minimum elevation and external to each macro-node. If such a saddle does not exist, then the VM is connected to the first complex maximum area, otherwise the ERG is a trivial graph connecting the VM to the only simple maximum existing. Then, using the notion of *influence zone*, the Reeb graph arcs are partially computed from the adjacency relationships between influence zones. In this manner the arcs connected to the terminal nodes of the ERG are identified. To complete the ERG construction the links between saddles and, in general, complex areas, have to be determined. Intuitively, arcs correspond to ascending paths between critical areas which are determined by expanding the associated influence zones, following free directions in the outmost boundary component (see figure 5). Free directions are those which do not correspond to an already identified arc.

In the following, the construction algorithm is described using a C pseudo-code. At each step, if the critical region is multiply-connected, then a macro-node is defined, with as many arcs as the inner components of the critical region. Note that the definition of influence zone guarantees that the Reeb graph is correctly constructed.

```
ERG_Construction(N,A)
  /* The ERG is defined by the set of nodes, N, and of arcs, A */
{ N=CriticalAreasRecognition(tin, contours);
  /* Identify critical areas and initialize the virtual minimum */
OrderAreas(N);     /* Order the Critical Areas by elevation */
InfluenceZoneDetection(N);
  /* Influence zones are associated to saddle areas */
     and then to maximum and minimum critical areas */
LinkNodesbyInfluenceZone(N,A);
  /* Create a subset of A directly by checking the adjacencies
     between Influence Zones */
CompleteArcSet(N,A); }
```

For the sake of clarity the function “CompleteArcSet” is here expanded:

```
CompleteArcSet(N,A)
        /* N=nodes, A=arcs */
{for (each node in N)
   {if (IsGrowingArea(node))
      {for ( each non visited growing direction node)
         {while ((not(findBoundarySurface)) or
                                       (not(findOtherInfluenceZone)))
```

```
        ExpandToUpperLevel(node);
        if (R=OtherAreaReached)
          ConnectWithArc(node, R);
          } } } }
```

The function "IsGrowingArea(node)" returns a boolean value which is TRUE if the critical area has at least one growing direction which has not been visited yet.

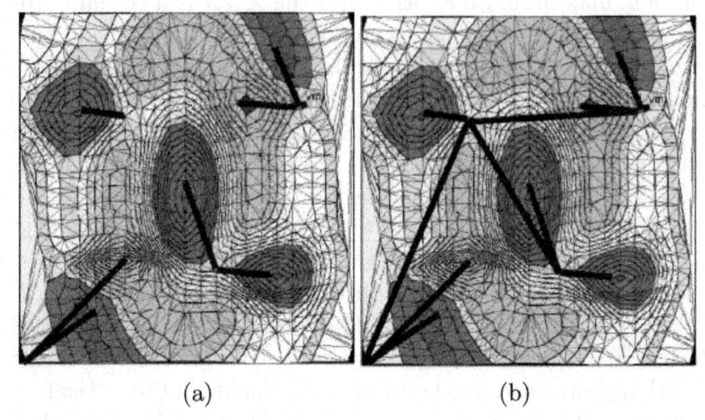

(a) (b)

Fig. 5. Two steps of the Reeb graph reconstruction process

Some examples are given: in figure 6(a), a test surface is depicted with the critical areas identified on the triangulation; its Reeb graph is depicted in 6(b) and 6(c). The surface that can be trivially reconstructed by triangulating only the critical sections of the ERG is shown in 6(d): even if rough, the reconstruction still faithfully reproduces the original surface morphology. Finally, in figure 7, another example of our characterisation method applied to a natural surface (a) is shown, then the reconstructed Reeb graph is depicted in (b) and (c) and the surface reconstructed considering only the critical sections is shown in (d).

4 Conclusions and Future Developments

In this paper a model has been defined which provides a high-level description of the surface shape. Morse theory concepts and Reeb graph properties have been used and translated to the discrete domain in order to define a sketch of the surface usable for classification or compression and decompression purposes [18].

With respect to previous work in this field, the proposed Extended Reeb Graph is able to faithfully represent the surface shape, even in case of degenerate critical points. The algorithm for constructing the ERG of scalar fields, represented as contour lines, is efficient and allows a topologically correct reconstruction of the surface shape.

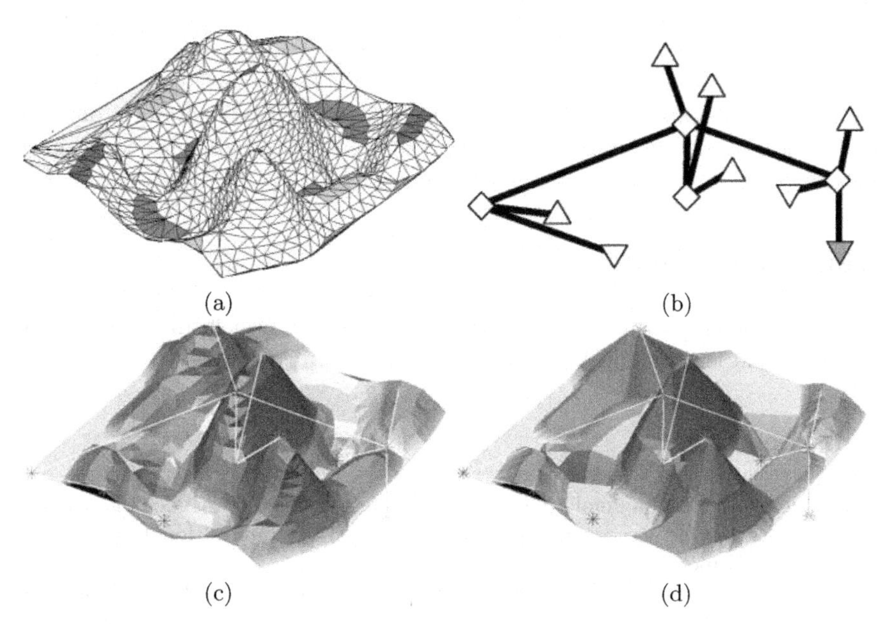

Fig. 6. A test surface (a), its Reeb graph (b) and (c) and the reconstructed model (d)

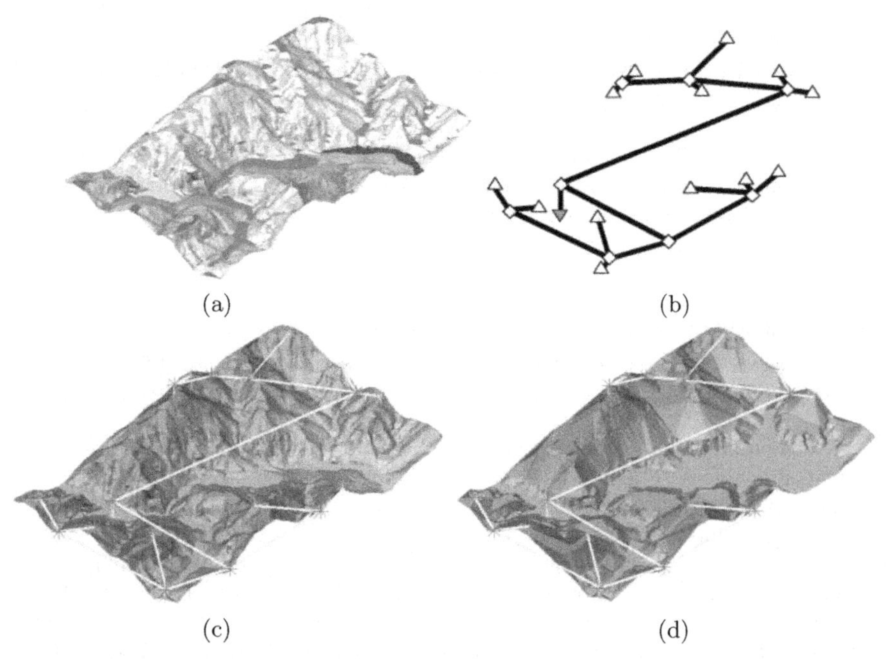

Fig. 7. The critical areas of a terrain (a), its Reeb graph (b), (c) and the recostruction from the critical sections (d)

Future developments are currently under consideration, mainly for the extension of the method to three dimensional surfaces, with or without boundary. Moreover, the shape decompression process is also being developed, taking into account several shape reasoning steps which allow a better definition of shape-based generation of intermediate sections between critical sections, coded in the Reeb graph [18].

Acknowledgements

Thanks are given to Prof. Tosiyasu Kunii for the many helpful discussions on this topic, and to Corrado Pizzi for his helpful technical support.

References

1. B. Falcidieno, M. Spagnuolo. Shape Abstraction Paradigm for Modelling Geometry and Semantics. In Proceedings of Computer Graphics International, pp. 646-656, Hannover, June 1998. 185, 186
2. W. D'Arcy Thompson. On growth and form. MA: University Press, Cambridge, second edition, 1942 185
3. P. Pentland. Perceptual organization and representation of natural form. Artificial Intelligence, Vol.28, pp. 293- 331, 1986. 185
4. L. R. Nackman. Two-dimensional Critical Point Configuration Graphs. IEEE Transactions on Pattern Analysis and Machine Intelligence, Vol. PAMI-6, No. 4, p. 442-450, 1984. 186
5. Y. Shinagawa, T. L. Kunii, Y. L. Kergosien. Surface Coding Based on Morse Theory. IEEE Computer Graphics & Applications, pp 66-78, September 1991. 186, 187, 189
6. J. L. Pfaltz. Surface Networks. Geographical Analysis, Vol. 8, pp. 77-93, 1990. 186
7. C. Bajaj, D. R. Schikore. Topology preserving data simplification with error bounds. Computer & Graphics, 22(1), pp. 3-12, 1998. 186, 189
8. Y. Shinagawa, T. L. Kunii. Constructing a Reeb graph automatically from cross sections. IEEE Computer Graphics and Applications, 11(6), pp 44-51, 1991. 186, 189
9. S. Takahashi, T. Ikeda, Y.Shinagawa, T. L.Kunii, M. Ueda. Algorithms for Extracting Correct Critical Points and Construction Topological Graphs from Discrete geographical Elevation Data. Eurographics '95, Vol. 14, Number 3, 1995. 186, 189
10. A. Fomenko. Visual Geometry and Topology. Springer- Verlag, 1994. 186
11. V. Guillemin, A. Pollack. Differential Topology. Englewood Cliffs, NJ: Prentice-Hall, 1974. 186
12. J. Milnor. Morse Theory. Princeton University Press, New Jersey, 1963. 186
13. G. Reeb. Sur les points singuliers d'une forme de Pfaff completement integrable ou d'une fonction numérique. Comptes Rendus Acad. Sciences, Paris, 222:847-849, 1946. 187
14. S. Biasotti. Rappresentazione di superfici naturali mediante grafi di Reeb. Thesis for the Laurea Degree, Department of Mathematics, University of Genova, September 1998. 189, 191

15. G. Aumann, H. Ebner, L. Tang. Automatic derivation of skeleton lines from digitized contours. ISPRS Journal of Photogrammetry and Remote Sensing, 46, pp. 259-268, 1991. 190, 192
16. M. De Martino, M. Ferrino. An example of automated shape analysis to solve human perception problems in anthropology. International Journal of Shape Modeling, Vol. 2 No. 1, pp 69-84, 1996. 190, 192
17. C. Pizzi, M. Spagnuolo. Individuazione di elementi morfologici da curve di livello. Technical Report IMA, No. 1/98, 1998. 190, 191, 192
18. S. Biasotti, M. Mortara, M. Spagnuolo. Surface Compression and Reconstruction using Reeb graphs and Shape Analysis. Spring Conference on Computer Graphics, Bratislava, May 2000. 194, 196

Euclidean Nets: An Automatic and Reversible Geometric Smoothing of Discrete 3D Object Boundaries

Achille J.-P. Braquelaire and Arnault Pousset

LaBRI – Laboratoire Bordelais de Recherche en Informatique - UMR 5800, Université Bordeaux 1
351 cours de la Libération, F-33405 Talence, France
{Braquelaire,Arnault.Pousset}@labri.u-bordeaux.fr

Abstract. In this work we describe a geometric method to smooth the boundary of a discrete 3D object. The method is reversible in the sense that the discrete boundary can be retrieved by digitizing the smoothed one. To this end, we propose a representation of the boundary of a discrete volume that we call Euclidean net and which is a generalization to the three-dimensional space of Euclidean Path introduced by Braquelaire and Vialard [4]. Euclidean nets can be associated either to voxel based boundaries or to inter-voxel based boundaries. In this paper we focus on the first approach.

Keywords: Digital Surfaces. Discrete boundary representation. Smoothing.

1 Introduction

Processing three-dimensional discrete images such as X-ray computer tomography images or MRI brings up the problem of defining and representing three-dimensional discrete objects [11]. Usually a *three-dimensional discrete image* (in short 3D image) is a parallelepipedic matrix of voxels and a *three-dimensional discrete object* (in short 3D object) of a 3D image is defined as a set of connected component of voxels of this image. In this work, we call *discrete object*, or simply *object*, any 6-connected set of voxels. For instance a raw visualization of a discrete sphere (a maximal object which voxel centers are inside a Euclidean sphere), also called cuberille representation [10], can be seen in Fig 1 (a).

There are several motivations to build an effective representation of the boundary of a 3D object. One of them is that a boundary representation fits with conventional rendering techniques. Another one is that the encoding of a 3D object with a matrix of labels does not provide a convenient framework for image analysis applications such as the extraction of geometric and topological feature or object recognition.

A first approach, known as the *marching cubes algorithm* [13], provides a method for extracting a polygonal surface mesh from the object. Each polygon

G. Borgefors, I. Nyström, and G. Sanniti di Baja (Eds.): DGCI 2000, LNCS 1953, pp. 198–209, 2000.
© Springer-Verlag Berlin Heidelberg 2000

is defined according to a local configurations of height adjacent centers of voxels. This method is easy to implement but has several drawbacks both from the geometrical and the topological points of view. A second approach known as *active surfaces* consists in moving the control points of a parametric surface enclosing the digital object in order to minimize an energy function [20,21]. With both these approaches the boundary of the discrete object is described by a continuous surface defined in the Euclidean space $\mathbb{R}^3$.

An alternative method is to use the *digital surface* (or *discrete surface*) of the object. Two approaches are commonly used : the *voxel based approach* [15,7,14,22] and the *inter-voxel based approach* [17,8,4]. It has been shown that the inter-voxel surface can be seen as a marching-cube algorithm define with an alternative interpolation scheme [12].

Several methods have been proposed to associate a surface mesh with the discrete surface of an object. For instance the set of centers of surface voxels can be triangulated by linking together some adjacent centers [24,22]. In Fig.1 (b) is displayed a wire-frame representation of the surface mesh computed by the method proposed by Thuermer [22], and in Fig.1 (c) is displayed a surface rendering of this surface mesh. The aspect of the rendered surface mesh is smoother than the cuberille representation but is still far from the aspect of a continuous sphere. A usual approach to improve the visualization of a digital surface is to estimate a normal at each point of the discrete surface and to perform a shading (for instance a usual Gouraud shading [9]).

Here we propose an alternative approach based on a *geometrical smoothing* of the digital surface. In general a discrete 3D image comes from the digitizing of a continuous 3D scene. In this context a discrete object is also the result of the digitizing of a continuous one. It is thus natural to try to retrieve from the boundary of a discrete object an approximation of the boundary of the underlying continuous object. The goal of the geometric smoothing is to approximate this underlying continuous boundary by performing local movings of the surface points. An example of geometric smoothing of a discrete sphere is shown in Fig.1 (d). Each point of the triangulated discrete surface has been moved according to a method described in the following of the paper.

This method is a generalization to the three-dimensional case of the method of Euclidean paths introduced by Braquelaire and Vialard [27,4] to smooth the discrete boundary of a discrete 2D region. A Euclidean path is associated to a discrete 4-connected or 8-connected path by moving each point of the discrete path from its position to any position of the open unit square centered on it. A method has been proposed to compute the moving of each point of the discrete path laying on the computation of a discrete tangent at each of these points. It has been shown that so called *tangent driven Euclidean paths* highly improve the visual aspect of a discrete region boundary and provides a convenient representation for geometric transformations and a good estimation of geometric features (tangent and normal, perimeter) [25]. The smoothing preserve the topology of the set of boundaries of a segmented image [1]. This method has been successfully used to solve some image processing problems [3,26].

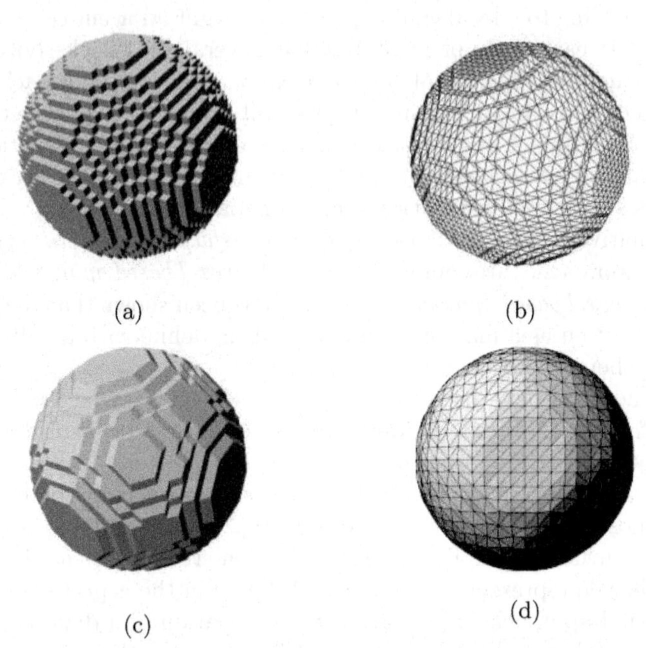

Fig. 1. Four representations of a discrete sphere: (a) cuberille representation, (b) wire-frame representation of a surface mesh associated with the discrete surface, (c) surface rendering of the surface mesh, (d) geometric smoothing of the discrete surface

The aim of this work is to propose a similar method to smooth the discrete surface of a 3D object. A geometrical analysis is performed at each point of the discrete surface providing an estimation of a *tangent plane.* The point is then projected onto this plane. This method is *reversible* in the sense that the discrete surface can be retrieved by digitizing the smoothed one. Moreover, this method computes a canonical result and does not need any manual parameterization.

In the following section we introduce discrete and Euclidean nets which are the 3D analogous of 2D discrete and Euclidean paths. In Section 3 we describe the model of tangent driven Euclidean nets. In Section 4 we address the problem of computing both a tangent plane an a normal vector to a point of a discrete surface. Finally in Section 5 we present both qualitative and quantitative results of experiments of geometric smoothing.

2 Discrete and Euclidean Nets

Let us now give some basic definitions used in the following. The coordinates of a point P are denoted by the tuple (x_P, y_P, z_P). The set $\mathbb{Z}^3$ of points with integer coordinates is called the *discrete space* and its elements are called *discrete points*. We denote by upper case letters the points of the discrete space $\mathbb{Z}^3$ and in lower case letters the points of the Euclidean space $\mathbb{R}^3$.

A *voxel* is a colored unit cube the center of which is a discrete point. The coordinates of a voxel are the coordinates of its center. An *image* is a set of voxels and the *image domain* is the set of voxel centers. In the following we consider images the domain of which is a parallelepipedic set of discrete points. The voxel based approach has been chosen to define the boundary of an object. More precisely, the *discrete surface* of an object V is the subset $S(V)$ of V such that each voxel of $S(V)$ is 6-connected to at least a voxel of the complement of V in the image. A voxel of $S(V)$ is called a *surface voxel* of V.

The following definitions are extensions for the three dimensional case of the definitions used in the model of Euclidean paths [4].

Definition 1. *Let $P = (x_P, y_P, z_P)$ be a discrete point. The* cell *of P is the set of points p of $\mathbb{R}^3$ verifying: $|x_P - x_p| < \frac{1}{2}$, $|y_P - y_p| < \frac{1}{2}$, $|z_P - z_p| < \frac{1}{2}$.*

The cell of a discrete point P is thus the open unite cube centered on P.

Definition 2. *Let P be a discrete point. We say that p is a Euclidean point associated with P if p belongs to the cell of P.*

In two dimensions it is natural to define the boundary of a discrete region by a discrete path (the nodes of the path being either pixel or inter-pixel positions). In three dimensions the analogous of a discrete path is a graph of discrete point which edges link neighboring elements of the boundary according to the considered topology. This graph describes the polygonal decomposition of the boundary. Since we consider here voxel based boundaries the edges of this graph link centers of adjacent boundary voxels. We do not focus in the following on the definition and the construction of this graph (any topological consistent definition can be used). Our experiments have been done by using the decomposition described by Thuermer [22].

Definition 3. *A* discrete net *is a graph associated which a connected component of the polygonal decomposition of the boundary of a 3D object. A* Euclidean net *associated with a discrete net N is the graph obtained by replacing in N each discrete point by an associated Euclidean point.*

The transform from a discrete point to an associated Euclidean one can be reversed simply by rounding the coordinates of the Euclidean point. The transform from discrete nets to Euclidean nets is thus also reversible. Since there is an infinity of Euclidean nets that can be associated with a discrete one, we must now define a strategy to associate a discrete point with to a Euclidean one. Recall that the goal of the geometric smoothing is to approximate the underlying continuous boundary. Good results were obtained in the 2D case by projecting the discrete point on the support of an arithmetic line *tangent* to the boundary. In the following section we describe a analogous 3D method based on arithmetic digital planes introduced by Reveilles [16].

3 Tangent Plane Driven Euclidean Nets

Let us first recall the following definition:

Definition 4. [16] *A discrete naive plane (in short discrete plane) is a set of points* (x, y, z) *of* $\mathbb{Z}^3$ *satisfying the double inequality* $\mu \leq ax + by + cz < \mu + \omega$, *with* $a, b, c, \mu, \omega \in \mathbb{Z}$ *and* $\omega = max(|a|, |b|, |c|)$.

The coefficients (a, b, c, μ) are called the *characteristics* of the the discrete plane. The vector of coordinates (a, b, c) is the normal vector of the discrete plane. The coefficient μ describes the position of the discrete plane.

Definition 5. *Let* $\mathcal{P}$ *be a discrete plane of characteristic* (a, b, c, μ). *The* lower *and* upper leaning planes *of* $\mathcal{P}$ *are the real planes defined respectively by the equations* $ax + by + cz = \mu$ *and* $ax + by + cz = \mu + \omega - 1$.

Definition 6. *Let* $\mathcal{P}$ *be a discrete plane of characteristic* (a, b, c, μ). *The* centered plane *of* $\mathcal{P}$ *is the real plane* $\overline{\mathcal{P}}$ *defined by the equation* $ax+by+cz = \mu+\frac{\omega-1}{2}$.

The upper and lower leaning planes of a discrete plane $\mathcal{P}$ bound the set of Euclidean points which can be associated with $\mathcal{P}$. Thus the centered plane is an average plane among all the real planes associated with $\mathcal{P}$. We suppose for the rest of this paper that $0 \leq a \leq b \leq c$ and $0 < c$. Thus we have $\omega = c$. The normal vector of such a discrete plane belongs to a 48th of space called *standard simplex* [6]. The other cases can be obtained by symmetry.

The strategy used to associate a Euclidean point p with a point P of a discrete net can be summarized as follows :

1. Select a discrete plane $\mathcal{P}(P)$ containing P.
2. Project P onto the centered plane of $\mathcal{P}(P)$ such that the projected point p satisfies the conditions of Definition 1.

There are several ways to select the plane $\mathcal{P}(P)$. It seams natural to try to find the discrete plane which is the better approximation of the Euclidean plane tangent to the underlying real boundary. We address the problem of the selection of a *discrete tangent plane* in the following section. For now we call tangent plane at P any discrete plane that contains P.

Definition 7. *Let* P *be a discrete point and let* $\mathcal{P}$ *be a discrete plane of characteristics* (a, b, c, μ) *and containing* P. *Let* $\overline{\mathcal{P}}$ *the centered plane of* $\mathcal{P}$. *The* canonical projection *of* P *according to* $\mathcal{P}$ *is the point* p *of* $\overline{\mathcal{P}}$ *satisfying* $\overrightarrow{Pp} = \alpha \begin{pmatrix} a \\ b \\ c \end{pmatrix}$.

From Definition 7 we have $x_p = \alpha a$, $y_p = \alpha b$, and $z_p = \alpha c$. Since $p \in \overline{\mathcal{P}}$ and by considering that P is the origin of the coordinate system we have $ax_p+by_p+cz_p = \mu+\frac{c-1}{2}$, and thus $\alpha a^2+\alpha b^2+\alpha c^2 = \mu+\frac{c-1}{2}$. Since we have supposed that $c > 0$, we finally get:

$$\alpha = \frac{\mu + \frac{c-1}{2}}{a^2 + b^2 + c^2} \tag{1}$$

Theorem 1. *According to the previous definitions the canonical projection of a point P is a Euclidean point associated with P.*

Proof. See appendix A

4 Selection of the Tangent Plane

The previous result states that the canonical projection of P on the centered plane of a discrete plane containing P is a Euclidean point associated with P. We have now to define a strategy to select a plane $\mathcal{P}(P)$ at each point of the discrete net. Several method have been proposed to estimate a *tangent plane* at a point of a discrete surface. In this work we have retained an approach based on the analysis of local configurations called *tricubes*. This choice is motivated by the fact that tricubes provide a simple and robust way to estimate a discrete tangent plane. Of course it would have been interesting to consider more general configurations like for instance (n,m)-cubes [29]. But in this work we have focused on the problem of providing a framework for reversible geometric smoothing in which any tangent plane method can be used. Improvement of methods of determination of tangent plane is out of the scope of this work.

Tricubes were introduced by Debled [6] and are the topic of several recent works [5,18,29,28]. A point P of a digital plane $\mathcal{P}$ has exactly height neighbors. The tricube of P is then the set of P and of its height neighbors. The point P is the center of the tricube. In the standard simplex there are exactly 40 configurations of tricubes.

For each point P of the discrete net we search for local configurations of tricubes centered at P and included in the net. When considering not only the standard simplex but the whole discrete space, some points of the discrete net can be associate with more than one center of tricube. There is an example of such a case in Fig. 2. When such a case arises we compute separately a Euclidean point for each tricube and we retain as Euclidean point the average of these points. Conversely there are configurations where some discrete points cannot be associated with any tricubes. For instance, there are only 58% surface voxels of a discrete sphere of radius 3 that can be associated with a tricube. This amount is of 84% for a sphere of radius 9, and is close to 100% for spheres which radius greater than 10 voxels. When such a case arises, the point may be either left unmoved or treated during a second step. If the second option is retained the associated Euclidean point is computed by interpolating Euclidean points associated with neighbors of the unmoved point.

The example shown in Fig.1(d) has been computed with this method. Of course, the quality of the smoothing highly depends on the quality of the method used to compute the tangent plane. The method used in this work is based on the analysis of small neighborhoods. The geometrical smoothing can be improved by improving the tangent plane recognition method.

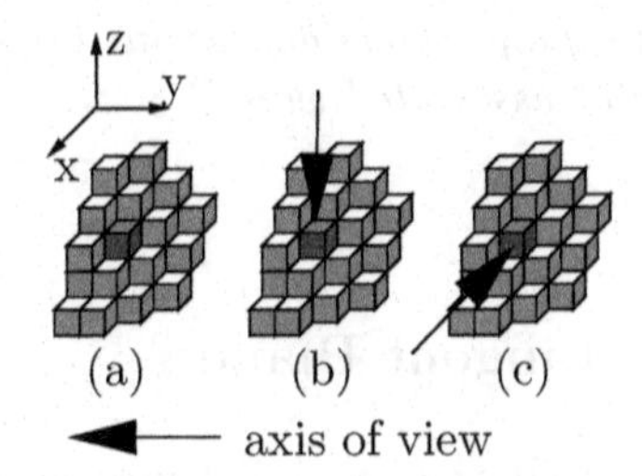

Fig. 2. The dark voxel of the discrete surface (a) can be associated with two centers of tricube (b) and (c)

5 Experiments

In this section we give some quantitative and qualitative results of experiments performed with the method of tangent plane estimation described in the previous section.

Let σ be a Euclidean sphere of radius r and S the associated digitized sphere (the set of voxels which the center is located inside the Euclidean sphere). For each point P of S we denote by $\rho(P)$ the distance from P to the center of the sphere (we suppose that the center C of the sphere σ is a discrete point). We define the *maximal geometrical error* (resp. *average geometrical error*) by the maximum (resp. average) of $\rho(P)$ for all the discrete points P of the boundary of S. In Fig. 3 (a) are displayed maximal and average errors for discrete spheres from radius 2 to radius 45 before and after smoothing. Discrete points that do not fit with a tricube are smoothed by interpolation. Both maximal and average errors are reduced by half.

In Figure 3 (b) is given the result of experiments of normal estimation. The normals have been estimated according to both the discrete net and the Euclidean associated net. The normal at a point of the net is computed by averaging the normals vectors of the adjacent facets. For each point p of either the discrete net or the Euclidean associated net the error on the normal is the angle between the vector $\overrightarrow{Cp}$ and the estimated normal vector. Both maximal and average errors are given for spheres from radius 2 to radius 45. The average error with the Euclidean net is lower than three degrees. The maximal error is about ten degrees for small radius and tends toward fifteen degrees when the radius increases.

Naturally these results depend on the method used to determine the tangent plane and can be improved by using a more global method than tricubes. For instance an idea to improve the estimation of the tangent plane would be to consider stochastic methods using disconnected configuration of (n, m)-cubes ((n, m)-cubes [29] are configurations that generalize tricubes).

Some examples of geometrical smoothing can be seen in Fig.4. For each example the image of the first column is a flat shading of the facets of the discrete net and the image of the second column is a flat shading of the facets of

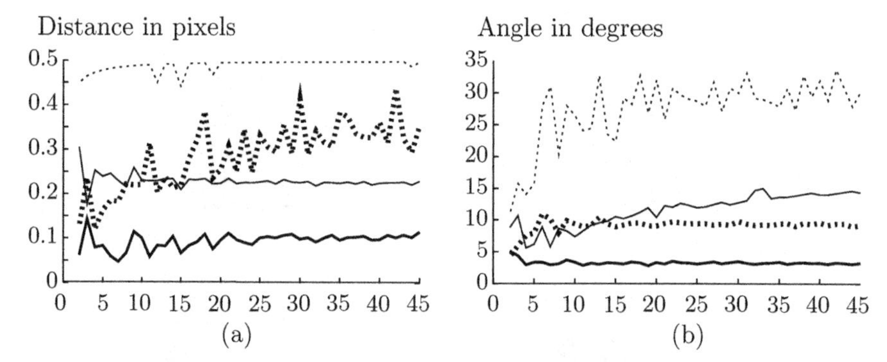

Fig. 3. (a) Maximal (dotted lines) and average (plain lines) geometrical error computed on discrete (thin lines) and smoothed (thick lines) spheres. (b) Maximal and average error of the estimated normal vectors of discrete and smoothed spheres

the associated Euclidean net. The three first rows show scenes made of regular objects (respectively a cube slightly rotated around one of its axis, a sphere of radius 20 and a combination of a plane with a sphere. The examples of the two last rows are computed from digitized 3D objects provided by the *Laboratoire des Sciences de l'Image, d'Informatique et de Télédétection* of Strasbourg University.

There are still visible staircase on the smoothed images that could be attenuated by improving the selection of the tangent plane. Nevertheless it is important to emphasize that the images of the second column show the result of geometric smoothing without rendering improvement. The displayed surface is exactly the smoothed surface and can be explicitly used for extract geometric features. Rendering methods based on a normal estimation [23,19] can give better visual results. On the other hand they does not provide an explicit geometric smoothing of the boundary.

6 Conclusion

In this work we have described a new approach to perform a discrete geometric smoothing of the boundary of a 3D object. This method extends to the three-dimension case the method of Euclidean paths. Each center of voxel P of the boundary is associated with a Euclidean point p located inside the open unit cube centered on P. Given a naive plane containing P, the Euclidean point p is the canonical projection of P on the centered plane of the naive plane. By this way a Euclidean net of Euclidean points is associated with the discrete boundary.

Any definition of *tangent* plane can be used to computed canonical projections. Thus the accuracy of the Euclidean net construction (and of the smoothing) highly depends on the accuracy of the method used to compute the tangent plane. In this work the tangent plane at a boundary point is estimated by an-

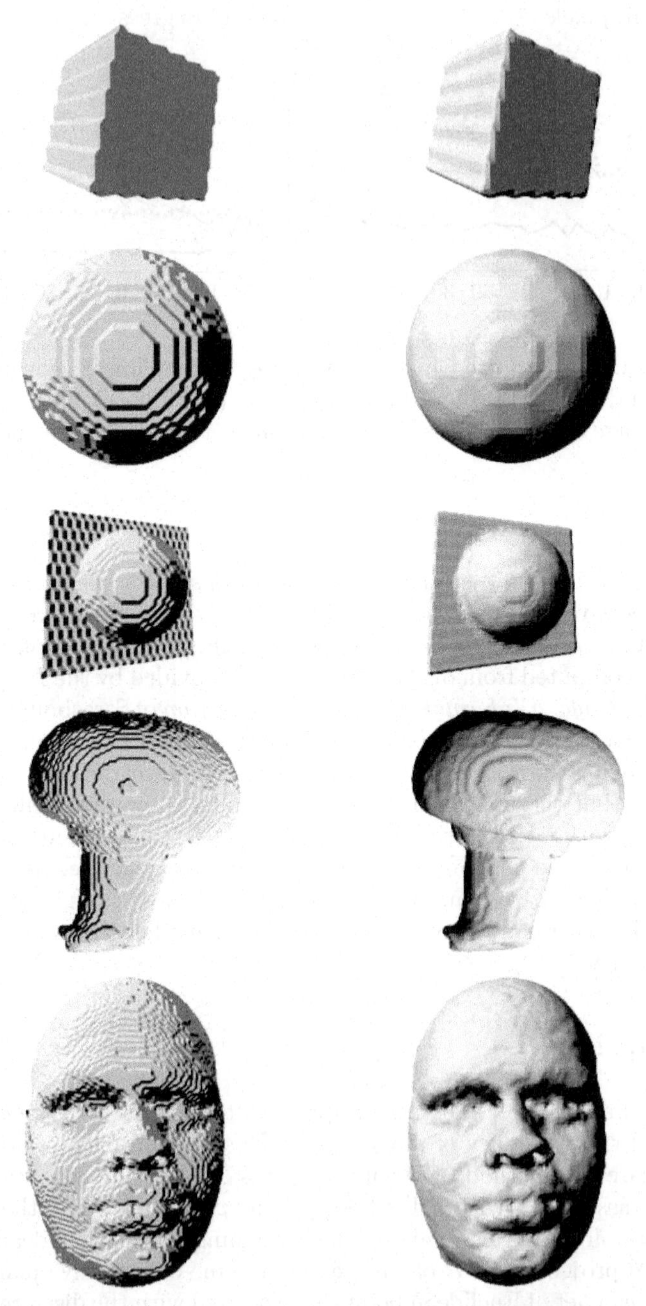

Fig. 4. Flat shading of the discrete net (first column) and of the associated Euclidean net (second column) of 3D discrete objects : a cube (first row), a sphere (second row), a combination of a plane and a sphere (third row), and digitized real objects provided by the LSIIT (fourth and fifth rows)

alyzing tricubes configurations centered on this point. Thought this method is limited to a local analysis of the boundary around the point it gives promising results. The geometric smoothing must now be experimented with other methods of determination of tangent plane. Both quantitative and qualitative results could then be improved.

We are currently working on Euclidean nets of inter-voxel boundaries. An advantage of inter-voxel boundaries is that they provide a simple and consistent representation of the topology of the object they bounds (the discrete net is trivially obtained from the data structure). On the other hand from a geometrical point of view the inter-voxel boundaries provide a quite poor representation of the shape of the object. In that case the relevance of the Euclidean net approach is thus increased.

The method has also to be adapted to the case of a non-manifold boundary resulting from the segmentation of an aggregate of 3D discrete objects [2]. Finally we think that Euclidean nets will provide and efficient framework to extract geometrical parameters from a 3D object as the method of Euclidean path does for 2D regions [25].

References

1. J. P. Braquelaire, L. Brun, and A. Vialard. Inter-pixel euclidean paths for image analysis. In *Discrete Geometry for Computer Imagery*, Lecture Notes in Computer Science 1176, pages 193–204. Springer Verlag, November 1996. 199
2. J. P. Braquelaire, P. Desbarats, J. P. Domenger, and C. A. Wüthrich. A topological structuring for aggregates of 3D discrete objects. In *Proc. of the 2nd IAPR-TC-15 Workshop on Graph-based Representation*, pages 193–202. Österreichische Computer Gesellschaft, 1999. ISBN 3-85804-126-2. 207
3. J. P. Braquelaire and A. Vialard. A new antialiasing approach for image compositing. *The Visual Computer*, 13(5):218–227, 1997. 199
4. J. P. Braquelaire and A. Vialard. Euclidean paths : a new representation of boundary of discrete regions. *Graphical Models and Images Processing*, 61:16–43, 1999. 198, 199, 201
5. J. M. Chassery and J. Vittone. Coexistence of tricubes in digital naive plane. In *Lecture Notes in Computer Science*, volume 1347, pages 99–110, December 1997. 203
6. Isabelle Debled. *Etude et reconnaissance des droites et plans discrets.* Phd thesis, Université Louis Pasteur, Strasbourg, France, 1995. 202, 203
7. T. J. Fan, G. Medioni, and R. Nevata. Recognising 3D objects using surface descriptions. *IEEE Transactions on Pattern Analisys and Machine Intelligence*, 1111:1140–1157, 1989. 199
8. J. Françon. Discrete combinatorial surfaces. *Grapical Models and Image Processing*, 57(1):20–26, 1995. 199
9. Henri Gouraud. Continuous shading of curved surfaces. *IEEE Transactions on Computers*, C-20(6):623–629, June 1971. 199
10. G. T. Herman and H. K. Liu. Three-dimensionnal display of uman organs from computed tomograms. *Computer Graphics and Image Processing 9*, 1:1–21, 1979. 198

11. K. H. Hohne, M. Bomans, A. Pommert, M. Riemer, C. Schiers, U. Tiede, and G. Wiebecke. 3D visualization of tomographic volume data using the generalized voxel model. *The Visual Computer*, 6(1):28–36, 1990. 198
12. J.-O. Lachaud and A. Montanvert. Deformable meshes with automated topology changes for coarse-to-fine 3D surface extraction. *Medical Image Analysis*, 3(2):187–207, 1999. 199
13. W. E. Lorenson and H. E. Cline. Marching cubes: A high resolution 3D surface construction algorithm. *Computer Graphics (SIGGRAPH'87)*, 21(4):111–118, 1987. 198
14. R. Malgouyres. A definition of surfaces of $\mathbb{Z}^3$: A new discrete jordan theorem. *Theoretical Computer Sciences*, 186:1–41, 1997. 199
15. FD. G. Morgenthaler and A. Rosenfeld. Surfaces in three-dimensional digital images. *Information and Control*, 51:227–247, 1981. 199
16. J. P. Reveilles. *Géométrie discrète, Calcul en nombres entiers et algorithmique.* PhD thesis, Université de Strasbourg, 1991. 201, 202
17. A. Rosenfeld, T.Yung Kong, and A. Y. Wu. Digital surfaces. *Grapical Models and Image Processing*, 53(4):305–312, 1991. 199
18. J. M. Schramm. Coplanar tricubes. In Springer-Verlag, editor, *Lecture Notes in Computer Science*, volume 1347, pages 87–98, 1997. 203
19. P. Tellier and I. Debled-Rennesson. 3d discrete normal vectors. In Springer-Verlag, editor, *Lecture Notes in Computer Science*, volume 1568, pages 447–458, 1999. 205
20. D. Terzopoulos, A. Witkin, and M. Kass. Symmetry-seeking models and 3D object reconstruction. *International Journal of Computer Vision*, 1(3):211–221, 1987. 199
21. D. Terzopoulos, A. Witkin, and M. Kass. Constraints on deformable models: Recovering 3D shape and nonrigid motion. *Artificial Intelligence*, 36(1):91–123, 1988. 199
22. G. Thürmer. *Surfaces in Three-Dimensional Discrete Space.* PhD thesis, Fakultät Medien der Bauhaus-Universität Weimar, 1998. Shaker Verlag. 199, 201
23. G. Thürmer and C. A. Wüthrich. Normal computation for discrete surfaces in 3D space. *Computer Graphics Forum*, 16(3):15–26, 1997. Proceedings of Eurographics'97. 205
24. G. Thürmer and C. A. Wüthrich. Polygon mesh generation for discrete surfaces in 3d space. In W. Lefer and M. Grave, editors, *Proc. of the Eighth Eurographics Workshop on Visualization in Scientific Computing*, pages 117–126, Laboratoire d'Informatique du Littoral, Boulogne sur Mer (France), 1997. 199
25. A. Vialard. Geometrical parameters extraction from discrete paths. In *Discrete Geometry for Computer Imagery*, Lecture Notes in Computer Science 1176, pages 24–35. Springer Verlag, November 1996. 199, 207
26. A. Vialard. Euclidean paths for representing and transforming scanned characters. In *Proceedings of GREC'97.* Second IAPR Workshop on Graphics Recognition, 1997. To be published in Lecture Notes in Computer Science, Springer Verlag. 199
27. Anne Vialard. *Chemins euclidiens : Un modèle de représentation des contours discrets.* Phd thesis, Université Bordeaux 1, 1996. 199
28. J. Vittone. *Caractérisation et reconnaissance de droites et de plans en géométrie discrète.* Phd thesis, Université de Grenoble, 1999. 203
29. J. Vittone and J. M. Chassery. (n,m)-cubes and farey nets for digital naive planes understanding. In *Lecture Notes in Computer Science*, volume 1568, pages 76–87, 1999. 203, 204

A Proof

We have to verify that the canonical projection p satisfies the condition of Definition 2. We can suppose that the origin of the coordinates system is the discrete point P. Since $a \leq b \leq c$, in order to show that the point

$$p = \left(a\frac{\mu + \frac{c-1}{2}}{a^2+b^2+c^2}, b\frac{\mu + \frac{c-1}{2}}{a^2+b^2+c^2}, c\frac{\mu + \frac{c-1}{2}}{a^2+b^2+c^2} \right)$$

belongs to the unit cell centered at P it suffices to verify that:

$$\left| c\frac{\mu + \frac{c-1}{2}}{a^2+b^2+c^2} \right| < \frac{1}{2}$$

The discrete tangent plane $\mathcal{P}$ is the set of points satisfying $\mu \leq ax+by+cz \leq \mu + c - 1$. From the hypothesis the point of coordinates $(0,0,0)$ belongs to $\mathcal{P}$. Thus we have from the definition of discrete plane $\mu \leq 0$ and thus

$$\begin{aligned}
\mu &\leq 0, \\
2\mu &< 1, \\
2\mu + c - 1 &< c, \\
c(2\mu + c - 1) &< c^2 \quad \text{since } c > 0, \\
c\frac{\mu + \frac{c-1}{2}}{a^2+b^2+c^2} &< \frac{c^2}{2(a^2+b^2+c^2)} \quad \text{since } a^2+b^2+c^2 > 0.
\end{aligned}$$

Finally, since $a^2+b^2+c^2 \geq c^2$ we have $c\frac{\mu+\frac{c-1}{2}}{a^2+b^2+c^2} < \frac{c^2}{2c^2}$ and thus we get:

$$c\frac{\mu + \frac{c-1}{2}}{a^2+b^2+c^2} < \frac{1}{2}.$$

By a symetrical argument from $0 \leq \mu + c - 1$ we get:

$$c\frac{\mu + \frac{c-1}{2}}{a^2+b^2+c^2} > -\frac{1}{2}. \qquad \square$$

Object Discretization in Higher Dimensions

Valentin E. Brimkov[1], Eric Andres[2], and Reneta P. Barneva[1]

[1] Eastern Mediterranean University,
Famagusta, P.O. Box 95, TRNC, Via Mersin 10, Turkey
{valentin.brimkov,reneta.barneva}@emu.edu.tr
[2] SIC-IRCOM, Université de Poitiers, Bat. SP2MI – BP. 179,
F–86960 Futuroscope Cédex – France
andres@sic.sp2mi.univ-poitiers.fr

Abstract. In this paper we study discretizations of objects in higher dimensions. We introduce a large class of object discretizations, called k-discretizations. This class is natural and quite general, including as special cases some known discretizations, like the standard covers and the naive discretizations. Various results are obtained in the proposed general setting.

Keywords: Object discretization, Supercover, Standard cover, Naive discretization, k-Discretization

1 Introduction

With the present paper we undertake an investigation aimed at providing a general framework for developing a discrete geometry of surfaces and bodies of *arbitrary* dimension. We define and study certain discretizations that generalize some well-known discretization schemes, applicable mainly to linear objects (e.g., hyperplanes). Among them, one can list some recent studies on supercovers and standard covers of hyperplanes and affine subspaces [3,1,2], as well as earlier discretization schemes (see, e.g., [18,20]).

One may expect that the theoretical and practical worth of multidimensional discrete geometry will be increasing. An argument in support of such an expectation is provided by some recent applications. In fact, 4-dimensional imaging is already existing, due to, e.g., PET scans and other dynamic medical images. Further applications are to be expected in multimedia where hypertexts together with images, motion video and sound are manipulated. Another motivation is inspired by mathematical arguments. For every Euclidean notion we may have numerous corresponding discrete notions, depending on the applications and the Euclidean property we want to preserve. The great diversity of possible discretizations is due to their discrete nature. If we consider objects and properties (especially topological properties) only in 2D or 3D, we will not necessarily choose definitions that would have natural extensions in higher dimensions later on, in case we need to go to higher dimensions (for animation, for instance). In other words, a geometry that is not coherent in all dimensions cannot serve as a

G. Borgefors, I. Nyström, and G. Sanniti di Baja (Eds.): DGCI 2000, LNCS 1953, pp. 210–221, 2000.
© Springer-Verlag Berlin Heidelberg 2000

good ground for future developments. More general results in higher dimensions will give insight to the practically important 2D and 3D versions of the theory, and may facilitate the deeper understanding of the matter. Thus, developing a reasonable n-dimensional discrete geometry is not simply an academic exercise, but is dictated by real needs.

We rely implicitly on the existing knowledge in the digital topology (see, e.g., [15,16,14,8]). In particular, we conform to certain widely adopted terminology, which we briefly recall in Section 2. In Section 3 we define a *minimal cover* of a *body* and study some basic properties of such covers. In Section 4 we introduce the fundamental concept of *k-discretization* and *k-cover* of *surfaces*. We also expose a relation between the k-discretizations and some other discretizations based on Hausdorff distance [21,19]. In Section 4.3 we present some results about connectivity of k-discretizations, while in Section 4.4 we illustrate some of the new notions and results in the case of hyperplanes and halfspaces. In particular, the 0-discretizations of hyperplanes appear to be *standard* discrete planes. The latter have been studied in the framework of the arithmetic geometry [2,17].

2 Preliminaries

We recall some basic notions to be used in the sequel. To obtain more details about some of them, the reader is referred to [9].

A *discrete* (resp. *Euclidean*) *point* p is an element of Z^n (resp. R^n). Throughout we will assume that $n \geq 2$. By p_i we denote the ith coordinate of p. A *discrete* (resp. *Euclidean*) *object* is a set of discrete (resp. Euclidean) points.

Let $p \in Z^n$. A *voxel* with center p is the unit cube $V(p) = [p_1 - \frac{1}{2}, p_1 + \frac{1}{2}] \times \ldots \times [p_n - \frac{1}{2}, p_n + \frac{1}{2}]$. A j-dimensional facet of $V(p)$ will be referred as to a *j-facet*, $0 \leq j \leq n$[1]. We also define $V(A) = \cup_{p \in A} V(p)$, i.e., $V(A) = \{x \in R^n : x \in V(p)\ for\ some\ p \in A\}$.

The Euclidean points $p = (p_1, \ldots, p_n)$ and $q = (q_1, \ldots, q_n)$ are said to be *k-neighbors*, with $0 \leq k \leq n$, if $|p_i - q_i| \leq 1$ for $1 \leq i \leq n$, and $k \leq n - \sum_{i=1}^{n} |p_i - q_i|$. A *k-path* in a discrete object A is a sequence of discrete points from A such that every two consecutive points are k-neighbors. A is called *k-connected* if there is a k-path connecting any two points of A. A *k-component* is a maximal k-connected subset of A.

Let D be a subset of a discrete object A. If $A - D$ is not k-connected then the set D is said to be *k-separating* in A.

Now we introduce some other notions. Let $M \subseteq R^n$ be a Euclidean object. For technical convenience throughout the paper we will assume that M is a closed set. A point $p \in R^n$ is called *boundary point* for M if it is neither internal nor external for M. The set of all boundary points for M is called *boundary* of M and denoted $Bd(M)$. Since M is closed, $Bd(M) \subseteq M$.

A *cover* of M is any set of voxels $C(M)$ whose union contains M. The *supercover* of M is the set $S(M)$ of all voxels which are intersected by M. Clearly,

[1] Thus the 0-facets are the voxel vertices, the 1-facets – its edges, the $(n-1)$-facets are the voxel faces, and the n-facet is the voxel itself.

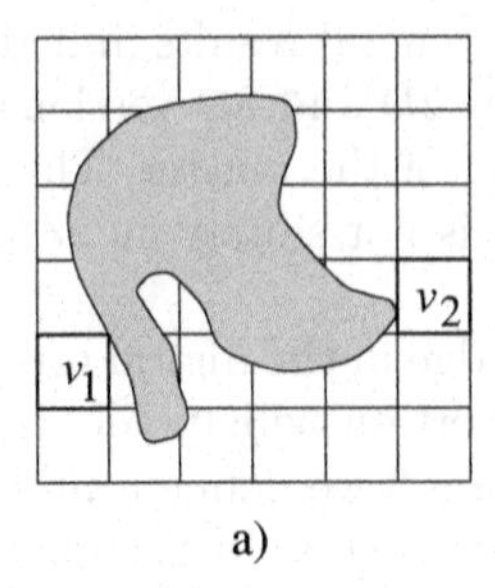

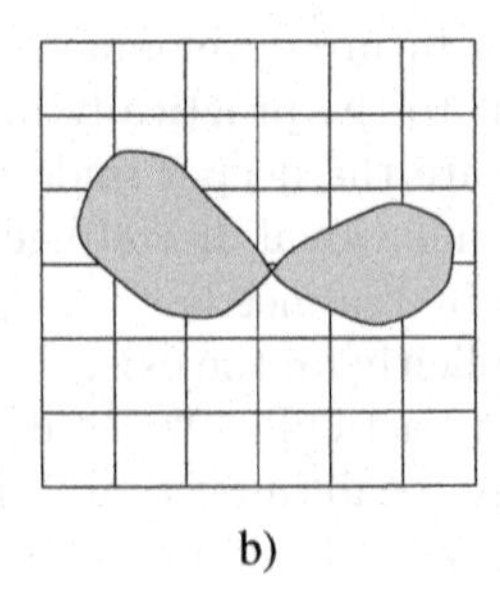

a) b)

Fig. 1. a) Example of an object that is a body in R^2. The voxels v_1 and v_2 are osculant for the object. b) Example of an object that is not a body. It has no osculant voxels

each of the following inclusions is possible for a cover $C(M)$ and a supercover $S(M)$: (i) $S(M) \subseteq C(M)$; (ii) $C(M) \subseteq S(M)$. In our study we will be interested in covers of Type (ii), which are "tighter" than or identical to supercovers. We will consider covers and other discretizations that are not necessarily covers, of Euclidean sets with topological dimension h, where $0 \leq h \leq n$. In particular, we will consider covers of bodies. We will say that a closed set $B \subseteq R^n$ is a *body* if it is simply connected[2], has a topological dimension $dim(B) = n$, and for every point $p \in Bd(B)$, $B - \{p\}$ is a simply connected set. For example, the set in Figure 1a is a body in R^2, while the one on Figure 1b is not. We will also study discretizations of surfaces of different dimension, in particular, $(n-1)$-dimensional surfaces. We will suppose that the surface is given in explicit form $\Gamma = \{x \in R^n : x_n = f(x_1, \ldots, x_{n-1})\}$, where f is a continuous function defined on R^{n-1}.

We will call a voxel $V(p)$ *osculant* for a Euclidean object M if $V(p) \cap M$ is contained in a j-facet of $V(p)$, for $0 \leq j < n$ (Figure 1a). An *n-bubble* in R^n is a set of 2^n voxels sharing a common vertex Q. Equivalently, an n-bubble can be considered as the supercover of a half-integer point, namely the point Q in the alternative definition.

We conclude this section with the following lemma, which will be used in the next section.

Lemma 1. *Let $M \subseteq \mathbf{R}^n$ be an arbitrary connected Euclidean set. Then $S(M)$ is $(n-1)$-connected.*

A proof can be obtained as a straightforward generalization of the proof for the case $n = 3$ [9]. For a detailed accounting on various other properties of supercovers we refer to [1]. In particular, it is well-known that a supercover is not, in general,

[2] We recall that a set $M \subseteq R^n$ is called *simply connected* if it is homeomorphic to the unit ball in R^n.

3 Minimal Covers

In this section we consider a natural and important class of covers.

Definition 1. *Let M be a Euclidean object. We call its cover $C(M)$* minimal *if there is no other cover of M that is a proper subset of $C(M)$.*

Next we study some properties of the minimal covers.

Proposition 1. *If a cover of an object M does not contain any osculant voxel, then it is minimal.*

Proof Let $C(M)$ be a cover of M that does not contain osculant voxels, and let v be an arbitrary voxel from $C(M)$. By definition of osculant voxel, the intersection $M_v = v \cap M$ is not contained in a facet of v, and therefore is also not contained in a facet of any other voxel. Let us remove v from $C(M)$. The remaining set of voxels $C(M) - \{v\}$ does not contain M_v, since a set cannot be contained in two distinct voxels at the same time, unless it is contained in a common facet of theirs. Thus $C(M) - \{v\}$ is not a cover of M. Hence the cover $C(M)$ is minimal. □

In the case of bodies, the condition of the above theorem completely characterizes the minimal covers, as the following theorem suggests.

Proposition 2. *A cover of a body (not necessarily bounded) is minimal if and only if it does not contain any osculant voxels.*

Proof One direction of the proof follows from Proposition 1. We prove the other direction. Let B be a body and $C(B)$ its minimal cover. Assume that $C(B)$ contains a voxel v which is osculant for B. Then we have $B_v = v \cap B \subseteq F$, where F is a facet of v.

If $C(B)$ contains another voxel $v' \supseteq B_v$, then $C(B) - \{v\}$ is still a cover of B and thus $C(B)$ is not minimal. Let us assume that B_v is not contained in any voxel of $C(B)$ other than v. Note that B_v cannot be a single point. The opposite would be possible either if B_v belongs also to another voxel (one containing the rest of B_v or a part of it) or if $B = B_v$, which is impossible since B is a body.

We distinguish between two cases. Let us first assume that $B_v = B$. Since B_v is contained in a facet of v with topological dimension less than or equal to $n-1$, the same follows for the dimension of B, which contradicts the fact that B is a body. The other possibility is $B = B_v \cup B'$ for some set $B' \neq \emptyset$, such that $B_v \cap B' = \emptyset$. Since B_v and B' are disjoint and B is simply connected, B_v is simply connected, as well. Since $0 \leq \dim(B_v) \leq n-1$, every point of B_v belongs to $Bd(B)$. Then there will be a point $p \in B_v$ such that the set $B_v - \{p\}$ is not simply connected, which contradicts the definition of body. □

Proposition 3. *A body B has a unique minimal cover.*

Proof Consider the supercover $S(B)$ of B, and let us remove from it all voxels which are osculant for B. Having in mind the proof of Proposition 2, we can conclude that the obtained set of voxels is a minimal cover, which is clearly unique. □

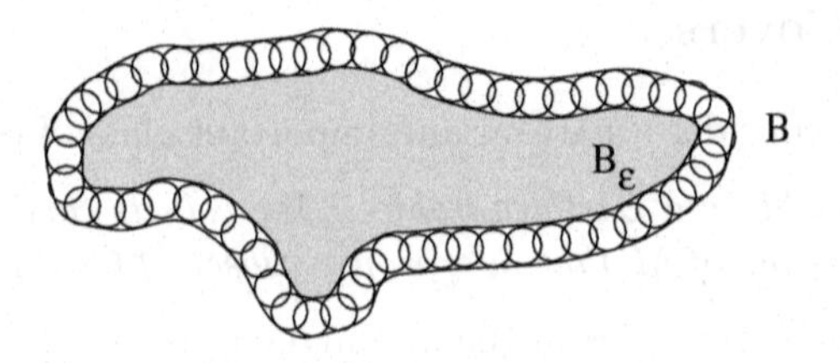

Fig. 2. Illustration to the proof of Theorem 1

Theorem 1. *The minimal cover of a body B is $(n-1)$-connected. It can be obtained from the supercover of B by removing all osculant voxels for B.*

Proof Let $S(B)$ be the supercover of a body B, and let us remove from it all the voxels that are osculant for B. Denote the resulting set of voxels by $C(B)$. By Propositions 2 and 3, $C(B)$ is the unique minimal cover of B. We will show that $V(B)$ is $(n-1)$-connected. For this, let us consider the set $Cl(B_\varepsilon) \subset B$, which is the closure of the set B_ε obtained from B by excluding the union of all the closed balls with diameter ε which are contained in B and have a single common point with the boundary of B (see Figure 2). Since B is a body, for ε small enough (e.g., $\varepsilon < 1/m$ for enough large integer m) B_ε will be a body as well, and thus a connected subset of R^n. It is also clear that for the so defined ε, the set $C(B)$ will be a supercover of B_ε. Then, by Lemma 1, $C(B)$ is $(n-1)$-connecded as a supercover of a connected set. □

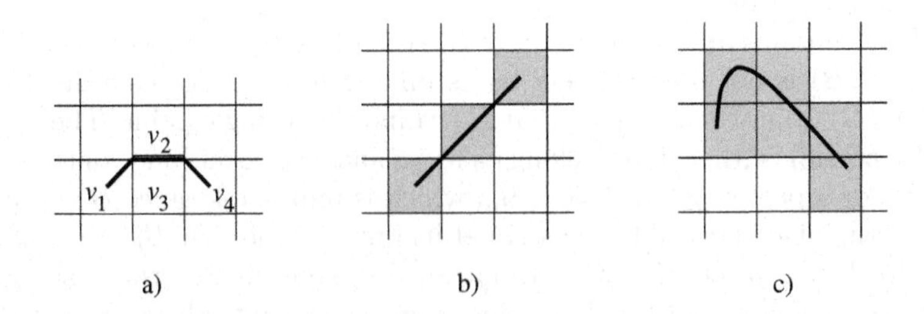

Fig. 3. a) The curve has two minimal covers: $\{v_1, v_2, v_4\}$ and $\{v_1, v_3, v_4\}$. Each of them contains a voxel that is osculant for the curve (v_2 and v_3, respectively). b) The segment on the figure has a minimal cover that is not $(n-1)$-connected. c) The minimal cover of the curve contains a 2-bubble

With surfaces, Propositions 2, 3, and Theorem 1 do not hold, in general, as illustrated in Figures 3a and b. A minimal surface cover may contain osculant voxels (Figure 3a), as well as bubbles (Figure 3c).

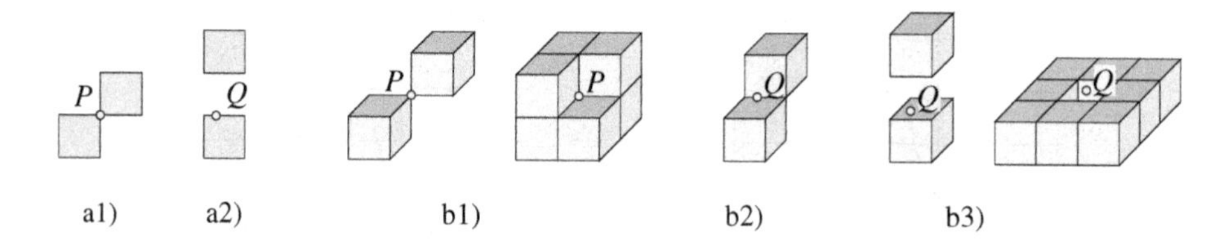

Fig. 4. Illustration to Definition 2. In dimension 2: a1) 0-tunnel; a2) 1-tunnel. In dimension 3: b1) 0-tunnel; b2) 1-tunnel; b3) 2-tunnel

4 Surface Discretizations

In this section we are concerned with discretizations of surfaces of different dimensions. We introduce and study a large class of discretizations called k-discretizations. For this purpose, we first introduce a definition of tunnels in discrete objects.

4.1 Tunnels in Discrete Objects

Bellow we suggest a definition of tunnels, applicable to a broad class of discrete objects. Let $A \subseteq R^n$ be a discrete object, such that the set $R^n - V(A)$ is connected. (This condition ensures that there are no "caves" in $V(A)$.)

Definition 2. *If the set $V(A)$ is not simply connected, we postulate that A has a j-tunnel, for every $0 \leq j \leq n-1$.*

0-tunnels*: If $n = 2$, then A has a 0-tunnel if A contains two voxels whose intersection is their common vertex and no other voxel from A shares the same vertex (Figure 4 a1). If $n > 2$, then A has a 0-tunnel either if A contains two voxels whose intersection is their common vertex and no other voxel from A shares the same vertex, or if there is a voxel $v \in A$ and a vertex P of v with $P \in Bd(V(A))$, such that $V(A) - \{P\}$ is not a simply connected set (Figure 4 b1).*

j-tunnels, $1 \leq j \leq n-1$*: A has a j-tunnel if there is a voxel $v \in A$ and a j-facet F of v, such that $F \subseteq Bd(V(A))$, and $V(A) - \{Q\}$ is not a simply connected set, for every point Q belonging to the relative interior*[3] *of F. Thus if A is disconnected, it has $(n-1)$-tunnels. (See Figure 4 a2,b2,b3.)*

A discrete object without k-tunnels is called *k-tunnel-free.*

4.2 k-Discretizations and k-Covers

We start with recalling the definition of Hausdorff distance. Let E be a metric space with metric d, and $\mathcal{E}$ a family of closed non-empty subsets of E. For every $x \in E$ and every $A \in \mathcal{E}$ let $d(x, A) = \inf\{d(x, y) : y \in A\}$. Then, given two sets

[3] *Relative interior* of a j-facet F is the interior of F, considered as a subset of R^j.

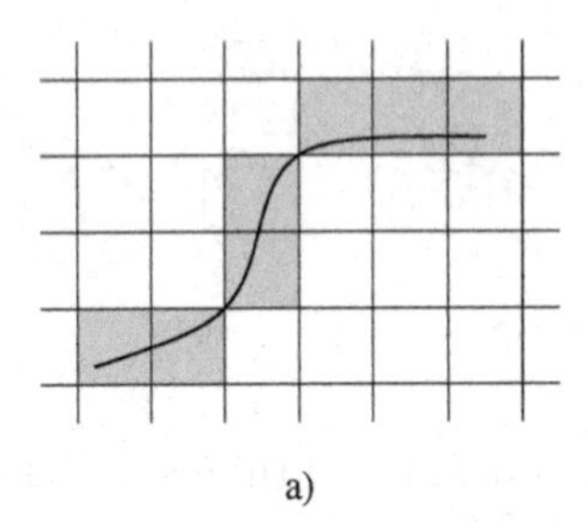

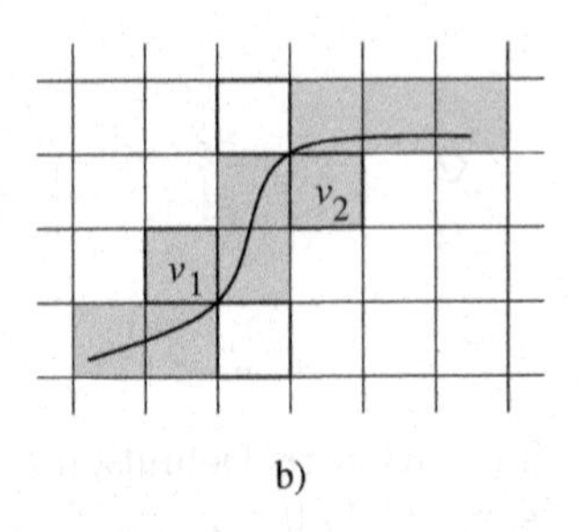

Fig. 5. Illustration to Definition 3. a) 1-Discretization. b) 0-Discretization. It can be obtained from the 1-discretization by adding the voxels v_1 and v_2

$A, B \in \mathcal{E}$, $H_d(A, B) = \max\{\sup\{d(a, B) : a \in A\}, \sup\{d(A, b) : b \in B\}\}$ is called the *Hausdorff distance* between A and B.

Let a surface $\Gamma \subset R^n$ be given, with topological dimension h, $0 \leq h \leq n-1$,[4] i.e., $\Gamma = \{x \in R^n : x_i = f_i(t_1, \ldots, t_h), 1 \leq i \leq n, t_j \in R, 1 \leq j \leq h\}$ where f_i are continuous functions of the variables $t_1, \ldots, t_h$. In a very general sense, we will consider every subset of the supercover $S(\Gamma)$ as a *discretization* of Γ. Of interest for us are discretizations that are $(n-1)$-tunnel-free (i.e., without "holes"). Let $D(\Gamma) \subseteq S(\Gamma)$ be a k-tunnel-free discretization of Γ. We will call a discrete point $p \in D(\Gamma)$ *k-simple*, if $D(\Gamma) - \{p\}$ is still k-tunnel-free. If a k-tunnel-free discretization $D(\Gamma)$ does not contain k-simple points, we will call it *k-minimal*[5]. Now we are able to give the following definition.

Definition 3. *Let $S(\Gamma)$ be the supercover of Γ, and $\mathcal{D}_k(\Gamma)$ the family of all subsets of $S(\Gamma)$ that are k-minimal, for some $0 \leq k \leq n-1$. We will call a set of voxels $D_k(\Gamma) \in \mathcal{D}_k(\Gamma)$ a* k-discretization *of Γ if it is $(n-1)$-tunnel-free and the Hausdorff distance $H_d(\Gamma, V(D_k(\Gamma)))$ is minimal, over all the elements of $\mathcal{D}_k(\Gamma)$.*

Further we consider mainly the case when d is the Euclidean metric in R^n. Some illustrations to Definition 3 are presented in Figure 5. We remark that in this definition the supercover of a surface is used as a natural upper bound for the amount of voxels included in a k-discretization.

Clearly, for $1 \leq k \leq n-1$, a k-discretization of a surface is not necessarily a cover of its, while a 0-discretization always is. If $\Gamma \subseteq D_k(\Gamma)$, we call the discretization $D_k(\Gamma)$ a *k-cover* of Γ. One can observe that the maximal k for

[4] For example, if $h = 0$, Γ is a Euclidean point, whereas if $h = 1$, Γ is a curve in R^n.

[5] Traditionally, the following definitions are used. Let B be k-separating in a discrete object A and $A - B$ have exactly two connected components. A *k-simple point* of B is a point $p \in B$ such that $B - p$ is still k-separating. A k-separating discrete object is called *k-minimal* if it does not contain any k-simple point. The advantage of our version of the definition of k-minimality is that it applies also to an object that is not k-separating in another object.

which a k-discretization of a surface appears to be a cover, corresponds to a minimal cover as defined in the previous section.

Of a special interest are the two extreme cases $k = 0$ and $k = n - 1$. As a matter of fact, 0- and $(n - 1)$-discretizations have been studied in the special cases when Γ is a straight line or plane. Among the earliest works is the one of Bresenham who proposes an algorithm to discretize a straight line in the plane [5]. Reveillès [17] considers Bresenham lines in the framework of the arithmetic geometry. He shows that the Bresenham line, called also *naive* line, admits an analytical description. 3D naive planes and lines are also defined and studied [10,12,13,11]. The naive lines/planes are proved to be the thinnest possible discretizations that approximate best the corresponding Euclidean lines/planes. It is easy to observe that they are $(n - 1)$-discretizations. Naive discretizations have also been studied in the case of hyperplanes. Stojmenovic et al. [18] propose a scheme for discretizing hyperplanes. Further related results within the Stojmenovic scheme are obtained in [20]. In [3] it is shown that the Stojmenovic discretization is an nD analog of naive discretization. As in the 2D and 3D cases, a naive discretization of a hyperplane appears to be an $(n - 1)$-discretization.

0-Discretizations of 2D lines and of planes are also well-studied. Such sort of discretizations have been defined and investigated in analytical setting [17,10], and are known as *standard* discretizations. In higher dimensions, standard discretizations are considered in [2].

Definition 3 provides a structural framework for studying naive $(n - 1)$-discretizations and standard 0-discretizations for *arbitrary* surfaces. It also gives a more general view on the matter, considering the naive and the standard discretizations as (extreme) particular cases within the class of k-discretizations. One can expect that various k-discretizations with $k \neq 0, n-1$ may be interesting for certain applications.

We remark that discretizations based on the concept of Hausdorff distance have been proposed in [21,19]. More precisely, a *Hausdorff discretization* of a compact set $K \subset R^n$ is defined as the union of all subsets of Z^n minimizing the Hausdorff distance to K. One can observe that all the naive $(n-1)$-discretizations that are in a minimal Hausdorff distance from a Euclidean object contribute to its Hausdorff discretization.

The Hausdorff and the k-discretization elucidate discretization issues from different angle. The former one is mathematically elegant and has some interesting properties, see [19, Sections 3,4]. Besides, interesting relations hold between the Hausdorff discretization and other well-known discretizations. For instance, under certain conditions, the Hausdorff discretization is equivalent to the supercover discretization defined above [19, Theorem 2].

On the other hand, with the study of k-discretizations we pursue deeper understanding of the structure of some natural discretizations (like the naive and the standard discretizations) within a more general setting. The k-discretizations can be considered as the thinnest possible discretizations without tunnels of certain type, while the Hausdorff discretizations, in general, can be seen to be thicker. In this sense, our approach is much in the spirit of some recent work [7,6,4] on thin polyhedral discretizations.

4.3 Connectivity of k-Discretizations

In this section we report our basic result about connectivity of k-discretizations of surfaces.

Theorem 2. *Let $\Gamma \subset R^n$ be an h-dimensional surface, $0 \le h \le n-1$. A k-discretization of Γ with $0 \le k \le n-1$ is $\max\{h-1, n-k-1\}$-connected.*

Idea of the Proof Let $D_k(\Gamma)$ be a k-discretization of Γ. Let v' and v'' be arbitrary two voxels from $D_k(\Gamma)$. Since $D_k(\Gamma)$ is connected, there is an ordered set of voxels $\{v_1, v_2, \ldots, v_l\}$ with $v_1 = v'$, $v'' = v_l$, such that any two consecutive elements v_i, v_{i+1}, $1 \le i \le l-1$, are at least 0-connected. To obtain a proof of the theorem, it suffices to show that v_i and v_{i+1} are $(h-1)$- and $(n-k-1)$-connected.

1. $(h-1)$-*connectivity.* Let F be the facet of maximal dimension f, $0 \le f \le n-1$, so that F is common for v_i and v_{i+1}. If $f \ge h-1$, then v_i and v_{i+1} are $(h-1)$-connected. Suppose that $f < h-1$. Let W be the set of all voxels that share the facet F. One can show that there is an ordered set of voxels $W' = \{v_1, v_2', \ldots, v_m', v_l\}$, $m < l$, $W' \subseteq W$, so that any two consecutive voxels in W' are $(h-1)$-connected. If one assumes the opposite, one can show that there must be an $(n-1)$-tunnel in $D_k(\Gamma)$, which contradicts the definition of k-discretization. The proof is based on a detailed analysis of the structure of the multidimensional discrete space, and will be included in the full journal version of the paper.

2. $(n-k-1)$-*connectivity.* By definition, if $D_k(\Gamma)$ is k-minimal, then it is k-tunnel-free. Then one can show that if $D_k(\Gamma)$ is not $(n-k-1)$-connected, it will contain k-tunnels. The details will be included in the full paper. □

The above theorem implies a number of corollaries about certain important special classes of surfaces.

Corollary 1. *Let Γ be an $(n-1)$-dimensional surface in R^n.*

(a) For any $1 \le k \le n-1$, a k-discretization of Γ is $(n-2)$-connected. Thus, in particular, any naive $(n-1)$-discretization is $(n-2)$-connected.

(b) Any standard cover (0-discretization) of Γ is $(n-1)$-connected.

For the case $h = 1$, i.e., when the surface is a curve, we obtain the following corollary.

Corollary 2. *A k-discretization of a curve in R^n is $(n-k-1)$-connected. (In particular, a standard 0-discretization is $(n-1)$-connected, and a naive $(n-1)$-discretization is 0-connected, see Figure 6a.)*

In the case of a naive $(n-1)$-discretization we have the following corollary.

Corollary 3. *A naive $(n-1)$-discretization of a h-dimensional surface is $(h-1)$-connected. (Thus a naive $(n-1)$-discretization of an $(n-1)$-dimensional surface is $(n-2)$-connected, in accordance with Corollary 1, see Figure 6b.)*

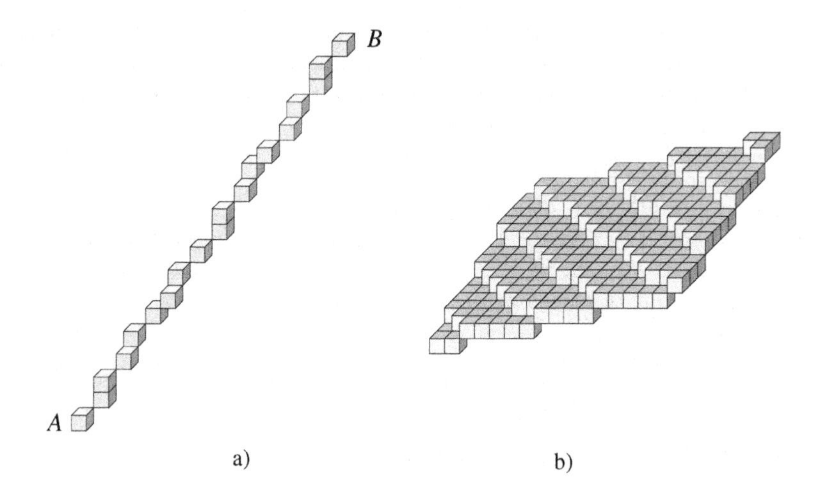

Fig. 6. Illustrations to Corollaries 2 and 3. a) A portion of naive 2-discretization of the Euclidean line between the points $A = (0, 0, 0)$ and $B = (11, 13, 18)$. It has 0-tunnels and is 0-connected. b) A portion of naive 2-discretization of the Euclidean plane $-4x - 9y + 18z = 0$. It has 1-tunnels and is 1-connected

4.4 Applications: k-Discretizations of Hyperplanes and Halfspaces

In this section we illustrate the concept of k-discretizations in the important case when the surface is a hyperplane. We recall that an *arithmetic plane*[6] $P = P(\beta, \alpha_1, \alpha_2, \ldots, \alpha_n, \omega)$ is defined as the set of integer points which satisfy the double linear Diophantine inequality $0 \leq \beta + \sum_{i=1}^{n} \alpha_i x_i < \omega$.

Let $H : b + \sum_{i=1}^{n} a_i x_i = 0$ be a hyperplane. Let j be the least index for which $a_j \neq 0$, and let $a_j > 0$[7]. Then a *standard cover* of H has been defined as

$$St(H) = \left\{ x \in Z^n \ : \ -\frac{\sum_{i=1}^{n} |a_i|}{2} \leq b + \sum_{i=1}^{n} a_i x_i < \frac{\sum_{i=1}^{n} |a_i|}{2} \right\}. \tag{1}$$

It is proved [2, Proposition 13] that $St(H)$ is indeed a cover, which is $(n-1)$-connected, 0-minimal, and n-bubble-free. One can show that the Hausdorff distance between H and $St(H)$ is the minimal possible over all 0-tunnel-free discretizations of H. This, together with the 0-minimality, implies that the above definition is in accordance with the global definition of a 0-discretization (standard cover) given in the previous section. More precisely, we can state the following theorem.

Theorem 3. *The set of voxels determined by Condition (1) is a 0-discretization (standard cover) of the hyperplane* $H : b + \sum_{i=1}^{n} a_i x_i = 0$.

[6] Sometimes also called *discrete analytical plane.*

[7] Then the hyperplane H is said to have a *standard* orientation [2].

This result can be extended to arbitrary k-discretizations. As a first step we reveal which arithmetic planes are k-discretizations. In [3] the connection between the tunnels of arithmetic hyperplanes and their thickness is investigated, and the following basic result is obtained[8].

Theorem 4. *[3, Proposition 9] Let* $P = P(b, a_1, a_2, \ldots, a_n, \omega) = \{x \in Z^n | 0 \leq b + \sum_{i=1}^n a_i x_i < \omega\}$ *be a discrete hyperplane, where* $b \geq 0$, $a_i \geq 0$ *for all* i, *and* $a_i \leq a_{i+1}$ *for* $1 \leq i \leq n-1$. *Then, if* $\omega < a_n$, *the hyperplane has* $(n-1)$-*tunnels; For* $0 < k < n$, *if* $\sum_{i=k+1}^n a_i \leq \omega < \sum_{i=k}^n a_i$, *the hyperplane has* $(k-1)$-*tunnels and is* k-*separating; If* $\omega \geq \sum_{i=1}^n a_i$, *the hyperplane is tunnel-free.*

The next theorem characterizes the arithmetic planes that are k-discretizations.

Theorem 5. *Let* P *be a discrete hyperplane as in Theorem 4. If* $\omega = \sum_{i=k+1}^n a_i$, *then* P *is a* k-*discretization.*

The proof will be included in the full paper. The notion of surface k-discretization or k-cover can be extended to subsets of R^n with topological dimension n. For this, one can use the corresponding discretization of the boundary of the given set. Let us consider, for instance, a half-space E determined by a linear inequality $a_1x_1 + a_2x_2 + \ldots + a_nx_n \leq b$. One can obtain a k-discretization of E as $D_k(E) = \{x \in Z^n : b + \sum_{i=1}^n a_i x_i < \omega\}$, where $\omega = \sum_{i=k+1}^n a_i$, for a standard discretization convention (see [2] for getting acquainted with the notion of discretization convention). In particular, the set of voxels $St(E) = \{x \in Z^n : b + \sum_{i=1}^n a_i x_i < \omega\}$, with $\omega = \sum_{i=1}^n a_i$, is a 0-discretization (standard cover) of E.

5 Concluding Remarks

In this paper we have studied minimal covers of bodies and k-discretizations of surfaces of arbitrary dimension. A further direction of research can be seen in studying minimal covers and k-discretizations of some important special classes of Euclidean objects, like affine subspaces of arbitrary dimension $0 \leq m \leq n-1$ (m-flats), simplexes or some other interesting classes of polytopes, spheres, etc.

It would also be interesting to study the properties of k-discretizations for different choices of a metric d in the definition of Hausdorff distance.

Acknowledgments

We thank the Referees for several valuable comments and suggestions.

[8] In the special case of discrete hyperplanes, a j-tunnel can be defined as follows. A discrete plane $P = P(\beta, \alpha_1, \alpha_2, \ldots, \alpha_n, \omega)$ has a j-tunnel if there are two j-neighboring voxels $v = (v_1, \ldots, v_n)$ and $w = (w_1, \ldots, w_n)$ such that $\alpha_1 v_1 + \ldots + \alpha_n v_n + \beta < 0$ and $\alpha_1 w_1 + \ldots + \alpha_n w_n + \beta \geq 0$ [17]. It is not difficult to see that this definition is in accordance with the more general Definition 2.

References

1. Andres, E.: An m-Flat Supercover is a Discrete Analytical Object. TR 1998-07, IRCOM-SIC, Université de Poitiers, Bat. SP2MI – BP. 179, F–86960 Futuroscope Cédex – FRANCE (1998) 210, 212
2. Andres, E.: Standard Cover: a New Class of Discrete Primitives. TR 1998-02, IRCOM-SIC, Université de Poitiers, Bat. SP2MI – BP. 179, F–86960 Futuroscope Cédex – FRANCE (1998) 210, 211, 217, 219, 220
3. Andres, E., Acharya, R., Sibata, C.: Discrete Analytical Hyperplanes. Graphical Models and Image Processing, **59** (5) (1997) 302–309 210, 217, 220
4. Barneva, R. P., Brimkov, V. E., Nehlig, P.: Thin Discrete Triangular Meshes. Theoretical Computer Science **264** (1-2) (2000) 73–105 218
5. Bresenham, J. E.: Algorithms for Computer Control of a Digital Plotter. IBM Syst. J., **4** (1) (1965) 25–30 217
6. Brimkov, V. E., Barneva, R. P.: Graceful Planes and Lines, Theoretical Computer Science, Elsevier, to appear 218
7. Brimkov, V. E., Barneva, R. P., Nehlig, P.: Minimally Thin Discrete Triangulations. In: Volume Graphics, M. Chen, A. Kaufman, R. Yagel (Eds.), Springer Verlag (2000) 51–70 218
8. Bertrand, G., Couprie, M.: A Model for Digital Topology. Lecture Notes in Computer Science, Vol. 1568, Springer-Verlag, Berlin (1999) 229–241 211
9. Cohen, D., Kaufman, A.: Fundamentals in Surface Voxelization. CGVIP Graphical Models and Image Processing, **57** (6) (1995) 211, 212
10. Debled-Rennesson, I., Reveillès, J-P.: A new approach to digital planes. In: SPIE Vision Geometry III, Boston **2356** 1994 217
11. Figueiredo, O., Reveillès, J-P.: A Contribution to 3D Digital Lines. In: Proc. the 5th Internat. Workshop DGCI, Clermont-Ferrand (1995) 187–198 217
12. Françon, J.: Discrete Combinatorial Surfaces. CGVIP Graphical Models and Image Processing, **57** (1995) 217
13. Kaufman, A., Shimony, E.: 3D Schan-Conversion Algorithms for Voxel-Based Graphics. In: Proc. 1986 Workshop on Interactive 3D Graphics, Chapel Hill, NC, ACM, New York (1986) 45–75 217
14. Kong, T. Y., Kopperman, R., Meyer, P. R.: A Topological Approach to Digital Topology. American Mathematical Monthly, **38** (1991) 901–917 211
15. Kong, T. Y., Rosenfeld, A.: Digital Topology: Introduction and Survey. Computer Vision, Graphics, and Image Processing **48** (1989) 357–393 211
16. Kovalevsky, V. A.: Finite Topology as Applied to Image Analysis. Computer Vision, Graphics, and Image Processing **46** (1989) 141–161 211
17. Reveillès, J.-P.: Géométrie Discréte, Calcul en Nombres Entiers et Algorithmique. Thèse d'État, Université Louis Pasteur, Strasbourg, France (1991) 211, 217, 220
18. Stojmenovic, I., Tosic, R.: Digitization Schemes and the Recognition of Digital Straight Lines, Hyperplanes and Flats in Arbitrary Dimensions. Digital Geometry. Contemporary Math. Series, Amer. Math. Soc. Providence, RI **119** (1991) 197–212 210, 217
19. Tajine, M., Wagner, D., Ronse, C.: Hausdorff Discretizations and its Comparison to Other Discretization Schemes. Lecture Notes in Computer Science, Vol. 1568, Springer-Verlag, Berlin (1999) 399–410 211, 217
20. Veerlaert, P.: On the Flatness of Digital Hyperplanes. J. Math Imaging Vision **3** (1993) 205–221 210, 217
21. Wagner, D., Tajine, M., Ronse, C.: An Approach to Discretization Based on Hausdorff Metric. ISMM'98, Kluwer Academic Publisher (1998) 67–74 211, 217

Strong Thinning and Polyhedrization of the Surface of a Voxel Object*

Jasmine Burguet and Rémy Malgouyres

GREYC, ISMRA
6 bd Maréchal Juin 14000 Caen, France
{Burguet,Remy.Malgouyres}@llaic.u-clermont1.fr

Abstract. We first propose for digital surfaces a notion analogous to the notion of strong homotopy which exists in 3D [1]. We present an associated parallel thinning algorithm. The surface of an object composed of voxels is a set of surfels (faces of voxels) which is the boundary between this object and its complementary. But this representation is not the classical one to visualize and to work on 3D objects, in frameworks like Computer Assisted Geometric Design (CAGD). For this reason we propose a method for passing efficiently from a representation to the other. More precisely, we present a three-step algorithm to polyhedrize the boundary of a voxel object which uses the parallel thinning algorithm presented above. This method is specifically adapted to digital objects and is much more efficient than such existing methods [12]. Some examples are shown, and a method to make the reverse operation (discretization) is briefly presented.

Keywords: digital surface, thinning, strong homotopy, parallel algorithm, polyhedrization.

Introduction

To study a three-dimensional object, its surface is often used because it is less costy (in particular concerning the memory cost) than studying the whole 3D volume [2,3,4,7,8]. For this reason, the discrete representation of digital surfaces of 3D objects is of practical interest. Indeed, it allows us to compute easily the union, intersection and difference (i.e. boolean set operations) of two discrete objects represented by their surfaces. On the other hand, classical modeling techniques use a continuous representation for surfaces. More precisely, to encode surfaces, a representation is often used to vizualize and to work on 3D objects: polyhedrons. This representation has a lot of advantages, particulary in Computer Assisted Geometric Design (CAGD) applications. However, floating points computations are much less relevant to perform boolean set operations. Indeed, there are some problems when the objects we want to work on are tangeant to

* The authors' new permanent address is: LLAIC1 - IUT Département Informatique, BP 86, 63172 AUBIERE Cédex

G. Borgefors, I. Nyström, and G. Sanniti di Baja (Eds.): DGCI 2000, LNCS 1953, pp. 222–234, 2000.
© Springer-Verlag Berlin Heidelberg 2000

each other. This leads us to find a way to pass efficently and quickly from a discrete representation to a polyhedron representation, and conversely.

The existing methods to polyhedrize the surfaces, except in [12], are not especially adapted to our initial data structure for the discrete surfaces, namely a set of surfels.

The purpose of this paper is to introduce an algorithm which enables us to compute a *polyhedrized surface* from a *discrete surface.*

The classical methods, using for exemple the Voronoï diagram and the Delaunay triangulation [17], are not effective for the non-euclidien spaces like surfaces. So, we use a topological approach for the discrete surfaces (see [5] and [6]). This leads us to define a "topological Voronoí diagram" on these surfaces which allows us to built the triangular faces of our polyhedrized surface.

In the first part, we present a parallel thinning algorithm to compute the skeleton of a set of surfels. This algorithm is based on an analogue for the surfaces of the notion of *strong homotopy* existing in 3D and introduced by G. Bertrand ([1,13]).

Then, we present a polyhedrization method which is based on the use of a germ-obtained skeleton. The germs are particular surfels chosen, with respect to the maximal curvature, in different locations of our surface; the higher is the curvature, the more numerous are the germs. Then, using the parallel algorithm previously presented, we compute the skeleton of the set of surfels composed of our surface from which the germs are removed. The obtained skeleton enables us to draw triangles which will be the faces of the polyhedrized surface. These triangles are built during a cover of the skeleton. Since the germs are more numerous where the surface is very curved, the number of faces is higher in such locations. Some examples of the polyhedrization algorithm results are given.

The presented method, which can also be used for (non conservative) data compression of binary 3D images, is much faster than the method introduced in [12]. Moreover, it produces more regular faces, which are all triangles, which makes it much more convenient for CAGD applications.

A method to make the reverse operation (discretization) is briefly presented. It is based on a particular data structure: an array the cells of which contain chained lists of surfels. Note that the boolean set operations can easily be performed using this data structure.

1 Notations and Basic Notions

Let X and E be two sets such as $X \subset E$. We denote by $\overline{X}$ the complement of X in E and by $card(X)$ the number of elements of X. In the sequel, an *adjacency relation* on X is an antireflexive symmetric binary relation on X. Given α an adjacency relation, an *α-path* in X with length l is a sequence $(x_0, x_1, \ldots, x_l)$ in X and such as x_{i-1} is α-adjacent to x_i, for $i = 1, \ldots, l$. An α-path is called *closed* if $x_0 = x_l$ and *simple* if the x_i are pairwise distinct. Two elements y and y' of X are said to be α-connected in X if there exists an α-path $(x_0, x_1, \ldots, x_l)$ contained in X such as $y = x_0$ and $y' = x_l$. The α-connectedness is an equivalence relation and its equivalence classes are called *α-connected components.* We denote by $C_{alpha}(X)$ the set of connected components of X and X is said to be *α-connected* if $card(C_\alpha(X)) = 1$.

1.1 Surface of a 3D Object

A 3D object O is a set of points in $\mathbb{Z}^3$. These points, which we can represent as unit cubes centered at a point $(i, j, k) \in \mathbb{Z}^3$, are called *voxels.* Two voxels are said to be *6-adjacent* if they share a face, *18-adjacent* if they share a face or an edge. We derive from this as above the classical notions of an *n-path* and *n-connected component*, with $n \in \{6, 18\}$, and we denote by $C_n(O)$ the set of n-connected components of the object O. We define the n-neighborhood of a voxel x denoted by $N_n(x)$ by $N_n(x) = \{y \in \mathbb{Z}^3 / y \text{ is } n\text{-adjacent to } x\}$.

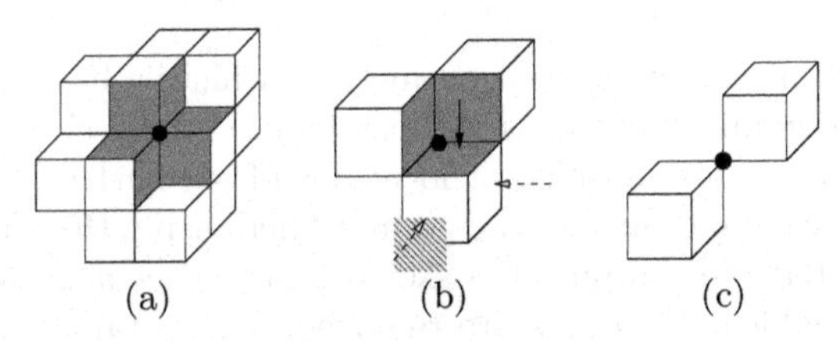

Fig. 1. Examples

Now, a *surfel* is a couple (v_1, v_2) of 6-adjacent voxels v_1 and v_2. A surfel may be seen as an unit square shared by two 6-adjacent voxels, and then we call *surface* of an object $O \subset \mathbb{Z}^3$ the set of all surfels (v_1,v_2) such as $v_1 \in O$ and $v_2 \in \overline{O}$. We can define adjacency relations between surfels, and these relations depend on the one we consider on voxels. We consider the surface Σ of an object O. Now, we call *e-adjacency* the adjacency relation on surfels considering the edges. Under this one, a surfel has exactly four neighbors, one per edge. For example, let call x the grey surfel pointed by a full arrow on the object of the figure 1, (b). If we consider the 6-adjacency, the object is composed of three 6-connected sets (the three voxels), and the e-neighbors of x are the four surfels of the lower voxel which share an edge with x. By contrast, if we consider the object with the 18-adjacency, the three voxels are 18-connected and the four e- neighbors of x are the two other grey surfels and the two surfels pointed by the dotted arrow. We are now going to define an adjacency relation associated with the vertices of the surfels. First, we define a *loop* in Σ as an e-connected component of the set of all surfels in Σ which share a given vertex (figure 1). One vertex may define two loops (see figure 1, b) and c)), so the loops are a way to duplicate formaly vertices. Thus, two surfels are called *v-adjacent* if they belong to a common loop. Then, e and v are adjacency relations on Σ from which follow the classic notions of *n-paths*, *n-neighborhoods*, and *e-connected components*, with

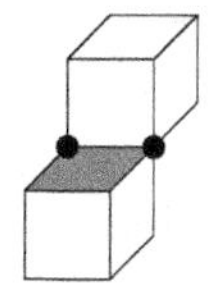

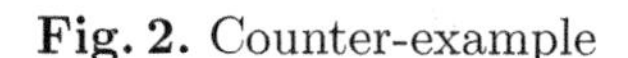

Fig. 2. Counter-example

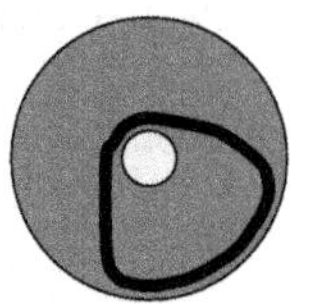

Fig. 3. A decentred skeleton

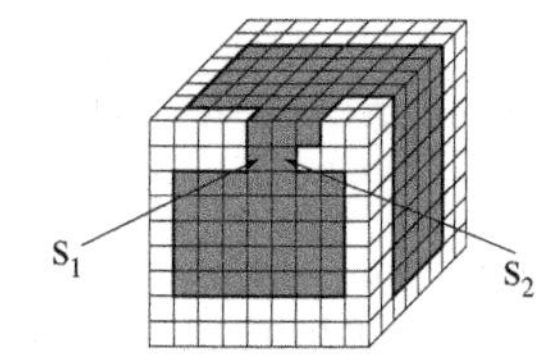

Fig. 4. No parallel strategy

$n \in \{e, v\}$. The kind of surfaces defined above satisfies the Jordan property [11], namely a e-connected component of a surface separates the space into two parts: a 6-connected and a 18-connected one [16].

Afterwards, we will assume that each loop of our surface is a topological disk. In other words, two v-adjacent surfels which are not e-adjacent cannot belong simultaneously to two distinct loops. The object of the figure 2 does not satisfy our hypothesis because the two loops defined by the marked vertices contain surfels which are v-adjacent but not e-adjacent. (see [9] for a short discution on this hypothesis).

1.2 Neighborhood Graph of Surfels

Let $x \in \Sigma$, $n \in \{e, v\}$ and $X \subset \Sigma$ be a set of surfels. We introduce the *neighborhood graph of surfels* of x as follows. We have assumed that each loop in Σ is a topologic disk, but there exist some v-neighborhoods $N_v(x)$ which are not topological disks (see the surfels of the two loops defined by the marked vertex on figure 1 (b). Then we have to define an other "topology" under which $N_v(x) \cup \{x\}$ is a topological disk. Two surfels s_1 and s_2 in $N_v(x) \cup \{x\}$ are said to be *e_x-adjacent* (respectively *v_x-adjacent*) if they are e-adjacent (respectively *v-adjacent*) and are contained in a common loop which contains x. Then, we denote by $G_e(x,X)$ (respectively $G_v(x,X)$) the graph whose vertices are the surfels of $N_v(x) \cap X$ and whose edges are the pairs of e_x-adjacent (respectively v_x-adjacent) surfels of $N_v(x) \cap X$. The set of all connected components of $G_n(x,X)$ which are n-adjacent to x is denoted by $C_n^x(G_n(x,X))$. Note that $C_n^x(G_n(x,X))$ is a set of sets of surfels and not a set of surfels.

A surfel x is said to be *n-isolated* in X if $N_n(x) \cap X = \emptyset$ and to be *n-interior* in X if $N_{\overline{n}}(x) \cap \overline{X} = \emptyset$.

2 Topology Preservation and Thinning

2.1 P-Simple Surfels

First, we introduce the notion of *n-simpleness* of surfels [10]:

Definition 1. *[9]. Let $X \subset \Sigma$ be a set of surfels of a digital surface Σ. A surfel $x \in X$ is said to be* n-simple *in X if and only if x is not n-interior in X and if $card(C_n^x(G_n(x, X)) = 1$.*

Intuitively, a surfel is n-simple in X if its deletion from X does not change the topology of X.

Definition 2. *Let $Y \subset X \subset \Sigma$. The set Y is said to be (lower) n-homotopic to X if and only if Y can be obtained from X by sequential deletion of n-simple surfels.*

Thus, we can obtain a skeleton from a set of surfels by a sequential erosion analogous to classical 2D sequential thining [10,9]. As in the planar 2D case [10], this method presents some drawbacks : there is no parallel way to test the n-simplicity of surfels (on figure 4, the surfels s_1 and s_2 of the set X composed of grey surfels are n-simple in X, but we cannot remove simultaneously these surfels without changing the topology). Moreover, the shape of the skeleton depends on the order with which the n-simplicity of surfels is tested (see figure 3 if we use the lexicographical order to test the n-simplicity of surfels). Finally, the strategy using end surfels (surfels which have only one n-neighbor) causes the apparition of a lot of undesirable branches (see figure 6).

For these reasons, we are now interested in a new thinning method of sets of surfels using a parallel strategy. This method is an analogue of a parallel thinning algorithm existing in 3D introduced by G. Bertrand [1], and using the notion of *P-simple points.*

First we must state some theoretical definitions and results.

Definition 3. *Let $Y \subset X \subset \Sigma$. The set Y is called* strongly n-homotopic to X *if for any Z such that $Y \subset Z \subset X$, the set Z is n-homotopic to X. If Y is strongly n-homotopic to X, then we call the set $S = X \backslash Y$ a* strongly n-simple set *in X.*

Note that if Y is strongly n-homotopic to X, then for any set $S \subset X \backslash Y$, the set $X \backslash S$ is always n-homotopic to X.

Definition 4. *Let $X \subset \Sigma$, $P \subset X$ and $x \in X$. The surfel x is called P_n-simple in X if, for any set S such that $S \subset P \backslash \{x\}$, the surfels x is n-simple for $X \backslash S$. We denote by $S_n(P)$ the sets of the P_n-simple surfels. A set $E \subset X$ is said* P_n-simple *in X if $E \subset S_n(P)$.*

The P_n-simple surfels satisfy a strong constraint, namely a P_n-simple surfel can be removed whatever the set contained in P already removed. Now, we establish a relationship between the notions of strongly n-simplicity and P_n-simplicity which is directly deduced from the definitions.

Theorem 1. *Let $Y \subset X \subset \Sigma$. The set Y is strongly lower n-homotopic to X if and only if the set $P = X \backslash Y$ is P_n-simple for X.*

Now, we propose a local caracterization of the P_n-simple surfels, which makes effective the notion of a P_n-simple surfel.

Definition 5. *We denote by $A_n(x, X)$ the set of surfels defined by*

$$y \in A_n(x, X) \Leftrightarrow \exists \text{ a } n_x\text{-path in } N_v(x) \cap X \text{ from } y \text{ to an } n\text{-neighbor of } x$$

and we denote by $T_n(x, X)$ the number of n_x-connected components of $A_n(x, X)$.

In other words, $A_n(x, X)$ is the union of all connected components of $G_n(x, X)$ which are n-adjacent to x.

Theorem 2. *Let $P \subset X \subset \Sigma$ and $x \in P$. We denote $R = X \backslash P$. Then the surfel x is P_n-simple in X if and only if the four following properties are satisfied:*

- $T_n(x, R) = 1$;
- $T_{\overline{n}}(x, \overline{X}) = 1$;
- $\forall y \in N_n(x) \cap P$, *$y$ is n-adjacent to* $A_n(x, R)$;
- $\forall y \in N_{\overline{n}}(x) \cap P$, *$y$ is $\overline{n}$-adjacent to* $A_{\overline{n}}(x, \overline{X})$.

This theorem enables us to test the P_n-simpleness of surfels by an algorithm using only local considerations, and enables us to define an algorithm using a parallel strategy.

2.2 Parallel Algorithm and Results

First, we present a sequential thinning algorithm of a subset X of a surface Σ, implemented by A. Lenoir [9,4]. The two-step principle is the following one: while there exist some n-simple surfels in X,

- for all x in X, if x is n-simple, mark x with REMOVABLE;
- for all surfels in X, if x is marked with REMOVABLE, and x is n-simple in the remaining set, remove x from X.

To preserve the general shape of our initial set of surfels, we impose on the end surfels to be unremovable (i.e. surfels of X with exactly one neighbor in X).

Now let us present our parallel algorithm (see figure 5). Only the surfels $\overline{n}$-adjacent to X can be be removed from a set X at each step. So, to test the P_n-simplicity of the surfels in X, our set P will be the set of the surfels in X which are $\overline{n}$-adjacent to $\overline{X}$. Now, a surfel which is not P_n-simple may be n-simple. So, after using the parallel algorithm, we will use the sequential one to remove all n-simple surfels which are not end surfels. The parallel algorithm is presented on figure 5.

The results of this algorithm (see figure 6) are, as in the 3D case, quite better than the ones obtained by the sequential algorithm. First, there are less undesirable branches on the final squeleton obtained by the parallel algorithm. Next, the symmetry of this skeleton is closer to the symmetry of the initial set of surfels.

Parallel Algorithm

```
M = array of all surfels;
P = set of the surfels on the border;
R = set of the interior surfels;
Repeat
    made := FALSE;
    For all x in P
            If (x is P-simple)
              Then mark x with REMOVABLE;
                   made := TRUE;
    End For
    For all x in P
            If (x is marked with REMOVABLE)
                Then remove x from M;
    End For
    Update P;
While made;
```

Fig. 5. Parallel algorithm

3 Polyhedrization

The method of polyhedrization presented here consists of three steps.

3.1 Step One: The Choice of the Germs

The first step consists in choosing some special surfels on the surface of the object we want to polyhedrize. These surfels, called *germs*, are distributed according to the curvature of the surface: the more curved is the surface, the more numerous are the germs. First, using the method of A. Lenoir [3,4], we compute the maximal curvature of all the surfels of the surface, and we choose a surfel at random, which is our first germ, on the surface. Then:

1. determination of a maximal width l depending on the maximal curvature of the surfel;
2. breadth-first exploration (using a FIFO structure) of the e-adjacency graph from the germ until the width l is reached; we mark the covered surfels;
3. if there are some non-marked surfels, choice of a new germ among the non-marked neighbors of the marked surfels, and back to the step 1. Otherwise, STOP.

At the third step, for better results, the next germ is the one which has the greater maximal curvature. Indeed, with this choice, we obtain more faces at the more curvated places.

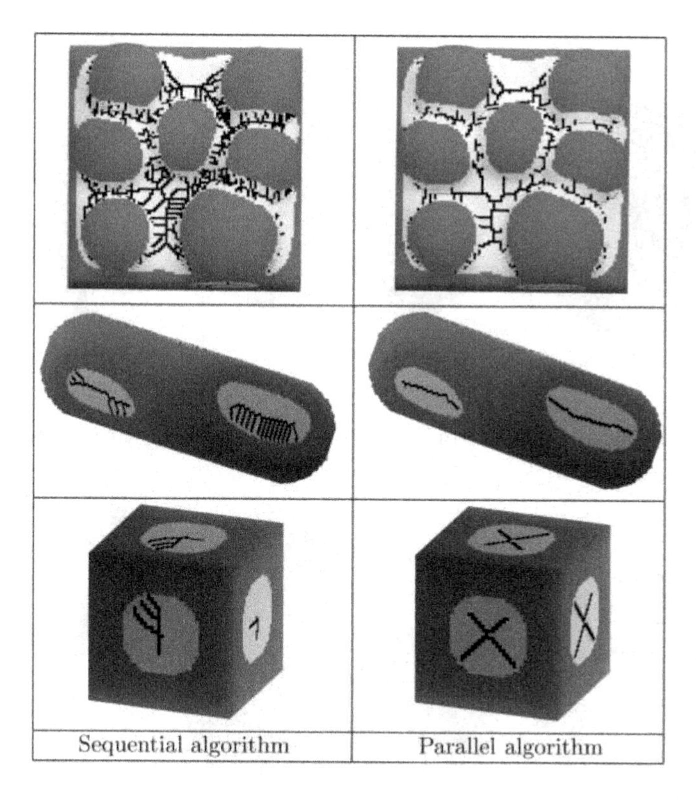

Fig. 6. Comparison of results

An interesting point is that we can choose a limit width L for the breadth-first exploration: if the computed width l is larger than L, then we replace l by L. Moreover, we use a slowdown strategy at the more curved places during the breadth-first exploration. So we can have germs more or less dense on the less curved places of the surface.

We denote by G the set of all germs thus obtained. Some results of this strategy are presented on figure 7.

3.2 Step 2: Obtaining of a "Topological Voronoï Diagram"

Let Σ be the surface of the object. To obtain the "topological Voronoï diagram" (i.e. a skeleton in which there are no more simple surfels), we use the thinning algorithm presented in the first section on the set $\Sigma \backslash G$, without the end surfels conditions. We obtain the skeletons visible on figure 8 for the three previous presented objects.

The skeletons are obviously more dense where the germs are more numerous.

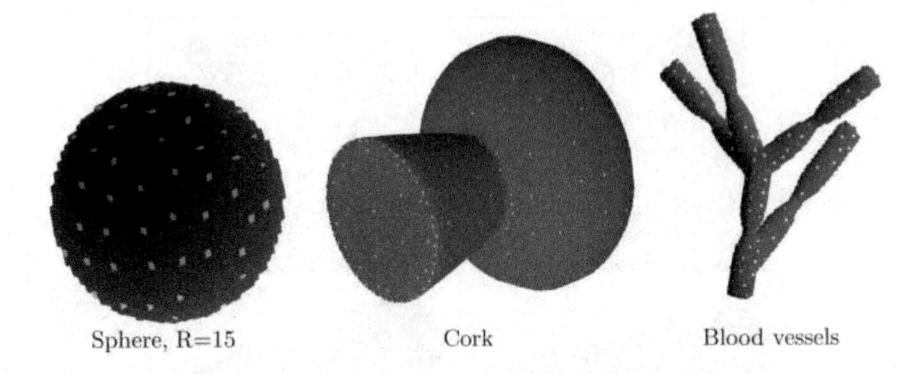

Fig. 7. The choosen germs (in white) on the surfaces of three voxels objects

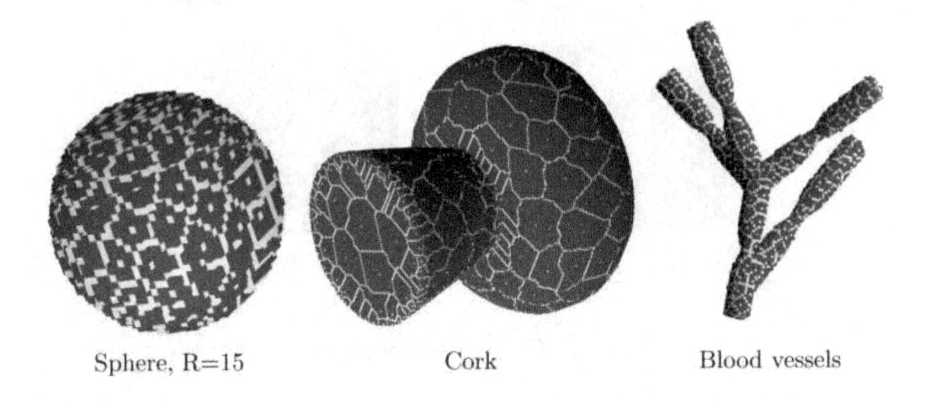

Fig. 8. The chosen germs and the "topological Voronoï diagrams"

3.3 Step 3: The Obtaining of the Faces

Now, from the skeleton and the germs, we are going to cover the triangular faces which will form our polyhedrized surface.

Let consider the figure 9, (a). The germs are represented by the black points, and the squeleton by the thick black lines. There exist two kinds of surfels on the skeleton: the surfels of the branches (surfels which have exactly two neighbors in the skeleton) and the surfels of the intersections (surfels which have strictly more than two neighbors). So, we cover the skeleton by covering of the branches from an intersection to an other one. Let I_1 and I_2 be the barycenters of two intersections connected by a branch. The segment I_1I_2 is the edge of two triangles, and the other vertices of these triangles are the germs which the branch separates (see figure 9, (b)).

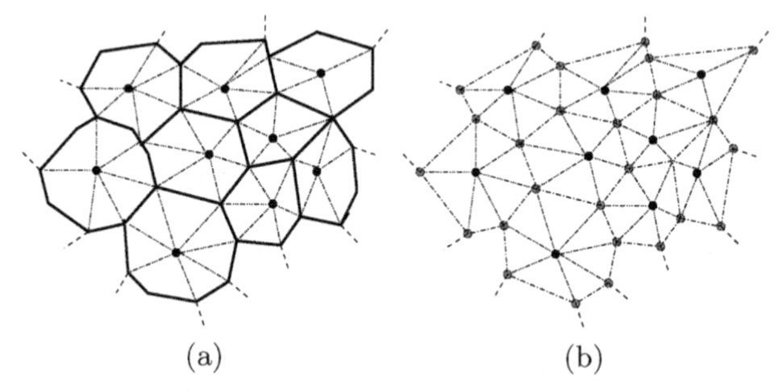

Fig. 9. Obtaining of triangular faces from the skeleton

3.4 Results

The results of the algorithm for the sphere with radius 15, the cork and the blood vessels are presented on figure 10. The initial numbers of surfels of these surfaces are respectively 4254, 81066, 10792 and the chosen limit widths L 4, 8 and 4. The numbers of triangular faces obtained by the algorithm are 504, 3914 and 1316. It takes a few seconds to compute the polyhedrized surfaces (for the bigger example, the cork, it takes about 30 seconds on Ultra Sparc Sun^{tm} work stations with bi-processors).

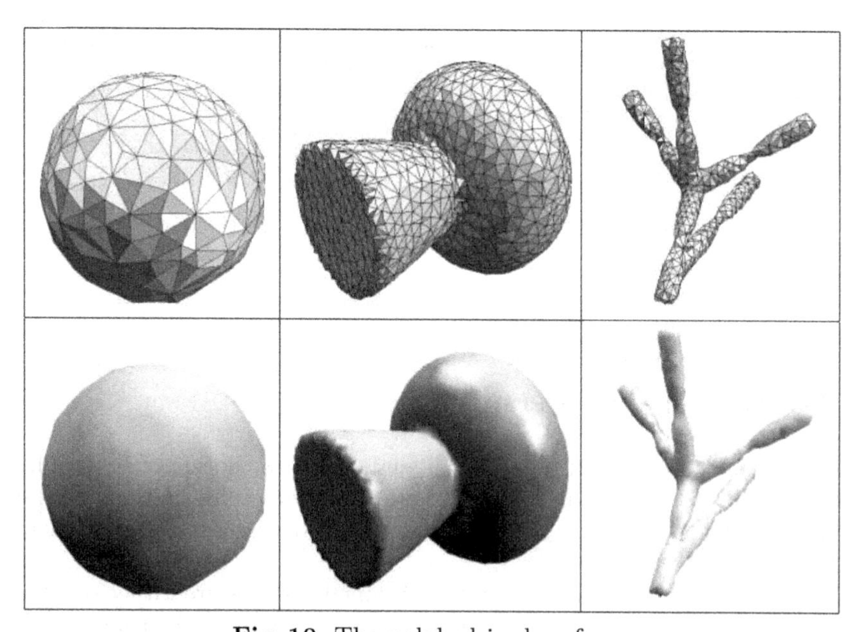

Fig. 10. The polyhedrized surfaces

As we can see on the figure, these faces are regular, well located and one gets good quality surfaces, even on thin objects (see the result on the blood vessels). The worst results are near the angles (see base of the cork), but they remain satisfying.

4 Discretization

Now, we have an effective tool to obtain a polyhedrized surface from a discrete one. Here, we present briefly a method to realize the reverse operation.

First, we choose a step of discretization P (smaller than the size of the surface S we want to discretize). Then, we are going to work according to the first axis of the classical coordinates space. We define a 2-dimensionnal array $(kP) * (kP)$, with $k \in \mathbb{N}*$ according to the plane containing the second and the third axis. Each cell of this array contains a chained list of surfels (figure 11). The method to fill up this array is to study the intersection between the line D_{ij} (the line coming from the cell (i, j) and parallel to the first axis) and S (figure 12). During the recover of D_{ij}, for each intersection we have to determine by a parity rule if we enter into the surface or if we get out. Then, we can compute the coordinates of the deduced surfels and obtain the discretized surface.

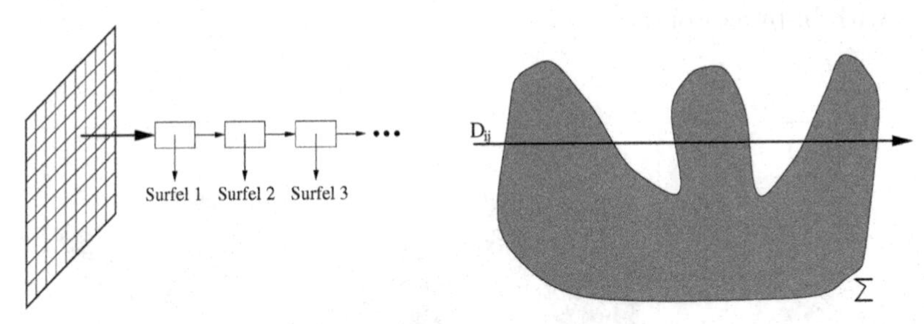

Fig. 11. Data structure **Fig. 12.** Method

5 Conclusions and Perspectives

Following section 2, we have obtained an analogue of the notion of P_n-simplicity existing in 3D for discrete surfaces. Moreover, the results of thinning using the parallel strategy are far better than the ones deriving from the sequential method.

Next, as for the polyhedrization, the obtained surfaces are rather satisfying. The computing time for results is quite short (a few seconds against several days for the method of [12]) and the obtained faces are much more regular, in comparison to what would be obtained by any other pre-existing method [12].

Note that the boolean set operations between the objects surrounded by the discretized surface can easily be performed using the obtained discrete data structure.

In future works, we shall include this work into a *multi-scale* context. Indeed, if we have to work with a set of multi-scaled objects, it is more relevant to consider different scales related to these objects. Thus, we will be able to represent the whole multi-scaled set with one method.

Acknowledgment

We thank A. Lenoir for providing us the source code of his computer program, including curvature computation and display for digital surfaces.

References

1. G. BERTRAND. *On P-simple points*, no. 321, C.R Académie des Sciences, 1995. 222, 223, 226
2. T. J. FAN, G. MEDIONI, R. NEVATA. *Recognising 3D Objects Using Surface Description*, IEEE Trans. Pattern Anal. Mach. Intelligence **11**(11),1140-1157, 1989. 222
3. A. LENOIR. *Fast Estimation of Mean Curvature on the Surface of a 3D Discrete Object*, Proceedings of DGCI'97, Lecture Notes in Computer Science, Vol. 1347, pp. 213-222, 1997. 222, 228
4. A. LENOIR. *Des Outils pour les Surfaces Discrètes*, PhD Thesis, 1999. 222, 227, 228
5. R. MALGOUYRES. *Presentation of the fundamental group in digital surfaces*, Proceedings of DGCI'99, Lecture Notes in Computer Science, Vol. 1568, pp. 136-150, March 1999. 223
6. S. FOUREY, R. MALGOUYRES. *Intersection number of paths lying on a digital surface and a new Jordan theorem*, Proceedings of DGCI'99, Lecture Notes in Computer Science, Vol. 1568, pp. 104-117, March 1999. 223
7. T. J. FAN, G. MEDIONI, R. NEVATA. *Recognising 3D Objects Using Surface Descriptions*, IEEE Trans. Pattern Anal. Mach. Intelligence 11(11), pp. 1140-1157, 1989. 222
8. G. FARIN. *Computer Assisted Geometric Design (CGAD)*. Academic Press, Inc. 222
9. R. MALGOUYRES, A. LENOIR. *Topology Preservation Within Digital Surfaces* Graphical Models (GMIP) 62, 71-84, 2000. 225, 226, 227
10. T. Y. KONG, A. ROSENFELD. *Digital Topology: Introduction and Survey*. Computer Vision, Graphics, and Image Processing 48, 357-393, 1989. 226
11. G. T. HERMAN. *Discrete Multidimensional Jordan Surfaces*, CVGIP: Graph. Models Image Process. 54, 507-515, 1992. 225
12. J. FRANÇON, L. PAPIER. *Polyhedrization of the Boundary of a Voxel Object*, Proceedings of DGCI'99, Lecture Notes in Computer Science, Vol. 1568, pp. 425-434, Mars 1999. 222, 223, 232
13. G. BERTRAND. *Simple points, topological numbers and geodesic neighborhoods in cubics grids*, Patterns Recognition Letters, 15:1003-1011, 1994. 223

14. R. MALGOUYRES. *Homotopy in 2-Dimensional Digital Images* Theoretical Computer Science. Vol. 230, pp. 221-233, 2000.
15. R. MALGOUYRES, S. FOUREY. *Intersection Number and Topology Preservation Within Digital Surfaces*, Proceedings of DGCI'99, Lecture Notes in Computer Science, Vol. 1568, pp. 104-117, Mars 1999.
16. J. K. UDUPA, V. G. AJJANAGADDE. *Boundary and Object Labelling in Three-Dimensonal Images*, Computer Vision, Graphics, and Image Processing 51, 355-369, 1990. 225
17. R. KLEIN. *Concrete and Abstract Voronoï Diagrams* . Lecture Notes in Computer Science, Springer Verlag Ed., 1989. 223

Deformable Modeling for Characterizing Biomedical Shape Changes

Matthieu Ferrant[2], Benoit Macq[2], Arya Nabavi[1], and Simon K. Warfield[1]

[1] Surgical Planning Laboratory, Brigham and Women's Hospital
Harvard Medical School, Boston, USA
{warfield,arya}@bwh.harvard.edu

[2] Telecommunications Laboratory, Université catholique de Louvain, Belgium
{ferrant,macq}@tele.ucl.ac.be

Abstract. We present a new algorithm for modeling and characterizing shape changes in 3D image sequences of biomedical structures. Our algorithm tracks the shape changes of the objects depicted in the image sequence using an active surface algorithm. To characterize the deformations of the surrounding and inner volume of the object's surfaces, we use a physics-based model of the objects the image represents. In the applications we are presenting, our physics-based model is linear elasticity and we solve the corresponding equilibrium equations using the Finite Element (FE) method. To generate a FE mesh from the initial 3D image, we have developed a new multiresolution tetrahedral mesh generation algorithm specifically suited for labeled image volumes. The shape changes of the surfaces of the objects are used as boundary conditions to our physics-based FE model and allow us to infer a volumetric deformation field from the surface deformations. Physics-based measures such as stress tensor maps can then be derived from our model for characterizing the shape changes of the objects in the image sequence. Experiments on synthetic images as well as on medical data show the performances of the algorithm.

Keywords: Deformable models, Active surface models, Finite elements, Tetrahedral mesh generation

1 Introduction

Today, there is a growing need for physics-based image analysis of deformations in image sequences (e.g. real-time MRI of the heart, image sequences showing brain deformation during neurosurgery, etc.). The subject has recently lead to considerable interest in the medical image analysis community [1,2,3,4,5,6,7,8].

Medical image analysis has in the past relied heavily upon qualitative description. Today, modern applications can be enabled by providing to the clinician quantitative data derived from these images. For example, rather than simply observing erratic heart beat with real-time MRI, clinicians want to measure ejection fraction and estimate stress in the heart muscle quantitatively.

G. Borgefors, I. Nyström, and G. Sanniti di Baja (Eds.): DGCI 2000, LNCS 1953, pp. 235–248, 2000.
© Springer-Verlag Berlin Heidelberg 2000

Shape- and surface-based image analysis is being increasingly used in the bio-medical image analysis community, e.g., for pathological analysis [9] and for tracking deformations [10]. Shape-based models are also being used for image segmentation [11,12,13] to constrain active surface models [1,14]. Such active surface models do not allow any physical interpretation of the deformation the surfaces undergo. Also, no volumetric deformation field is available.

In an attempt to overcome these problems, several authors have proposed to use a physics-based model to infer a volumetric deformation field (e.g. [15,16]) from surface-based deformations. But the used parameters were determined heuristically, and one could therefore not exploit the information generated by the model to extract biomechanical properties. Other authors have also proposed to use physical deformation models to constrain a volumetric deformation field computed from image data using elastic [17] or even viscous fluid deformation models [18,19]. But in these applications, the models did not account for actual material characteristics, because the matching is done minimizing an energy measure that consists of a weighted sum of an image similarity term and a relaxation term representing the potential energy of a physical body (e.g., elastic). Therefore, the actual physics of the phenomenon cannot be properly captured by these models. In order to capture a physics-based deformation field, one needs to use biomechanical models and image-derived forces for deforming them.

In the context of brain shift analysis, there recently has been a significant amount of work directed towards simulation [7] using models driven by physics-based forces such as gravity. Skrinjar et al. [4] have proposed a model consisting of mass nodes interconnected by Kelvin models to simulate the behavior of brain tissue under gravity, with boundary conditions to model the interaction of the brain with the skull. Miga et al. [3] proposed a Finite Element (FE) model based on consolidation theory where the brain is modeled as an elastic body with an interstitial fluid. They also use gravity induced forces, as well as experimentally determined boundary conditions.

Even though these models are very promising, it remains difficult to accurately estimate all the forces and boundary conditions that interact with the model.

The cardiac image analysis community has been using physics-based models – mainly FE models – they deform with image-derived forces. These models then provide quantitative, and physically interpretable 3D deformation estimates from image data. Papademetris et al. [20] derive the forces they apply to the FE model from Ultrasound (US) using deformable contours they match from one image to the next one using a shape-tracking algorithm. Metaxas et al. [1] derive their forces from MRI-SPAMM data for doing motion analysis of the left or right ventricle [21,22].

In the context of deformable brain registration, Hagemann et al. [5] use a biomechanical model for registering brain images, but they enforce correspondances between landmark contours manually. Moreover, the basic elements of their FE model are pixels, which causes the computations to be very slow. Kyriacou et al. [23] study the effect of tumor growth in brain images for doing atlas registration. They use a FE model and apply concentric forces to the tumor boundary to shrink it. In these two studies, the experiments were performed in 2D, thereby limiting the clinical utility and the possibility to efficiently assess the accuracy of the methods.

We propose to merge the prior physical knowledge physicians have about the object that is being imaged with the information that can be extracted from the image sequence to obtain quantitative measurements. We extract shape information of the objects in the image sequence using an active surface model, and characterize the changes the objects undergo using a physics-based model.

The idea is similar to that used for cardiac analysis; we track boundary surfaces in the image sequence, and we use the boundary motion as input for a FE model. The boundary motion is used as a boundary condition for the FE model to infer a volumetric deformation field, as proposed in [5].

The main contribution of this paper is that instead of using a generic FE model that is fitted to the image data as it is done for cardiac image analysis [1], or using the pixels (or voxels) as basic elements of the FE model, we propose an algorithm for generating patient-specific tetrahedral FE models from the initial 3D image in the sequence, with locally adaptable resolution, and integrated boundary surfaces. Also, this enables us to perform computations in 3D, without manual interaction, and moreover, on a limited number of elements, with equivalent precision, and in a reasonable amount of time on a common workstation, thanks to an efficient implementation of the FE deformation algorithm.

2 Description of the Algorithm

There are two important points for doing physics-based modeling of the deformation in 3D image sequences. One first needs to have a prior bio-mechanical model of the object represented by the image, i.e., the constitutive equations of the bodies (elastic, fluid, viscous fluid, etc.) represented in the image. On the other hand, one also needs a way of applying forces and boundary conditions to the model using the image information.

In this work, we have chosen to model image structures as elastic bodies. More elaborate models can of course very easily be integrated into our algorithm. Thus, we assume that the objects that are being imaged have an elastic behavior during deformation. The deformations will be tracked using the boundary information of the objects in the image sequence. The boundary surfaces are deformed towards the boundaries of the next 3D image in the sequence using an active surface algorithm. The deformation field of the boundary surfaces is then used as a boundary condition for our bio-mechanical model, that will be used to infer the deformation field throughout the entire volume.

This will provide us with physically realistic and interpretable information (such as stress tensors, compression measures, etc.) of the imaged objects during the whole sequence.

3 Mathematical Formulation

Assuming a linear elastic continuum with no initial stresses or strains, the potential energy of an elastic body submitted to externally applied forces can be expressed as [24] [1]:

$$E = \frac{1}{2}\int_{\Omega} \boldsymbol{\sigma}^T \boldsymbol{\epsilon}\, d\Omega + \int_{\Omega} \mathbf{F}\mathbf{u}\, d\Omega \tag{1}$$

where $\mathbf{u} = \mathbf{u}(x, y, z)$ is the displacement vector, $\mathbf{F} = \mathbf{F}(x, y, z)$ the vector representing the forces applied to the elastic body (forces per unit volume, surface forces or forces concentrated at the nodes), and Ω the body on which one is working. $\boldsymbol{\epsilon}$ is the strain vector, defined as

$$\boldsymbol{\epsilon} = \left(\frac{\partial \mathbf{u}}{\partial x}, \frac{\partial \mathbf{u}}{\partial y}, \frac{\partial \mathbf{u}}{\partial z}, \frac{\partial \mathbf{u}}{\partial x} + \frac{\partial \mathbf{u}}{\partial y}, \frac{\partial \mathbf{u}}{\partial y} + \frac{\partial \mathbf{u}}{\partial z}, \frac{\partial \mathbf{u}}{\partial x} + \frac{\partial \mathbf{u}}{\partial z}\right)^T = \mathbf{L}\mathbf{u} \tag{2}$$

and $\boldsymbol{\sigma}$ the stress vector, linked to the strain vector by the constitutive equations of the material. In the case of linear elasticity, with no initial stresses or strains, this relation is described as

$$\boldsymbol{\sigma} = \left(\sigma_x, \sigma_y, \sigma_z, \tau_{xy}, \tau_{yz}, \tau_{xz}\right)^T = \mathbf{D}\boldsymbol{\epsilon} \tag{3}$$

where $\mathbf{D}$ is the elasticity matrix characterizing the properties of the material [24].

This equation is valid whether one is working with a surface or a volume. We model our active surfaces, which represent the boundaries of the objects in the image, as elastic membranes, and the surrounding and inner volumes as 3D volumetric elastic bodies.

Within a finite element discretization framework, an elastic body is approximated as an assembly of discrete finite elements interconnected at nodal points on the element boundaries. This means that the volumes to be modeled need to be meshed, i.e. divided into elements. Our meshing algorithm will be described in the next section.

The continuous displacement field $\mathbf{u}$ within each element is a function of the displacement at the element's nodal points $\mathbf{u}_i^{el}$ weighted by its shape functions $N_i^{el} = N_i^{el}(x, y, z)$ (4).

$$\mathbf{u} = \sum_{i=1}^{N_{nodes}} N_i^{el}\mathbf{u}_i^{el} \tag{4}$$

The elements we use are tetrahedra ($N_{nodes} = 4$) for the volumes and triangles for the membranes ($N_{nodes} = 3$), with linear interpolation of the displacement field. Hence, the shape function of node i of tetrahedron el is defined as:

$$N_i^{el}(\mathbf{x}) = K\left(a_i^{el} + b_i^{el}x + c_i^{el}y + d_i^{el}z\right) \tag{5}$$

[1] Superscript T designs the transpose of a vector or a matrix

where $K = \frac{1}{6V^{el}}$ for a tetrahedron, and $K = \frac{1}{2S^{el}}$ for a triangle. The computation of V^{el}, S^{el} (volume, surface of el) and the other constants is detailed in [24].

For every node i of each element el, we define the matrix $\mathbf{B}_i^{el} = \mathbf{L}_i N_i^{el}(\mathbf{x})$. The function to be minimized at every node i of each element el can thus be expressed as :

$$E(\mathbf{u}_i^{el}) = \int_{\Omega} \sum_{j=1}^{N_{nodes}} \mathbf{u}_i^{el^T} \mathbf{B}_i^{el^T} \mathbf{D} \mathbf{B}_j^{el} \mathbf{u}_j^{el} + \mathbf{F}(\mathbf{x}) N_i^{el}(\mathbf{x}) \mathbf{u}_i^{el} \, d\Omega \tag{6}$$

We seek the minimum of this function by solving for $\frac{dE(u_i^{el})}{du_i^{el}} = 0$. Equation (6) then becomes :

$$\int_{\Omega} \sum_{j=1}^{N_{nodes}} \mathbf{B}_i^{el^T} \mathbf{D} \mathbf{B}_j^{el} \mathbf{u}_j^{el} \, d\Omega = -\int_{\Omega} \sum_{j=1}^{N_{nodes}} \mathbf{F}(\mathbf{x}) N_i^{el}(\mathbf{x}) \, d\Omega \tag{7}$$

This last expression can be written as a matrix system for each finite element:

$$\mathbf{K}^{el} \mathbf{u}^{el} = -\mathbf{F}^{el} \tag{8}$$

Matrices $\mathbf{K}^{el}$ and vector $\mathbf{F}^{el}$ are defined as follows: $\mathbf{K}_{i,j}^{el} = \int_{\Omega} \mathbf{B}_i^{el^T} \mathbf{D} \mathbf{B}_j^{el} \, d\Omega$, $\mathbf{F}_j^{el} = \int_{\Omega} \mathbf{F} N_i^{el} \, d\Omega$; where every element i, j refers to pairs of nodes of the element el (i and j range from 1 to 4 for a tetrahedron – 1 to 3 for a triangle). $\mathbf{K}_{i,j}^{el}$ is a 3 by 3 matrix, and $\mathbf{F}_j^{el}$ is a 3 by 1 vector. The 12 by 12 (9 by 9 for a triangle) matrix $\mathbf{K}^{el}$, and the vector $\mathbf{F}^{el}$ are computed for each element and are then assembled in a global system $\mathbf{Ku} = -\mathbf{F}$, the solution of which will provide us with the deformation field corresponding to the global minimum of the total energy.

We now have constitutive equations that model surfaces as elastic membranes and volumes as elastic bodies.

3.1 Finite Element Mesh Generation

In [5], Hagemann et al. propose to use the pixels of the image as basic elements of his FE mesh. This approach does not take advantage of the intrinsic formulation of FE modeling, which assumes that the mechanical properties are constant over the element, suggesting that one can use elements covering several image pixels. Also, when performing computations in 3D, which is eventually what is needed for medical applications, the amount of degrees of freedom will be far too large (for a typical 256x256x60 MR image, this means about 12 million degrees of freedom at worst case !) to perform efficient computations in a reasonable time, even on high performance computing equipment.

Most available meshing software packages do not allow meshing of multiple objects (e.g., [25,26]), and are usually designed for regular and convex objects, which is often not the case for anatomical structures. Therefore, we have implemented a tetrahedral mesh generator specifically suited for labeled 3D medical images. The mesher can be seen as the volumetric counterpart of a marching tetrahedra surface generation algorithm, the only difference being that the initial tetrahedralization we use can have an adaptive resolution with sizes of tetrahedra depending on the underlying image content.

The labeled 3D image from which the mesh needs to be computed is first divided into cubes of a given size, which are further divided into 5 tetrahedra with an alternating pattern so as to avoid diagonal crossings on the shared quadrilateral faces of neighboring cubes. The initial cube size determines the size of the largest tetrahedra the mesh will contain. Each tetrahedron is checked for subdivision according to the underlying image content. In our case, we decided to only subdivide tetrahedra that lie across boundaries of given objects, so as to have a detailed description of their boundaries. The edges of those tetrahedra to be subdivided are labeled for subdivision, and a new vertex is inserted at their middle point. This process is executed iteratively until the smallest edges have reached a specified minimum size.

At each iteration, the mesh is re-tetrahedrized given the required edge subdivisions for each tetrahedron. The main problem is to re-mesh tetrahedra that lie next to tetrahedra that are being split. For those tetrahedra, only some edges have been split. The mesh is therefore re-tetrahedrized using a case table with the $2^6 = 64$ possible edge splitting configurations. There are 10 basic configurations, the others are symmetrical to those presented on Figure 1 (the gray coloring and the different node labelings are represented only to facilitate visualization of the tetrahedra's subdivisions). From upper left to lower right, Figure 1 successively presents the tetrahedrization if one edge is split, if two edges are split (2 possible configurations), if three edges are split (3 possible configurations), if 4 edges are split (3 possible configurations), if 5 edges are split and finally if all edges of the tetrahedron are split.

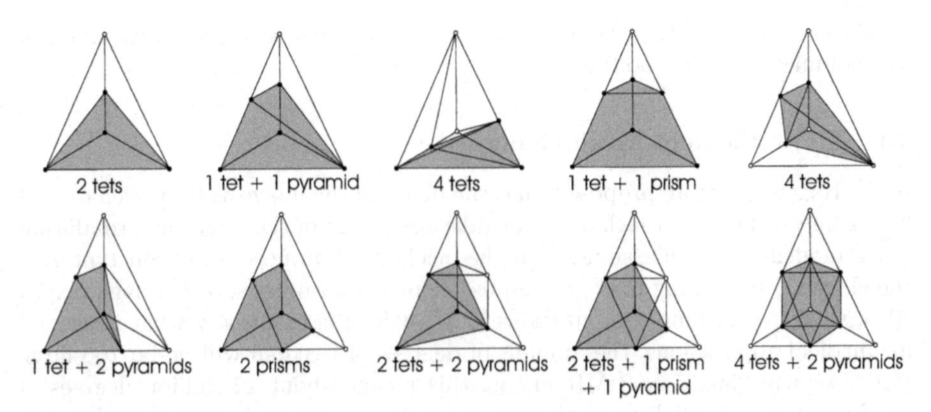

Fig. 1. Different subdivisions of a tetrahedron given edge splittings

The resulting mesh contains tetrahedra, but also pyramids and prisms, which need to be further tetrahedrized. The main problem is to ensure consistency [2] between the diagonals of quadrilateral faces shared by 2 elements (pyramids or prisms). We split the quadrilateral faces along the shortest diagonal so as to have better shaped tetrahedra. The subdivision of a pyramid into two tetrahedra is straightforward given the diagonal of the quadrilateral face. For a prism, there are eight possible configurations for tetrahedralization given the diagonal configuration. Figure 2 presents the different possible tetrahedralizations of a prism given the diagonal's configuration. If no straight tetrahedralization is possible (cases 1 and 8), a vertex is inserted in the middle of the prism which is then divided into 8 tetrahedra.

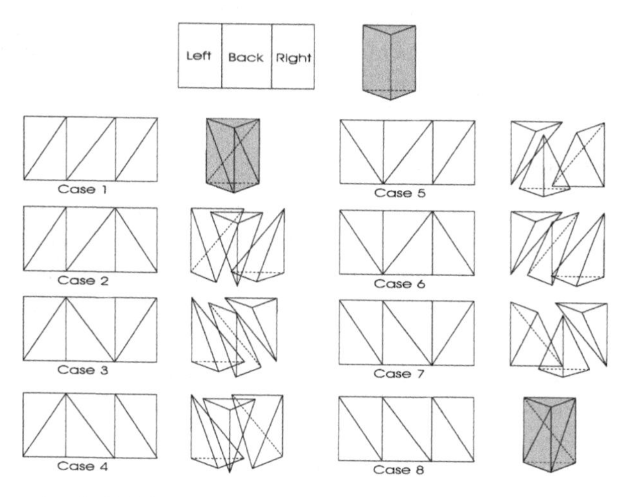

Fig. 2. Different subdivisions of a prism given the quadrilateral faces' diagonals

Finally, we apply a marching tetrahedra-like approach to generate the actual tetrahedral mesh with accurately represented boundary surfaces. For each tetrahedron, the image labels at its nodes are checked. A case table draws the elements to be added to the mesh. If all 4 nodes have non-object labels, no tetrahedron is added to the mesh. If all nodes have an object label, the tetrahedron is added to the mesh as is. If the tetrahedron lies across two objects (i.e. all nodes do not have the same label), the subdivision of the original tetrahedron is looked up in the case table.

Figure 3 shows the 5 basic cases. There are actually 16 cases, but the remaining cases are symmetric to cases 2, 3, and 4. The resulting prisms are divided into tetrahedra using the same approach as presented above.

The resulting mesh structure is built such that for images containing multiple objects, a fully connected and consistent tetrahedral mesh is obtained for every cell, whith a given label corresponding to the object the cell belongs to.

[2] A consistent tetrahedral mesh is built such that every (non-boundary) triangular face is shared by exactly 2 tetrahedra.

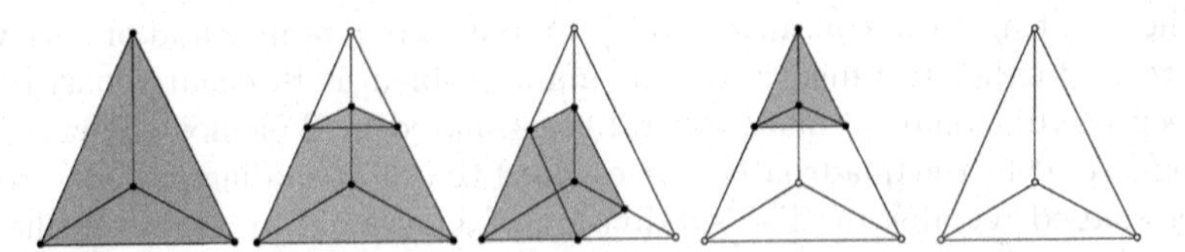

Fig. 3. Different tetrahedral cases depicted from left to right. Case 1: all nodes belong to structure; case 2: 3 nodes belong to structure; case 3: 2 nodes belong to structure; case 4: 1 node belongs to structure; case 5: no nodes belong to structure

Therefore, different biomechanical properties and parameters can easily be assigned to the different cells or objects composing the mesh. Boundary surfaces of objects represented in the mesh can be extracted from the mesh as triangulated surfaces, which is very convenient for running an active surface algorithm.

3.2 Active Surface Algorithm

The active surface algorithm deforms the boundary surface of an object in one volumetric scan of the sequence towards the boundary of the same object in the next scan of the sequence. This is done iteratively by applying image-derived forces $\mathbf{F}^{\mathbf{u}^t}$ (forces computed using the surface's nodal positions $\mathbf{u}$ at iteration t) to the elastic membrane. The temporal variation of the surface can be discretized using finite differences, provided the time step τ is small enough. This yields the following semi-implicit iterative equation [3]:

$$\frac{\mathbf{u}^t - \mathbf{u}^{t-1}}{\tau} + \mathbf{K}\mathbf{u}^t = -\mathbf{F}^{\mathbf{u}^{t-1}} \tag{9}$$

which can be rewritten as :

$$\left(\mathbf{I} + \tau\mathbf{K}\mathbf{u}^t\right) = \mathbf{u}^{t-1} - \tau\mathbf{F}^{\mathbf{u}^{t-1}} \tag{10}$$

The external forces driving the elastic membrane toward the edges of the structure in the image are integrated over each element of the mesh and distributed over the nodes belonging to the element using its shape functions (see Eqn. 4). Classically, the image force $\mathbf{F}$ is computed as a decreasing function of the gradient so as to be minimized at the edges of the image [14,27]. A potential weakness of active surface methods is that for correct convergence, the surfaces need to be initialized very close to the edges of the object to be segmented. In [14], Cohen et al. proposed to use inflation or deflation forces (so-called balloon forces) to circumvent that problem. To increase the robustness and the convergence rate of the surface deformation, we compute our forces as a gradient descent on a distance map of the edges in the target image. The distance map is computed very efficiently using our fast distance transformation algorithm [28]. To prevent the surface from sticking on a wrong edge, or to prevent two sides of a thin

[3] Superscript t refers to the current iteration.

surface from sticking together on the same edge , we have included the expected gradient sign of the structure to be segmented in the force expression. Also, we have signed the distance map for improved convergence. More details about our active surface algorithm can be found in [29].

3.3 Inferring Volumetric Deformations from Surface Deformations

The deformation field obtained for the boundary surfaces is then used in conjunction with the volumetric model to infer the deformation field inside and outside the boundary surfaces.

The idea is to apply forces to the boundary surfaces that will produce the same displacement field at the boundary surfaces that was obtained with the active surface algorithm. The volumetric biomechanical model will then compute the deformation of the surrounding nodes in the mesh.

Let $\widetilde{\mathbf{u}}$ be the vector representing the displacement of the boundary nodes to be imposed. Hence, the equilibrium equation of the elastic body (Eqn. 1) needs to be rewritten with the following external forces to impose these displacements to the volume :

$$\mathbf{F} = \mathbf{K}\widetilde{\mathbf{u}} \tag{11}$$

The solution of the global equilibrium system (see Eqn. 8) will provide us with the displacement at all the nodes in the volumetric mesh with the imposed displacements at the nodes of the boundary surfaces delimiting the objects represented in the mesh. This volumetric displacement field is then interpolated back onto the image grid using the shape functions of every element of the mesh (see Eqn. 4).

Biomechanical parameters such as the stress tensors can then be derived from the displacements at the nodes using the stress-strain relationship (Eqn. 3) :

$$\boldsymbol{\sigma}_i = \sum_{Tet|i \in Tet} \mathbf{D}\epsilon_i = \sum_{Tet|i \in Tet} \mathbf{D}\mathbf{L}_i N_i^{Tet} \mathbf{u}_i \tag{12}$$

4 Experiments

4.1 Synthetic Image Sequence

We have tested the algorithm on a sequence of two 3D images of an elastic sphere being squeezed in a given direction. The object is surrounded by another elastic object.

The original active surface extracted from the volumetric tetrahedral mesh is shown in Figure 4b. Note that for this experiment, the initial tetrahedralization from which the mesh was computed was not multi-resolution, it had constant tetrahedral sizes. When running the active surface algorithm, the surface readily converges to the boundary of the ellipsoid in the target image.

Figure 5 shows 3D views of the mesh associated to the initial image and of the mesh after deformation, while Figure 6 shows cuts through the original

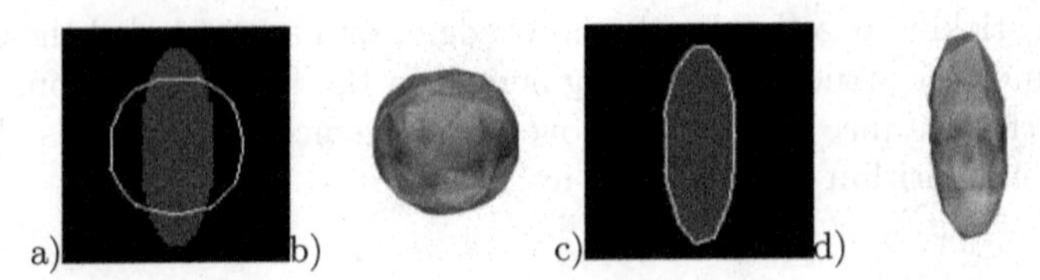

Fig. 4. a) Slice 30 of the target image with a cut through the initial surface of the object overlayed. b) 3D surface rendering of the initial surface. c) The same slice with a cut through the deformed surface. d) 3D surface rendering of the deformed surface

(a) and deformed mesh (b) and the deformation field (c) interpolated back onto the image grid (downsampled for clarity) overlayed on a cut through the target volume. One can very well observe the physical squeezing of the sphere onto the ellipsoid, also deforming the surrounding elastic medium.

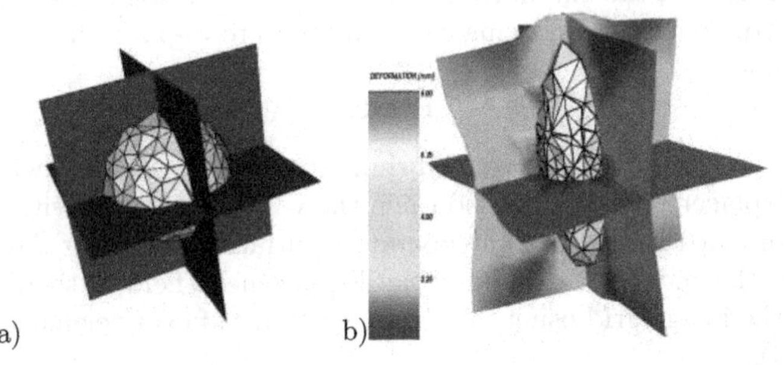

Fig. 5. a) Orthogonal cuts through the initial volumetric mesh with the sphere extracted and b) the same with deformed mesh

4.2 Brain Shift Analysis

In this experiment, we wanted to characterize the deformation the brain undergoes during neurosurgery after craniotomy. Two 3D MR volumetric scans were taken before and after craniotomy and partial tumor resection, and a significant shift could be observed. Figure 7 shows cuts through a sample tetrahedral mesh of the brain overlayed on the corresponding initial image.

The active surface is extracted from the intraoperative scan at start of surgery, before opening the dura mater (see Figure 8a), and deformed towards the brain in a later intraoperative image (see Figure 8b). Figure 8c shows the 3D surface deformation field the brain has undergone. One can observe that the deformation of the cortical surface is happening in the direction of gravity and is mainly located where the dura was removed. Part of the shift (especially on the left of the picture) is also due to the tumor resection that was done between the two scans.

The deformation field obtained with the active surface algorithm is then used as a boundary condition for our biomechanical FE model, and allows us to infer

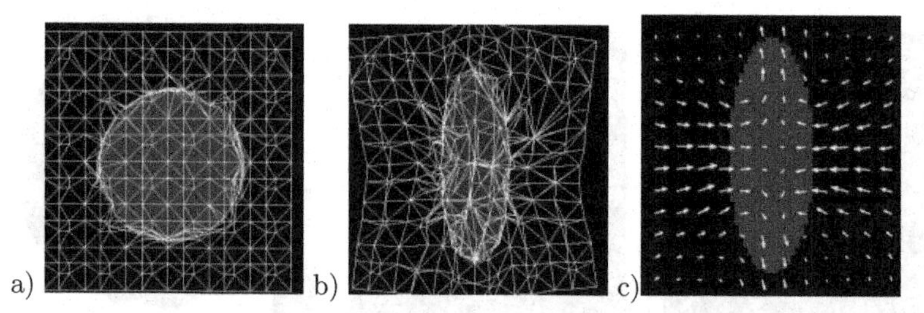

Fig. 6. a) Axial cut through original mesh overlayed on slice of original image and b) the same with deformed mesh on target image. c) The deformation field overlayed on slice of target image

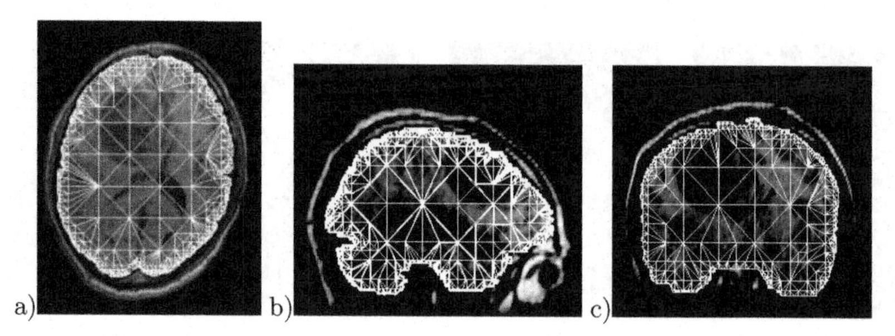

Fig. 7. Axial (a), sagittal (b), and coronal (c) cuts through tetrahedral mesh of the brain overlayed on corresponding cuts through preoperative image

a volumetric deformation field. Elasticity parameters were chosen according to in-vivo studies carried out by Miga et al. [3]. Figure 9a shows the obtained deformation field overlayed on a slice of the initial scan, and Figure 9b shows the same slice of the initial scan deformed with the obtained deformation field. Figure 9 also presents the same slice of the target scan and the magnitude of the difference with the initial scan showing the closeness of the alignment of the brain. The gray-level mean square difference between the target scan and the deformed original scan on the image regions covered by the mesh went down from 181 to 88. The remaining difference is due to the fact that the model we used did not incorporate the ventricular thinning and the tumor resection that occurred between both scans.

5 Conclusions

We have presented a new algorithm for tracking and characterizing shape changes in 3D image sequences of physics-based objects. The algorithm incorporates a biomechanical model of the deforming objects and uses image-based information to drive the deformation of our model through an active surface algorithm. One of the main contributions of this paper is an improved algorithm for generating multi-resolution patient-specific FE meshes from labeled 3D images.

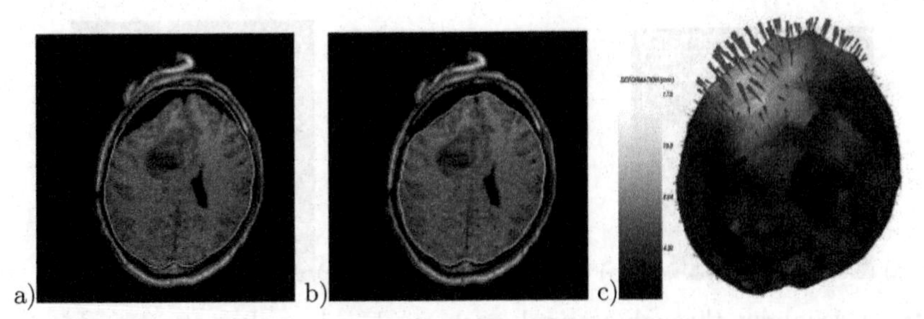

Fig. 8. Axial cut through active surface a) initial, and b) deformed, overlayed on corresponding slice of intraoperative MR image. c) 3D surface rendering of active surface with colorcoded intensity of the deformation field

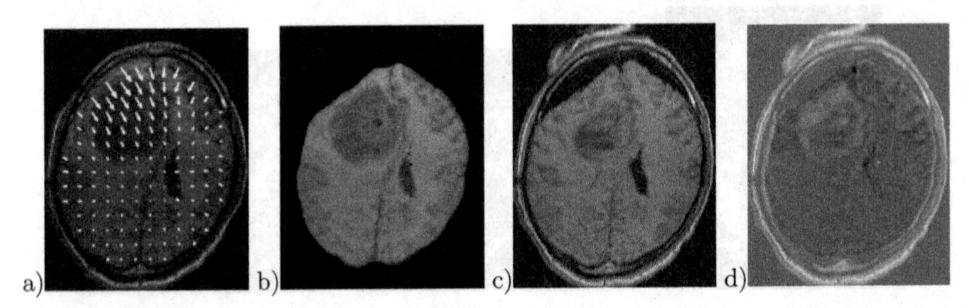

Fig. 9. a) Volumetric deformation field overlayed on initial intraoperative image slice. b) The same slice of initial image deformed using deformation field. c) Same slice of target image. d) The difference between target and deformed images with same slice

The algorithm is a promising tool for the analysis of 3D medical image sequences. It will provide physicians with a tool for measurement and physical interpretation of deformation in 3D image sequences, and can thus be of great aid in in the interpretation and diagnosis of these images.

References

1. D. M. Metaxas. *Physics-Based Deformable Models: Applications to Computer Vision, Graphics and Medical Imaging.* Kluwer Academic Publishers, 1997. 235, 236, 237
2. D. L. G. Hill, C. R. Maurer, A. J. Martin, S. Sabanathan, W. A. Hall, D. J. Hawkes, D. Rueckert, and C. L. Truwit. Assessment of intraoperative brain deformation using interventional mr imaging. In Berlin Springer-Verlag, editor, *MICCAI '99*, pages 910–919, 1999. 235
3. M. I. Miga, K. D. Paulsen, P. J. Hoopes, F. E. Kennedy, A. Hartov, and D. W. Roberts. In vivo quantification of a homogeneous brain deformation model for updating preoperative images during surgery. *IEEE Transactions on Medical Imaging*, 47(2):266–273, February 2000. 235, 236, 245
4. O. M. Skrinjar and J. S. Duncan. Real time 3d brain shift compensation. In *IPMI '99*, 1999. 235, 236

5. A. Hagemann, Rohr K., H. S. Stiel, U. Spetzger, and Gilsbach J. M. Biomechanical Modeling of the Human Head for Physically Based, Non-Rigid Image Registration. *IEEE Transactions on Medical Imaging*, 18(10):875–884, October 1999. 235, 236, 237, 239
6. Paul M. Thompson, Jay N. Giedd, Roger P. Woods, David MacDonald, Alan C. Evans, and Arthur W. Toga. Growth patterns in the developing brain detected by using continuum mechanical tensor maps. *Nature*, (404):190–193, 2000. Macmillan Publishers Ltd. 235
7. H. Delinguette. Toward Realistic Soft-tissue Modeling in Medical Simulation. *Proceedings of the IEEE*, 86(3):512–523, March 1998. 235, 236
8. Pengcheng Shi, Albert J. Sinusas, R. Todd Constable, and James S. Duncan. Volumetric deformation using mechanics-based data fusion: Applications in cardiac motion recovery. *International Journal of Computer Vision*, November 1999. 235
9. J. Martin, A. Pentland, S. Sclaroff, and R. Kikinis. Characterization of Neuropathological Shape Deformations. *IEEE Transactions on Pattern Analysis and Machine Intelligence*, 20(2):97–112, February 1998. 236
10. Jean-Philippe Thirion and Guillaume Calmon. Deformation analysis to detect and quantify active lesions in 3d medical image sequences. Technical Report 3101, INRIA Sophia Antipolis, February 1997. 236
11. Stephen M. Pizer, Christina A. Burbeck, Daniel S. Fritch, Bryan S. Morse, Alan Liu, Shobha Murthy, and Derek T. Puff. Human Perception and Computer Image Analysis of Objects in Images. In *DICTA-93 Digital Image Computing: Techniques and Applications*, volume 1, pages 19–26, 1993. 236
12. G. Stetten and S. M. Pizer. Medial Node Models to identify and Measure Objects in Real-Time 3D Echocardiography. *IEEE Transactions on Medical Imaging*, 18(10):1025–1034, 1999. 236
13. S. M. Pizer, D. S. Fritsch, P. Yushkevich, V. Johnson, and E. Chaney. Segmentation, registration, and measurement of shape variation via image object shape. *IEEE Transactions on Medical Imaging*, pages 851–865, 1996. 236
14. L. D. Cohen and Cohen I. Finite Element Methods for Active Contour Models and Balloons for 2D and 3D Images. *IEEE Transactions on Pattern Analysis and Machine Intelligence*, 15:1131–1147, 1993. 236, 242
15. C. Davatzikos. Spatial Transformation and Registration of Brain Images Using Elastically Deformable Models. *Computer Vision and Image Understanding*, 66(2):207–222, May 1997. 236
16. P. Thompson and A. W. Toga. A Surface-Based Technique for Warping Three-Dimensional Images of the Brain. *IEEE Transactions on Medical Imaging*, 15(4):402–417, 1996. 236
17. R. Bajcsy and S. Kovacic. Multi-resolution Elastic Matching. *Computer Vision, Graphics, and Image Processing*, 46:1–21, 1989. 236
18. G. E. Christensen, S. C. Joshi, and M. I. Miller. Volumetric Transformation of Brain Anatomy. *IEEE Transactions on Medical Imaging*, 16(6):864–877, December 1997. 236
19. M. Bro-Nielsen and C. Gramkow. Fast Fluid Registration of Medical Images. In *Visualization in Biomedical Computing (VBC '96)*, pages 267–276, 1996. 236
20. X. Papademetris, A. J. Sinusas, D. P. Dione, and J. S. Duncan. 3D Cardiac Deformation from Ultrasound Images. In *MICCAI 1999 : Medical Image Computing and Computer Assisted Intervention*, pages 420–429. Springer, September 1999. 236
21. J. Park, D. Metaxas, and L. Axel. Analysis of left ventricular wall motion based on volumetric deformable models and MRI-SPAMM. *Medical Image Analysis*, 1(1):53–71, 1996. 236

22. E. Haber, D. N. Metaxas, and L. Axel. Motion Analysis of the Right Ventricle from MR images. In *MICCAI 1998 : Medical Image Computing and Computer Assisted Intervention*, pages 177–188. Springer, October 1998. 236
23. S. K. Kyriacou, C. Davatzikos, S. J. Zinreich, and R. N. Bryan. Nonlinear elastic registration of brain images with tumor pathology using a biomechanical model. *IEEE Transactions on Medical Imaging*, 18(7):580–592, july 1999. 236
24. O. C. Zienkewickz and R. L. Taylor. *The Finite Element Method.* McGraw Hill Book Co., 1987. 238, 239
25. Will Schroeder, Ken Martin, and Bill Lorensen. *The Visualization Toolkit: An Object-Oriented Approach to 3D Graphics.* Prentice Hall PTR, New Jersey, second edition, 1998. 239
26. B. Geiger. Three dimensional modeling of human organs and its application to diagnosis and surgical planning. Technical Report 2105, INRIA, 1993. 239
27. M. Kass, A. Witkin, and D. Terzopoulos. Snakes: Active Contour Models. *International Journal of Computer Vision*, 1(4):321–331, 1988. 242
28. O. Cuisenaire and B. Macq. Fast Euclidean Distance Transformation by Propagation Using Multiple Neighborhoods. *Computer Vision and Image Understanding*, 76(2):163–172, November 1999. 242
29. M. Ferrant, O. Cuisenaire, and B. Macq. Multi-Object Segmentation of Brain Structures in 3D MRI Using a Computerized Atlas. In *SPIE Medical Imaging '99*, volume 3661-2, pages 986–995, 1999. 243

Naive Planes as Discrete Combinatorial Surfaces

Yukiko Kenmochi[1] and Atsushi Imiya[2]

[1] School of Information Science, Japan Advanced Institute of Science and Technology
kenmochi@jaist.ac.jp
[2] Department of Information and Image Sciences, Chiba University
imiya@ics.tj.chiba-u.ac.jp

Abstract. An object of interest is digitized if we acquire its 3-dimensional digital images by using techniques such as computerized tomographic imaging. For recognition or shape analysis of such digitized objects, we need the study of 3-dimensional digital geometry and topology. In this paper, we focus on one of the simplest geometric objects such as planes and study their geometric and topological properties which are expressed by using an algebraic method.

1 Introduction

In this paper, we deal with the geometric and topological properties of digitized objects which are expressed by using an algebraic method. An object of interest is digitized if we acquire its 3-dimensional digital images by using techniques such as computerized tomographic imaging. For recognition and shape analysis of such digitized objects, topological constraints such as topological equivalence in use of Euler characteristics [1], skeletons [2], combinatorial manifolds [4], etc. are used as well as geometric constraints which are given by algebraic equations/inequations. In this paper, we focus on one of the simplest geometric objects such as planes and study not only their digital geometry but their digital topology for 3-dimensional computer imagery.

In the context of digital geometry, an algebraic approach for the study of geometric properties of planes in an integer lattice space has been proposed in [3,4,5,6,7]. They have defined a naive/standard plane which is a set of integer lattice points and proposed the theory using algebraic properties of a lattice space. The algebraic properties such as local configurations of points in naive/standard planes have been derived by their algebraic approach.

On the other hand, we have proposed an approach based on combinatorial topology [8,9] for the definition and construction of topological planes in an integer lattice space; they are called discrete planar surfaces and constructed by applying our boundary extraction algorithm [8]. In this paper, we clarify the relations between our discrete planar surfaces and naive planes such that our discrete planar surface for 18- or 26-neighborhood system is a triangulation of a naive plane in the aspect of combinatorial topology [10]; in [4], a triangulation has been introduced in a similar approach for a standard plane, but not for a naive plane. By seeing the topological structures of naive planes as discrete

G. Borgefors, I. Nyström, and G. Sanniti di Baja (Eds.): DGCI 2000, LNCS 1953, pp. 249–261, 2000.
© Springer-Verlag Berlin Heidelberg 2000

planar surfaces, we show their geometric and topological properties in an integer lattice space, which can be derived from the algebraic properties of naive planes.

2 Definition of Naive Planes

Let $\mathbf{R}$ be the set of real numbers; $\mathbf{R}^3$ denotes the 3-dimensional Euclidean space. A plane $\mathbf{P}$ in $\mathbf{R}^3$ is defined by

$$\mathbf{P} = \{(x, y, z) \in \mathbf{R}^3 \; : \; ax + by + cz + d = 0\} \tag{1}$$

where a, b, c, d are real numbers. Let $\mathbf{Z}$ be the set of real numbers; $\mathbf{Z}^3$ denotes the set of lattice points whose coordinates are all integers. We introduce the definition of planes in $\mathbf{Z}^3$ based on algebraic approach [5]. The naive plane is defined with respect to $\mathbf{P}$ by

$$\mathbf{NP} = \{(x, y, z) \in \mathbf{Z}^3 \; : \; 0 \leq ax + by + cz + d < w\} \tag{2}$$

where $w = max\{|a|, |b|, |c|\}$. The parameter w is called the width of $\mathbf{NP}$.

3 Definition of Discrete Planar Surfaces

3.1 Definition of Discrete Combinatorial Surfaces

In this subsection, we introduce the definition of surfaces in $\mathbf{Z}^3$ based on the approach of combinatorial topology [10]. In $\mathbf{Z}^3$ we define three different neighborhoods of a lattice point $\boldsymbol{x} = (i, j, k)$ as

$$\mathbf{N}_m(\boldsymbol{x}) = \{(p, q, r) \in \mathbf{Z}^3 : (i-p)^2 + (j-q)^2 + (k-r)^2 \leq t\} \tag{3}$$

for $m = 6, 18, 26$ corresponding to $t = 1, 2, 3$. They are called 6-, 18- and 26-neighborhoods, respectively. Depending on each neighborhood, we define elements of 1-dimensional curves and 2-dimensional surfaces in $\mathbf{Z}^3$. These elements are called 1- and 2-dimensional discrete simplexes and abbreviated as 1- and 2-simplexes, respectively. Suppose we define 0-dimensional discrete simplexes, which are called 0-simplexes, as isolated points in $\mathbf{Z}^3$. Then 1- and 2-simplexes are defined recursively as follows.

Definition 1. *An n-simplex for $n = 1, 2$ is defined as a set of k points in $\mathbf{Z}^3$, $[\boldsymbol{x}_1, \boldsymbol{x}_2, \ldots, \boldsymbol{x}_k] = \{\boldsymbol{x}_1, \boldsymbol{x}_2, \ldots, \boldsymbol{x}_k\}$ so that the closed convex hull of $\boldsymbol{x}_1, \boldsymbol{x}_2, \ldots, \boldsymbol{x}_k$ is one of n-dimensional minimum nonzero regions in $\mathbf{R}^3$ which are bounded by the closed convex hulls of $(n-1)$-simplexes.*

According to Definition 1, a 1-simplex consists of two neighboring points in $\mathbf{Z}^3$. The configurations of those two neighboring points depend on the neighborhood systems as shown in the first line of Table 1. A 2-simplex is defined as a set of points whose closed convex hull is bound by a set of the closed convex

Table 1. Each n-simplexes for $n = 1, 2$ is defined as a set of points in $\mathbf{Z}^3$ for the 6-, 18- and 26-neighborhood systems. All n-simplexes are obtained by rotation and translation of those in the table

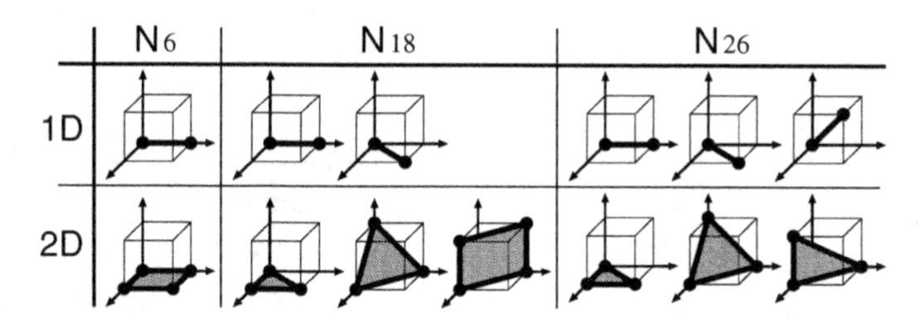

hulls of 1-simplexes and holds a 2-dimensional minimum nonzero area. All 2-simplexes for each neighborhood system are shown in the second line of Table 1. The constructive definitions of 1- and 2-simplexes are presented in [8].

If an n_1-simplex is a subset of an n_2-simplex where $n_1 < n_2$, the n_1-simplex is called a face of the n_2-simplex; it is also called an n_1-face. For instance, a 2-simplex for the 26-neighborhood system has three 0-faces and three 1-faces. A set of all faces included in a discrete simplex $[a] = [\boldsymbol{x}_1, \boldsymbol{x}_2, \dots, \boldsymbol{x}_k]$ is denoted by $face([a])$. Let the closed convex hull of k points, $\boldsymbol{x}_1, \boldsymbol{x}_2, \dots, \boldsymbol{x}_k$, be denoted by $\mathbf{CH}(\{\boldsymbol{x}_1, \boldsymbol{x}_2, \dots, \boldsymbol{x}_k\})$. The embedded discrete simplex is defined as

$$\|a\| = \mathbf{CH}([a]) \setminus (\bigcup_{[b] \in face([a])} \mathbf{CH}([b])) \quad (4)$$

for any n-simplex $[a]$, and $\|a\|$ is called the embedded n-simplex of $[a]$. An n-simplex and the embedded n-simplex are clearly different since $[a]$ and $\|a\|$ are defined as sets of points in $\mathbf{Z}^3$ and $\mathbf{R}^3$, respectively.

Definition 2. *A finite set* $\mathbf{K}$ *of discrete simplexes is called a discrete complex if it satisfies the following conditions: if* $[a] \in \mathbf{K}$, $face([a]) \subseteq \mathbf{K}$*; if* $[a], [b] \in \mathbf{K}$ *and* $\|a\| \cap \|b\| \neq \emptyset$*, then* $[a] = [b]$.

The dimension of $\mathbf{K}$ is equal to the maximum dimension of discrete simplexes which belong to $\mathbf{K}$. Hereafter, we abbreviate n-dimensional discrete complexes to n-complexes as well as n-simplexes. Suppose that $\mathbf{K}$ is an n-complex. If there exist at least one n-simplex $[a] \in \mathbf{K}$ for every s-simplex $[b] \in \mathbf{K}$ such that $[b] \in face([a])$ and $s < n$, $\mathbf{K}$ is called pure. If we can find a chain of discrete simplexes between two arbitrary elements $[c], [d] \in \mathbf{K}$, $[c_1] = [c], [c_2], \dots, [c_k] = [d]$, such that $[c_i]$ and $[c_{i+1}]$, $i = 1, 2, \dots, k-1$, has a common face in $\mathbf{K}$, $\mathbf{K}$ is called connected.

Definition 3. *If a 2-complex* $\mathbf{K}$ *is pure and connected,* $\mathbf{K}$ *is a discrete combinatorial surface.*

More discussion on discrete combinatorial surfaces in the sense of combinatorial topology is given in [8]. Note that discrete complexes and discrete combinatorial surfaces are constructed with respect to each neighborhood system.

Table 2. A set $\partial \mathbf{I}_m^+(i,j,k)$ for each of eight possible configurations of black and white points in $\mathbf{C}(i,j,k)$, $m = 6, 18, 26$. In the table, we consider $\mathbf{P}$ such that $0 \leq a \leq b \leq c$, $c > 0$. The configurations within parentheses are ignored for the construction of $\partial \mathbf{I}_m^+$ because black points in such $\mathbf{C}(i,j,k)$ are regarded as 0- or 1-faces of 2-simplexes in the adjacent cubes of $\mathbf{C}(i,j,k)$

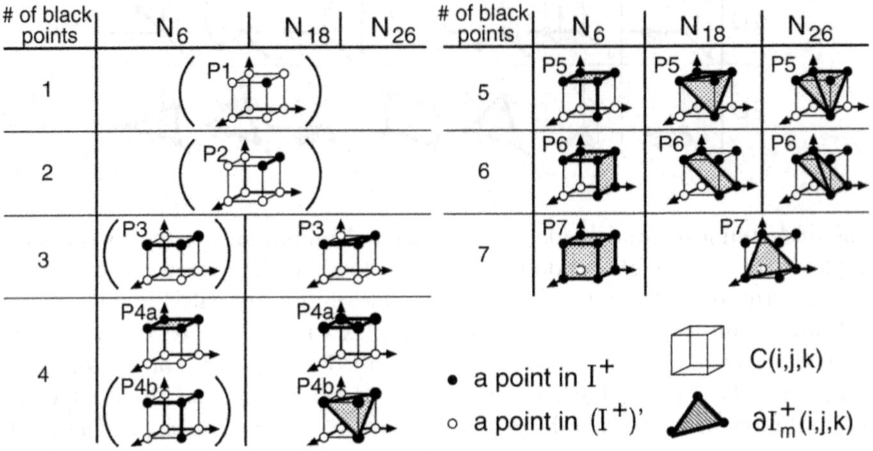

# of black points	N_6	N_{18}	N_{26}
1	(P1)		
2	(P2)		
3	(P3)	P3	
4	P4a, (P4b)	P4a, P4b	
5	P5	P5	P5
6	P6	P6	P6
7	P7	P7	

• a point in I^+
○ a point in $(\mathrm{I}^+)'$
C(i,j,k)
$\partial \mathrm{I}_m^+(i,j,k)$

3.2 Construction of Discrete Planar Surfaces

The following two regions in $\mathbf{R}^3$ are separated by $\mathbf{P}$:

$$\mathbf{H}^- = \{(x,y,z) \in \mathbf{R}^3 \ : \ ax + by + cz + d \leq 0\} \ , \tag{5}$$

$$\mathbf{H}^+ = \{(x,y,z) \in \mathbf{R}^3 \ : \ ax + by + cz + d \geq 0\} \ . \tag{6}$$

Obviously, we have

$$\mathbf{H}^- \cap \mathbf{H}^+ = \mathbf{P} \ . \tag{7}$$

Just as $\mathbf{H}^-$ and $\mathbf{H}^+$ in $\mathbf{R}^3$, there are two regions in $\mathbf{Z}^3$, which are separated by $\mathbf{P}$ as follows:

$$\mathbf{I}^- = \{(x,y,z) \in \mathbf{Z}^3 \ : \ ax + by + cz + d \leq 0\} \ , \tag{8}$$

$$\mathbf{I}^+ = \{(x,y,z) \in \mathbf{Z}^3 \ : \ ax + by + cz + d \geq 0\} \ . \tag{9}$$

We say that $\mathbf{I}^-$ and $\mathbf{I}^+$ are the digitization of $\mathbf{H}^-$ and $\mathbf{H}^+$, respectively. For both $\mathbf{I}^-$ and $\mathbf{I}^+$, we can construct the boundaries which are discrete combinatorial surfaces with the m-neighborhood system for $m = 6, 18, 26$, denoted by $\partial \mathbf{I}_m^-$ and $\partial \mathbf{I}_m^+$, using the similar algorithm for boundary extraction [8]. Both $\partial \mathbf{I}_m^-$ and $\partial \mathbf{I}_m^+$ are considered to be the digitization of $\mathbf{P}$ and called discrete planar surfaces with respect to $\mathbf{P}$. In this subsection, we present how to generate $\partial \mathbf{I}_m^+$ from $\mathbf{I}^+$. The same procedure can be applied to generate $\partial \mathbf{I}_m^-$ if $\mathbf{I}^+$ and $\partial \mathbf{I}_m^+$ are replaced by $\mathbf{I}^-$ and $\partial \mathbf{I}_m^-$, respectively.

Algorithm 1.
input: $\mathbf{I}^+$.
output: $\partial \mathbf{I}_m^+$.
begin

1. *Points in* $\mathbf{I}^+$ *and the complement* $(\mathbf{I}^+)' = \mathbf{Z}^3 \setminus \mathbf{I}^+$ *are assigned black and white points, respectively;*
2. *in any unit cubic region*

$$\mathbf{C}(i,j,k) = \{(x,y,z) \in \mathbf{Z}^3 : i \leq x \leq i+1, j \leq y \leq j+1, k \leq z \leq k+1\} \quad (10)$$

   such that $\mathbf{C}(i,j,k) \cap \mathbf{I}^+ \neq \emptyset$ *and* $\mathbf{C}(i,j,k) \cap (\mathbf{I}^+)' \neq \emptyset$, *the black and white points have either of eight different configurations as shown in Table 2;*
3. *for each* $\mathbf{C}(i,j,k)$, $\partial \mathbf{I}_m^+(i,j,k)$ *is obtained as a set of 2-simplexes and their faces by referring to Table 2;*
4. *obtain*

$$\partial \mathbf{I}_m^+ = \bigcup_{(i,j,k) \in \mathbf{Z}^3} \partial \mathbf{I}_m^+(i,j,k) \; . \quad (11)$$

end

The next theorem is derived; the proof is given in [9].

Theorem 1. *Each* $\partial \mathbf{I}_m^+$ *(resp.* $\partial \mathbf{I}_m^-$*) for* $m = 6, 18, 26$ *obtained from* $\mathbf{I}^+$ *(resp.* $\mathbf{I}^-$*) by Algorithm 1 is a discrete combinatorial surface and called a discrete planar surface.*

4 Relations between Naive Planes and Discrete Planar Surfaces

Let $\mathbf{B}_m^+$ be the set of all lattice points in $\partial \mathbf{I}_m^+$ for $m = 6, 18, 26$, such that

$$\mathbf{B}_m^+ = \bigcup_{[a] \in \partial \mathbf{I}_m^+} [a]. \quad (12)$$

Then, the following lemma is derived.

Lemma 1. *For any plane* $\mathbf{P}$, *the inclusion and equality relations*

$$\mathbf{B}_6^+ \supseteq \mathbf{B}_{18}^+ = \mathbf{B}_{26}^+ \quad (13)$$

hold.

Proof. Using $\mathbf{C}(i,j,k)$ of (10), for each m, we define

$$\mathbf{B}_m^+(i,j,k) = \mathbf{B}_m^+ \cap \mathbf{C}(i,j,k) \quad (14)$$

which is a subset of $\mathbf{B}_m^+$. Let us compare a triplet of $\mathbf{B}_m^+(i,j,k)$, $m = 6, 18, 26$ for every $\mathbf{C}(i,j,k)$ in $\mathbf{Z}^3$. If we make a comparison between $\mathbf{B}_6^+(i,j,k)$ and $\mathbf{B}_{18}^+(i,j,k)$ in Table 2, we see

$$\mathbf{B}_6^+(i,j,k) \supset \mathbf{B}_{18}^+(i,j,k) \quad (15)$$

Table 3. The classification of all points in $\mathbf{I}^+$ into two types with respect to each configuration of Table 2: a set of black and gray points which are included in $\mathbf{B}_{26}^+$ and a set of white points which are not included in $\mathbf{B}_{26}^+$

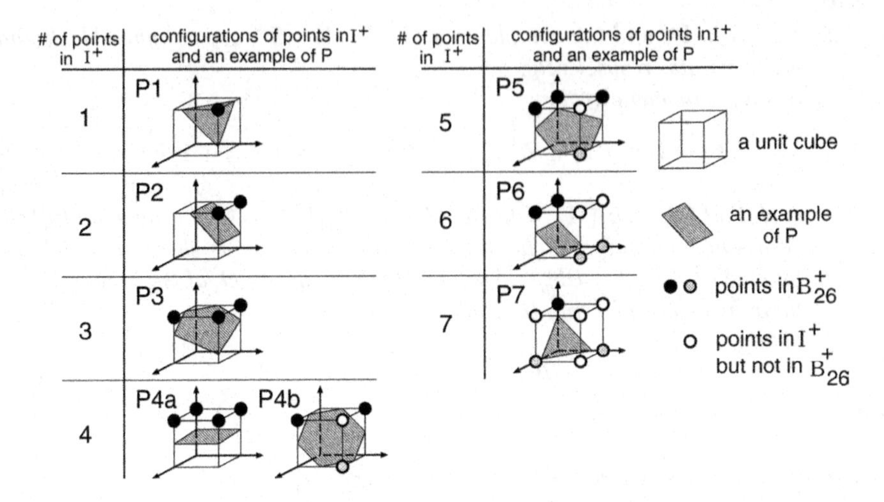

for configurations P4b, P5, P6 and P7, otherwise we obtain

$$\mathbf{B}_6^+(i,j,k) = \mathbf{B}_{18}^+(i,j,k). \tag{16}$$

Between $\mathbf{B}_{18}^+(i,j,k)$ and $\mathbf{B}_{26}^+(i,j,k)$, we see

$$\mathbf{B}_{18}^+(i,j,k) = \mathbf{B}_{26}^+(i,j,k) \tag{17}$$

for any configuration in Table 2, even if $\partial\mathbf{I}_{18}^+(i,j,k)$ and $\partial\mathbf{I}_{26}^+(i,j,k)$ are different for P5 and P6. From (15), (16) and (17), we see that (13) holds. □

We then prove the next theorem.

Theorem 2. *For any* $\mathbf{P}$*, the equality relations*

$$\mathbf{NP} = \mathbf{B}_{18}^+ = \mathbf{B}_{26}^+ \tag{18}$$

hold.

Proof. Let us consider $\mathbf{P}$ such that $0 \le a \le b \le c$, $c > 0$. In this case $w = c$. From (2) we obtain

$$\mathbf{NP} = \{(x,y,z) \in \mathbf{Z}^3 \;:\; -\frac{a}{c}x - \frac{b}{c}y - \frac{d}{c} \le z < -\frac{a}{c}x - \frac{b}{c}y - \frac{d}{c} + 1\}. \tag{19}$$

For every point $\boldsymbol{x} = (x,y,z)$ in $\mathbf{NP}$, if we set a point $\mathbf{c}_{\boldsymbol{x}} \in \mathbf{P}$ such that

$$\mathbf{c}_{\boldsymbol{x}} = (x, y, -\frac{a}{c}x - \frac{b}{c}y - \frac{d}{c}), \tag{20}$$

then we see that

$$0 \leq |\boldsymbol{x} - \boldsymbol{c_x}| < 1 \tag{21}$$

from (19). Since $\mathbf{B}_{18}^+ = \mathbf{B}_{26}^+$ from Lemma 1, we will show that every $\boldsymbol{x} \in \mathbf{B}_{26}^+$ satisfies (21) and no point in $\mathbf{Z}^3 \setminus \mathbf{B}_{26}^+$ satisfies (21) to prove this theorem. Let us consider a cubic region $\mathbf{C}(i,j,k)$ of (10). Table 2 gives all configurations of points in $\mathbf{I}^+$ and $(\mathbf{I}^+)'$ for a $\mathbf{C}(i,j,k)$. All black points in Table 2 are classified into black, gray and white points in Table 3. All black points in Table 3 apparently satisfy (21). Let us consider the case $a > 0$. For each gray point $\boldsymbol{x} = (x,y,z)$, if we set two points in $\mathbf{P}$ such as

$$\boldsymbol{a_x} = (-\frac{b}{a}y - \frac{c}{a}z - \frac{d}{a}, y, z) \tag{22}$$

and

$$\boldsymbol{b_x} = (x, -\frac{a}{b}x - \frac{c}{b}z - \frac{d}{b}, z)\ , \tag{23}$$

we obtain

$$|\boldsymbol{x} - \boldsymbol{a_x}| \geq |\boldsymbol{x} - \boldsymbol{b_x}| \geq |\boldsymbol{x} - \boldsymbol{c_x}| \tag{24}$$

since $|\boldsymbol{x} - \boldsymbol{a_x}| : |\boldsymbol{x} - \boldsymbol{b_x}| : |\boldsymbol{x} - \boldsymbol{c_x}| = 1/a : 1/b : 1/c$ from Lemma 2 in Appendix and $0 < a \leq b \leq c$. We then see in Table 3 that every gray point $\boldsymbol{x}$ satisfies

$$|\boldsymbol{x} - \boldsymbol{a_x}| < 1 \quad \text{or} \quad |\boldsymbol{x} - \boldsymbol{b_x}| < 1\ , \tag{25}$$

and from (24) we obtain (21). In the case $a = 0$, we will have only P2, P4a and P6 for configurations of $\mathbf{C}(i,j,k)$ in Table 3, and gray points exist only in P6. If $b > 0$, we set $\boldsymbol{b_x}$ and have the second inequation of (25). Thus, we also obtain (21). If $b = 0$, we will have only P4a in which no gray point exists. Obviously, no white point in Table 3 satisfies (21). From a comparison between Tables 2 and 3, we see that a set of black and gray points in Table 3 is equal to a set of points of $\mathbf{B}_{26}^+$ in Table 2. Thus, we have (18). □

Theorem 2 indicates that either of $\partial\mathbf{I}_{18}^+$ or $\partial\mathbf{I}_{26}^+$ is a triangulation of $\mathbf{NP}$ in the aspect of combinatorial topology. From Lemma 1 and Theorem 2, we obtain the next corollary.

Corollary 1. *For any* $\mathbf{P}$*, the inclusion relation*

$$\mathbf{NP} \subseteq \mathbf{B}_6^+ \tag{26}$$

holds.

If $\mathbf{NP} = \mathbf{B}_6^+$, we say that $\partial\mathbf{I}_6^+$ is also a triangulation of $\mathbf{NP}$, but if $\mathbf{NP} \subset \mathbf{B}_6^+$, it is not obviously. If we define a naive plane such that

$$\mathbf{NP}^- = \{(x,y,z) \in \mathbf{Z}^3 \ : \ -w < ax + by + cz + d \leq 0\} \tag{27}$$

instead of $\mathbf{NP}$ and set

$$\mathbf{B}_m^- = \bigcup_{[a] \in \partial\mathbf{I}_m^-} [a] \tag{28}$$

for $m = 6, 18, 26$, then the following corollary is derived.

Corollary 2. *For any* $\mathbf{P}$, *the relations*

$$\mathbf{NP}^- = \mathbf{B}_{18}^- = \mathbf{B}_{26}^- \tag{29}$$

and

$$\mathbf{NP}^- \subseteq \mathbf{B}_6^- \tag{30}$$

hold.

Equation (29) in Corollary 2 indicates that either of $\partial\mathbf{I}_{18}^-$ or $\partial\mathbf{I}_{26}^-$ is a triangulation of $\mathbf{NP}^-$ in the aspect of combinatorial topology.

5 Properties of Naive Planes as Discrete Planar Surfaces

In this section, we discuss the local configurations of discrete simplexes in $\partial\mathbf{I}_6^+$, $\partial\mathbf{I}_{18}^+$ and $\partial\mathbf{I}_{26}^+$ with respect to $\mathbf{P}$. From Theorem 2 and the properties of local point configurations of $\mathbf{NP}$ which have been introduced in [4,5,6,7], we can derive the combinatorial properties of $\partial\mathbf{I}_6^+$, $\partial\mathbf{I}_{18}^+$ and $\partial\mathbf{I}_{26}^+$; they are summarized in Propositions 1 to 5. Let us consider the configurations of discrete simplexes in the parts of $\partial\mathbf{I}_m^+$ for $m = 6, 18, 26$ which project on the coordinate plane $z = 0$ as a rectangle whose sizes are $\lambda \times \mu$.

Proposition 1. *In the case of* $\lambda = \mu = 2$, *there exist five different configurations in* $\mathbf{NP}$*s and the corresponding configurations of discrete simplexes for* $\partial\mathbf{I}_m^+$, $m = 6, 18, 26$, *are shown in Fig. 1 with respect to any* $\mathbf{P}$ *such that* $0 \leq a \leq b \leq c$, $c > 0$.

Proposition 2. *At most four different configurations of discrete simplexes for* $\lambda = \mu = 2$ *are contained in a* $\partial\mathbf{I}_m^+$, $m = 6, 18, 26$.

Proposition 3. *In the case of* $\lambda = \mu = 3$, *there exist 40 different configurations in* $\mathbf{NP}$*s and the corresponding configurations of discrete simplexes for* $\partial\mathbf{I}_m^+$, $m = 6, 18, 26$, *are shown in Figs. 2, 3 and 4, respectively, with respect to any* $\mathbf{P}$ *such that* $0 \leq a \leq b \leq c$, $c > 0$.

Proposition 4. *At most nine different configurations of discrete simplexes for* $\lambda = \mu = 3$ *are contained in a* $\partial\mathbf{I}_m^+$, $m = 6, 18, 26$.

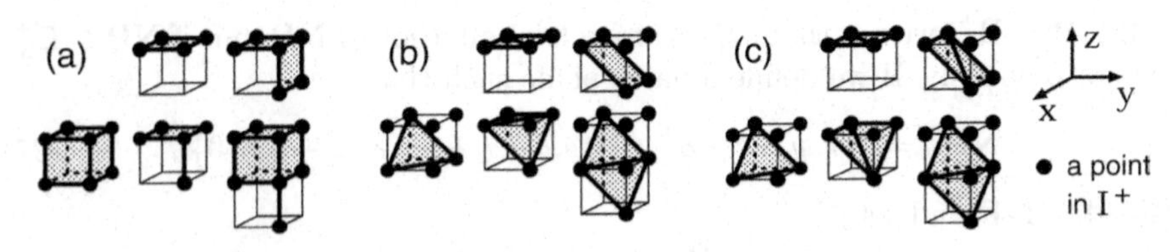

Fig. 1. All five configurations of discrete simplexes in $\partial\mathbf{I}_6^+$ (a), $\partial\mathbf{I}_{18}^+$ (b) and $\partial\mathbf{I}_{26}^+$ (c), whose projections on plane $z = 0$ lie on the 2×2 square grids

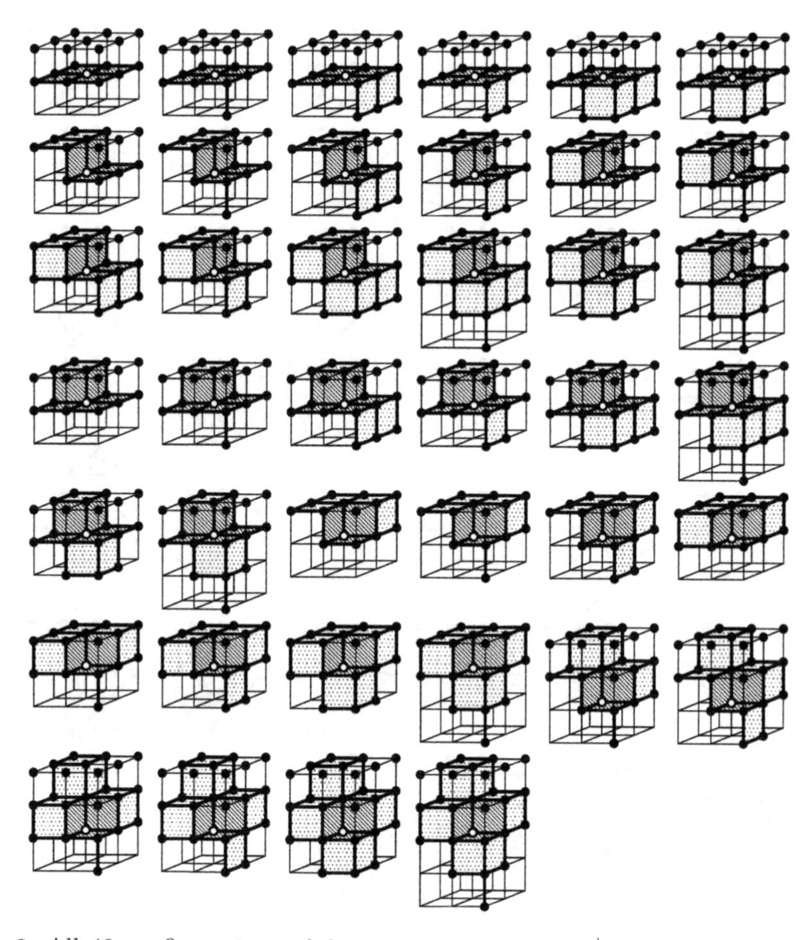

Fig. 2. All 40 configurations of discrete simplexes in $\partial\mathbf{I}_6^+$ whose projections on plane $z = 0$ lie on the 3×3 square grids. The star of each white point is also shown as discrete simplexes with diagonal lines in the figures

Propositions 1 to 4 give the coexistence of adjacent 2-simplexes in a $\partial\mathbf{I}_m^+$, $m = 6, 18, 26$. There are actually two simplicial configurations for P4a in $\partial\mathbf{I}_{18}^+$ and $\partial\mathbf{I}_{26}^+$, and for P6 in $\partial\mathbf{I}_{26}^+$ as shown in Fig. 5. For each 0-simplex $[\boldsymbol{x}] \in \partial\mathbf{I}_m^+$, we can define the star such that

$$\sigma([\boldsymbol{x}] : \partial\mathbf{I}_m^+) = \{[a] \in \partial\mathbf{I}_m^+ \ : \ [\boldsymbol{x}] \in face([a])\} \ . \tag{31}$$

In Figs. 3 and 4, we choose one of the configurations in Fig. 5 for P4a or P6 so that the number of 2-simplexes in the star of a white point $[\boldsymbol{x}]$ becomes as small as possible. The projection of $\sigma([\boldsymbol{x}] : \partial\mathbf{I}_m^+)$ on the coordinate plane $z = 0$ is in a square whose size is 3×3 if $0 \le a \le b \le c$ and $c > 0$. From this fact, we also derive the following proposition.

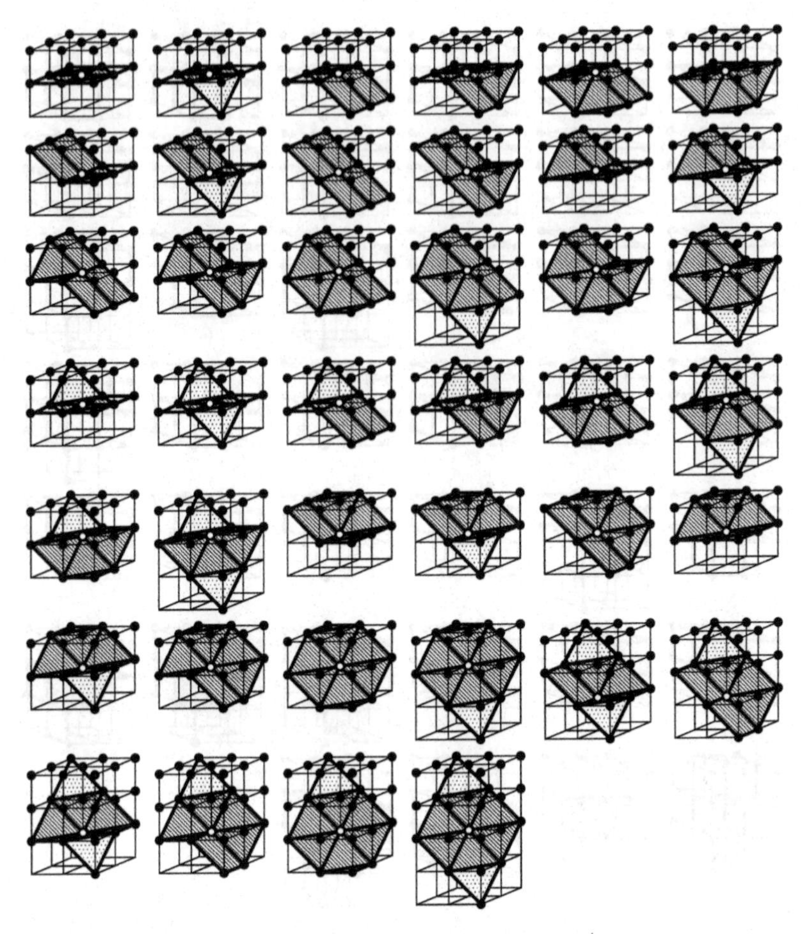

Fig. 3. All 40 configurations of discrete simplexes in $\partial \mathbf{I}_{18}^{+}$ whose projections on plane $z = 0$ lie on the 3×3 square grids. The star of each white point is also shown as discrete simplexes with diagonal lines in the figures

Proposition 5. *Any discrete planar surface* $\partial \mathbf{I}_{m}^{+}$ *for* $m = 6, 18, 26$ *consists of 2-simplexes and their faces so that every* 0*-simplex* $[\boldsymbol{x}] \in \partial \mathbf{I}_{m}^{+}$ *has one of the stars whose configurations are illustrated in Figs. 2, 3 and 4, respectively.*

We see that the equivalent simplicial configurations of a star can appear in different simplicial configurations each of which projects on the coordinate plane $z = 0$ as a 3×3 square in Figs. 2, 3 and 4. Thus, the total number of different configurations of discrete simplexes of a star will be less than 40, *i.e.* 4, 18 and 23 configurations for the 6-, 18- and 26-neighborhood systems, respectively. The similar results of local configurations for $\partial \mathbf{I}_{6}^{+}$ are also presented in [1,4].

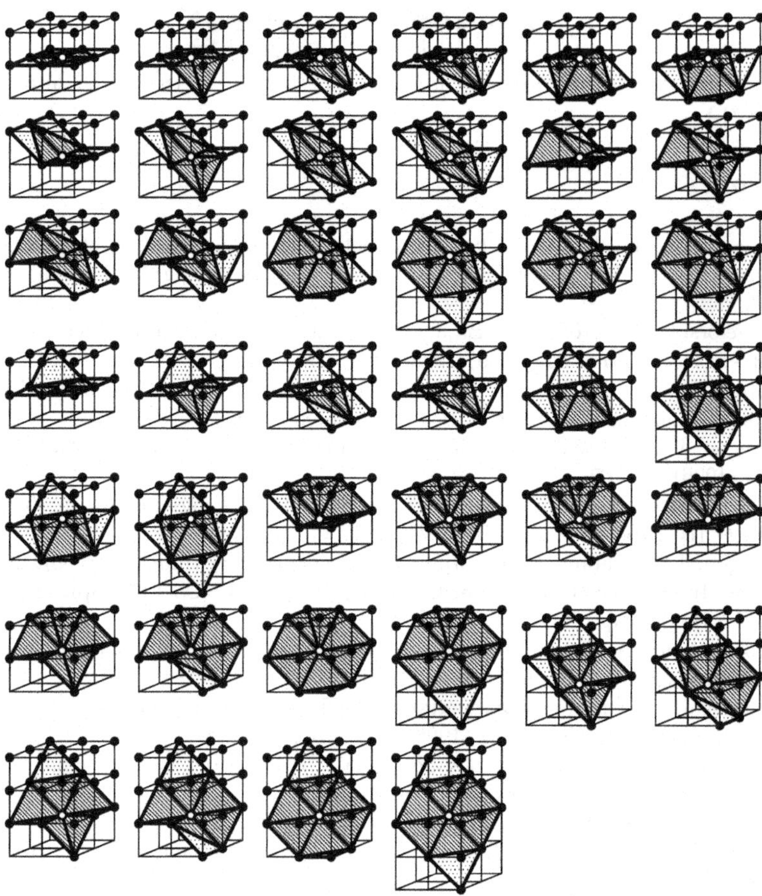

Fig. 4. All 40 configurations of discrete simplexes in $\partial\mathbf{I}_{26}^{+}$ whose projections on plane $z = 0$ lie on the 3×3 square grids. The star of each white point is also shown as discrete simplexes with diagonal lines in the figures

6 Conclusions

In this paper, we first proved that $\mathbf{B}_{18}^{+}$ and $\mathbf{B}_{26}^{+}$ are equal to **NP**. From the equality relations, we see that $\partial\mathbf{I}_{m}^{+}$ for $m = 18, 26$ is a triangulation of **NP** in the sense of combinatorial topology. Since $\partial\mathbf{I}_{m}^{+}$ consists of discrete simplexes, we described the local properties of $\partial\mathbf{I}_{m}^{+}$ by using configurations of discrete simplexes instead of those of lattice points, such as the coexistence of adjacent 2-simplexes and the configuration of discrete simplexes of a star in a $\partial\mathbf{I}_{m}^{+}$. If we set $w = |a| + |b| + |c|$ in (2), we obtain standard planes with respect to **P** of (1), instead of naive planes [4]. It is our future work to clarify the relations between standard planes and $\partial\mathbf{I}_{m}^{+}$, $m = 6, 18, 26$. A part of this work was supported by JSPS Grant-in-Aid for Encouragement of Young Scientists (12780207).

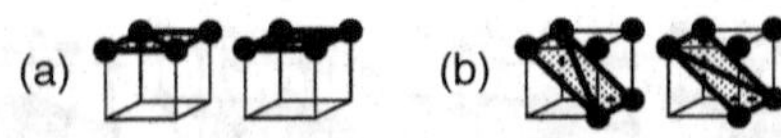

Fig. 5. Two simplicial configurations of P4a for $m = 18, 26$ (a) and P6 for $m = 26$ (b)

References

1. Imiya, A., Eckhardt, U.: The Euler Characteristics of Discrete Objects and Discrete Quasi-Objects. Computer Vision and Image Understanding **75 3** (1999) 307–318 249, 258
2. Borgefors, G., Nyström, I., di Baja, G. S.: Computing skeletons in three dimensions. Pattern Recognition **32** (1999) 1225–1236 249
3. Andres, E.: Le Plan Discret. In Proceedings of 3e Colloque Géométrie discète en imagerie: fondements et applications. Strasbourg (1993) 45–61 249
4. Françon, J.: Sur la topologie d'un plan arithmétique. Theoretical Computer Science **156** (1996) 159–176 249, 256, 258, 259
5. Reveillès, J. P.: Combinatorial Pieces in Digital Lines and Planes. In Vision Geometry III. Proceedings of SPIE, Vol. 2573 (1995) 23–34 249, 250, 256
6. Françon, J., Schramm, J. M., Tajine, M.: Recognizing arithmetic straight lines and planes. In Discrete Geometry for Computer Imagery. LNCS **1176** Springer-Verlag, Berlin, Heidelberg (1996) 141–150 249, 256
7. Debled-Renesson, I.: Etude et reconnaissance des droites et plans discrets. PhD thesis, University of Louis Pasteur (1995) 249, 256
8. Kenmochi, Y.: Discrete Combinatorial Polyhedra: Theory and Application. Doctoral thesis, Chiba University (1998) 249, 251, 252
9. Kenmochi, Y., Imiya, A.: On Combinatorial Properties of Discrete Planar Surfaces. In Proceedings of International Conference on Free Boundary Problems: Theory and Applications. Vol. 2. Gakkotosho, Tokyo (2000) 255–272. 249, 253
10. Aleksandrov, P. S.: Combinatorial Topology. Vol. 1. Graylock Press, Rochester, New York (1956) 249, 250

Appendix: Lemma 2

Let us consider $\mathbf{P}$ of (1) such that $a, b, c > 0$. For each point $\boldsymbol{p} \in \mathbf{I}^+ \setminus \mathbf{P}$ such that $\boldsymbol{p} = (s, t, u)$, we set three planes such as

$$\mathbf{S} = \{(x, y, z) \in \mathbf{R}^3 \ : \ x = s\} , \tag{32}$$

$$\mathbf{T} = \{(x, y, z) \in \mathbf{R}^3 \ : \ y = t\} , \tag{33}$$

$$\mathbf{U} = \{(x, y, z) \in \mathbf{R}^3 \ : \ z = u\} . \tag{34}$$

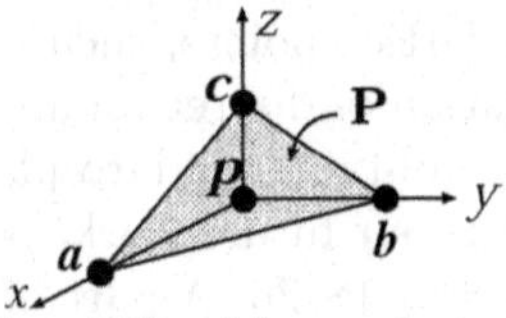

Fig. 6. Three points $\boldsymbol{a}$, $\boldsymbol{b}$ and $\boldsymbol{c}$ defined for a plane $\mathbf{P}$ and a point $\boldsymbol{p}$ which is not in $\mathbf{P}$

Let $\boldsymbol{a}$, $\boldsymbol{b}$ and $\boldsymbol{c}$ be the intersection points of $\mathbf{P}$, $\mathbf{T}$ and $\mathbf{U}$, $\mathbf{P}$, $\mathbf{S}$ and $\mathbf{U}$, and $\mathbf{P}$, $\mathbf{S}$ and $\mathbf{T}$, respectively, as illustrated in Fig. 6. Then the next lemma is derived.

Lemma 2. *For any* $\boldsymbol{p} \in \mathbf{I}^+ \setminus \mathbf{P}$, *we obtain*

$$|\boldsymbol{p} - \boldsymbol{a}| : |\boldsymbol{p} - \boldsymbol{b}| : |\boldsymbol{p} - \boldsymbol{c}| = 1/a : 1/b : 1/c \tag{35}$$

where $a, b, c > 0$.

Proof. The equation of the line which is the intersection of $\mathbf{P}$ and $\mathbf{U}$ is given by $ax + by + cu + d = 0$. Thus, the slope of the line in $\mathbf{U}$ is given by

$$\frac{|\boldsymbol{p} - \boldsymbol{b}|}{|\boldsymbol{p} - \boldsymbol{a}|} = \frac{a}{b} \,. \tag{36}$$

Similarly, the slopes of the intersection lines between $\mathbf{P}$ and $\mathbf{T}$, and $\mathbf{P}$ and $\mathbf{S}$, are respectively given by

$$\frac{|\boldsymbol{p} - \boldsymbol{c}|}{|\boldsymbol{p} - \boldsymbol{b}|} = \frac{b}{c} \quad \text{and} \quad \frac{|\boldsymbol{p} - \boldsymbol{a}|}{|\boldsymbol{p} - \boldsymbol{c}|} = \frac{c}{a} \,. \tag{37}$$

From (36) and (37), we obtain (35). □

Surface Digitizations by Dilations Which Are Tunnel-Free

Christoph Lincke and Charles A. Wüthrich

Computer Graphics, Visualization, Man-Machine Communication Group
Faculty of Media, Bauhaus University Weimar
99421 Weimar, Germany

Abstract. In this article we study digital topology with methods from mathematical morphology. We introduce reconstructions by dilations with appropriate continuous structural elements and prove that notions known from digital topology can be defined by continuous properties of this reconstruction. As a consequence we determine the domains for tunnel-free surface digitizations. It will be proven that the supercover and the grid-intersection digitization of every surface with or without boundary is always tunnel-free.

1 Introduction

Various approaches have been made to study geometrical and topological properties of binary digital images. Discussing the advantages and disadvantages of them would be far beyond this paper. However, there is a growing interest in relating these approaches to each other in order to develop a foundation for a mathematically consistent theory.

The most well-known approach, known as *digital topology* [KR89], is derived from graph theory. Elements of $\mathbb{Z}^n$ are interpreted as vertices. Edges are defined by different adjacency relations between object and background points. This approach serves well for two-dimensional image analysis. The 3D case [MR81] is far more complicated and a generalization to higher dimensions has not been made yet.

A *cellular approach* has been applied by Kovalevsky [Kov89] in 2D and by Herman et al. [HW83] in 3D. Voss [Vos88] studied a dual cell-structure in $\mathbb{Z}^n$ and Khalimsky [KKM90] developed a topological approach based on *connected ordered topological spaces*. The structure studied in these approaches is the *discrete* or *Alexandrov topology*. It is equivalent to a tessellation of $\mathbb{R}^n$ by n-dimensional unit cubes. Each approach maps $\mathbb{Z}^n$ onto different elements of the structure. In the first case an element of $\mathbb{Z}^n$ is associated with an open n-dimensional unit-cube, whereas Voss interprets $\mathbb{Z}^n$ as the set of the vertices of these cubes. In the third approach, every element of that structure is associated with an element of $\mathbb{Z}^n$.

Bertrand and Couprie [BC99] proposed a model for digital geometry that associates two orders to each subset of $\mathbb{Z}^n$. These order relations correspond to

G. Borgefors, I. Nyström, and G. Sanniti di Baja (Eds.): DGCI 2000, LNCS 1953, pp. 262–271, 2000.
© Springer-Verlag Berlin Heidelberg 2000

the different adjacencies as used in digital geometry. Moreover, the authors have proven that the notion of surfaces and simple points in their model correspond exactly to that very notions in digital topology. This justifies the original graph based approach for 2D and 3D space.

Digital images can also be investigated using a *digitization approach* [Ser82]. A discrete object has a certain property if it is a digitization of an appropriate continuous object with that property. Dual to this is the *embedding approach* in which continuous analogs [KR85] of discrete objects are studied. Both approaches define properties of discrete objects by well-known continuous, usually Euclidean, notions.

Reveillès [Rev92] introduced the *arithmetical geometry* approach. He defined *discrete analytical objects* as discrete objects which are the integer solution of a finite set of inequalities. Recently, Andres studied the supercover digitization of m-flats in the context of discrete analytical objects [And99]. In [LW00b] we generalized these results to linear analytical objects.

The digitization approach has been related to digital topology. Various researchers [Pav82,Ser82,GL95] studied the preservation of topological features of continuous objects under digitization. Bærentzen et al. [BŠC00] presented a criterion for determining whether a 3D solid is suitable for digitization at a given resolution. Although these articles make different assumptions, the common idea is to consider objects that are morphologically open and closed by a closed ball whose radius depends on the grid resolution.

We applied a similar approach to study the digitizations of surfaces without boundary [LW00a]. Since the opening of a simple surface is the empty set, we employed morphologically closed surfaces with respect to a ball of a radius r. We called these objects r*-surfaces.* Contrary to Bærentzen et al. [BŠC00] our approach did not take a reconstruction kernel into account, but this article explains that properties from digital geometry, such as connectivity and separability, can defined by a dilation of the discrete object with a continuous structural element.

An extension of these results for surfaces without boundary to surfaces with boundary is essential, because many real-life objects can be described as the union of surface patches. To evaluate the quality of digitizations of surfaces with boundary Cohen-Or et al. introduced the notion *tunnel-free* [CK95]. This notion has been applied successfully to polygons and polyhedra [ANF97]. In this paper we develop a theoretical framework for digitizations of surfaces with and without boundary which is based on mathematical morphology [Ser82,Hei94]. The same theoretical background has been used by Schmitt to study digitizations and connectivity [Sch98].

This article is outlined as follows: Section 2 states the basic definitions from differential geometry, digital topology and mathematical morphology. In section 3 important results about surface digitizations and digitizations by dilation will be recalled. In the following section *reconstructions by dilation* will be introduced and in section 5 this notion is employed to prove a condition under which a surface digitization is tunnel-free. We conclude with a summary and remarks on future work.

2 Basic Definitions

2.1 Differential Geometry

In differential geometry continuous objects are studied as their parametrization. Curves are basically 1-dimensional and surfaces are $(n-1)$-dimensional parametrizations in $\mathbb{R}^n$. Thus, in $\mathbb{R}^2$ curves can be considered as surfaces.

A set of points $C \subseteq \mathbb{R}^n$ $(n \geq 2)$ is said to be a *$C^r(I)$-curve* $(r \geq 1)$ if there exists an open interval $I \subseteq \mathbb{R}$ and an r times continuously differentiable function $\gamma : I \to \mathbb{R}^n$ such that $C = \gamma(I)$. The function γ is called *parametrization.* A curve γ is *smooth* if, for all $t \in I$, the first derivative exists and is non-zero. A curve is *simple* if it has no self-intersection.

From now on all curves will be considered to be smooth and simple. A curve $C = \gamma([a,b])$ with end points $\gamma(a)$ and $\gamma(b)$ is a subset of curve $\gamma(I)$ that is defined on an appropriate open interval I that contains $[a,b]$. Let $C = \gamma((a,b))$ be a simple curve and let $\gamma(a) = \gamma(b)$, then $C = \gamma([a,b])$ is a *simple closed curve.*

A set of points $S \subseteq \mathbb{R}^n$ $(n \geq 2)$ is said to be a *$C^r(U)$-surface $(r \geq 1)$* if there exists a non-empty open set $U \subseteq \mathbb{R}^{n-1}$ and an r-times continuously differentiable function $f : U \to \mathbb{R}^n$ such that $S = f(U)$.

Again, only *simple, smooth* surfaces with or without boundary are considered. A simple *surface without boundary* is either a closed surface, such as a sphere, or an infinite object homeomorphic to a hyperplane. These notions are intuitively clear and similar to those for curves. For a detailed definition the reader is referred to text books on differential geometry such as [LV85].

2.2 Digital Topology

We define a *discrete object* A as a subset of $\mathbb{Z}^n$. Its complementary set $A^C = \mathbb{Z}^n \setminus A$ is called the *background.* We think of $\mathbb{Z}^n$ as a subset of n-dimensional Euclidean space $\mathbb{R}^n$. An element $z \in \mathbb{Z}^n$ is called a *grid point.*

There are various equivalent ways to introduce the basic notions of digital topology. We define the neighborhood of grid points through Voronoi sets [Kle85,Wüt98]. Other definitions are based on distances or differences in the coordinates of these points.

The *Voronoi set* $\mathbb{V}(z)$ of a grid point z is the set of all points in $\mathbb{R}^n$ which are at least as close to z as to any other grid point. $\mathbb{V}(z)$ is a closed axes-aligned n-dimensional unit cube with center z. The Voronoi sets of a 2D and 3D grid point are known as *pixel* and *voxel,* respectively. Neighboring n-dimensional Voronoi sets can share a point, a straight line segment, up to an $(n-1)$-dimensional cube.

Two grid points $z, z' \in \mathbb{Z}^n$ are said to be *k-neighbors* $(0 \leq k \leq n-1)$ if their Voronoi sets share a point set of dimension k or higher, i.e. if $\dim(\mathbb{V}(z) \cap \mathbb{V}(z')) \geq k$.

A sequence $(z_0, \ldots, z_l)$ of points of an object $A \subseteq \mathbb{Z}^n$ is said to be a *k-arc* from z_0 to z_l in A if successive elements are k-neighbors. $K \subseteq \mathbb{Z}^n$ is a *(simple closed) k-curve* if each point of K has exactly two k-neighbors.

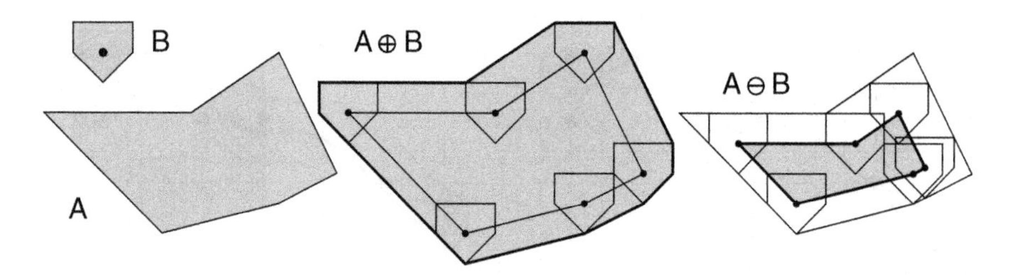

Fig. 1. Dilation $A \oplus B$ and erosion $A \ominus B$ of a set A by a structuring Element B

An object $A \subseteq \mathbb{Z}^n$ is k-*connected* if there exists a k-arc in A from z to z' for any points $z, z' \in A$. A k-*component* of $A \subseteq \mathbb{Z}^n$ is defined as a maximal k-connected non-empty subset of A.

A discrete object $A \subseteq \mathbb{Z}^n$ is said to be k-*separating* if the background $\mathbb{Z}^n \setminus A$ consists of exactly two k-components. A k-separating object A is called k-*minimal* if for any $z \in A$ $A \setminus \{z\}$ is not k-separating. A k-*separating surface* (without boundary) is a minimal k-separating object.

To avoid pathological situations in 2D, a 1-curve must consist of at least 8 points and an 0-curve of at least 4 points [KR89]. A discrete surface should have no touching points. Traditionally [KR89], in $\mathbb{Z}^2$ 1- and 0-neighbors are called 4-neighbors and 8-neighbors, respectively, and in $\mathbb{Z}^3$ 26-, 18- and 6-neighbors are common notions.

2.3 Morphological Definitions

In this article, morphological operations [Ser82,Hei94] on point sets will be required. Let A and B be two subsets of $\mathbb{R}^n$. Since $\mathbb{Z}^n \subseteq \mathbb{R}^n$, the following operations can be applied to continuous as well as to discrete point sets.

$A \oplus B = \{a + b : a \in A, b \in B\}$ is called *Minkowski addition* and $A \ominus B = \{p : b + p \in A \text{ for all } b \in B\}$ is the *Minkowski subtraction* of A and B. In mathematical morphology, $A \oplus B$ and $A \ominus B$ are known *dilation* and *erosion* of A by the *structuring element* B, respectively. Fig. 1 shows an example.

The operation $A \circ B = (A \ominus B) \oplus B$ is called the *opening* and $A \bullet B = (A \oplus B) \ominus B$ is called the *closing* of A with respect to B. A is *morphologically open (closed)* with respect to B if $A \circ B = A$ (resp. $A \bullet B = A$).

Finally, $A_z = A \oplus \{z\}$ is the translate of A by z and $\check{A} = \{-a : a \in A\}$ denotes the reflected set of A.

3 Surface Digitizations

In our previous work [LW00a] we studied a class of digitizations, commonly known as digitizations by dilations [Hei94]. The grid-intersection [Kle85] and the supercover [CK95] digitization schemes, which are common for surfaces, are special cases of digitizations by dilations.

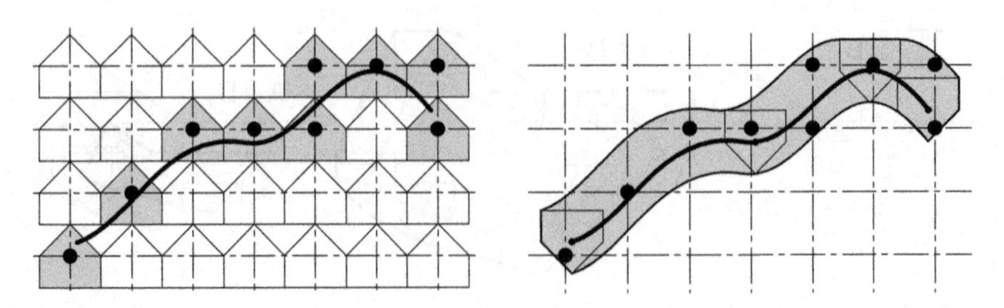

Fig. 2. Digitization of a curve as the set of translated basic domains D_z hit by A (left) and as the set of grid points contained in $A \oplus \check{D}$

A *digitization by dilation* with *domain* $D \subseteq \mathbb{R}^n$ is a function $\Delta_{\oplus}^{D} : \wp(\mathbb{R}^n) \to \wp(\mathbb{Z}^n)$ that is defined as $\Delta_{\oplus}^{D}(A) = \{z \in \mathbb{Z}^n : A \cap D_z \neq \emptyset\}$ for every continuous object $A \subseteq \mathbb{R}^n$.

By virtue of this definition, a grid point z belongs to digitization $\Delta_{\oplus}^{D}(A)$, if and only if D_z, the domain translated to $z \in \mathbb{Z}^n$, hits the continuous object A. $\Delta_{\oplus}^{D}(A)$ is called digitization by dilation because it is the set of grid points contained in the dilation of A by the reflected domain $\check{D}$, i.e. $\Delta_{\oplus}^{D}(A) = (A \oplus \check{D}) \cap \mathbb{Z}^n$ [Hei94].

If the domain of a digitization is not specified then $\Delta(A)$ is any subset of $\mathbb{Z}^n$ that is intended to serve as a discrete approximation of a continuous object $A \subseteq \mathbb{R}^n$. It does not need to be a digitization by dilation when we focus on criteria for the quality of these approximations.

The criterion "k-separating" can only be applied to discrete surfaces without boundaries. To overcome this limitation the notion of a k-tunnel-free digitization of a surface has been introduced [CK95].

Let $(z_0, \ldots, z_l)$ be a k-*arc*. Then the continuous polygonal arc consisting of the straight line segments $[z_0, z_1], [z_1, z_2], \ldots, [z_{l-1}, z_l]$ is called a *continuous* k-*path*. A continuous path π *hits* a surface $S \subseteq \mathbb{R}^n$ in a point $p \in S$ if $p \in \pi \cap S$. A continuous path π *crosses* a surface $S \subseteq \mathbb{R}^n$ in $p \in S$ if there exists an $\epsilon > 0$ such that π hits two different components of $B_\epsilon(p) \setminus S$. $B_\epsilon(p)$ denotes the closed ball of radius ϵ with center p.

A digitization $\Delta(S) \subseteq \mathbb{Z}^n$ of a continuous surface $S \subseteq \mathbb{R}^n$ is k-*tunnel-free* $(0 \leq k \leq n-1)$ if every continuous k-path in $(\Delta(S))^C = \mathbb{Z}^n \setminus \Delta(S)$ does not cross S. A continuous k-path in $(\Delta(S))^C$ that crosses S is called k-*tunnel*.

If $\Delta(S) \subseteq \mathbb{Z}^n$ is a k-tunnel-free digitization of S then $\Delta(S) \cup A$ is also k-tunnel-free for every $A \subseteq \mathbb{Z}^n$. As illustrated in Fig. 3, a k-tunnel-free digitization of a continuous surface without boundary is not necessarily a k-separating discrete object.

4 Reconstructions by Dilation

The foundation of digital topology is the notion "k-neighbor". In Section 2 k-neighborhood of grid points was defined by means of their Voronoi sets. In this

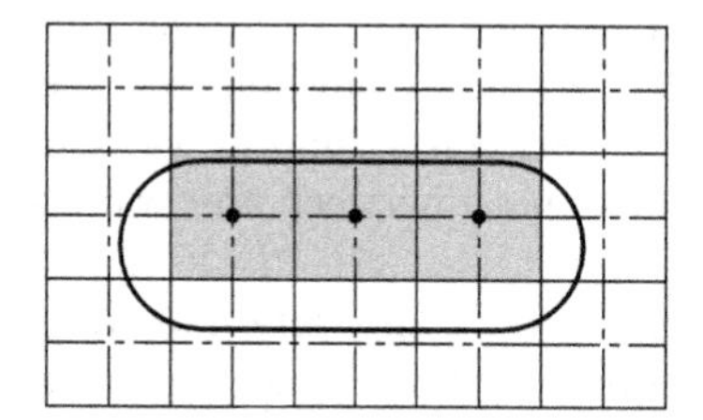
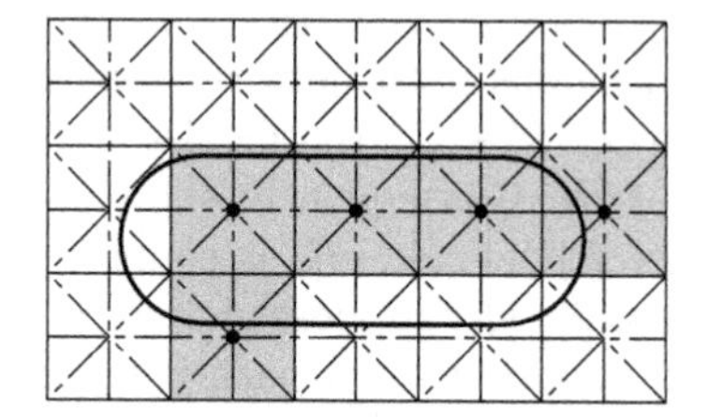

Fig. 3. A 1-tunnel-free (left) an 0-tunnel-free (right) digitization of a simple closed curve, that are no discrete curves

section will be shown that two point $z, z' \in \mathbb{Z}^n$ are neighbored if $R_z \cup R_{z'} = (\{z\} \oplus R) \cup (\{z'\} \oplus R)$ is a connected set in $\mathbb{R}^n$ for an appropriate structural element $R \subseteq \mathbb{R}^n$. Consequently, if $A \in \mathbb{Z}^n$ is a discrete object, we can think of the continuous set $(A \oplus R) \subseteq \mathbb{R}^n$ as its *reconstruction by dilation.*

Let $x = (x_1, \ldots, x_n)$ and $y = (y_1, \ldots, y_n)$ be two points in $\mathbb{R}^n$ then $[x, y]$ and $[(x_1, \ldots, x_n), (y_1, \ldots, y_n)]$ denotes the straight line segment between these points. $R_k^{(n)} \in \mathbb{R}^n$ is the union of all line segments $[(0, \ldots, 0), (x_1, \ldots, x_n)]$ with the property that at least k $(0 \leq k < n)$ of the coordinates x_i are 0, while the others are either $\frac{1}{2}$ or $-\frac{1}{2}$.

For simplicity the index for the dimension will be omitted if it is clear by the context. Fig. 4 shows the sets R_0 and R_1 in $\mathbb{R}^2$ and R_0, R_1 and R_2 in $\mathbb{R}^3$. The following lemma is a simple conclusion of the definitions of R_k and k-neighbors.

Lemma 1. *Two grid points $z, z' \in \mathbb{Z}^n$ are k-neighbors $(0 \leq k \leq n-1)$, if and only if the reconstruction $\{z, z'\} \oplus R_k$ is a connected set in $\mathbb{R}^n$.*

There exists a straight line segment between k-neighbored two grid points in the reconstruction by R_k and, conversely, if there is a straight line segment between two grid points z and z' in $\{z, z'\} \oplus R_k$ then these points are k-neighbors. Notice that there may exist other continuous paths between these points. Only for $(n-1)$-neighbors this path is unique. A k-arc from z to z' in $A \subseteq \mathbb{Z}^n$ exists, iff there exists a continuous path between z and z' in $A \oplus R_k$.

Lemma 2. *An object $A \subseteq \mathbb{Z}^n$ is k-connected if and only if the reconstruction $A \oplus R_k$ is a connected set in $\mathbb{R}^n$.*

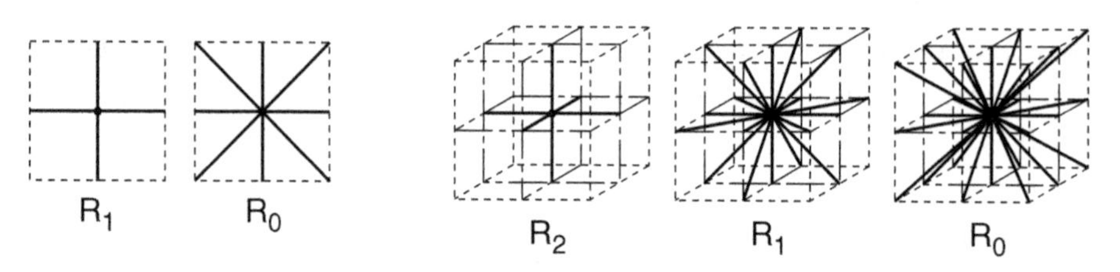

Fig. 4. R_1 and R_0 in $\mathbb{R}^2$ and R_2, R_1 and R_0 in $\mathbb{R}^3$

Proof. Assume $A \subseteq \mathbb{Z}^n$ is k-connected. Then, for any $z, z' \in A$, there exists a continuous path between z and z' in $A \oplus R_k$. R_k is a connected set. Hence, for every $z'' \in \mathbb{Z}^n$ there exist a continuous path between z'' and every point of $\{z''\} \oplus R_k$. Consequently, there exists a continuous path between any two points in $A \oplus R_k$, i.e. $A \oplus R_k$ is connected in $\mathbb{R}^n$.

Conversely, suppose $A \oplus R_k$ is connected in $\mathbb{R}^n$. By construction of R_k there exist a continuous k-path between any grid points in A and A is k-connected. q.e.d.

As a result of this lemma k-components can be defined as continuous components of the reconstruction by R_k. The proofs of the following two lemmas will be omitted. They are similarly simple.

Lemma 3. *A discrete object $A \subseteq \mathbb{Z}^n$ is k-separating iff $(\mathbb{Z}^n \setminus A) \oplus R_k$ consists of exactly two continuous components.*

Lemma 4. *A k-separating object A is k-minimal iff $((\mathbb{Z}^n \setminus A) \cup \{z\}) \oplus R_k$ is not separating for all $z \in A$.*

Let us conclude this section with some remarks. The reconstruction by dilation with a structural element R_k represents only the number of components of a discrete object. For example the reconstruction of a discrete curve or a surface is not a continuous curve or surface. It also does not reconstruct the number of background components.

Instead of R_k we could have used other structural elements such as the convex hull of R_k or the smallest closed ball that contains R_k. For these examples all results of this section would still hold.

We have chosen the term "reconstruction by dilation" to represent a dual or opposite operation of digitization by dilation. It should be pointed out that this term is also used in mathematical morphology with a different meaning. In the context of geodesic transformations there is a notion of "reconstruction by dilation of a mask image from a marker image" [Soi99] which is not related to our definition.

5 Tunnel-Free Surface Digitizations

In this section tunnel-free digitizations by dilations will be studied for surfaces with or without boundary. The first theorem establishes a sufficient condition such that a digitization by dilation is tunnel-free.

Theorem 1. *Let $\Delta(S)$ be a digitization of a surface $S \in \mathbb{R}^n$ such that the reconstruction of the background $(\Delta(S))^C \oplus R_k$ does not hit S. Then $\Delta(S)$ is k-tunnel-free.*

Proof. Let us assume $\Delta(S)$ is a digitization of $S \in \mathbb{R}^n$ such that $(\Delta(S))^C \oplus R_k$ does not hit the surface S. Suppose that there exists a k-tunnel in $\Delta(S)$. Then there exists a continuous path in $(\Delta(S))^C \oplus R_k$ that hits the surface S, which is a contradiction to our assumptions.

Note that the condition "the reconstruction of the background by R_k does not hit the surface" is not necessary for a k-tunnel-free digitization. Using this theorem, a relationship between existence of k-tunnels a digitization by dilation $\Delta_\oplus^D$ and the choice of the domain D can be proven.

Theorem 2. *Let $\Delta_\oplus^D$ be a digitization by dilation with the domain $D \supseteq R_k$. Then $\Delta_\oplus^D(S)$ is k-tunnel-free for every surface $S \subseteq \mathbb{R}^n$.*

Proof. Let $\Delta_\oplus^D$ be a digitization by dilation with the domain $D \supseteq R_k$. For every $S \subseteq \mathbb{R}^n$, $\Delta_\oplus^{R_k}(S)$ is a subset of $\Delta_\oplus^D(S)$. The digitization $\Delta_\oplus^{R_k}(S)$ contains all points $z \in \mathbb{Z}^n$ such that $z \oplus R_k$ hits S. The construction of its background $(\Delta_\oplus^{R_k}(S))^C \oplus R_k$ does not hit S. Hence, $\Delta_\oplus^{R_k}(S)$ is k-tunnel-free and so is $\Delta_\oplus^D(S)$.

This theorem justifies the grid-intersection [Kle85] and supercover digitization [CK95] as appropriate digitizations schemes for surfaces with or without boundary.

The grid-intersection digitization is a digitization by dilation whose domain is R_{n-1}. As a consequence of Theorem 2 every grid-intersection digitization of surfaces is always $(n-1)$-tunnel-free. The domain of the supercover digitization is $\mathbb{V}(0)$, which is a superset of R_0. Hence, the supercover of every surface is 0-tunnel-free.

6 Summary and Future Work

In this article we investigated digital topology with methods from mathematical morphology. We introduced reconstructions by dilations with a structural element R_k. We have proven that important notions from digital topology, such as k-neighbors, k-connected and k-separating objects, can be defined by continuous properties of the reconstruction dilation with R_k.

As a consequence the new notions have been used to prove that every digitization by dilation whose basic domain is a superset of R_k is k-tunnel-free. In particular the grid-intersection digitization and the supercover of every surface is always 0-tunnel-free and $(n-1)$-tunnel-free, respectively.

Currently we are relating our work on r-surfaces [LW00a] to the results of this paper in order to obtain a theoretical framework for the digitization of surfaces with boundary. Our future work includes also an algebraic study of the relationship between digitizations and reconstructions by dilations on the abstraction level of windowing functions.

References

And99. E. Andres. An m-flat supercover is a discrete analytical object. submitted to Theoretical Computer Science, 1999. 263

ANF97. E. Andrés, P. Nehlig, and J. Françon. Tunnel-free supercover 3D polygons and polyhedra. *Computer Graphics Forum*, 16(3):3–14, August 1997. ISSN 1067-7055. 263

BC99. G. Bertrand and M. Couprie. A model for digital topology. *Lecture Notes in Computer Science*, 1568:229–241, 1999. 262

BŠC00. J. A. Bærentzen, M. Šrámek, and N. A. Christensen. A morphological approach to the voxelization of solids. In V. Skala, editor, *Proceedings of the WSGG 2000*, pages 44–51, Plzen, Czech Republic, 2000. University of West Bohemia. 263

CK95. D. Cohen-Or and A. Kaufman. Fundamentals of surface voxelization. *GMIP*, 57(6):453–461, November 1995. 263, 265, 266, 269

GL95. Ari Gross and Longin Latecki. Digitizations preserving topological and differential geometric properties. *Computer Vision and Image Understanding: CVIU*, 62(3):370–381, November 1995. 263

Hei94. H. J. A. M. Heijmans. *Morphological Image Operators*. Academy Press, Boston, 1994. 263, 265, 266

HW83. G. T. Herman and D. Webster. A topological proof of a surface tracking algorithm. *CVGIP*, 23:92–98, 1983. 262

KKM90. E. Khalimsky, R. Kopperman, and P. R. Meyer. Computer graphics and connected topologies on finite ordered sets. *Topology and its Applications*, 36:1–17, 1990. 262

Kle85. R. Klette. The m-dimensional grid point space. *Computer Vision, Graphics, and Image Processing*, 30:1–12, 1985. 264, 265, 269

Kov89. V. A. Kovalevsky. Finite topology as applied to image analysis. *Computer Vision, Graphics, and Image Processing*, 46:141–161, 1989. 262

KR85. T. Y. Kong and A. W. Roscoe. Continuous analogs of axiomatized digital surfaces. *Computer Vision, Graphics, and Image Processing*, 29(1):60–86, January 1985. 263

KR89. T. Y. Kong and A. Rosenfeld. Digital topology. introduction and survey. *Computer Vision, Graphics, and Image Processing*, 48(3):357–393, December 1989. 262, 265

LV85. Walter Ledermann and Steven Vajda. Combinatorics and geometry. In Walter Ledermann, editor, *Handbook of applicable mathematics*, volume V. John Wiley & Sons, 1985. 264

LW00a. C. Lincke and C. A. Wüthrich. Properties of surface digitizations. Media Systems Research Report RR00-01 – paper submitted for publication, 2000. 263, 265, 269

LW00b. C. Lincke and C. A. Wüthrich. Towards a unified approach between digitization of linear objects and discrete analytical objects. In V. Skala, editor, *Proceedings of the WSGG 2000*, pages 124–131, Plzen, Czech Republic, 2000. University of West Bohemia. 263

MR81. D. G. Morgenthaler and A. Rosenfeld. Surfaces in three-dimensional digital images. *Information And Control*, 51:227–247, 1981. 262

Pav82. T. Pavlidis. *Algorithms for Graphics and Image Processing*. Computer Science Press, 1982. 263

Rev92. J.-P. Reveillès. *Geometrie Discrète, Calcul en Nobres Entiers et Algorithmique*. Thèse d'Etat. Université Louis Pasteur, Strasbourg, F, 1992. 263

Sch98. M. Schmitt. Digitization and connectivity. In H. Heijmans and J. Roerdink, editors, *Mathematical Morphology and its applications to Signal Processing*, volume 12 of *Computational Imaging and Vision*, pages 91–98. Kluwer Academic Publishers, Dordrecht, 1998. 263

Ser82. J. Serra. *Image Analysis and Mathematical Morphology*. Academic Press, London, 1982. 263, 265

Soi99. P. Soille. *Morphological Image Analysis*. Springer-Verlag, Berlin, 1999. 268

Vos88. K. Voss. *Theoretische Grundlagen der Digitalen Bildverarbeitung*. Akademie-Verlag, 1988. 262

Wüt98. Ch. A. Wüthrich. A model for curve rasterization in n-dimensional space. *Computer & Graphics*, 22(2–3):153–160, 1998. 264

Delaunay Surface Reconstruction from Scattered Points

Angel Rodríguez[1], José Miguel Espadero[1], Domingo López[2], and Luis Pastor[3]

[1] U. Politécnica de Madrid, Dep. de Tecnología Fotónica,
Campus de Montegancedo s/n, 28660 Boadilla del Monte, Madrid, Spain
{arodri,jespa}@dtf.fi.upm.es
[2] Visual Tools S.A., Dep. de I+D.,
C. Xaudaró, 13 bis, 2ª planta. 28034, Madrid, Spain.
dlopez@vtools.es
[3] U. Rey Juan Carlos, Dep. de Ciencias Experimentales y Tecnológicas,
C. Tulipán, s/n, 28933 Móstoles,Madrid, Spain.
lpastor@escet.urjc.es

Abstract. The use of three-dimensional digitizers in computer vision and CAD systems produces an object description consisting of a collection of scattered points in $\mathbb{R}^3$. In order to obtain a representation of the objects' surface it is necessary to establish a procedure that allows the recovering of their continuity, lost during the data acquisition process. A full automatic $O(n^2)$ algorithm is presented. Such algorithm obtains surface representations of free genus objects described from a set of points that belong to the original surface of the object. The only information available about each point is its position in $\mathbb{R}^3$. The achieved surface is a Delaunay triangulation of the initial cloud of points. The algorithm has been successfully applied to three-dimensional data proceeding from synthetic and real free shape objects.

Keywords: Automatic surface reconstruction, 3D Delaunay triangulation, 3D Modeling

1 Introduction

This article deals with the problem of tessellating real objects' surface representations in an automatic way. The free genus objects are exclusively defined by a set of three-dimensional points that belong to their surface. This problem can be found in areas such as computer vision or CAD, that accept object descriptions based on 3D scattered points. The input data is composed by a set of unstructured points which represent the samples digitized from the surface of the objects using passive or active acquisition techniques [23]. Furthermore, they could be obtained from the segmentation of 3D volumetric images, although this approach is more frequently used in medical environments because of its high cost [20]. From this set of points, represented by their 3D coordinates, we are intending to automatically build a mesh of triangular facets produced by the proximity

G. Borgefors, I. Nyström, and G. Sanniti di Baja (Eds.): DGCI 2000, LNCS 1953, pp. 272–283, 2000.
© Springer-Verlag Berlin Heidelberg 2000

relationships between the points of the surface. Another desirable property of the mesh is to be as regular as possible. So, the extracted mesh under these conditions will be a Delaunay triangulation.

1.1 Related Work

The use of 3D digitizers to capture the shape of the objects is more and more frequent in computer vision and CAD systems. This kind of devices make a discrete sampling over the surface of the object obtaining a description based on a cloud of points. However, we know few works in $\mathbb{R}^3$ that deal with the automatic recovering of the surface continuity without more information that the points' coordinates. Most of the existing techniques manage some type of additional information in order to characterize the cloud of points. The $\alpha-$shapes of Edelsbrunner *et al.* [10] are a good effort to formalize the concept of "shape" for the discussed descriptions. Several solutions require to use some kind of manipulation from a human being, making the process highly interactive [5,21]. In other cases, some limits to the data sampling process are fixed [22,8,1]. For example, Amenta and Bern restrict the sampling distance to avoid the problems that appear with the existence of creases and corners. Other methods need to know the orientation or the position of the sensor during the sampling stage in order to recover the connectivity between the points of the surface [12,26]. Another approach is to use the orientation of the normals associated to each point to establish neighbor relationships between the samples [4]. Hoppe *et al.* [19] and Bajaj *et al.* [3] define a signed distance function to compute its zero-set, considering the surface as an implicit function defined over the input data. Another alternative approach used by some authors consists to perform a deformation over a reference mesh by means of an iterative process which adjust the mesh to the cloud of points [25,27,18]. The only known work in 3D that deals with the stated problem is the Attali's one [2], where the boundary and surface extraction in two and three dimensions is intended to be formalized. In 2D she extracts the boundary from a Delaunay triangulation and uses the existence of topological relationships between neighbor points with the so-called *τ-regular shapes*. However, the extrapolation of this problem to 3D can not be formalized in the same way so she offers a heuristic solution to extract the surfaces, but limited to closed surfaces.

The contents of this paper is as follows. Section 2 describes the algorithm implemented to obtain the Delaunay tetrahedralization. Section 3 presents the algorithm that allows the computation of the objects' surface Delaunay triangulation in an automatic way. Section 4 shows some meshes extracted with the proposed algorithm. Last, section 5 summarizes the paper's main conclusions.

2 3D Delaunay Tetrahedralization

It is well known that there is an equivalence relationship between a Voronoi diagram and the corresponding Delaunay triangulation that allows changing from one type of representation to its dual by means of a $O(n)$ process [12].

For our purposes, we have chosen the incremental algorithm described by Faugueras in [12] to obtain a Delaunay tetrahedralization of the cloud of points, although there are other methods to compute the Delaunay representation [17,11,6,9]. It is a $O(n^3)$ algorithm that can be summarized in the next steps:

1. Compute a Delaunay tetrahedralization of the vertex of a cube $\{V_i, i = 1, \ldots, 8\}$ that contains all the points $\{M_j \mid M_j \in \mathbb{R}^3 \quad j = 1, \ldots, p\}$ belonging to the surface of the object, and the centers $\{C_i, i = 1, \ldots, 8\}$ and the radius $\{R_i, i = 1, \ldots, 8\}$ of the circumscribed spheres of the computed tetrahedra. The centers of the spheres are the Voronoi points.
2. If M_j is the current point to be inserted in the tetrahedralization and k is the current number of tetrahedra, each new point M_j will belong, at least, to one of the current spheres. If it would belong to p spheres, all the tetrahedra of those spheres must be marked in order to be deleted from the tetrahedralization, because the Voronoi region $R(p_i)$ is a convex polyhedron that contains the corresponding p_i generator inside it. E.g., if $d^2(M_j, C_i) - R_i^2 \leq 0 \; i = 1, \ldots, k$ then mark the i-th tetrahedron T_i.
3. For all the marked tetrahedra, extract the list of their faces.
4. Remove from this list all the faces that appear twice.
5. For each non-removed face, create a new tetrahedron with M_j and insert it in the tetrahedralization.
6. Delete from the tetrahedralization all the marked tetrahedra and return to 2 until there are no more points to process.

One of the main advantages of this algorithm is the local nature of the operations performed from steps 3 to 6, because the point M_j deals with a reduced number of Delaunay tetrahedra. This allows a local and simple update of the tetrahedralization.

To solve the degeneration cases of the Voronoi diagram computed with this algorithm, produced by numerical errors or by the own location of the points, it is possible to lightly modify the coordinates of the new point until the degeneration disappears and then insert it in the usual way. Another option is to discard this point from the cloud and not to include it in the tetrahedralization.

3 Surface Extraction Algorithm

The purpose of the surface extraction algorithm is first to remove the tetrahedra of the simplicial complex resulting from the Delaunay tetrahedralization described in section 2 that are outside of the object, and then extract the external surface as the list of facets of the simplicial complex that only belong to one tetrahedron. In this process it is necessary to use a heuristic to determine those tetrahedra that are inside of the object, because there is not a deterministic procedure to do this only knowing the Cartesian coordinates of the sampling points. Our proposed heuristic is to start with the tetrahedra placed inside of the convex hull of the objects' cloud and remove those that offers to the outside any facet greatest than some given threshold, repeating the process until the size of all the exterior facets is under the threshold. There are several ways to measure the size of a facet, for example, the perimeter, the area or the radius of the circumscribed circle. We have chosen the perimeter for two main reasons:

1. From a computational point of view, it is faster and easier to compute than other measures.
2. The tests performed have shown a better response of this measure than the others mentioned. We think that this is due to the fact that our starting simplicial complex is a Delaunay triangulation. As it is well known, the Delaunay triangulation has the property that the partition obtained maximizes the minimum angle of the triangular facets achieved. Although this property approximates the triangular facets to equilateral triangles, sometimes the position of the samples makes very irregular triangles, so the area or the radius are not very good measures to clean the tetrahedralization. With this in mind, we are interested to compute a mesh where the distance between the samples, remember that the samples are the vertices of the triangles, is minimum.

Following is the description of the algorithm that we propose.

Algorithm

1. Begin using the Delaunay tetrahedralization computed from the input cloud of points with the algorithm described in section 2.
2. Discard all the tetrahedra having a vertex $\{V_i, i = 1, \ldots, 8\}$ that belongs to the external cube defined in step 1 of the tetrahedralization algorithm (section 2). The result is a simplicial complex of the convex hull of the input data.
3. Compute the external surface of the complex as the list of facets that only belong to one tetrahedron.
4. Compute the threshold v_u as the mean of the perimeters of the facets in the external surface list. Sometimes it will be interesting to multiply this value by a factor, as will be explained below.
5. If a facet in the external surface list exceeds the threshold v_u, remove from the complex the tetrahedron who owns the facet and insert the remainder three facets in the surface list.
6. Repeat step 5 until all the perimeters of the facets in the surface list are below v_u.
7. Recompute the mean perimeter of the external surface. If the difference with respect the old value is up 1%, return to step 4.

The most time consuming step of this algorithm is, by far, the calculus of the Delaunay tetrahedralization in step 1. The main disadvantage of the method is that it employs a global threshold for all the surface, so if the input points are not registered in a uniform way and there is too much variability in the perimeter measures along the surface, some holes will appear over the areas with less density of points. This could lead to the loss of some of the points from the original input representation. Sometimes, instead of computing the surface in a fully automatic way, it is possible to achieve better results if we multiply the mean perimeter value by a factor between 1.0 and 2.0 in the step 4 of the algorithm.

The complexity of the algorithm from steps 2 to 7 is $O(n^2)$, where n is the number of tetrahedra of the tetrahedralization. Each iteration treats n tetrahedra, and in the worst case, we need to loop $n - 1$ times.

4 Experimental Results

The main purpose of this section is to present some representative objects used in the experiments and the results achieved with the algorithm proposed. The meshes shown in this section have been generated on a Silicon Graphics Origin2000. The main hardware features of this multiprocessor are 2 GB of RAM and eight 250 MHz R10000 MIPS processors, although our actual implementation computes only over one of the eight available processors. The application is implemented in C++ and has been ported successfully to several UNIX based platforms (Linux). Although the development has been performed on a proprietary computer, we have used free development distribution tools to assure the portability of the code. The development tools have been those that provide GNU [16], and the 3D visualization of the objects has been carried out with Geomview [14].

The clouds of points presented with more detail in this paper belong to two synthetic objects, the goblet and the sea shell, proceeding from the GTS Library [24]. GTS is an Open Source Free Software Library intended to provide a set of useful functions to deal with 3D surfaces meshed with interconnected triangles. A third object with higher genus than the previous has been considered too with a similar detail: a human skull extracted from 3D tomographic real data. A fourth object proceeding from the Large Geometric Models Archive of the Georgia Institute of Technology is presented: a human skeleton hand [15]. In this case, the number of available samples is very higher, but the hand presents areas where the density of points is very variable too. The reason to show these examples is that the fourth objects present very different geometric features that allow to illustrate how the algorithm works. The selection of the first two synthetic shapes is motivated by the fact that the available sensors can not capture the internal geometry of the objects, like the details recovered inside the sea shell. Table 1 summarizes some statistics of the examples shown, together with the data of the other meshes extracted from some of the real objects used in the tests. It includes the number of points of the clouds describing the geometry of the objects, the number of vertices and facets of the final meshes and the execution time of each one of the most demanding stages of the whole process: the 3D tetrahedral Delaunay computation, and the final triangular facet extraction. Although there is some loss in the number of vertices with respect the number of input samples, the examples show that the shape, globally and locally, is cor-

Table 1. Statistic data about some of the objects managed in the experiments

Object	No. of input points	No. of output vertices	No. of output facets	Exec. time(s): 3D Delaunay Tetrahedra	Exec. time(s): Triang. mesh Extraction	Total Execution Time(s)
Bowl	1093	685	1412	6.800	0.988	8.00
Cup	3051	2791	5630	14.773	1.087	16.391
Amphora	6287	6041	16880	45.757	4.781	51.48
Human skull	6391	5707	14158	63.220	3.570	68.172
Tip of an arrow	7609	5405	12342	85.305	5.838	92.561
Sea shell	7782	5379	10906	38.298	2.728	42.259
Hand	52704	47342	144166	1090.171	41.158	1132.884

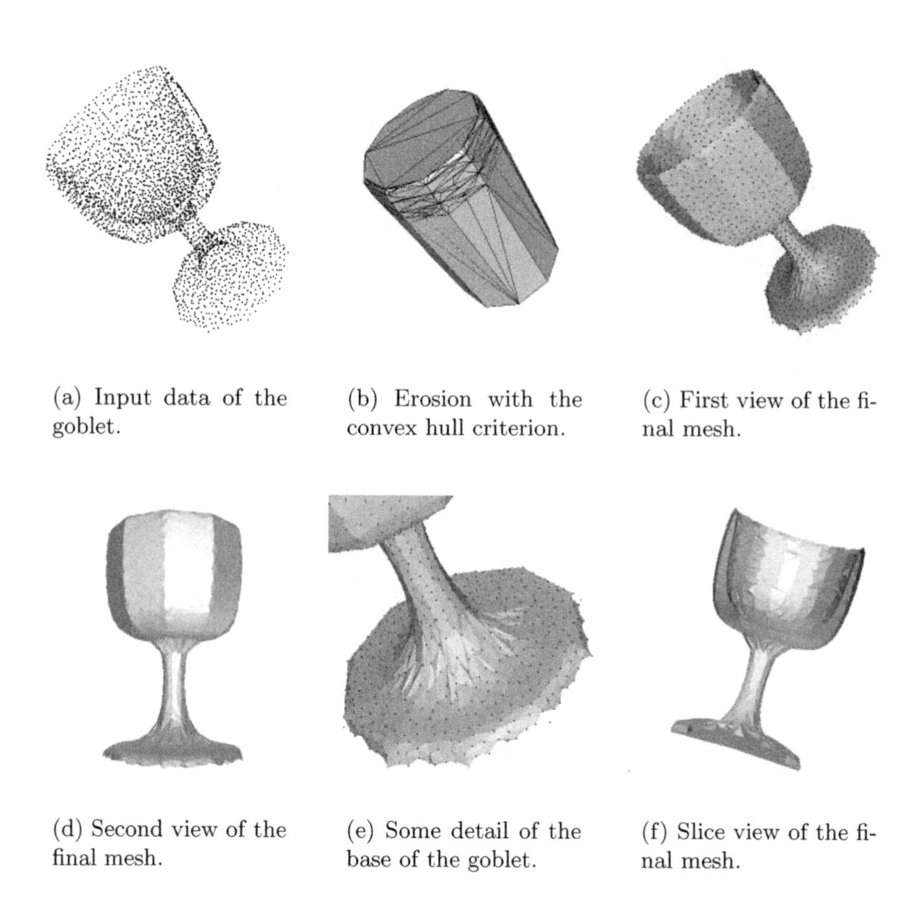

(a) Input data of the goblet.

(b) Erosion with the convex hull criterion.

(c) First view of the final mesh.

(d) Second view of the final mesh.

(e) Some detail of the base of the goblet.

(f) Slice view of the final mesh.

Fig. 1. The goblet. No. of points: 3051. No. of facets: 5630

rectly recovered. The most of the points removed are discarded in the Delaunay tetrahedralization step due to numerical errors.

The first object, Fig. 1, is a goblet that proceeds from [24]. The surface description of the goblet is a set of 3051 3D points. Figure 1(a) shows the input

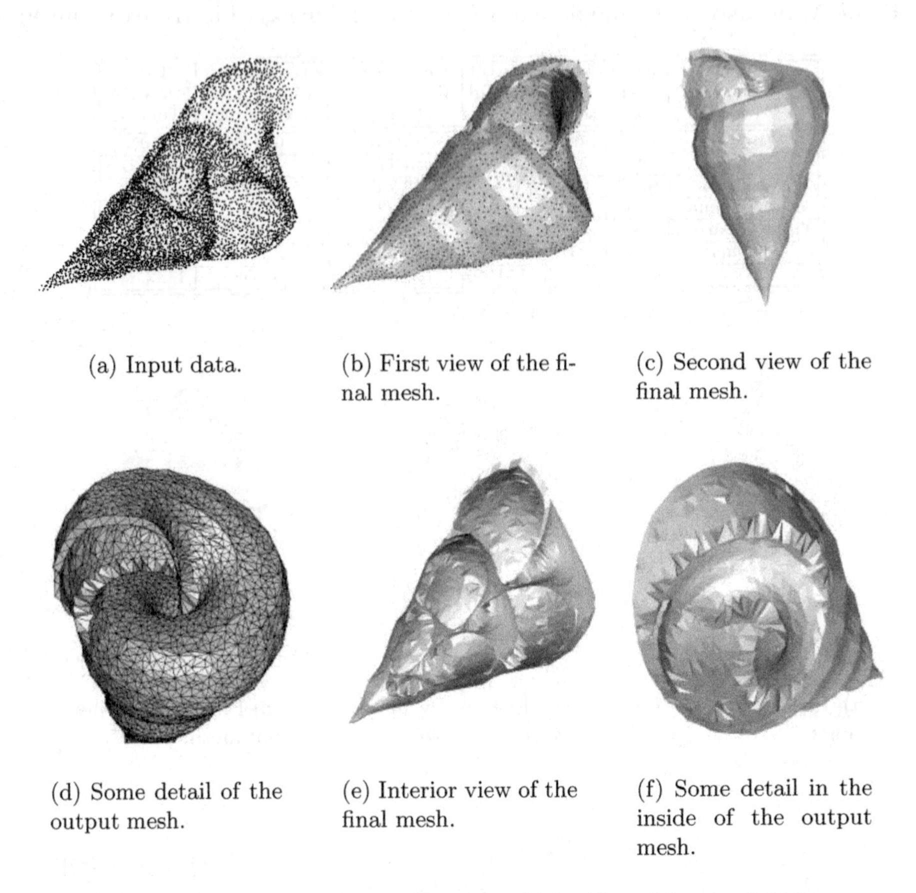

(a) Input data.

(b) First view of the final mesh.

(c) Second view of the final mesh.

(d) Some detail of the output mesh.

(e) Interior view of the final mesh.

(f) Some detail in the inside of the output mesh.

Fig. 2. The sea shell. No. of points: 7782. No. of facets: 10906

cloud of points that describes the geometry of the goblet. The mesh obtained after the first stage of the algorithm is shown on figure 1(b). This is the stage where the algorithm makes an erosion of the tetrahedra whose circumscribed spheres have centers outside of the convex hull. Figures 1(c)–1(f) show several views of the final mesh. Figure 1(c) overlaps the cloud of points with the output mesh, like Fig. 1(e), but in this one, we have performed a zoom over the base of the goblet to take a closer look at the extracted mesh and the input data. The black dots represent the 3D input points. The extracted mesh is a Delaunay triangulation of a subset of the input data. As it is shown in this and in the next figures, it can be considered that the corners, holes and cavities are satisfactory recovered.

The second object, Fig. 2, is a synthetic sea shell taken from the GTS Library [24]. Fig. 2(a) shows the input data; Fig. 2(b) shows the mesh achieved after applying the algorithm described in section 3 with the superposition of the

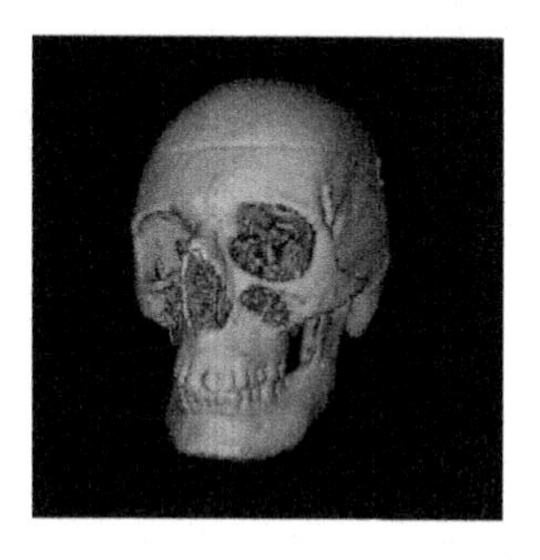

(a) Volumetric data of the human skull.

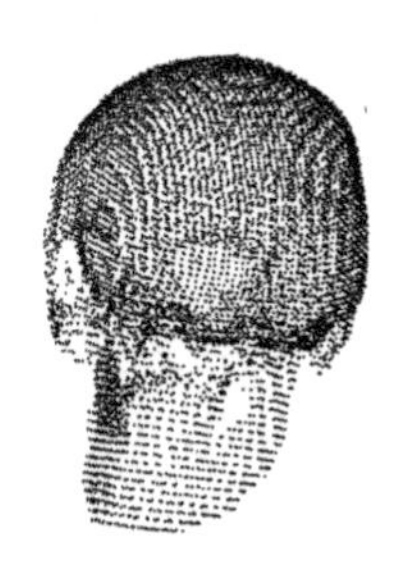

(b) Cloud of points extracted from the volumetric data of Fig. 3(a).

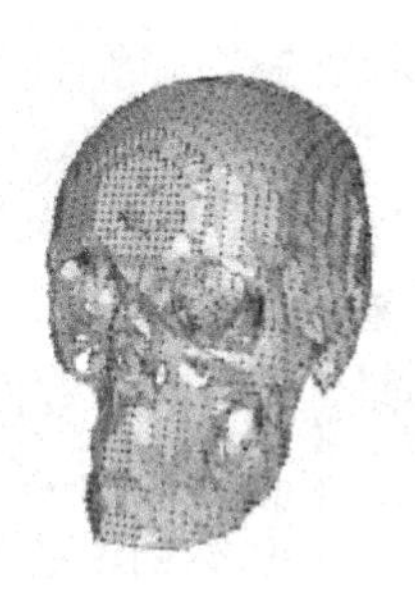

(c) First view of the human skull.

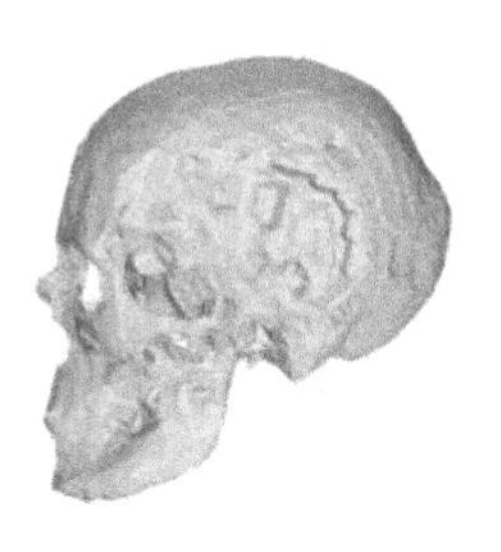

(d) Second view of the human skull.

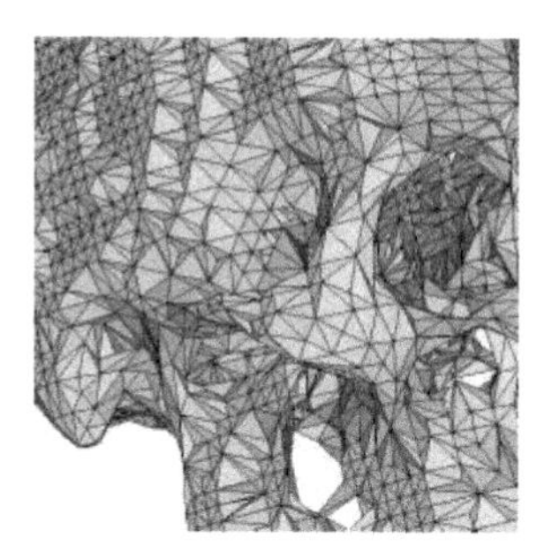

(e) A closer view showing some detail of the human skull.

Fig. 3. The human skull. No. of points: 6391. No. of facets: 14158

input cloud of points; Fig. 2(c) presents another view of the final mesh; we have applied a zoom process over the shell's opening in Fig. 2(d) for a better view of the details of the sea shell hole; in Fig. 2(e) we see a cut of the final mesh in order to see the inside and the cavities of the output mesh; finally, in Fig. 2(f), we present a zoom over the half-bottom interior of the sea shell.

The third object is a human skull (Figure 3). The main problems of this shape are the high complexity and the variable density of the cloud of points, presenting zones where there are very few points, for example in the eye holes, and others where the distribution is more uniform. Like in the previous figures, we present the input data, Fig. 3(b) and several views of the final mesh, Fig. 3(c)–3(e).

Figure 4 shows the surface extracted from a human hand. It presents some difficulties, like the inter-finger space or the finger tips, but in all the cases the surface recovered can be considered very close to the ideal.

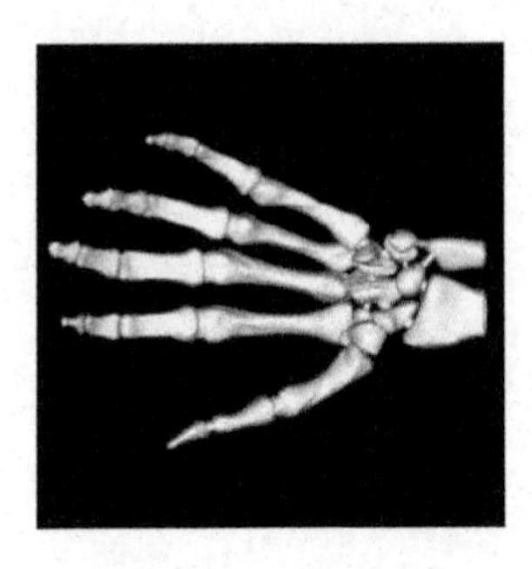

(a) A photograph taken from the original human skeleton hand.

(b) Cloud of points extracted from the human skeleton hand.

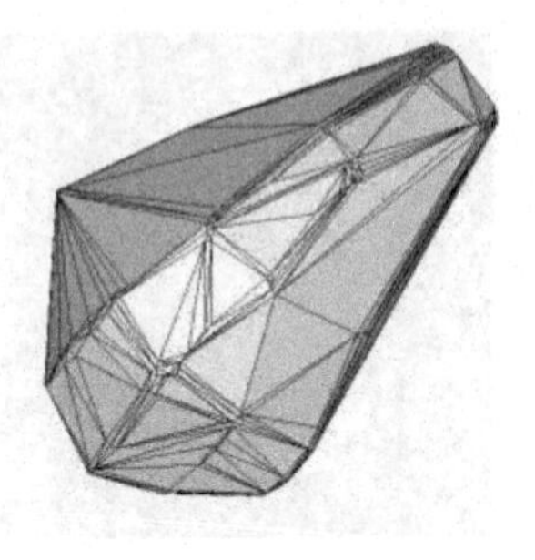

(c) Erosion with the convex hull criterion.

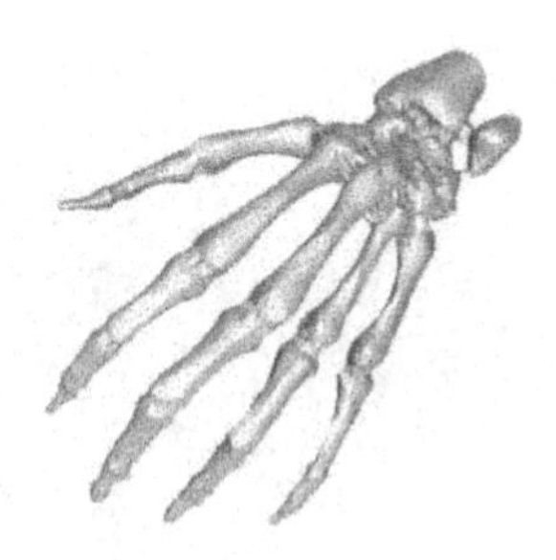

(d) First view of the human hand.

(e) Second view of the human hand.

Fig. 4. The human hand. No. of points: 47342. No. of facets: 144166

Last, in Fig. 5 we present the shape of the other objects mentioned in Table 1. In this case, the digitization of the bowl and the tip of an arrow has been carried out with a hand-held 3D digitizer. This device is a tactile low-price sensor manipulated by a human operator, which allows the simulation of any other kind of sensor that provides scarce and irregularly sampled measurements. It provides exclusively geometric information (only the points' coordinates are kept). The amphora has been digitized with a laser-based sensor, which provides a more regular sampling than the hand-held digitizer, but its prize is considerably higher too.

The current version of the implemented algorithm allows the specification of a threshold defined by the user. Sometimes, the geometry of the object does not allow to reach satisfactory results in a fully automatic way and the operator has the possibility to decide his preference: reducing the value of the threshold achieves a finer resolution mesh, but may produce some undesirable holes on the

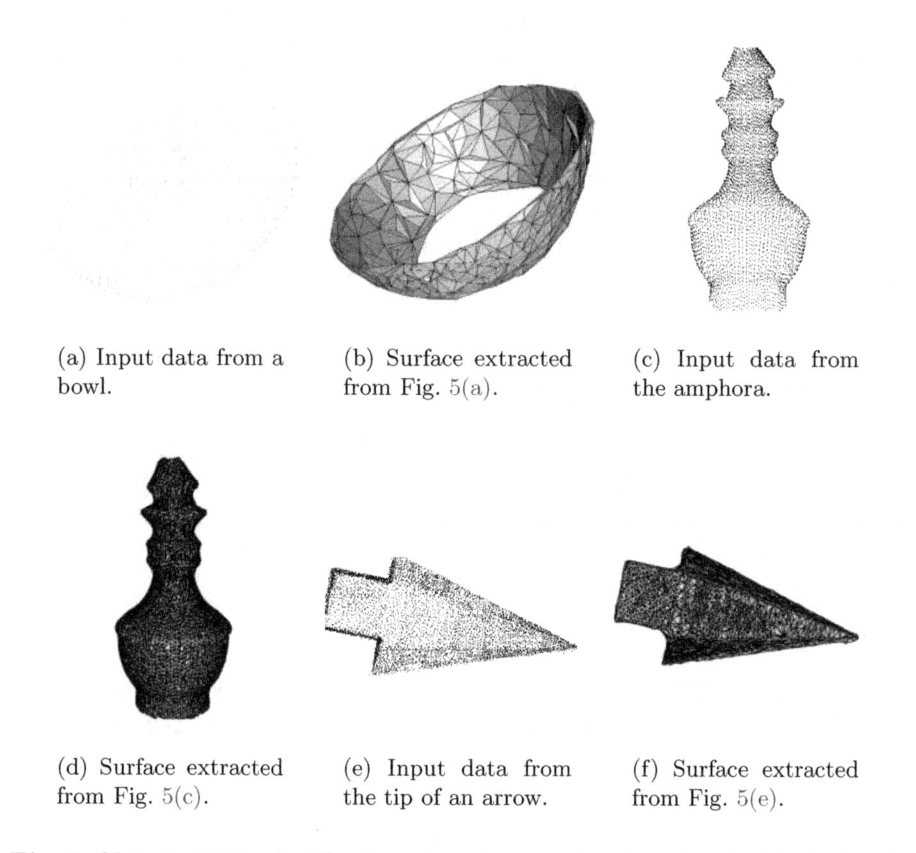

(a) Input data from a bowl.

(b) Surface extracted from Fig. 5(a).

(c) Input data from the amphora.

(d) Surface extracted from Fig. 5(c).

(e) Input data from the tip of an arrow.

(f) Surface extracted from Fig. 5(e).

Fig. 5. Objects digitized with a laser-based sensor (amphora) and with the hand-held digitizer (bowl and tip of an arrow) and meshes obtained

final mesh. Usually, this defect is not present in the meshes generated with a more uniform distribution, but it may be noticeable if the density of points over the object's surface is very variable.

We must notice that all the examples shown have been refined by the user specifying a threshold very similar to that computed in the fully automatic process.

5 Conclusions and Future Work

It has been presented a $O(n^2)$ algorithm that allows the automatic extraction of 3D Delaunay triangulations from free genus real objects. The only input information about the geometry of the objects is a set of 3D points irregularly distributed over the whole objects' surfaces. We must remark that n is the number of tetrahedra achieved by the spatial partition algorithm stated by Faugueras [12], and n depends not only on the number of considered points, but also on their positions.

The use of a topological criteria is not enough to obtain a mesh that correctly fits the shape of the objects without a human interaction, and it has been necessary to include additional restrictions to the process in order to become determinist. In our case, the heuristic chosen has been the perimeter of the triangular facets, mainly, because of its simplicity and good results achieved over other measures. The algorithm has been tested successfully with numerous free shape real objects, as we have shown with some examples presented above.

Thinking in future works, it should be desirable an optimization of the algorithm to reduce the response time by means of a parallel implementation, taking advantage of the local nature of the operations [7].

Acknowledgments

This work has been partially funded by the Spanish Commission for Science and Technology (grants CICYT TIC98-0272-C02-01 and TIC99-0947-C02-01). The authors gratefully acknowledge the help provided by Jaime Gómez (3D amphora data).

References

1. Nina Amenta and Marshall Bern. Surface reconstruction by Voronoi filtering. In 14th *ACM Symposium on Computational Geometry*, pages 39–48. ACM, June 1998. 273
2. D. Attali. τ-regular shape reconstruction from unorganized points. *Computational Geometry*, 10:239–247, 1998. 273
3. C. L. Bajaj, F. Bernardini, and G. Xu. Reconstructing surfaces and functions on surfaces from unorganized three-dimensional data. *Algorithmica*, 19:243–261, 1997. 273
4. Jean-Daniel Boissonnat and Frédéric Cazals. Smooth surface reconstruction via natural neighbour interpolation of distance functions. In *Proc.* 16th *Annu. ACM Symposium on Computational Geometry*, 2000. 273
5. S. Campagna and H.-P. Seidel. Parameterizing meshes with arbitrary topology. In H.Niemann, H.-P. Seidel, and B. Girod, editors, *Image and Multidimensional Signal Processing'98*, pages 287–290, 1998. 273
6. P. Cignoni, C.Montani, R.Perego, and R.Scopigno. Parallel 3D Delaunay triangulation. *Computer Graphics Forum*, 12(3):129–142, 1993. 274
7. P. Cignoni, D. Laforenza, C. Montani, R. Perego, and R. Scopigno. Evaluation of parallelization strategies for an incremental Delaunay triangulator in E^3. *Concurrency: Practice and Experience*, 7(1):61–80, 1995. 282
8. T. K. Dey, K. Mehlhorn, and E. R. Ramos. Curve reconstruction: Connecting dots with good reason. In *Proc.* 15th *Annu. ACM Sympos. Comptuational Geometry*, pages 197–206, 1999. 273
9. R. Dwyer. A faster divide and conquer algorithm for constructing Delaunay triangulations. *Algorithmica*, 2:137–151, 1989. 274
10. H. Edelsbrunner and E. P. Mücke. Three-dimensional alpha shapes. *ACM Trans. on Graphics*, 12(1):43–72, 1994. 273

11. T. Fang and L. Piegl. Delaunay triangulation using a uniform grid. *IEEE Computer Graphics and Applications*, pages 36–47, 1993. 274
12. Olivier Faugueras. *Three-Dimensional Computer Vision. A Geometric Viewpoint*, chapter 10 Interpolating and Approximating Three-Dimensional Data, pages 403–482. In Olivier Faugueras [13], 1993. 273, 274, 281
13. Olivier Faugueras. *Three-Dimensional Computer Vision. A Geometric Viewpoint.* The Massaachusetts Institute of Technology, 1993. 283
14. The Geometry Center. Geomview. University of Minnesota. www.geom.umn.edu/docs/software/download/geomview.html 276
15. Georgia Institute of Technology. Large Geometrical Models Archive http://http://www.cc.gatech.edu/projects/large_models 276
16. GNU. www.gnu.org 276
17. L. J. Guibas, D. E. Knuth, and M. Sharir. Randomized incremental construction of Delaunay and Voronoi diagrams. *Algorithmica*, 7:381–261, 1992. 274
18. A. Hilton and J. G. M. Gonçalves. 3D scene representation using a deformable surface. In IEEE Computer Society Press, editor, *Proc. on Physics Based Modeling Workshop in Computer Vision*, volume 1, pages 24–30, June 1995. 273
19. H. Hoppe, T. DeRose, T. Duchamp, J. McDonal, and W Stuetzle. Surface reconstruction from unorganized points. In *SIGGRAPH'92*, pages 71–78, July 1992. 273
20. Andrew E. Johnson and Martial Hebert. Control of polygonal mesh resolution for 3-D computer vision. *Graphical Models and Image Processing*, 60:261–285, 1998. 272
21. Leif Kobbelt, Jens Vorsatz, and Hans-Peter Seidel. Multiresolution hierarchies on unstructured triangle meshes. *Computational Geometry*, 14:5–24, November 1999. 273
22. C. Oblonsek and N. Guid. A fast surface-based procedure for object reconstruction from 3D scattered points. *Computer Vision and Image Understanding*, 2(69):185–195, 1998. 273
23. Michael Petrov, Andrey Talapov, Timothy Robertson, Alexeis Lebedev, Alexander Zhilayaev, and Leonid Polonsky. Optical 3D digitizers: Bringing life to the virtual world. *IEEE Computer Graphics and Applications*, 18(3):28–37, May 1998. 272
24. Stéphan Popinet. The GNU Triangulated Surface Library. http://gts.sourceforge.net 276, 277, 278
25. H. Shum, M. Hebert, K. Ikeuchi, and R. Reddy. An integral approach to free-form object modeling. *IEEE Transactions on Pattern Analysis and Machine Intelligence*, 12(19):1366–1370, 1997. 273
26. Hiromi T. Tanaka. Accuracy-based sampling an reconstruction with adaptive meshes for parallel hierarchical triangulation. *Computer Vision and Image Understanding*, 61(3):335–350, May 1995. 273
27. D. Terzopoulos and D. Metaxas. Dynamic 3D models with local and global deformations: Deformable superquadrics. *IEEE Trans. on Pattern Analysis and Machine Intelligence*, 13(7):703–714, 1991. 273

Go Digital, Go Fuzzy

Jayaram K. Udupa

Medical Image Processing Group, Department of Radiology University of Pennsylvania
423 Guardian Drive 4th Floor Blockley Hall, Philadelphia, PA 19104-6021
jay@mipg.upenn.edu

Abstract. In many application areas of imaging sciences, object information captured in multi-dimensional images needs to be extracted, visualized, manipulated, and analyzed. These four groups of operations have been (and are being) intensively investigated, developed, and applied in a variety of applications. In this paper, after giving a brief overview of the four groups of operations, we put forth two main arguments: (1) Computers are digital, and most image acquisition and communication efforts at present are toward digital approaches. In the same vein, there are considerable advantages to taking an inherently digital approach to the above four groups of operations rather than using concepts based on continuous approximations. (2) Considering the fact that images are inherently fuzzy, to handle uncertainties and heterogeneity of object properties realistically, approaches based on fuzzy sets should be taken to the above four groups of operations. We give two examples in support of these arguments.

1 Introduction

In imaging sciences, particularly medical, there are many sources of multidimensional images [1].

2D: digital/digitized radiographic images, tomographic slices.

3D: a time sequence of 2D images of a dynamic object, a stack of slice images of a static object.

4D: a time sequence of 2D images of a dynamic object for a range of imaging parametric values (e.g., MR spectroscopic images of a heart), a time sequence of 3D images of a dynamic object.

5D: a time sequence of 3D images of a dynamic object for a range of parametric values.

Three- and higher-dimensional images may also be generated by a computational step. For example, a 5D binary image is produced if we apply a "lifting" operation [2] to a 4D gray-level image to represent the 4D image intensity distribution as a surface in a 5D space. Columns of volume elements in this higher dimensional space represent the height of the surface.

G. Borgefors, I. Nyström, and G. Sanniti di Baja (Eds.): DGCI 2000, LNCS 1953, pp. 284-295, 2000.
© Springer-Verlag Berlin Heidelberg 2000

For ease of further reference, we will refer to a multidimensional image simply as a *scene* and represent it by a pair $\mathcal{C}$ = (C, *f*), where C, the *scene domain,* is a multidimensional rectangular array of spatial elements (*spels* for short), and *f*, the *scene intensity,* is a function that assigns to every spel a vector whose component elements are from a set of integers.

There is usually an *object* of study for which the scene is generated. This may be a physical object such as an organ/tissue component, a tumor/fracture/lesion/an abnormality, a prosthetic device, or a phantom. It may also be a conceptual object such as an isodose surface in a radiation treatment plan, an activity region highlighted by PET or functional MRI, or a mathematical phantom. The objects may be rigid, deformable, static, or dynamic. For example, the bones at a moving joint represent a dynamic, rigid object assembly. The heart muscle and the blood pool in the various chambers constitute a dynamic, deformable object assembly. We will refer to any operation, which, for a given set of multimodality scenes of an object of study, produces information about the object of study, as *3D imaging*. The information may be qualitative - for example, a 3D rendition of a static object, an animation sequence of a dynamic object or an animation sequence of a static object corresponding to different viewpoints. It may be quantitative - for example, the volume of a static object, and the rate of change of volume of a dynamic, deformable object, the extent of motion of a rigid dynamic object. The purpose of 3D imaging is, therefore, given multiple scenes as input, to output qualitative/quantitative information about the object of study.

In this exposition, we will give a quick overview of the various 3D imaging operations that are commonly used and identify the challenges currently faced. We argue with strong evidence that, since the scenes are inherently digital, we should take entirely digital approaches to realize all 3D imaging operations. We also suggest that, since object information in scenes is always fuzzy, we should take approaches that are based on fuzzy set concepts. The need to fulfill these strategies opens numerous problems that are inherently digital in both hard and fuzzy settings. The reader is referred to [2] for a detailed account of 3D imaging operations and their medical applications.

Although the examples considered in this exposition are all medical, the principles and the arguments are also applicable to other application areas.

2 An Overview of 3D Imaging Operations

3D imaging operations may be classified into four groups: *preprocessing, visualization, manipulation*, and *analysis*. All preprocessing operations aim at improving or extracting object information from the given scenes. The purpose of visualization operations is to assist humans in perceiving and comprehending the object structure/characteristics/function in two, three and higher dimensions. The manipulation operations allow humans to interactively alter the object structures (mimic surgery, for example). The analysis operations enable us to quantify object structural/morphological/functional information.

In the rest of this section, we will cursorily examine these four classes of operations. The references given here are only samples from a literature, which has over a thousand publications excluding pure application–directed papers.

2.1 Terminology

Body region: Region of the space, for example, of the human body, that is imaged.
Pixel: Image element in a 2D scene.
Spel: Image element in a 2D, 3D, 4D, . . ., scene.
Scalar scene: A scene wherein each spel has one value assigned to it.
Vector scene: A scene wherein each spel has two or more values assigned to it. For example, T2 and proton density value in an MR scene.
Scene intensity: The value(s) assigned to the spel.
Scene domain: The set of all spels in a scene.
Pixel size: The size of the spel within the natural slice of the scene.
Slice spacing: The distance between the centers of two successive slices. In case of isotropic resolution, the spels have equal length in all dimensions.
Scene coordinate system: A coordinate system affixed to the scene.
Object: An object of study in the body region such as an organ or a pathology such as a tumor.
Object system: A collection of objects.
Structure: A computer representation of an object. It may be hard (crisp or binary) or fuzzy. In the hard case, a spel in the scene domain is considered to be either in the object or not in the object. In the fuzzy case, each spel has a value between 0 and 100 that indicates the degree of objectness assigned to the spel.
Structure system: A collection of structures.
Structure coordinate system: A coordinate system affixed to the structure/structure system.
Rendition: A 2D scene created by the computer that depicts some aspects of the object information contained in a scene. For example, the display of a slice of the scene, a shaded-surface display of a structure computed from the scene.
Display coordinate system: A coordinate system affixed to the display device.
Imaging device coordinate system: A coordinate system affixed to the imaging device. Information about the various coordinate systems is useful in registering scenes and structures.

2.2 Preprocessing Operations

The input to these operations is a set of scenes of a body region and the output is either a set of scenes or a structure system. Typically, the structure system contains only one structure corresponding to the object of interest.

Volume of Interest (VOI) [3]:

The input scene $\mathcal{C}_i = (C_i, f_i)$ is converted to another smaller scene $\mathcal{C}_o = (C_o, f_o)$ such that $C_o \subset C_i$ and f_o is a restriction of f_i to C_o. C_o is specified, usually interactively, to be the set of spels in a rectangular box, a sphere, or an ellipsoid (and their equivalents in higher dimensions). The purpose here is to make C_o just contain the objects of interest and to reduce the amount of data to be handled in other expensive operations. In large, routine applications, to automate this operation is a challenge. For this, clearly, some object knowledge is essential.

Filtering:

In these operations, the output scene C_o is obtained from the input scene C_i by modifying the intensity values in such a way as to either *enhance* object information [4] or to *suppress* unwanted non-object information such as noise [5] in C_i. Usually for these operations $C_o = C_i$. Various types of filters are available under these two categories. Unfortunately, usually some non-object information is also enhanced by the former methods and object information is also suppressed by the latter methods. The challenge in filtering is how to incorporate object knowledge into the filter design so as to achieve optimum performance.

Interpolation:

The purpose of interpolation is to change the level and orientation of discretization of C_i to a desired entity. For these operations, usually $C_o \neq C_i$. These operations are needed for converting a non-isotropically sampled scene into a scene with isotropic resolution, or for obtaining scenes at desired level and orientation of discretization. Two classes of operations are available: scene-based and object-based. In *scene-based* methods [6], $f_o(c)$ is estimated from $f_i(c')$ for spels c' lying in the close vicinity of c. In *object-based* methods [7], some object information is used in guiding interpolation. Object-based methods have been shown repeatedly in the literature to produce better results than scene-based methods. The challenge here is how to incorporate object knowledge specific to the application into the interpolation process.

Registration:

The purpose of registration is to represent structures obtained from multiple scenes in a common coordinate system. These are needed for combining object information obtained from multiple modalities (such as MRI, PET and CT) for the same body region, and for determining change, growth, motion, and displacement of objects over time. Again two classes of methods are available: scene-based and object-based. In *scene-based* methods [8], scene intensity patterns are matched, while in *object-based* methods [9] structures derived from scenes are matched. Both rigid and deformable methods of registration under both categories are available. As in other operations, the challenge here is to incorporate specific object knowledge into the registration process in both object-based and scene-based methods.

Segmentation:

This operation, the most crucial among all 3D imaging operations, outputs a structure system from a given set of scenes. It consists of two related tasks - recognition and delineation. *Recognition* is the process of determining roughly the whereabouts of the objects in the scene, and *delineation* is the process of determining precisely the spatial extent and composition of the objects in the scene. Knowledgeable humans usually outperform computer algorithms in the high-level task of recognition, while computer algorithms can outperform humans in the precise, accurate, and efficient delineation of objects.

There are two classes of approaches to recognition – *automatic* (knowledge-based/model-based) [10] and *human-assisted.* By far the latter is the most commonly used in practice. At the outset, two classes of approaches to delineation may be identified - *boundary-based* [11] and *region-based* [12]. In the former, the output is a set of boundaries of the object, and in the latter, it is a set of object regions. Each of these two strategies can be further divided into subgroups - *hard* [11] and *fuzzy* [12] - depending on whether the output is a hard or a fuzzy set. Combined with the two strategies of recognition, therefore, we may identify eight classes of approaches to segmentation.

As seen from the outline of other preprocessing operations, object knowledge usually facilitates all 3D imaging operations, including segmentation. This implies that segmentation is helpful/needed for all 3D imaging operations, ironically also for segmentation itself. To devise generic segmentation methods with high precision, accuracy, and efficiency, that can be quickly adapted to a given application, is indeed the greatest challenge in 3D imaging and in image analysis.

2.3 Visualization

The input to these operations is a set of scenes or a structure system and the output is a rendition.

Two classes of methods are available: *scene-based* wherein renditions are created directly from given scenes, and *object-based,* wherein renditions are created from (hard or fuzzy) structures extracted from the scene. Scene-based methods may be further divided into two groups: slice mode and volume mode. In the former, 2D "slices" of different orientations - natural, orthogonal, oblique and curved - are extracted from the scene and displayed as a montage or in a roam through fashion using gray scale, color or overlay [2]. In the latter, surfaces, interfaces and intensity distributions are displayed with a variety of 3D cues using *surface rendering* and *volume rendering* techniques [13, 14]. In object-based methods, the hard or fuzzy structures extracted from the scene are displayed with 3D cues using surface rendering and volume rendering techniques [15, 16]. The challenges in visualization are the realistic display of objects including color, texture and surface properties, speeding up volume rendering, and objective evaluation of the large number of rendering schemes that are available.

2.4 Manipulation

These operations are needed for editing structures (for simulating surgery in medical applications) for unobscured visualization, and for developing aids for interventional procedures. They take as input a structure system and output another structure system by altering structures or their relationship. Two classes of operations are being developed: rigid [15, 17] with operations to cut, separate, add, subtract, move and mirror structures and their components, and deformable [18]. The main challenge in manipulation is to realize the manipulative operations realistically utilizing object material properties.

2.5 Analysis

The purpose of these operations is, given a set of scenes or a structure system, to generate a quantitative description of the morphology or function of the objects in the body region captured in the scene.

Two classes of operations are available: *scene-based*, wherein quantities based on scene intensities such as intensity statistics within object regions, tissue density, activity, perfusion, flow are estimated; and *object-based*, wherein quantities measured from segmented structures and how they change with time are estimated including distance, length, width, curvature, area, volume, kinematics and mechanics. The main challenge here is the assessment of the accuracy of the estimated measures.

3 Go Digital, Go Fuzzy

Any scene of any body region exhibits the two following important characteristics of the objects contained in the body region.

Graded Composition: The spels in the same object region exhibit heterogeneity of scene intensity due to the heterogeneity of the object material, and noise, blurring, and background variation introduced by the imaging device. Even if a perfectly homogeneous object were to be imaged, its scene will exhibit graded composition.

Hanging-togetherness (Gestalt): In spite of the graded composition, knowledgeable humans do not have any difficulty in perceiving object regions as a whole (Gestalt). This is a fuzzy phenomenon and should be captured through a proper theoretical and computational framework.

There are no binary objects or acquired binary scenes. Measured data always have uncertainties. Additionally, scenes are inherently digital. As seen from the previous section, no matter what 3D imaging operation we consider, we cannot ignore these two fundamental facts – fuzziness in data and their digital nature.

We have to deal with essentially two types of data - scenes and structures. We argue that, instead of imposing some sort of a continuous model on the scene or the structure in a hard fashion, taking an inherently digital approach, preferably in a fuzzy setting, for all 3D imaging operations can lead to effective, efficient and practically viable methods. Taking such an approach would require, in almost all cases, the development of the necessary mathematical theories and algorithms from scratch. We need appropriate theories and algorithms for topology, geometry and morphology, all in a fuzzy, digital setting, for dealing with the concept of a "structure" in scenes. Note that almost all the challenges we raised in the previous section relate to some form of object definition in scenes. Therefore, they all need the above mathematical and algorithmic developments. Additionally, since we deal with deformable objects (all soft-tissue organs, and often even bone, as in distraction osteogenesis – the process of enlarging or compressing bones through load applied over a long period of time), we also need fuzzy digital mechanics theories and algorithms to handle structure data realistically for the operations relating to registration, segmentation, manipulation and analysis. Because of the digital setting, almost all challenges raised previously lead to discrete problems. Since we are dealing with n-dimensional scenes, the mathematics and algorithms need to be developed for hard and fuzzy sets of spels defined in n-dimensions.

In the rest of this section, we shall give two examples of digital/fuzzy approaches that motivated us to the above argument. The first relates to modeling an object as a 3D surface and visualizing it. The second relates to the fuzzy topological concept of connectedness and its use in segmentation.

3.1 Polygonal versus Digital Surfaces

In hard, boundary-based 3D scene segmentation (e.g., thresholding approaches), the output structure is often a 3D surface. The surface is represented usually either as a set of polygons (most commonly triangles [19]), or in a digital form [20] as a set of cubes or as a set of oriented faces of cubes. We shall describe in some detail how both the representation and rendering of such surfaces is vastly simpler and more efficient using digital approaches than using polygonal or other continuous approximations for them.

Let us consider the representation issues first. We shall subsequently consider the rendering issues. It is very reasonable to expect these surfaces to satisfy the following three properties since surfaces of real objects possess these properties.

(1) The surface is connected.
(2) The surface is oriented. This means that it has a well defined inside and outside.
(3) The surface is closed; that is, it constitutes a Jordan boundary. The latter implies that the surface partitions the 3D space into an "interior" set and an "exterior set such that any path leading from a point in the interior to a point in the exterior meets the surface.

The definitions of connectedness, orientedness, and Jordan property are much simpler, more natural and elegant in the digital setting using faces of cubes (or cubes) than using any continuous approximations such as representations via triangles. These global concepts can be arrived at using simple local concepts for digital surfaces [21, 22]. For example, orientedness can be defined by thinking of the faces to be oriented. That is, a face with a normal vector pointing from inside of the surface to its outside in the $-x$ direction is distinguished from a face at the same location with a face normal in exactly the opposite direction. Usually the surface normals at various points p on these surfaces (in both the polygonal and the digital case) are estimated independent of the geometry of the surface elements, for example, by the gradient at p of the intensity function f of the original scene from which the surface was segmented [23]. The gradient may also be estimated from other derived scenes such as a Gaussian smoothed version of the segmented binary scene [24]. Since the surface normals dictate the shading in the rendering process, the surface elements are used here only as a geometric guide rather than as detailed shape descriptors of the surface. Therefore the digital nature of the geometry introducing "staircase" effects in renditions can be more or less completely eliminated. Since the surface elements in the digital case are simple *and* all of identical size and shape, they can be stored using clever data structures that are typically an order of magnitude more compact than their polygonal counterparts [25]. Finally, there is a well-developed body of literature (e.g., [26], Chapters 6, 7) describing the theory of digital surfaces that naturally generalizes to n-dimensions for any n. Such a generalization is very difficult for polygonal representations.

Let us now come to the rendering issues. Rendering of digital surfaces is considerably simpler and more efficient than that of polygonal surfaces, the main reason being the simplicity of the surface elements and of their spatial arrangement in the former case [15, 27]. There are mainly two computational steps in any rendering algorithm: hidden surface removal and shading. Both these steps can be considerably simplified exploiting the special geometry of the surface elements, reducing most expansive computations to table lookup operations. For example, the faces in a digital surface can be classified into six groups based on their face normals (corresponding to the directions ($+x$, $+y$, $+z$, -x, $-y$, $-z$). For any viewpoint, all faces from at least three of these groups are not visible and hence can be discarded without doing any computation per face [28]. There are other properties that allow the rapid projection of discrete surfaces in a back-to-front or front-to-back order by simply accessing the surface elements in some combination of column, row, and slice order from a rectangular array. Triangular and other continuous approximations to surfaces simply do not possess such computationally attractive properties. It was shown in [25] based on 10 objects of various sizes and shapes that, digital surface rendering entirely in software on a 300 MHz Pentium PC can be done about 4-17 times faster than the rendering of the same objects, whose surfaces are represented by triangles, on a Silicon Graphics Reality II hardware rendering engine for about the same quality of renditions. Note that the Reality II workstation is vastly more expensive than the Pentium PC.

The design of rendering engines such as Reality II was motivated by the need to visualize, manipulate, and analyze structure systems representing human-made objects such as automobiles and aircrafts. Early efforts in modeling in computer graphics took, for whatever reason, a continuous approach. Digital approaches to modeling have originated mostly in medical imaging [29] and have recently been applied also to the modeling of human-made objects [30] with equal effectiveness. Digital approaches are more appropriate for modeling natural objects (such as internal human organs) than continuous approaches. Natural objects tend to be more complex in shape and morphology than human-made objects. The applications of digital approaches to human-made objects [30] have demonstrated that the latter are equally appropriate even for human-made objects.

3.2 Fuzzy Connected Object Definition

Although much work has been done in digital geometry and topology based on hard sets, analogous work based on fuzzy sets is rare [12, 31]. As pointed out earlier, the uncertainties about objects inherent in scenes must be retained as realistically as possible in all operations instead of making arbitrary hard decisions. Although many fuzzy strategies have been proposed particularly for image segmentation, none of them has considered the spatio-topological relationship among spels in a fuzzy setting to formulate the concept of hanging-togetherness. (We note that the notion of hard connectedness has been used extensively in the literature [31], Chapter 1. But hard connectedness already assumes segmentation and removes the flexibility of utilizing the strength of connectedness in segmentation itself.) This is a vital piece of information that can greatly improve the immunity of object definition (segmentation) methods to noise, blurring and background variation.

Given the fuzzy nature of object information in scenes, the frameworks to handle fuzziness in scenes should address questions of the following form: How are objects to be mathematically defined in a fuzzy, digital setting taking into account the graded composition and hanging-togetherness of spels? How are fuzzy boundaries to be defined satisfying a Jordan boundary property? What are the appropriate algorithms to extract these entities from scenes in such a way as to satisfy the definitions? These questions are largely open. We will give one example below of the type of approaches that can be taken. This relates to fuzzy connected object definition [12]. This framework and the algorithms have now been applied extensively on 1000's of scenes in several routine clinical applications [32-34] attesting to their strength, practical viability, and effectiveness.

We define a fuzzy *adjacency* relation α on spels independent of the scene intensities. The strength of this relation is in [0, 1] and is greater when the spels are spatially closer. The purpose of α is to capture the blurring property of the imaging device.

We define another fuzzy relation κ, called *affinity*, on spels. The strength of this relation between any two spels c and d lies in [0, 1] and depends on α as well as on how similar are the scene intensities and other properties derived from scene intensities in the vicinity of c and d. Affinity is a local fuzzy relation. If c and d are far apart, their affinity is 0.

Fuzzy *connectedness* is yet another fuzzy relation on spels, defined as follows. For any two spels c and d in the scene, consider all possible connecting paths between them. (A path is simply a sequence of spels such that the successive spels in the sequence are "nearby".) Every such path has a strength which is the smallest affinity of successive pairwise spels along the path. The strength of fuzzy connectedness between c and d is the largest of the strength of all paths between c and d.

A fuzzy connected object of a certain strength θ is a pool O of spels together with an objectness value assigned to each spel in O. O is such that for any spels c and d in O, their strength of connectedness is at least θ, and for any spels e in O and g not in O, their strength is less than θ.

Although the computation of a fuzzy connected object in a given scene for a given κ and θ appears to be combinatorially explosive, the theory leads to solutions for this problem based on dynamic programming. In fact, fuzzy connected objects in 3D scenes (256×256×60) can be extracted at interactive speeds (a few seconds) on modern PCs such as a 400 MHz Pentium PC.

4 Concluding Remarks

In this article, we have first given an overview of the operations available for 3D imaging - a discipline wherein, given a set of multidimensional scenes, the aim is to extract, visualize, manipulate, and analyze object information captured in the scenes. We have also raised numerous challenges that are encountered in real applications. Computers are digital. Current attempts in image acquisition, storage, and communication are completely digital or are proceeding in that direction. We have presented an argument with evidences that there are considerable advantages in taking an inherently digital approach to realizing all 3D imaging operations. We have also argued that, since object information in scenes is fuzzy, the digital approaches should be developed in a fuzzy framework to handle the uncertainties realistically. This calls for the development of topology, geometry, morphology and mechanics, all in a fuzzy and digital setting, all of which are likely to have a significant impact on imaging sciences such as medical imaging and their applications.

Acknowledgments

The authors' research is supported by an NIH grant NS37172, a grant from the Department of Army DAMD 179717271, and a contract from EPIX Medical, Inc. He is grateful to Mary A. Blue for typing the manuscript.

References

[1] Cho, Z.H., Jones J.P. and Singh, M.: *Foundations of Medical Imaging*, New York, New York: John Wiley & Sons, Inc. (1993).

[2] Udupa, J. and Herman, G. (eds.): *3D Imaging in Medicine*, 2nd Edition, Boca Raton, Florida: CRC Press (1999).

[3] Udupa, J.K.: "Interactive Segmentation and Boundary Surface Formation for 3-D Digital Images," *Computer Graphics and Image Processing*, 18 (1982), 213-235.

[4] Perona, P. and Malik, J.: "Scale Space and Edge Detection Using Anisotropic Diffusion," *IEEE Transactions on Pattern Analysis and Machine Intelligent* 12 (1983), 629-639.

[5] Chin, R.T. and Yeh, C.L.: "Quantitative Evaluation of Some Edge-Preserving Noise Smoothing Techniques," *Computer Graphics and Image Processing* 23 (1983), 67-91.

[6] Herman, G.T., Rowland, S.W. and Yau, M.-M.: "A Comparative Study of the Use of Linear and Modified Cubic Spline Interpolation for Image

Reconstruction," *IEEE Transactions on Nuclear Science* NS-26 (1979), no. 2, 2879-2894.

[7] Raya, S.P. and Udupa, J.K.: "Shape-Based Interpolation of Multidimensional Objects," *IEEE Transactions on Medical Imaging* 9 (1990), no. 1, 32-42, 1990.

[8] Woods, R.P. Cherry, S.R. and Mazziotta, J.C.: "Rapid Automated Algorithm for aligning and Reslicing PET Images," *Journal of Computer Assisted Tomography* 17 (1992), no. 4, 620-633.

[9] Pelizzari, C.A., Chen, G.T.Y., Spelbring, D.R., Weichselbaum, R.R. and Chen, C.-T.: "Accurate Three-Dimensional Registration of CT, PET, and/or MR Images of the Brain," *Journal of Computer Assisted Tomography* 13 (1989), 20-26.

[10] Gong, L. and Kulikowski, C.: "Composition of Image Analysis Processes Through Object-Centered Hierarchical Planning," *IEEE Transactions on Pattern Analysis and Medicine Intelligence* 17 (1995), no. 10, 997-1008.

[11] McInerney, T. and Terzopoulos, D.: "A Dynamic Finite Element Surface Model for Segmentation and Tracking in Multidimensional Medical Images With Application to Cardiac 4D Image Analysis," *Computerized Medical Imaging and Graphics* 19 (1995), no. 1, 69-83.

[12] Udupa, J.K. and Samarasekera, S.: "Fuzzy Connectedness and Object Definition: Theory, Algorithms, and Applications in Image Segmentation," *Graphical Models and Image Processing* 58 (1996), no. 3, 246-261.

[13] Goldwasser, S. and Reynolds, R.: "Real-time Display and Manipulation of 3-D Medical Objects: The Voxel Machine Architecture," *Computer Vision, Graphics, and Image Processing* 39 (1987), no.4, 1-27.

[14] Lacroute, P.G.: "Fast Volume Rendering Using a Shear-Warp Factorization of the Viewing Transformation," Technical Report CSL-TR-95-678, Ph.D. Thesis, Departments of Electrical Engineering and Computer Science, Stanford University, Stanford, California, September (1995).

[15] Udupa, J.K. and Odhner, D.: "Fast Visualization, Manipulation, and Analysis of Binary Volumetric Objects," *IEEE Computer Graphics and Applications* 11 (1991), no. 6, 53-62.

[16] Udupa, J.K. and Odhner, D.: "Shell Rendering," *IEEE Computer Graphics and Applications* 13 (1993), no. 6, 58-67.

[17] Odhner, D. and Udupa, J.K.: "Shell Manipulation: Interactive Alteration of Multiple-Material Fuzzy Structures," *SPIE Proceedings* (1995) 2431:35-42.

[18] Waters, K.: "A Muscle Model for Animating Three Dimensional Facial Expression," *Proceedings of SIGGRAPH'87* 21 (1987), no. 3, 17-24.

[19] Lorensen, W. and Cline, H.: "Marching Cubes: A High Resolution 3D Surface Construction Algorithm," *Computer Graphics* 21(1987), 163-169.

[20] Udupa, J.K., Srihari, S.N. and Herman G.T.: "Boundary Detection in Multidimensions," *IEEE Transactions on Pattern Analysis and Machine Intelligence* PAMI-4 (1982), 41-50.

[21] Artzy, E., Frieder, G. and Herman, G.: "The Theory, Design, Implementation and Evaluation of a Three-Dimensional Surface Detection Algorithm, " *Computer Graphics and Image Processing* 15 (1981), 1-24.

[22] Gordon, D. and Udupa, J.: "Fast Surface Tracking in Three-Dimensional Binary Images, " *Computer Vision, Graphics, and Image Processing* 45 (1989), 196-214.

[23] Höhne, K.H. and Bernstein, R.: "Shading 3D Images from CT Using Gray-Level Gradients," *IEEE Transactions on Medical Imaging* 5 (1986), 45-47.

[24] Udupa, J. and Goncalves, R.: "Imaging Transforms for Visualizing Surfaces and Volumes," *Journal of Digital Imaging* 6 (1993), 213-236.

[25] Grevera, G.J. and Udupa, J.K.: "Order of Magnitude Faster Surface Rendering Via Software in a PC Than Using Dedicated Hardware," *SPIE Proceedings* 3658 (1999), 202-211.

[26] Kong, T.Y. and Rosenfeld, A. (eds.): *Topological Algorithms for Digital Image Processing*, New York, New York: Elsevier Science B.V. (1996).

[27] Frieder, G, Gordon, G. and Reynolds, R.: "Back-to-Front Display of Voxel-Based Objects," *IEEE Computer Graphics and Applications* 5 (1985), 52-60.

[28] Artzy, E.: "Display of Three-Dimensional Information in Computed Tomography, " *Computer Graphics and Image Processing* 9 (1979), 196-198.

[29] Herman, G. and Liu, H.: "Three-Dimensional Display of Human Organs from Computed Tomograms, " *Computer Graphics and Image Processing* 9 (1979), 679-698.

[30] Kaufman, A.: "Efficient Algorithms for 3-D Scan Conversion of Parametric Curves, Surfaces, and Volumes," *Computer Graphics* 21 (1987), 171-179.

[31] Rosenfeld, A.: "Fuzzy Digital Topology," *Information and Control* 40 (1979), 76-87.

[32] Udupa, J.K., Wei, L. Samarasekera, S., Miki, Y., van Buchem, M.A. and Grossman, R.I.: "Multiple Sclerosis Lesion Quantification Using Fuzzy Connectedness Principles," *IEEE Transactions on Medical Imaging* 16 (1997), 246-261.

[33] Rice, B.L. and Udupa, J.K.: "Clutter-Free Volume Rendering for Magnetic Resonance Angiography Using Fuzzy Connectedness," *International Journal of Imaging System and Technology* 11 (2000), 62-70.

[34] Saha, P.K., Udupa, J.K., Conant, E.F. and Chakraborty, D.P.: "Near-Automatic Segmentation and Quantification of Mammographic Glandular Tissue Density," *SPIE Proceedings* 3661 (1999), 266-276.

Recognition of Digital Naive Planes and Polyhedrization

Joëlle Vittone[1] and Jean-Marc Chassery[2]

[1] LSIIT - Louis Pasteur University, Pole API
Bd Sébastien Brandt, 67400 Illkirch-Graffenstaden, France
Joelle.Vittone@dpt-info.u-strasbg.fr
http://dpt-info.u-strasbg.fr/~vittone/
[2] LIS - ENSIEG, Domaine universitaire
B.P.46, 38402 Saint-Martin d'Hères, France
Jean-Marc.Chassery@inpg.fr

Abstract. A digital naive plane may be seen as a repetition of (n, m)-cubes, set composed of $n \times m$ adjacent voxels or more generally sets of p voxels. In a previous works [VC99a], we have shown how to link the parameters of a naive plane to the different configurations of voxels sets by the construction of the associated Farey net. We propose an algorithm to recognize any set of coplanar voxels. This algorithm will be used for the polyhedrization of voxel objects. This is an original contribution offering a new method for digital plane recognition.

Keywords: Digital naive plane - Polyhedrization - Recognition - Equivalence classes

1 Introduction

The characterization of digital naive planes is now solved. Effectively, naive planes have been studied through their configurations of tricubes [Sch97,VC97], of (n, m)-cubes [VC99b] and connected or not connected voxels set [VC99a,Gér99]. The link between the normal equation of a plane and configuration of voxels set has been studied by the construction of the corresponding Farey net [VC99a].

We can find many references about the recognition of digital planes. Some algorithms were related to the construction of the convex hull of the studied voxels set [KS91,KR82]. Other approaches use linear programming[ST91], mean square approximation [BF94] or Fourier-Motzkin transform [FP99,FST96,Vee94].

The first algorithms entirely discrete were to recognize rectangular pieces of naive planes [Deb95,DRR94,VC99b]. In this paper, we propose an incremental algorithm to recognize any coplanar voxels set.

2 Definitions

Let a, b, c, r be four integers such as a, b, c are not null all together and verify $\gcd(a, b, c) = 1$.

G. Borgefors, I. Nyström, and G. Sanniti di Baja (Eds.): DGCI 2000, LNCS 1953, pp. 296–307, 2000.
© Springer-Verlag Berlin Heidelberg 2000

The digital naive plane $\mathcal{P}(a,b,c,r)$, where (a,b,c) is its normal vector and r its translation parameter, is the set of points (x,y,z) in $\mathbb{Z}^3$ verifying:

$$0 \leq ax + by + cz + r < \max(|a|,|b|,|c|)$$

We will limit our study to naive planes $\mathcal{P}(a,b,c,r)$ in the 48th part of space such as $0 \leq a \leq b \leq c$ and $c \neq 0$. These planes are `functional` in $0xy$. For each point (x,y) in $\mathbb{Z}^2$, we have only one point (x,y,z) in $\mathbb{Z}^3$ belonging to the naive plane. Let us notice $f(a,b,c,r)$ the function from $\mathbb{Z}^2$ to $\mathbb{Z}$ defined by:

$$f(a,b,c,r)(u,v) = -\left[\frac{au+bv+r}{c}\right]$$

where $[w]$ denotes the integer part of the real number w, then $z = f(a,b,c,r)(x,y)$.

The points (x,y,z) of the naive plane $\mathcal{P}(a,b,c,r)$ which verify $ax+by+cz+r=0$ (resp. $ax+by+cz+r=c-1$) are the `lower` (resp. `upper`) `leaning points` of the naive plane.

3 Equivalence Class of a Voxels Set

Let n in $\mathbb{N}^*$ and $V = \{(i_1,j_1),\cdots,(i_n,j_n)\}$ a set of n points of $\mathbb{Z}^2$.

The cluster of voxels $S(a,b,c,r)(x,y)$ of the naive plane $\mathcal{P}(a,b,c,r)$ indexed by V and with origin (x,y) is defined by:

$$S(a,b,c,r)(x,y)$$
$$=$$
$$\bigcup_{q=1}^{n}\{(x+i_q,y+j_q,z+k_q) \mid (i_q,j_q) \in V,\ k_q = f(a,b,c,r)(x+i_q,y+j_q) - z\}$$

where $z = f(a,b,c,r)(x,y)$.

$S(a,b,c,r)(x,y)$ is the part of the naive plane going through the n voxels $(x+i_q,y+j_q,z+k_q)$, $q=1,\cdots,n$.

Let (x_l,y_l,z_l) be a lower leaning point from $\mathcal{P}(a,b,c,r)$. It verifies $ax_l + by_l + cz_l + r = 0$. The point (x,y,z) can be written as (x_l+u, y_l+v, z_l+w) with (u,v,w) in $\mathbb{Z}^3$. As it belongs to the naive plane $\mathcal{P}(a,b,c,r)$, it verifies $0 \leq a(x_l+u)+b(y_l+v)+c(z_l+w)+r < c$. Let us notice r' equal to $au+bv+cw$. For $q=1,\cdots,n$, the point $(x+i_q,y+j_q,z+k_q)$ verifies $0 \leq ai_q+bj_q+ck_q+r' < c$. Consequently, $S(a,b,c,r)(x,y)$ can be written as:

$$S(a,b,c,r)(x,y) = \{(x,y,z)\} \oplus S(a,b,c,r')(0,0)$$

where $A \oplus B$ is the Minkowski sum between the sets A and B.

All real planes for which the discretization by the *object boundary quantization* method on the set $\{(x,y)\} \oplus V$ is the set $S(a,b,c,r)(x,y)$ have to go through the point (x_l,y_l,z_l). Moreover on the point (x,y) the discretization must be the voxel (x,y,z). A first equivalence class of $S(a,b,c,r)(x,y)$ is the set of parameters (α,β) with $0 \leq \alpha \leq \beta \leq 1$ of the real plane $\alpha(x-x_l)+\beta(y-y_l)+z-z_l=0$ verifying $0 \leq \alpha u + \beta v + w < 1$. We notice this equivalence class by the set:

$$\overline{S}(r)(x,y) = \bigcap_{q=1}^{n} \{(\alpha,\beta) \mid 0 \le \alpha \le \beta \le 1,$$
$$(x+i_q, y+j_q, z+k_q) \in S(a,b,c,r)(x,y), -1 < i_q\alpha + j_q\beta + k_q < 1\}$$

For each integer point (i,j), we have $f(a,b,c,r)(x_l+i, y_l+j) = f(a,b,c,r)(x_l,y_l)$ $+f(a,b,c,0)(i,j)$. For $q = 1,\cdots,n$, the integer k_q satisfies $k_q = f(a,b,c,0)(u+i_q, v+j_q) - f(a,b,c,0)(u,v)$. So $\overline{S}(r)(x,y)$ is equal to $\overline{S}(0)(u,v)$.

Let $\mathcal{D}(i,j,k)$ be the line in the parametric space $W = \{(\alpha,\beta),\ 0 \le \alpha \le \beta \le 1\}$ with equation $i\alpha + j\beta + k = 0$. Let $B(i,j,k)$ be the open-band of this space limited by the two parallel lines $\mathcal{D}(i,j,k+1)$ and $\mathcal{D}(i,j,k-1)$.

The equivalence class becomes:

$$\overline{S}(r)(x,y) = W \bigcap \left(\bigcap_{q=1}^{n} B(i_q,j_q,k_q) \right)$$

In a previous work [VC99a], we proved that the voxels set $\mathcal{E}$ centered on the lower leaning point (x_l,y_l,z_l) and defined by:

$$\mathcal{E} = \bigcup_{q=1}^{n} S(a,b,c,r)(x_l - i_q, y_l - j_q)$$

is a complete system. It is representative of the different configurations of voxels sets defined on V which generate the naive planes with normal (a,b,c). The equivalence class $\overline{\mathcal{E}}$ of that set is the intersection of the open bands $B(i,j,k)$ for (i,j) belonging to $V \ominus V$ ($\ominus$ designs the Minkowski difference between two sets) and k verifying $k = f(a,b,c,r)(x_l+i, y_l+j) - f(a,b,c,r)(x_l,y_l)$. The equivalence classes of the different configurations appearing around leaning points split the space W in polygonal areas called *Farey net associated to voxels sets defined on V*.

Example 1. Let A be the voxels set of a naive plane defined on $V = \{(0,0), (1,-1), (2,0), (2,1)\}$ and illustrated in figure 1(a). The equivalence class $\overline{A}$ of that set is the intersection of the four bands $B(0,0,0)$, $B(1,-1,0)$, $B(2,0,-1)$ and $B(2,1,-2)$ (cf. Fig. 1(b)). Each rational point in that area corresponds to the parameters of a naive plane containing that configuration of voxels set. In figure 1(c), we have the Farey net associated to the voxels set defined on V.

Now if we look for real planes $\alpha x + \beta y + z + \gamma = 0$ with $0 \le \alpha \le \beta < 1$ for which the discretization by the object boundary quantization method on the point (x,y) is the voxel (x,y,z) then the parameter γ has to verify $\gamma = \gamma' - (\alpha x + \beta y + z)$ with $0 \le \gamma' < 1$. Moreover if the dicretization on $\{(x,y)\} \oplus V$ is the set $S(a,b,c,r)(x,y)$, the parameters of that planes belong to the set:

$$\overline{S}'(x,y) = \bigcap_{q=1}^{n} \{(\alpha,\beta,\gamma' - (\alpha x + \beta y + z)) \mid 0 \le \alpha \le \beta \le 1, 0 \le \gamma' < 1,$$
$$(x+i_q, y+j_q, z+k_q) \in S(a,b,c,r)(x,y), 0 \le i_q\alpha + j_q\beta + \gamma' + k_q < 1\}$$

This second equivalence class will be used in the recognition algorithm.

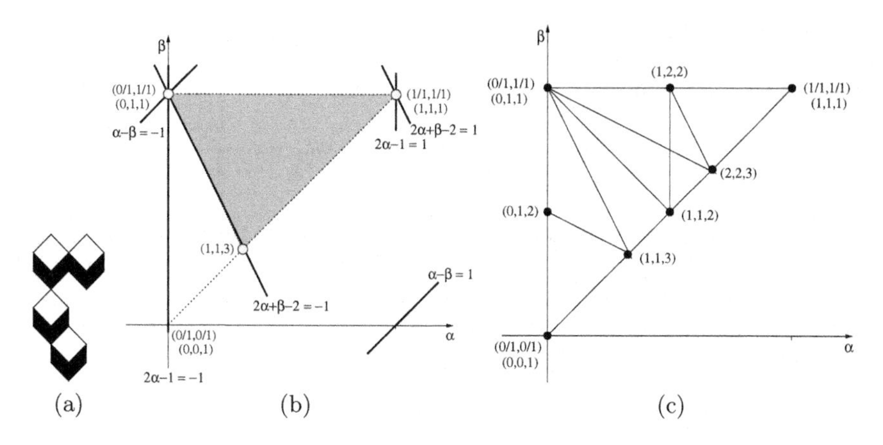

Fig. 1. (a) Set $A = \{(0,0,0),(1,-1,0),(2,0,-1),(2,1,-2)\}$; (b) Equivalence class of A; (c) Farey net associated to voxels set defined on $V = \{(0,0),(1,-1),(2,0),(2,1)\}$

4 Recognition Algorithm

Let $S = \{(x_q, y_q, z_q),\ q = 1, \cdots, n\}$ be a set of n voxels.

We are going to establish an incremental algorithm to identify the parameters of the naive planes going through the n points of S.

The naive planes solutions are the planes $\mathcal{P}(a,b,c,r-(ax_1+by_1+cz_1))$ for which the parameters $\left(\frac{a}{c},\frac{b}{c},\frac{r}{c}\right)$ belong to the set:

$$\overline{S} = \left\{(\alpha,\beta,\gamma) \in [0,1]^2 \times [0,1[\ |\ \forall q \in \{1,\cdots,n\} \quad 0 \le i_q\alpha + j_q\beta + \gamma + k_q < 1\right\}$$

where (i_q, j_q, k_q) is defined as to be the integer points $(x_q - x_1, y_q - y_1, z_q - z_1)$ for $q = 1, \cdots, n$.

Let $\mathcal{B}_q$ in $\mathbb{N}^4$ be the set of vectors (a,b,c,r) such that the projection in the plane $(c = 1)$ are the vertices of the convex hull of the space containing the parameters (α,β,γ) of real planes for which the dicretization on the point (i_p, j_p) is the point (i_p, j_p, k_p) for p varying from 1 to q.

We are going to construct the sets $\mathcal{B}_q$ for $q = 1, \cdots, n$. The following algorithm gives at step q the set $\mathcal{B}_q$ or the empty set if there is no solution. In the first case, the solution with the minimal periodicity can correspond to a vertex of the convex-hull, the median point between the projection of two vertices of $\mathcal{B}_q$ or the median point of the area limited by the projection of the vectors of $\mathcal{B}_q$.

As the discretization of all real planes of the working space goes through the origin (here the origin is taken at point (x_1, y_1, z_1)), the algorithm starts with

$$\mathcal{B}_1 = \{(0,0,1,0),(0,1,1,0),(1,1,1,0),(0,0,1,1),(0,1,1,1),(1,1,1,1)\}$$

composed by the six vectors (a,b,c,r) such that the projection in the plane $(c = 1)$ are the vertices of the convex-hull limiting the solution space of (α,β,γ).

Algorithm at step q, $q \geq 2$:
We introduce the point (i_q, j_q, k_q).
Let L_q and L_q^+ the functions from $\mathbb{N}^4$ to $\mathbb{Z}$ defined by:

$$L_q(a,b,c,r) = ai_q + bj_q + ck_q + r$$
$$L_q^+(a,b,c,r) = L_q(a,b,c,r) - c$$

The naive plane of parameters (a,b,c,r) goes through the voxel (i_q, j_q, k_q) if and only if $0 \leq L_q(a,b,r,c) < c$. Consequently the vectors (a,b,c,r) of $\mathcal{B}_q$ verify $L_p(a,b,c,r) \geq 0$ and $L_p^+(a,b,c,r) \leq 0$ for $p = 1, \cdots, q$.

Initialization: $\mathcal{B}_q = \emptyset$.

For all the vectors V_i in $\mathcal{B}_{q-1}$, $i = 1, \cdots, \#(\mathcal{B}_{q-1})$ do:

1. Process

   Step 1 : If $L_q(V_i) \geq 0$ and $L_q^+(V_i) \leq 0$ then the projection of V_i is still on the convex hull of the domain solution. We insert V_i in $\mathcal{B}_q$. More particularly, if $L_q(V_i) = 0$ (resp. $L_q^+(V_i) = 0$) we can say that the voxel (i_q, j_q, k_q) (resp. $(i_q, j_q, k_q - 1)$) is a lower leaning point of the naive plane of parameters V_i.

   Step 2 : If $L_q(V_i) < 0$ (resp. $L_q^+(V_i) > 0$), we are going to search the point P such as $L_q(P) = 0$ (resp. $L_q^+(P) = 0$). To do that, we use an algorithm based on the notion of median point [Far16,Gra92].

   For each vectors V_j, $j > i$, belonging to $\mathcal{B}_{q-1}$ and verifying $L_q(V_j) > 0$ (resp. $L_q^+(V_j) < 0$) **do**
   $P_1 = V_i$ and $P_2 = V_j$
   While $L_q(P_1) + L_q(P_2) \neq 0$ (resp. $L_q^+(P_1) + L_q^+(P_2) \neq 0$) **do**
   if $L_q(P_1) + L_q(P_2)$ and $L_q(V_i)$ (resp. $L_q^+(P_1) + L_q^+(P_2)$ and $L_q^+(V_i)$) have the same sign **then**
   $P_1 = P_1 + P_2$ and $P_2 = V_j$
   else
   $P_1 = V_i$ and $P_2 = P_1 + P_2$
   End While
   End For
   The solution is given by $P = P_1 + P_2$.

   The point (i_q, j_q, k_q) (resp. $(i_q, j_q, k_q - 1)$) is a lower leaning point of the naive plane of parameters P.
   We insert the point P in $\mathcal{B}_q$.

2. Validation of $\mathcal{B}_q$

   (a) For each vector $V = (a,b,c,r)$ in $\mathcal{B}_q$ we verify if the projection in the plane $(c = 1)$ is a vertex of the convex hull. If V can be written as a combination of vectors of $\mathcal{B}_q$ then the projection of V is on the convex hull but it is not a vertex. So we suppress that point from the list.

(b) If $\#(\mathcal{B}_q) \leq 2$, there is no solution and we suppress all the vertices from the list.
(c) If $\#(\mathcal{B}_q) = 3$, we verify that the points $(a/c, b/c)$ corresponding to the vectors (a, b, c, r) of $\mathcal{B}_q$ are not alined otherwise we suppress the vertices from the list.

Example 2. We are going to illustrate this algorithm on an example. We want to know if the set of voxels in figure 2 belong to a naive plane of the studied 48th part of space. We start with a first point defined as the origin of the voxels

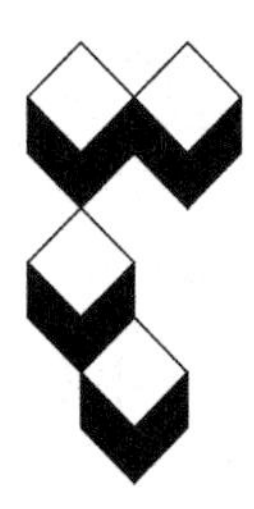

Fig. 2. Set of voxels to recognize

set (cf. Fig 3(a)). The parameters set of naive planes including the origin are the rational points (α, β, γ) contained in the domain limited by the projection $(a/c, b/c, r/c)$ of the vectors (a, b, c, r) of $\mathcal{B}_1$. The set $\mathcal{B}_1$ is composed by the six vectors:

$$\mathcal{B}_1 = \{(0,0,1,0), (0,1,1,0), (1,1,1,0), (0,0,1,1), (0,1,1,1), (1,1,1,1)\}$$

As it was previously mentioned, it is equivalent to say that parameters (α, β) belong to the area limited by the points $(a/c, b/c)$ (cf. Fig 3(b)). We introduce the second point $(1, -1, 0)$ (cf. Fig 4(a)). Let us compute the value $L_2(V)$ and $L_2^+(V)$ on the different vectors V from $\mathcal{B}_1$:

(a,b,c,r)	$L_2(a,b,c,r) = a - b + r$	$L_2^+(a,b,c,r) = a - b + r - c$
$(0,0,1,0)$	0	-1
$(0,1,1,0)$	-1	-2
$(1,1,1,0)$	0	-1
$(0,0,1,1)$	1	0
$(0,1,1,1)$	0	-1
$(1,1,1,1)$	1	0

The vectors $(0,0,1,0)$, $(1,1,1,0)$, $(0,0,1,1)$, $(0,1,1,1)$ and $(1,1,1,1)$ verify the property indiced by step 1 of the algorithm. Ve insert these vectors in $\mathcal{B}_2$. As $L_2(0,1,1,0) < 0$, we apply the algorithm presented in step 2. We introduce the new vectors $(0,1,2,1)$ and $(1,2,2,1)$. But $(0,1,2,1)$ is the sum of vectors

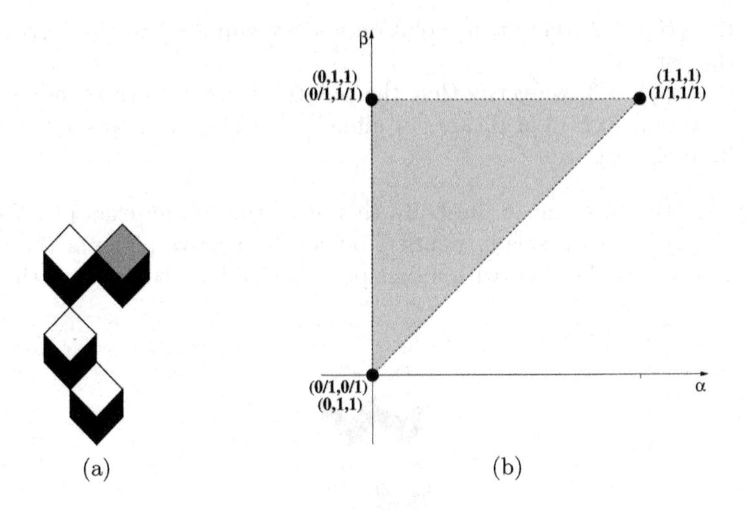

Fig. 3. (a) Set $\{(0,0,0)\}$; (b) Equivalence class

$(0,0,1,0)$ and $(0,1,1,1)$. Similarly, the vector $(1,2,2,1)$ is the sum of vectors $(0,1,1,1)$ and $(1,1,1,0)$. Consequently, the vectors $(0,1,2,1)$ and $(1,2,2,1)$ are not present in $\mathcal{B}_2$.
Finally, we have:

$$\mathcal{B}_2 = \{(0,0,1,0),(1,1,1,0),(0,0,1,1),(0,1,1,1),(1,1,1,1)\}$$

Every naive planes for which the projection $(a/c, b/c)$ of the normal (a,b,c) is contained in the area limited by the points $(a'/c', b'/c')$ with (a',b',c',r') belonging to $\mathcal{B}_2$ are solutions (cf. Fig 4(b)). We introduce the third point $(2,0,-1)$ (cf. Fig 5(a)). We compute the value $L_3(V)$ and $L_3^+(V)$ on the different vectors V of $\mathcal{B}_2$:

(a,b,c,r)	$L_3(a,b,c,r) = 2a - c + r$	$L_3^+(a,b,c,r) = 2a - 2c + r$
$(0,0,1,0)$	-1	-2
$(1,1,1,0)$	1	0
$(0,0,1,1)$	0	-1
$(0,1,1,1)$	0	-1
$(1,1,1,1)$	2	1

The vectors $(1,1,1,0)$, $(0,0,1,1)$ and $(0,1,1,1)$ verify the property indiced by step 1 of the algorithm. We insert these vectors in $\mathcal{B}_3$. As $L_3(0,0,1,0) < 0$ and $L_3^+(1,1,1,1) > 0$, we apply for these vectors the algorithm presented in step 2. We make appear the new vectors $(1,1,2,0)$, $(1,1,3,1)$, $(2,2,3,2)$, $(1,1,2,2)$ and $(1,2,2,2)$. But we have: $(1,1,3,1) = (1,1,2,2) + (1,1,1,0)$ and $(2,2,3,2) = (1,1,2,2) + (1,1,1,0)$. These two vectors are not inserted.
Finally, we have:

$$\mathcal{B}_3 = \{(1,1,1,0),(0,0,1,1),(0,1,1,1),(1,1,2,0),(1,1,2,2),(1,2,2,2)\}$$

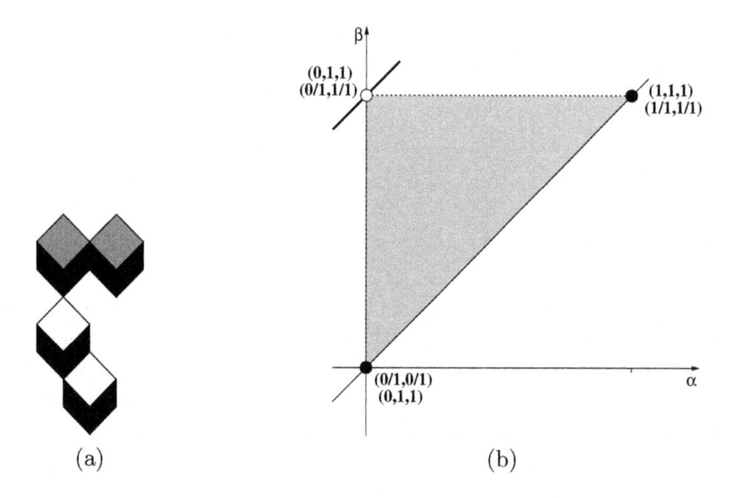

Fig. 4. (a) Set $\{(0,0,0),(1,-1,0)\}$; (b) Equivalence class

Every naive planes for which the projection $(a/c, b/c)$ of the normal (a, b, c) is contained in the area limited by the points $(a'/c', b'/c')$ with (a', b', c', r') belonging to $\mathcal{B}_3$ are solutions (cf. Fig 5(b)). More particularly, the naive planes with normal $(1, 1, 2)$ or $(1, 2, 2)$ contains the initial configuration of 3 voxels. Finally,

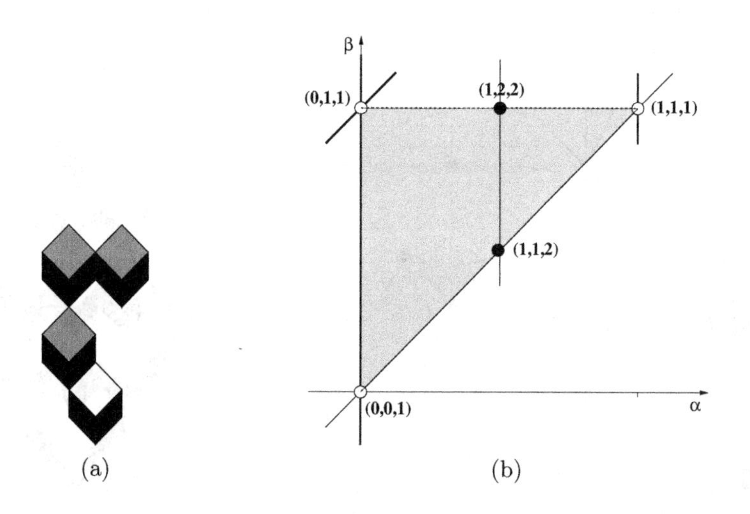

Fig. 5. (a) Set $\{(0,0,0),(1,-1,0),(2,0,-1)\}$; (b) Equivalence class

we introduce the last point $(2, 1, -2)$ (cf. Fig 6(a)). We compute the value $L_4(V)$ and $L_4^+(V)$ on the different vectors V of $\mathcal{B}_3$:

(a,b,c,r)	$L_4(a,b,c,r) = 2a+b-2c+r$	$L_4^+(a,b,c,r) = 2a+b-3c+r$
$(1,1,1,0)$	1	0
$(0,0,1,1)$	-1	-2
$(0,1,1,1)$	0	-1
$(1,1,2,0)$	-1	-3
$(1,1,2,2)$	1	-1
$(1,2,2,2)$	2	0

The vectors $(1,1,1,0)$, $(0,1,1,1)$, $(1,1,2,2)$ and $(1,2,2,2)$ verify the property indiced by step 1 of the algorithm. We insert these vectors in $\mathcal{B}_4$. As the value L_4 is negative for the vectors $(0,0,1,1)$ and $(1,1,2,0)$, we applied for these vectors the algorithm presented in step 2. We make appear the new vectors $(1,1,2,1)$, $(1,1,3,3)$, $(2,2,3,0)$, $(1,2,4,4)$ and $(3,4,6,2)$. But we have: $(1,2,4,4) = (1,1,3,3) + (0,1,1,1)$ and $(3,4,6,2) = (1,1,2,1) + (0,1,1,1) + (2,2,3,0)$.These two vectors are not inserted.
Finally, we have:

$$\mathcal{B}_4 = \{(1,1,1,0),(0,1,1,1),(1,1,2,2),(1,2,2,2),(1,1,2,1),(1,1,3,3),(2,2,3,0)\}$$

Every naive planes for which the projection $(a/c, b/c)$ of the normal (a,b,c) is contained in the area limited by the points $(a'/c', b'/c')$ with (a',b',c',r') belonging to $\mathcal{B}_4$ are solutions (cf. Fig 6(b)). More particularly, the naive planes with normal $(1,1,2)$ or $(1,2,2)$ or $(2,2,3)$ contains the set of 4 voxels. We can verify in figure 6(c) that this configuration of voxels set is contained in the naive plane with normal $(1,1,2)$.

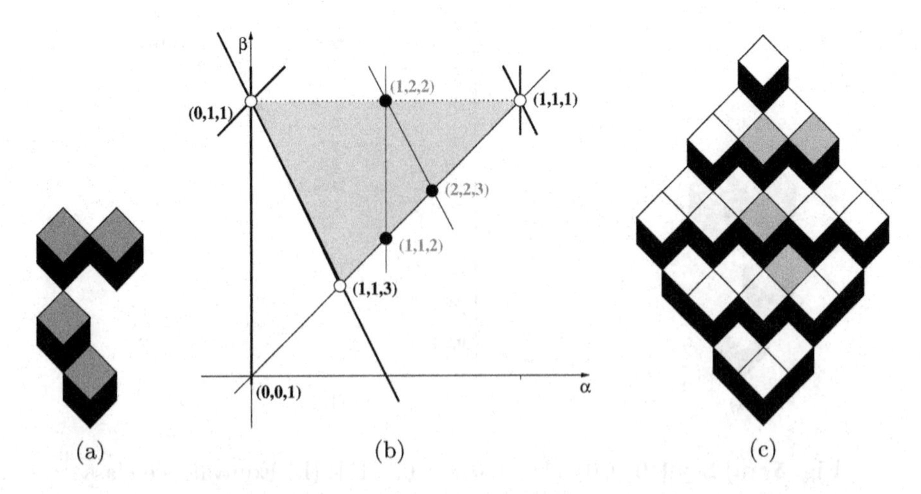

Fig. 6. (a) Set $\{(0,0,0),(1,-1,0),(2,0,-1),(2,1,-2)\}$; (b) Equivalence class; (c) Part of the naive plane with normal $(1,1,2)$l

5 Polyhedrization of a Voxels Object

The polyhedrization of a voxelized object has been studied using mainly approximation approaches [BF94]. Here we propose an algorithm which fully works on the discrete representation of digital naive planes. The polyhedrization can be made in different ways in order to obtain straight or smooth angles between adjacent planes.

The present algorithm is an iterative process based on the recognition algorithm.

Let V_0 be a voxel of the object such that his surfel with normal N belongs to the boundary of the object. We considere this voxel as the origin of the plane. By applying the algorithm of recognition, we verify that V_0 is center of a tricube such as all the surfels of the voxels with normal N belong to the boundary of the object. If V_0 is not center of a tricube we color its surfel in white. Otherwise, we verify that its neighbours are centers of tricubes of the same naive plane. If a neighbour belongs to that plane we mark the surfel as belonging to that plane and we can analyze in a same way its neighbours. Otherwise, this voxel will be treated as the origin of a new plane.

We analyze in the same way all surfels of the object boundary until they are colored in white or marked.

The obtained results are shown on a chanfrein cube and a cube cut by three naive planes (cf. Fig 7).

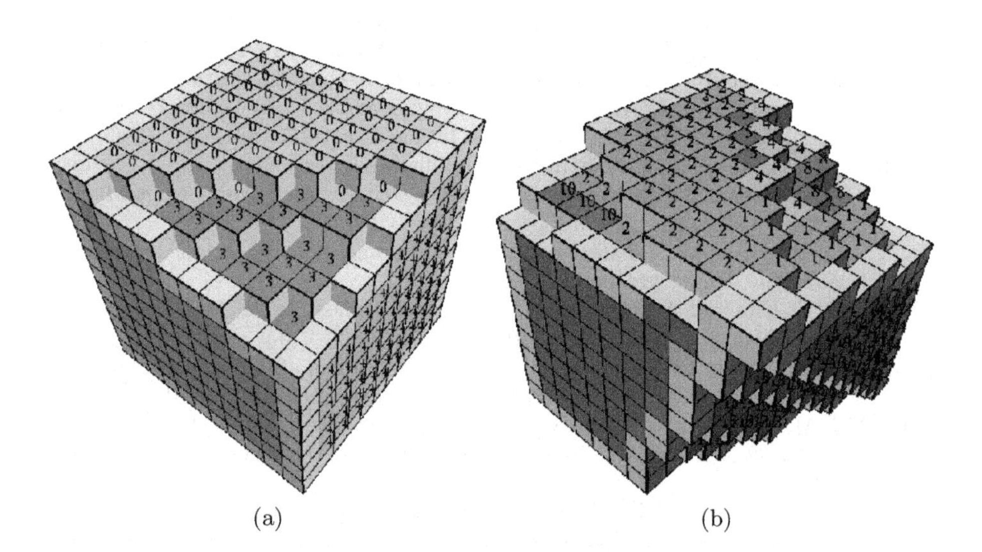

Fig. 7. Chanfrein cube and polytope: Voxels with the same gray level belong to a same naive plane; white voxels belong to the boundaries between adjacent planes

6 Conclusion

A generic algorithm for coplanar voxels recognition has been presented. This algorithm analyses any configuration of voxels set either connected or not connected. It is fully discrete working in the dual space issued from Farey net representation of the normal equation of a digital naive plane. This algorithm has been used for polyhedrization of the boundary of voxelized objects. As perspective, a lot of work has to be achieved precisely in computation of the exact area evaluation of such a boundary using the Pick theorem well known in 2D [Sta86,CM91].

References

BF94. Ph. Borianne and J. Françon. Reversible polyhedrization of discrete volumes. In *DGCI'94*, pages 157–168, Grenoble, France, September 1994. 296, 305

CM91. J. M. Chassery and A. Montanvert. *Géométrie discrète en analyse d'images.* Hermès, Paris, 1991. 306

Deb95. I. Debled. *Etude et reconnaissance des droites et plans discrets.* PhD thesis, Louis Pasteur University, Strasbourg, France, 1995. 296

DRR94. I. Debled-Renesson and J. P. Reveillès. An incremental algorithm for digital plane recognition. In *DGCI'94*, pages 207–222, Grenoble, France, September 1994. 296

Far16. J. Farey. On a curious property of vulgar fractions. *Phil. Mag. J. London*, 47:385–386, 1816. 300

FP99. J. Françon and L. Papier. Polyhedrization of the boundary of a voxel object. In M. Couprie G. Bertrand and L. Perroton, editors, *DGCI'99, Lect. Notes Comput. Sci.*, volume 1568, pages 425–434. Springer, 1999. 296

FST96. J. Françon, J. M. Schramm, and M. Tajine. Recognizing arithmetic straight lines and planes. In A. Montanvert S. Miguet and S. Ubéda, editors, *DGCI'96, Lect. Notes Comput. Sci.*, volume 1176, pages 141–150. Springer, 1996. 296

Gér99. Y. Gérard. Local configurations of digital hyperplanes. In M. Couprie G. Bertrand and L. Perroton, editors, *DGCI'99, Lect. Notes Comput. Sci.*, volume 1568, pages 65–75. Springer, 1999. 296

Gra92. D. J. Grabiner. Farey nets and multidimensional continued fractions. *Monatsh. Math.*, 114(1):35–61, 1992. 300

KR82. C. E. Kim and A. Rosenfeld. Convex digital solids. *IEEE Trans. on Pattern Anal. Machine Intell.*, PAMI-4(6):612–618, 1982. 296

KS91. C. E. Kim and I. Stojmenović. On the recognitioon of digital planes in three dimensional space. *Pattern Recognition Letters*, 12:612–618, 1991. 296

Sch97. J. M. Schramm. Coplanar tricubes. In E. Ahronovitz and C. Fiorio, editors, *DGCI'97, Lect. Notes Comput. Sci.*, volume 1347, pages 87–98. Springer, 1997. 296

ST91. I. Stojmenović and R. Tošić. Digitization schemes and the recognition of digital straight lines, hyperplanes and flats in arbitrary dimensions. *Contemporary Mathematics*, 119:197–212, 1991. 296

Sta86. R. Stanley. *Enumerative Combinatorics*, volume 1. WadsworthandBrooks, 1986. 306

VC97. J. Vittone and J. M. Chassery. Coexistence of tricubes in digital naive plane. In E. Ahronovitz and C. Fiorio, editors, *DGCI'97, Lect. Notes Comput. Sci.*, volume 1347, pages 99–110. Springer, 1997. 296

VC99a. J. Vittone and J. M. Chassery. Digital naive plane understanding. In *Vision Geometry VIII*, volume 3811, pages 22–32. SPIE - The International Society for Optical Engineering, 1999. 296, 298

VC99b. J. Vittone and J. M. Chassery. (n, m)-cubes and farey nets for naive planes understanding. In M. Couprie G. Bertrand and L. Perroton, editors, *DGCI'99, Lect. Notes Comput. Sci.*, volume 1568, pages 76–87. Springer, 1999. 296

Vee94. P. Veelaert. Recognition of digital algebraic surfaces by large collections of inequalities. In *Vision Geometry III*, volume 2356, pages 2–11. SPIE - The International Society for Optical Engineering, 1994. 296

Shape Representation

Topological Encoding of 3D Segmented Images

Yves Bertrand[1], Guillaume Damiand[2], and Christophe Fiorio[2]

[1] IRCOM-SIC, SP2MI
BP 179, F-86960 Futuroscope Cedex
bertrand@sic.sp2mi.univ-poitiers.fr
[2] LIRMM
161 rue Ada, F-34392 Montpellier Cedex 5
{damiand,fiorio}@lirmm.fr

Abstract. In this paper we define the *3d topological map* and give an optimal algorithm which computes it from a segmented image. This data structure encodes totally all the information given by the segmentation. More, it allows to continue segmentation either algorithmically or interactively. We propose an original approach which uses several levels of maps. This allows us to propose a reasonable and implementable solution where other approaches don't allow suitable solutions. Moreover our solution has been implemented and the theoretical results translate very well in practical applications.

1 Introduction

An early stage of an image analysis process is to group related information in the image into regions: it is segmentation. This problem has been widely studied and efficent algorithms which perform segmentation are well known. A related problem concerns the data structure used to represent the segmentation. Indeed, after segmentation, more complex algorithms (such as pattern recognition) need to be run in order to give useful results. These algorithms need a structuration of the data which represents completely all the information given by the segmentation. Moreover this structuration need to be efficiently computable. For example we may have to deal with the topological characteristics of the volume found, as well as to work interactively on the objects modelled.

Numerous data structures have been already proposed. For example, a simple one is to associate each pixel to the region it belongs to [15]. But this solution is not efficient if we have to cut or merge regions. There is also a solution using quadtrees or octrees [8,16] which are based on a recursive decomposition of the image. But this structure is difficult to modify and doesn't allow to check easily adjacency relations between regions. Same drawbacks can be reproached to the pyramidal structures [2,12].

Other structures focus on information related to the regions, for example the region adjacency graph [14], which codes the adjacency relation between regions. But it is not topologically consistent and doesn't relate the inclusion or multiple adjacency, so it doesn't allow a precise analysis.

G. Borgefors, I. Nyström, and G. Sanniti di Baja (Eds.): DGCI 2000, LNCS 1953, pp. 311–324, 2000.
© Springer-Verlag Berlin Heidelberg 2000

Some other solutions try to use several of the formers in order to gather the advantages offered by each of them. But this lead to complex solutions, often non efficiently computable.

This is this lack of adequate solutions which has lead [10] to define the topological graph of frontiers. The interest of such a structure is to be consistent with the topology of the objects it represents. Moreover, this structure can be efficiently computed thanks to a linear algorithm in the number of pixels which proceeds in one scan of the image [9]. One can remark that this solution is close to the one of [3,4,5] which defines the 2d topological map, the difference being in the extraction algorithm.

But these solutions are only defined in dimension 2. The need to work in dimension 3 has lead [1] to propose the border map. This solution can easily be extended to dimension n. Moreover, they have proposed a very simple algorithm which extract this structure from a 3d segmented image. But the border map is not homogeneous. It mixes topological and geometrical information. Furthermore, the optimal algorithm proposed is not practicable in dimension 3 since it requires the definition of more than 4000 differents cases, which is impossible to achieve.

We then propose an evolution of the border map: the 3d topological map. This data structure is completely homogeneous and requires less memory than the border map. We also give an extraction algorithm which uses a limited number of cases and so allows an implementation. The topological and the geometrical information has been completely separated allowing to modify one type of characteristic on one object independently to the other.

In order to present the topological map, we define several maps levels, the border map becoming one of this levels. We start Sect. 2 by a short introduction on combinatorial maps. Then Sect. 3 we come back on the basis of border map. Section 4 we show the notion of topological map and the principle of the extraction algorithm in dimension 2. Then Sect. 5 we extend the precedent results to the dimension 3. At last Sect. 6 we show and analyse some results and we conclude this paper Sect. 7.

2 Combinatorial Maps

Combinatorial maps are a mathematical model of representation of space subdivisions in any dimension. They are consistent with the topology of the space as they also code the adjacency relation between all the subdivisions. For a complete definition see for example [6,7,11].

A combinatorial map of dimension n is called a *n-map*. The *n-map* allows to represent subdivisions of orientable n-manifolds. The notion of maps has been generalized in [13] for the representation of quasi-manifold cellular of dimension n, orientable or not.

In dimension n, a n-map is an $(n+1)$-tuple $M = (D, \beta_1, \beta_2, \ldots, \beta_n)$ such that D is the set of darts, β_1 is a permutation on D and the other β_i are involutions[1]. Each β_i links two cells of dimension i. The β_i are called sews, and a dart b is said *i-sewn* to a dart c if $\beta_i(b) = c$. For each dimension the sets of darts corresponding to the i-cells are a partition of the set of all darts. Maps code the topology of objects, but it is not sufficient in order to represent them completely. Information on how drawing them are necessary: it's the notion of embedding. For a cells subdivision of space, this embedding associates to each i-cell a geometric object of dimension i.

For example in dimension 3, the embedding of a surface consists in relating a point to each vertex, an open curve to each edge and an open surface to each face. An open face is associated to each cell in order to avoid duplication of information. Indeed, border of an i-cell is an $(i-1)$-cell to which an embedding is already associated.

But according to our needs or constraints, we can choose not to embed each cell and compute the missing embedding. For example, in dimension 2, we can embed only the edge with closed curves. In this case, the embedding of the vertices can be retrieved by taken the extremities of one incident edge. It is then necessary to ensure the coherence between topological and geometrical information: if two edges are incident to the same vertex, there embeddings have to have the same extremity.

3 Border Maps

We give, here, a brief recall on the border map defined in [1]. Note that in this paper we use combinatorial map instead of generalized map, this, in order to save memory.

Definition 1. *The* border map *is a combinatorial map which codes the interpixel contours of a segmented image. Each edge of the map corresponds to a segment of the interpixel contours of the image.*

The border map codes the topology of the contours but also the geometry. So it contains all the information given by the segmentation about the objects found.

A level 0 map is the complete map obtained by sewing $n_1 \times n_2$ squares between them. Then level i map is the map obtained from i different sorts of merging. For dimension 2, we get the level 1 map by merging adjacent faces of the same region. Level 2 map is then obtained by merging adjacent edges on the same line incident to a degree 2 vertex[2]. This level 2 map is in fact the border map.

We can see Fig. 1, the border map of an image and also the two intermediate levels of maps. We can see on this example, that after the merging of faces, the map can be disconnected. This problem is solved by adding an inclusion tree to the connected components of the map. This tree can easily be computed by a linear algorithm in the number of darts of the map [1].

[1] An application f is an involution if and only if $f \circ f$ is the identity application.

[2] A vertex is of degree 2 if and only if it's incident to exactly two edges.

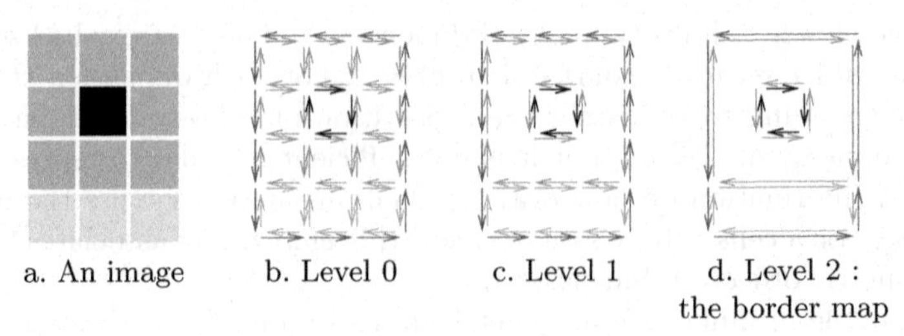

Fig. 1. The border map

For dimension 3, we proceed on a similar way to obtain the border map. But an additionnal level is necessary (see Algo. 1).

Algorithm 1: The border map in dimension 3

Input: An image I of $n_1 \times n_2 \times n_3$ pixels
Output: The border map of I.

1 Build a 3-map C of $n_1 \times n_2 \times n_3$ cubes sewn in between them;
2 Merge all couple of adjacent volumes having the same region;
3 Merge all couple of coplanar faces incident to a degree 2 edge;
4 Merge all couple of aligned edge incident to a degree 2 vertex;

The extraction algorithm in dimension 3 is simply the extension of the naive extraction algorithm of the border map in dimension 2. We now have three sorts of merging:

- Merging of adjacent volumes which consists in deleting the faces in between the volumes and to update the sews in order to have only one volume (see Fig. 2.*b*): we get the Level 1.
- Merging couple of coplanar faces which is equivalent to two merging of faces in dimension 2 (see Fig. 2.*a*): we get the Level 2.
- At last, the merging of edges on the same line which is equivalent to do, one by one, several merging of edges in dimension 2: we then get the Level 3 which is also the border map.

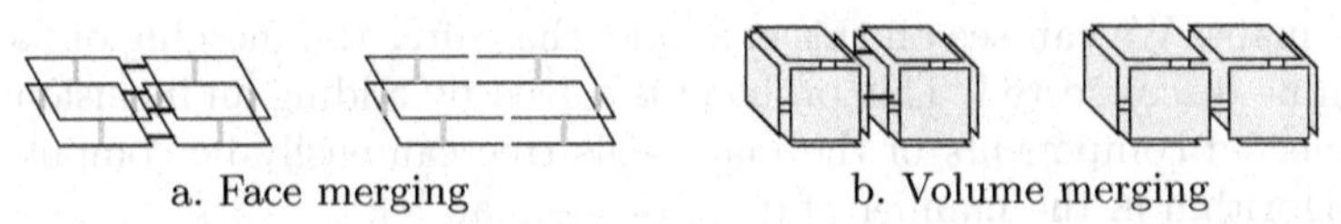

Fig. 2. Face and volume merging in dimension 3

As for dimension 2, an inclusion tree of the regions is needed. But it is not sufficient since we can also have non-connected faces. Indeed when the merging of coplanar faces is done, the border of faces can be disconnected and so the map. In order to solve this new problem we can add an inclusion tree of faces which, for each face, gives the list of faces included.

In order to embed a border map in dimension 2 or 3, it is sufficient to give the coordinates of each vertex of the map. Indeed only elements on the same plane or line are merged, so edges are only straight segments and faces are coplanar to these edges. So the embedding of the vertices allows to compute the embedding of an edge from the embeddings of the two incident vertices, and the embedding of a face from the embedding of all incident edges. This choice of embedding facilitates process and extraction of the border map.

But this leads to an important problem: the same topological information can be coded differently according to the geometry of the image : two topologically equivalent objects could have two different border maps.

The border map mixes topological and geometrical information. The notion of topological map has been introduced to solve this problem. We first present it in dimension 2 in order to help us to understand the idea of the algorithms and the relation in the differents precodes of our different levels. But the goal remains the definition of a 3d topological map and this is presented in Sect. 5.

4 2d Topological Map: Last Level of an Hierarchy of Maps

In the border map, some merging are done accordingly to the embedding of the associated cells. So, two topologically equivalent images, but with different embeddings will have two different border maps.

So we have to do merging only on topological criteria. Practically for dimension 2, in addition to the merging of edges on the same line and incident to degree 2 vertices, we also have to merge edges incident to degree 2 vertices but not on the same line. Edges embeddings are no longer line segment and embed only vertices as for the border map is no longer sufficient.

Definition 2. *The* topological map *is a combinatorial map which codes the interpixel contours of a segmented image. Each edge of the map corresponds to an entire interpixel frontier of the image.*

If we forget for a moment the problem of embedding, computing the topological map is not really a problem, there is just another sort of merging to add after the computation of the border map: the merging of each couple of edges incidents to a degree 2 vertex. This Level 3 map is the topological map. If we take the example of Fig. 1 showing maps of Level 0 to 2, the topological map is then the one of Fig. 3. On this figure, the darts are drawn without taking into account their embedding. In fact the embedding of each edge should be coded by an open curve. An interesting way of doing this is to use a 1-map representing the

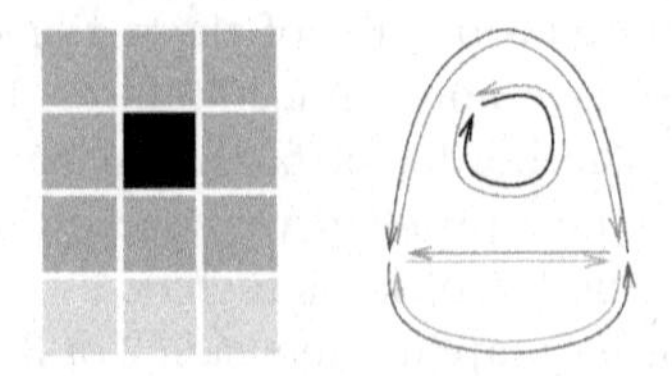

Fig. 3. Topological map without embedding

curve corresponding to the embedding of the edge. We then get an hierarchical structure where i-maps are used to code the embedding of i-cells.

We can see on Fig. 3 that there is exactly one edge for one frontier between two regions, so a frontier is always coded by exactly two darts. This imply that the topological map requires less memory than the border map. This can be very important to deal with 3d images. In the topological map only topological information darts have been kept. So two topologically equivalent images will give the same topological map, except for the embedding.

We can easily deduce from the definition of the topological map a first extraction algorithm: building the different levels, starting from level 0 and then each level is built from the former by doing the supplementary mergings. Indeed deducing one level from the former level is very simple. Moreover, the same method can easily be applied for the definition of the topological map and its associated extraction algorithm for any dimension.

But this algotihm is not optimal since it needs several image scans and it creates many darts which are in the following deleted. In order to improve this algorithm, [1] gives an algorithm based on precodes which computes directly the border map. We have extended this algorithm in order to compute the map of any level directly and in one scan.

4.1 Extraction of the Level 1 Map in One Scan

This algorithm scans the image with a 2×2 window and does the operations given by the configuration of the frontiers (called a precode) in the 2×2 window. So the map is built in one scan and no more than the exact number of darts are created.

The idea of the algorithm is that during a scan from top to bottom and from the left to the right, the face corresponding to the current pixel can only be merged with the one above or the one on the left. Then, We only have four possible cases which corresponds to all the combinations of merging this face with the two others. Figure 4 shows this four face merging precodes.

The algorithm uses two invariants which say that before processing one pixel:

- The map corresponding to the image already scanned is built.
- The dart corresponding to the right border of the last processed pixel is known.

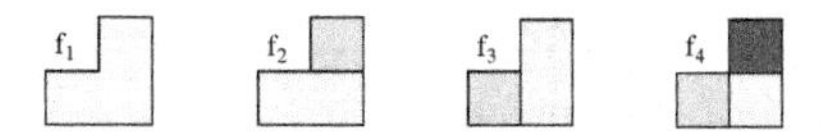

Fig. 4. The four precodes of face fusion

These two invariants insure us that we can proceed by modifying the map. Indeed we have to know how to link the map associated to the current pixel to the global map. So before scanning the image we have to build a map done with n_1 horizontal darts, corresponding to the upper border of the image, and 1 vertical dart to the left, all these darts being sewn by β_1. So the invariant is also verified for the first pixel and for the first line of the image.

During the scanning, we consider that the image is on a cylinder. Thus, when the algorithm treats the last pixel of a line, it can create the dart for the left border of the next line. Special treatments for borders of the image are so avoided. We only have to define the treatment associated for each of the four precode and the Level 1 map extraction algorithm will be defined. In fact this is easily done as we can see on the example of Fig. 5 and on Algo. 2 associated.

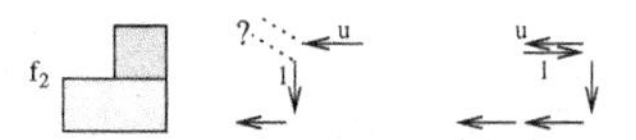

Fig. 5. Precode f_2: map before processing this precode and map to be obtained

---

Algorithm 2: Processing of precode f_2

$t_1 \leftarrow$ a new dart; $t_2 \leftarrow$ a new dart;
$\beta_1 - sew(t_2, \beta_1(l))$; $\beta_1 - sew(t_1, t_2)$;
$\beta_1 - sew(l, t_1)$; $\beta_2 - sew(l, u)$;
$(x, y) \leftarrow$ the embedding of vertex incident to u;
Embedding of vertex incident to $t_2 \leftarrow (x + 1, y + 1)$;
return *t1*

We can see on this example that not all the local configuration is known (the dart sewn to the dart u is not known), but it is not a problem since the operations to be done are independent of the fourth pixel. This precode returns the dart which will be the dart to the left for the next precode. At last we can see that the treatment is very simple, this is the same thing for the other three precodes.

Extraction algorithm of Level 1 map in one scan is very simple and is described in Algo. 3. This algorithm is in fact similar for all maps level, it is just necessary to define the precodes corresponding to the level concerned. So we will now focus on the precodes for the other levels of the map.

Algorithm 3: Extraction algorithm in one scan

Entrée: An image I of $n_1 \times n_2$ pixels
Sortie: Level i map of I.

1 Build the upper border of the map;
2 **for** $j = 1$ $n_2 + 1$ **do**
  for $i = 1$ *to* $n_1 + 1$ **do**
    Process the precode (i, j);

4.2 Extraction of Level 2 Map in One Scan

For level 2 map, we have to add the merging of edges incident to a degree 2 vertex and on the same line. We can easily check that only two precodes are concerned by this merging: these are the precodes of Fig. 6.*a*.

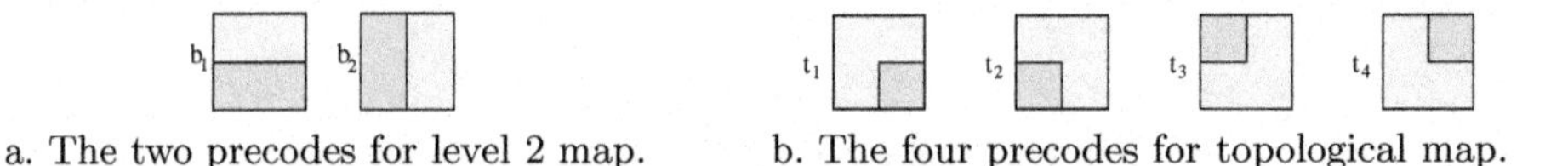

a. The two precodes for level 2 map. b. The four precodes for topological map.

Fig. 6. Precodes for level 2 and 3 maps

Finally we have six precodes for the Level 2 map: the four of the level 1 and these two of Fig. 6.*a*. The extraction algorithm is similar, we just have to look at the six precodes.

4.3 Extraction of the Topological Map in One Scan

For the topological map, in addition to the precedent merging, we have to merge the edges incident to a degree 2 vertex. These edges are necessary not on the same line, else they have been processed by one of the two precedent precodes for level 2 map. It is easy to see that there are only four precodes concerned by this sort of merging. These precodes are shown on Fig. 6.*b*.

In addition, this level requires an embedding for the edges, which can be represented by a 1-map. The general algorithm is also similar to Algo. 3 but ten precodes have to be checked. Nevertheless, this algorithm is still simple and can easily be implemented. It is also optimal and built directly the topological map.

Here we see the advantage of having defined several map levels: we have a relation between the number of precodes and the level of the map we want to extract in one scan. It allows to choose to extract directly Level i map accordingly to the need or development constraints, without paying more in processing time. This approach is particularly interesting in the 3d case where the number of precodes is huge.

5 3d Topological Map

As for dimension 2, we have to do merging according to topological criteria and without taking into account the embedding associated to the cells. We have seen, Sect. 3 that the border map of dimension 3 is the Level 3 map where adjacent volumes of the same region, coplanar faces incident to degree 2 edge and same line edges incident to degree 2 vertex have been merged. The level 4 map is then the map obtained from the Level 3 map where incident to degree 2 edge faces are merged.

As in dimension 2 where an edge embedding was added, this merging of non coplanar faces will require to add a face embedding. We can also use here a map for coding the embedding. Of course this map will be a 2d topological map where their edge embedding will be 1d map.

The topological map of dimension 3 (Level 5 map) is defined algorithmitically by merging in the Level 4 map each couple of edges incident to a degree 2 vertex: all the redundant topological information has been merged.

The simple algorithm which builds the map of Level i consists now to build a complete map (Level 0 map) of $n_1 \times n_2 \times n_3$ voxels sewn, then to build level after level the different maps. The Level $i+1$ map is obtained from the Level i map by doing the adequate additional merging.

[1] have tried to apply the same method as for dimension 2 in order to bring a more efficient algorithm. But they have calculated, thanks to the Stirling numbers, that it could be necessary to define 4140 different precodes in order to be able to compute optimally and in one scan the border map! In fact this number denotes the total number of possible precodes in dimension 3.

But we are interested only in manifolds, so in a first time, we have tried to calculate the number of manifold precodes. But we didn't manage to achieve a formula giving us this result. We finally computed it by the help of a program and found 958 manifold precodes. This number is more encouraging but the implementation of such a number of precodes is not yet easy. In fact the decomposition of the problem in different levels of map will bring us a reasonable solution.

5.1 8 Precodes for Level 1 Map

The scanning of the image starts from up to down, from behind to front of and from left to right. So a voxel can only be merged with the upper voxel, the voxel behind and the voxel to the left. We have then 8 possibility of volume merging and we get the 8 precodes of Fig. 7. More Generally, in dimension n, we get 2^n precodes for the Level 1 map. Indeed, the current element has exactly n neighbours already scan.

The extraction algorithm requires a similar invariant as for the dimension 2 version which ensure us that there exists in the map a face to the left, a face behind and a face with the top of the current voxel. The definition of the operations for each precode is not a lot more difficult that in dimension 2. The difficulties provide from the number of darts in one precodes, and from the fact that this

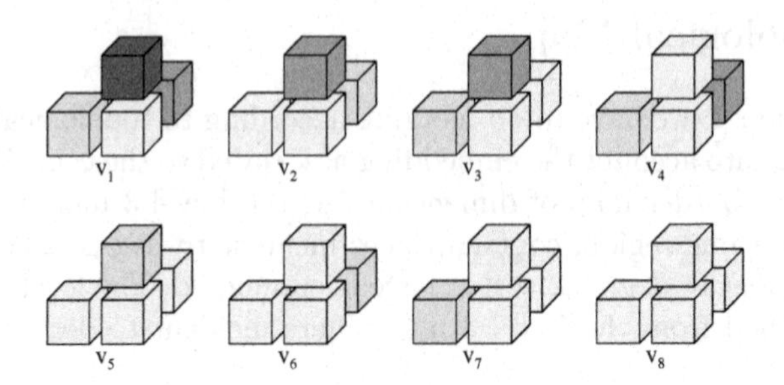

Fig. 7. 8 Precodes of Level 1 Map

is not easy to represent and visualize the object in dimension 3. The general algorithm is similar as the one of dimension 2, it requires only an additional loop for the additional dimension.

5.2 18 Precodes for Level 2 Map

At this level we have to do the merging of coplanar faces incident to degree 2 edge. Intuitively, in order to merge coplanar faces, the two voxels "containing" this faces are to be of the same region, and that the two front of voxels are of another region but the same for the two. We can see Fig. 8.*a* some examples of such precodes.

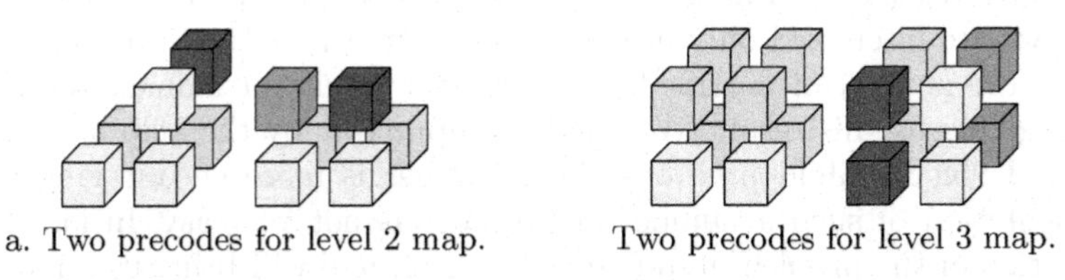

a. Two precodes for level 2 map. Two precodes for level 3 map.

Fig. 8. Examples of precodes for Level 2 and 3 Map

The low number of cases allows us to find exhaustively the 18 precodes for this sort of merging. We, then, immediately get the optimal extraction algorithm in one scan of Level 2 map: it's enough to check for each voxel in which of the 26 (the 18 for the coplanar face merging plus the 8 volume merging precodes) cases (precodes) it is and then to apply to corresponding operations on the map.

5.3 27 Precodes for the Border Map

For this level, we have to look at the merging of edges incident to a degree 2 vertex and on the same line. Intuitively, this merging can only be done if the vertex in the center of the precode is of degree 2. It means that all the voxels are two by two of the same region. Fig. 8.*b* shows some examples of such precodes.

In this case also, the low combinatorial allows us to find easily all the 27 precodes involved. So the total number of precodes to extract directly and in one scan the border map is of 53. We obtain a number which is far away from the 4140 first claimed in [1], and far away from the 958 cases of manifold. This last result is obtained thanks to our new approach which considers different levels of map.

But the topological map, which is our final objective, requires the implementation of Level 4 and 5.

5.4 Level 4 Map and Topological Map

Until the Level 3, only on the same line or on the same plane elements are merged. So the number of precodes remains limited and we have been able to determine all the precodes. But from level 4, things become more complicated and finding all the precodes is a lot more difficult and a long work. So we have just tried to determine the number of precodes for each level. This was done by the help of a program which, for each precode among the 958, builds the corresponding map, then does the merging required by the Level i considered. It is then sufficient to group the precodes with the same merging and count the number of equivalence classes.

The numbers of 8, 18 and 27 precodes for the first three levels of map have been confirmed. Then we get 98 precodes for the Level 4 map and 216 precodes more for the topological map. We recall that these numbers are the numbers of precodes to add to the precedent level. So the total number for an algorithm which would compute the topological map directly and in one scan would require the definition of 367 precodes! Our approach allows us to divide by half the total number of precodes required. But even if we are still far away from the 958 or better from the 4140 initially claimed, there are still too many precodes for a human.

But an intermediate solution exists. It consists in computing directly the border map by the way of the 53 precodes, and then to proceed as for the general algorithm and compute Level 4 and Level 5 (the topological map) by doing the additional merging on the border map. This solution is the one we have implemented and is an interesting compromise between the running time, the development time and the memory space required.

6 Experiments and Analysis

In dimension 2, we have implemented the extraction of maps for each level with the precode based algorithm. Results about memory needed can be seen on Table 1. The running time doesn't change a lot for the different levels and is about 0.20 seconds for a 512×512 on a 600 MHz PIII with 256 Mb of memory, included the computation of the inclusion tree.

In dimension 3, we can see on Table 2, for each level, the differences between the number of darts and the memory space. The method implemented is the

Table 1. Comparison of the different levels in dimension 2

	Level 1	Level 2	Level 3
Lena, 256 region	58 724 2d darts - 1 650 004 Bytes	31 648 2d darts - 891 876 Bytes	1 160 2d darts 15 244 1d darts 350 948 Bytes
Lena, 14218 regions	220 900 darts 2d - 6 478 564 Bytes	139 848 darts 2d - 4 209 108 Bytes	67 112 darts 2d 36 368 darts 1d 3 095 524Bytes

Table 2. Comparison of the different levels in dimension 3

	Level 1	Level 2	Level 3	Level 5	Unit
Heart,	245 296	33 936	23 900	1 408	3d darts
$64 \times 64 \times 64$	-	-	-	16 435	2d darts
12 regions	-	-	-	225	1d darts
	7 998 920	1 235 400	854 032	598 068	Bytes
Skeleton,	10 109 680	1 354 412	922 072	44 300	3d darts
$512 \times 512 \times 177$	-	-	-	632 396	2d darts
4 regions	-	-	-	14 828	1d darts
	329 647 688	49 479 112	33 050 192	22 905 824	Bytes
Legs,	17 869 952	5 152 672	3 282 702	710 518	3d darts
$512 \times 512 \times 55$	-	-	-	1 667 613	2d darts
13926 regions	-	-	-	184 554	1d darts
	584 564 284	177 611 324	117 772 284	82 155 560	Bytes

one described at the end of Sect. 5.The computation of the Level 5 map takes about 3 seconds for the heart (Fig. 9.a), 470 seconds for the skeleton (Fig. 9.b) and about 200 seconds for the legs. The running time of the other levels is a little more important, due to much more dart creations and so much more memory allocation calls.

We can see on Table 2 that the gain in memory is very significant. These first results confirm that the 3d topological map is well suited for applications on 3d images. Indeed in dimension 3 the big amount of memory required could limit the size of the images processed. For example, for the legs where all regions are considered, only 55 slices have been processed and the level 1 map requires too much memory space to be used in a basic computer. In the contrary the topological map only needs about $80Mb$ which is reasonable and can be used on a classical computer.

7 Conclusion

In this paper we have presented a data structure and an algorithm which reconstruct a 3d segmented image. This structure, the 3d *topological map* represents

a. Heart b. Part of a skeleton

Fig. 9. Topological maps extracted from a segmented image

totally the objects found by the segmentation. It allows to continue the segmentation either algorithmically or interactively, since based on combinatorial maps used in geometric modeling. This structure can be used to compute geometric properties as well as topological properties of the objects represented.

The approach proposed which uses different levels of simplification has allowed to define on a simple manner the structure and to propose a simple, linear (but not optimal) extraction algorithm. In order to achieve an optimal solution, we have introduced a new algorithm based on precodes. This algorithm builds the different levels of map in one scan of the image and is then very efficient. Nevertheless, it remains very simple since it proceeds by checking a configuration, called a precode, and realizes the operations associated to the met configuration.

In order to implement this algorithm, it is then sufficient to define these operations. A first study gave 4140 possibles cases, but we have proven here that there are only 958 precodes suitable for the non-manifold case which we are interested in, and moreover that our approach gives the possibility to study only one half, i.e. 367 precodes, for the last level of map. But the intermediate levels require only 8, 18, 27 and 53 precodes and then can easy be implemented.

This level approach allows to choose an hybrid solution in order to avoid the 367 precodes: compute directly an intermediate level of map and then build the last levels from this one. This is the solution we have implemented and tested. The border map have been computed directly with the 53 precodes and then, level 4 and the topological map has been built from the border map by doing appropriate merging of elements of the map.

At last our experiments prove that the topological map requires less memory than the other map levels, which is an important point for 3d images. Moreover in order to apply segmentation algorithm on a structure it is important that it doesn't code redundant information. In fact, adjacency relation, that is topological information, is the key point of such algorithms. Our structure is then well-adapted, not only to reconstruction and modeling, but also to image analysis .

Acknowledgements

We wish to thank Lionel Gral for useful comments and careful reading of this paper.

References

1. Y. Bertrand, C. Fiorio, and Y. Pennaneach. Border map : a topological representation for nd image analysis. In G. Bertrand, M. Crouprie, and L. Perroton, editors, *Discrete Geometry for Computer Imagery*, number 1568 in Lecture Notes in Computer Science, pages 242–257, Marne-la-vallée, France, March 1999. 312, 313, 316, 319, 321
2. M. Bister, J. Cornelius, and A. Rosenfeld. A critical view of pyramid segmentation algorithms. *Pattern recognition Letters*, 9(11):605–617, 1990. 311
3. J. P. Braquelaire and L. Brun. Image segmentation with topological maps and inter-pixel representation. *Journal of Visual Communication and Image Representation*, 1(9):62–79, 1998. 312
4. J. P. Braquelaire and J. P. Domenger. Representation of segmented images with discrete geometric maps. *Image and Vision Computing*, pages 715–735, 1999. 312
5. L. Brun. *Segmentation d'images couleur à base Topologique.* Thèse de doctorat, Université Bordeaux I, décembre 1996. 312
6. R. Cori. *Un code pourles graphes planaires et ses applications.* PhD thesis, Université Paris VII, 1973. 312
7. R. Cori. Un code pour les graphes planaires et ses applications. In *Astérisque*, volume 27. Soc. Math. de France, Paris, France, 1975. 312
8. R. C. Dyer, A. Rosenfeld, and H. Samet. Region representation : boundary codes for quadtrees. In *ACM 23*, pages 171–179, 1980. 311
9. C. Fiorio. *Approche interpixel en analyse d'images : une topologie et des algorithmes de segmentation.* PhD thesis, Université Montpellier II, 1995. 312
10. C. Fiorio. A topologically consistent representation for image analysis: the frontiers topological graph. In S. Miguet, A. Montanvert, and S. Ubeda, editors, *6th International Workshop, DGCI'96*, number 1176 in Lecture Notes in Computer Sciences, pages 151–162, Lyon, France, November 1996. 312
11. A. Jacques. Constellations et graphes topologiques. In *Combinatorial Theory and Applications*, pages 657–673, Budapest, 1970. 312
12. W. G. Kropatsch. Building irregular pyramids by dual graph contraction. In *IEE Proceedings : Vision, Image and Signal Processing*, volume 142(6), pages 366–374, 1995. 311
13. P. Lienhardt. Subdivision of n-dimensional spaces and n-dimensional generalized maps. In 5th *Annual ACM Symposium on Computational Geometry*, pages 228–236, Saarbrücken, Germany, 1989. 312
14. A. Rosenfeld. Adjacency in digital pictures. *Inform. and Control*, 26:24–33, 1974. 311
15. A. Rosenfeld and A. C. Kak. *Digital Picture Processing*, volume 2. Academic Press, New York, 1982. 311
16. H. Samet. Region representation : quadtrees from boundary codes. In *ACM 23*, pages 163–170, 1980. 311

Some Weighted Distance Transforms in Four Dimensions

Gunilla Borgefors

Centre for Image Analysis, Swedish University of Agricultural Sciences
Lägerhyddvägen 17, SE-752 37 Uppsala, Sweden
gunilla@cb.uu.se

Abstract. In a digital distance transform, each picture element in the shape (background) has a value measuring the distance to the background (shape). In a weighted distance transform, the distance between two points is defined by path consisting of a number of steps between neighbouring picture elements, where each type of possible step is given a length-value, or a weight. In 4D, using $3\times3\times3\times3$ neighbourhoods, there are four different weights. In this paper, optimal real and integer weights are computed for one type of 4D weighted distance transforms. The most useful integer transform is probably $\langle 3,4,5,6\rangle$, but there are a number of other ones listed. Two integer distance transform are illustrated by their associated balls.

1 Introduction

Results regarding the 4D digital space, $\mathbf{Z}^4$, are being found more and more in literature, both regarding theory and emerging applications. Examples where 4D is used are: when processing 3D grey-level images, just as some 2D problems are solved using temporary 3D images; for volume data sequences, as ultrasound volume images of a beating heart; or for the discretisation of the parameter space of a robot or robot arm. Some examples are [6,7,8,9].

In a Distance Transform (denoted DT), each element in the shape (background) has a value measuring the distance to the background (shape). DTs have proven to be an excellent tool for many different image operations. Therefore, distance transforms (DT) in 4D are moving from being a theoretical curiosity, [1], to becoming a useful tool.

The basic idea, utilised for most DTs, is to approximate the global Euclidean distance by propagation of local distances, i.e., distances between neighbouring pixels. This idea was probably first presented by Rosenfeld and Pfaltz in 1966, [10]. This approach is motivated by ease of computation. In sequential computation only a small area of the image is available at the same time. In massively parallel computation (if such an approach still exists) each pixel has access only to its immediate neighbours.

Weighted or chamfer distance transforms, denoted WDT. The local steps between neighbouring pixels are given different weights. In 2D, the most common WDTs are $\langle 2,3\rangle$ and $\langle 3,4\rangle$, were the first number is the local distance between

G. Borgefors, I. Nyström, and G. Sanniti di Baja (Eds.): DGCI 2000, LNCS 1953, pp. 325–336, 2000.
© Springer-Verlag Berlin Heidelberg 2000

edge-neighbours and the second number is the local distance between point-neighbours, [2]. Weighted DTs can be computed in arbitrary dimensions by two raster scans through the image, where, at each point, the image values in a small neighbourhood of the point are used to compute the new point value [1,2]. A first, not very good, effort of discovering WDTs in higher dimensions is found in [1].

Important theoretical results on general DTs in higher dimensions have been published [8] a few years ago. In this paper, necessary conditions for an nD DT to be a metric are presented. In [3], WDTs in 3D, fulfilling these criteria were exhaustively investigated for $3 \times 3 \times 3$ neighbourhoods. There proved to be two types of such DTs, the "obvious" one and one less intuitive. In 4D, the situation is even more complex, and there are at least eight different cases of WDTs. In this paper, the most "natural" case will be investigated, and optimal real and integer weights for this case will be presented.

In Section 2, the geometry and general equations are developed. In Section 3, optimal weights are computed, where optimality is defined as minimising the maximum difference from the Euclidean distance in an $M \times M \times M \times M$ image. In Section 4, the optimal real and integer WDTs are listed and two integer DTs are illustrated by their associated balls.

2 Geometry and Equations

Denote a digital shape on a hypercubic grid F, and the complement of the shape $\overline{F}$, where the sets F and $\overline{F}$ are not necessarily connected. A distance transformation converts the binary image to a distance image, or Distance Transform (DT). In the DT each *hyxel* (hypervolume picture element) has a value measuring the distance to the closest hyxel in $\overline{F}$.

A good underlying concept for all digital distances is the one proposed by Yamashita and Ibaraki, [11]:

Definition 1 *The distance between two points* $\boldsymbol{x}$ *and* $\boldsymbol{y}$ *is the length of the shortest path connecting* $\boldsymbol{x}$ *to* $\boldsymbol{y}$ *in an appropriate graph.*

They proved that any distance is definable in the above manner, by choosing an appropriate neighbourhood relation and an appropriate definition of path length.

In 4D hypercubic space, each hyxel has four types of neighbours: 8 volume neighbours, 24 face neighbours, 32 edge neighbours, and 16 point neighbours. A path between two hyxels in the 4D image can thus include steps in 80 directions, if only steps between immediate neighbours are allowed.

The DT(i, j, k, l) of a hyxel in F is the minimum length of a path connecting (i, j, k, l) to any hyxel in $\overline{F}$, where steps between volume neighbours have length a, steps between face neighbours have length b, steps between edge neighbours have length c, and steps between point-neighbours have length d, and no other steps are allowed. Due to symmetry, it is enough to consider distances from the origin to a hyxel (x, y, z, w), where $0 \leq w \leq z \leq y \leq x \leq M$ and M

is the maximal dimension of the image when computing optimal a, b, c, and d. The distance to be minimised then becomes $D(x, y, z, w)$. A $3 \times 3 \times 3 \times 3$ WDT will be denoted $\langle a, b, c, d \rangle$.

As the length of any minimal path is defined only by the numbers of steps of different types in it, the *order* of the steps is arbitrary. Therefore, we can always assume a minimal path where the steps are arranged in a number of straight line segments, equal to the number of different directions of steps used.

Not all combinations of local distances a, b, c, d result in useful distance transforms. The DT should have the following property.

Definition 2 *Consider two picture elements that can be connected by a straight line, i.e., by using only one type and direction of local step. If that line defines the distance between the pixels, i.e., is a minimal path, then the resulting DT is* **semi-regular**. *If there are no other minimal paths, then the DT is* **regular**.

From [8] we have the following result.

Theorem 1 *A distance transform in $\mathbf{Z}^n$ that is a metric is semi-regular. A semi-regular distance transform in $\mathbf{Z}^2$ is a metric.*

Thus, all suggested DTs should be semi-regular as this is a necessary but, in higher dimensions, not sufficient condition for being metrics.

As there are four types of steps, there are four types of straight paths possible in the hypercubic grid. To find the conditions for 4D regularity we must investigate all the ways these four straight paths can be approximated by paths using other steps and find the conditions for the straight path being shortest. The result is that a 4D WDT is semi-regular if the following inequalities hold (see [3] for a complete description of the method of computation):

$$a \leq b, \quad b \leq 2a, \quad b \leq c, \quad c \leq \frac{3}{2}b, \quad c \leq d, \quad d \leq \frac{4}{3}c. \tag{1}$$

These inequalities define a hyperpolyhedron in a, b, c, d-parameter space. A cut through this polyhedron at $d = 2$ and with a as a scale factor is shown in Fig. 1.

The conditions in (1) may seem restrictive, but they are *not* sufficient to determine unique expressions for the WDTs. If we compute the distances from the origin, choosing the shortest paths and assuming that the local distances have the properties in (1), we discover (at least) eight different, equally valid cases. For example, the hyxel $(2, 2, 1, 1)$ can be reached either as $(1, 1, 1, 1) + (1, 1, 0, 0) = d + b$ or as $(1, 1, 1, 0) + (1, 1, 0, 1) = 2c$. The inequalities in (1) do not determine which is the shorter path. In 3D there is the same phenomenon, but there are only two cases [3].

In Fig. 1, the eight cases discovered are marked by thin lines. In each of the Cases, expressions could be found for the distance transforms, and the local distances could be optimised. However, the area marked "Case I" is the most interesting and easiest to handle, as there is only one expression valid for all hyxels, and that expression, moreover, is the one we would expect, as it is an extensions of the equation in 2D and in Case I 3D. In the triangular hypercone

$$b \leq 2a, \quad c \leq d, \quad a + c \leq 2b, \quad b + d \leq 2c. \tag{2}$$

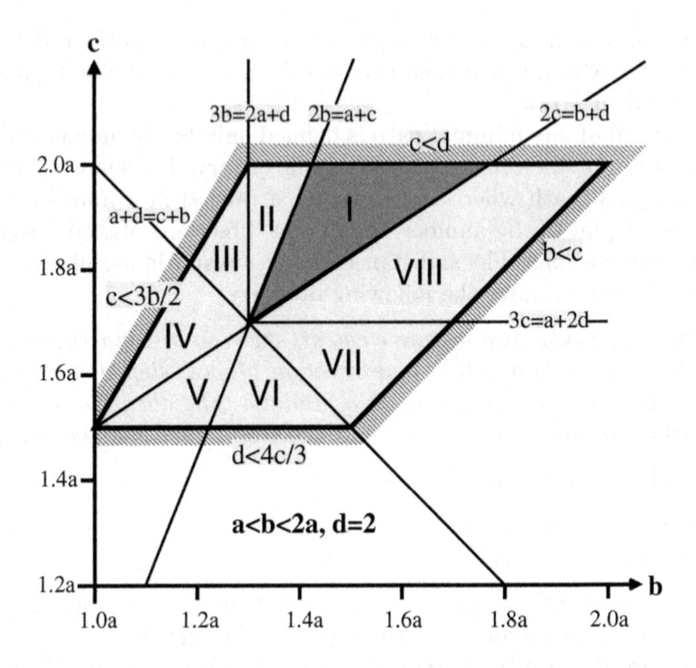

Fig. 1. The hyperpolyhedron in a, b, c, d-space that results in semi-regular 4D weighted distances (thick lines). The thin lines separates different cases. The grey area is Case I, which will be covered in this paper

the distance between the origin and (x, y, z, w) is

$$\text{Case I: } D = wd + (z - w)c + (y - z)b + (x - y)a, \text{ for } 0 \leq w \leq z \leq y \leq x \quad (3)$$

This equation, without the limitation of parameter space (2), is found in [1].

The 4D DT analogous to the chessboard DT in 2D is $\langle 1, 1, 1, 1 \rangle$, or D^{80}, where the distance to all 80 neighbours is set to 1. The 4D DT analogous to the city block DT in 2D is $\langle 1, 2, 3, 4 \rangle$, or D^8 (not to be confused with chessboard DT in 2D), where the distance to the eight volume neighbours is set to 1. Both D^8 and D^{80} are semi-regular (but not regular) according to the inequalities in (1). The equations for the distances can be expressed as in (3), still with $0 \leq w \leq z \leq y \leq x$:

$$D^8 = x + y + z + w, \quad (4)$$
$$D^{80} = x. \quad (5)$$

3 Optimality Computations

In this Section, the optimal local distances for the Case I 4D WDT will be computed. Optimality is defined as minimising the maximal difference between the WDT and the Euclidean distance in an image of size $M \times M \times M \times M$. The choice of optimality criterion is somewhat arbitrary. However, this one has the advantage that is does not depend on any non-digital structure, such as an imbedded Euclidean sphere, which would be necessary to, e.g., minimise the average error.

The maximum of a particular type of function will often have to be computed. The following Lemma is used.

Lemma 2 *The function* $f(\xi) = \alpha\xi + \beta - \lambda\sqrt{\gamma + k\xi^2}$, *where* $|\alpha| < \sqrt{k}|\lambda|$ *and* $|\gamma| > 1$ *has the maximum value*

$$f_{\max} = \beta - \sqrt{\lambda^2 - \frac{\alpha^2}{k}} \cdot \sqrt{\gamma} \quad \textit{for} \quad \xi = \frac{\alpha\sqrt{\gamma}}{\sqrt{k^2\lambda^2 - k\alpha^2}}$$

Proof: The extremal value is found by setting the derivative of $f'(\xi)$ to zero, solving for ξ and simplifying the resulting expressions. □

The difference between the computed distance, see (3), and the Euclidean distance is

$$E(x, y, z, w) = (d - c)w + (c - b)z + (b - a)y + xa - \sqrt{x^2 + y^2 + z^2 + w^2}, \quad (6)$$

where $0 \leq w \leq z \leq y \leq x \leq M$. This difference is to be minimised in an $M \times M \times M \times M$ image. Put the origin in a corner of the image. We can then assume that the maximum difference, denoted maxdiff, occurs for $x = M$. As $0 \leq w \leq z$, the maximum of E(w) occurs for $w = 0$, $\partial/\partial w E(y, z, w) = 0$, or $w = z$. The difference in these three cases are found by simple insertion or by using Lemma 2 with $\xi = w, \alpha = (d - c), \beta = (c - b)z + (b - a)y + Ma, \lambda = k = 1$, and $\gamma = M^2 + y^2 + z^2$. We get:

$$\begin{aligned}
E_1(y, z) &= (c - b)z + (b - a)y + aM - \sqrt{M^2 + y^2 + z^2} \quad &&\text{for } (M, y, z, 0),\\
E_2(y, z) &= (c - b)z + (b - a)y + aM - \sqrt{1 - (d - c)^2}\sqrt{M^2 + y^2 + z^2} \\
& &&\text{for } (M, y, z, w_{\max}),\\
E_3(y, z) &= (d - b)z + (b - a)y + aM - \sqrt{M^2 + y^2 + 2z^2} &&\text{for } (M, y, z, z).
\end{aligned}$$

For each of these three expressions the maximum can occur for $z = 0$, $\partial/\partial z E(y, z) = 0$, or $z = y$, as $0 \leq z \leq y$. We get the following nine difference expressions, using insertion and Lemma 2.

$$\begin{aligned}
E_{11}(y) &= (b-a)y + aM - \sqrt{M^2+y^2} && \text{for } (M,y,0,0),\\
E_{12}(y) &= (b-a)y + aM - \sqrt{1-(c-b)^2}\sqrt{M^2+y^2} && \text{for } (M,y,z_{\max},0),\\
E_{13}(y) &= (c-a)y + aM - \sqrt{M^2+2y^2} && \text{for } (M,y,y,0),\\
E_{21}(y) &= \emptyset, \quad \text{as } 0 = z < w = w_{\max}, &&\\
E_{22}(y) &= (b-a)y + aM - \sqrt{1-(d-c)^2-(c-b)^2}\sqrt{M^2+y^2} &&\\
& && \text{for } (M,y,z_{\max},w_{\max}).\\
E_{23}(y) &= (c-a)y + aM - \sqrt{1-(d-c)^2}\sqrt{M^2+2y^2} && \text{for } (M,y,y,w_{\max}),\\
E_{31}(y) &\equiv E_{11}(y), &&\\
E_{32}(y) &= (b-a)y + aM - \sqrt{1-\tfrac{1}{2}(d-b)^2}\sqrt{M^2+y^2} && \text{for } (M,y,z_{\max},z_{\max}),\\
E_{33}(y) &= (d-a)y + aM - \sqrt{M^2+3y^2} && \text{for } (M,y,y,y).
\end{aligned} \tag{7}$$

For each of these seven expressions the maximum can occur for $y = 0$, $\partial/\partial y E(y) = 0$, or $y = M$, as $0 \le y \le x = M$. We get the following 21 expressions, again using insertion and Lemma 2.

$$\begin{aligned}
E_{111} &= (a-1)M && \text{for } (M,0,0,0),\\
E_{112} &= (a-\sqrt{1-(b-a)^2})M && \text{for } (M,y_{\max},0,0),\\
E_{113} &= (b-\sqrt{2})M && \text{for } (M,M,0,0),\\
E_{121} &= \emptyset, \quad \text{as } 0 = y < z = z_{\max}, &&\\
E_{122} &= (a-\sqrt{1-(c-b)^2-(b-a)^2})M && \text{for } (M,y_{\max},z_{\max},0),\\
E_{123} &= (b-\sqrt{2}\sqrt{1-(c-b)^2})M && \text{for } (M,M,z_{\max},0),\\
E_{131} &\equiv E_{111}, &&\\
E_{132} &= (a\sqrt{1-\tfrac{1}{2}(c-a)^2})M && \text{for } (M,y_{\max},y_{\max},0),\\
E_{133} &= (c-\sqrt{3})M && \text{for } (M,M,M,0),\\
E_{221} &= \emptyset, \quad \text{as } 0 = y < z = z_{\max} &&\\
E_{222} &= (a-\sqrt{1-(d-c)^2-(c-b)^2-(b-a)^2})M &&\\
& && \text{for } (M,y_{\max},z_{\max},w_{\max}),\\
E_{223} &= (b-\sqrt{2}\sqrt{1-(d-c)^2-(c-b)^2})M && \text{for } (M,M,z_{\max},w_{\max}),\\
E_{231} &= \emptyset, \quad \text{as } 0 = y = z < w = w_{\max} &&\\
E_{232} &= (a-\sqrt{1-(d-c)^2-\tfrac{1}{2}(c-a)^2})M && \text{for } (M,y_{\max},y_{\max},w_{\max}),\\
E_{233} &= (c-\sqrt{3}\sqrt{1-(d-c)^2})M && \text{for } (M,M,M,w_{\max}),\\
E_{321} &= \emptyset, \quad \text{as } 0 = y < z = z_{\max}, &&\\
E_{322} &= a-\sqrt{1-\tfrac{1}{2}(d-b)^2-(b-a)^2})M && \text{for } (M,y_{\max},z_{\max},z_{\max}),\\
E_{323} &= (b-\sqrt{2}\sqrt{1-\tfrac{1}{2}(d-b)^2})M && \text{for } (M,M,z_{\max},z_{\max}),
\end{aligned}$$

$$\begin{aligned}
E_{331} &\equiv E_{111}, \\
E_{332} &= (a - 2\sqrt{1-(d-a)^2})M \quad \text{for } (M, y_{\max}, y_{\max}, y_{\max}), \\
E_{333} &= (d-2)M \quad \text{for } (M, M, M, M).
\end{aligned}$$

Thus 15 difference expressions E_{ijk} remain. The maximum of these 15 expressions should now be minimised by varying a, b, c, and d. Numerical experimentation show that $\max(E_{ijk})$ is minimal for $E_{222} = -E_{111} = -E_{113} = -E_{133} = -E_{333}$. Solving these equations yields

$$\begin{aligned}
a_{opt} &= 1 - \mathcal{R} && \approx 0.9048, \\
b_{opt} &= \sqrt{2} - \mathcal{R} && \approx 1.3191, \\
c_{opt} &= \sqrt{3} - \mathcal{R} && \approx 1.6369, \\
d_{opt} &= 2 - \mathcal{R} && \approx 1.9048, \\
\text{with} \quad \text{maxdiff} &= \mathcal{R}M && \approx 0.0951M,
\end{aligned} \tag{8}$$

$$\text{where} \quad \mathcal{R} = \frac{1}{2}(1 - \sqrt{2}\sqrt{\sqrt{6} + 2\sqrt{3} + \sqrt{2} - 7}).$$

The optimal solutions for $a \equiv 1$ are needed when computing integer DTs, as then a becomes a scale factor. In this case, we solve $E^*_{222} = -E^*_{113} = -E^*_{133} = -E^*_{333}$, where the star denotes that $a = 1$ has been substituted in the expressions. The solutions are

$$\begin{aligned}
a^*_{opt} & && = 1, \\
b^*_{opt} &= \sqrt{2} - \mathcal{S} && \approx 1.2796, \\
c^*_{opt} &= \sqrt{3} - \mathcal{S} && \approx 1.5975, \\
d^*_{opt} &= 2 - \mathcal{S} && \approx 1.8654, \\
\text{with} \quad \text{maxdiff}^* &= \mathcal{S}M && \approx 0.1346M,
\end{aligned} \tag{9}$$

$$\text{where} \quad \mathcal{S} = \frac{1}{2}\sqrt{2} - \sqrt{\sqrt{6} + 2\sqrt{3} + \sqrt{2} - 7}.$$

In both situations, free a and $a \equiv 1$, it is easy to check that the optimal solutions fulfil the inequalities (2), and thus are in the allowed hypercone in parameter space. Also, in both situations the maximum difference from the Euclidean distance is a fraction of the size of the image, as can be expected.

Using real valued local distances in digital images is generally not desirable. Integer local distances are preferable. Candidate integer approximations of the optimal values, denoted A, B, C, and D, are found by multiplying the optimal local distances by an integer scale factor and rounding to the nearest integer. Then the maximal differences are computed (all expressions are available from the computations of the optimal local distances), to check the approximations. The smallest local distance, a, will act as a scale factor, therefore the resulting WDT will become $\langle 1, B/A, C/A, D/A \rangle$. It is of course important to check that (2) are fulfilled, otherwise the difference expressions are invalid. The best approximation result possible is maxdiff* and the optimal local distances to be multiplied by the scale factor are b^*_{opt}, c^*_{opt}, and d^*_{opt}. Good integer $3 \times 3 \times 3 \times 3$ WDTs are listed in the next Section.

Table 1. Integer $3 \times 3 \times 3 \times 3$ distance transformations

Case	a	b	c	d	maxdiff
D^8	1	-	-	-	2.00000
D^{80}	1	1	1	1	1.00000
real	a_{opt}	b_{opt}	c_{opt}	d_{opt}	0.09515
real	1	b^*_{opt}	c^*_{opt}	b^*_{opt}	0.13456
integer	2	3	4	4	0.29289
integer	3	4	5	6	0.18350
integer	6	8	10	11	0.16667
integer	6	9	10	11	0.16667
integer	7	10	12	13	0.15485
integer	8	11	13	15	0.14304
integer	15	20	24	28	0.13590

4 Results

In this section the results of the optimality computations are summarised and illustrated. Table 1 lists a number of distance transforms. First, the simple D^8 and D^{80} are listed, with their associated maxdiff. These are easily computed from the expressions (4) and (5). Next in Table 1 comes the optimal values for Case I, both for free a and $a \equiv 1$.

After the real valued WDTs, the best integer approximations, using scale factors ($= A$) up to 20, are listed in Table 1. For practical purposes, $\langle 3, 4, 5, 6 \rangle$ is probably the best choice. The maxdiff is reasonably good, with a small scale factor. Note that it is hard to improve on $\langle 15, 20, 24, 28 \rangle$. Note also that all integer DTs are on the border of the allowed hypercone defined by (2), except $\langle 7, 10, 12, 13 \rangle$, which should thus exhibit the most "typical" traits of Case I DTs in 4D.

It must be remembered that even if Case I is the "natural" one of the Cases for 4D DTs, there is no guarantee that it is the best Case. In 3D, the analogous Case I_3 gave the best maxdiff_3 *but* the other case, Case II_3, gave the best maxdiff^*_3 (with $a \equiv 1$), so better integer approximations could be found for Case II_3 than for Case I_3, see [3]. The same may well be true in 4D.

A good way to characterise a DT is the shape of its associated ball, defined as all pixels/voxels/hyxels with a distance less than or equal to the radius from a single central element. In 2D, the city block and chessboard distance balls are a diamond and a square, respectively. In 4D, the D^8 and D^{80} balls are a hyper-octahedron and a hypercube, respectively (the tetrahedron, the octahedron, and the cube are the only "Platonic" solids that exist in any dimension, see [4]). In 4D, the Case I $3 \times 3 \times 3 \times 3$ WDT balls are hyperpolyhedra.

Illustrating hyperpolyhedra is, however, not very easy. One way of doing this was presented in [5]. A 4D digital image is created, with a single object hyxel in the middle. The DT is then computed from this object into the background, in the standard way. If this image is thresholded at a suitable level, a ball with the radius of the threshold value is created. The threshold should be as large as possible while the ball created is still completely within the image. We now have a binary 4D image containing the ball we wish to visualise. If we fix the w-level in this image, we will get a 3D image with a "hyperslice" of the 4D ball, which is in itself a polyhedron. This 3D object can be visualised using a simple binary 3D imaging technique. Ideally, the consecutive hyperslices can be shown as a sequence, a "movie," but here we are constrained to show a few sample hyperslices. In Figs. 2 and 3, we show the $\langle 3,4,5,6 \rangle$ and the $\langle 7,10,12,13 \rangle$ balls with radius 46. The six "hyperslices" were chosen so that the different shapes the ball has at different levels are shown. They are *not* equally spaced in 4D, see the Figures for the chosen w-values, where $w = 1$ denotes the slice with the "first" ball hyxel. These two WDTs were chosen, as the $\langle 3,4,5,6 \rangle$ is the one most probable to be used and the $\langle 7,10,12,13 \rangle$ is the only one exhibiting all the faces that a general 4D Case I WDT can have. The "mid-slice", $w = 46$, is a $\langle 3,4,5 \rangle$ ball and a $\langle 7,10,12 \rangle$ ball, respectively (see [3]).

5 Conclusions

Optimal weighted distance transforms in 4D using $3 \times 3 \times 3 \times 3$ neighbourhoods have been investigated. The best possible such DT, using real-valued weights, has a maximal difference from the Euclidean distance of 9.51%. The best possible integer valued DT is proven to have a maximal difference of 13.46%. The most useful WDT is probably $\langle 3,4,5,6 \rangle$ with a maximal difference of 18.35%. This was, in fact, what was suggested already in [1], but there the motivation was much weaker. A number of other integer WDTs with higher scale factors, but with smaller maximal differences are also listed.

Acknowledgement

Thanks to Stina Svensson for computing and visualising the two balls in Figs 2 and 3.

References

1. G. Borgefors, Distance transformations in arbitrary dimensions, *Computer Vision, Graphics, and Image Processing* **27**, 1984, pp. 321–345. 325, 326, 328, 333
2. G. Borgefors, Distance transformations in digital images *Computer Vision, Graphics, and Image Processing* **34**, 1986, pp. 344–371. 326
3. G. Borgefors, On digital distance transforms in three dimensions, *Computer Vision and Image Understanding,* Vol. **64**, No. 3, (1996), pp. 368–376. 326, 327, 332, 333
4. H. S. M. Coxeter, Regular polytopes, Dover Publications, Inc., New York, 1973. 332
5. G. Borgefors, H. Guo, Weighted distance transform hyperspheres in four dimensions, Proc. SSAB Symposium on Image Analysis 1997, Stockholm, Sweden, March 1997, pp. 71–76. 332

6. M. Fidrich: Iso-surface extraction in 4D with applications related to scale space, In Miguet, Montanvert, Ubéda, Eds., Discrete Geometry for Computer Imagery, Springer 1996 (LNCS 1176), pp. 257–268. 325
7. P. P. Jonker and O. Vermeij, On skeletonization in 4D images, In Perner, Wang, Rosenfeld, Eds., Advances in Structural and Syntactical Pattern Recognition, Springer 1996 (LNCS 1121), pp. 79–89. 325
8. C. O. Kiselman, Regularity properties of distance transformations in image analysis, *Computer Vision and Image Understanding,* Vol. **64**, No. 3, (1996), pp. 390–398. 325, 326, 327
9. T. Y. Kong, Topology-preserving deletion of 1's from 2-, 3-, and 4-dimensional binary images, In Ahronovitz and Fiorio, Eds., Discrete Geometry for Computer Imagery, Springer 1997 (LNCS 1347), pp. 3–18. 325
10. A. Rosenfeld and J. Pfaltz, Sequential operations in digital picture processing, *Journal of the ACM* **13** (4), 1966, pp. 471–494. 325
11. M. Yamashita and T. Ibaraki: Distance defined by neighborhood sequences, *Pattern Recognition* **19**, 1986, pp. 237–246. 326

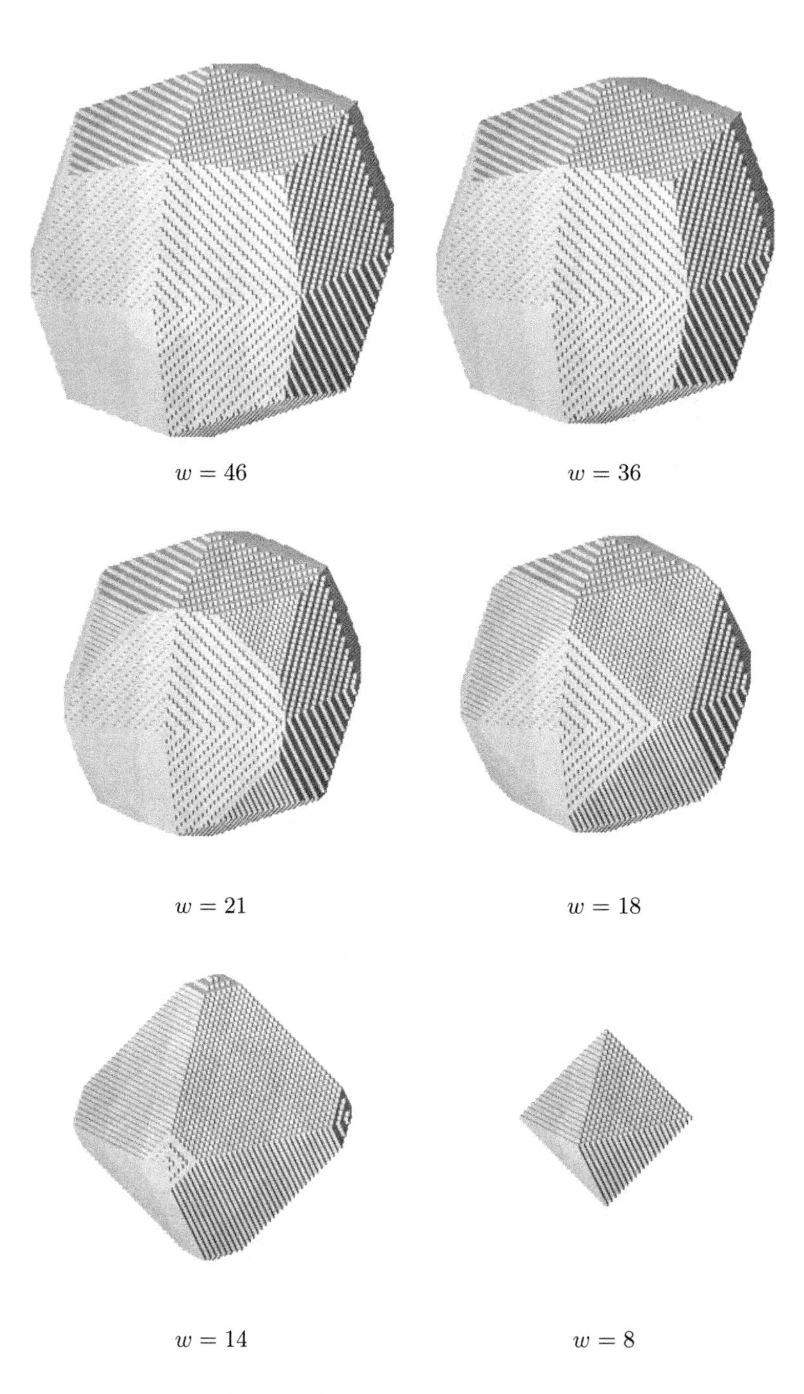

Fig. 2. The ball of the $\langle 3, 4, 5, 6 \rangle$ distance transform, shown as 3D cuts through 4D space at six different levels

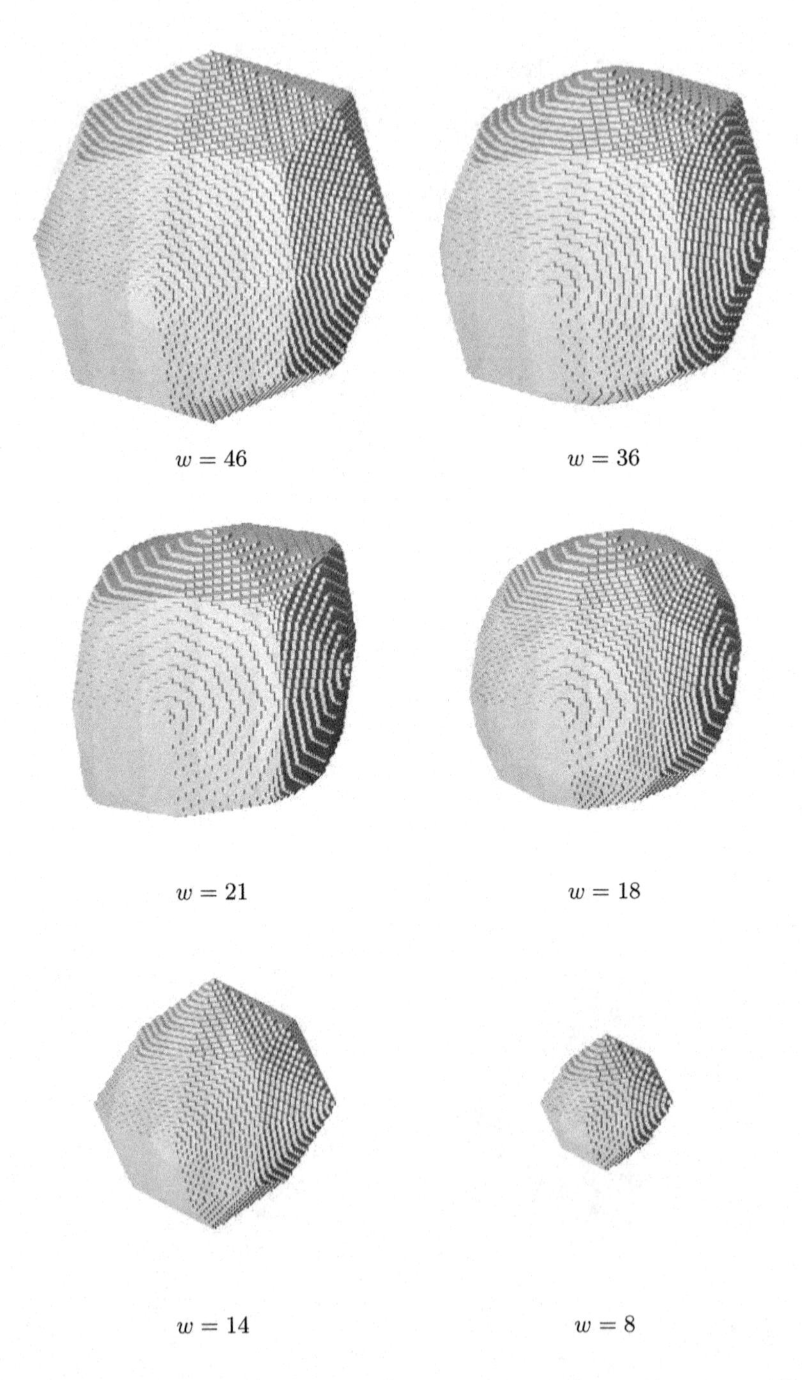

Fig. 3. The ball of the $\langle 7, 10, 12, 13 \rangle$ distance transform, shown as 3D cuts through 4D space at six different levels

Representing 2D Digital Objects

Vito Di Gesù and Cesare Valenti

Dipartimento di Matematica ed Applicazioni, University of Palermo
Via Archirafi 34 - 90123 Palermo, Italy
{digesu,cvalenti}@math.unipa.it

Abstract. The paper describes the combination a multi-views approach to represent connected components of $2D$ binary images. The approach is based on the *Object Connectivity Graph* (OCG), which is a sub-graph of the connectivity graph generated by the *Discrete Cylindrical Algebraic Decomposition*($DCAD$) performed in the $2D$ discrete space. This construction allows us to find the number of connected components, to determine their connectivity degree, and to solve visibility problem. We show that the CAD construction, when performed on two orthogonal views, supply information to avoid ambiguities in the interpretation of each image component. The implementation of the algorithm is outlined and the computational complexities is given.

Keywords: shape representation, shape decomposition, shape description, digital topology

1 Introduction

The combination of multiple views of the same object can be used to improve its reconstruction. In principle, it is possible to recover the whole object information from partial ones; the number of views that are necessary to fully recover an object-representation depends also on the feature space. For examples in [1] two views are considered to reconstruct convex polyminoes; computerized tomography [2] is another example of reconstruction from projection. In [3] the symmetry transform, performed on multiple views, is used for object representation and classification.

The paper describes how to combine the information of two orthogonal views to represent the connected components in $2D$ binary images. The combination method uses the *Connectivity Graph* (CG) as it is derived from the *Cylindrical Algebraic Decomposition* (CAD)introduced by Collins [4] for the decomposition of the Euclidean space, E^d. In [5] the CAD construction has been extended to the analysis of components of a set of points of a $2D$ discrete space, $D \subseteq N^2$, where N is the set of natural numbers and it has been applied for the solution of geometry problems of the first and the second order [6,7], to computer aided design, and shape analysis.

The CAD-algorithm performs a cellular decomposition [8,9] of the projective plane in open ball, also named i-cells, with $i \geq 0$. For example in E^2 a 0-cell is a point, 1-cell is an open arc, and 2-cell is an open region. The cellular

G. Borgefors, I. Nyström, and G. Sanniti di Baja (Eds.): DGCI 2000, LNCS 1953, pp. 337–347, 2000.
© Springer-Verlag Berlin Heidelberg 2000

decomposition of E^2 may be defined as a nested sequence $X^0 \subset X^1 \subset X^2 \subset E^2$ of closed subspaces, such that X^0 consists of finitely many 0-cells, X^1 consists of finitely many disjoint 1-cells, X^2 consists of finitely many disjoint 2-cells.

Let us describe the classical CAD-algorithm with an example. Let C be an algebraic curve in the real projective plane (see Fig.1a):

$$y^4 - 2xy^3 - x^2y^2 + y^2 + 2x^3y + x^2 - 1 = 0$$

then a cellular decomposition can be performed as in Fig.1b, by intersecting C with linear varieties. The decomposition is arbitrary, however after that it is performed the i-cells are determined, and they may be used to study C, and its connected components ($\leq \binom{n}{2} - 1$, where n is the degree of C). Two distinct cells

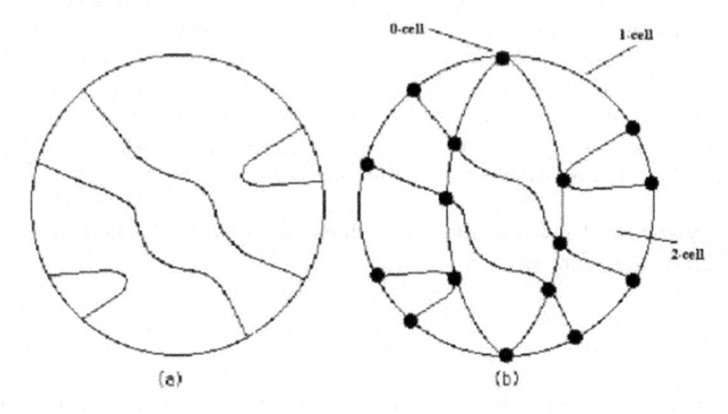

Fig. 1. a) Algebraic curve C in the projective real plane; b) The *Cylindrical Algebraic Decomposition*

are adjacent if they touch each other. Clearly adjacency is a symmetric relation. Therefore the CG nodes are cells, and the arcs represent the adjacency between cells.

The extension of the CAD decomposition to a digital spaces is named *Discrete CAD* ($DCAD$) [5]. Its definition is straightforward by providing the proper definitions of $i - cell$, and connectivity relation on D [10,11].

The $DCAD$ provides unambiguous topological information (connected components, holes); however, the structural description of a given component could be ambiguous. This ambiguity can be avoided by using two or more views. In the paper we show that two views are sufficient to avoid shape ambiguities that are on the component borders. An algorithm for combining two orthogonal views of the $DCAD$ construction is also given and its computational complexity outlined.

The paper is organized as follows. Section 2 describes the $DCAD$-algorithm. The algorithm to combine the CG is described in Section 3. Section 4 reports results and implementation notes. Final remarks are given in Section5.

2 The Discrete CAD Algorithm

In the following the internal borders of connected components in E^2 will play the role of the algebraic curves. In this case the CAD-algorithm consists in the partition of the space by means of parallel straight lines that intersect the borders of connected components. The i-cells, generated by the decomposition, are points (p), bend-points (s), open lines representing borders (b), open lines representing cylinders (c) and open $2D$ regions (r).

Starting from this decomposition a labeled not oriented connectivity graph (CG) is derived, nodes of which are of type p, s, b, c, and r. The arcs of a CG connect only nodes of different type that are adjacent. Fig.s 2a,b show an example of CAD, and the corresponding CG in E^2. An ordered pair of integer number, (L, R), is assigned to each node of the CG; where L and R are named the left and right labels respectively. This labeling allows us to identify univocally nodes of type c, r, and b, while nodes of type p and s have the same kind of pair, as stated below.

R/L	*Odd*	*Even*
Odd	c	r
Even	p,s	b

This labeling rule is determined by scanning the plane left to right (bottom to up), and increasing by 1 L (R) each time a new element of the decomposition is intersected (initially $L = R = 0$). For example the region r1 in Fig.2a is labeled $(0, 1)$, the line $c2$ is labeled $(1, 3)$ the line $b1$ is labeled $(2, 2)$ and the point $p2$ is labeled $(3, 2)$, see Fig.2b. In E^2 an infinity number of CAD's can be performed; in fact, it is possible to consider all possible directions of straight lines. Here, only straight lines, parallel to the $Y(X)$-axis that intersect concavity, convexity, and bend-points are selected.

The $DCAD$-decomposition is obtained by considering as cylinders digital straight paths in D that are parallel to the $Y(X)$-axis. Moreover, the cylinders selected cross the internal borders of the components in D, where one of the configurations of pixels shown in Fig.3 is present. These configurations correspond to concavities, convexities, and bends of digital borders.

The $DCAD$ construction may generate *empty regions* between two adjacent cylinders. In the algorithm these regions are treated as *virtual regions*, and their labeling is determined by considering the labels of adjacent cylinders. Each of these configurations became a single vertex in the construction of the CG. The labeling of CG is determined as in the continuous space, by scanning D left-right and bottom-up. The arcs represent the adjacency relation among points, paths and regions.

Note that the 4-adjacency holds between points, paths, and internal regions; while 8-adjacency holds between points and external regions.

2.1 Properties of the CG

The CG allows to retrieve the connected components of binary images, and to derive information about the visibility-problem. that is stated as follows: *for*

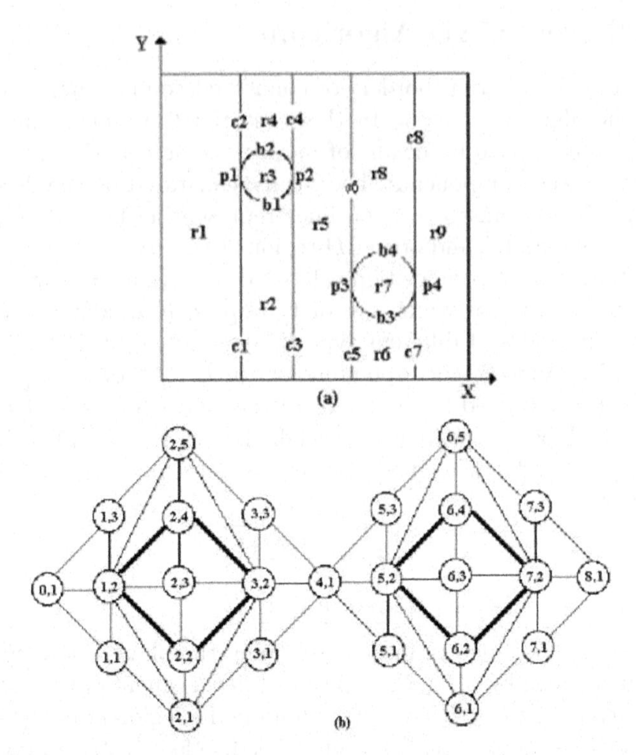

Fig. 2. a) Example of CAD in E^2; b) corresponding Connectivity Graph of the CAD

a given illumination of a scene, with more then one component, decide which component is totally or partially hidden.

The CG derived is planar graph three colorable; its faces are triangles, nodes of which, represent points, bend-points, lines, and $2D$-regions. As said before, each node has three labels $(\lambda, L_\lambda, R_\lambda)$, with $\lambda \in \{p, s, b, c, r\}$, that represent, in the order, its type, left, and right labels.

Proposition 1. *A simple cycle, C, of CG, such that nodes of type p (s) ($L_{p(s)}$ is odd and $R_{p(s)}$ is even), are connected to nodes of type b (L_b and R_b are both even), represents an internal border of a connected component.*

The internal regions of a component of D are obtained as follows:

Proposition 2. *The* internal *region of a component, represented by the cycle C, is the union of the regions, such that their labeling $(\lambda, L_\lambda, R_\lambda)$, $\lambda \in \{p, s\}$, satisfies the condition: L is even and $R \bmod 4 = 3$ and are connected to some nodes of type p (s) of C. An analogous statement can be formulated for the external regions of a cycle (component), by replacing the condition $R \bmod 4 = 3$ with $R \bmod 4 = 1$.*

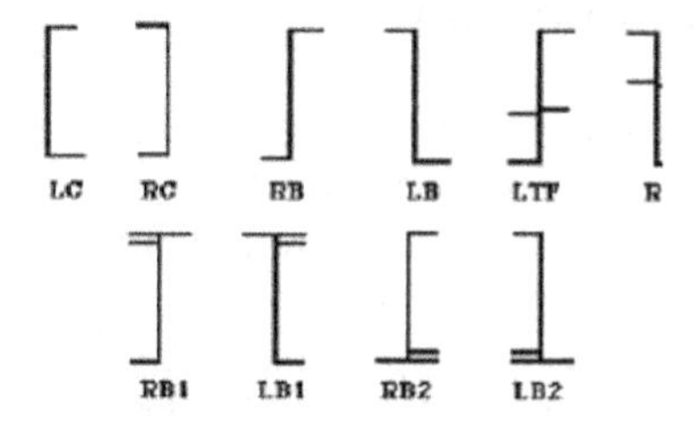

Fig. 3. Basic configurations: left concavity (LC); right concavity (RC); right bend (RB) and left bend (LB); left double point (LTF); right double point (RTF); right bend of type 1 ($RB1$); left bend of type 1 ($LB1$); right bend of type 2 ($RB2$); left bend of type 2 ($LB2$)

Proposition 1 and 2 are straightforward; they follow from the ordering that is settled in the nodes of CG, during the computation of the right and left labels.

In the case of simply connected components the number of cycles is equal to the number of connected components in D. For example, the cycle (bold arcs) in Fig.2b represent the connected components in Fig.2a. Connected components may have N-holes ($N \geq 0$) and the number of components and their holes are retrieved by searching cycles in CG. In this case it is necessary to determine the relation of *inclusion* between cycles. The algorithm generate a graph is a forest of rooted oriented trees, nodes of which represents borders. Oriented arcs represent the relation *to be directly included*. Nodes of the forest represent a component *iff* it belongs to an *even* level. A node of an even level, with N-sons, represents a component with N-holes.

The inclusion relation is easily determined by comparing the highest, and lowest left/right labels of their nodes.

Proposition 3. *Given two simple cycles, C_i, C_j , of CG, representing internal borders of connected components in D, then $C_i \subset C_j$ iff the inequalities:*

$$Lm(C_i) > Lm(C_j) \quad LM(C_i) < LM(C_j)$$

and

$$Rm(C_i) > Rm(C_j) \quad RM(C_i) < RM(C_j)$$

are all satisfied. Where $Lm(C_i)$, $Lm(C_j)$, $LM(C_i)$, $LM(C_j)$, $Rm(C_i)$, $Rm(C_j)$, $RM(C_i)$, $RM(C_j)$ are the lowest and highest left/right labels of the nodes in C_i, and C_j.

Here, we omitted the labeling information to light the notation. This proposition is easily proved by considering again the ordering of the nodes in CG, and by the fact that two cycles, representing regular components in D, cannot have common nodes.

3 Two Views Combination

In the following we will consider the sub-graph of the CG, named *Object CG* (OCG), that is obtained deleting from the CG all nodes and the corresponding arcs representing components of the background of the digital scene (the value 0 is assigned to the background by convention). The label of each node is preserved therefore all results, obtained in the previous section, still hold. The connected components of the OCG represent objects in the scene. The OCG allows us to store less space memory and makes easier the visualization of the results. Moreover, the λ-label will not be explicitly shown in OCG.

In the following we denote by $DCAD_{Y(X)}$ the $DCAD$ along the $Y(X)$-axis and by $OCG_{Y(X)}$ the corresponding graphs. The $DCAD_X$ is performed by scanning the image in a top down and left-right order (see Fig.4).

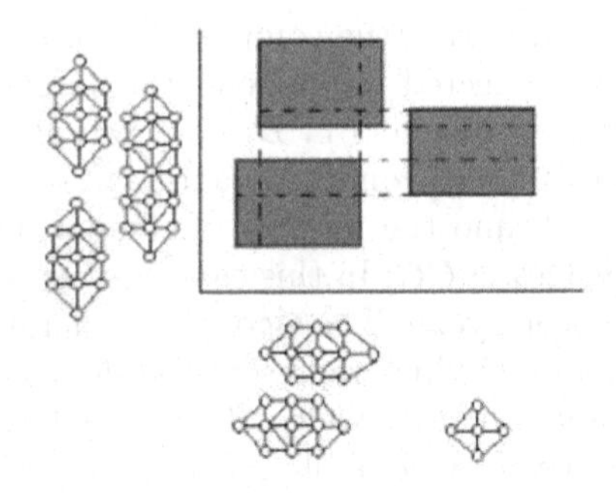

Fig. 4. An example of $DCAD_Y$ and $DCAD_X$ and related graphs

The $DCAD$ of a $2D$ digital scene and the related OCG includes by its definition all topological information regarding the connected components in the scene. However, in the case on non-convex components structural information can be lost or could be ambiguous if we consider only one view.

In the case of binary images most of the shape information is on the borders. The ambiguities, we want to eliminate, regard the positions and orientation of bend points and concavities (these one can be considered the combination of two bend points.

For example, the $DCAD_Y$ of the left and right images in Fig.5 generate the same OCG_Y (see Fig.6), even if the bottom components have a different structure (the bend-points are in the opposite position). Note that the ordering of the λ-labels remains the same in both configurations.

Fig.7 shows the $DCAD_X$ of the previous images. The OCG_X in Fig.s 8a,b corresponds to the left and right images in Fig.7 respectively; note hat in this case the two OCG_X's are different; because the ordering of the λ-labels is not preserved. In the following we will show how to combine the two OCG's to solve this ambiguity.

Fig. 5. $DCAD_Y$ of the left and right images

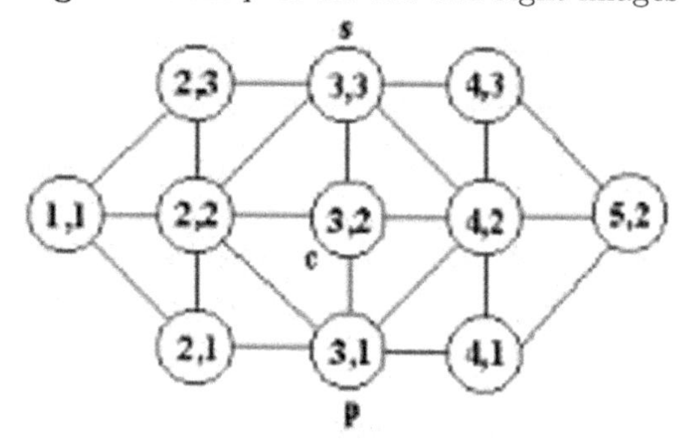

Fig. 6. The OCG_Y of the images in Fig.5

The following properties are used in the algorithm for combining the OCG_Y and the OCG_X graphs.

Proposition 4. *A connected component of an OCG starts with at least one node with label (p, L, R) with L odd and R even, which is connected to nodes $(\lambda, L+1, R1)$, $(\lambda, L+1, R1+1)$, $(L+1, R1+2)$ with $R1 \geq R$ even.*

This proposition follows from the scanning rules.

Proposition 5. *A sequence of nodes labeled (p, L, R), $(c, L, R+1)$, $(p, L, R+2)$ of a connected component of an OCG can be deleted with the arcs and the nodes on the left and right of the triple can be merged as follows: $\{(\lambda, L-1, x), (\lambda, L+1, x)\} \longrightarrow (\lambda, L-1, x)$ for $x \equiv R, R+1, R+2$. The reduced connected components maintain both topological and structural properties.*

Proof. In fact, the cylinder represented by the nodes (p, L, R), $(c, L, R+1)$, $(p, L, R+2)$ makes a partition of a rectangle into two sub-rectangles that merged do not change neither the topology neither the structure of the image component. The components obtained are said *normalized.*

In the following the sketch of the combining algorithm is given. The time complexity of each step is also evaluated as a function of the linear size n of D and the number of components N_c in the image.

Fig. 7. $DCAD_X$ of the same images

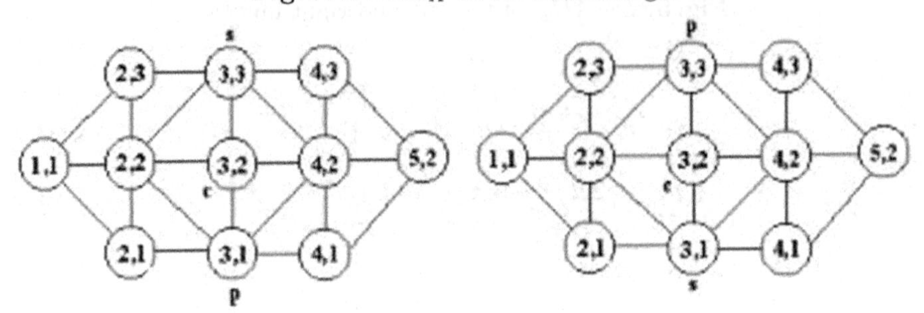

Fig. 8. a) The OCG_X of the left image; b) the OCG_X of the right image

- Algorithm *COMPOUND*

1 *Input* OCG_Y, OCG_X

2 *Components association*
In this step the labeling, generated by the scanning rule, is used. Components in OCG_X are selected by searching for starting nodes that satisfy Proposition 4 that are yet marked and with the lowest labels L and R. The association is made with the corresponding not marked component in OCG_Y that has the highest Y spatial value. The associated components are marked. The association stops after that all components in OCG_X and OCG_Y have been marked. The time complexity of this step is $O(N_c \times N_c)$.

3 *Component normalization*
Using iteratively the reduction rule stated in Proposition 5 performs the normalization of the $OCG_Y(X)$. The time complexity of this step is $O(n \times N_c)$.

4 *Component combination*
The combination of the associated normalized components is performed by considering that for a given odd value of $L > 1$ the sequence of labels is of the form

$$pc(sc)^*p \cup pc(sc)^*s \cup (sc)^*p \cup (sc)^*s$$

Moreover, by using the 9 correspondences between the sequence OCG_Y and OCG_X (see Fig.9 and Table 1) it is possible to set the position of bend-points in each component eliminating the related ambiguities.

5 *Output structural and shape information about each component*
– *end COMPOUND*

The time complexity of this step is $O(n \times N_c)$.

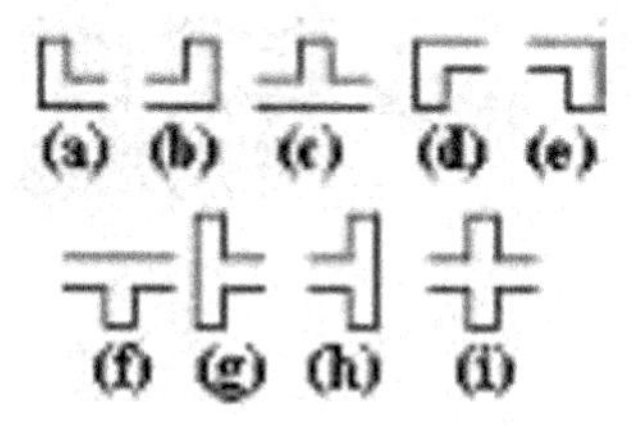

Fig. 9. The 9 correspondences between OCG_Y and OCG_X

Table 1. Words corresponding to configurations in Fig.9

Y	X	$Config.$	Y	X	$Config.$	Y	X	$Config.$
pcs	pcs	$Fig.9a$	pcs	scp	$Fig.9b$	pcs	scs	$Fig.9c$
scp	pcs	$Fig.9d$	scp	scp	$Fig.9e$	scp	scs	$Fig.9f$
scs	pcs	$Fig.9g$	scs	scp	$Fig.9h$	scs	scs	$Fig.9i$

Proposition 6. *The combination phase of the algorithm COMPOUND eliminates ambiguities due to position and orientation of bend points.*

Proof. For binary images, represented in a squared lattice, configurations, given in Fig.9, cover all possible cases. Ambiguties are then solved by parsing left to right the words corresponding to the same object in OCG_X and OCG_Y and testing the occurrencies given in Table 1.

4 Implementation Notes

The serial version of the $DCAD$-algorithm has been implemented in C++ under Linux. The whole computation time of the $COMPOUND$ algorithm is, as we have seen, of the order $O(n \times N_c)$. Note that in most applications $N_c \ll n$. The computation time to perform the $DCAD$ algorithm is of the order $O(n^2)$ and it is the more expensive part of the whole computation. In [12] a parallel version of the $DCAD$-algorithm is given.

The CPU time to run $DCAD$ and $COMPOUND$ algorithms on a PENTIUM III 400MHz was of about $T_{DCAD} = 0.5sec + T_{COMPOUND} = 30msec$ for 256×256 synthetic images containing at most 10 components. The maximum

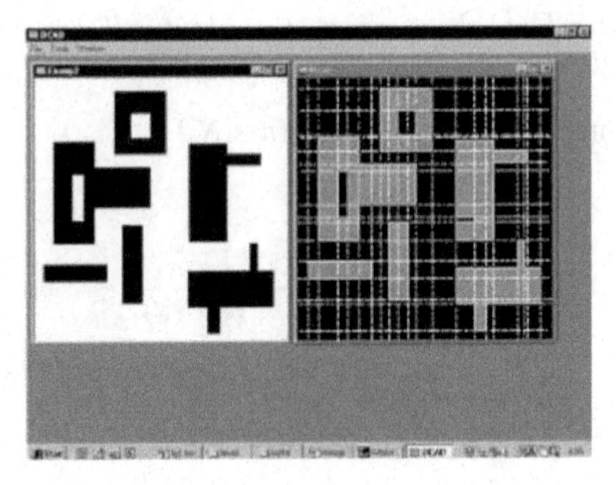

Fig. 10. The $DCAD$ construction of a binary image

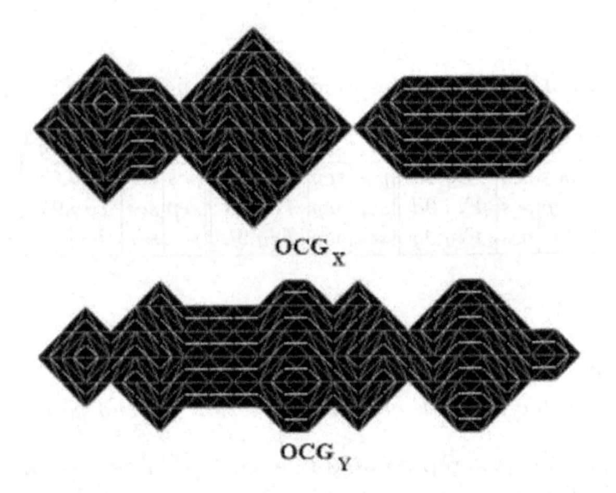

Fig. 11. The OCG_X and OCG_Y of the binary image in Fig.10

number of nodes to be explored was 530 in the case of ten components. The right side of Fig.10 shows the $DCAD$ of a binary image on the left in both views (X, Y). The top and the bottom of Fig.11 show the corresponding OCG_X and OCG_Y respectively.

Either the discrete nature of D or the noise background may generate dummy bend-nodes. They have been removed by using heuristic arguments that are related to the image size; for example discard bend-points with length less than $0.05 \times n$ (e.g. in our case length< 10).

5 Final Remarks

The paper shows a new approach to the analysis of connected components of binary images. The *DCAD* algorithm allows us to analyse topological properties of binary images. In the paper it is show how to combine the *DCAD* along two views in order to provide a structural description of digital images. The ordering of the *OCG*'s is at the basis of the search and combining algorithms, allowing them to be tractable and faster. The extension of the *DCAD* on $3D$ spaces is under study.

References

1. Barcucci, E., Del Lungo, A., Nivat, M., Pinzani, R.: Reconstructing convex polyminoes from horizontal and vertical projections. Theoretical Computer Science **155** (1996) 321–347. 337
2. Herman, G. T.: Image Reconstruction from Projections: The Fundamentals of Computerized Tomography. Academic Press (New York) (1980). 337
3. Chella, A., Di Gesù, V., Infantino, I., Intravaia, D., Valenti, C.: Cooperating Strategy for Objects Recognition. in Lecture Notes in Computer Science book "Shape, contour and grouping in computer vision", Springer Verlag, **1681** (1999) 264–274. 337
4. Collins, G. E.: Quantifier elimination for real closed fields by cylindrical algebraic decomposition. Proc.of the Second GI Conference on Automata Theory and Formal Languages, Springer Lect.Notes Comp. SCi. **33** (1975) 515–532. 337
5. Di Gesù, V. and Renda, R.: An algorithm to analyse connected components of binary images. Geometrical Problems of Image Processing **4** (1991) 87–93. 337, 338
6. Arnon, D. S., McCallum, S.: A polynomial-time algorithm for the topological type of a real algebraic curve. Journal Symb. Comp. **5** (1988) 213–236. 337
7. Arnborg, S., Feng, H.: Algebraic decomposition of regular curves. Journal Symb. Comp. **5** (1988) 131–140. 337
8. Françon, J. M.: Sur la topologie d'un plan arithmétique. Theoretical Computer Science **156** (1996) 31–40. 337
9. Khalimsky, E. D., Kopperman, R., Meyer P. R.: Computer graphics and connected topologies on finite ordered sets. Topology and Applications **36** (1990) 1–17. 337
10. Chassery, J. M. and Montanvert, A.: Geómetrie discrete en analyse d'images. Hermeś, Paris (1991). 338
11. Kovalevsky, V. A.: Finite Topology as Applied to Image Analysis. Computer Vision, Graphics, and Image Processing **45** (1989) 141–161. 338
12. Chiavetta, F., Di Gesù, V., and Renda, R.: A Parallel Algorithm to analyse connected components on binary images. International Journal of Pattern Recognition and Artificial Intelligence **6**(2,3) (1992) 315–333. 345

Plane Embedding of Dually Contracted Graphs*

Roland Glantz and Walter G. Kropatsch

Pattern Recognition and Image Processing Group 183/2
Institute for Computer Aided Automation, Vienna University of Technology
Favoritenstr. 9, A-1040 Vienna, Austria phone ++43-(0)1-58801-18358, fax ++43-(0)-1-58801-18392
{glz,krw}@prip.tuwien.ac.at

Abstract. The use of plane graphs for the description of image structure and shape representation poses two problems : (1) how to obtain the set of vertices, the set of edges and the incidence relation of the graph, and (2) how to embed the graph into the plane image. Initially, the image is represented by an embedded graph G in a straight forward manner, i.e. the edges of G represent the 4-connectivity of the pixels. Let $\overline{G}$ denote a (planar) abstract dual of G. Dual graph contraction is used to reduce the pair $(\overline{G}, G)$ to a pair $(\overline{H}, H)$ of planar abstract duals. Dual graph contraction is unsymmetric due to an extra condition on the choice of the contraction kernels in G. This condition is shown to be necessary and sufficient for H to be embedded onto G. The embedding is applied to the description of image structure and to shape representation.

1 Introduction

A key concept in combinatorial topology is the separation of topology (the graph) and embedding (of the graph) [Fra96]. In a hierarchical representation of an image by a sequence of plane graphs on increasing levels of abstraction and scale, however, the hierarchy should also be reflected by the embedding. Thus, the embedding of a lower level graph induces constraints on the embedding of higher level graphs. Let a high and a low level graph be denoted by $H = (V_H, E_H)$ and $L = (V_L, E_L)$ respectively. The hierarchy is reflected by the embedding of H and L (Fig. 1), if

1. $V_H \subset V_L$,
2. each embedded edge from E_H (as subset of the plane) is a union of embedded edges from E_L,
3. each region from the embedding of H (as subset of the plane) is a union of regions from the embedding of L.

Note the transitivity of the above conditions. The high level graph is a connected *topological minor* of the low level graph [Die97]. This paper is devoted to

* This work has been supported by the Austrian Science Fund (FWF) under grant S7002-MAT and grant P14445-MAT.

G. Borgefors, I. Nyström, and G. Sanniti di Baja (Eds.): DGCI 2000, LNCS 1953, pp. 348–357, 2000.
© Springer-Verlag Berlin Heidelberg 2000

hierarchies of graphs as obtained by dual graph contraction [KM95]. It will be shown that an embedding of a higher level graph may always be derived from the embedding of a lower level graph such that the above conditions are fulfilled. Note, however, that dual graph contraction yields two hierarchies of graphs: on a planar graph and on its (planar) abstract dual. It will turn out that, in general, it is impossible to obtain a proper embedding of **both** hierarchies. Proper embedding of graph hierarchies is crucial for document image analysis [BK99] and the description of image structure [GEK99]. In this paper we propose a new application for dual graph contraction in which embedding is crucial, i.e. shape representation. The outline of the paper is as follows: In Section 2 dual graph

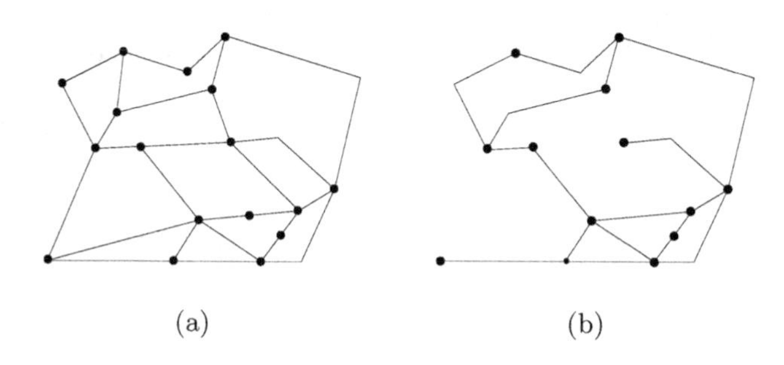

Fig. 1. Hierarchical embedding. (a) Low level. (b) High level

contraction is described in terms of equivalence relations. In Section 3 the embedding of high level graphs onto low level graphs is related to the unsymmetry of dual graph contraction. Section 4 demonstrates the use of the embedding for the description of image structure. Section 5 is devoted to shape representation by means of embedded graphs that describe the structure of distance transforms. We conclude in Section 6.

2 Dual Graph Contraction by Equivalence Relations

Throughout the paper we refer to the following definition of a graph.

Definition 1 (Graph). *A graph $G = (V, E)$ is given by a finite set V of elements called vertices, a finite set E of elements called edges with $E \cap V = \emptyset$ and an incidence relation ι which associates with each edge $e \in E$ a subset of V with one or two elements. The vertices in $\iota(e)$ are called the end vertices of e.*

Note that the definition includes graphs with self-loops (i.e. edges with only one end vertex) and multiple edges (i.e. several edges with identical sets of end vertices).

Let $(\overline{G}, G)$ denote a pair of planar graphs, where $\overline{G}$ is an abstract dual [Die97] of G. We write $\overline{G} = (\overline{V}, \overline{E})$ and $G = (V, E)$. *Dual edge contraction* [Kro95] [KM95] of the pair $(\overline{G}, G)$ is specified by contraction kernels which form a spanning forest $\overline{F}$ of $\overline{G}$. The result of dual edge contraction consists of a pair denoted by $(\overline{G}/\overline{F}, G \setminus F)$, in which

- $\overline{G}/\overline{F}$ is obtained from $\overline{G}$ by contracting the edges of $\overline{F}$ as shown below.
- F denotes the subgraph of G that is induced by the edges which are dual to the edges of $\overline{F}$.
- $G \setminus F$ is obtained from G by the removal of all edges from F.

In the following, the vertex set, the edge set and the incidence relation of $\overline{G}/\overline{F}$ are defined by means of an equivalence relation on the vertex set $\overline{V}$ of $\overline{G}$.

The spanning forest $\overline{F}$ partitions $\overline{V}$: each vertex belongs to exactly one connected component of $\overline{F}$. Hence, the binary relation $\sim_{\overline{F}}$ defined as

$$\overline{v} \sim_{\overline{F}} \overline{w} :\Leftrightarrow \overline{v} \text{ and } \overline{w} \text{ belong to the same connected component of } \overline{F} \tag{1}$$

is an equivalence relation on $\overline{V}$. Let the equivalence class of $\overline{v}$ be denoted by $[\overline{v}]_{\overline{F}}$. The vertex set of $\overline{G}/\overline{F}$ equals the *quotient of* $\overline{V}$ *by* $\sim_{\overline{F}}$, defined as

$$\overline{V} \setminus \sim_{\overline{F}} := \{[\overline{v}]_{\overline{F}} \mid \overline{v} \in \overline{V}\}. \tag{2}$$

If $\overline{E}_{\overline{F}}$ denotes the set of edges in $\overline{F}$, the edge set of $\overline{G}/\overline{F}$ equals $\overline{E} \setminus \overline{E}_{\overline{F}}$. The incidence relation $\iota_{\overline{G}/\overline{F}}$ of $\overline{G}/\overline{F}$ is derived from the incidence relation $\iota_{\overline{G}}$ of $\overline{G}$ in the following way: For a non-loop $\overline{e} \in \overline{E} \setminus \overline{E}_{\overline{F}}$ with $\iota_{\overline{G}}(\overline{e}) = \{\overline{v}, \overline{w}\}$ set

$$\iota_{\overline{G}/\overline{F}}(\overline{e}) := \{[\overline{v}]_{\overline{F}}\} \cup \{[\overline{w}]_{\overline{F}}\}. \tag{3}$$

For a loop $\overline{e} \in \overline{E} \setminus \overline{E}_{\overline{F}}$ with $\iota_{\overline{G}}(\overline{e}) = \{\overline{v}\}$ set

$$\iota_{\overline{G}/\overline{F}}(\overline{e}) := \{[\overline{v}]_{\overline{F}}\}. \tag{4}$$

The second step of dual graph contraction, i.e *dual face contraction*, consists in the contraction of edges from the graph $G \setminus F$ and the removal of the corresponding dual edges in $\overline{G}/\overline{F}$. The contraction kernels in $G \setminus F$ are to form a spanning forest F' of $G \setminus F$. However, each connected component of F' is required to contain at most one vertex whose degree in $G \setminus F$ is larger than two. In the following, this constraint will be referred to as *degree constraint.* It assures that the removal of the corresponding dual edges in $\overline{G}/\overline{F}$ is restricted to so called *redundant edges*, i.e. parallel edges, or loops that do not surround a subgraph of $\overline{G}/\overline{F}$ (Fig. 2).

The result of dual graph contraction is a pair $(\overline{H}, H)$ of planar graphs, in which $\overline{H}$ is an abstract dual of H. If $\overline{F'}$ denotes the subgraph of $\overline{G}$ that is induced by the edges which are dual to the edges in F', then

$$H = (G \setminus F)/F' \quad \text{and} \quad \overline{H} = (\overline{G}/\overline{F}) \setminus \overline{F'}. \tag{5}$$

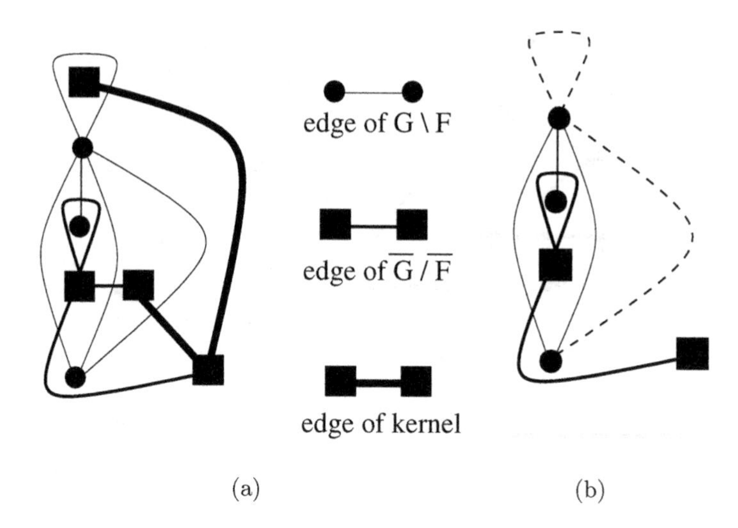

Fig. 2. Dual face contraction. (a) Contraction kernel in $G \setminus F$. (b) Removal of redundant edges in $\overline{G}/\overline{F}$ (dotted)

The degree constraint ensures that the degree of a vertex $[v]_{F'}$ in H cannot exceed the degree of vertex v in $G \setminus F$. Furthermore, the degree of vertex v in $G \setminus F$ is restricted by the degree of the vertex v in G. Hence, if the degree of vertex $[v]_{F'}$ in H is denoted by $deg_H([v]_{F'})$, the following inequation holds:

$$deg_H([v]_{F'}) \leq deg_{G \setminus F}(v) \leq deg_G(v) \quad \forall v \in V. \tag{6}$$

In particular, the maximal vertex degree in H is restricted by the maximal vertex degree in G.

3 Embedding of Contracted Graphs

Let $G = (V, E)$ denote a plane graph and assume F to be a spanning forest of G that fulfills the degree constraint. An embedding of G/F can be obtained from the embedding of G in the following way (Fig. 3):

1. Interpret the connected components of F as *rooted* trees: If the connected component has a (unique) vertex r, whose degree **in G** is larger than two, declare r to be the root. Otherwise declare any vertex of the connected component to be the root.
2. Let R denote the set of all roots from step 1. For the vertex set of G/F, i.e. the quotient $V \setminus \sim_F$ of V by $\sim_F$ the following holds

$$V \setminus \sim_F = \{[r]_F \mid r \in R\}. \tag{7}$$

   For each $r \in R$ set the location of $[r]_F$ to the location of r in G.

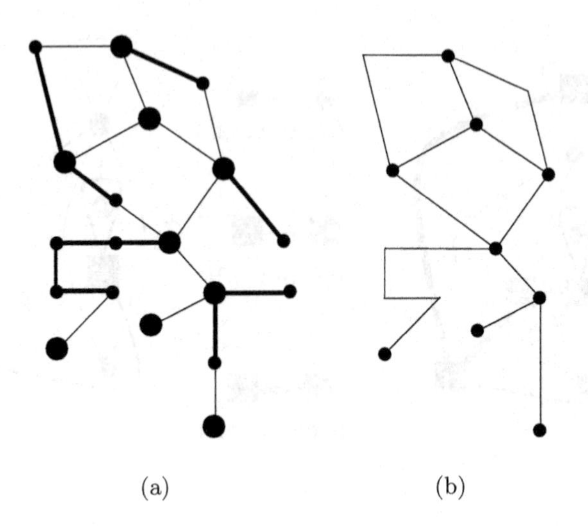

Fig. 3. (a) Graph G. Bold edges belong to spanning forest F. F fulfills the degree constraint. Roots are enlarged. (b) Embedding of G/F

3. Let E_F denote the set of edges in F. The edge set of G/F equals $E \setminus E_F$. Let e be an edge from $E \setminus E_F$ and let the connected components of F, where the end vertices of e **in G** belong to, be denoted by CC_1 and CC_2. Since the roots r_1 and r_2 of CC_1 and CC_2 are the only vertices in $CC_1 \cup CC_2$ that may have a degree greater than two **in G**, there is a unique path $\Pi = \Pi(e)$ from r_1 to r_2 in G which contains e. Furthermore, for any two edges e, e' the paths $\Pi(e)$ and $\Pi(e')$ may only intersect at vertices from R. The embedding of e **in G / F** is given by the union of all (embedded) edges that belong to $\Pi(e)$.

Note that the embedding of the paths, as described in step 3, fails whenever F does not fulfill the degree constraint. Hence, the degree constraint is necessary and sufficient for the graph G/F to be embedded onto G.

4 Image Structure

Gray level images may be interpreted as digital elevation models [DEM], in which the altitudes are given by the gray levels. In [KD94] the structure of an image is defined via the crest lines of the corresponding DEM. In this section, a plane graph is constructed, which describes the crest lines in a DEM. It is referred to as *crest graph*. The crest graph for the gray level image in Fig. 4(a) is depicted in Fig. 5(a). In contrast to the graph constructed in [GEK99], the crest graph has the following properties:

– The crest graph is embedded such that the edges describe the run of the crest lines in the DEM.

– The vertex set of the crest graph may include vertices that do not represent a hill of the DEM. The extra vertices represent branching points of crest lines that are not located on the top of hills.

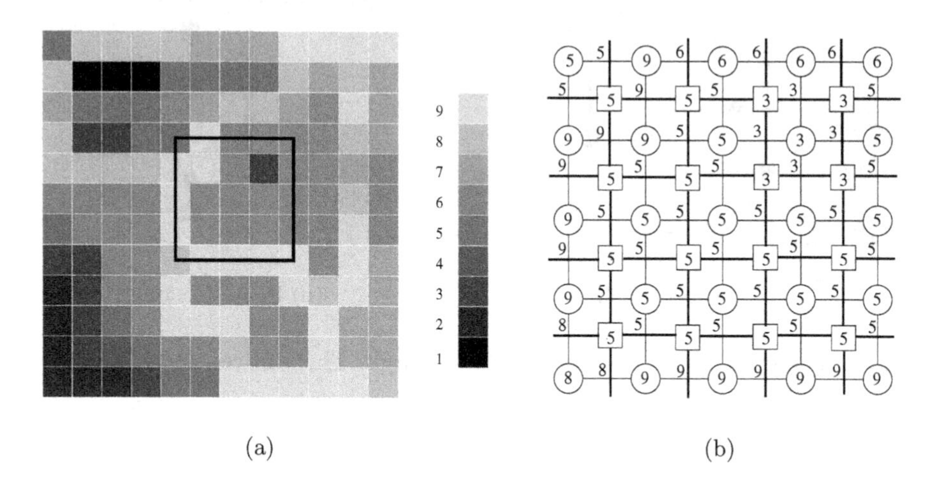

(a) (b)

Fig. 4. (a) The gray levels of the pixels. (b) The pair $(\overline{G}, G)$ restricted to the square region marked in (a). The vertices of $\overline{G}$ [G] are depicted as squares [circles]

The crest graph is constructed in the following way:

1. The DEM is transformed into a pair $(\overline{G}, G)$ of attributed plane graphs, in which G and $\overline{G}$ are plane duals [Die97]. The pair $(\overline{G}, G)$ will later be transformed into a pair $(\overline{G_c}, G_c)$, in which G_c is the crest graph. An example of $(\overline{G}, G)$ is given in Fig. 4(b). The vertices of G represent the pixels centers, while the edges of G indicate the 4-neighborhood of the pixels. The vertex value is set to the altitude associated with the corresponding pixel. Throughout the transformation of $(\overline{G}, G)$ into $(\overline{G_c}, G_c)$, the value of an edge from G is to indicate the lowest altitudes along the embedded edge. Hence, the value of an edge e with end vertices u and v is initialized to the minimum of the vertex values of u and v. The vertices of $\overline{G}$ represent the regions of G and are located at the crossings of the pixel borders. The edges of $\overline{G}$ are straight line segments that reflect the 4-neighborhood of the regions from G. Their values are initialized to the values of the corresponding dual edges in G. A vertex $\overline{v}$ of $\overline{G}$ is initialized to the minimal value of all edges to surround the region (basin) represented by $\overline{v}$, i.e. the minimal value of all edges in $\overline{G}$ that are incident to $\overline{v}$. Throughout the construction of $(\overline{G_c}, G_c)$ from $(\overline{G}, G)$, the value of a vertex $\overline{v}$ from $\overline{G}$ is to indicate the lowest altitude in the region (basin) represented by $\overline{v}$.

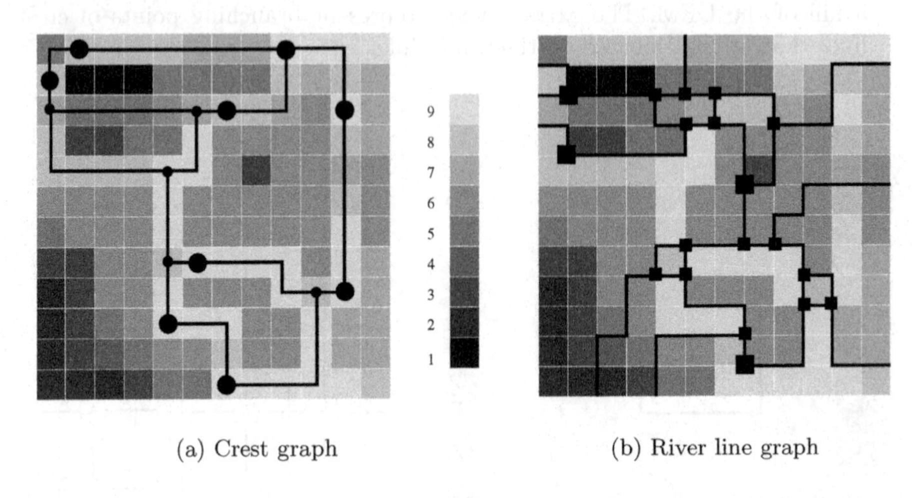

(a) Crest graph (b) River line graph

Fig. 5. The crest graph and the river line graph. The vertices which represent hills respectively basins are drawn larger

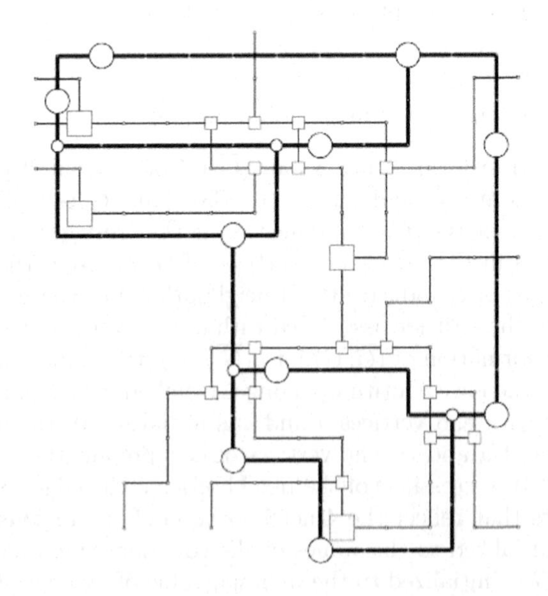

Fig. 6. Overlay of the crest graph (thick lines) and the river line graph (thin lines) from Fig. 5

2. The pair $(\overline{G_c}, G_c)$ is generated from $(\overline{G}, G)$ by a sequence of monotonic dual graph contractions [GEK99]: Besides the general requirements described in Section 2, each contraction kernel of G [$\overline{G}$] has to contain exactly one local maximum [minimum] of vertex values. The vertex value in the contracted graph is set to the maximum [minimum] of the vertex values in the corresponding contraction kernel. Monotonic dual graph contraction stops when no contraction kernel with more than one vertex can be chosen.
3. The crest graph G_c is embedded into the plane as described in Section 3.

If monotonic dual graph contraction is applied to the pair $(G, \overline{G})$ instead of $(\overline{G}, G)$, the graph $\overline{G}$ will be transformed into a graph that describes the river lines of a DEM. The river line graph of Fig. 4(a) is depicted in Figure 5(b)). An overlay of the the crest graph and the river line graph is shown in Fig. 6. Note that, in general, the crest graph and the river line graph are not even abstract duals. However, embedding the two graphs into the DEM, two basins separated by a crest line are always connected by at least one river line that crosses the crest line at a local minimum (saddle point).

The calculation of the crest graph was implemented in C^{++} based on LEDA [MN99]. In order to describe gray level edges in images, we calculated the crest graph on the image transformed by the Sobel operator (Fig. 7). Note that the crest graph is always connected.

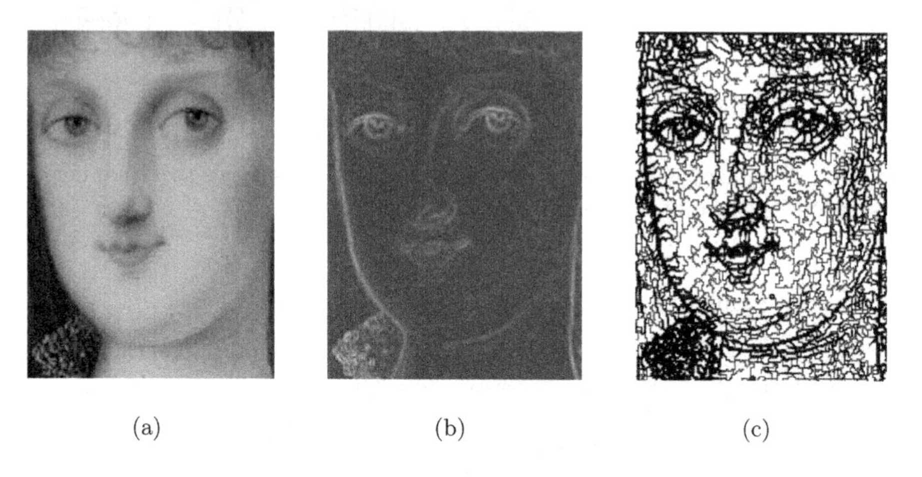

Fig. 7. (a) Gray level image. (b) Input image for the calculation of the crest graph (Sobel operator on (a)). (c) Crest graph of (b). The widths of the edges correspond to the edge values, i.e. the altitudes of the saddle points

5 Shape from the Structure of Distance Transforms

In this section the shape of a 4-connected set S of square pixels is represented by an embedded graph. The embedded graph is the crest graph calculated from a distance transform of S, i.e. the gray values indicate the distance of the pixel to the outside of S. The calculation of the crest graph is done exactly as explained in the previous section. Note that this concept is not restricted to a special grid or to a special distance.

As an example consider the 4-connected set of pixels in Fig. 8(b). The numbers in Fig. 8(b) indicate the *chamfer-3-4* distance [CM91] from the pixels to the outside. A sphere with respect to the chamfer-3-4 distance is shown in Fig. 8(a). The crest graph of the distance transform is depicted in Fig. 8(c). The advantages

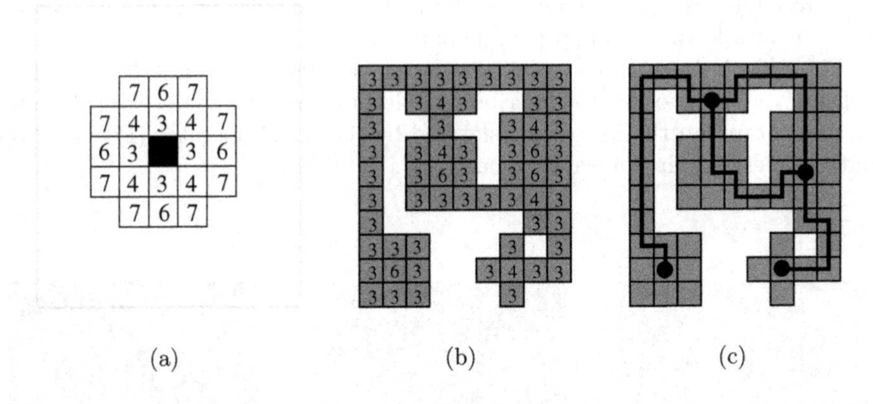

Fig. 8. (a) Sphere with respect to the chamfer-3-4 distance. (b) Chamfer-3-4 distance transform of an object. (c) Crest graph of (b)

of the crest graph compared to the medial axis [Ser82] are as follows.

- Whatever structuring elements are taken for the computation of the distance transform, the crest graph is always connected.
- No pruning is needed [Ogn94].
- The structure of the crest graph, i.e. the crest graph without its embedding, yields a more compact and more abstract description of the shape.

In contrast to the medial axis, shape representation by means of the crest graph is lossy. In particular, the crest graph does not represent narrowing dead ends. In the future we will try to overcome this drawback by a modification of the distance transform: the pixel values at the edge of the image are raised such that they form a new summit. The edges which connect this summit with the summits in the inner part of the image will run through the narrowing dead ends.

6 Conclusion

The hierarchy of planar graphs, as obtained by dual graph contraction, has been combined with a hierarchical plane embedding of the graphs. It was shown that, in general, it is impossible to find such an embedding also for the corresponding hierarchy of the dual graphs. The proposed hierarchical embedding is very useful for the description of image structure by monotonic dual graph contraction and improves the method proposed in [GEK99]. If the image structure is computed on gray values coming from a distance transform of a binary image, monotonic dual graph contraction yields a skeleton-like plane graph. In contrast to the skeletons formed by pixels no pruning is necessary. Future work will focus on the proper representation of narrowing dead ends.

References

BK99. Mark J. Burge and Walter G. Kropatsch. A minimal line property preserving representation of line images. *Computing*, 62:355 – 368, 1999. 349

CM91. Jean Marc Chassery and Annick Montanvert. *Géometrie Discrète en Imagery*. Hermes, Paris, 1991. 356

Die97. Reinhard Diestel. *Graph Theory*. Springer, New York, 1997. 348, 350, 353

Fra96. Jean Francon. On Recent Trends in Discrete Geometry in Computer Science. In Serge Miguet, Michel Montanvert Annick, and Stéphane Ubéda, editors, *Discrete Geometry for Computer Imagery, DGCI'96*, volume Vol. 1176 of *Lecture Notes in Computer Science*, pages 3–16, Lyon, France, 1996. Springer, Berlin Heidelberg, New York. 348

GEK99. Roland Glantz, Roman Englert, and Walter G. Kropatsch. Dual image graph contractions invariant to monotonic transformations of image intensity. In *Proc. of the 2nd IAPR Workshop on Graph-based Representation*, 1999. 349, 352, 355, 357

KD94. J. Koenderink and A. van Doorn. Image structure. In E. Paulus and F. Wahl, editors, *Mustererkennung 1997*, pages 401–408. Springer, 1994. 352

KM95. Walter G. Kropatsch and Herwig Macho. Finding the structure of connected components using dual irregular pyramids. In *Cinquième Colloque DGCI*, pages 147–158. LLAIC1, Université d'Auvergne, ISBN 2-87663-040-0, September 1995. 349, 350

Kro95. Walter G. Kropatsch. Building Irregular Pyramids by Dual Graph Contraction. *IEE-Proc. Vision, Image and Signal Processing*, 142(6):366 – 374, 1995. 350

MN99. K. Mehlhorn and S. Näher. *The LEDA Platform of Combinatorial and Geometric Computing*. Cambridge University Press, Cambridge, U. K., 1999. 355

Ogn94. R. L. Ogniewicz. A Multiscale MAT from Voronoi Diagrams: The Skeleton-Space and its Application to Shape Description and Decomposition. In Carlo Arcelli, Luigi P. Cordella, and Gabriella Sanniti di Baja, editors, *2nd Intl. Workshop on Visual Form*, pages 430–439, Capri, Italy, June 1994. 356

Ser82. Jean Serra. *Image and Mathematical Morphology*, volume 1. Academic Press, London, G. B., 1982. 356

A New Visibility Partition for Affine Pattern Matching

Michiel Hagedoorn*, Mark Overmars, and Remco C. Veltkamp

Department of Computer Science, Utrecht University
Padualaan 14, 3584 CH Utrecht, The Netherlands
{mh,markov,Remco.Veltkamp}@cs.uu.nl

Abstract. Visibility partitions play an important role in computer vision and pattern matching. This paper studies a new type of visibility, *reflection-visibility*, with applications in affine pattern matching: it is used in the definition of the *reflection metric* between two patterns consisting of line segments. This metric is affine invariant, and robust against noise, deformation, blurring, and cracks. We present algorithms that compute the reflection visibility partition in $O((n+k)\log(n)+v)$ randomised time, where k is the number of visibility edges (at most $O(n^2)$), and v is the number of vertices in the partition (at most $O(n^2+k^2)$). We use this partition to compute the *reflection metric* in $O(r(n_A+n_B))$ randomised time, for two line segment unions, with n_A and n_B line segments, separately, where r is the complexity of the overlay of two reflection-visibility partitions (at most $O({n_A}^4+{n_B}^4)$).

1 Introduction

The visibility from a particular viewpoint in a pattern gives a local description of the pattern. The visibilities from all possible viewpoints give a complete representation of the pattern. This insight led to the use of visibility for pattern matching. Visibility has been used for object-recognition as early as 1982, see Chakravarty and Freeman [6]. Visibility is defined in terms of affine geometry, the concept does not depend on Euclidean distances. Therefore, visibility can be used as a tool for affine invariant shape recognition and affine pattern matching. We use a strong type of visibility to compute an affine invariant pattern metric, called the reflection metric. This metric is affine invariant, and robust against noise, deformation, blurring, and cracks, see Section 2.

A well-studied structure related with visibility is the visibility graph. For a collection of n planar line segments, the visibility graph is the graph having the endpoints of the line segments as vertices, and edges between vertices for which the corresponding endpoints can be connected by an open line segment disjoint with all segments in the collection. A trivial algorithm, consisting of three nested loops, computes the visibility graph in $O(n^3)$ time. Lee [18] was the first to improve this, by giving an $O(n^2\log(n))$ time algorithm. Optimal $O(n^2)$

* supported by Philips Research

G. Borgefors, I. Nyström, and G. Sanniti di Baja (Eds.): DGCI 2000, LNCS 1953, pp. 358–370, 2000.
© Springer-Verlag Berlin Heidelberg 2000

time algorithms were found by Welzl [23], and Asano et al. [3]. If the number of visibility edges is k, an algorithm by Pocchiola and Vegter [22] computes the visibility graph in $O(n \log(n) + k)$ time and $O(n)$ space.

A plausible approach to using visibility for pattern matching would be visibility graph recognition, see Ghosh [9], and Everett [8]. However, visibility graph recognition is a computationally expensive problem, see Lin and Skiena [19]. Moreover, visibility graphs depend heavily on the topology of the pattern, and are therefore by themselves not suitable for robust pattern matching.

Our approach uses a special form of visibility, called reflection-visibility, for defining a similarity measure on line patterns. This measure, the reflection metric, is affine invariant by definition. That is, the distance between affine transformed patterns $t(A)$ and $t(B)$ equals the distance between the original A and B. The reflection metric is robust. It responds proportionally when lines are deformed a little, slightly translated copies of existing lines are added, small cracks are made in lines, and small new lines are added.

To compute the reflection metric we need to know the structure of the visibility partition corresponding to a special type of visibility. Visibility partitions consists of equivalence classes with constant combinatorial visibility. Plantinga and Dyer [21] call this structure the *viewpoint space partition.* The dual of the visibility partition is called the *aspect graph,* see Kriegman and Ponce [17], Bowyer and Dyer [5], and Gigus et al. [11]. The number of possible views, the size of the visibility partition, was investigated, under varying assumptions, by Agarwal and Sharir [1], and de Berg et al. [7]. For polygons, results about visibility partitions were found by Guibas et al. [13], Aronov et al. [2], and Bose et al. [4].

In this paper, we focus on the structure of visibility partitions as two-dimensional arrangements. We will investigate an alternative visibility partition, called reflection-visibility. As a start, we consider the standard visibility partition. We use an alternative way to define and describe it that is useful for the other types of visibility. After that, the analysis becomes more interesting as we proceed to the reflection-visibility partition. Let k be the number of visibility edges (at most quadratic in n) and v be the number of vertices in the partition (at most quadratic in $n + k$). We present randomised algorithms that compute the partitions in $O((n+k)\log(n)+v)$ time. We use this to compute the reflection distance. We assume a model of computation in which the absolute value of any rational function, a quotient of polynomials of degree at most d, can be integrated over any triangle in $\Theta(d)$ time. Let A and B be unions of n_A and n_B segments, respectively. If the overlay of the two corresponding reflection-visibility partitions has complexity r, the reflection distance between A and B can be computed in $O(r(n_A + n_B))$ time.

2 The Reflection Metric

The reflection metric, introduced in [16,14], defines a distance between finite unions of algebraic curve segments in the plane. If A and B are such unions, the reflection distance is denoted as $d_{\mathtt{R}}(A, B)$. The reflection metric turns the patterns A and B into functions $\rho_A, \rho_B : \mathbb{R}^2 \to \mathbb{R}$, after which the integrated absolute difference of these functions is normalised. The functions are defined using a strong form of visibility, called reflection-visibility. We defer the exact definition of the reflection metric until Section 5.

Figure 1 shows a two-dimensional pattern A consisting of a finite number of straight line segments. Figure 2 shows the corresponding function evaluated on a discrete lattice, represented as a grey-scale image in which black corresponds with value 0. The example pattern is hieroglyphic 'A1' obtained from the hieroglyphics sign list, see [12].

Fig. 1. A straight line pattern

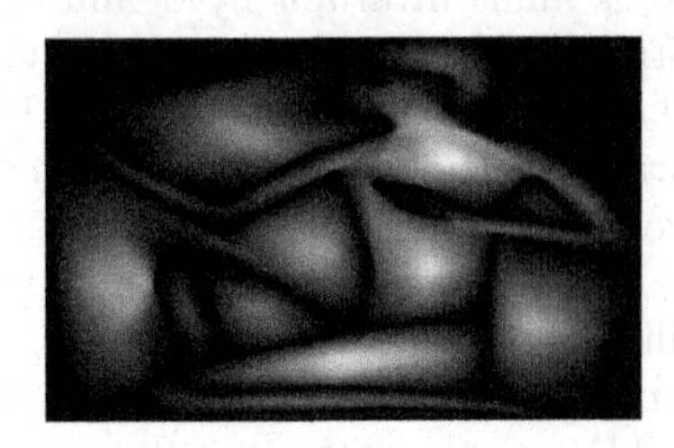

Fig. 2. The function ρ_A evaluated on a discrete lattice

Let T be the group of affine transformations on $\mathbb{R}^2$. The reflection metric is invariant for T, meaning that $d_{\mathbb{R}}(t(A), t(B)) = d_{\mathbb{R}}(A, B)$ for any affine transformation $t \in T$. As a result, $d_{\mathbb{R}}$ can be used to construct a metric on *affine shapes*, patterns modulo affine transformation:

$$D_R(T(A), T(B)) = \inf_{t \in T} d_{\mathbb{R}}(t(A), B).$$

The reflection metric is robust for various types of effects caused by discretisation and unreliable feature extraction. Slight *deformations* of patterns only increase or decrease distances slightly. Introducing *blur*, by adding new lines near existing lines in a pattern, only causes a proportional change in the reflection distance. Making *cracks* in the interior of lines, splitting them up into multiple smaller ones, only changes the distance proportional to the length of these cracks. Adding *noise* in the form of new lines, far away from other lines, only increases the distance proportionally to the length of the added line. Figure 3 illustrates the effects of deformation, blur, cracks, and noise. The left pattern is the "original", the right pattern is "affected" by the four types of distortion.

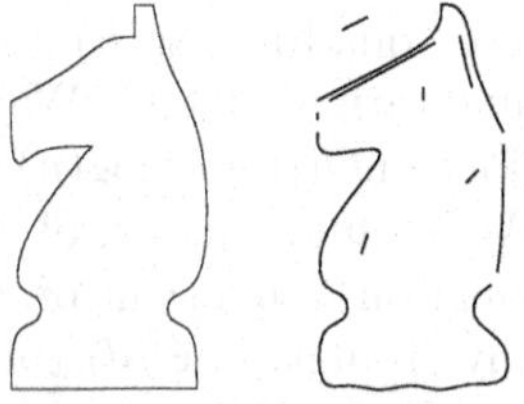

Fig. 3. Deformation, blur, cracks and noise

In Section 5, we show that the reflection metric can be computed by constructing and traversing an arrangement, which is the overlay of two reflection-visibility partitions. In the following sections, we characterise the visibility and reflection visibility partitions. We will give a randomised algorithm that is optimal in the number of segments n.

3 Visibility Partitions

Let $\mathcal{S} = \{ S_1, \dots, S_n \}$ be a collection of closed line segments and let $P = \{ p_1, \dots, p_m \}$ be the corresponding set of endpoints. In all that follows, we assume that the endpoints are in general position. For convenience, set $A = \bigcup \mathcal{S}$. We say a point $y \in \mathbb{R}^2$ is visible from $x \in \mathbb{R}^2$, if the open line segment $\overline{xy}$ is disjoint with A. For any viewpoint $x \in \mathbb{R}^2$, define the *visible part* of A as the subset $\mathrm{Vp}_A(x) \subseteq A$ given by:

$$\mathrm{Vp}_A(x) = \{ a \in A \mid A \cap \overline{xa} = \varnothing \}.$$

This set is sometimes called the *visibility polygon* of x in A, see [3]. The *visibility star* $\mathrm{Vst}_A(x)$ is the union of all open line segments connecting the viewpoint x with the visible part of A:

$$\mathrm{Vst}_A(x) = \bigcup_{a \in \mathrm{Vp}_A(x)} \overline{xa}.$$

Visibility stars are similar to *view zones*, see [20] pp. 383–391. A view zone is a visibility star extended with all infinite rays from x disjoint with A. Figure 4 shows a visibility star for an example consisting of eight line segments with thirteen distinct endpoints. The visible part $\mathrm{Vp}_A(x)$ is drawn thick. The visibility star $\mathrm{Vst}_A(x)$ is the light grey region, including the dotted lines.

Consider the endpoints and segments bounding the visibility star $\mathrm{Vst}_A(x)$ ordered by slope with respect to x. This describes the structure of the visibility star. The visibility star is a finite union of triangles. Each triangle is an intersection of three half-planes. Two of the half-planes are bounded by lines through x and a point in P. The third half-plane is bounded by the line through a segment of $\mathcal{S}$. If a segment $S_i \in \mathcal{S}$ has a visible endpoint p_j, and x is collinear with S_i, then the triangle "degenerates" to the open line segment $\overline{xp_j}$. We are interested in the regions in the plane in which the structure of the visibility star is constant.

We simplify the presentation by introducing an additional "line segment". Let D be an open rectangle containing the union of segments A, and let S_0 be its boundary. We will simply call S_0 a segment. This gives an extended collection of segments $\mathcal{S}' = \{ S_0 \} \cup \mathcal{S}$. Let $A' = S_0 \cup A$. Each ray starting from any point in D, intersects A'.

We need a compact description of the structure of the visibility star $\mathrm{Vst}_A(x)$, for any viewpoint $x \in D$. For this purpose, we define a collection of identifiers, referring either to segments or endpoints. An identifier is an (integer) index subscripted with a p or an s, indicating an endpoint or a segment, respectively. We order the identifiers linearly as follows: $0_s < 1_s < \cdots < n_s < 1_p < \cdots < m_p$. Each point $a \in A'$ is assigned an identifier $\mathrm{id}(a)$ as follows. If $a = p_i$, then set $\mathrm{id}(a) = i_p$. If $a \in S_i - P$, then set $\mathrm{id}(a) = i_s$.

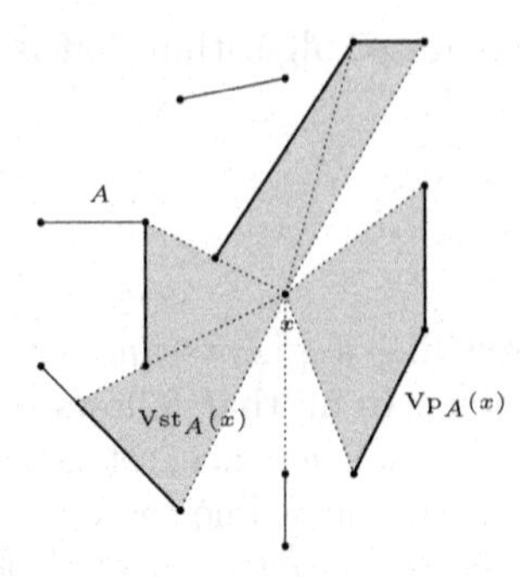

Fig. 4. A visibility star

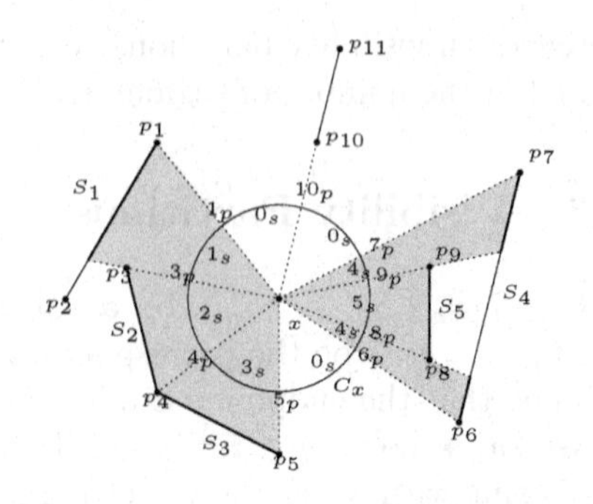

Fig. 5. A view map

We represent the structure of the visibility star by a tuple identifiers. Choose any closed disc centred at x disjoint with A. The boundary of such a disc is called the *view circle*, denoted by C_x. We give each point $c \in C_x$ a label $l(c)$, an identifier, as follows. For each point $a \in A'$ visible from x, compute the intersection $c \in C_x \cap \overline{xa}$ and set $l(c) = \mathrm{id}(a)$.

The view circle C_x is a disjoint union of inverse images $l^{-1}(d)$, for each identifier d. Each non-empty inverse image $l^{-1}(d)$, the subset of points in C_x with label d, can be decomposed into its connected components. These components are either single points or open arcs. The *view map* of x, denoted with $\mathrm{Vmp}_{\mathcal{S}}(x)$, is a labeled circuit graph whose vertices are the components with their constant labels. We call vertices labeled with endpoint identifiers (i_p) *p-vertices.* We call vertices labeled with segment identifiers (i_s) *s-vertices.* Edges of $\mathrm{Vmp}_{\mathcal{S}}(x)$ are defined by pairs of vertices, constant-label components, with intersecting closures. Figure 5 shows the labeled view circle, inducing the view map, for a collection of four closed line segments having nine distinct endpoints. Labels of p-vertices are indicated on the dotted lines on the outside of C_x. Labels of s-vertices are indicated inside the view circle between successive dotted lines.

The view map $\mathrm{Vmp}_{\mathcal{S}}(x)$ can be represented using a tuple of labels encountered when traversing all edges, starting with some initial vertex and some incident edge. Of all possible tuples, the lexicographically smallest one represents the view map. This representation does not depend on the direction (clockwise or counter-clockwise) in which the labels occur on the view circle. We identify the view map with this unique tuple. In the situation of Figure 5 this gives $\mathrm{Vmp}_{\mathcal{S}}(x) = (0_s, 1_p, 1_s, 3_p, 2_s, 4_p, 3_s, 5_p, 0_s, 6_p, 4_s, 8_p, 5_s, 9_p, 4_s, 7_p, 0_s, 10_p)$.

The view map $\mathrm{Vmp}_{\mathcal{S}}(x)$ is a function of points $x \in D$. Define points $x, y \in D$ to be equivalent if their view maps (labeled graphs) are isomorphic, that is, $\mathrm{Vmp}_{\mathcal{S}}(x) = \mathrm{Vmp}_{\mathcal{S}}(y)$. This equivalence relation results in a partition of D into equivalence classes. This *visibility partition* is denoted by $\mathcal{Q}_v(\mathcal{S})$. If x and y lie in the same class $Q \in \mathcal{Q}_v(\mathcal{S})$ of the partition, the visibility stars $\mathrm{Vst}_A(x)$ and $\mathrm{Vst}_A(y)$ have the same structure.

The visibility partition is affine invariant: the partition for the affine transformed set $\mathcal{S}$ equals the affine transformed partition (including the labels). The reflection-visibility partition also has this property. A single class in the visibility partition can have more than one connected component. In the example of a single segment $\mathcal{S} = \{ S_1 \}$ with endpoints $P = \{ p_1, p_2 \}$, the open half-planes left and right of S_1 are the connected components of the equivalence class in $\mathcal{Q}_v(\mathcal{S})$ having view map $(0_s, 1_p, 1_s, 2_p)$.

The visibility partition has the structure of an arrangement induced by a finite union of closed line segments. Each cell in this arrangement is a connected component of an equivalence class in the partition. As the viewpoint x moves continuously within D, changes occur in $\mathrm{Vmp}_{\mathcal{S}}(x)$. Each time such a change occurs, the set of vertices visible from x changes. The sets of viewpoints x on which changes in the viewmap occur form one-dimensional boundaries in the arrangement describing the visibility partition.

We construct a collection of "event segments" for the view map. Let $E_{\mathcal{S}}$ be the collection of (directed) edges in the visibility graph. That is, $E_{\mathcal{S}}$ consist of all pairs of endpoint-indices (i, j) such that p_j is visible from p_i. We extend $E_{\mathcal{S}}$ to a collection $E'_{\mathcal{S}}$ by also including the endpoint-index pairs of each segment in $\mathcal{S}$. Given an endpoint p_i, sort all endpoints p_{j_k}, with $(i, j_k) \in E'_{\mathcal{S}}$, on clockwise angle. This results in a list of endpoint identifiers $j_1, \ldots, j_c$. Let s_k be the segment-identifier of the segment visible from p_i inbetween the angles of p_{j_k} and $p_{j_{k+1}}$ relative to p_i (where $k+1$ is modulo c). We construct event segments bounding the set of points in D from which p_i is visible.

First, we define the collection of event segments $\mathcal{P}_i$. For each $k = 1, \ldots, c$, we include in $\mathcal{P}_i$ the closure of the visible part of segment S_{s_k} (visible from p_i). This includes parts of the special segment with index 0_s.

Second, we construct a segment collection $\mathcal{B}_i$ connecting pairs of segments in $\mathcal{P}_i$. For each $k = 1, \ldots, c$, we construct a closed segment between the two intersections of $\mathrm{ray}(p_i, p_{j_k})$ with segments in $\mathcal{S}'$. If these two intersection coincide, we include no segment in $\mathcal{B}_i$, for that particular k.

The third and last types of segments $\mathcal{X}_i$ are extensions of segments in $\mathcal{S}$. Consider each segment S in $\mathcal{S}$ having endpoint p_i, and having another endpoint p_j. Include in $\mathcal{X}_i$, the closed segment having endpoints p_i, and the intersection of $\mathrm{ray}(p_i, p_j)$ with $\bigcup \mathcal{P}_i$ that is closest to p_i.

The three types of segments result in the arrangement describing the visibility partition. Let $\mathcal{P}$, $\mathcal{B}$, and $\mathcal{X}$, denote the unions of $\mathcal{P}_i$, $\mathcal{B}_i$, and $\mathcal{X}_i$, over all $i = 1, \ldots, m$, respectively.

Theorem 1. *The boundaries in the visibility partition are formed by the event segments:* $\bigcup_{Q \in \mathcal{Q}_v(\mathcal{S})} \partial Q = \bigcup \mathcal{P} \cup \bigcup \mathcal{B} \cup \bigcup \mathcal{X}$.

Figure 6 shows the visibility partition for four line segments having seven distinct endpoints. The union of $\mathcal{P}$ coincides with A. The segments of $\mathcal{B}$ are drawn dashed. The segments of $\mathcal{X}$ are drawn coarse dashed. The points where event segments meet are indicated as dots. The rectangle containing the segments is the "segment" S_0.

For a derivation of the worst-case complexity of the visibility partition, see [15]. The visibility partition $\mathcal{Q}_v(\mathcal{S})$ corresponding to $\mathcal{S}$ can be computed as follows. First, we compute the visibility graph for $\mathcal{S}$, in $O(n\log(n)+k)$ time using algorithms by Ghosh and Mount [10] or Pocchiola and Vegter [22]. Using the visibility graph, the view map can be computed for each endpoint. A simple algorithm discovers the segment visible between successive visibility edges incident to each endpoint by performing $\log(n)$ iterations over the visibility edges. Include the segments of $\mathcal{S}$ as visibility edges. For each (directed) visibility edge $e=(p,q)$, such that no segment adjacent to q is visible to the left of q relative to p, store a pointer $\mathrm{r}(e)$ to the visibility edge incident to q that turns right relative to e. If a segment adjacent to q is visible to the left of q (relative to p), we store a pointer to this segment in $\mathrm{r}(e)$. In each iteration we consider all edges e in the visibility graph. If $\mathrm{r}(e)$ is not a segment blocking the view to the left we replace $\mathrm{r}(e)$ by $\mathrm{r}(\mathrm{r}(e))$. Analogous, we maintain pointers $\mathrm{l}(e)$, where the roles of left and right exchange. After $O(\log(k))=O(\log(n))$ iterations, we have found the segments that block the view directly to the left and the right of each directed visibility edge. This takes a total of $O((n+k)\log(n))$ time. Using this information, we can generate the total collection of event segments in $O(n+k)$ time. Let v be the number of intersections in the collection of event segments thus generated. Using randomised incremental construction we construct a trapezoidal decomposition of this collection, see Mulmuley [20] pp. 84-94, in $O((n+k)\log(n)+v)$ time. The arrangement defined by the event segments can be obtained by merging together trapezoids into polygonal cells. Thus, the visibility partition, represented as an arrangement, can be computed using randomised techniques in $O((n+k)\log(n)+v)$. This results in the following theorem.

Theorem 2. *The visibility partition of n segments has worst-case complexity $\Theta(n^4)$. Using randomisation, it can be computed in $O((n+k)\log(n)+v)$ time, where $k=\Theta(n^2)$ is the number of visibility edges, and $v=O(n^2+k^2)$ is the number of vertices in the arrangement.*

4 Reflection Partitions

In this section, we consider a stronger notion of visibility, resulting in different stars and partitions. We say that a point $y\in\mathbb{R}^2$ is *trans-visible* from a point $x\in\mathbb{R}^2$ if y is visible from x and both $\mathrm{ray}(x,y)$ and $\mathrm{ray}(y,x)$ intersect A. The *trans-visible part* of A is given by:

$$\mathrm{Tp}_A(x)=\{\,a\in\mathrm{Vp}_A(x)\mid A\cap\mathrm{ray}(a,x)\neq\varnothing\,\}.$$

The *trans-visibility star* $\mathrm{Tst}_A(x)$ is the union of all open line segments between x and the trans-visible part of A:

$$\mathrm{Tst}_A(x)=\bigcup_{a\in\mathrm{Tp}_A(x)}\overline{xa}.$$

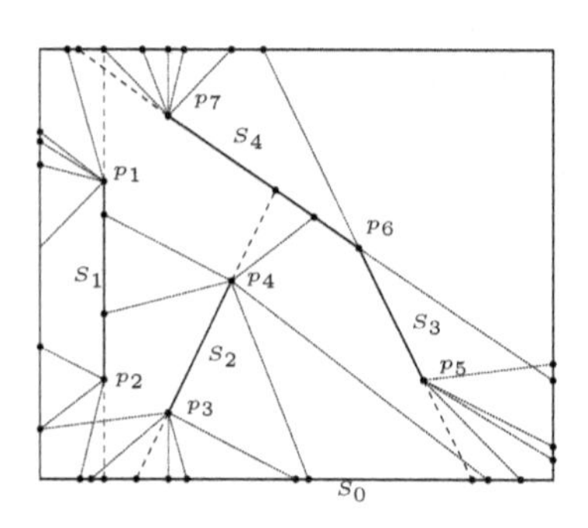

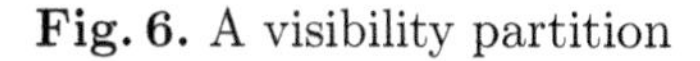
Fig. 6. A visibility partition

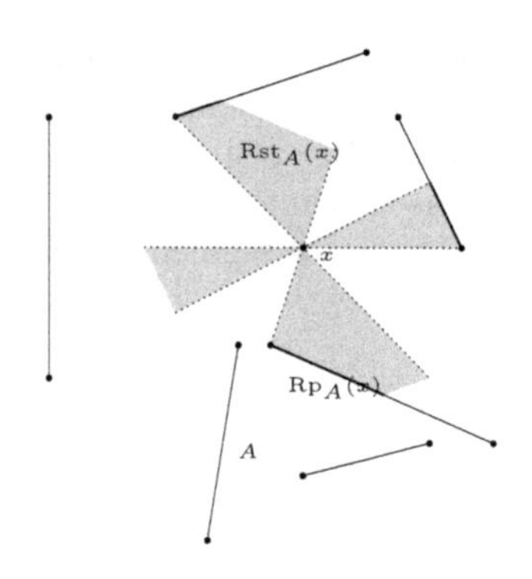

Fig. 7. A reflection star

Based on trans-visibility, we can define a trans-view map, and a trans-visibility partition, see [15].

We say a point $y \in \mathbb{R}^2$ is *reflection-visible* from a point x, if y is trans-visible from x and the open segment between y and the reflection of y in x is disjoint with A. Observe that in contrast with visibility and trans-visibility, reflection-visibility is not symmetric. Define the *reflection-visible part* of A as follows:

$$\mathrm{Rp}_A(x) = \{\, x + v \in \mathrm{Tp}_A(x) \mid \overline{(x+v)(x-v)} = \varnothing \,\}.$$

Define the *reflection-visibility star* as:

$$\mathrm{Rst}_A(x) = \bigcup \{\, \overline{(x+v)(x-v)} \mid x + v \in \mathrm{Rp}_A(x) \,\}.$$

The reflection-visibility star equals the intersection of a trans-visibility star $\mathrm{Tst}_A(x)$ with its reflection around x. Figure 7 shows a reflection-visibility star for six disjoint segments. The grey area, including the dotted segments, forms the reflection-visiblity star.

Consider a view circle C_x centred at x. We label the view circle to describe the structure of the reflection-visibility star at x. We define a labeling l of a view circle C_x, that represents the structure of the reflection visibility star. We use polar coordinates, such that each point $c \in C_x$, is represented as $c = (\alpha, r)$, where r is the radius of C_x. Let $L(x, \alpha)$ be the line through x and (α, r). Let $\epsilon > 0$ be smaller than the angle between any two endpoints. There are three cases:

1. If $L(x, \alpha)$ intersects a visible point of S_0, then $l(c) = 0_s$.
2. If $L(x, \alpha)$ intersects a visible endpoint $p \in P$ and $L(x, \alpha - \epsilon)$ or $L(x, \alpha + \epsilon)$ intersects a visible point of S_0, then $l(c) = \mathrm{id}(p)$.
3. In all other cases, set $l(c) = \mathrm{id}(a)$, where a is the visible point in $A \cap L(x, \alpha)$ closest to x.

Overlaps in the rules are resolved by choosing the minimum-index identifier.

The labeling l defines a labeled circuit graph called the *reflection view map*, denoted by $\mathrm{Rmp}_{\mathcal{S}}(x)$. Figure 8 shows the view circle along with the labels of the reflection-view map. The reflection-visible part is shown thick. The dashed lines are reflections of segments in the view point.

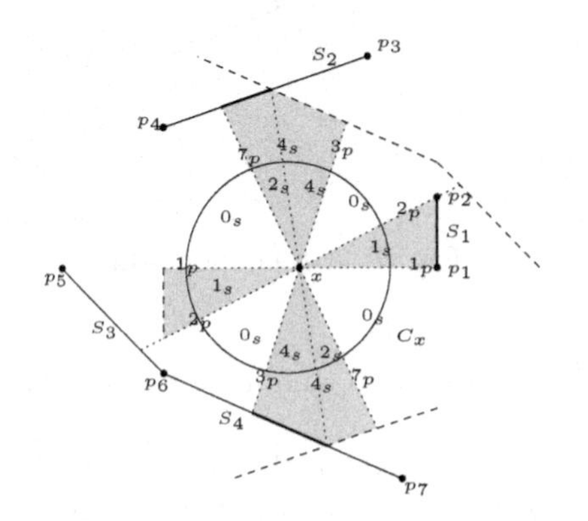

Fig. 8. A reflection-view map

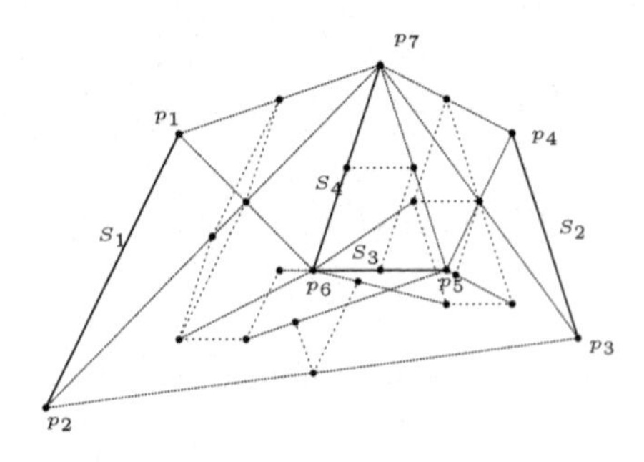

Fig. 9. A reflection-visibility partition

The reflection-view map $\mathrm{Rmp}_{\mathcal{S}}(x)$ at a point x contains the structure of the reflection-visibility star $\mathrm{Rst}_A(x)$ at a given point. Starting at some vertex and some initial edge, we obtain a tuple of labels representing the reflection-view map. Since this tuple repeats itself, we only take the first half. We choose the lexicographically smallest half-tuple as a unique representation of $\mathrm{Rmp}_{\mathcal{S}}(x)$. In the situation of Figure 8 this gives $\mathrm{Rmp}_{\mathcal{S}}(x) = (0_s, 1_p, 1_s, 2_p, 0_s, 3_p, 4_s, 2_s)$.

We obtain a reflection-visibility partition by identifying points $x, y \in D$ if their reflection-view maps $\mathrm{Rmp}_{\mathcal{S}}(x)$ and $\mathrm{Rmp}_{\mathcal{S}}(y)$ are equal. When moving x, if the reflection-view map changes, then the set of reflection-visible endpoints (relative to x) changes: if we move x, either an endpoint identifier, a segment identifier, or an intersection identifier, appears or disappears. In case of an endpoint identifier, the result is immediate. In case of a segment identifier, a change occurs only if one endpoint starts to occlude another one. If an intersection identifier disappears it is replaced by an endpoint identifier. Since a change in the set of endpoints also implies a change in the reflection view map, we can conclude the following. Each class in the reflection-visibility partition is a maximal connected subset of D in which a fixed set of endpoints is reflection-visible.

We construct a collection of event segments for the reflection-view map. Consider an endpoint p_i. Consider $\mathcal{P}_i$ as for visibility. Replace each segment in $\mathcal{P}_i$ that lies in S_0 by a "degenerate" segment consisting of the point p_i. Now build $\mathcal{B}_i$ and $\mathcal{X}_i$ using $\mathcal{P}_i$ just as in the visibility case.

Consider a scaling transformation f that leaves p_i fixed and which has a scaling factor of $1/2$. That is, all coordinates relative to p_i are multiplied by

1/2. To obtain the desired reflection-view map, replace all segments in $\mathcal{P}_i$, $\mathcal{B}_i$ and $\mathcal{X}_i$ by their images under f. Like before, we construct unions of the segment collections over $i = 1, \ldots, m$.

Theorem 3. *The boundaries in the reflection-visibility partition are formed by the event segments:* $\bigcup_{Q \in \mathcal{Q}_r(\mathcal{S})} \partial Q = \bigcup \mathcal{P} \cup \bigcup \mathcal{B} \cup \bigcup \mathcal{X}$.

Figure 9 shows a reflection-visibility partition. The complexity of the reflection-visibility partition is at most $O(n^4)$. For the $\Omega(n^4)$ worst-case lower bound, see [15]. To compute the reflection-visibility, we can use the same basic techniques to compute the normal visibility partition.

Theorem 4. *The reflection-visibility partition of n segments has worst-case complexity $\Theta(n^4)$. Using randomisation, it can be computed in $O((n+k)\log(n)+v)$ time, where $k = \Theta(n^2)$ is the number of visibility edges, and $v = O(n^2+k^2)$ is the number of vertices in the arrangement.*

5 Computing the Reflection Metric

As an application of the previously defined structures, we use them in computing the reflection metric. Let $\rho_A(x)$ be the area of the reflection star $\mathrm{Rst}_A(x)$ for each $x \in \mathbb{R}^2$, see Figure 2 for an example. Observe that for points x outside the convex hull of A, this area is always zero. If we have two finite unions of line segments A and B, the reflection metric d_{R} is defined as:

$$d_{\mathrm{R}}(A,B) = \frac{\int_{\mathbb{R}^2} |\rho_A(x) - \rho_B(x)|\, dx}{\int_{\mathbb{R}^2} \max(\rho_A(x), \rho_B(x))\, dx}.$$

The reflection metric can be generalised to finite complexes of $d-1$ dimensional algebraic hyper-surface patches in d dimensions. For this, we refer to [16]. Here, we focus at the computation of the reflection metric for finite unions of segments in the plane.

In Section 2, we emphasised the fact that the reflection metric is robust for deformation, blur, cracks, and noise. This property can be derived from the definition without much difficulty. Given a fixed viewpoint, the change in the area of the visibility star caused by each of the above effects is proportional. The area of the reflection-visibility star, $\rho_A(x)$, changes at most twice as much as the area of the visibility star. This pointwise behaviour of the ρ_A-function is preserved as we integrate it, showing robustness of the reflection metric.

Now, we apply the results from the previous sections to compute the reflection metric. We assume A and B are unions of n_A and n_B line segments, respectively. having at most k edges in their visibility graphs. The reflection metric can be rewritten as follows:

$$d_{\mathrm{R}}(A,B) = \frac{2\int_{\mathbb{R}^2} |r(x)|\, dx}{\int_{\mathbb{R}^2} |p(x)|\, dx + \int_{\mathbb{R}^2} |q(x)|\, dx + \int_{\mathbb{R}^2} |r(x)|\, dx},$$

where $p(x) = \rho_A(x)$, $q(x) = \rho_B(x)$, and $r(x) = p(x) - q(x)$. The functions p, q and r are piecewise rational functions in two variables. With piecewise we mean that there is a finite number of triangles covering the support of the function, such that the restriction of the function to each such triangle is a rational function in two variables. The functions p and q are quotients of polynomials of degrees $O(n_A)$ and $O(n_B)$, respectively. The function r is a quotient of polynomials having degree $O(n_A + n_B)$. We adopt a model of computation in which the absolute value of a rational function in two variables, can be integrated over a triangular domain in $\Theta(d)$ time, where d is the maximum degree of the polynomial numerator and denominator.

The computation of the integrals of the rational functions p, q and r proceeds as follows. Let k_A and k_B denote the number of visibility edges corresponding to A and B respectively. First, we compute the visibility graphs of A and B, taking times $O(n_A \log(n_A) + k_A)$ and $O(n_B \log(n_B) + k_B)$, respectively. Using the algorithm sketched in Section 3, the event segments that correspond to the reflection visibility partition, can be found in $O(s_A \log(n_A))$ and $O(s_B \log(n_B))$ time, where $s_A = \Omega(k_A)$ and $s_B = \Omega(k_B)$ are the number of event segments for A and B, respectively. Then, we compute a trapezoidal decomposition for the union of both event segment collections in time $O((s_A + s_B)\log(n_A + n_B) + v)$, where v is the number of intersections. We integrate the absolute values of p, q, and r by summing the partial integrals over all trapezoids (each trapezoid is a union of two triangles). In our model of computation, this takes $\Theta(n_A + n_B)$ time for each trapezoid. Since the summation of partial integrals dominates the overall complexity, we arrive at the following result.

Theorem 5. *Let A and B each be unions of n_A and n_B segments, respectively. Using randomisation, the reflection distance $d_{\mathbb{R}}(A, B)$ can be computed in $O(r(n_A + n_B))$ time, where r is the complexity of the overlay of the reflection-visibility partitions of A and B.*

6 Conclusion

We presented a new metric for pattern matching, the reflection metric. This metric is invariant under the group affine transformations and can therefore be used for affine shape recognition. The reflection metric is robust for pattern-defects such as deformation, blur, cracks, and noise. It can be generalised to finite unions of algebraic hyper-surface patches in any dimension.

The reflection metric is defined in terms of reflection visibility. Trans-visibility and reflection-visibility are stronger than visibility. Reflection-visible points are always trans-visible, and trans-visible points are always visible. We analysed the partitions corresponding to the visibilities, starting at normal visibility, and proceeding with reflection-visibility. New types of events emerged, making the resulting partitions more complex.

Constructions show that the worst-case combinatorial complexity for each of the types of partitions is $\Omega(n^4)$. Using randomised incremental construction, each of the corresponding arrangements can be built in $O((n + k)\log(n) + v)$

time, where k is the number of visibility edges, and v is the number of intersections in the arrangement. The structure of reflection-visibility partitions can be used to compute the reflection metric for two collections of segments in $O(r(n_A + n_B))$ randomised time, where r is the complexity of the overlay of two reflection-visibility partitions. Limiting the endpoints of segments to $\mathbb{Z}^2$ would not change the complexity of the partitions, since any finite arrangement in $\mathbb{R}^2$ can be transformed into an arrangement in $\mathbb{Z}^2$. Limiting the segment slopes to set of allowed slopes does not affect the worst-case complexity of the visibility partition, but does affect the complexity of computing the metric, since the maximal polynomial degree of the piecewise function ρ is linear in the number of different segment slopes.

References

1. P. K. Agarwal and M. Sharir. On the number of views of polyhedral terrains. *Discrete & Computational Geometry*, 12:177–182, 1994. 359
2. B. Aronov, L. J. Guibas, M. Teichmann, and L. Zhang. Visibility queries in simple polygons and applications. In *ISAAC*, pages 357–366, 1998. 359
3. T. Asano, T. Asano, L. Guibas, J. Hershberger, and H. Imai. Visibility of disjoint polygons. *Algorithmica*, 1:49–63, 1986. 359, 361
4. P. Bose, A. Lubiw, and I. Munro. Efficient visibility queries in simple polygons. In *4th Canadian Conference on Computational Geometry*, 1992. To appear in International Journal of Computational Geometry and Applications. 359
5. K. W. Bowyer and C. R. Dyer. Aspect graphs: An introduction and survey of recent results. *Int. J. of Imaging Systems and Technology*, 2:315–328, 1990. 359
6. I. Chakravarty and H. Freeman. Characteristic views as a basis for three-dimensional object recognition. In *Proc. SPIE: Robot Vision*, pages 37–45, 1982. 358
7. M. de Berg, D. Halperin, M. H. Overmars, and M. van Kreveld. Sparse arrangements and the number of views of polyhedral scenes. *Int. J. Computational Geometry & Applications*, 7(3):175–195, 1997. 359
8. H. Everett. *Visibility graph recognition.* PhD thesis, Department of Computer Science, University of Toronto, 1990. 359
9. S. K. Ghosh. On recognizing and characterizing visibility graphs of simple polygons. *Discrete & Computational Geometry*, 17(2):143–162, 1997. 359
10. S. K. Ghosh and D. M. Mount. An output-sensitive algorithm for computing visibility graphs. *SIAM Journal on Computing*, 20:888–910, 1991. 364
11. Z. Gigus, J. Canny, and R. Seidel. Efficiently computing and representing aspect graphs of polyhedral objects. *IEEE Transactions on Pattern Analysis and Machine Intelligence*, 13:542–551, 1991. 359
12. N. Grimal, J. Hallof, and D. van der Plas. Hieroglyphica, sign list. http://www.ccer.theo.uu.nl/hiero/hiero.html, 1993. 360
13. L. J. Guibas, Rajeev Motwani, and Prabhakar Raghavan. The robot localisation problem. *SIAM J. Computing*, 26(4), 1997. 359
14. M. Hagedoorn. Pattern matching using similarity measures PhD thesis, Department of Computer Science, Utrecht University, ISBN 90-393-2460-3, 2000. 359
15. M. Hagedoorn, M. H. Overmars, and R. C. Veltkamp. New visibility partitions with applications in affine pattern matching. Technical Report UU-CS-1999-21,

Utrecht University, Padualaan 14, 3584 CH Utrecht, the Netherlands, July 1999. http://www.cs.uu.nl/docs/research/publication/TechRep.html. 364, 365, 367

16. M. Hagedoorn and R. C. Veltkamp. Measuring resemblance of complex patterns. In L. Perroton G. Bertrand, M. Couprie, editor, *Discrete Geometry for Computer Imagery*, Lecture Notes in Computer Science 1568, pages 286–298, 1999. Springer. 359, 367
17. D. J. Kriegman and J. Ponce. Computing exact aspect graphs of curved objects: Solids of revolution. *ACM Symp. Computational Geometry*, 5:119–135, 1990. 359
18. D. T. Lee. *Proximity and Reachability in the plane.* PhD thesis, University of Illinois at Urbana-Champaign, 1978. 358
19. Y. Lin and S. S. Skiena. Complexity aspects of visibility graphs. *International Journal of Computational Geometry and Applications*, 5:289–312, 1995. 359
20. K. Mulmuley. *Computational Geometry: An introduction through randomized algorithms.* Prentice Hall, 1994. 361, 364
21. W. H. Plantinga and C. R. Dyer. Visibility, occlusion and the aspect graph. *ACM Symp. Computational Geometry*, 5(2):137–160, 1990. 359
22. M. Pocchiola and G. Vegter. Topologically sweeping visibility complexes via pseudotriangulations. *Discrete & Computational Geometry*, 16(4):419–453, 1996. 359, 364
23. E. Welzl. Constructing the visibility graph for n-line segments in $O(n^2)$ time. *Inform. Process. Lett.*, 20:167–171, 1985. 359

Morphological Operations on 3D and 4D Images: From Shape Primitive Detection to Skeletonization

Pieter P. Jonker

Pattern Recognition Group, Faculty of Applied Sciences,
Delft University of Technology,
Lorentzweg 1, 2628 CJ Delft,The Netherlands
pieter@ph.tn.tudelft.nl

Abstract. This paper describes a practical approach to mathematical morphology and ways to implement its operations. The first chapters treat a formalism that has the potential of implementing morphological operations on binary images of arbitrary dimensions. The formalism is based on sets of structuring elements for hit-or-miss transforms whereas each structuring element actually describes a shape primitive. The formalism is applied to two and three dimensional binary images and the paper includes structuring elements for topology preserving thinning or skeletonization and various skeleton variants. The generation of shape primitive detecting masks is treated as well as their application in segmentation, accurate measurement and conditions for topology preserving. The formalism is expanded to four-dimensional images and elaborates on the extension of 3D skeletonization to 4D skeletonization. A short excursion was made to methods based on 3D and 4D Euler - cluster count methods.

1 Introduction

In this paper a practical introduction to Mathematical Morphology applied on binary images of arbitrary dimensions is presented. Mathematical Morphology applied on binary images is also known as Cellular Logic Processing. Whereas the Mathematical Morphology sprouts from a mathematical domain, Cellular Logic Processing has its roots in binary image processing on massively parallel Cellular Logic Machines. The approaches are, however, similar and have their own charm. The method and formalism as described have been developed in parallel with the design of an architecture for real-time image processing [1]. However, the method is (also) implemented in software. In the next chapter, the use of sets of structuring elements is explained and operations in 2D and 3D. In chapter three, the connectivity paradox and topology breaking properties are studied for N dimensional images, followed by skeletonization using shape primitives. The generation and verification of shape primitives is explained, and in chapter four, their extension to the fourth dimension.

G. Borgefors, I. Nyström, and G. Sanniti di Baja (Eds.): DGCI 2000, LNCS 1953, pp. 371-391, 2000.
© Springer-Verlag Berlin Heidelberg 2000

2 Sets of Structuring Elements

2.1 Hit-or-Miss Transformations

The field of cellular logic image processing and mathematical morphology is extensively described in several works (Golay 1969 [2]; Preston 1970 [3]; Serra 1982 [4], 1988 [5], Giardina 1988 [6], Heijmans 1994 [7], Soille 1999 [8], and many others). A basic operation in mathematical morphology is the Hit-or-Miss transformation, and as a starting point for this section we will quote its definition by Serra (1982):

> *The Hit-or-Miss transformation is a point-by-point transformation of a set X, that is performed in the following way:*
> *Choose and fix a structuring element S, the datum of S being two sets S^1 and S^2. Suppose S is centered at the point x of X, denoted by $S=\left(S_x^1,S_x^2\right)$, then a point belongs to the Hit-or-Miss transformation $Y \leftarrow X \otimes S$ of X, if and only if S_x^1 is included in X and S_x^2 is included in the complement X^c of X, or:*

$$Y \leftarrow X \otimes S \equiv \left\{x \middle| \left(S_x^1 \subset X, S_x^2 \subset X^c\right)\right\} \tag{1}$$

To show the similarities, this definition in the mathematical domain will be mirrored to the domain of binary image processing with its convolution kernel-like approach, by associating an origin and a vector structure with it.

We associate with the universe -the union of *X* and X^c -, a square tessellated image X of size m x n, and with the elements of the universe, the pixels of image X. We can then associate with the set *X*, the foreground of image X and with the set X^c its background. Likewise for the sets *Y* and Y^c, and image Y. As structuring element *S* consists of two disjunct sets S^1 and S^2, then its universe is formed by the union of the disjunct sub-sets S^1, S^2 and S^c, S^c being the complement of the union of S^1 and S^2. We associate with the union of S^1, S^2 and S^c, a square tessellated image S of size i x j, and with the elements of S^1, S^2 and S^c, the pixels in the image S. Then we can associate with the set S^1 the foreground pixels, with the set S^2 the background pixels, with the union of S^1 and S^2 the **do-care** pixels and with the set S^c the **don't care** pixels. In our case of binary image processing, the foreground is defined to have the value TRUE or 1, the background to have the value FALSE or 0, and the don't cares to have the value D. Hence, don't cares are used to implement the unrestricted set *S* with image S having a fixed size and shape.

Let x, y and s be the elements (pixels) of X, Y and S, and let the size of S e.g. be i = j = 3, then the 3 x 3 structuring element S consists of $s^0, s^1, ... s^8$ the east (E), north-east (NE), north (N), ... , south-east (SE) neighbours of the central (C) pixel s^8 itself, and an equivalent 3 x 3 neighbourhood M_k around pixel x_k, consists of x_k^0, $x_k^1 .... x_k^7$: the E, NE, N, ... ,SE neighbours of the pixel x_k and x_k the central (C) pixel x_k itself.

The point-by-point transformation $Y \leftarrow X \otimes S$ can now be implemented for binary image processing with the neighbourhood transformation $\mathrm{Y} \leftarrow \mathrm{X} \cong \mathrm{S}$, or $\{\forall \mathrm{k}{:}\mathrm{y}_\mathrm{k} \leftarrow \mathrm{M}_\mathrm{k} \cong \mathrm{S}\}$ with M_k and S centered around pixel x_k, and with $\cong$ the symbol for an inexact match. Informally, the inexact neighbourhood matches $\{\forall \mathrm{k}{:}\mathrm{y}_\mathrm{k} \leftarrow \mathrm{M}_\mathrm{k} \cong \mathrm{S}\}$ can be described as follows:

If for any pixel x_k *in an input image* X, *its neighbourhood* M_k *matches inexactly with a structuring element* S, *the pixel* y_k *of output image Y is set to one. If* M_k *doesn't match* S, y_k *is set to zero.*

In the inexact neighbourhood match the foreground pixels in S should match with foreground pixels in M_k at the same positions AND the background pixels in S should match with background pixels in M_k at the same positions, whereas in the don't care positions of S a match is not required, or:

$$\mathrm{y}_\mathrm{k} \leftarrow \bigwedge_{i=0}^{8} \{\mathrm{x}_\mathrm{k}^\mathrm{i} \cong \mathrm{s}^\mathrm{i}\} \text{ with:} \tag{2}$$

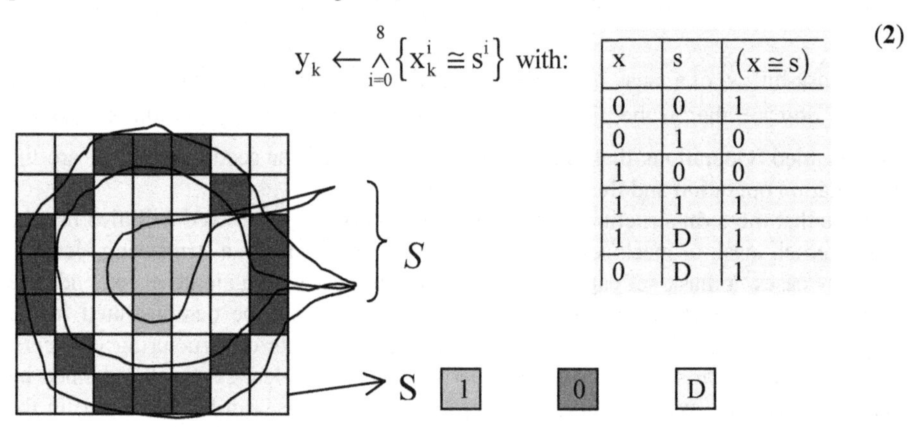

x	s	$(\mathrm{x} \cong \mathrm{s})$
0	0	1
0	1	0
1	0	0
1	1	1
1	D	1
0	D	1

Fig. 1. Structuring element *S* and its implementation with an image S with fixed size and shape

We now extend this inexact neighbourhood match in two ways, first to:

$$\mathrm{y}_\mathrm{k} \leftarrow \bigvee_{i=1}^{p} \{\mathrm{M}_\mathrm{k} \cong \mathrm{S}_\mathrm{i}^\mathrm{S}\} \tag{3}$$

or: *If for any pixel* x_k *in an input image* X, *its neighbourhood* M_k *matches one mask* $\mathrm{S}_\mathrm{i}^\mathrm{S}$ *from a set of masks* S^S *the pixel* y_k *of output image Y is set to one, else to zero.*

Meaning that the union of all mask matches is taken.

And further to:
$$\mathrm{y}_\mathrm{k} \leftarrow \left\{ \bigvee_{i=1}^{p} \{\mathrm{M}_\mathrm{k} \cong \mathrm{S}_\mathrm{i}^\mathrm{S}\} \right\} \vee \left\{ \bigvee_{i=1}^{q} \{\overline{\mathrm{M}_\mathrm{k} \cong \mathrm{S}_\mathrm{i}^\mathrm{R}}\} \right\} \tag{4}$$

or: *If for any pixel* x_k *in an input image* X*, its neighbourhood* M_k *matches any mask* S_i^S *from a given set of masks* S^S *the pixel* y_k *of output image* Y *is set to one, else set to zero, OR if its neighbourhood matches any mask* S_i^R *from a given set of masks* S^R*, the pixel* y_k *is set to zero, else to one.*

The image transformation $Y \leftarrow X \cong S$ is now implemented with a set of masks S as structuring element, consisting of a subset S^S (the SET-masks) and a subset S^R (the RESET-masks), one of which may be empty. This means that either a pixel y_k is set to zero, if one of the RESET masks fits, or the pixel is set to one, if one of the SET mask fits, where the SET masks are chosen to dominate over the RESET masks.

Finally, a second input image Z can be used to locally enable/disable the transformation, yielding the dyadic operation $Y \leftarrow f(\{X, Z\} \cong S)$, or:

$$y_k \leftarrow \left(\left\{ \bigvee_{i=1}^{p} \left\{ M_k \cong S_i^{S,z} \right\} \right\} \vee \left\{ \bigvee_{i=1}^{q} \overline{\left\{ M_k \cong S_i^{R,z} \right\}} \right\} \right) \tag{5}$$

If a "mask-bit" z of a mask of the set S is set don't care, the transformation is enabled. If z is do-care, then if the pixel z_k of Z matches with z, the operation is disabled, else enabled. Operations that use Z to locally mask-off or seed operations, are the propagation operation and the anchor-skeleton.

Note that the term structuring element is used in broad sense. We will also use the word kernel, mask or mask set (all indicated with S) to indicate a structuring element. In many cases, a mask set can be transformed into a set with a single mask, and vice versa, without effecting its functional properties. This will be demonstrated in the sequel. A theoretical treatment of the manipulation of sets of structuring elements, finding the minimal set (the basis) and deriving the equivalent set that is spanned by scanning an image with a sequence of structuring elements, can be found e.g. in the article of Jones and Svalbe (1994)[9], which starts from Matheron's theorem (1975)[10] that "a simple algebra of union intersection and translation suffices to implement all general translation invariant and increasing set mappings".

An operation on an image X can be done by performing the Hit-or-Miss transformation with a mask set S once (6), twice or more (7), or infinite (8: the n-pass form):

$$Y \leftarrow (X \cong S) \tag{6}$$

$$Y \leftarrow \left(\left(\left((X \cong S) \cong S \right) \cong S \right) .... \cong S \right) \tag{7}$$

$$X_0 \leftarrow X;\ \{\forall i{:}X_{i+1} \leftarrow (X_i \cong S)\};\ Y \leftarrow X_\infty \tag{8}$$

In the latter case an iterative scheme is used, which can be implemented by transforming the image until it is idempotent under the transformation.

In many cases practical use can be made of the intermediate results in the output image. This is called **(spatial) recursion**. For instance, if the transformation is performed by a raster scan over the image, i.e. from top-left to bottom-right, the fact that the neighbours NE, N, NW, W of the pixel y_k have already obtained a new value can be utilized. For this purpose, the neighbourhood M_k of pixel x_k^i will be extended by $y_k^{r1}, y_k^{r2}, y_k^{r3}, y_k^{r4}$, the newly obtained values and the structuring element S will be extended by: $S^{r1}, S^{r2}, S^{r3}, S^{r4}$.

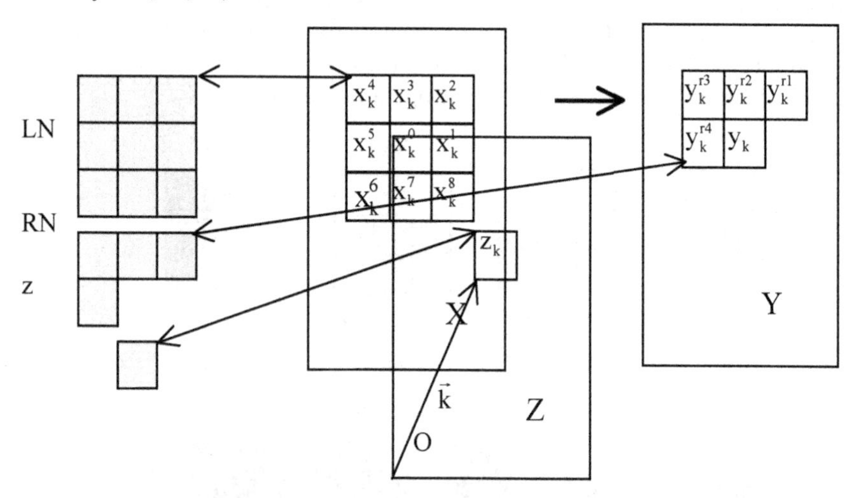

Fig. 2. Drawing conventions for $Y \leftarrow f(\{X, Y, Z\} \cong S)$ Here shown for 2D images and 3 x 3 neighbourhoods. The LN is taken from input image X, the RN from output image Y. For dyadic operations a second input image Z is used

So now both the Local Neighbourhood (LN: the original value x_k and its neighbourhood $x_k^0, x_k^1, x_k^2, \ldots, x_k^7$), and the Recursive Neighbourhood (RN: the newly obtained values $y_k^{r1}, y_k^{r2}, y_k^{r3}, y_k^{r4}$), can be used in the neighbourhood matching procedure. The class of cellular logic operations performed with this extended neighbourhood will be referred to as *recursive neighbourhood operations* (RNO) in contrast with the *local neighbourhood* operations (LNO) that use the normal neighbourhood only. Note that both classes have a dyadic form for which S is extended with an element z and M_k with a pixel z_k of a second image Z, yielding:

$$Y \leftarrow f(\{X, Y, Z\} \cong S) \tag{9}$$

Figure 2 pictures the matching process with a single mask from a set S, whereas the neighbourhood M_k is extracted from the three different images. Note that the Recursive Neighbourhood that can be realised is implementation dependent, and that when using a software raster scan over the image, it is beneficial for RNOs to scan in all odd scans from top-left to bottom-right and in all even scans from bottom-right to top-left. In this case, the RN should be transposed to match with the scan direction.

2.2 Basic Morphological Operations in X_N

The most frequently used known Local Neighbourhood Operations are: Erosion, Dilation, Contour extraction, Spot noise removal (pepper and salt removal) and Majority vote (binary rank filtering). The two most frequently used recursive *and* iterative cellular logic operations are: Connected component extraction or Propagation and Topology Preserving Thinning or Skeletonisation. Figures 3 and 4 show some structuring elements in the form of mask-sets for simple operations in X_2 and X_3.

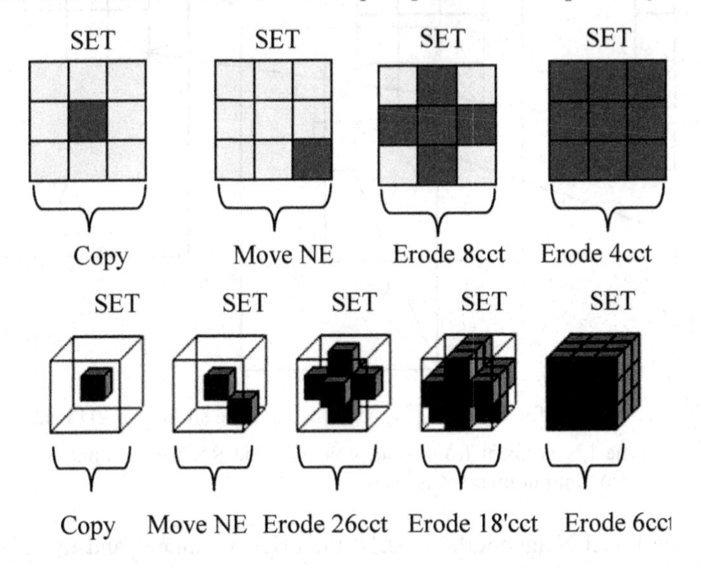

Fig. 3. Structuring elements for simple operations in X_2 and X_3

The structuring elements in Figure 3 are LNOs based on single masks of the SET type, meaning that the central element is set to foreground if the 3^N neighbourhood matches, else to background.

The operations for X_2 are: copy the image, move the image to North-East, erode an 8 connected contour from all objects in the image, erode a 4 connected contour from all objects in the image. Note that the "erode 8cct mask" is in fact a template that matches on foreground area in X_2, it does neither match on background nor places where it touches the contour of an object in any of the directions N,E,S,W. Erode 4cct does the same but in addition will not match on pixels in the NE, NW, SE, SW directions. Because of the template matching aspect of the masks, the principle can be expanded to X_N as the masks for X_3 show.

Figure 4 shows the dilation operation: It comprises a mask that matches on background area in X_2 and on background volume in X_3 while not touching objects. Where the mask matches, the background is reset to background, where the mask does not match, the background is set to foreground. From the contour detection mask set in X_2 and X_3 the first mask matches on foreground not touching the object contour, which is reset to background, and the second mask matches on all background which is also reset to background. Both masks do not match on the object contours, which will hence be set to foreground. A result is shown for X_3, where on a cubic volume with a ball shaped hole, de 26 connected contour surface was detected using the mask-set "detect 26cct". Note that for a better view of the result, the cube was cut open short behind its front face. Finally, the mask-set for Pepper and Salt removal for X_2 shows that the first mask repairs the pepper and the second mask repairs the salt noise.

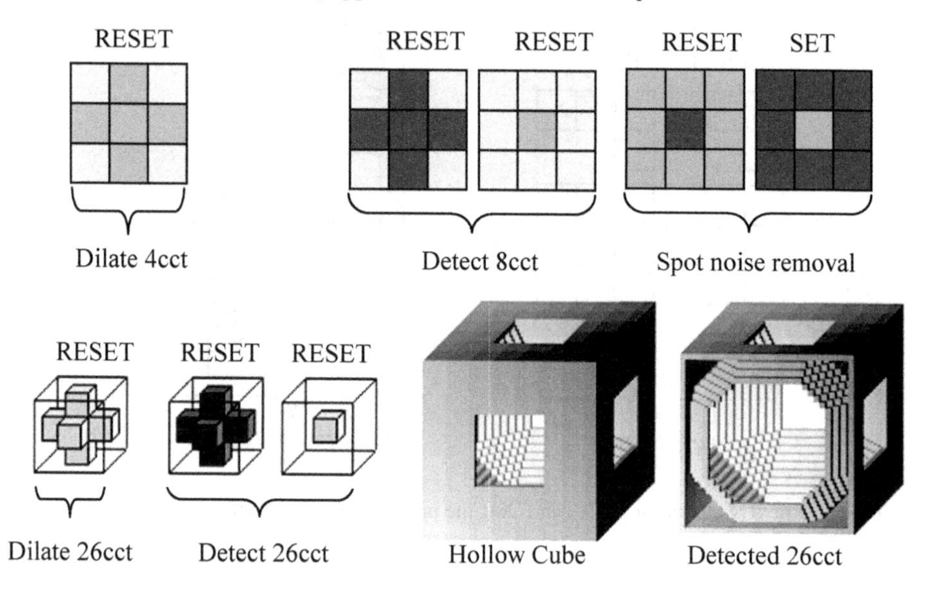

Fig. 4. RESET masks, masks sets and mixed SET / RESET sets. Opaque elements are don't cares, light grey are background elements and dark grey are foreground elements

The relation between the SET and RESET forms of the mask-sets is given by the Boolean algebra, i.e. the theorem of De Morgan (1847)[11]. Defining mask sets using only SET conditions is possible, although in some cases this is not very parsimonious. For example, the transformation of dilate_4cct from a conjunctive canonical form to a disjunctive canonical form is given by:

$$\overline{(\bar{C} \wedge \bar{N} \wedge \bar{W} \wedge \bar{S} \wedge \bar{E})} \equiv (C \vee N \vee W \vee S \vee E) \equiv \\ (\bar{C} \wedge N) \vee (\bar{C} \wedge W) \vee (\bar{C} \wedge S) \vee (\bar{C} \wedge E) \vee (C) \tag{10}$$

Dilate_4cct, first expressed as the "erosion of the background", expands to a set of five masks, expressing the dilation as: "copy the image in directions N, W, S, E and take the union of all results and original image", and finally to a set of 5 masks that indicate that "any background pixel adjacent to the foreground is set to foreground". The discussion above leads us to the observation, that mask sets can be minimized by transforming them from the disjunctive canonical form to the conjunctive canonical form, or vice versa. And, that mask-sets can be manipulated using Boolean algebra and minimisation.

Figure 5 shows an example of a dyadic Recursive Neighbourhood Operation in X_2 and X_3; the propagation operation. Objects in an image are recursively dilated (the first mask), wherever foreground in image Z is found and background in X was found (the second mask).

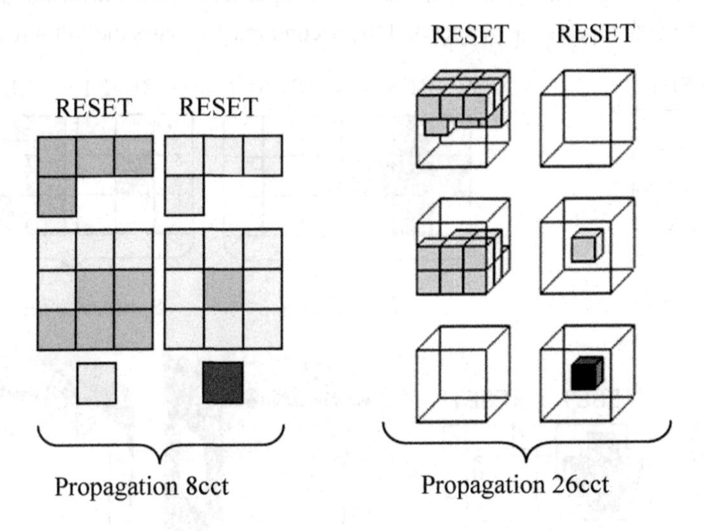

Fig. 5. Dyadic RN Operation in X_2 and X_3; the propagation operation

2.3 Connectivities in N Dimensional Images X_N

A set is connected if each pair of its point can be joined by a path along points that are all in the set. Within a square tessellated two-dimensional binary image X_2 with objects on a background, there are two possibilities for the connectivity of object and background pixels. The objects can be chosen to consist of pixels connected with one or more of their 8 neighbours at (E, NE, N, NW, W, SW, S, SE) or they can be chosen to be 4-connected, with one or more of their neighbours at (E, N, W, S). See also Rosenfeldt and Pfaltz (1966)[12] for the connectivity paradox. This paradox extends to higher dimensions. Note that we define the connectivity of an object as the highest connectivity between any two adjacent elements within that object.

Tessellation in higher dimensions is also known as honeycombing. Coxeter [13] proves in his book on regular polytopes, that the only regular honeycomb (a single regular identical cell on a lattice) that exist in all dimensions N is the cubic honeycomb. In addition in 2D the tessellation with hexagons exist and in 4D two types of honeycombs exist (each-others dual), however, not with their vertices on a lattice.

Although in 3D the Dodecahedron and Icosahedron exits, they cannot be used for honeycombing in 3D. C60 molecules or "bucky-balls" are truncated Icosahedrons, made from hexagons and pentagons. Above dimension 5 only the equivalents of the regular Tetrahedron, Octahedron and Cube exist as polytopes. Which leaves hypercubic honeycombing as the only way to extend our method to higher dimensions.

Let us define an N-dimensional binary image X_N as a bounded section of an equidistantly sampled Euclidean space of dimension N, with element values {0,1}. The elements of X_2 are referred to as pixels, the elements of X_3 as voxels, of X_4 and higher as hyper-voxels. Let M_N^n be an N-dimensional hyper-cubic neighbourhood with odd size n = 2k+1. For many morphological operations 3^N neighbourhoods are used (k=1, n=3), so M_N^n is abbreviated to M_2 for a 3x3 and M_3 for a 3 x 3 x 3 neighbourhood, etc. Connectivities can be derived by counting the number of image elements around a central element, within a hypersphere and within the neighbourhood. E.g. in X_2 there are 4 pixels within radius 1 and 8 pixels within radius $\sqrt{2}$. In [1] and [14] it was shown that the connectivity between image elements can be derived by counting E, the number of elements on a hyper sphere with radius *d* around an image element that lie within the neighbourhood M_N^n. E can be calculated by permuting the neighbourhood element co-ordinates, and is given by:

$$E=\frac{N!}{\prod_{j=0}^{k}(n_j!)}2^{(N-n_0)} \quad \textbf{(11)}$$

From E, we can derive V, the total number of elements within the hyper-sphere within M_N and hence the connectivity G = V–1. This yields as connectivities in a 3^N neighbourhood for X_2 : {4,8}, for X_3: {6, 18, 26}, for X_4 : {8, 32, 64, 80}, etc., in a 5^N neighbourhood for X_2 : {12, 20, 24}, for X_3 : {32, 56, 80, 92, 116, 124}, etc.

2.4 Connectivity Paradox in X_3

Connectivity rests on connected paths between pairs of points in a set. The paradox exists because with the sort of operations we do on the sets (like segmenting regions into objects) we demand that paths are boundaries. I.e. they cannot be crossed -nor tunneled- without breaking the path. In topology, the Jordan Curve theorem states that a simple closed curve lying in a plane divides the plane into precisely two regions and forms their common boundary. [15]. This theorem extends to X_N.

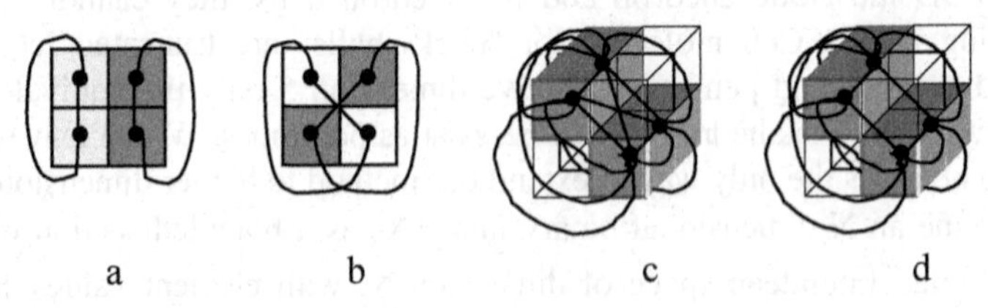

Fig. 6. Jordan Curves applied in X_2 and X_3

Consider a set of 4 points from a larger universe X_2 with two sets each of two points (white and grey), see figure 6. We will call this 2 x 2 neighbourhood a *tile*. Clearly when the grey set forms anywhere in the universe a *closed* boundary, incorporating the connection between the two grey points of the neighborhood, the white set cannot be connected. When we define that in a neighbourhood all points of a set directly adjacent to the boundary of the neighbourhood are assumed to be connected to any other point of the same set over that boundary, then for figure 6b this means that if the grey set is connected, the white set cannot be connected. If we assume closure over the tile boundary, we can say that two points each at another side of a boundary cannot be connected. So we must choose, either white is face and edge connected and grey is only face connected, or vice versa.

We can extend this to higher dimensions, but then tiles are 2^N neighbourhoods and closed boundaries are topologically equivalent to (hyper) spheres. Note that we consider connections over the tile boundaries now separately for each dimension. Figures 6c and d shows tiles in X_3. Figure 6c is a topological equivalent of a sphere when we assume that the grey set is 26 connected and the white set is not. The white set must then be 6 connected. Figure 6d is a topological equivalent of a sphere when we assume that the grey set is 18 connected and the white set not. The white set must then be 6 connected, while we don't care about the vertex connections. Vice versa, when we are sure that the white set is 6 connected, we might as well say that the grey set is 18 connected, as we don't care about the vertex connections. We can now make the observations that:

- The lowest (face) connectivity in any image X^N is always a perfect boundary, it never leaks, all higher connectivities do. The reason is that a pair of face connected elements has unit extends perpendicular to its connection axis in all remaining dimensions.
- Usually we are interested in the properties of the foreground objects, the curvature of objects being one of them. The higher the connectivity, the smaller radii can be made, as can be imagined by observing the radii of 8 and 4 connected curves in X_2 $(1, \sqrt{2})$ and 26, 18 and 6 connected rings in X_3 $(1, \sqrt{2}, \sqrt{3})$. If we assume that that the centers of the image elements represent sampling points of a physical phenomenon, then, with the same sampling density, smaller structures can be represented more accurately using a higher connectivity. This gives a preference for taking the highest possible connectivity as the foreground connectivity.

Consequently: *A good choice is to take the lowest connectivity as background as it never leaks and the highest connectivity as foreground as it provides us with the highest bending radius.*

3 Skeletonization in X_N

In the previous sections, we have described a general principle for morphological operations on cubic tessellated binary images X_N, based on the scanning of the image with a set of masks that comprise the structuring element. The masks contain foreground, background and don't-cares. On each element of the image an inexact match between all masks of the set and the neighbourhood extracted from the image is performed, whereas the result written to the output image is the union of all matches.

So far, only erosion of foreground and background and matches on single elements (pixels, voxels, ...) have been used. Erosions can be described as a match on foreground area in X_2 and on foreground volume in X_3. In this chapter, skeletonization in X_N is treated. Skeletonization can be seen as conditional erosion [16]. Figure 7 shows that in X_3 a volume is eroded to a curved surface -the surface skeleton-, where after the surface is eroded to a space curve -the curve skeleton-. Apparently, for skeletonization the condition for the erosion is that surfaces, or curves, should not be eroded. For those conditions, we need the notion of shape primitives in X_N.

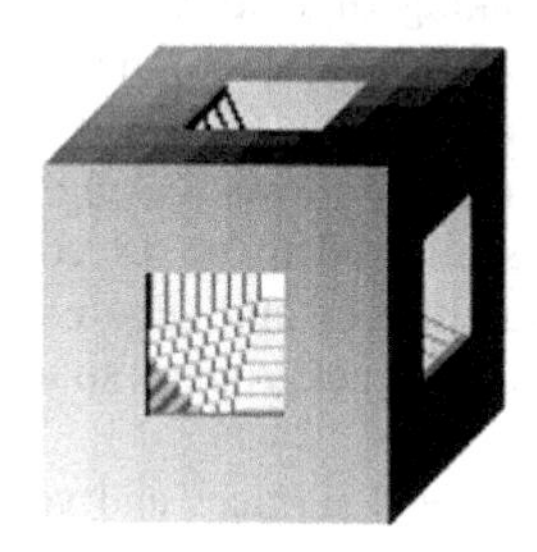
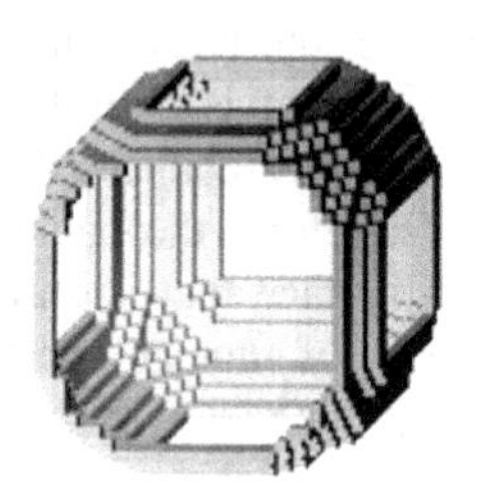
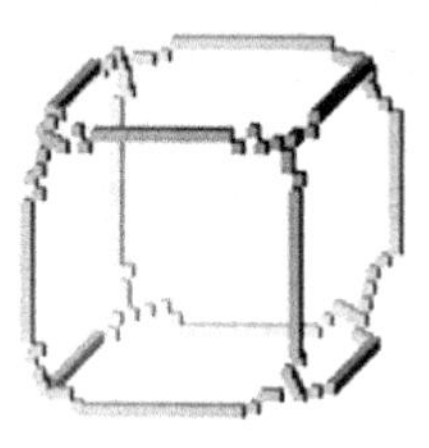

Fig. 7. Original, contour, surface-, and curve skeleton of a cube with hole in X_3

3.1 Shape Primitives in X_N

In X_N we can identify shape primitives with a certain intrinsic dimension $\widetilde{N}$. For example, in X_2 we can distinguish points ($\widetilde{N}$ = 0), curves ($\widetilde{N}$ = 1) and surfaces ($\widetilde{N}$ = 2), in X_3 points, space curves, curved surfaces and flat volumes ($\widetilde{N}$ = 3), and in X_4 we can distinguish points, space curves, curved surfaces, curved volumes and flat hyper-volumes ($\tilde{N} = 4$). Objects in X_N can now be considered as made of any arbitrary combination of shape primitives with intrinsic dimensions ($0 \leq \tilde{N} \leq N$).

Figure 7 shows objects only consisting of volume, or surface, or space curve. It was already remarked that the (6, 18', 26 connected) erosion conditions for X_3 match on volumes except for the outer boundary of those volumes. Consequently, they are the shape primitive detectors for volumes in X_3. The contour detection operations "Detect 6cct", "Detect 18'cct", and "Detect 26cct" were applied on a hollow volume (obtained by dilating a single voxel with 26 connected surface contours obtaining a sphere, and XOR-ing this with a cube). See figure 8.

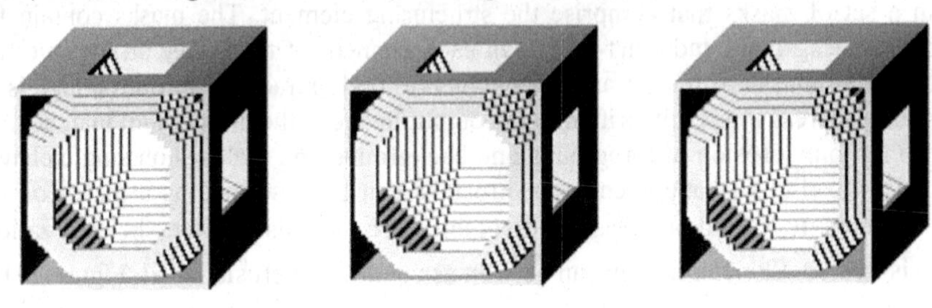

Fig. 8. 6, 18' and 26 connected surface contours in X_3

(The cube's front slices have been removed for better insight). All resulting surfaces prevent the leaking of 6 connected background. The first result shows a surface that is entirely made of face connected voxels, which prevents leaking of face, edge and vertex connected voxels. Therefore, the background is allowed to be 26 connected. The second result shows a surface that is entirely made of face connected voxels, which prevents leaking of face and edge connected background but allows leaking of vertex connected background. Therefore, the background is allowed to be 18 connected. The third result shows a surface that is entirely made of face and edge connected voxels, which allows leaking of edge and vertex connected background. So the background is should be face (6) connected.

In order to detect primitive shapes such as space curves, curved surfaces and volumes, we need shape primitives that match on all possible incarnations of those primitives shapes.

3.2 The Generation of Shape Primitives in X_3

Shape primitives for space curves can be found by remarking that an open (non circular) space curve in a 3^3 neighbourhood consists of a central voxel and two mutually non touching neighbours. Figure 9 shows all possible configurations for 6 and 26 connected space curves generated by this rule.

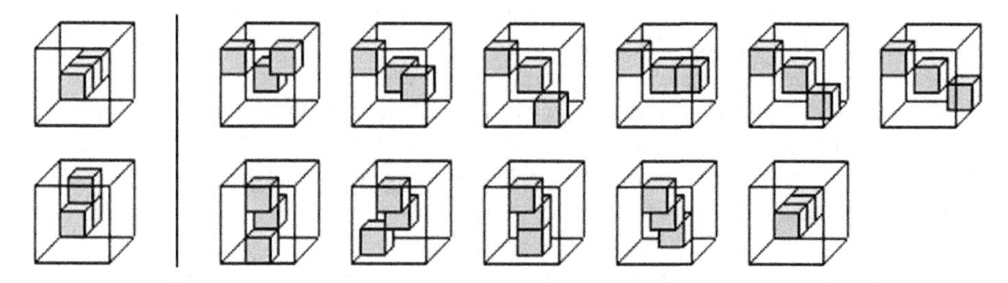

Fig. 9. The 6 connected space curve primitives and the 26 connected space curve primitives

Note that we omitted all rotated and mirrored versions and that non touching is defined for a certain connectivity. Hence, the second mask from the 6 connected set is not present in the 26 connected set. The set of 18 connected space curves is a subset of the 26 connected set.

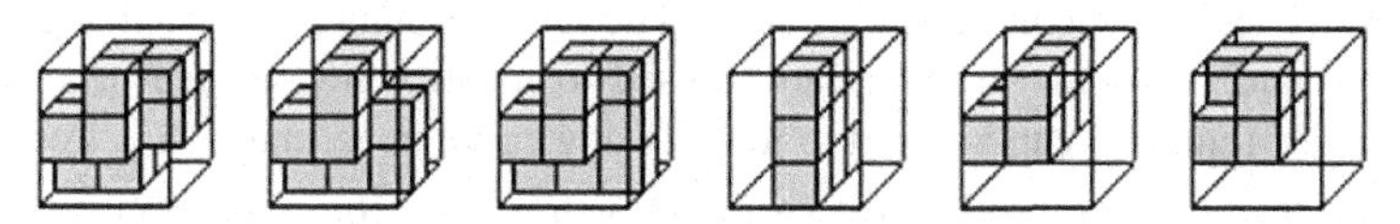

Fig. 10. The 6 Connected curved surface primitives

In a similar way, the surface primitives can be found. They can be generated by a central voxel encircled by a non-touching closed space curve. For a 6 connected surface set, the central voxel should be encircled by a 6 connected space curve (figure 10), for a 26 connected set by a 26 connected space curve (figure 11).

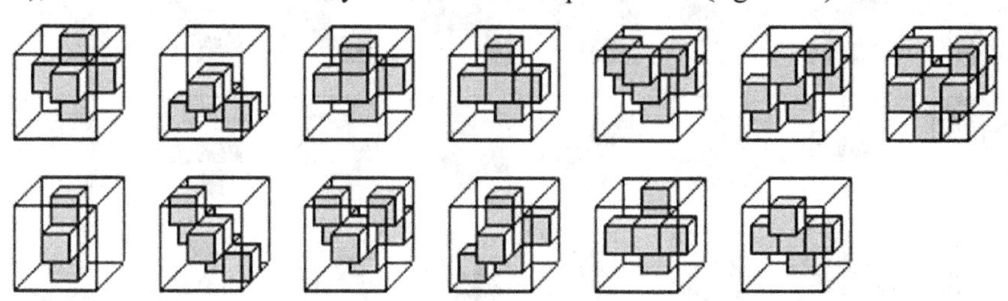

Fig. 11. The 26 connected curved surface primitives

Again non touching is defined for a certain connectivity. The generation procedure is a recursive search that generates all possible occasions. After this, all identical, mirrored and rotated variants are removed and the set is logically reduced to a minimal covering set [17]. Note when generating the surface primitives, their voxels stay in the 18 connected neighbourhood of the central voxel in both cases. The vertex connected positions of the central voxel are apparently don't care. Observe, that in the 6BG - 26FG system, the space curves roam over the 26 positions, the curved surfaces over the 18 connected positions, and the voxels of the volume detection mask are found on the 6 connected positions. Consequently, we can state that in a 6BG-26FG system the space curves are 26 connected, the surfaces are 18 connected and the volumes are 6 connected. In a 26BG-6FG system the space curves are 6 connected, the surfaces 18 connected and the volumes 26 connected.

If we have compound (bifurcating) objects, such as forking space curves, splitting surfaces, surfaces with sprouting curves etc., we must introduce don't cares. We can do this, because we can be more precise in our foreground-background paradox. We can state this paradox now as follows:

1. *A foreground curved surface is threatened to be pierced by a background space curve.*
2. *A foreground space-curve is threatened to be broken by a background surface, breaking the connectivity of the curve.*

We can now merge the sets. Rule 1 leads to figure 12, a merging of the sets from figures 11 and the first set of figure 9. Rule 2 leads to figure 13, a merging of the sets from figure 10 and the second set of figure 9. The set from figure 12 (when augmented with rotated and mirrored variants) can be used to identify any surface in an image X_3. The set of figure 13 can be used to identify any curve in an image X_3. For all the masks hold, that if the central voxel swaps its value, e.g. from foreground to background, the connectivity of the foreground is broken and the topology changes. As each mask hits individually, it can be tagged with a contribution to the length of a curve or to the area of a surface. Hence, surface area and curve length's of objects can be measured more accurately then simply counting voxels. [18].

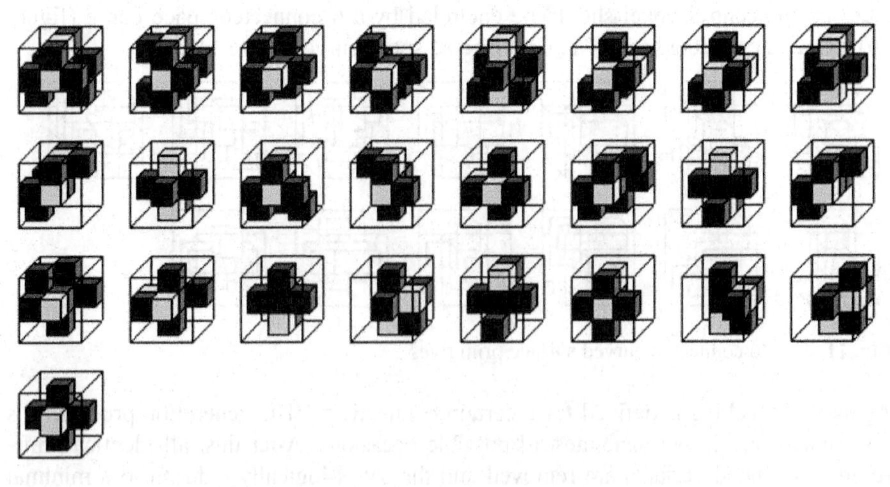

Fig. 12. Shape detectors for curved surfaces "Surf26"

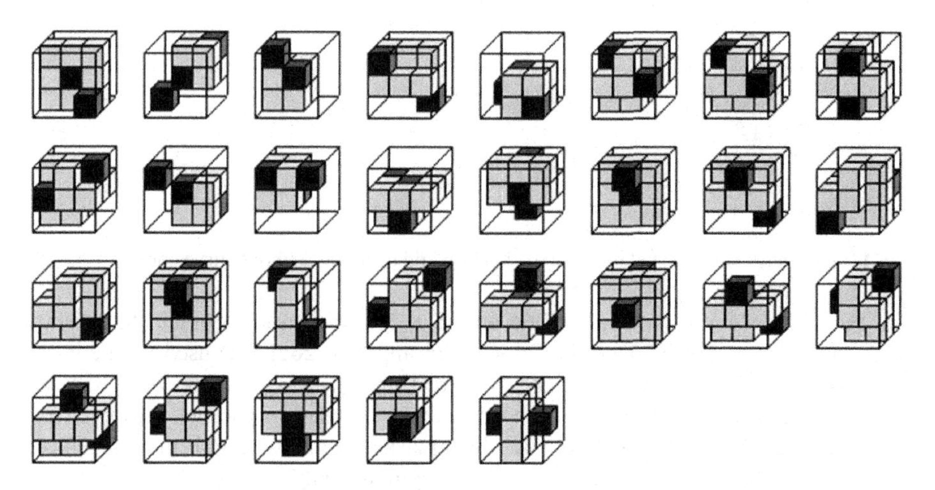

Fig. 13. Shape detectors for space curves "Curv26"

3.3 Erosion with Shape Primitives as Conditions: Skeletonization

The main application of the shape detectors is that they can be used as conditions for topology preserving erosion, or skeletonization. The method of chapter 2, iterating over an image X_3 with the erosion mask "Erode26cct" of figure 3 and the sets of figure 12 (Surf26) and 13 (Curv26), yields a skeleton. Erode26cct erodes surfaces from volumes (it hits only on the core of the volumes, not on their boundaries). The set Curv26 prevents the erosion of surfaces (the masks only hit on the core of the surface, not on the surface boundaries). The set Surf26 prevents the erosion of curves (the masks only hit on the kernel of the curves). Consequently, as volumes, surfaces and curves cannot be eroded from their kernels, they are eroded from their boundaries. So only closed surfaces and curves will remain.

To prevent the erosion of boundaries, from the set Curv26 a new set Curv26e is made that contains all surface edge situations. Similarly, a set Curve26e can be made from the set CURV26 containing all curve end situations. A simple procedure can be followed. A surface extends in 4 directions. Systematically one and two directions can be disabled by setting them to background. A curve extends in two directions, one of which can be disabled by setting it to background. This yields 87 unique configurations for the surface edges and 3 possibilities for curve ends.

A final remark should be made on the detection of two element thick structures. They cannot be detected properly in a 3^N neighbourhood. To circumvent this, the recursive neighbourhood must be used to detect foreground-background changes. However, only for the Surf 26 and Curv26 sets and not for the erosion mask and the Surf26e and Curv26e masks!

A variety in skeletons can now be made (see figures 14 and 15)

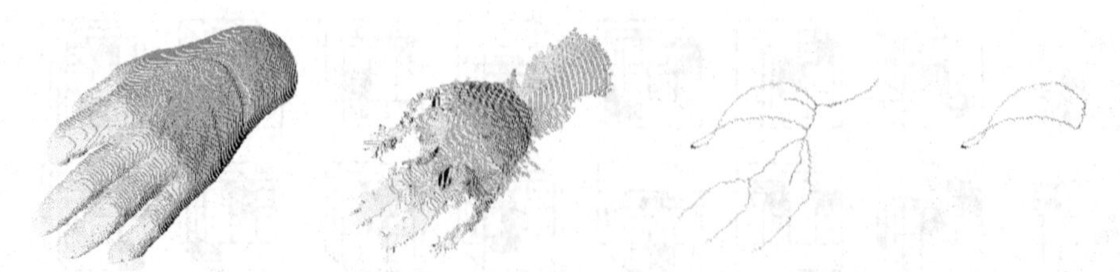

Fig. 14. Original[1], surface skeleton, curve skeleton and skeleton without curve ends

The set {Erode26cct, Surf26, Surf26e and Curv26} was used to obtain the surface skeleton. The set {Erode26cct, Surf26, Curv26 and Curv26e} was used to obtain the curve skeleton. The set {Erode26cct, Surf26, Curv26} was used to obtain the last skeleton of figure 14. All masks from all sets can be used in parallel. However, the local situation in the image determines which masks are active. E.g. if the image does not contain voluminous and surface objects anymore, only the Curv26 set is active, preventing the breaking of curves that are eroded from their ends.

An important observation is that the masks that prevent the breaking of the topology for a certain intrinsic dimension $\widetilde{N}$, are the erosion conditions for objects with that intrinsic dimension $\widetilde{N}$. Consequently, the mask Erode26cct prevents breaking the topology of a (flat) volume and the set Surf26 incorporate the erosion conditions and metric (!) for the curved surfaces. These metrics, due to the mask Erode26cct and set Surf26, are not Euclidean, which results in deviations of the skeleton from the medial axes of the original object. When the skeleton is needed for measurement purposes, the sets Surf26, Surf26e, Curv26, Curv26e cannot be used in parallel, but must be applied sequentially in the following way. A constrained Euclidean distance transform [19] should be performed on the original volume, where-after the surface skeleton should be made, eroding voxels in an order that depends on their distance to the volume boundary. Subsequently, on the surface skeleton, a constrained distance transform should be made, after which the curve skeleton should be made eroding voxels in the order of their distance to the surface edge.

Figure 15 shows the use of a curve skeleton {Erode26cct, Surf26, Curv26} in path-finding through a maze. Start and endpoint are inserted in image Z^3 and using an extra mask anchor the skeleton onto those two points.

For Robot Soccer [20] a similar anchor skeleton was used in the "Dutch Team" to quickly plan the collision free path from start to goal of an attacker robot. The image X_3 represented the universe of the robot with two place dimensions (x, y) and one time dimension (t). Manipulating both the erosion metric and the mask sets made it possible to prevent the time from running backwards and setting a speed maximum for the robot. If the orientation of the robot should be taken into account also, the problem should be solved in X_4 (x, y, θ, t).

---

[1] The real original image (from Toriwaki) contained two holes, giving rise to two closed surfaces at the skeleton near the wrist. We closed the holes beforehand, for didactical reasons.

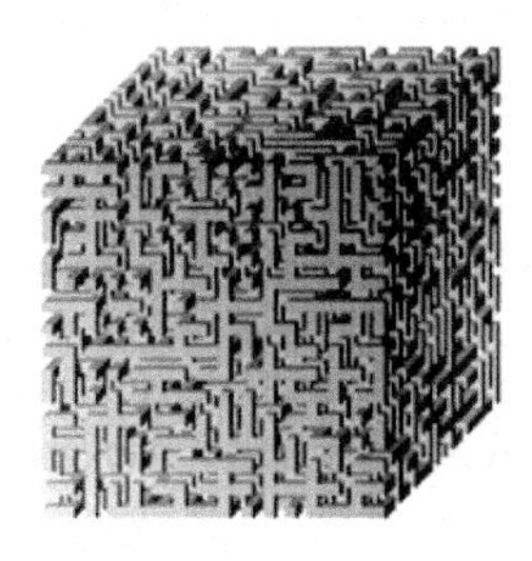

Fig. 15. An anchor skeleton is used to solve a path finding problem

3.4 Verification

The method described in this paper lends itself very well for extension into higher dimensions. Provided the approach is valid. This was verified by comparing the outcome of the masks for X_2 with the Hilditch [21] skeleton and outcome of the masks for X_3 with the approaches of Lobregt[22] / Toriwaki [23] and Malandain [24]. They were checked by comparing and found to have an identical outcome in all 2^9 and 2^{27} possibilities of the 3^2 and 3^3 neighbourhood. Note, however, that in all three methods only the curve skeleton could directly be made, as the conditions can not be split over surface and a space curve sets. Hence, no accurate Euclidean variants could be established with those methods.

To prepare for verification in X_4 we extended the Euler count (Lobregt) / cluster count (Toriwaki) method to 4D [25]. As the Euler-sum[26] in 4D is always zero, we solved this by embedding 4D objects in X_5 to calculate the 5D Euler number. This number changes due to 3 events: 1D tunnels ($\widetilde{N}=1$), 2D tunnels ($\widetilde{N}=2$) and 3D tunnels ($\widetilde{N}=3$). As the latter cannot occur in X_4, with Euler-, and fore- and background cluster counting, the number of equations just equals the number of unknowns and a topology change can be detected. As the Euler sum applied to objects in X_5 counts also the possible 3D tunnels, fore- and background cluster counting is now not enough to detect the *cause* of change in the Euler sum. Consequently, as fore- and background cluster counting is all we can do in any dimension, the Euler/cluster count approach is only applicable in X_3 and (using X_5) in X_4, making X_4 the limit.

We generated test images in X_4 based on circular closed objects of various intrinsic dimension, such as 4D equivalents of hollow balls, circles, etc. and contaminated them with foreground noise. The 4D skeleton showed to be robust.

4 Shape Primitives in X_4

The methods described in this paper extend straightforward to X_N, especially for the basic operations such as erosion, dilation, contour extraction, propagation etc. involving only a few masks. To get an impression how the interaction between foreground and background shapes primitives and their intrinsic dimension works out in X_4 we show in figure 16 how the paradox of foreground and background shape primitives works out. We define:

Topology breaking of a foreground shape primitive of dimension $\tilde{N}$ *is done by a background shape primitive of intrinsic dimension* $N-\tilde{N}$ *(and vice versa).*

For example, a space curve cannot be broken by a point. As we only have foreground and background, the background point must be connected to other background. This cannot be a line, because the foreground curve leaves two degrees of freedom c.q. intrinsic dimensions open. For a simplified view on the matter in X_4, we assumed in figure 16 lowest connectivity for fore- and background and the existence of flat objects only. Figure 16 shows five sets of shape primitives: Hyper-volume threatened to be broken by a point, volume threatened by a line, plane threatened by a plane, line threatened by a volume an point threatened by a hyper-volume. It indicates that because we consider the topology breaking as the union of a number of individual events, we do not run into the problem of the Euler-Cluster count method, that cannot make a distinction between the occurrences that lead to a change in the Euler number.

On top of the paradox of the shape primitives we find the connectivity paradox. Table 1 indicates the relation between a dimension, its shape primitives and their associated connectivities. We distinguish CE: Connectivity to the central Element, NN: number of neighbours required for this shape primitive, RC: Recursive connectivity.

The generation of the various sets follows the same lines as was done for X_3. A shape primitive of intrinsic dimension $\tilde{N}$ is made by encircling the central element over the positions CE, by shape primitives of intrinsic dimension $\tilde{N}$-1 and connectivity RC. This is a recursive procedure. E.g. a foreground hyper-volume is made by encircling it by a closed foreground surface, etc.

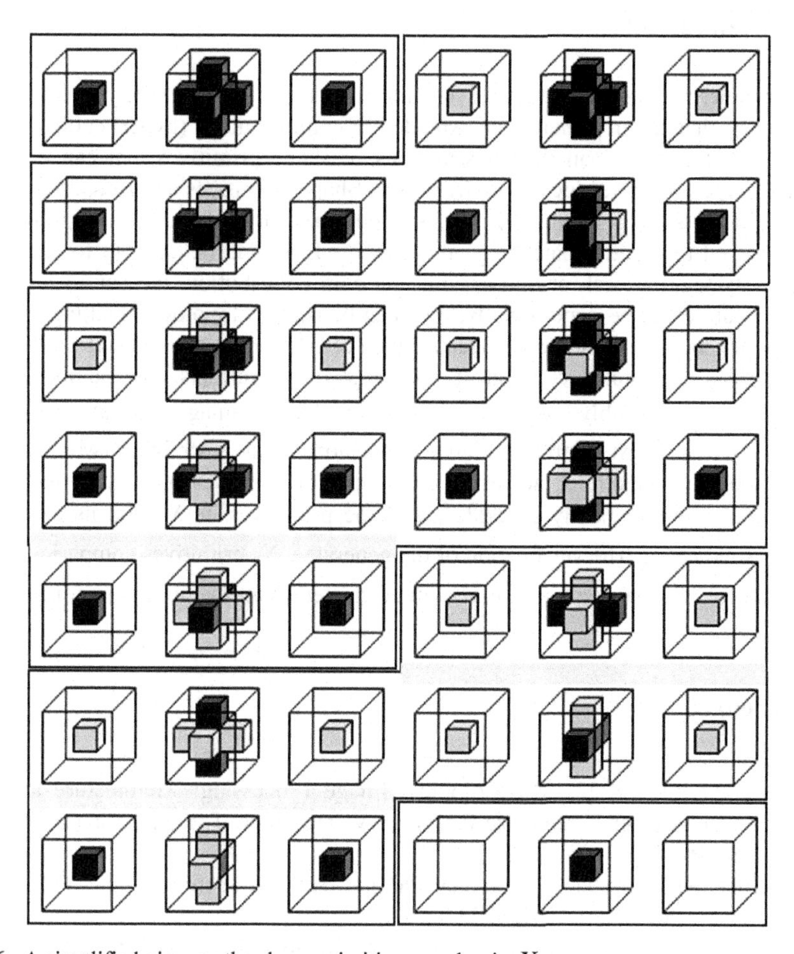

Fig. 16. A simplified view on the shape primitive paradox in X_4

Table 1. Dimension, intrinsic dimension and connectivity for various shapes

$N, \widetilde{N}$	Foreground shape primitive	C E	N E	R C	Background shape primitive	C V	N E	R C
(2,2)	Flat surface	4	4	-	Point	-	-	-
(2,1)	Curve	8	2	-	Curve	4	2	-
(3,3)	Flat volume	6	6	-	Point	-	-	-
(3,2)	Curved surface	18	≥4	26	Space curve	6	2	6
(3,1)	Space curve	26	2	-	Curved surface	18	≥4	6
(4,4)	Flat hyper-volume	8	8	-	Point	-	-	-
(4,3)	Curved volume	32	≥6	64	Space curve	80	2	8
(4,2)	Curved surface	64	≥4	80	Curved surface	64	≥4	8
(4,1)	Space curve	80	2	-	Curved volume	32	≥6	8

5 Conclusions

In this paper we described a framework for mathematical morphology or cellular logic operations that can in principle be extended to N dimensional (hyper) cubic tessellated images. We focussed on the constructions of shape primitives that can be used to detect shape properties locally in the image. Shape primitives are grouped in intrinsic dimension sets, such as volume, surface and curve primitives. The shape properties can be used to identify parts (e.g. all surface edges) and hence help to segment the image. By tagging each primitive with a certain contribution, measurements that are more accurate can be done. Finally, they can be used as topology breaking conditions for the generation of skeletons. Because they are the union of a sub-set of conditions, this enables the smooth creation of various skeleton variants. The number of shape primitives is remarkably low in X_3. A skeleton of a 200^3 image takes about 5 minutes.

The method is extendable to higher dimensions. This is straightforward for simple morphological operations, however, not simple for shape primitives. A method was presented to automatically generate the shape primitives in X_3 and its extension to X_4 was indicated. The verification of the generated X_4 primitives comparing it with a skeleton for X_4 based on an Euler count / cluster count method, is ongoing research [27].

References

1 Jonker P.P. (1992) Morphological Image Processing: Architecture and VLSI design. Kluwer Technische Boeken BV, Deventer, The Netherlands. ISBN 90-201-2766-7

2 Golay M.J.E. (1969) Hexagonal Parallel pattern transformations. IEEE transactions on computers C-18, 8:733-740

3 Preston K. Jr. (1970) Feature extraction by golay hexagonal pattern transforms. IEEE Symposium on feature extraction and selection in pattern recognition, Argonne, III

4 Serra J. (1982) Image analysis and mathematical morphology. Academic Press, Inc. London.

5 Serra J. (1988) Image analysis and mathematical morphology. Volume II: Theoretical Advances. Academic Press, Inc. London.

6 Giardina, C.R. (1988) Morphological methods in image and signal processing. Prentice Hall, Englewood Cliffs-NJ. ISBN 0-13601295-7

7. Heijmans H.J.A.M., (1994) Morphological Image Operators. Academic Press. Boston

8 Soille P., (1999) Morphological Image Analysis, Springer Verlag, Berlin

9 Jones R., and Svalbe, I.D., (1994) Basis Algorithms in Mathematical Morphology, Advances in Electronics and Electron Physics, Vol.89, Academic Press, Inc. ISBN 0-12-014731-9

10 Matheron, G. (1975) Random Sets and Integral geometry. Wiley, New York

11 McClusky E.J. (1965) Introduction to the theory of Boolean functions, mcCraw-Hill, New York
12 Rosenfeld A., Pfalts J.L. (1966) Sequential operations in Digital Image Processing. Journal of the ACM, 471-494.
13 Coxeter H.S.M. (1974) Regular Polytopes (3rd edition) Dover Publications, Inc. New York, ISBN 0-486-61480-8
14 Jonker P.P., Vossepoel A.M. (1994) Connectivity in high dimensional images, Proc. MVA 94, IAPR Workshop on Machine Vision Applications, Kawasaki, Japan, Dec.13-15.
15 P. Alexandroff (1961)Elementary concepts of topology. Dover Publications, Inc. New York, ISDN 0-486-60747-X
16 Jonker P.P., Vossepoel A.M. (1995) On skeletonization algorithms for 2, 3 .. N dimensional images, in: D. Dori, A. Bruckstein (eds.), Shape, Structure and Pattern Recognition, World Scientific, Singapore, 71-80. ISBN 981-02-2239-4
17 Brayton R.K., Hachtel G.D., McMullen C.T., Sangiovanni-Vincetelli A.L. (1984) Logic minimization algorithms for VLSI synthesis. Kluwer, Dordrecht / Boston
18 Mullikin, J. J.C., Verbeek P.W.(1993), Surface area estimation of digitized planes, BioImaging, vol. 1, no. 1, 6-16.
19 Verwer B.J.H. (1991), Local distances for distance transformations in two and three dimensions, Pattern Recognition Letters, vol. 12, no. 11, 1991, 671-682.
20 www. robocup.org, www.robocup.nl
21 Hilditch C.J (1969) Linear Skeletons from square cupboards. in B. Meltzer and D. Mitchie (eds.) Machine Intelligence Vol. 4. Edinburgh: University Press, 404-420.
22 Lobregt S., Verbeek P.W, Groen F.C.A, (1980) Three dimensional skeletonization: Principle and algorithm IEEE Trans. Patt. Anal. Machine Intell. vol. 2, pp. 75-77
23 Toriwaki J., Yokoi S., T. Yonekura T, Fukumura F (1982) Topological properties and topological-preserving transformation of a three dimensional binary picture. in Proc. Int. Conf. Patt. Recogn., Munich 414-419
24 Malandain C, Bertrand B (1992) Fast Characterixation of 3D Simple Points. Proceedings of the 11th ICPR, Vol III, The Hague.
25 Jonker P.P., Vermeij O. (1996), On skeletonization in 4D images, in: P. Perner, P. Wang, A. Rosenfeld (eds.), Advances in Structural and Syntactical Pattern Recognition, Lecture Notes in Computer Science, vol. 1121, Springer Verlag, Berlin, 79-89.
26 Hilbert D, Cohn-Vossen S (1932) Anschauliche Geometrie, Springer Verlag, Berlin.
27 Tools for generating, manipulating and executing mask sets for 2, 3 and 4 dimensional images can be found on ftp.ph.tn.tudelft.nl/pub/clop

Efficient Algorithms to Implement the Confinement Tree

Julian Mattes[1,2] and Jacques Demongeot[1]

[1] TIMC-IMAG, Faculty of Medicine
38700 La Tronche, France
Julian.Mattes@imag.fr,Jacques.Demongeot@imag.fr
[2] iBioS, DKFZ Heidelberg
69120 Heidelberg, Germany
J.Mattes@dkfz.de

Abstract. The aim of this paper is to present a new algorithm to calculate the confinement tree of an image – also known as component tree or dendrone – for which we can prove that its worst-case complexity is $O(n \log n)$ where n is the number of pixels. More precisely, in a first part, we present an algorithm which separates the different kinds of operations – which we call scanning, fusion, propagation, and attribute operations – such that we can separately apply complexity analysis on them and such that all operations except propagation stay in $O(n)$. The implementation of the propagation operations is presented in a second part, first in $O(n_n^2)$, where n_n is the number of nodes in the tree ($n_n \leq n$). This is sufficient if the number of pixels is much larger than the number of nodes ($n_n << n$). Else, we show how to obtain $O(n_n \log n_n)$ complexity for propagation. We construct two example images to investigate the behavior of two known algorithms for which we can show worst-case complexity of $O(n^2 \log n)$ and $O(n^2)$, respectively, and we compare it to our algorithm. Finally, a practical evaluation will be opposed to the theoretical results. Several variations of the implementation will show which operations are time consuming in practice.

1 Introduction

In the last years, the confinement tree proved to be a useful tool in image processing. This summary of an image in which each node corresponds to an image region (called a *confiner*) has been (re)discovered by several authors according to various applications in image processing, but its origins lie in statistics [14,6]. Applications have been image filtering, segmentation [5,4,3,7], matching [10,9,12,11], and classification [10], object detection [5] and recognition [10,12]. Structures similar to the confinement tree are investigated in [13]. Note that especially when we are concerned with 3D imaging the time to calculate the tree (or the trees for each of the 2D slices forming the 3D image) becomes prohibitive. For 3D reconstruction in X-ray audio-graph imaging we have to align (match) up to 1000 2D image slices and to calculate for each one, as well as for

G. Borgefors, I. Nyström, and G. Sanniti di Baja (Eds.): DGCI 2000, LNCS 1953, pp. 392–405, 2000.
© Springer-Verlag Berlin Heidelberg 2000

the whole 3D image, the confinement tree for image matching [11]. Fast execution time is also required for segmentation of cell regions in high-throughput scanning [15].

In this paper we want to present and investigate the difficulties that appear when looking for an efficient implementation of the confinement tree and to propose possible solutions. There is a need for such investigations because even if for some applications very fast execution times are required, in a first approach, people focus more on the ability of the algorithm to solve the studied problem than on its optimization. For instance, Guillataud [4] proposed an implementation which determines the confiners successively and independently at each level and [5,3] made only suggestions for an algorithm which "may be in almost-linear time" [3] based on immersion simulation. Jones [7,8] made a first attempt to consider also the optimization problem. He obtained in general much faster execution times but for certain images (as shown on theoretical examples in section 3) we can considerably improve the fastness. In worst-case we can improve the complexity by passing from $O(n^2 \log n)$ to $O(n \log n)$. Salembier et al. [13] presented already before an efficient algorithm to calculate the max-tree which is a tree related to the confinement tree (see section 3). We apply similar basic techniques as Jones in [8] but our global strategy is different. In addition we will give more details (sections 3.3 and 3.2) and provide a complexity analysis in section 4. Our experimental evaluation in section 5 will show where to achieve more efficiency in practice. Before presenting our algorithm in section 3 we give first definitions and notations in section 2.

2 Definitions and Notations

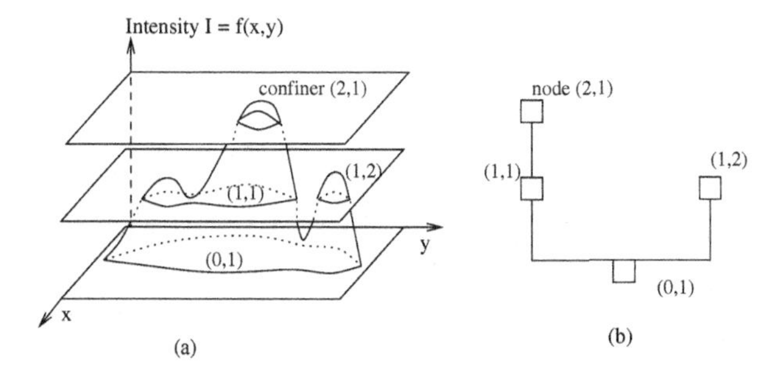

Fig. 1. Illustration after [4] for (a) the definition of the confiners and (b) the confinement tree

Given an intensity function $f : \mathbb{R}^d \rightarrow \mathbb{R}_{\geq 0}, d = 2, 3$, the *confiners* are defined as the maximal connected subsets (i.e., *components*) $\mathcal{C}_l$ of the level sets $\mathcal{L}_l = \{x \in \mathbb{R}^d | f(x) \geq l\}, l \in \mathbb{R}_{\geq 0}$ (see Figure 1(a)). Considering them taken on several levels $l_k, k = 1, ..., r$ (r from resolution) including the 0 level, they define obviously a tree (by "set inclusion"; see Figure 1(b)) called *confinement tree.* In practice, the levels are those of the image's grey intensity function and we take all available grey levels into account. If we do not represent identical confiners in the tree but only the highest among them (see Figure 2) we call this the *unique representation.* We choose the discrete $d4-distance$ for defining the connectivity between pixels (see [7] for illustrations).

We found these components (confiners) and tree first in the classification domain in a paper by Wishart [14] and further investigations by Hartigan [6], where the confiners are called *high density clusters.* In image processing these sets and this structure appear first in [5]. See [14,5,10,11] for the invariance properties of the confinement tree. Another kind of components, induced in a natural way by the grey level function, are the connected components of the sets $\{x \in \mathbb{R}^d | f(x) = l\}$ [13]. They are called *flat regions.* We will use throughout this paper the notations n for the number of pixels in the image, n_n for that of nodes in the tree, and r for the number of grey levels. We assume $r \leq n$.

3 Implementation

When designing an efficient algorithm implementing the confinement tree, there arise some difficulties from the necessity to scan at each grey level only pixels bringing a new knowledge about the tree and to propagate information about connectivity of regions in the image. Hereby, we will distinguish in the following *attribute operations*, applied to update the tree attributes, and *connectivity operations* which are necessary to find the connection from a pixel to a confiner and to establish the father-son relationship in the tree. They include *scanning, fusion*, and *propagation operations*, the latter are needed to propagate neighborship information through the already updated part of the tree.

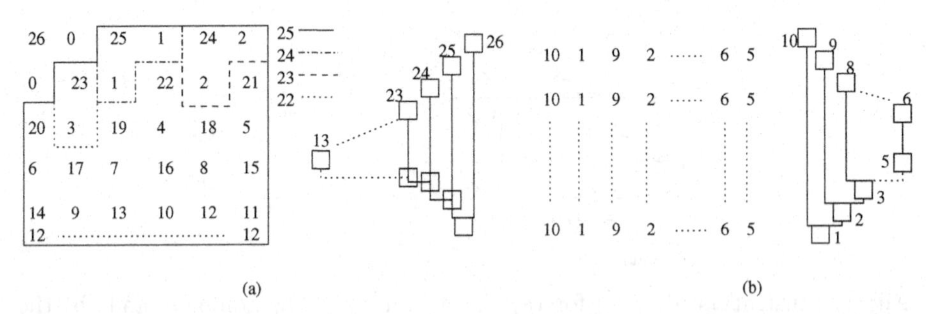

Fig. 2. Example images: (a) for worst case analysis for the algorithms of Jones [8] and Salembier et al. [13]; (b) for comparison with our algorithm. On the corresponding confinement trees we have chosen unique representation of identical confiners. We assume 4-connectivity

Algorithms to calculate confiners for a given grey level in linear time with respect to n are well known [1]. However, when we look for an algorithm which calculates the confiners for *all* grey levels but stays within $O(n)$ attribute operations and $O(n \log n)$ connectivity operations there arise some complications making it worth to be detailed.

A first attempt at designing an efficient algorithm tackling this problem was made by Jones [7,8]. His algorithm calculates first all local maxima of the image's grey level function (i.e., the leaves of the tree). After it detects, starting from each local maxima in sequential order, the confiners at the different grey levels on the path (called *branch* [8]) between the leaf and the root which are not already updated. For each leaf m the k_m pixels of these confiners are updated using a priority data queue in which they are ordered according to their grey value. In Figure 2(a) this will be done, for a given leaf m, in $O(k_m \log k_m)$ time. The pixels associated to the leaf at level 25 are delimited by the continuous 25-line. Pixels in not updated confiners on the branch between leaf 24 and the root are confined by the 24-line and the longer part of the 25-line, etc. Therefore, if – as in Figure 2 – many isolated local maxima with high grey level appear in the image the algorithm examines multiple times large parts of it, even if n_n is small as depicted in Figure 2(b) (see also section 5). The worst case behavior (Figure 2(a)) is $O(n^2 \log n)$. If we try to overcome the problem by stopping this detection process for a given leaf as soon as the grey levels increase we have to pay attention to the fact that in this case two higher branches (as 23 and 24 in 2(a)) can be separated by lower ones (as 19 and 22). We would need procedures as we will propose in steps (B2 l) and (B3 l).

Another approach for calculating the confinement tree would be to add attribute operations to the max-tree creation procedure and then to derive the confinement tree from the max-tree. Salembier et al. [13] presented an elegant algorithm to calculate the max-tree and achieved very fast results in practice. Applied to the image of Figure 2(a) however, their algorithm runs in $O(n^2)$ time. As their paper did not address principally to the optimization problem, a detailed evaluation is not given.

In the following, we describe first globally the algorithm, after we give detailed information for procedures implementing it, as presented in the Figures 4, 5, 6.

3.1 Global Strategy

The discrete connectivity model we have chosen is that of 4-connectivity (for illustration see [7]). The pseudo-code used below is based on ANSI-C.

Data Structure of the Result to be Determined Our algorithm aims to calculate the components of the following data structure: for each confiner (node) we keep in a structure (`struct`) (1) its surface, (2) its mass, (3) the X and the Y coordinates of its gravity center, (4) the eastmost among its northmost pixels (to be able to recalculate easily its contour), and (5) a number for identifying the node of its father in the array `Tree` defined next. The tree is implemented then as a 2 dimensional (dynamic) array `Tree`. It associates to a given level l (in the tree) and to the given number i of a node at this level the structure of the corresponding confiner (node) C_{il}. C_{il} is also at level l in the image `F` (in the procedures we will use the notation `L` for the level). In an additional array `NodeNumb` we keep the number of nodes at the different levels. In the description of our implementation we will first represent identical confiners (see Figure 3(c), confiner 4 at the levels 10, 11, 12) at each new level again. After, we will show what is to change when representing only the confiner at highest level among identical confiners ("unique representation"). This is necessary for worst case analysis.

Principal Steps of the Algorithm The algorithm proceeds in two steps: first (A) the pixels are sorted in decreasing order according to their grey level. The order at the same level is arbitrary. This is done by a standard sort algorithm called "counting sort" [2] in $O(n+r) = O(n)$ (we assume $r \leq n$) time. In the second step (B) we calculate the confiners for all grey levels l (step B l) in decreasing order, beginning with the highest grey level, and we update the tree structure up to the given level. Step (B l) consists of four sub-steps (B1 l), ..., (B4 l). Step (B1 l) (Figures 3(a), 4) implements the *scanning* operations and determines the flat regions at level l. For each pixel the number of the corresponding flat region is provisionally kept in memory (in `Pix2cf`, see below). Step (B1 l) determines also which already updated confiners (obtained using `Pix2cf`) – and even their ancestors at level $l+1$ using the data structure maintained in step (B3 m), $m > l$ – are adjacent to such a region. This information is kept in memory (in `Fusion`, see below). We deduce from it in step (B2 l) (Figures 3(b), 5) which confiners at higher levels and which flat regions at level l are fusing for obtaining the new confiners at level l (stored then in `Pix2cf`). Step (B2 l) implements the *fusion* operations.

That we have to know (in constant time if possible) for a given confiner its ancestor at level $l+1$ (Figure 3(a),(c)), in order to fuse the confiners and flat regions, is one of the major difficulties (with respect to worst-case behavior) of the confinement tree construction. To do this, in step (B3 l), detailed in section 3.3, we define and maintain (by the *propagation* operations) a data structure associated to the tree. It is called `Node2ances` and replaced in the optimized versions of (B3 l) (cf. section 3.3) by `Leaves2ances` or by `SubtreeL2ances`. Finally we pass in step (B4 l) once more through all pixels at level l to update (by the *attribute* operations) the attribute values of the now identified confiners (Figure 6).

Step (B3 l) is crucial for worst-case analysis. In section 3.3 we present an algorithm (Figure 7) to execute step (B3 l) in $T_3(l)$ time such that $\sum_l T_3(l) = O(n_n \log n_n) = O(n \log n)$ time, assuming unique representation of identical confiners. Steps (Bi l), $i = 1, 2, 4$, are executed in $\sum_l T_i(l) = O(n)$ time by the procedures presented in Figures 4, 5 and 6. However, as we have in imaging practice often $n_n << n$, step (B3 l) will not in general be the most consuming one (cf. section 5).

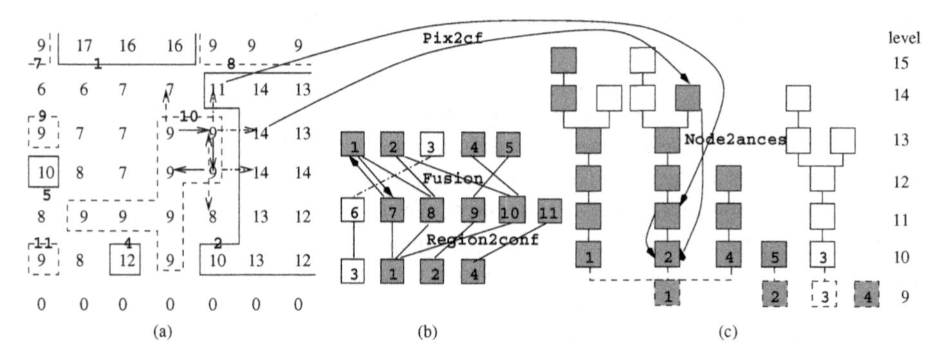

Fig. 3. (a) Part of an image to illustrate the detection of flat regions and confiners for a given grey level: dashed lines delimit flat regions at the current level 9 and continuous lines delimit confiners at level 10; the numbers on the lines are that of the corresponding confiners or regions respectively; different arrow types correspond to different cases in Figure 4 (see section 3.2). (b) Boxes contain confiner numbers at level 10 (upper row) and 9 (lower row) and flat region numbers (middle row). Broken lines and (as in (c)) white boxes correspond to not shown parts of the image. Arrows illustrate (as in (c)) the association made by the stated data structures; we omit in `Region2conf` implicitly given associations between the confiners at levels 9 and 10. (c) Confinement tree updated up to level 9; broken lines illustrate information added in step (B 9)

Four Principal Data Structures The sub-steps (B1 l), ..., (B4 l) are related by four data structures essential for our algorithm. There are (cf. Figure 3) `Pix2cf`, `Fusion`, `Region2conf`, and, in a first approach, `Node2ances` which will be replaced by `Leaves2ances` or by `SubtreeL2ances` in section 3.3 when showing how to pass from $O(n^2)$ complexity to $O(n \log n)$ for (B3 l). `Fusion` is a dynamic array of stacks (see below); the other structures are dynamic arrays of integer values. Let us detail now, how this structures are defined and cooperate during the sub-steps (B1 l), ..., (B4 l).

To each flat region at level l, determined in step (B1 l), we allocate a number $R_{i'l}$ which we store provisionally in each of its pixels (in an array `Pix2cf`). Equally, we store for each (already discovered) confiner $C_{j(l+1)}$ at level $l+1$ adjacent to such a region this number $R_{i'l}$ in a stack `Fusion`($C_{j(l+1)}$) and vice versa $C_{j(l+1)}$ in a stack `Fusion`($R_{i'l}$). See in step (B3 l) and in Figure 3(a),(c) how we find the number of $C_{j(l+1)}$. The stacks in `Fusion` permit (as illustrated in Figure 3(b)) to connect in step (B2 l) all flat regions $R_{i'l}$ and all confiners $C_{j(l+1)}$ at the previous level which belong now to the new confiner C_{il} at the current level l. Therefore, these connections enable us also to establish the father-son relationships between the confiners (nodes) at two successive levels. `Fusion` is implemented as an array containing after step (B1 l) the stacks `Fusion`($C_{j(l+1)}$) for all confiners $C_{j(l+1)}$ at level $l+1$ and the stacks `Fusion`($R_{i'l}$) for all flat regions $R_{i'l}$ at level l. Using `Fusion` we merge in step (B2 l) all connected regions in order to obtain the confiners at level l. Here, we associate in the array `Region2conf` (see Figure 3(b)) each number identifying one of the confiners $C_{j(l+1)}$ at level $l+1$ and each flat region number $R_{i'l}$ with the number of the new confiner C_{il} containing the given confiner or region. Using `Region2conf`, we replace in the array `Pix2cf` the flat region number $R_{i'l}$ by that of the confiner C_{il} enclosing it. This is done by the procedure in Figure 5.

Step (B3 l) operates on the tree only. We update an array `Node2ances` (in a first approach: more sophisticated approaches to implement step (B3 l) are proposed in section 3.3; they do not require changes in the other parts of the algorithm). `Node2ances` maps (Figure 3(c)), before step (B3 l), each already determined node to its ancestor at level $l+1$ and to its ancestor at level l, after (B3 l). This will be necessary for the steps (B1 $l-1$) and (B2 $l-1$): we stored (in the array `Pix2cf`) in the steps (B2 m), $m \geq l$, for each pixel at level m the number of the node/confiner at level m containing it but we will need its ancestor node at level l. In step (B4 l), finally, we examine once more all pixels at level l. We add to each new node first the attribute information coming from its sons in the tree by the procedure `addNode(node,prevNode)` and after the attribute information coming from the pixels by `addPixel(node,pixel)`. The corresponding procedure is given in Figure 6.

3.2 Our Algorithm in Detail

Let us give now some more detailed information for the procedures presented in the Figures 4, 5, 6 and show some techniques to speed up the algorithm.

```
1. init(Fusion);    /* initial. Fusion as array of empty stacks */
2. RegNumber = NodeNumb[L+1];
3. for each regElement at level L do
4.    if (Pix2cf[regElement] == 0)               /* new flat region */
5.       RegNumber++;        /* Number Ri'l of this new flat region */
6.       Pix2cf[regElement] = RegNumber;
7.       push(pixS, regElement); /* init. of stack pixS to scan it*/
8.       while (pop(pixS, pixel) != 0)
9.          for all nghb of pixel
10.            if((F[nghb] == L)&& !Pix2cf[nghb])/*arrow: cont. line*/
11.               Pix2cf[nghb] = RegNumber;
12.               push(pixS, nghb);
13.            else if (F[nghb] > L)          /* arrow: dots + dashes */
14.               node = Pix2cf[nghb];/*node containing nghb and its*/
15.               anc = Node2ances[F[nghb]][node]; /* ances. at L+1 */
16.               push(Fusion[anc], RegNumber);
17.               push(Fusion[RegNumber], anc);
```

Fig. 4. Procedure to detect the flat regions and to establish the inter-region connexions (step B1 L)

Variables As a general convention we use capitals at the beginning of a variable name only if the variable appears also outside the procedure. These variables are F, Tree, and NodeNumb as described above, RegNumber ($= R_{i'l}$ and $=$ number of confiners at level $l+1$ plus flat regions at l, after (B1 l)), L= l, and R= r, and the four data structures Pix2cf, Fusion, Node2ances, and Region2conf. These four variables, described already above and illustrated in Figure 3, are internal variables of (B) but Pix2cf and Node2ances are input and output for each step (B l) whereas Fusion and Region2conf are cleared after step (B2 l). RegNumber reappears indirectly in line 12 of (B2 l) but is an internal variable of (B l). Outside the procedures, after (A), Pix2cf is initialized to zero for each pixel, and we set NodeNumb[R+1]=0. All variables beginning with a small letter are internal variables of the respective procedures.

```
1. nodeNumber = 0; init(Region2conf);/*init. all array el. to 0*/
2. for each prevNode at level L+1 do
3.    if (Region2conf[prevNode] == 0)        /* new confiner at L*/
4.       nodeNumber++;
5.       Region2conf[prevNode] = nodeNumber;
6.       push(regS, prevNode); /* regS: stack of regions to fuse */
7.       while (pop(regS, reg1) != 0)
8.          while (pop(Fusion[reg1],reg2) != 0)
9.             if(Region2conf[reg2] == 0)
10.               Region2conf[reg2] = nodeNumber;
11.               push(regS, reg2);
12.for each flatReg at level L do
13.   if (Region2conf[flatReg] == 0)   /* new leaf in the tree */
14.      nodeNumber++;
15.      Region2conf[flatReg] = nodeNumber;
16.NodeNumb[L] = nodeNumber;
17.for each regElement at level L do
18.   Pix2cf[regElement] = Region2conf[Pix2cf[regElement]];
```

Fig. 5. Procedure to associate the flat regions and the confiners at level L+1 with the new confiners (nodes). The association is kept in the array Region2conf (step B2 l, l=L)

Sub-steps (B1 l) and (B2 l) In the procedure in Figure 4, the flat regions are represented by the current value of RegNumber ($= R_{i'l}$). They are initialized in line 5-7 with regElement which corresponds in the example in Figure 3(a) to the encircled pixel in the middle at level 9. Pixels at the border are supposed to be set at 0 (see lowest row in Figure 3(a)). The procedure (Figure 4) detects each pixel in each flat region following a recursive scheme. It implements this scheme iteratively using the data stack pixS. We will remark here that we do not need a priority queue as in [7], where the pixels in the queue had to be ordered according to their grey level. This is because all pixels which put in the queue have the same grey level `L`. The auxiliary function `push(pixS, element)` places `element` at the top of the stack `pixS` as `push(Fusion[number1], number2)` places the number `number2` at the top of the stack `Fusion[number1]`. `pop(pixS, element)` removes the element at the top of the stack and stores it in `element`. It returns 1 if there was an element in the stack, else 0. For a `pixel` fetched from `pixS` (line 8) we will examine all neighbors `nghb` (line 9): (i) if the neighbor is in the same flat region and not treated up to now (this corresponds to the continuous arrows in Figure 3(a)), it will be treated in lines 11, 12; (ii) if the neighbor is in a confiner at a previous level (arrow with dashes and dots), lines 14-17 are executed; (iii) else, if the neighbor is at a lower level or already updated (dashes), nothing is done. In case (ii) the connexions between the lowest updated confiner containing the neighbor (represented by `anc`) and the current flat region (represented by `RegNumber`) are stored in Fusion (line 16, 17; Figure 3(b)).

The array `Region2conf` in Figure 5 is of length `RegNumber` and its indices correspond to the confiners/nodes at the previous level $l+1$ followed by the flat regions at level l as represented by $R_{i'l}$ (Figure 3(b), line 2 and 5 in Figure 4). The $R_{i'l}$ are stored in the `Fusion` stacks. The procedure in Figure 5 implements step (B2 l) following a very similar scheme up to line 11 as that implementing (B1 l). The differences are that only the number of the new confiner has to be kept (line 10 which corresponds to line 11 in Figure 4 and lines 9-11 against lines 10-17 in Figure 4) and that the number of connexions changes for different regions and confiners (line 8 against line 9 in Figure 4). Lines 12-15 detect the new leaves. Figure 6 implements step (B4 l) as already described above. The remaining step (B3 l) is detailed in section 3.3.

Ways to Speed Up the Algorithm To improve in practice the calculation time, first, we will implement a way to store a connexion between a confiner and a flat region just once in the corresponding stack of `Fusion` during step (B1 l) (Figure 4). We use an array which marks for each confiner at the previous level if the current flat region has already been treated. We refer to this in section 5 as "`mark_neighConf`". We achieve in a similar way, to deal with a tree implementation using unique representation of identical confiners where we mark the confiners adjacent to a flat region at the current level.

The second improvement for which we will present timings in section 5 is an attempt to implement all stacks in the `Fusion` array in one big array. It will be referred below as "`arrayFusion_Proc`".

3.3 Propagating Information through the Tree

Let be given a rooted tree T known from its highest level down to a level l where the levels l decrease iteratively from the highest to the lowest one. This subsection addresses the *problem* of maintaining a data structure associated to T permitting, at each iteration, to find in constant time for any node at any level $l' > l$ its ancestor at level l.

```
init(Tree[L]); /* initial. Tree[L][node] to 0 for all nodes at L */
for each prevNode at level L+1 do
  node = Region2conf[prevNode];
  addNode(Tree[L][node],Tree[L+1][prevNode]);
for each regElement at level L do
  addPixel(Tree[L][Pix2cf[regElement]],regElement);
```

Fig. 6. Procedure updating the tree with the attributes of the confiners at level L (step B4 L)

In a first approach, we implement this data structure using the array `Node2ances` (see Figures 3(c), 4) associating a node to its ancestor at the required level l. We update `Node2ances` for each new level $l-1$ by replacing for each already updated node the number of its ancestor at level l by that at level $l-1$. This is illustrated in Figure 3(c). We call this procedure "`all_nodes_Proc`".

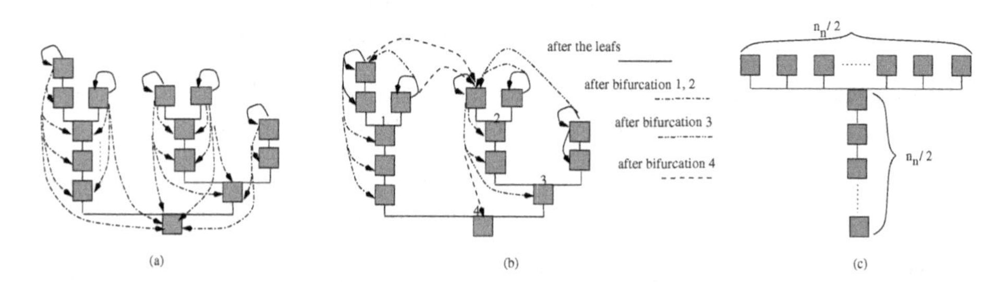

Fig. 7. (a), (b) Two ways to maintain a data structure permitting to find for each leaf in constant time its ancestor at level l; (c) example for worst-case analysis of `all_leaves_Proc`

We can reduce the operations by associating to each node one of its descendant leaves. Then, we can reduce the above stated problem to leaves instead of nodes by implementing the required data structure as an array `Leaves2ances` and proceeding as in the procedure before, but updating the ancestor associations only for each leaf. This is illustrated in Figure 7(a). We call this procedure "`all_leaves_Proc`". However, as we will see in section 4 the worst-case complexity of `all_leaves_Proc` stays $O(n^2)$ as for the procedure `all_nodes_Proc`.

We show now, how to stay – as proven in section 4 – within a worst-case complexity of $O(n \log n)$. We have to proceed as illustrated in Figure 7(b). For a given subtree, only for one leaf m we set an association to the subtree root. We can implement this association either directly as a new field of a leaf or as a separate list or array `SubtreeL2ances`. The other leaves in the subtree point to m and get the ancestor value from m. If a bifurcation appears in a node b, we have to change the pointers of the leaves in the new subtree rooted in b to point all to one single leaf of this subtree. The number of operations for this step is limited by the number of leaves in the subtree (see the proof of proposition 2 in section 4, where we call this procedure "`subtreeL_Proc`").

4 Complexity Analysis

Lemma 1. *If n is the number of pixels of an image and if n_n denotes the number of nodes in the corresponding confinement tree, we have $n_n \leq n$.*

Proof The lemma follows from the unique representation of identical confiners.

Given a rooted tree with n_n nodes (leaves included), how many nodes have to be passed through in order to go from all leaves to the root? One could think, the answer is $O(n_n \log n_n)$ as in two extremal cases – the complete binary tree and the tree with $n_n - 1$ leaves – at most $O(n_n \log n_n)$ passes are necessary, even only $O(n_n)$ in the latter case. But Figure 7(c) illustrates a case[1] where $O(n_n^2)$ passes are required. This is also the worst-case as there are at most $n_n - 1$ leaves and at most n_n nodes on the path between a leaf and the root. This proves the

Proposition 1. *The procedure `all_leaves_Proc`, given in section 3.3, updates the array `Leaves2ances` for all levels in a total time $O(n_n^2)$.*

Proposition 2. *The procedure `subtreeL_Proc`, given in section 3.3, updates the array `SubtreeL2ances` for all levels in a total time $O(n_n \log n_n)$.*

Proof We will prove that we have at most $n_n \log n_n$ operations without counting the $O(n_n)$ initialization operations. This, we will prove by induction on the number of nodes of the tree. The following logarithms are at base 2. For $n_n = 1$ (initialization of the induction) there is nothing to show. We will consider the root of the tree and suppose that it has k ($k \geq 2$; if $k = 1$ we have $(n_n - 1) \log (n_n - 1) + 1 \leq n_n \log n_n$) sons defining k subtrees for which the induction hypothesis is true. Therefore, and as we need less than n_n operations when passing from one level to the next one (see section 3.3), we can reduce the problem to the following statement: $\sum_{i=1}^{k} n_i \log n_i + n_n \leq n_n \log n_n, n_n = 1 + \sum_{i=1}^{k} n_i, n_i \geq 1$. As we are in the convex part of the function $x \log x$ (as $n_i \geq 1$) this follows directly from Jensen's inequality for convex functions: we have $\frac{1}{k} \sum_{i=1}^{k} n_i \log n_i \leq (\frac{1}{k} \sum_i n_i) \log (\frac{1}{k} \sum_i n_i)$ from which follows $\sum_{i=1}^{k} n_i \log n_i + n_n \log k \leq n_n \log(n_n - 1) \leq n_n \log n_n$. As $1 \leq \log k$ ($k \geq 2$) this proves our statement and therefore our proposition 2.

Theorem 1. *The algorithm calculates the confinement tree in $O(n \log n)$ time using $O(n)$ scanning, fusion, attribute and $O(n_n \log n_n)$ propagation operations.*

[1] An image corresponding to this tree is possible, even for $n_n = n/2$

Proof We will just prove that the total time to execute step (B2 l) for all levels l is $O(n)$. The rest follows either from lemma 1, proposition 2 or is clear. In step (B2 l), for a given level l, each flat region number at l and each number of a confiner at level $l+1$ is put exactly once in the data stack `regS`. For each element in the data stack the number of operations is the number of its connexions with flat regions or confiners. The number of all this connexions at a level l is limited by 4 times the number of pixels (forming the flat regions) at grey level l as each pixel has at most 4 neighbors. Therefore, the total number of connexions for all levels l is within $O(n)$ and our theorem is proven.

5 Experimental Evaluation

We measured the execution times for the algorithm and its parts as presented in the procedures of the Figures in section 3.2 for several images on a Intel Pentium III, 450 MHz. The timings are measured with the `clock()`-function under LINUX operation system and the results are summarized in figure 8 as well as information about the images (they have all 8 bit grey values; the MRI brain image is presented and treated in [12,11] and the microscopic images A, B in [11]). We observe stable behavior of our algorithm for approximately the same number of leaves and nodes but also increasing importance of step (B3 l) for larger images with larger trees: the propagation operations are about 10 times more time consuming for the microscopic images C and D with respect to A and B, whereas the other operations increase only 3-4 times. However, as $n_n << n$ the propagation operations does not become dominant. At the contrary, Figure 2(b) illustrates that the execution time of the algorithm of Jones [8] can become prohibitive *even if* $n_n << n$: for an increasing number of maxima (5 in 2(b)), all operations are multiplied by a factor depending linearly on this number. Figure 2(b) presents a synthetical image, however, the timings measured by Jones [8] on real images allow to conclude that our algorithm can be 10 up to 50 times faster in practice.

6 Conclusion

We presented a new algorithm for calculating the confinement tree and could prove that its worst-case complexity is $O(n \log n)$. Moreover, the algorithm separates different kinds of operations such that all operations except propagation stay in $O(n)$ time. We showed how to achieve complexity $O(n_n \log n_n)$ for propagation operations where n_n is the number of nodes in the tree (in a first approach, we implement propagation in $O(n_n^2)$, sufficient if $n_n << n$). Two example images illustrate $O(n^2 \log n)$ complexity for a known algorithm, prohibitive computation time for this known algorithm even if $n_n << n$, and $O(n^2)$ for an algorithm calculating a tree related to the confinement tree. Practical evaluation allows to conclude that our improvement leads to much faster calculation times on standard medical images.

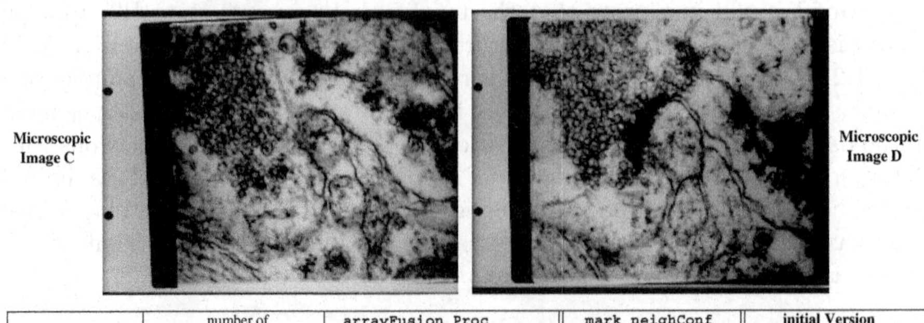

		number of			arrayFusion_Proc					mark_neighConf					initial Version				
		pixels	nodes	leaves	total algorithm	step B1	B2	B3	B4	total	B1	B2	B3	B4	total	B1	B2	B3	B4
MRI Brain Image		256x256	14217	7963	0.17	0.07	0.06	0.01	0.03	0.20					0.28				
Microscopic Images	A	512x480	28425	8984	0.68	0.31	0.19	0.07	0.07	0.77	0.34	0.26	0.07	0.07	1.10	0.40	0.53	0.07	0.07
	B	512x480	30128	8494	0.69	0.31	0.20	0.06	0.08										
	C	850x714	127263	26284	2.75	0.94	0.76	0.67	0.29										
	D	850x714	128983	26448	2.69	0.95	0.74	0.62	0.28										

Fig. 8. Test images and execution times in seconds for the implementation of our algorithm

Acknowledgments

We would like to thank Prof. M. Couprié for helpful bibliographic remarks.

References

1. Chassery, J.-M., Montanvert, A.: Géométrie discrète en analyse d'images. Hermes (1991) 395
2. Cormen, T. H., Leiserson, L. E., Rivest, R. L.: Introduction to algorithms. MIT Press (1998) 396
3. Couprié, M., Bertrand, G.: Topological grey scale watershed transformation. In: SPIE Vision Geometry V Proceedings, Vol. 3168 Bellingham, WA (1997) 136–146 392, 393
4. Guillataud, P.: Contribution à l'analyse dendronique des images. PhD thesis, Université de Bordeaux I (1992) 392, 393
5. Hanusse, P., Guillataud, P.: Sémantique des images par analyse dendronique. In: AFCET, 8th RFIA, Vol. 2. Lyon (1992) 577–588 392, 393, 394
6. Hartigan, J. A.: Statistical theory in clustering. J. of Classification **2** (1985) 63–76 392, 394
7. Jones, R.: Connected Filtering and Segmentation Using Component Trees. Computer Vision and Image Understanding **75** (1999) 215–228 392, 393, 394, 395, 399
8. Jones, R.: Connected Filtering and Segmentation Using Component Trees: Efficient Implementation Algorithms. http://www.dms.CSIRO.AU/ ronaldj/pseudocode (1999) 393, 394, 395, 403
9. Kok-Wiles, S.L, Brady, J. M., Highnam, R.: Comparing mammogram pairs for the detection of lesions. In: Karssemeijer, N. (ed.): 4th Int. Workshop of Digital Mammography, Nijmegen, Netherlands, June 1998. Kluwer,Amsterdam, (1998) 392

10. Mattes, J., Demongeot, J.: Dynamic confinement, classification, and imaging. In: 22nd Ann. Conf. GfKl, Dresden, Germany, March 1998. Studies in Classification, Data Analysis, and Knowledge Organization. Springer-Verlag (1999) 205–214 392, 394
11. Mattes, J., Demongeot, J.: Tree representation and implicit tree matching for a coarse to fine image matching algorithm. In: MICCAI'99, C. Taylor, A. Clochester (Eds.). LNCS. Springer-Verlag (1999) 646–655 392, 393, 394, 403
12. Mattes, J., Richard, M., Demongeot, J.: Tree representation for image matching and object recognition. In: DGCI'99, G. Bertrand and M. Couprié and L. Perroton (Eds.). LNCS. Springer-Verlag (1999) 298–309 392, 403
13. Salembier, P., Oliveras, A., Garrido, L.: Antiextensive Connected Operators for Image and Sequence Processing. IEEE Trans. on Image Processing **7** (1998) 555–570 392, 393, 394, 395
14. Wishart, D.: Mode analysis: A generalization of the nearest neighbor which reduces chaining effects. In: Cole, A. J. (Ed.): Numerical Taxonomy. Academic Press, London (1969) 282–319 392, 394
15. Zuck, P., Lao, Z., Skwish, S., Glickman, J. F., Yang, K., Burbaum, J., Inglese, J.: Ligand-receptor binding measured by laser-scanning imaging. PNAS **96** (1999) 11122–7 393

A 3D 3–Subiteration Thinning Algorithm for Medial Surfaces

Kálmán Palágyi

Department of Applied Informatics, University of Szeged,
H–6701 Szeged P.O.Box 652, Hungary
palagyi@inf.u-szeged.hu

Abstract. Thinning on a binary picture is an iterative layer by layer erosion to extract a reasonable approximation to its skeleton. This paper presents an efficient 3D parallel thinning algorithm which produces medial surfaces. Three–subiteration directional strategy is proposed: each iteration step is composed of three parallel subiterations according to the three deletion directions. The algorithm makes easy implementation possible, since deletable points are given by matching templates containing twentyeight elements. The topological correctness of the algorithm for $(26, 6)$ binary pictures is proved.

1 Introduction

Skeletonization provides shape features that are extracted from binary image data. A very illustrative definition of the skeleton is given using the prairie–fire analogy: the object boundary is set on fire and the skeleton is formed by the loci where the fire fronts meet and quench each others [3]. In discrete spaces, the thinning process is a frequently used method for producing an approximation to the skeleton in a topology–preserving way [6]. It based on digital simulation of the fire front propagation: border points of a binary object that satisfy certain topological and geometric constraints [21] are deleted in iteration steps. The entire process is repeated until only the "skeleton" is left. Therefore, a thinning algorithm can be regarded as a reduction operation that changes some 1's (object elements or black points) to 0's (white points) but does not alter 0's.

A reduction operation does *not* preserve topology if

- any object in the input picture is split (into two or more) or completely deleted,
- any cavity in the input picture is merged with the background or another cavity, or
- a cavity is created where there was none in the input picture.

There is an additional concept called hole in 3D pictures. A hole (that doughnuts have) is formed by 0's, but it is not a cavity [6]. Topology preservation implies that eliminating or creating holes is not allowed.

G. Borgefors, I. Nyström, and G. Sanniti di Baja (Eds.): DGCI 2000, LNCS 1953, pp. 406–418, 2000.
© Springer-Verlag Berlin Heidelberg 2000

A simple point is an object point whose deletion does not alter the topology of the picture [13]. Sequential thinning algorithms delete simple points which are not end points, since preserving end–points provides important information relative to the shape of the objects. Curve thinning (i.e., a thinning process for extracting medial line) preserves line–end points while surface thinning (i.e., a thinning process for extracting medial surface) does not delete surface–end points.

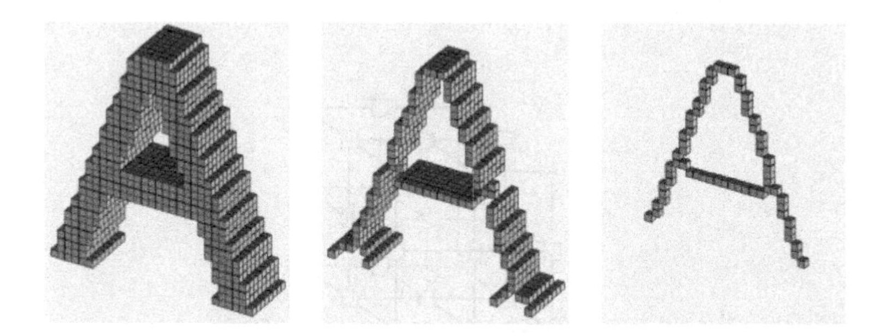

Fig. 1. A 3D synthetic picture containing a character "A" (left), result of a surface thinning process (centre), and result of a curve thinning process (right). (Cubes represent black points.)

Parallel thinning algorithms delete a set of simple points. A possible approach to preserve topology is to use directional strategy; each iteration step is composed of a number of parallel subiterations where only border points of certain kind can be deleted in each subiteration. There are six kinds of border points in 3D pictures on cubic grid, therefore, 6–subiteration directional thinning algorithms were generally proposed.

In this paper, a 3–subiteration directional algorithm is proposed for surface thinning. Some experiments are made on synthetic objects and the topology preservation for (26,6) binary pictures [6] is proved. Our approach demonstrates a possible way for constructing non–conventional directional thinning algorithms.

2 Basic Notions and Results

Let p be a point in the 3D digital space $\mathbb{Z}^3$. Let us denote $N_j(p)$ (for $j = 6, 18, 26$) the set of points *j–adjacent* to point p (see Fig. 2). The sequence of distinct points $\langle x_0, x_1, \ldots, x_n \rangle$ is a *j-path* of length $n \geq 0$ from point x_0 to point x_n in a non–empty set of points X if each point of the sequence is in X and x_i is j–adjacent to x_{i-1} for each $1 \leq i \leq n$. (Note that a single point is a j–path of length 0.) Two points are *j–connected* in the set X if there is a j–path in X between them. A set of points X is *j–connected* in the set of points $Y \supseteq X$ if any two points in X are j–connected in Y.

The *3D binary (m,n) digital picture* $\mathcal{P}$ is a quadruple $\mathcal{P} = (\mathbb{Z}^3, m, n, B)$ [6]. Each element of $\mathbb{Z}^3$ is called a *point* of $\mathcal{P}$. Each point in $B \subseteq \mathbb{Z}^3$ is called a *black point* and value 1 is assigned to it. Each point in $\mathbb{Z}^3 \setminus B$ is called a *white point* and value 0 is assigned to it. Adjacency m belongs to the black points and adjacency n belongs to the white points. A *black component* (or *object*) is a maximal m–connected set of points in B. A *white component* is a maximal n–connected set of points in $B \subseteq \mathbb{Z}^3$.

We are dealing with (26,6) pictures. It is assumed that any picture contains finitely many black points.

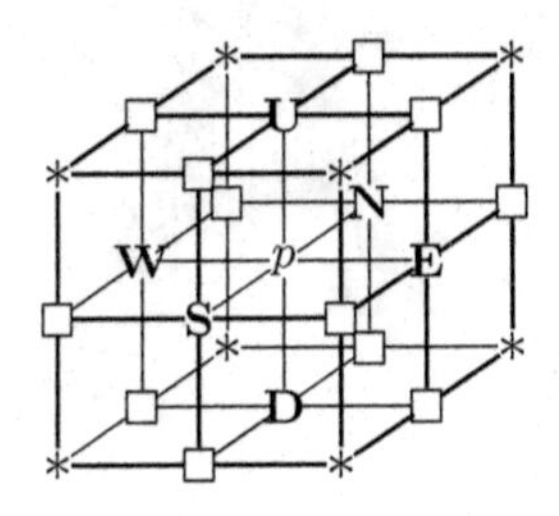

Fig. 2. The frequently used adjacencies in $\mathbb{Z}^3$. The set $N_6(p)$ contains the central point p and the 6 points marked **U**, **D**, **N**, **E**, **S**, and **W**. The set $N_{18}(p)$ contains the set $N_6(p)$ and the 12 points marked "□". The set $N_{26}(p)$ contains the set $N_{18}(p)$ and the 8 points marked "$*$"

A black point is called *border point* if it is 6–adjacent to at least one white point. (Note that this definition is correct only for the special cases $m = 26$ and $m = 18$.) A border point p is called **U**–*border point* if the point marked by **U** in Fig. 2 is white. We can define **N**–, **E**–, **S**–, **W**–, and **D**–border points in the same way.

A black point is called *simple point* if its deletion does not alter the topology of the picture. We make use of the following result for (26,6) pictures:

Theorem 1. [12,19] *Black point p is simple in picture $(\mathbb{Z}^3, 26, 6, B)$ if and only if all of the following conditions hold:*

1. *the set $(B \setminus \{p\}) \cap N_{26}(p)$ contains exactly one 26–component; and*
2. *the set $(\mathbb{Z}^3 \setminus B) \cap N_6(p)$ is not empty and it is 6–connected in the set $(\mathbb{Z}^3 \setminus B) \cap N_{18}(p)$.*

Theorem 1 shows that the simplicity in $(26, 6)$ pictures is a local property; it can be decided using the $3 \times 3 \times 3$ neighborhood of a given point.

We need to consider what is meant by topology preservation when a number of black points are deleted simultaneously. Ma [8] and Kong [5] gave sufficient conditions for parallel reduction operations of 3D (26,6) pictures. We use the following more general sufficient conditions:

Theorem 2. [17,18] *Let $\mathcal{T}$ be a parallel reduction operation on* (26, 6) *pictures. Then $\mathcal{T}$ is topology preserving, if for all picture $\mathcal{P} = (\mathbb{Z}^3, 26, 6, B)$, all of the following conditions hold:*

1. *for all points $p \in B$ that are deleted by $\mathcal{T}$ and for all sets $Q \subseteq (N_{18}(p) \backslash \{p\}) \cap B$ that are deleted by $\mathcal{T}$, p is simple in the picture $(\mathbb{Z}^3, 26, 6, B \backslash Q)$; and*
2. *no black component contained entirely in a unit lattice cube (i.e., a $2 \times 2 \times 2$ configuration in $\mathbb{Z}^3$) can be deleted completely by $\mathcal{T}$.*

We propose a surface thinning algorithm. The deletable points of the algorithm are border points of certain types and not *surface end–points* (i.e., which are not extremities of surfaces). The proposed algorithm uses the following characterization of the the surface end–points.

Definition 1. [17,18] *The set $N_6(p)$ is subdivided into three kinds of opposite pairs of points* (**U**,**D**), (**N**,**S**), *and* (**E**,**W**) (see Fig. 2).
A black point p is a surface end–point in a picture if the set $N_6(p)$ contains at least one opposite pair of white points.

3 Existing Parallel Thinning Algorithms

Most of the existing thinning algorithms are parallel, since the fire front propagation is by nature parallel. Those algorithms delete a set of simple points simultaneously that can alter the topology. There are three major strategies to overcome this problem:

- **Fully parallel algorithms:**
  Algorithms from this group do not divide an iteration step into subiterations. In order to preserve topology, the known three fully parallel 3D thinning algorithms investigate larger neighborhood than the $3 \times 3 \times 3$ one: Ma proposed an algorithm, in which the new value of a black point depends on 30 points (and a parallel rechecking pass is required) [9], the fully parallel algorithm of Ma and Sonka uses a special neighborhood containing 50 points [10], and Manzanera et al. developed an algorithm using a symmetric neighborhood consisting of 81 points [11].
- **Directional (or border sequential) algorithms:**
  Iteration steps are divided into a number of successive subiterations, where only border points of certain kind can be deleted in parallel in each subiteration. Consequently, each subiteration uses different deletion rule. Since there are six kinds of major directions in 3D pictures, 6–subiteration directional thinning algorithms were generally proposed [2,4,7,14,15,22]. Note that Palágyi and Kuba developed 8–subiteration [17] and 12–subiteration [18] directional thinning algorithms, too. Each existing directional algorithm examines the $3 \times 3 \times 3$ neighborhood of each border point.

– **Subfield sequential algorithms:**
The 3D digital space $\mathbb{Z}^3$ is subdivided into more disjoint subfields that are alternatively activated. At a given iteration step, only border points in the active subfield are designated to be deleted. Each subiteration is executed in parallel (i.e., all border points in the actual subfield satisfying the deletion condition are simultaneously deleted). Two subfield sequential 3D thinning algorithms working in cubic grid has been proposed so far [1,20]. Both algorithms investigate the $3 \times 3 \times 3$ neighborhood and use eight subfields, therefore, each iteration step contains eight successive subiterations. Note that Palágyi and Kuba proposed a hybrid thinning algorithm [16]. It uses both subfield sequential and directional approaches (with two subfields and eight deletion directions).

The algorithm proposed in this paper follows the directional strategy. It requires only three subiterations (corresponding to the three kinds of opposite pair of points) in each iteration step, but an additional point not in the $3 \times 3 \times 3$ neighborhood is examined in each subiteration.

4 The New Thinning Algorithm

In this section, a new algorithm is presented for extracting medial surfaces from 3D $(26, 6)$ pictures.

Each conventional 6–subiteration directional thinning algorithm uses the six deletion directions that can delete certain **U**–, **D**–, **N**–, **E**–, **S**–, and **W**–border points, respectively [2,4,7,14,15,22]. In our 3–subiteration approach, two kinds of border points can be deleted in each subiteration. The three deletion directions correspond to the three kinds of opposite pairs of points, and are denoted by **UD**, **NS**, and **EW**.

Suppose that the 3D $(26, 6)$ picture to be thinned contains finitely many back points. Reduction operations associated with the three subiterations are called `deletion_from_UD`, `deletion_from_NS`, and `deletion_from_EW`. We are now ready to present the 3–subiteration approach formally:

Input: picture $\mathcal{P} = (\mathbb{Z}^3, 26, 6, B)$
Output: picture $\mathcal{P}' = (\mathbb{Z}^3, 26, 6, B')$

```
3-subiteration_thinning(B,B')
  begin
    B' = B;
    repeat
      B' = deletion_from_UD(B');
      B' = deletion_from_NS(B');
      B' = deletion_from_EW(B');
    until no points are deleted;
  end.
```

The new value of a black point depends on 28 points in each subiteration. The three special neighborhoods assigned to the different subiterations are presented in Fig. 3.

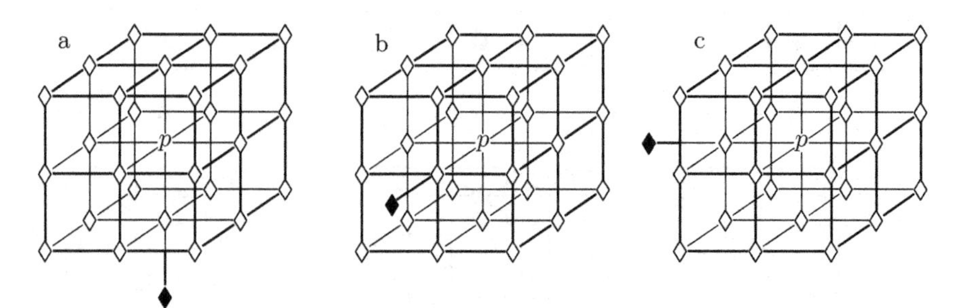

Fig. 3. The special local neighborhoods assigned to the deletion directions **UD** (a), **NS** (b), and **EW** (c), respectively. The new value of a black point p depends on the $3 \times 3 \times 3$ neighborhood of p (marked "◊") and an additional point (marked "♦") that is not in $N_{26}(p)$

Deletable points in a subiteration are given by a set of matching templates. A black point is deletable if at least one template in the set of templates matches it. Templates are usually described by three kinds of elements, "black", "white", and "don't care", where "don't care" matches either black or white point in a given picture. In order to reduce the number of masks we use additional notations (see Fig. 4).

The first subiteration assigned to the deletion direction **UD** can delete certain **U**– or **D**–border points; the second subiteration associated with the deletion direction **NS** attempt to delete **N**– or **S**–border points, and some **E**– or **W**–border points can be deleted by the third subiteration corresponding to the deletion direction **EW**. The set of templates $\mathcal{T}_{\mathbf{UD}}$ is given by Fig. 4. Note that Fig. 4 shows only the eight base templates **T1**–**T8**. Additionally, all their rotations around the vertical axis belong to $\mathcal{T}_{\mathbf{UD}}$, where the rotation angles are 90°, 180°, and 270°. This set of templates was constructed for deleting some simple points which are neither surface end–points (see Definition 1) nor extremities of surfaces. The deletable points of the other two subiterations (corresponding to deletion directions **NS** and **EW**) can be obtained by proper rotations of the templates in $\mathcal{T}_{\mathbf{UD}}$. Each template of our algorithm can be given by a Boolean condition that makes easy implementation possible.

Note that choosing another order of the deletion directions yields another algorithm. The proposed algorithm terminates when there are no more black points to be deleted. Since all considered input pictures are finite, it will terminate.

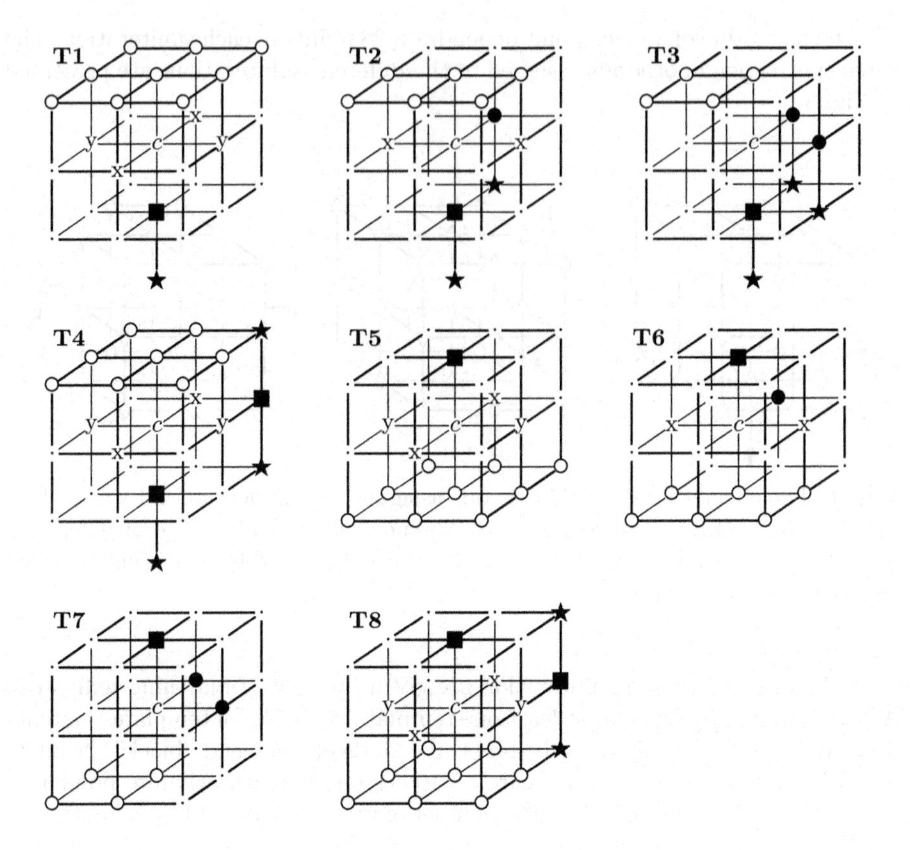

Fig. 4. Base templates **T1**–**T8** and their rotations around the vertical axis form the set of templates $\mathcal{T}_{\mathbf{UD}}$ assigned to the deletion direction **UD**. This set of templates belongs to the first subiteration. Notations: each position marked "c", "●", "■", and "★" matches a black point; each position marked "○" matches a white point; each "·" ("don't care") matches either a black or a white point; at least one position marked "**x**" matches a black point; at least one position marked "**y**" matches a black point. Emphasis is to be put that "**x**" and "**y**" positions provide that surface end–points cannot be deleted. (Note that using different symbols for black template positions helps us to prove the topological correctness of the algorithm)

5 Discussion

The proposed algorithm has been tested on objects of different shapes. Here we present only four examples (see Figs. 5–8).

The proposed 3–subiteration thinning algorithm is topology preserving for $(26, 6)$ pictures. It is sufficient to prove that reduction operation given by the set of templates $\mathcal{T}_{\mathrm{UD}}$ is topology preserving. If the first subiteration of the algorithm is topology preserving, then the other two are topology preserving, too, since the applied rotations of the deletion templates do not alter the topological properties. Therefore, the entire algorithm is topology preserving, since it is composed of topology preserving reductions.

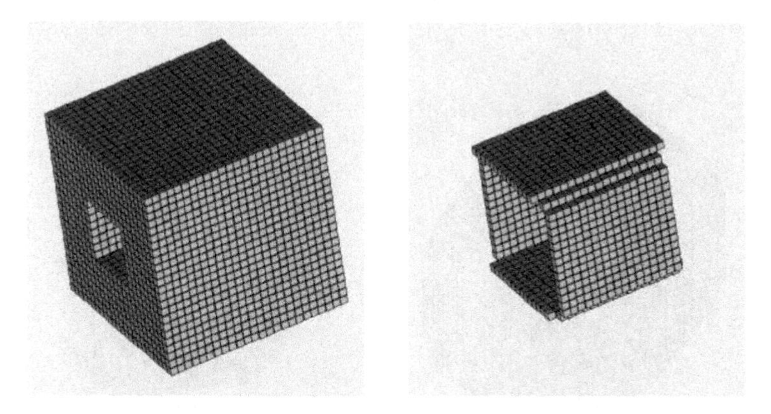

Fig. 5. A synthetic object containing a cube with a hole (left) and its medial surface produced by the proposed algorithm (right)

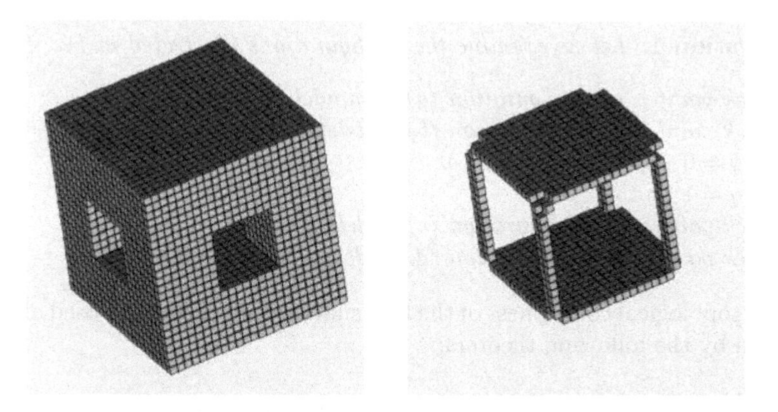

Fig. 6. A synthetic object containing a cube with two holes (left) and its medial surface produced by the proposed algorithm (right)

In order to prove both conditions of Theorem 2, we classify the elements of templates and state some properties of the set of templates $\mathcal{T}_{\mathrm{UD}}$. The element in the very centre of a template is called *central* (marked by "*c*" in Fig. 4). A noncentral template element is called *black* if it is always black (marked by "●", "■", and "★" in Fig. 4). A noncentral template element is called *white* if it is always white (marked by "○" in Fig. 4). Any other noncentral template element which is neither white nor black, is called *potentially black* (marked by "**x**", "**y**", and "·" in Fig. 4). A black or a potentially black noncentral template element is called *nonwhite*. A black point p is *deletable* if it can be deleted by at least one template in $\mathcal{T}_{\mathrm{UD}}$; p is *nondeletable* otherwise.

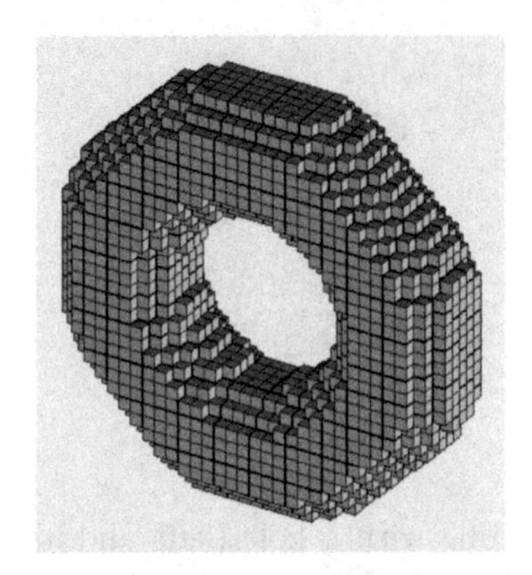

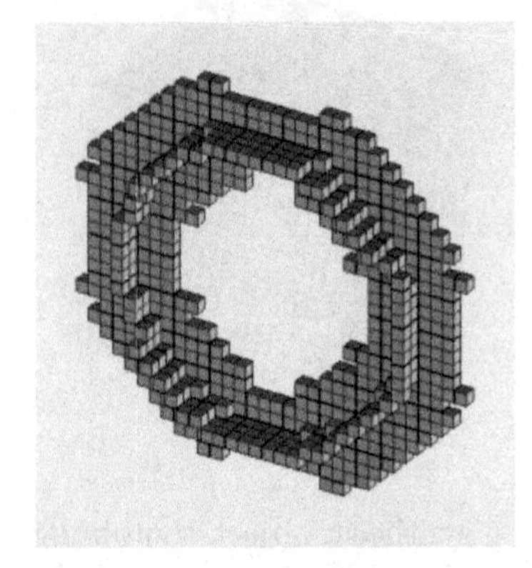

Fig. 7. A synthetic object containing a doughnut (left) and its medial surface produced by the proposed algorithm (right)

Observation 1. *Let us examine the configurations illustrated in Fig. 9.*

1. *Black point p in configuration (a) is nondeletable.*
2. *Black point p in configuration (b) is deletable if*
   - $q = 0$, $r = 1$, *and* $s = 1$, *or*
   - $q = 1$ *and* $r = 0$.
3. *Black point p in configuration (c) is deletable if $q = 1$ and $r = 1$.*
4. *Black point p in configuration (d) is deletable if $q = 1$.*

The topological correctness of the first subiteration of the proposed algorithm is stated by the following theorem:

Theorem 3. *Reduction operation given by the set of templates $\mathcal{T}_{\mathrm{UD}}$ is topology preserving for* $(26, 6)$ *pictures.*

Proof. (sketch) It is easy to see that each template in $\mathcal{T}_{\mathrm{UD}}$ deletes only simple points of (26,6) pictures.

The first point is to verify that there exists a 26–path between any two nonwhite positions (condition 1 of Theorem 1). It is sufficient to show that any

Fig. 8. A synthetic picture containing four characters (top) and their medial surface produced by the proposed algorithm (bottom)

potentially black position is 26–adjacent to a black position and any black position is 26–adjacent to another black position. It is obvious by careful examination of the templates in $\mathcal{T}_{\mathrm{UD}}$.

To prove that condition 2 of Theorem 1 holds, it is sufficient to show for each template in $\mathcal{T}_{\mathrm{UD}}$ that:

1. there exists a white position 6–adjacent to the central position,
2. for any two white positions 6–adjacent to the central position p are 6–connected in the set of white positions 18–adjacent to p,
3. and for any potentially black position 6–adjacent to the central position p, there exists a 6–adjacent white 18–neighbor which is 6–adjacent to a white position 6–adjacent to p.

The three points are obvious by a careful examination of the templates in $\mathcal{T}_{\mathrm{UD}}$.

We know that each deletable point p is simple. It can be stated that the value of any point coinciding with a potentially black template position does not alter the simplicity of p. We can state that the simplicity of a point p does not depend on the points that coincide with a template position marked "★", "x", or "y" (see Fig. 4). In addition, black points that coincide with template positions

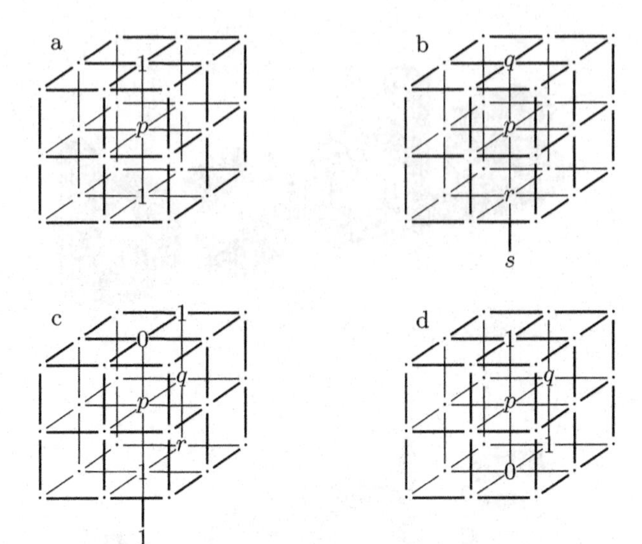

Fig. 9. Configurations assigned to Observation 1. Note that Observation 1 holds for their rotations around the vertical axis, too. (The rotation angles are 90°, 180°, and 270°.)

marked "■" are nondeletable (by Observation 1/1). Therefore, it is sufficient to deal with deletable points that coincide with template positions marked "●". Note that base templates **T1**, **T4**, **T5**, and **T8** (and their rotated versions) do not contain any positions marked "●". Therefore, only base templates **T2**, **T3**, **T6**, and **T7** (and their rotated versions) are to be investigated. It is easy to see with the help of Observation 1 that deletion of points coinciding with template positions marked "●" do not alter the simplicity of point p. Therefore, Condition 1 of Theorem 2 is satisfied.

Condition 2 of Theorem 2 can be seen with the help of Observation 1/2, too. Let us consider a unit lattice cube containing an upper set of four points $U = \{u_1, u_2, u_3, u_4\}$ and a lower set of four points $L = \{l_1, l_2, l_3, l_4\}$. Let $C \subseteq U \cup L$ be a black component contained in the unit lattice cube. If $C \cap L$ contains a deletable point then $C \cap U \neq \emptyset$ by Observation 1/2. It is easy to see, that any point in $C \cap U$ is nondeletable by Observation 1/2. Therefore, black component C cannot be deleted completely. □

Acknowledgements

The author is grateful to Attila Kuba and László G. Nyúl for their valuable suggestions. This work was supported by OTKA T023804 and FKFP 0908/1997 Grants.

References

1. Bertrand, G., Aktouf, Z.: A 3D thinning algorithms using subfields. In: Proc. SPIE Conference on Vision Geometry III, Vol. 2356 (1994) 113–124
2. Bertrand, G.: A parallel thinning algorithm for medial surfaces. Pattern Recognition Letters **16** (1995) 979–986
3. Blum, H.: A transformation for extracting new descriptors of shape. Models for the Perception of Speech and Visual Form, MIT Press, (1967) 362–380
4. Gong, W. X., Bertrand, G.: A simple parallel 3D thinning algorithm. In: Proc. 10th International Conference on Pattern Recognition (1990) 188–190
5. Kong, T. Y.: On topology preservation in 2–d and 3–d thinning. International Journal of Pattern Recognition and Artifical Intelligence **9** (1995) 813–844
6. Kong, T. Y., Rosenfeld, A.: Digital topology: Introduction and survey. Computer Vision, Graphics, and Image Processing **48** (1989) 357–393
7. Lee, T., Kashyap, R. L., Chu, C.: Building skeleton models via 3–D medial surface/axis thinning algorithms. CVGIP: Graphical Models and Image Processing **56** (1994) 462–478
8. Ma, C. M.: On topology preservation in 3D thinning. CVGIP: Image Understanding **59** (1994) 328–339
9. Ma, C. M.: A 3D fully parallel thinning algorithm for generating medial faces. Pattern Recognition Letters **16** (1995) 83–87
10. Ma, C. M., Sonka, M.: A fully parallel 3D thinning algorithm and its applications. Computer Vision and Image Understanding **64** (1996) 420–433
11. Manzanera, A., Bernard, T. M., Preteux, F., Longuet, B.: Medial faces from a concise 3D thinning algorithm. In: Proc. 7th IEEE Int. Conf. Computer Vision, ICCV'99 (1999) 337–343
12. Malandain, G., Bertrand, G.: Fast characterization of 3D simple points. In: Proc. 11th IEEE International Conference on Pattern Recognition (1992) 232–235
13. Morgenthaler, D. G.: Three–dimensional simple points: Serial erosion, parallel thinning and skeletonization. Technical Report TR–1005, Computer Vision Laboratory, Computer Science Center, University of Maryland (1981)
14. Mukherjee, J., Das, P. P., Chatterjee, B. N.: On connectivity issues of ESPTA. Pattern Recognition Letters **11** (1990) 643–648
15. Palágyi K., Kuba, A.: A 3D 6–subiteration thinning algorithm for extracting medial lines. Pattern Recognition Letters **19** (1998) 613–627
16. Palágyi K., Kuba, A.: A hybrid thinning algorithm for 3D medical images. Journal of Computing and Information Technology - CIT **6** (1998) 149–164
17. Palágyi K., Kuba, A.: Directional 3D thinning using 8 subiterations. In: Proc. 8th Int. Conf. on Discrete Geometry for Computer Imagery, DGCI'99, Lecture Notes in Computer Science, Vol. 1568. Springer (1999) 325–336
18. Palágyi K., Kuba, A.: A parallel 3D 12–subiteration thinning algorithm. Graphical Models and Image Processing **61** (1999) 199–221
19. Saha, P. K., Chaudhuri, B. B.: Detection of 3–D simple points for topology preserving transformations with application to thinning. IEEE Transactions on Pattern Analysis and Machine Intelligence **16** (1994) 1028–1032
20. Saha, P. K., Chaudhuri, B. B., Majumder, D. D.: A new shape–preserving parallel thinning algorithm for 3D digital images. Pattern Recognition **30** (1997) 1939–1955
21. Székely, G.: Shape characterization by local symmetries. Habilitationsschrift, ETH Zürich (1996)
22. Tsao, Y. F., Fu, K. S.: A parallel thinning algorithm for 3–D pictures. Computer Graphics and Image Processing **17** (1981) 315–331

Computing 3D Medial Axis for Chamfer Distances

Eric Remy and Edouard Thiel

Laboratoire d'Informatique de Marseille
LIM, Case 901, 163 av Luminy, 13288 Marseille Cedex 9, France
{Eric.Remy,Edouard.Thiel}@lim.univ-mrs.fr

Abstract. Medial Axis, also known as Centres of Maximal Disks, is a representation of a shape, which is useful for image description and analysis. Chamfer or Weighted Distances, are discrete distances which allow to approximate the Euclidean Distance with integers. Computing medial axis with chamfer distances has been discussed in the literature for some simple cases, mainly in 2D. In this paper we give a method to compute the medial axis for any chamfer distance in 2D and 3D, by local tests using a lookup table. Our algorithm computes very efficiently the lookup tables and, very important, the neighbourhood to be tested.

Keywords: Medial axis, Centres of Maximal Disks, Chamfer Distances, Lookup Table Transform, Shape Representation.

1 Introduction

A digital image is a set of colored points in a space, for instance the voxels on the cubic grid. A shape in a binary image, is a subset of the points of the image, sharing the same color. The points of the shape, taken individually, do not provide any global information for the shape description.

Given a family of disks and a shape, we consider the disks completely included in the shape. If the smallest disk in the family is a single point, then the shape can be completely covered by such disks. Coding their positions and sizes is sufficient to reconstruct the original shape.

Now consider a covering of the shape by maximal disks, defined as follows:

Definition 1. A disk is maximal in the shape if the disk is not included in any single other disk in the shape.

The maximal disks are centred in the shape. Note that a maximal disk can be included in the union of other maximal disks; so the covering by maximal disk, which is unique by construction, is not always a minimal covering.

Definition 2. The Medial Axis (MA) of a shape is the set of centres of maximal disks in the shape.

G. Borgefors, I. Nyström, and G. Sanniti di Baja (Eds.): DGCI 2000, LNCS 1953, pp. 418–430, 2000.
© Springer-Verlag Berlin Heidelberg 2000

The medial axis is a reversible coding of the shape; it is a global representation, centred in the shape, which allows shape description, analysis, simplification or compression. The medial axis is often not thin and disconnected. It is a step for weighted skeleton computation [12] and implicit surface reconstruction [7].

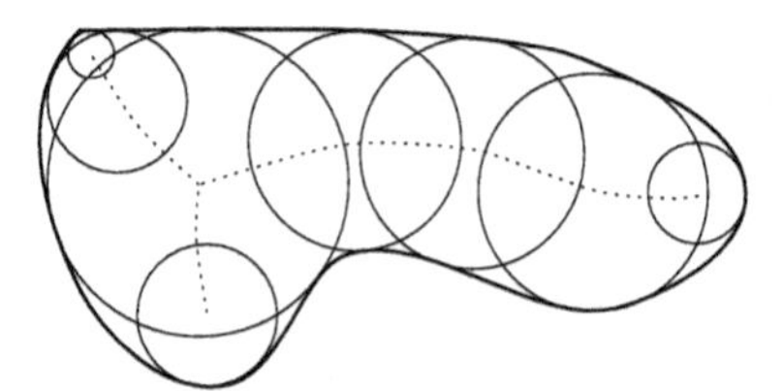

Fig. 1. Medial Axis with circles

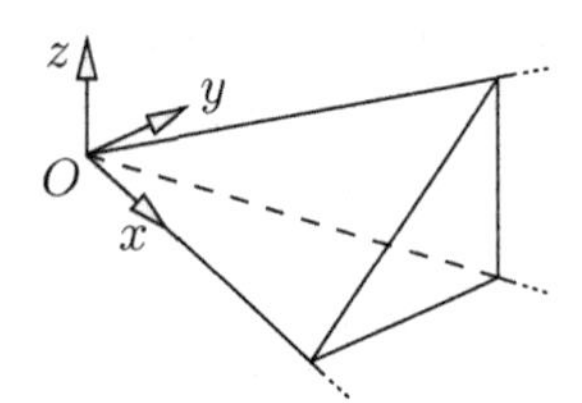

Fig. 2. The 48th of space $\mathcal{S}$

The aspect of the medial axis and its computation depend on the geometry of the family of disks, which can be circles, rectangles, distance balls, etc. For instance, figure 1 schematizes a medial axis covering a shape with circles.

The medial axis is especially attractive when the disks are distance balls, for computation cost and for description purpose. A distance is positive defined, symetric and respects the triangular inequality. Given a distance d, the distance ball B_d of centre p and radius r, is the set of points

$$B_d(p, r) = \{ q : d(p, q) \leq r \} \tag{1}$$

which is centre-symetric. There exists innumerable distances functions; since digital images are the most often encoded in integer arrays, we focus on discrete distances, which are distances working with integers.

2 Chamfer Distances

Chamfer distance d_C can be defined in the following way: a chamfer mask $\mathcal{M}_C$ is a set of legal displacements in a neigbourhood, each displacement being weighted by an integer cost; the chamfer distance d_C between two points is the cost of the path of least cost joining them, formed with legal displacements in the mask.

Borgefors popularizes chamfer distances in [2], in any dimension. Afterwards many optimization methods have been proposed, to approximate the Euclidean distance d_E; the major contribution is due to Verwer in 2 and 3 dimensions [13]. One can find in [12] a complete history of chamfer distance, the comparison of different optimization methods and crossing formulas between them. More recently, we propose in [8] new results in the 3D case, using Farey triangulations.

Chamfer distances have many advantages, which justify their success in applications. They are local distances, that is to say, which permit to deduce a distance from the distances of close neighbours, unlike d_E. All computations are done using integers and linear operations $\{+, -, <\}$. As we will see, the computation of the medial axis can also be done by local tests.

The major attraction is the high speed — and simplicity — of the distance transform algorithm, denoted DT, due to Rosenfeld et al. [10]. The DT consists in labeling each point of a shape to its distance to the complementary. The DT is global, and operates in 2 passes on the image, independently of the thickness of the shape in the image, and of the dimension. The Reverse Distance Transform (RDT) allows to recover a shape from it's medial axis, also in 2 passes.

a	$(1,0,0)$
b	$(1,1,0)$
c	$(1,1,1)$
d	$(2,1,0)$
e	$(2,1,1)$
f	$(2,2,1)$

g	$(3,1,0)$
h	$(3,1,1)$
i	$(3,2,0)$
j	$(3,2,1)$
k	$(3,2,2)$
l	$(3,3,1)$
m	$(3,3,2)$

Fig. 3. First visible points

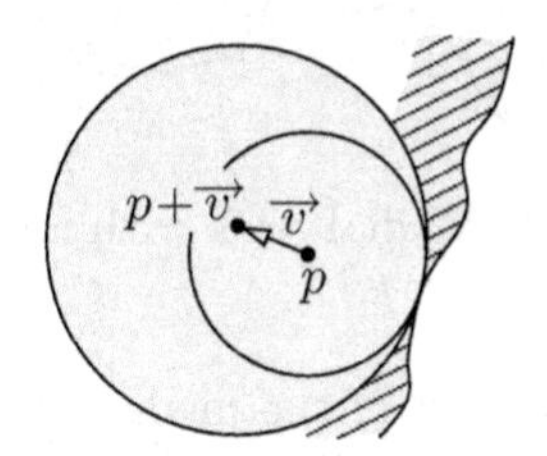

Fig. 4. Balls inside the shape

Our work space is the cubic grid, associated with the fundamental lattice Λ of $\mathbb{Z}^3$. The cubic grid implies the symmetry towards planes of axes and bissectrices, called 48-symmetry : it divides $\mathbb{Z}^3$ into 48 sub-spaces ($48 = 2^3.3!$ with 2^3 sign combinations and 3! coordinates permutations), versus 8 octants in $\mathbb{Z}^2$. We denote $\mathcal{S}$ the 48th of space represented in figure 2:

$$\mathcal{S} = \left\{ 0 \le z \le y \le x \,,\; (x,y,z) \in \mathbb{Z}^3 \right\} . \tag{2}$$

We call weighting $M(\vec{v}, w)$ a displacement $\vec{v}(x,y,z) \in \mathbb{Z}^3$ associated with a weight $w \in \mathbb{N}^*$, also denoted $W[\vec{v}]$. In our work space, a chamfer mask $\mathcal{M}_C$ is a 48-symmetric set of m weightings

$$\mathcal{M}_C = \{\, M_i(x_i, y_i, z_i, w_i) \,,\; 1 \le i \le m \,\} . \tag{3}$$

The generator $\mathcal{M}_C^g$ of a mask $\mathcal{M}_C$ is the part $\mathcal{M}_C \cap \mathcal{S}$, from which are deduced all other weightings by the 48-symmetry. The cardinal of $\mathcal{M}_C^g$ is denoted m_g. Given a displacement $\vec{v}$ in $\mathcal{M}_C$, we name $\vec{v}^g$ the corresponding displacement in $\mathcal{M}_C^g$ by the 48-symmetry.

A weighting (x,y,z,w) generates by translation the periods $(2x,2y,2z,2w)$, $(3x,3y,3z,3w)$, etc. For the sake of efficiency during DT, it is self-evident that $\mathcal{M}_C^g$ should only be formed of points such that $\gcd(x,y,z) = 1$. The points having this property are said visible (from the origin, see [5]). The set of visible points of $\mathcal{S}$ can be obtained with a sieve upon the periods of visible points, by scanning $\mathcal{S}$ on x, y, z. Visible points are named a, b, c, ... in the sieve order. We give figure 3 the cartesian coordinates of the first visible points in $\mathcal{S}$. Properties of the choice of visible point subsets in a chamfer mask are studied in [8,9].

3 Existing Methods to Extract MA

3.1 Local Maxima

After the DT, each shape point p is labeled to its distance $DT[p]$ to the complementary. Let $\overrightarrow{v}$ be a displacement of the mask $\mathcal{M}_C$. The point $p + \overrightarrow{v}$ is deeper inside the shape than p (see figure 4) if $DT[p + \overrightarrow{v}] > DT[p]$. Because of the definition of the chamfer distances, the greatest possible value of $DT[p + \overrightarrow{v}]$ is $DT[p] + W[\overrightarrow{v}]$. If this hapens, then the point p propagates to $p + \overrightarrow{v}$ the distance information during the DT. We deduce that the disk centered in $p + \overrightarrow{v}$ completely overlaps the disk centered in p (figure 4), thus $p \notin$ MA.

On the contrary, if p does not propagate any weighting, then p is called a Local Maximum. Such a point verifies

$$DT[p + \overrightarrow{v}] < DT[p] + W[\overrightarrow{v}] \ , \quad \forall \overrightarrow{v} \in \mathcal{M}_C \tag{4}$$

which we name Local Maximum Criterion (LMC). The set of points detected by the LMC, includes the MA by construction. Rosenfeld and Pfaltz showed in [10] that for the basic distances such that $\mathsf{a} = 1$ (d_4 and d_8 in 2D, d_6, d_{18} and d_{26} in 3D), the LMC set is exactly MA.

3.2 Equivalent Disks

This is no more true from the moment that $\mathsf{a} > 1$, since the LMC detects the MA plus erroneous points, which are not center of maximal disks. The LMC set is still reversible ; but the erroneous points are generally numerous, in particular close to the border of the shape, and they make the MA completely unusable for applications.

The trouble comes from the check in (4) in the difference between disk radii on DT. In fact, if $\mathsf{a} > 1$, then several radii may correspond to a same disk, ie. a given disk may have an interval of equivalent integer radii. The computation of the equivalence classes is related to the Frobenius problem [11,6]. For instance, the class of equivalence of the single pixel ball is $[1 \mathinner{..} \mathsf{a}]$. The corollary is that (4) is inadequate for $\mathsf{a} > 1$.

Arcelli and Sanniti di Baja showed in the 2D case for 3×3 masks that it is sufficient to bring down each value on the DT to the lowest term in its equivalence class ; the LMC is then exact on the modified DT. For instance, $d_{3,4}$ simply needs to bring the 3 down to 1 and the 6 down to 5 [1]. Their method is inappropriate for masks greater than 3×3 in 2D and $3 \times 3 \times 3$ in 3D, because of the appearance of influence cones in chamfer balls [12].

3.3 Lookup Tables

The most general and efficient solution is the method of the lookup tables, which stores the corrections to the LMC.

A shape point p is a maximal centre if there is no other shape point q such that the ball $B_d(q, DT[q])$ entirely overlaps the ball $B_d(p, DT[p])$. The presence of q forbids p to be a MA point. Suppose that (i) it is sufficient to search q in a local neigbourhood of p and (ii) that we know for each $DT[p]$ the minimal value $DT[q]$, stored in a lookup table Lut, which forbids p in direction $\overrightarrow{v} = \overrightarrow{pq}$.

(i) The local neighbourhood of vectors to be tested is denoted $\mathcal{M}_{Lut}$ and is 48-symmetric. The generator of $\mathcal{M}_{Lut}$ is denoted $\mathcal{M}^g_{Lut}$. Given $\overrightarrow{v} \in \mathcal{M}_{Lut}$, we name $\overrightarrow{v}^g$ the corresponding vector by the 48-symmetry in $\mathcal{M}^g_{Lut}$.

(ii) The minimal value for p and $\overrightarrow{v}$ is stored in $Lut[\overrightarrow{v}][DT[p]]$. Because of the 48-symmetry, it is sufficient to store only the values in $\mathcal{M}^g_{Lut}$; hence the minimal value for p and $\overrightarrow{v}$ is accessed using $Lut[\overrightarrow{v}^g][DT[p]]$.

Finally we have the following criterion:

$$p \in \text{AM} \iff DT[p + \overrightarrow{v}] < Lut[\overrightarrow{v}^g][DT[p]] \; , \; \forall \overrightarrow{v} \in \mathcal{M}_{Lut} \; . \tag{5}$$

The first use of lookup tables is due to Borgefors, Ragnemalm and Sanniti di Baja in [4], for d_E in 2D. The lookup tables are computed with exhaustive search; the combinatory is enormous, but the computations are done once for all. The tables are given for radii less than $\sqrt{80}$. Borgefors gives in [3] the lookup tables for the 2D distance $d_{5,7,11}$, which entries differs from the LMC for radii less than 60; but she does not generalize her lookup table computation method.

One of the authors proposes in [12] an efficient algorithm to compute the lookup tables for any chamfer mask in 2D, assuming that $\mathcal{M}_{Lut} = \mathcal{M}_C$. But he points out that for large masks, erroneous points may be detected: a contradiction, and a fondamental question about the validity of the whole method.

We have recently discovered that the assumption $\mathcal{M}_{Lut} = \mathcal{M}_C$ actualy is the error. In fact, the two masks often completely differ, and we propose in the following the correct and efficient algorithm which computes both $\mathcal{M}_{Lut}$ and Lut in 3D. The following method is immediately applicable to the 2D case by skipping any reference to z.

4 Proposed Method to Compute Lookup Tables

Let us come back to the meaning of the labels on the DT image. A point p is labeled to its distance $r = DT[p]$ to the complementary. If we apply the RDT on p, it generates the reverse ball denoted $B_d^{-1}(p, r)$, which is the set of points

$$B_d^{-1}(p, r) = \{ \, q \, : \, r - d(p, q) > 0 \, \} \tag{6}$$

and which is the greatest ball inside the shape, centred in p. From (1) and (6) it comes immediately the following lemma:

Lemma 1. $B_d(p, r) = B_d^{-1}(p, r + 1)$.

While the sets are the same, it is important to note that resulting labels have different values on DT and RDT, as demonstrated figure 5.

The computation of an entry $Lut[\overrightarrow{v}][r]$ in the lookup table for $r = DT[p]$ in a direction $\overrightarrow{v}$, consists in finding the smallest radius R of a ball $B_d^{-1}(p + \overrightarrow{v}, R)$

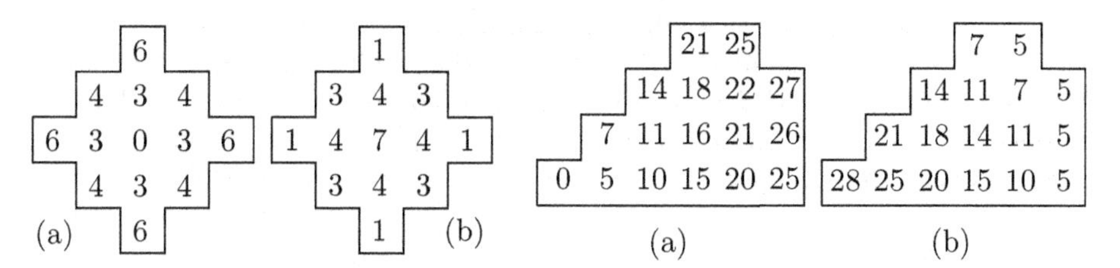

Fig. 5. Difference of labeling between (a) $B_{d_{3,4}}(6)$ and (b) $B^{-1}_{d_{3,4}}(6+1)$

Fig. 6. Difference between (a) $B_{d_{5,7,11}}(27) \cap \mathcal{S}$ produced by CT^g and (b) its DT^g

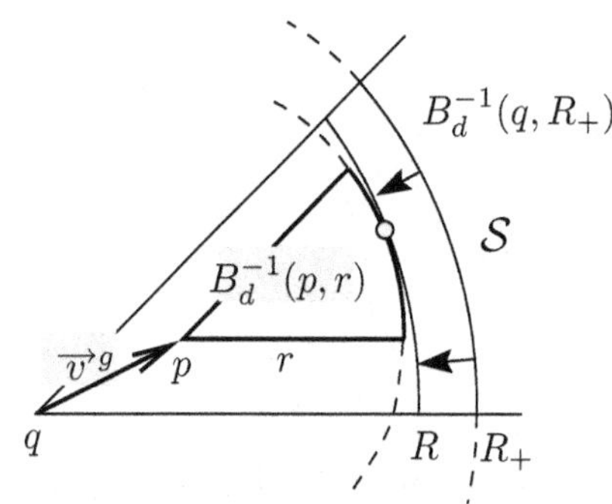

Fig. 7. Overlapping test on two balls restricted to $\mathcal{S}$

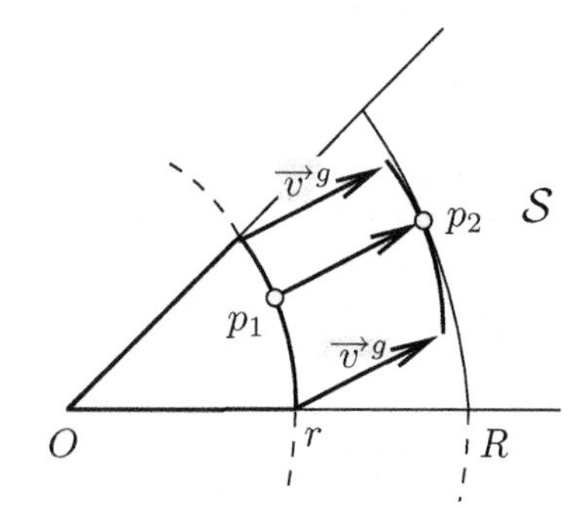

Fig. 8. Translated overlapping test on DT^g

which overlaps completely $B_d^{-1}(p, r)$ (see figure 4). As illustrated figure 7, a method to find R is to decrease the radius R_+ while the ball overlaps, but each step needs a RDT, and the cost is prohibitive. The lemma 1 is the starting point of our method: we will show that it is sufficient to compute the DT in $\mathcal{S}$ only a single time at the beginning; the lemma 1 allows to test the overlappings on $\mathcal{S}$.

4.1 Computing an Entry of the Lookup Table

We choose to restrict our study to balls which are convex and symetric, ie. to distances inducing a norm (see [8]). Therefore, we can limit the overlapping test by restricting the two balls to $\mathcal{S}$, which gives the figure 7. This loss of generality should not be annoying since non-norm distances are not homogeneous (the shortest path may zigzag over the grid). Note however that it is still possible not to restrict to $\mathcal{S}$ leading to a greater computationnal load.

Instead of reducing the radius R_+ while the ball overlaps in figure 7, we propose to translate both $B_d^{-1}(p, r)$ and $B_d^{-1}(q, R)$ to the origin as shown in figure 8. We scan each point p_1 of $B_d^{-1}(O, r) \cap \mathcal{S}$, which by translation of vector $\overrightarrow{v}^g$ gives p_2. Finally, we find

$$R = \max \left\{ d(O, p_2) \,:\, p_2 = p_1 + \overrightarrow{v}^g \,,\; p_1 \in B_d^{-1}(O, r) \cap \mathcal{S} \right\} \tag{7}$$

$$= \max \left\{ d(O, p_1 + \overrightarrow{v}^g) \,:\, p_1 \in B_d^{-1}(O, r) \cap \mathcal{S} \right\} . \tag{8}$$

This process can be very efficiently implemented, using a correspondence between figure 8 and the image CT^g, which is the result of the Cone Transform. It gives for any point of $\mathcal{S}$ its distance from the origin, see figure 6.a. For any chamfer distance that induces a norm (see [9,8]), a fast algorithm (given in figure 9) exists that computes CT^g in a single pass, only using $\mathcal{M}_C^g$.

```
     PROCEDURE ComputeCT^g ( L, M_C^g, Cone ) ;
 1   Cone[(0,0,0)] = 0 ;
 2   FOR x = 1 TO L-1 , FOR y = 0 TO x , FOR z = 0 TO y DO
 3   {
 4     min = +∞ ;
 5     FOREACH →v^g IN M_C^g DO
 6     {
 7       (x',y',z') = (x,y,z) - →v^g ;
 8       IF (x',y',z') ∈ S AND Cone[(x',y',z')] + W(→v^g) < min THEN
 9             min = Cone[(x',y',z')] + W(→v^g) ;
10     }
11     Cone[(x,y,z)] = min ;
12   }
```

Fig. 9. Fast Cone Distance Transform Algorithm. Input: L the side length, $\mathcal{M}_C^g$ the generator of the d_C mask. Output : *Cone* the L^3 distance image containing the cone

```
     PROCEDURE ComputeLutColumn ( Cone, L, →v^g, R_max, Lut[→v^g] ) ;
 1   FOR r = 1 TO R_max DO Lut[→v^g][r] = 0 ;          // Initializes Lut[→v^g] to 0
 2   FOR x = 1 TO L-1 , FOR y = 0 TO x , FOR z = 0 TO y DO
 3   {
 4     r1 = Cone[(x,y,z)] + 1 ;          // radius of the ball where p1 is located
 5     r2 = Cone[(x,y,z) + →v^g] + 1 ;                              // same for p2
 6     Lut[→v^g][r1] = MAX {Lut[→v^g][r1], r2} ;
 7   }
 8   r2 = 0 ;
 9   FOR r1 = 0 TO R_max DO
10     IF Lut[→v^g][r1] > r2 THEN r2 = Lut[→v^g][r1] ELSE Lut[→v^g][r1] = r2 ;
```

Fig. 10. *Lut* Computation Algorithm. Input : *Cone* the cone, L the side length, $\overrightarrow{v}^g$ the column of *Lut* to be computed, R_{max} the greatest radius value seekable in *Lut*. Output : $Lut[\overrightarrow{v}^g]$ is filled with the correct values

The implementation of *Lut* computation shown in figure 10, consists in a scan of $\mathcal{S}$ (line 2–7). For each point p_1, we look for the corresponding radius r_1 which is $CT^g[p_1] + 1$ because of lemma 1. We then look for the radius r_2 of the ball passing by point p_2; its value is $CT^g[p_2] + 1 = CT^g[p_1 + \overrightarrow{v}^g] + 1$, also because of lemma 1. During the scan, we keep the greatest value found for r_2, which at the end, is R by (8).

Due to the discrete nature of the balls, one can observe in Lut, cases where for instance $r_1 < r_2$ while $Lut[\overrightarrow{v}^g][r_1] > Lut[\overrightarrow{v}^g][r_2]$, which means that the ball overlapping $B_d(r_1)$ is bigger than the ball overlapping $B_d(r_2)$. These artefacts of discrete distances should not happen because it has been proved in [13], p. 20, that any d_C is a distance function, and thus

$$\forall r_1, r_2 \quad r_1 < r_2 \Rightarrow B_d(r_1) \subseteq B_d(r_2)\,. \tag{9}$$

We therefore have to correct the table by assuming that in this case, $Lut[\overrightarrow{v}^g][r_2]$ should at least equals $Lut[\overrightarrow{v}^g][r_1]$ (figure 10, lines 8–10).

4.2 Computing the Mask of the Lookup Table

We now focus on the computation of the set of weightings $\mathcal{M}^g_{Lut}$.

We assume that a given set $\mathcal{M}^g_{Lut}$ is sufficient to extract correctly the MA from any DT image which values does not exceed R_{Known}, ie. this $\mathcal{M}^g_{Lut}$ enables to extract from any ball $B_d(O, r)$ where $r \leq R_{Known}$, a medial axis which is (by definition 2) the sole point O (the origin). At the beginning, $\mathcal{M}^g_{Lut}$ is empty and $R_{Known} = 0$.

We propose to test each ball $B_d(O, r)$, where $r > R_{Known}$, each time extracting its DT and then its MA, until whether r reaches R_{Target}, or a point different from O is detected in the MA of $B_d(r)$. If r reaches R_{Target}, then we know that $\mathcal{M}^g_{Lut}$ enables to extract MA correctly, for any DT containing values lower or equal to R_{Target}.

On the contrary, if we reach the case where (at least) one extra point p is found in MA, it means that $\mathcal{M}^g_{Lut}$ is not sufficient to correctly extract MA. We propose to introduce the new weighting $(p, CT^g[p])$ in $\mathcal{M}^g_{Lut}$. If this new mask is sufficient to remove p then the process continues. If not, there is no possible Lut for this distance and the process stops, reporting the problem (this should never happen if d actualy induces a norm).

4.3 Related Algorithms

The full algorithm given figure 11, uses two other algorithms given figures 12 and 13. They are dedicated versions (limited to $\mathcal{S}$ and thus 48 times faster) of the classical Distance Transform and Medial Axis Extraction.

Note that the use of DT^g (figure 11, line 7) is mandatory, since the MA is extracted from the DT to the complementary (see §2 and §3). In fact, a simple threshold on image CT^g to the radius R gives only the $B_d(O, R) \cap \mathcal{S}$ set, but not the correct labels (see figure 6, where values of (a) differ from (b)).

```
   PROCEDURE ComputeAndVerifyLut ( L, M^g_Lut, R_Known, R_Target ) ;
 1 ComputeCT^g(L, M^g_C, Cone) ;
 2 R_max = MAX {Cone[p] : p ∈ Cone} ;
 3 FOREACH v^g IN M^g_Lut DO ComputeLUTColumn ( v^g, R_max, Lut ) ;
 4 FOR R = R_Known + 1 TO R_Target DO
 5 {
 6    ∀p, DT^g[p] = { 1 if Cone[p] ≤ R ; 0 else }  ;     // Copy B_d(R) ∩ S to DT^g
 7    ComputeDT^g(M^g_C, DT^g) ;
 8    FOR x = 1 TO L , FOR y = 0 TO x , FOR z = 0 TO y DO
 9       IF IsMA ( (x,y,z), M^g_Lut, Lut, DT^g ) THEN
10       {
11          M = (x,y,z,Cone[(x,y,z)]) ;                 // Build a new weighting M
12          M^g_Lut = M^g_Lut ∪ {M} ;                      // Add M to M^g_Lut
13          ComputeLUTColumn( (x,y,z), R_max, Lut[(x,y,z)] ) ;  // and Lut
14          IF IsMA ( (x,y,z), M^g_Lut, Lut, DT^g ) THEN ERROR ;
15       }
16 }
17 R_Known = R_Target ;
```

Fig. 11. Full *Lut* Computation Algorithm with determination of $\mathcal{M}^g_{Lut}$. Input: L the side length of the images, $\mathcal{M}^g_{Lut}$ the generator of the *Lut* neighbourhood, R_{Known} the last verified radius, R_{Target} the maximum radius to be verified. Output: *Lut* the lookup table, R_{Known} the new verified radius

5 Results

While the computation of the *Lut* array in figure 10 is very fast (less than a second[1]), the computation of $\mathcal{M}^g_{Lut}$ in figure 11, involving its verification, is rather slow, as shown below, and its result should hence be saved for further usage. It takes approximatively six minutes (370s[1]) to compute the $\mathcal{M}^g_{Lut}$ for distance $d_{11,16,19,\mathrm{j}=45}$ for $L = 100$ and from $R_{Known} = 0$ to $R_{Target} = 1100$ (the radius of the biggest $B_d \cap \mathcal{S}$ possible in the L^3 sized image CT^g). This load is explained by the systematic test of 1100 balls $B_d(r)$. Each of them involves computations (CT^g and *MA* extraction) on $o(r^3)$ points. It is therefore much more interesting to use chamfer distances with small values of $W[\mathrm{a}]$ since this gives fewer balls to test and thus a faster result.

In this scope, in most cases, it is interesting to compensate for the quality loss by more weightings in $\mathcal{M}^g_C$. For example, the extraction of $\mathcal{M}^g_{Lut}$ for distance $d_{7,10,12,\mathrm{d}=16,\mathrm{e}=17}$ for $L = 100$ from $R_{Known} = 0$ to $R_{Target} = 700$ is 33% faster, while achieving a better approximation of d_E.

We give in figure 14 some examples of *Lut* arrays where $\mathcal{M}^g_C = \mathcal{M}^g_{Lut}$. The tables show only the values which differ from the LMC (ie. $R + W[\overrightarrow{v}^g] \neq Lut[\overrightarrow{v}^g][R]$, see (4) and (5)), and thus represent the irregularities.

[1] On a SGI Octane/IRIX 6.5, Mips R10000 180 MHz

```
    PROCEDURE ComputeDT^g ( DT^g, L, M_C^g ) ;
 1  FOR z = L - 1 TO 0, FOR y = L - 1 TO z, FOR x = L - 1 TO y DO
 2    IF DT^g[(x,y,z)] ≠ 0 THEN
 3    {
 4      min = +∞ ;
 5      FOREACH v^g IN M_C^g DO
 6       {
 7         IF (x,y,z) + v^g ∈ S THEN w = DT^g[(x,y,z) + v^g] + W(v^g)
 8                              ELSE w = +∞ ;
 9         min = MIN {w, min} ;
10       }
11      DT^g[(x,y,z)] = min ;
12    }
```

Fig. 12. Fast Distance Transform in $\mathcal{S}$. Input: DT^g the shape (limited to $\mathcal{S}$), L the side length, $\mathcal{M}_C^g$ the generator of the d_C mask. Output: DT^g the distance map

```
    FUNCTION IsMA ( p, M_Lut^g, Lut, DT^g ) ;
 1  FOREACH v^g IN M_Lut^g DO
 2    IF DT^g[p + v^g] < Lut[v^g][DT^g[p]] THEN RETURN FALSE ;
 3  RETURN TRUE ;
```

Fig. 13. Fast extraction of MA points from $B_d \cap \mathcal{S}$. Input: p the point to test, $\mathcal{M}_{Lut}^g$ the generator of the Lut neighbourhood, Lut the lookup table, DT^g the distance transform of the section of the ball. Output: returns TRUE if point p is detected as MA in DT^g

The figure 15 shows a full example of both the computed mask $\mathcal{M}_{Lut}^g$ and the Lut array for distance $d_{11,16,19,\mathrm{j}=45}$. One must note the difference between $\mathcal{M}_C^g$ and $\mathcal{M}_{Lut}^g$ and the presence of point 3e which is not a visible point, and thus would not have appeared in $\mathcal{M}_C^g$ as seen in section 2. We finally give in figure 16 the $\mathcal{M}_{Lut}^g$ for distance $d_{19,27,33,\mathrm{d}=43,\mathrm{e}=47}$, which is as in figure 15, very different from its corresponding $\mathcal{M}_C^g$.

6 Conclusion

In this paper, we give the solution to the fundamental problem pointed out in [12], where the computed lookup tables detected erroneous MA points for large masks. We introduce a new mask $\mathcal{M}_{Lut}$, different from the mask $\mathcal{M}_C$ used to compute DT. We present and justify our algorithm to compute $\mathcal{M}_{Lut}$ and Lut. The correctness of the method is verified during the computation of $\mathcal{M}_{Lut}$ and thus ensures that the medial axis is free of the error points encountered in [12]. We use the symetry of the ball of chamfer norms to minimize computations. Finally, we give results for various chamfer norms given in [8].

$d_{3,4,5}$

R	a	b	c
3	4	5	6

$d_{4,6,7,d=9,e=10}$

R	a	b	c	d	e
4	5	7	8	10	11
6	9	10	11	14	15
7	10				
8	11				
9		14	15		
12	15	17	18	20	21
16	19				

$d_{19,27,33}$

R	a	b	c	R	a	b	c	R	a	b	c
19	20	28	34	76	93	101	107	111	129	137	143
27	39	47	53	79	96	104	110	114	132	140	146
33	47	55	61	81	99	107	113	117	134	142	148
38	53	61	67	84	101	109	115	122	140	148	154
46	58	66	72	87	105	113	119	125	143	151	157
52	66	74	80	90	107	115	121	130	148	156	162
54	72	80	86	92	110	118	124	135	153	161	167
57	74	82	88	95	113	121	127	141	159	167	173
60	77	85	91	98	115	123	129	144	162	170	176
65	80	88	94	103	120	128	134	149	167	175	181
71	86	94	100	106	124	132	138	168	186	194	200
73	91	99	105	108	126	134	140				

$d_{7,10,13,e=18}$

R	a	b	c	e
7	8	11	14	19
10	15	18	19	26
13	18	21	24	29
14	19			

R	a	b	c	e
17	22	25	27	33
18			29	
20	26	29	32	37
23	29	32	34	40

R	a	b	c	e
24	29			
25			37	
26			37	
27	33			

R	a	b	c	e
28			40	
30	36	39	42	47
33			45	
34	40			

R	c
35	47
36	47
38	50
43	55

R	c
46	58
48	60
56	68

Fig. 14. Examples of Lut of 4 d_C for which $\mathcal{M}^g_C = \mathcal{M}^g_{Lut}$

Acknowledgements

This work was performed within the framework of the joint incentive action "Beating Heart" of the research groups ISIS, ALP and MSPC of the French National Center for Scientific Research (CNRS).

References

1. C. Arcelli and G. Sanniti di Baja. Finding local maxima in a pseudo-euclidean distance transform. Computer Vision, Graphics and Image Processing, 43:361–367, 1988. 421
2. G. Borgefors. Distance transformations in arbitrary dimensions. Computer Vision, Graphics and Image Processing, 27:321–345, 1984. 419
3. G. Borgefors. Centres of maximal disks in the 5-7-11 distance transform. In 8th Scandinavian Conf. on Image Analysis, pages 105–111, Tromsø, Norway, 1993. 422
4. G. Borgefors, I. Ragnemalm, and G. Sanniti di Baja. The Euclidean Distance Transform : finding the local maxima and reconstructing the shape. In 7th Scandinavian Conf. on Image Analysis, volume 2, pages 974–981, Aalborg, Denmark, 1991. 422
5. G. H. Hardy and E. M. Wright. An introduction to the theory of numbers. Oxford University Press, fifth edition, October 1978. §3.1. 420

$$\mathcal{M}^g_C = \left\{ \begin{array}{l} \mathsf{a} = (1,0,0,11) \\ \mathsf{b} = (1,1,0,16) \\ \mathsf{c} = (1,1,1,19) \\ \mathsf{j} = (3,2,1,45) \end{array} \right\} \qquad \mathcal{M}^g_{Lut} = \left\{ \begin{array}{ll} \mathsf{a} = (1,0,0,11) & \mathsf{k} = (3,2,2,49) \\ \mathsf{b} = (1,1,0,16) & \mathsf{d} = (2,1,0,27) \\ \mathsf{c} = (1,1,1,19) & \mathsf{3e} = (6,3,3,90) \\ \mathsf{f} = (2,2,1,35) & \mathsf{j} = (3,2,1,45) \end{array} \right\}$$

R	a	b	c	f	k	d	3e	j
11	12	17	20	36	50	28	91	46
16	23	28	31	46	61	39	102	57
19	28	33	36	52	65	44	106	62
22	31	36	39	55	69	46	110	65
27	34	39	42	57	72	50	113	68
30	39	44	46	62	76	55	117	73
32	42	46	50	65	80	57	121	76
33					81		121	
35	45	50	53	68	83	61	124	79
38	46	52	55	71	84	62	125	81
41	50	55	58	74	88	66	129	84
43	53	57	61	76	91	68	132	87
44			62	78	91		132	
45				79				
48	57	62	65	81	95	73	136	91
49					97		136	
51	61	66	69	84	99	77	140	95
52	62				100	78	140	
54	64	68	72	87	102	79	143	98
55					103		144	
56				90				
57				91				
59	69	74	77	93	107	84	148	103
60			78	94	107		148	
61				95				
63	73	78	81	97	110	89	151	107
64		79		98		90		

R	a	b	c	f	k	d	3e	j
66			84	100	114		155	
67				101				
70	80	84	88	103	118	95	159	114
71					119		159	
72				106				
73			91	107	121		162	
74	84				122	100	163	
75		90		109		101		
76				110				
79				113				
80		95		114		106		
81					129		170	
82			100	116	129		170	
83				117				
85	95	100	103	119	133	111	174	129
86		101		120		112		
88				122				
89				123				
92			110	126	140		181	
93					141		182	
95				129				
96						122		
98				132				
99				133				
101				135				
102			121	136	151		192	
104			122	138	152		193	

R	k
105	139
108	142
111	145
114	148
117	151
118	152
120	154
121	155
124	158
127	161
130	164
133	167
137	171
140	174
143	177
146	180
149	183
156	190
159	193
162	196
165	199
175	209
178	212
194	228

Fig. 15. $\mathcal{M}^g_{Lut}$ and *Lut* for $d_{11,16,19,\mathsf{j}=45}$

6. M. Hujter and B. Vizvari. The exact solutions to the Frobenius problem with three variables. Ramanujan Math. Soc., 2(2):117–143, 1987. 421
7. J. L. Mari and J. Sequeira. Using implicit surfaces to characterize shapes within digital volumes. In RECPAD'00, pages 285–289, Porto, Portugal, May 2000. 419
8. E. Remy and E. Thiel. Optimizing 3D chamfer masks with distance constraints. In 7th IWCIA, Int. Workshop on Combinatorial Image Analysis, pages 39–56, Caen, July 2000. 419, 420, 423, 424, 427
9. E. Remy and E. Thiel. Structures dans les sphères de chanfrein. In 12ème RFIA, congrès Reconnaissance des Formes et I.A, volume 1, pages 483–492, Paris, Fev 2000. 420, 424

$$\mathcal{M}^g_C = \left\{ \begin{array}{l} \mathsf{a} = (1,0,0,19) \\ \mathsf{b} = (1,1,0,27) \\ \mathsf{c} = (1,1,1,33) \\ \mathsf{d} = (2,1,0,43) \\ \mathsf{e} = (2,1,1,47) \end{array} \right\} \qquad \mathcal{M}^g_{Lut} = \left\{ \begin{array}{ll} \mathsf{a} = (1,0,0,19) & \mathsf{d} = (2,1,0,43) \\ \mathsf{b} = (1,1,0,27) & \mathsf{i} = (3,2,0,70) \\ \mathsf{c} = (1,1,1,33) & v_{17} = (5,3,0,113) \\ \mathsf{e} = (2,1,1,47) & \end{array} \right\}$$

Fig. 16. $\mathcal{M}^g_{Lut}$ for $d_{19,27,33,\mathsf{d}=43,\mathsf{e}=47}$

10. A. Rosenfeld and J. L. Pfaltz. Sequential operations in digital picture processing. Journal of ACM, 13(4):471–494, 1966. 420, 421
11. J. Sylvester. Mathematicals questions with their solutions. Educational Times, 41:21, 1884. 421
12. E. Thiel. Les distances de chanfrein en analyse d'images : fondements et applications. PhD thesis, UJF, Grenoble, Sept 1994. `http://www.lim.univ-mrs.fr/~thiel/these`. 419, 421, 422, 427
13. J. H. Verwer. Distance transforms: metrics, algorithms and applications. PhD thesis, Technische Universiteit, Delft, 1991. 419, 425

Multiresolution Modelling of Polygonal Surface Meshes Using Triangle Fans*

José Ribelles, Angeles López, Inmaculada Remolar,
Oscar Belmonte, and Miguel Chover

Departamento de Informática, Universitat Jaume I,
E-12080 Castellón, Spain
{ribelles,lopeza,remolar,belfern,chover}@uji.es

Abstract. Multiresolution modelling of polygonal surface meshes has been presented as a solution for the interactive visualisation of scenes formed by hundreds of thousands of polygons. On the other hand, it has been shown that representing surfaces using sets of triangle strips or fans greatly reduces visualisation time and provides an important memory savings. In this paper we present a new method to model polygonal surface meshes. Like the previously explained *Multiresolution Ordered Meshes* (MOM), this method permits the efficient management of an ample range of approximations of the given model. Furthermore, this method utilises the triangle fan as its basic representation primitive. Experiments realised with data sets of varying complexity demonstrate reduced storage space requirements, while retaining the advantages of MOMs.

1 Introduction

One of the principle objectives of multiresolution modelling [1] is to permit interactive visualisation of surfaces formed by thousands of polygons. Given a mesh, M, a multiresolution model defines how to store and retrieve n different approximations or levels of detail (LOD), $M_0, M_1, ..., M_{n-1}$, in an efficient manner.

Several mechanisms have been developed to accelerate the process of visualising polygonal models. For example, strips and fans of triangles (see figure 1) appear as drawing primitives in some graphics libraries, such as *OpenGL*. This type of primitives allow for rapid visualisation. To draw a fan of n triangles, for example, it is only necessary to pass $n+2$ vertices, instead of $3n$, to the graphics processor. This not only reduces computation time due to a reduction in vertices, but also an important memory savings.

Obtaining the optimal set of strips or fans for a given surface is a process realised off-line when working with static models. However, when working with a multiresolution model, the surface connectivity changes with changes in level

* Supported by grant TIC1999-0510-C02-02 (CICYT, Ministerio de Educación y Ciencia)

G. Borgefors, I. Nyström, and G. Sanniti di Baja (Eds.): DGCI 2000, LNCS 1953, pp. 431–443, 2000.
© Springer-Verlag Berlin Heidelberg 2000

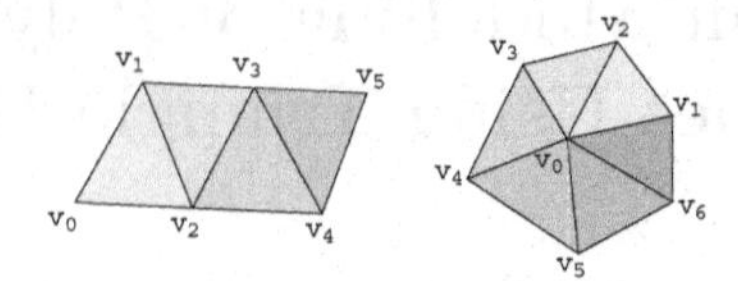

Fig. 1. Example of a strip and a fan of triangles. On the left, a strip defined by $v0, v1, v2, v3, v4, v5$ and on the right, a fan defined by $v0, v1, v2, v3, v4, v5, v6, v1$

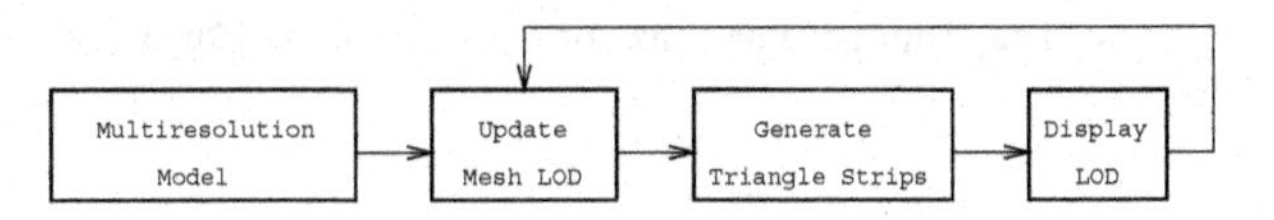

Fig. 2. Visualisation process of a LOD in *VDPM*

of detail, probably with each frame, therefore requiring the dynamic generation of the strips or fans.

This article presents a new multiresolution scheme permitting the recuperation of a level of detail, directly as a set of triangle fans. It is based on the *Multiresolution Ordered Meshes (MOM)* model presented earlier [2]. The new scheme, called *MOM-Fan*, defines a new data structure and a new traversal algorithm which optimises the triangle fans for visualisation at the LOD required.

Notation. The geometry of a triangulated model, M, is denoted as a tupla $\{\mathcal{V}, \mathcal{F}\}$, where $\mathcal{V}$ is a set of N positions $v_i = (x_i, y_i, z_i) \in \mathbb{R}^3$, $1 \leq i \leq N$, and $\mathcal{F}$ is a set of triples $\{j, k, l\}, j, k, l \in \mathcal{V}$, specifying positions of triangles faces.

2 Previous Work

Hoppe [3] presents a multiresoution model, *VDPM*, based on a hierarchical structure of vertices built from a sequence of contractions of edges. The level of detail is determined from a series of criteria based on the view frustum, surface orientation, etc. Changes in these conditions trigger changes in the required LOD, and it is proposed that triangle strips be generated using a greedy algorithm once the component triangles are determined for that LOD (figure 2).

El-Sana et al. [4] presents a data structure, *Skip-Strips*, that maintains triangle strips even though the LOD may change. A *Skip-Strip* is built at run time, from the multiresolution model. At the same time the triangle strips from the original model are obtained. Each time the level of detail changes, the *Skip-Strip* structure is updated based on the required LOD, which permits the update of the strips and their subsequent visualisation (figure 3). As the strips are generally quite short, it has been proposed that they be concatenated previous to visualisation.

The proposal in this article simplifies the LOD visualisation scheme. The multiresolution model itself is encoded using triangle fans, and therefore it is

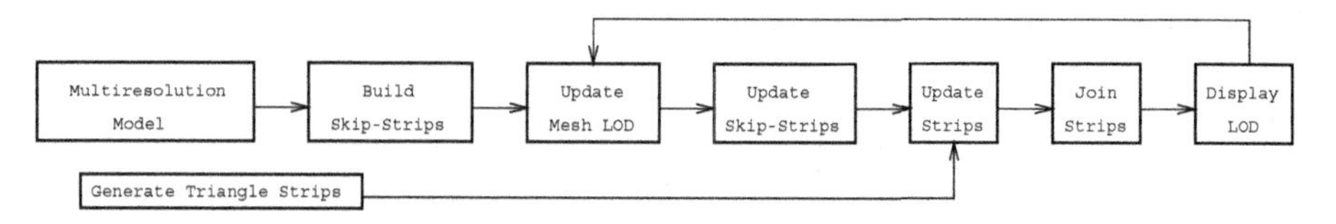

Fig. 3. Visualisation process of a LOD in *Skip-Strips*

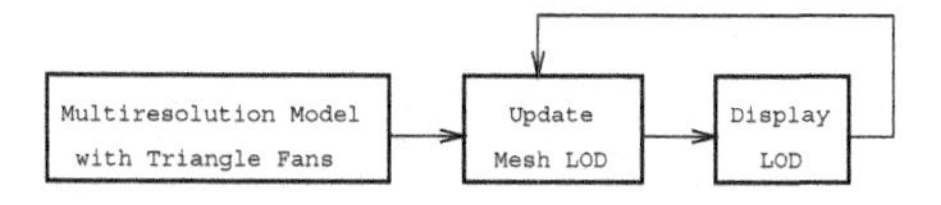

Fig. 4. Visualisation process of a LOD using *MOM-Fan*

necessary to make some adjustments to each fan according to the LOD (figure 4). Figure 9 illustrates three LODs of a *MOM-Fan* object, where the triangles composing each fan have been coloured alike and the boundaries have been highlighted.

Other multiresolution models exist which do not make use of either triangle fans or strips; see [1] for a recent survey of these.

2.1 Review of Multiresolution Ordered Meshes

Multitresolution Ordered Meshes was presented with the idea of improving the interactive visualisation of complex polygonal surfaces. Later, it was extended to exploit frame-to-frame coherence [5]. This permitted the acceleration of LOD recovery, while not affecting visualisation time. Finally, in [6] *MOM* was compared against *Progressive Meshes* [7].

Let M and M^r be the original and multiresolution meshes, respectively. M^r explicitly stores the vertices $\mathcal{V}^r$, and the faces $\mathcal{F}^r$, utilised to represent any resolution:

$$M^r = \{\mathcal{V}^r, \mathcal{F}^r\} \tag{1}$$

To build M^r with n levels of detail, we apply $n-1$ iterations of a simplification method. Each simplification S_i, $0 \leq i < n-1$, produces a new level of detail M_{i+1} and may be represented by the tuple $S_i = \{V_i, F_i, V_i', F_i'\}$ where V_i and F_i are the sets of vertices and faces which are eliminated from M_i, and V_i' and F_i' are the sets of vertices and faces which are added to M_i to gerenerate, finally, M_{i+1}. Therefore, we may express the resulting object, M_{i+1}, as:

$$M_{i+1} = (M_i - \{V_i, F_i\}) \cup \{V_i', F_i'\}, \; 0 \leq i < n-1 \tag{2}$$

Given that M^r stores all vertices and faces that can be used at any set level of detail, M^r can be defined as:

$$M^r = \bigcup_{i=0}^{n-1} M_i, \; n \geq 1 \tag{3}$$

From the equations 1 - 3, we derive that M^r can be expressed as the initial mesh, $M = M_0$, plus all vertices and faces generated in each iteration of the simplification process:

$$\mathcal{V}^r = \mathcal{V}_0 \cup V_0' \cup V_1' \cup ... \cup V_{n-2}' = \mathcal{V}_0 \cup \bigcup_{i=0}^{n-2} V_i' \tag{4}$$

$$\mathcal{F}^r = \mathcal{F}_0 \cup F_0' \cup F_1' \cup ... \cup F_{n-2}' = \mathcal{F}_0 \cup \bigcup_{i=0}^{n-2} F_i' \tag{5}$$

or also, as the mesh corresponding to the worst level of detail, M_{n-1}, plus the vertices and faces eliminated in each iteration of the simplification process:

$$\mathcal{V}^r = V_0 \cup V_1 \cup ... \cup V_{n-2} \; \cup \; \mathcal{V}_{n-1} = \bigcup_{i=0}^{n-2} V_i \; \cup \; \mathcal{V}_{n-1} \tag{6}$$

$$\mathcal{F}^r = F_0 \cup F_1 \cup ... \cup F_{n-2} \; \cup \; \mathcal{F}_{n-1} = \bigcup_{i=0}^{n-2} F_i \; \cup \; \mathcal{F}_{n-1} \tag{7}$$

The basic idea of *MOM* is based on the expressions of the two previous equations. That is, store in ordered form the sequences of vertices and faces eliminated, V_i and F_i, $0 \leq i < n-2$, plus the vertices and faces corresponding to the worst level of detail $M_{n-1} = \{\mathcal{V}_{n-1}, \mathcal{F}_{n-1}\}$. Each stored face is identified by a value representing its position in the face sequence ordered according to equation 5 above. *MOM-Fan* is based on the same idea, the difference being that here we store and manipulate triangle fans instead of isolated triangles. In section 3 we show how to store the fans in the data structure, and in section 4 how to recover those fans which form a given level of detail.

3 Data Structure

The data structure presented here is based on a list of lists similar to that described in [2]. The fundamental difference is the list which stores vertex sequences, which substitutes the anterior list of faces. In figure 5 the data structure of the model is shown. The data structure is formed of three lists:

- *Vertex list* (*vertexList* field). Stores the vertices of the mesh in an ordered fashion according to their elimination in the simplification process. Each represents the initial vertex of a fan and consists of its co-ordinates (*coord* field) and a pointer to the second vertex in the fan (*secondFanVertex* field).

```
struct Vertex
  float coord[3];
  int secondFanVertex;
end struct

struct FanVertex
  int id;
  struct Vertex *v;
end struct

struct ControlLod
  int deletedVertices;
  int generatedFaces;
end struct

struct MOM-Fan
  int nVertices, nFanVertices, nLods;
  struct Vertex vertexList[];
  struct FanVertex fanVertexList[];
  struct ControlLod controlLodList[];
end struct
```

Fig. 5. *MOM-Fan* Data Structure

- *Fan list* ($fanVertexList$ field). Stores the fans as vertex sequences (except the initial vertex, already stored in the vertex list). For each fan we store the vertices of the triangles which disappear when the common vertex is eliminated. Each stored vertex consists of a pointer to the vertex in the vertex list (v field) and an identifier (id field). The identifier references the face of the fan represented by that vertex. The fans are also ordered according to their elimination in the simplification process.
- *Control List* (*controlLodList* field). For each LOD, this list stores the information necessary to recover the data pertaining to that LOD.

3.1 Construction Process

The process of constructing the model is divided into 2 steps. The first involves the tasks which must be repeated during each iteration of the simplification. The second includes a group of operations which finish building the multiresolution model M^r. To simplify the explanation let us assume that each simplification realised eliminates only one vertex. In figure 6 we show the mesh to be utilised in explaining the construction, initialisation, the result of the first stage -eliminating vertices $v4$ and $v1$- and the result of the second stage.

Initialisation. Before beginning this process it is necessary to initialise M^r with M_0, as it is the first level of detail. For this it is necessary to update *controlLodList* with data from the mesh M_0, filling in the fields *deletedVertices= 0* and *generatedFaces=* $|\mathcal{F}_0|$. The vertex list and fan list are empty.

First stage. For each iteration in the simplification process the following tasks are realised:

1. Store in $fanVertexList$ the vertices which form the fan, except for the vertex eliminated (common vertex).
2. Store in *vertexList* the eliminated vertex and place the pointer at the first of the remaining vertices of the fan.
3. Update *controlLodList* with the deleted vertices and the total number of generated faces.

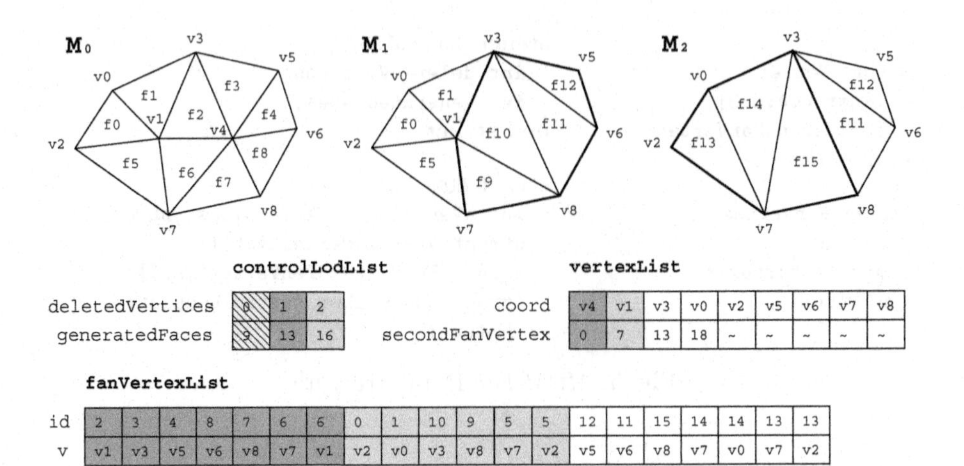

Fig. 6. Example of model construction. Shaded cells indicate initialisation. In dark grey, data added after the elimination of $v4$, and in light grey, data added after the elimination of $v1$, both in the first stage. In white, data added as a result of the second stage

Second stage. To complete the data structure:

1. Add the mesh containing the worst LOD, as a set of fans, in the same manner as in the first stage. The vertices that do not have an assigned fan are put at the end of the list.
2. Update the pointers of the fan vertices so that they point to the vertices in the new vertex list.

With this step we conclude the construction of the data structure, which we can generalise as being the elimination of more than one vertex in the simplification process similar to that in [2].

3.2 Storage Cost

Assuming that the cost of storing a real, an integer and a pointer is one word, and that we store three lists, one for each data type in the model (see figure 5), the total storage cost will be $4|\mathcal{V}^r| + 2|\mathcal{F}_a^r| + 2n$ words, being $|\mathcal{F}_a^r|$ the size of *fanVertexList*. The order of $|\mathcal{V}^r|$ (see eq. 4) in the worst case is quadratic with regard to $|\mathcal{V}_0|$. The best case appears when we use a simplification method based on vertex decimation [8], where $|\mathcal{V}^r| = |\mathcal{V}_0|$ and may reach $|\mathcal{V}^r| = 2|\mathcal{V}_0|$ in the case of simplification methods based on elimination of edges [9].

With regard to $|\mathcal{F}_a^r|$, in the worst case, the cost also is of quadratic order with respect to $|\mathcal{F}_0|$. If we use a simplification method based on vertex decimation and we assume that the number of faces around a given vertex to be an average of 6 and that $|\mathcal{V}_i| \approx |\mathcal{F}_i|/2, 0 \leq i < n-1$, we have $|\mathcal{F}_a^r| \approx 3.5|\mathcal{F}_0|$. With a method based on elimination of edges, $|\mathcal{F}_a^r| \approx 5.5|\mathcal{F}_0|$.

```
for each vertex v from DV and while there exist triangles to paint
  InterruptionFan <- true {force the a.1 case}
  for each vertex v_i of the fan associated with v, from the first to the penultimate
    if fanVertexList[v_i].id < GF then {case a) paint v_i}
      if InterruptionFan then
        paint(v) {case a.1) paint the initial vertex}
        InterruptionFan <- false {do not repeat case a.1)}
      end if
      paint(v_i)
      NF <- NF-1
    else {case b) do not visualise v_i}
      if no InterruptionFan then
        paint(v_i) {paint v_i to close triangle}
        InterruptionFan <- true {force case a.1)}
      end if
    end if
  end for
  if no InterruptionFan then
    paint(v_i) {last vertex v_i from the list, to close fan}
  end if
end for
```

Fig. 7. Visualisation algorithm

4 Algorithm for Visualising a LOD

The visualisation algorithm of a LOD, which is shown in figure 7, requires the initial calculation of the number of triangles to visualise at that LOD. Afterward, it traverses the data structure recovering the triangle fans pertaining to the required LOD.

Given a LOD k in the control list we store the number of vertices not pertaining to k, DV, and the number of generated faces, GF. Based on these data it is trivial to obtain the number of faces, NF, which pertain to k:

```
DV= controlLodList[k].deletedVertices;
GF= controlLodList[k].generatedFaces;
NF= GF-(vertexList[DV].secondVertexFan- DV)
```

Using the example in figure 6, suppose we wish to visualise the best LOD (M_0). The fans to be generated, ordered by the appearance of vertices in the data structrure, are: (v4 v1 v3 v5 v6 v8 v7 v1) which correspond to the faces (f2 f3 f4 f8 f7 f6); (v1 v2 v0 v3) and (v1 v7 v2), which correspond to the faces (f0 f1) and (f5), respectively. The first vertex of the fan always corresponds to the eliminated vertex. The remaining vertices are stored in *fanVertexList*, and are associated with the eliminated vertex. For each fan only the t first vertices of $t+1$ vertices composing it are processed, because each of them represents one of the t faces represented whereas the last vertex serves only to complete the fan.

However, it may occur that some vertices do not pertain to the required LOD. In the example, because f10 and f9, implicitly associated to v1, should not be visualised, there is a jump from vertex v3 to v7. To resolve these jumps without splitting fans, it is necessary to introduce the vertex v1 between v3 and v7, thus producing two degenerate triangles. While the first fan is resolved without degenerate triangles (v4 v1 v3 v5 v6 v8 v7 v1), the second fan (v1 v2 v0 v3 v1 v7 v2) includes two of them, (v1 v3 v1) and (v1 v1 v7). Upon processing each of the vertices two things may occur: a) the vertex identifier indicates that the triangle should be painted, or b) the vertex identifier indicates that the triangle should not be painted. When the second case occurs, this signifies an interruption in the fan, and if this continues further along it is necessary to insert the initial vertex. Therefore, the first time that case a) is encountered after an interruption the insertion should be realised (case a.1 in the algorithm).

The computational cost of the algorithm to extract LOD k depends on the total number of vertices not eliminated, which is at most $|\mathcal{V}^r| - k$, and the total number of vertices of the associated fans, which is at most $|\mathcal{F}_a^r| - 2k$. $|\mathcal{V}^r|$ and $|\mathcal{F}_a^r|$ are, in the worst case, of quadratic order with respect to $|\mathcal{V}_0|$ and $|\mathcal{F}_0|$, respectively, but this case differs substantially from the normal case. With the method of simplification by elimination of vertices the cost is $O(8|\mathcal{V}_k|)$ and with a method based on edge elimination, $O(13|\mathcal{V}_k|)$.

5 Results

The experiments were realised utilising a Silicon Graphics RealityEngine 2, with a MIPS R10000 at 194 MHz and 256 Mb RAM. Coding of the model was in *C++* and utilised *OpenGL* as its graphics library. The simplification method used to construct the multiresolution representations is that proposed by Garland and Heckbert [9] based on contraction of edges. The meshes come from the *Stanford University Computer Graphics Laboratory* (http://www-graphics.stanford.edu/data/3Dscanrep/) and *Cyberware* (http://www.cyberware.com/models/).

In table 1 we summarise the characteristics and storage costs of the objects used in the experiments. For each of them we indicate the number of vertices and faces of the original model, and its storage cost assuming a structure based on a vertex list and a triangle list [10]. Also it is assumed that a word (integer, real, or pointer) carries a set cost of 4 bytes. With regard to the multiresolution ordered mesh (MOM) representation, we indicate the number of levels of detail, the number of faces and the total storage cost. Regarding the new representation proposed in this article, we indicate the number of *fanVertex* and the total storage cost. It can be observed that the number of fan vertices stored in the new representation is higher than that of the faces in the MOM representation. However, given that the cost to store a fan vertex is less than for a face, the storage cost of the new list provides a memory savings of about 20% over the face list. The repercussion of this is that the total storage cost is reduced by an important amount, approximately 15%, due to the fact that the list which is reduced is the "heaviest" list in the model (compare the number of faces with the number of vertices and LODs).

Table 1. Characteristics and storage costs

	Original			MOM			MOM-Fan	
	Vertices	Faces	MB	Lods	Faces	MB	Fan V.	MB
Cow	2,905	5,804	0.100	2,803	14,982	0.237	17,852	0.202
Sphere	15,315	30,624	0.526	15,264	83,486	1.306	98,793	1.104
Bunny	34,835	69,451	1.193	33,990	182,192	2.876	216,759	2.445
Phone	83,045	165,963	2.850	81,668	441,181	6.940	523,844	5.887
Isis	187,871	375,736	6.450	187,370	993,559	15.667	1,181,373	13.309
Buda	543,653	1,085,636	18.646	543,106	2,754,083	43.957	3,297,593	37.598

In figures 8(a) and 8(d) the behavior of the new representation is shown, using the *Bunny* and *Buda* models (some views of them are shown in figures 10 and 11). On the X-axis the level of detail is represented, where 0 is the poorest and 1 the best. On the Y-axis we show the time spent (in seconds) for the model to recover the data pertaining to a given LOD and to visualise them. The behavior is similar to that obtained with the earlier MOM scheme and one can observe the linear response with respect to the number of triangles of the LOD visualised. The improvement gained by the use of triangle fans is diminished somewhat by the slight increase in data recovery time due to a more complex algorithm, the short length of fans (an average of 3.3 triangles per fan), the degenerate triangles, and the overhead caused by executing, for each fan, the instructions *glBegin* and *glEnd* in the *OpenGL* implementation. In figures 8(b) and 8(e) we show the number of non-degenerate triangles and the number of degenerates (around 23% of the total triangles) sent to the graphic subsystem per LOD. In figure 8(c) and 8(f) we show the number of vertices sent per LOD in the MOM representation, and those sent in the new representation (about 40% fewer).

6 Conclusions and Future Work

In this article we present a new multiresolution scheme which permits the storage and visualisation of the distinct levels of detail as triangle fans. The objective is double: to reduce the visualisation time and also the space (storage) cost.

The experimental results show a reduction of about 15% in storage cost with respect to the previous MOM representation, upon which the new approximation is based. However, the behavior of the new model regarding its visualisation time, is similar to its ancestor. A short average fan length, the high percentage of degenerate triangles, and the necessity to adjust the fans to the required LOD in real-time contribute to produce overall results which do not suppose a global improvement in visualisation time.

This work will proceed from this moment on, toward the utilisation of triangle strips which, in principle, will permit a higher average number of triangles per strip than has been obtained using fans. In this manner we expect the storage cost to be further improved, as well as visualisation times.

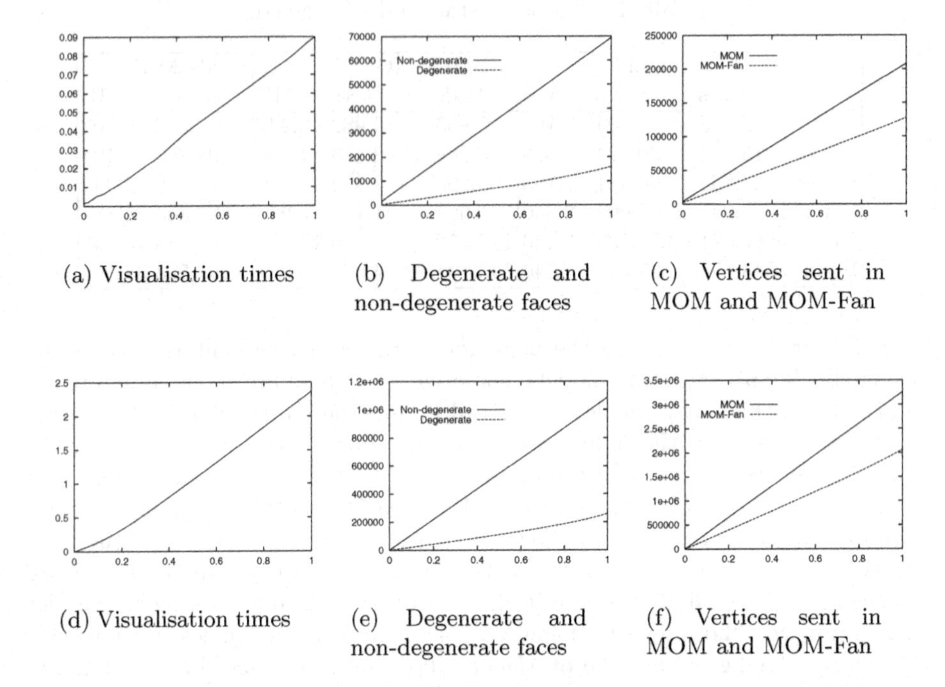

(a) Visualisation times

(b) Degenerate and non-degenerate faces

(c) Vertices sent in MOM and MOM-Fan

(d) Visualisation times

(e) Degenerate and non-degenerate faces

(f) Vertices sent in MOM and MOM-Fan

Fig. 8. Results: a), b) and c) *Bunny* model; d), e) and f) *Buda* model

References

1. Garland, M.: Multiresolution Modeling: Survey & Future Opportunities. State of the Art Reports of EUROGRAPHICS '99 (1999) 111–131 431, 433
2. Ribelles, J., Chover, M., Huerta, J., Quiros, R.: Multiresolution Ordered Meshes. Proc. of 1998 IEEE Conference on Information Visualization (1998) 198–204 432, 434, 436
3. Hoppe, H.: View-Dependent Refinement of Progresive Meshes. Proc. of SIGGRAPH'97 (1997) 189–198 432
4. El-Sana, J., Azanli, E., Varshney, A.: Skip Strips: Maintaining Triangle Strips for View-dependent Rendering. Proc. of IEEE Visualization 1999 (1999) 131–137 432
5. Ribelles, J., Chover, M., Lopez, A., Huerta, J.: Frame-to-frame Coherence of Multiresolution Ordered Meshes. Proc. of CEIG'99 (1999) 91–104 433
6. Ribelles, J., Chover, M., Lopez, A., Huerta, J.: A First Step to Evaluate and Compare Multiresolution Models. Short Papers and Demos of EUROGRAPHICS'99 (1999) 230–232 433
7. Hoppe, H.: Progressive Meshes. Proc. of SIGGRAPH'96 (1996) 99–108 433
8. Schroeder, W. J., Zarge, J. A., Lorensen, W. E.: Decimation of Triangle Meshes Proc. of SIGGRAPH'92 (1992) 65–70 436
9. Garland, M., Heckbert, P.: Surface Simplification Using Quadratic Error Metrics. Proc. of SIGGRAPH'97 (1997) 209–216 436, 438

10. Foley, J. D., van Dam, A., Feiner, S., Hughes, J., Phillips, R.: Computer Graphics. Principles and Practice. Addison-Wesley Publishing Company (1990) 438

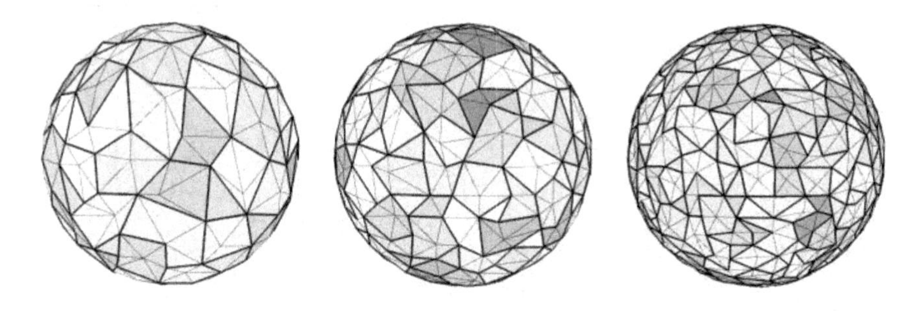

Fig. 9. Three levels of detail of the *Sphere* model visualised using fans

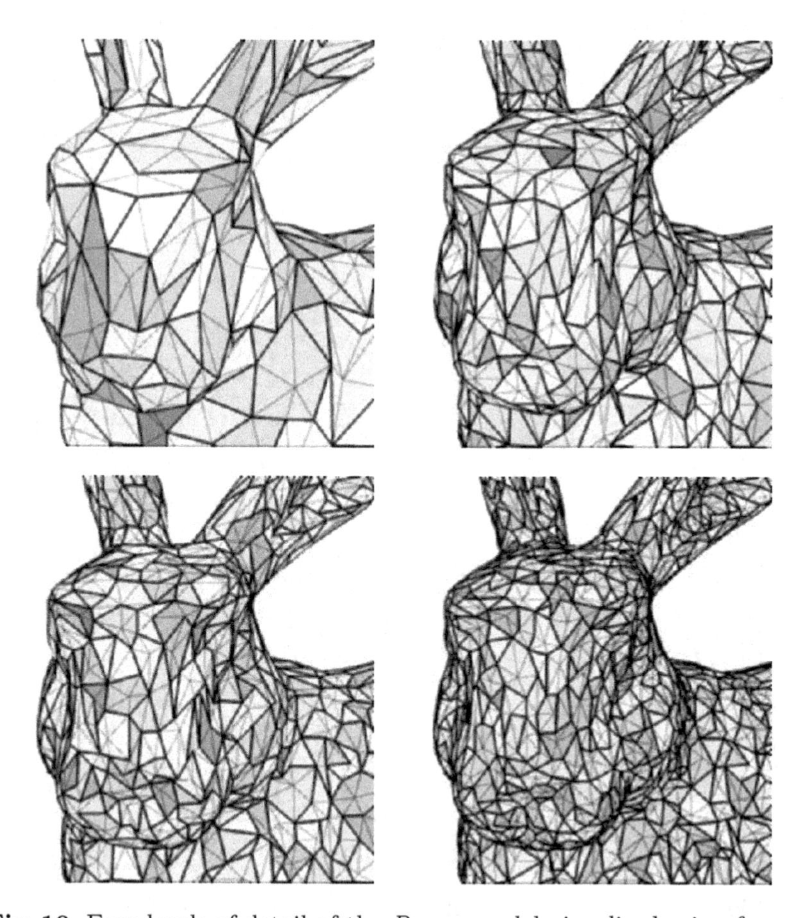

Fig. 10. Four levels of detail of the *Bunny* model visualised using fans

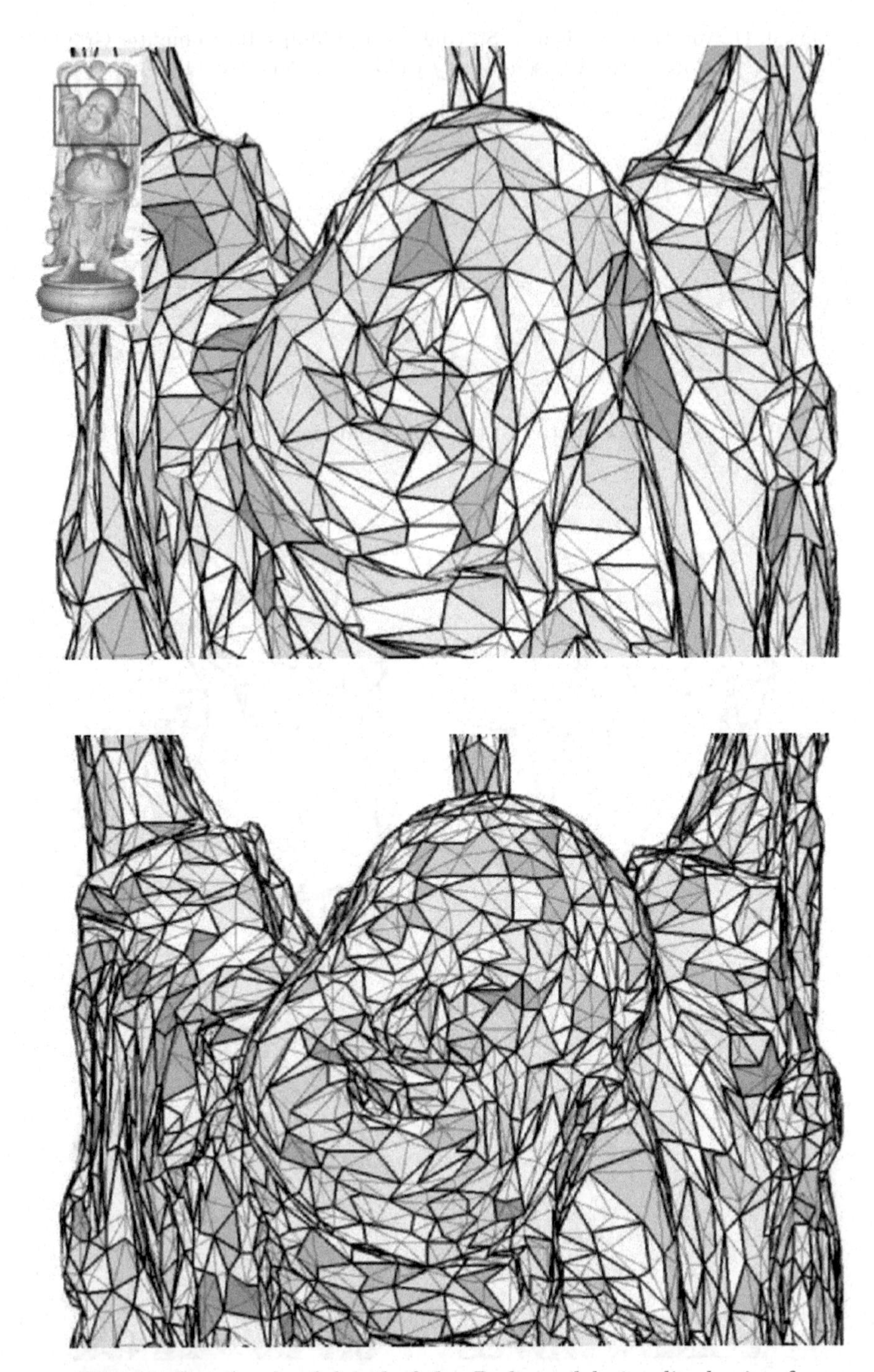

Fig. 11. Two levels of detail of the *Buda* model visualised using fans

Detecting Centres of Maximal Geodesic Discs on the Distance Transform of Surfaces in 3D Images

Gabriella Sanniti di Baja[1] and Stina Svensson[2]

[1] Istituto di Cibernetica, Italian National Research Council
Via Toiano 6, IT-80072 Arco Felice (Naples), Italy
gsdb@imagm.cib.na.cnr.it

[2] Centre for Image Analysis, Swedish University of Agricultural Sciences
Lägerhyddvägen 17, SE-75237 Uppsala, Sweden
stina@cb.uu.se

Abstract. We introduce the distance transform for surfaces in 3D images, i.e., the distance transform where every voxel in the surface is labelled with its geodesic distance to the closest voxel on the border of the surface. Then, the distance transform is used to identify the set of centres of maximal geodesic discs in the surface. The centres of maximal geodesic discs can be used to give a compact representation of any surface. In particular, they can provide a useful representation of the surface skeleton of solid volume objects.

1 Introduction

In [6], the *constrained distance transform* was introduced for computing the minimal path from any point in the image to a goal point, avoiding *obstacles*. This can be used to find optimal paths between obstacles (e.g., for robot navigation), measure perimeters, and determine convex hulls. A special case of the constrained distance transform is the distance transform of line patterns (DTLP), introduced in [12] and further investigated in [9]. In DTLP, distance information is propagated along the line pattern, starting from the end points of the line pattern itself (the background is used as an obstacle to the propagation). In presence of meeting lines that convey different distance information into their common meeting point, the minimum distance or the maximum distance can be used to continue distance propagation on the outgoing line. By selecting the maximum distance, the DTLP can be used, for instance, to find the longest path through a line pattern. The minimum distance can be used to guide hierarchical decomposition.

Here we extend the DTLP to surfaces in 3D images. In case of concurrent distinct distance values, we use the minimum distance to continue propagation. Every voxel in the surface is labelled with the distance to the closest voxel on the border of the surface. This implies that only *open* surfaces, i.e., surfaces that do not enclose any background component, can be considered. The first step in the algorithm is to detect the border of the surface. Then, once the distance transform of the surface is available, we identify therein the centres

G. Borgefors, I. Nyström, and G. Sanniti di Baja (Eds.): DGCI 2000, LNCS 1953, pp. 443–452, 2000.
© Springer-Verlag Berlin Heidelberg 2000

of the maximal geodesic discs (CMGDs). These, similarly to the centres of the maximal discs in the distance transform of a 3D object, can be used to provide a new representation of the surface. In fact, the CMGDs constitute a subset of the surface which, in turn, can be seen as the union of the maximal geodesic discs. The CMGDs can be suitably linked to each other to obtain a curve representation with as many components as there are components in the surface. If this is done, one can easily treat images containing more than one object component, without need of resorting to a preliminary connected component labelling. Moreover, connected curves can be traced, which is not possible for sparse voxels, and can be vectorized so saving memory occupation. Thus, the representation of the surface in terms of connected sets including the CMGDs is convenient. We remark that the new representation does not coincide with the skeleton of the surface. In fact, loops can be created which do not correspond to tunnels of the surface, so that the topology is not preserved. Nevertheless, it can be seen as a new linear shape descriptor of surfaces in 3D images, where the curves can be interpreted as the main symmetry axes of the surfaces.

To extract the new representation, it is convenient that the border of the surface is defined analogously to the contour of a 2D plane pattern. The approach to extract the new representation can then be seen as analogous to the one to extract the skeleton of the pattern from its distance transform. The contour of a pattern is the set of pattern pixels having a neighbour in the background. These pixels belong to protruding contour arcs, i.e., have no neighbours internal in the pattern, or belong to arcs delimiting the pattern itself, i.e., have neighbours internal in the pattern. To have a similar definition of border in case of a surface, we should identify there curves and edges, respectively corresponding to the protruding contour arcs and the arcs delimiting the pattern. Curves should be prevented from propagating distance information, as their 2D counterpart, the protruding arcs, consist exclusively of centres of maximal discs through which no distance information can be further more propagated. To this purpose, all curves have to be preliminarily identified and temporarily removed from the surface. It is then possible to identify only the border voxels (edges) from which distance information can be propagated to the surface. The curves are added again after the distance transform is computed. The distance label of their voxels is set to the value of the voxels in the edges of the surface, as also curves belong to the border.

The new representation can be applied to any open surface. In particular, it can be applied to the surface skeleton of a 3D solid object. This latter case is treated in this paper.

2 Definitions and Notions

In this paper, 3D bilevel images, i.e., volume images consisting of object and background, will be considered. Each voxel has three types of neighbours in its $3 \times 3 \times 3$ neighbourhood: six voxels sharing a face, twelve voxels sharing an edge, and eight voxels sharing a point with the central voxel. We will use 26-connectedness for the object and 6-connectedness for the background.

Each voxel in the object can be labelled with the distance to its closest background voxel. This labelling can be performed by computing the distance transform (DT) of the object, [2]. The distance depends on the chosen metric. Two simple metrics, commonly used, are the D^6 and the D^{26} metrics, where the distance between two voxels, v and w, is equal to the number of steps in a minimal 6- and 26-connected path, respectively, between v and w. The corresponding DTs, here called DT^6 and DT^{26}, can be computed by propagating distance information from the background in one forward and one backward scan over the image. Each voxel is assigned the minimum of the label of the voxel itself and the labels of its already visited neighbours increased by one. For DT^6 only face neighbours are considered.

DT^6 and DT^{26} give rough approximations to the Euclidean distance. To have a better approximation a *weighted distance transform* can be used. For the weighted DT, we use different weights, w_1, w_2, and w_3, for the distance to a face, an edge, and a point neighbour, respectively. Computation of the DT using weights $w_1 = 3$, $w_2 = 4$, and $w_3 = 5$, has been shown to give a good approximation to the Euclidean distance, [2]. The DT using $w_1 = 3$, $w_2 = 4$, and $w_3 = 5$ will here be denoted $3 - 4 - 5$ DT. When computing the DT, the contribution given by each already visited neighbour of a voxel is its corresponding label increased by the relative weight. To be consistent, we use weights also for DT^6 and DT^{26}. For DT^6, $w_1 = 1$, $w_2 = 2$, and $w_3 = 3$, while for DT^{26}, $w_1 = 1$, $w_2 = 1$, and $w_3 = 1$.

The distance label of a voxel v in the DT can be interpreted as the length of the radius of a ball, centred on the voxel and fully enclosed in the object. In the following, we will denote by v both the voxel and its associated distance label. A voxel is *centre of a maximal ball* (CMB) if the corresponding ball is not completely covered by any other single ball in the object, [1]. This can be checked by a suitable label comparison. A voxel v is a CMB if all its neighbours n_i, $i = 1, \ldots, 26$, with proper weights w_j, $j = 1, 2, 3$, satisfy

$$n_i < v + w_j.$$

For any voxel v, N^{26} is the number of 26-connected object components in the $3 \times 3 \times 3$-neighbourhood of v. $\overline{N}_f^{18}$ is the number of 6-connected background components, having v as a face-neighbour, in the $3 \times 3 \times 3$-neighbourhood of v where the point neighbours are disregarded (in other words only 18 neighbours of v are taken into account). For a deeper discussion, see [7,10,4]. Here we use the algorithm introduced in [3] for computing $\overline{N}_f^{18}$ and N^{26}.

3 Computing the Distance Transform of Surfaces

Before computing the DT of a surface, we should detect its border. In fact, the minimum distance will be propagated starting from it. The labels in the DT will indicate the distances to the background along paths in the surface, passing through the border. Therefore, we set the border voxels to w_1 instead of 0. The same convention is used for the end points when computing the DTLP, [12,9].

Surface skeletons can consist of both surfaces and curves, depending on the shape of the original object and the used metric. When computing the DT of the surface skeleton, only the voxels in the edges should be allowed to propagate distance information to the surface. Curves do not border any surface portion and we should prevent that any propagation occurs from them. Thus, we detect and temporarily remove the curves, before detecting the edge voxels from which propagation can be done. After the DT has been computed, the curves are added and the distance label of their voxels is set to w_1.

Our method can be applied to any surface skeleton of a solid 3D object. We prefer a skeleton that has a number of desirable properties, as it is the case for the surface skeleton introduced in [11], which will be used through the rest of the paper. The surface skeleton is centred within the original object with respect to the D^6 metric, topologically equivalent to the original object, almost symmetric, and allows complete object recovery. This last property, unfortunately, implies that the skeleton can be two-voxel thick. Thus, before detecting the border of the surface skeleton, this has to be reduced to a one-voxel thick surface. To this purpose, we use the local operations introduced in [4] to obtain a one-voxel thick surface skeleton starting from the two-voxel thick set. Thinning is achieved in six scans of the image, one for every face direction. During each scan, only configurations that are two-voxel thick in the scanning direction are taken into account, see Fig. 1. Voxels in these configurations are sequentially removed provided that their removal does not change the topology.

The above thinning is, however, not enough to obtain a one-voxel thick surface in presence of L-shaped regions, see Fig. 2. To remove voxels in such regions another step is necessary, based on the use of an L-shaped configuration detector. During this step, only voxels having $\overline{N}_f^{18} = 1$ (grey voxels in Fig. 2) and at least three face neighbours in the skeleton are considered. Among them, the voxels with two face neighbours in the skeleton, that are edge neighbours of each other and have both $\overline{N}_f^{18} = 1$ ($\bullet$) or have both $\overline{N}_f^{18} \neq 1$ ($\circ$), are marked as candidate to removal. Removal of marked voxels is then done sequentially, provided that topology is not altered. The resulting set is shown to the right in Fig. 2.

Once the one-voxel thick surface is obtained, the border is detected in two steps. In the first step, *curve voxels* are defined as voxels having $\overline{N}_f^{18} = 1$ and $N^{26} \geq 2$ (the case $N^{26} > 2$ occurs at curve junctions), [7,10]. Since propagation of distance information should not occur from curves, all voxels with $\overline{N}_f^{18} = 1$ and $N^{26} \geq 2$ are temporarily removed. Note that the tips of the curves, end points, are not detected by the previous criteria and hence remain in the surface skeleton. This does not create any problem, since they are voxels isolated from the rest of the surface. In the second step, voxels having $\overline{N}_f^{18} = 1$ and $N^{26} = 1$ are detected as *edge voxel.* With reference to Fig. 2, grey voxels are voxels having $\overline{N}_f^{18} = 1$ and $N^{26} = 1$, hence they would be identified as edge

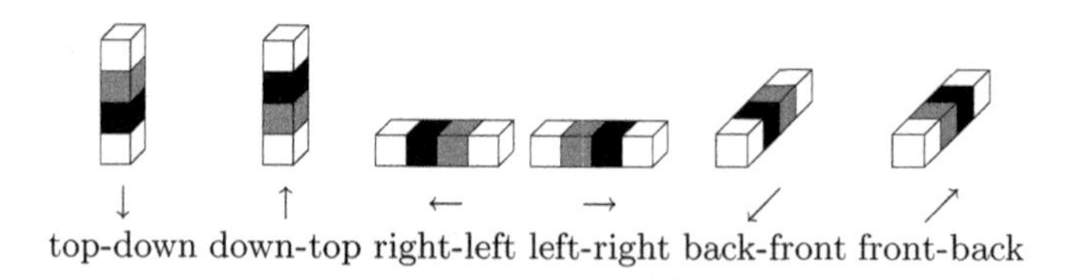

Fig. 1. Two-voxel thick configurations in the six scanning directions. The current voxel of the surface skeleton is shown in grey, the other voxel in the skeleton in black and background voxels in white

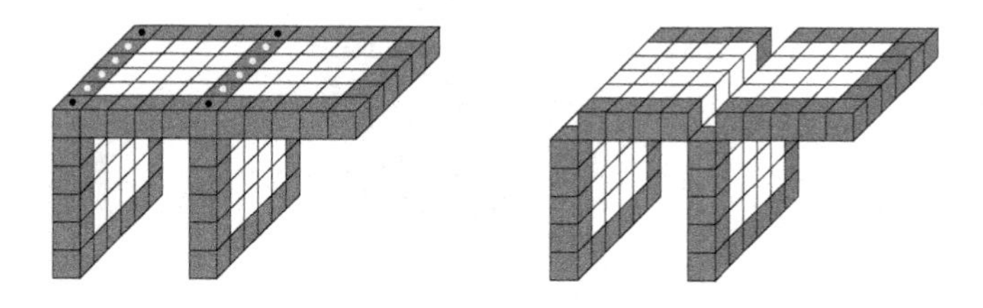

Fig. 2. Surface skeleton before, left, and after, right, removal of voxels in L-shaped configurations. Grey voxels satisfy $\overline{N}_f^{18} = 1$. See text

voxels. Removal of voxels in L-shaped configurations is indispensable to avoid that these voxels are wrongly classified.

To compute the DT of a surface, we use, analogously to the "regular" DT case, the weights w_1 for the distance to a face neighbour, w_2 for the distance to an edge neighbour, and w_3 for the distance to a point neighbour. Edge voxels are labelled w_1. Distance information is propagated from the edge over the surface. Each voxel in the surface is assigned the minimum of the label of the voxel itself and the labels of its already visited neighbours, also belonging to the surface, increased by the corresponding weight. The propagation is repeated, alternately in forward and backward fashion, until no more changes occur. The number of scans needed depends on the complexity of the surface. For discussion about the 2D case, [8]. The number of iterations of forward and backward scans needed for the examples shown in Figs. 5 and 6 are 6 and 2, respectively.

4 Representing the Surface

Any voxel in the DT of a surface can be seen as the centre of a geodesic disc with radius equal to the distance label of the voxel, and fully contained in the surface. A geodesic disc centred on a voxel v can be computed by applying the reverse distance transformation, [4], using v as a seed and the background of the surface as an obstacle. In Fig. 3, top, a surface is shown that consists of planes that meet each other along segments in face, edge, and point directions. The geodesic discs shown in Fig. 3, bottom, are centred at the intersection of the above segments and correspond to different DTs, DT^6, DT^{26}, and $3-4-5$ DT. The radius of the disc is the same for all DTs.

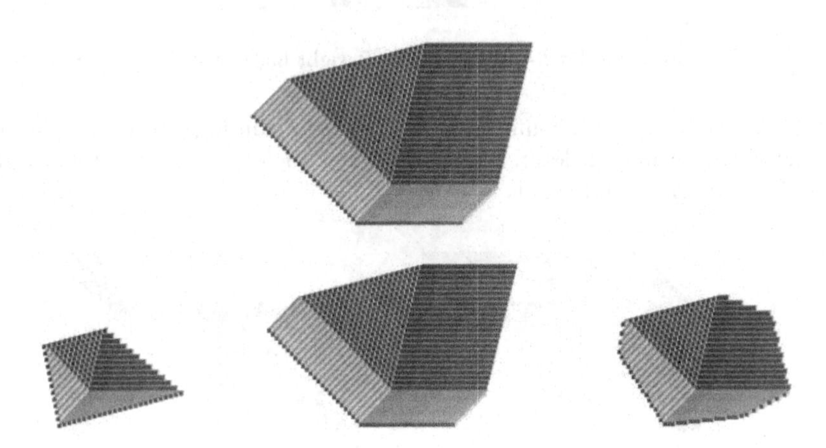

Fig. 3. Above, a surface consisting of four planes meeting each other along segments in face, edge, and point directions. Below (from left to right), geodesic discs centred at the intersection of the above segments and with radius 25, using DT^6, DT^{26} and $3-4-5$ DT, respectively

4.1 Centres of Maximal Geodesic Discs

A centre of a maximal geodesic disc can be defined in the same way as the centre of a maximal disc for 2D objects and the centre of a maximal ball for 3D objects, [1]. Thus, a *centre of a maximal geodesic disc* (CMGD) is a voxel whose corresponding geodesic disc is not completely covered by any other single geodesic disc. A voxel v in the DT is a CMGD if all its neighbouring voxels n_i, $i = 1, \ldots, 26$, in the surface, with the proper weights w_j, $j = 1, 2, 3$, satisfy

$$n_i < v + w_j.$$

In Fig. 4, the sets of CMGDs for the surface shown in Fig. 3, top, are shown for DT^6, DT^{26}, and $3-4-5$ DT, respectively.

The union of the geodesic discs corresponding to the set of CMGDs is equivalent to the original surface.

4.2 Connecting the Centres of Maximal Geodesic Discs

To have a representation of the surface with the same number of components as those of the surface, the CMGDs should be linked to each other. This could be

Fig. 4. From left to right, the set of CMGDs for the DT of the surface shown in Fig. 3, top, using DT^6, DT^{26}, and $3-4-5$ DT, respectively

done, e.g., by growing paths in the direction of the steepest gradient in the DT starting from each CMGD.

Indeed, besides the CMGDs, also other voxels, here called *saddle voxels* and *branch point voxels*, are important to gain the connectedness of the final representation. Saddle voxels are defined as voxels having in the surface skeleton more than one component of neighbours with higher label, or more than one component of neighbours smaller label. A branch point voxel is defined as a voxel having more than two components of neighbours with the same label as the voxel itself. In all cases, we use 26-connectedness.

The CMGDs, saddle voxels, and branch point voxels are called *intrinsic.* They are detected and marked in one scan of the DT, before starting path growing. During path growing, suitable voxels, here called *induced* are detected and marked on the DT. The DT is repeatedly inspected until no more voxels are marked. For each already marked voxel, v, either intrinsic or induced, its neighbourhood is inspected. The voxel(s) n_i, $i = 1, \ldots, 26$, on the DT, that by using the proper weights w_j, $j = 1, 2, 3$, maximise the gradient

$$grad(n_i) = \frac{n_i - v}{w_j}$$

are marked as belonging to a growing path. We denote the highest gradient $grad_{\max}$.

Due to the fact that we are growing paths on surfaces, where 18-connectedness is possible, we can have paths in certain directions that alternately include one single voxel and a pair of edge connected voxels. This might create diverging paths, as the steepest gradient could occur along non-connected, diverging directions from of the two voxels. We force, if possible, path growing to continue in only one direction, whenever two or more connected neighbours equally maximise the gradient. Thus, if a voxel can propagate on a number of neighbours n_i, all with $grad(n_i) = grad_{\max}$, and constituting a unique 26-connected component, we originate a single path by choosing, as the next voxel in the path, only one neighbour of the component of n_i's.

In Figs. 5 and 6, two examples of the set of CMGDs and the corresponding connected curve representations are shown. We started from 3D solid objects, a cylinder and a digital Euclidean ball, respectively, and computed the surface skeletons by using the algorithm introduced in [11]. These were reduced to one-voxel thick surfaces, as described in Section 3. The surface skeletons were then also simplified by pruning some non-significant peripheral curves by means of the algorithm introduced in [5]. The metric used is D^6.

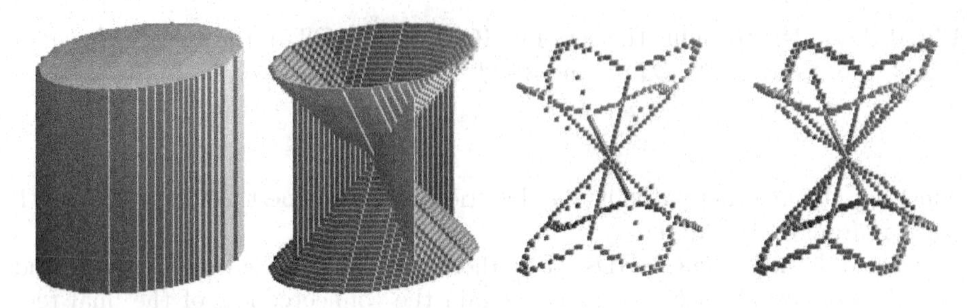

Fig. 5. From left to right: A cylinder with radius 25, and height 50, the simplified surface skeleton, the set of CMGDs, and the connected curve representation obtained after path growing

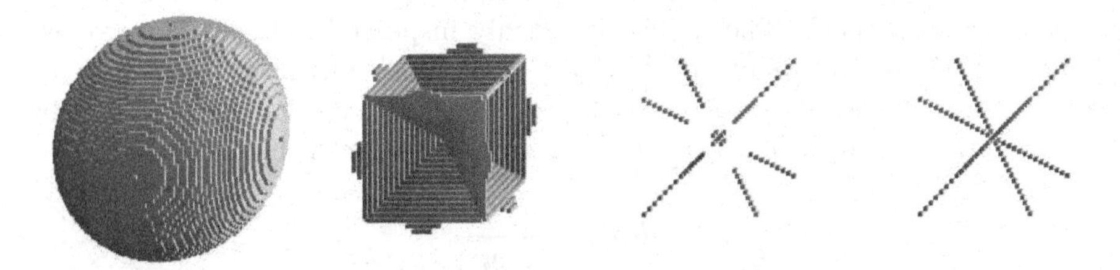

Fig. 6. From left to right: A digital Euclidean ball with radius 28, the simplified surface skeleton, the set of CMGDs, and the connected curve representation obtained after path growing

In Fig. 5, right, it is possible to see paths alternately including one voxel and a pair of edge connected voxels. We also note that the curve representation is not topologically equivalent to the surface skeleton and, hence, to the object. In fact, a number of *loops* characterise the curve representation, that do not correspond to tunnels in the surface skeleton and the object. The curve representation is not the curve skeleton of the object.

Note that it is generally preferable to remove the markers from the CMGDs with label w_1 (except for the case in which they belong to curves) before performing path growing. In fact, these CMGDs might cause the growth of a number of noisy paths. The marker has been removed from CMGDs with label w_1 in the examples shown in this paper.

5 Conclusion

We have introduced the distance transform of surfaces in 3D images and have used it to extract the centres of maximal geodesic discs. The centres of maximal geodesic discs provide a new representation of a surface in general, and of the surface skeleton of 3D solid objects in particular. To compute the distance transform, we detect the border. This is done in two steps. First the curves are identified and temporarily removed, then the edges are identified and used for propagating distance information.

The set of centres of maximal geodesic discs is connected by a path growing process, to obtain a richer representation of the surface consisting of a set of connected curves. The representation has the same number of object components as the surface, but might have loops. Hence the resulting representation is not topologically equivalent to the surface, (which is topologically equivalent to the original object when the surface is indeed the surface skeleton of the object).

Future work will be to further investigate the surface representation here introduced and, in particular, to identify the border starting directly from two-voxel thick surfaces. This is important to make our algorithm more general. We will also investigate the distance transform for surfaces where we use the maximum distance in case of concurrent distinct distance values. This will need a more complicated algorithm but could give interesting results.

Acknowledgements

Prof. Gunilla Borgefors and Dr. Ingela Nyström, both Centre for Image Analysis, Uppsala, Sweden, are gratefully acknowledged for their scientific support.

References

1. C. Arcelli and G. Sanniti di Baja. Finding local maxima in a pseudo-Euclidean distance transform. *Computer Vision, Graphics and Image Processing*, 43(3):361–367, Sept. 1988. 445, 448
2. G. Borgefors. On digital distance transforms in three dimensions. *Computer Vision and Image Understanding*, 64(3):368–376, 1996. 445
3. G. Borgefors, I. Nyström, and G. Sanniti di Baja. Connected components in 3D neighbourhoods. In M. Frydrych, J. Parkkinen, and A. Visa, editors, *Proceedings of 10th Scandinavian Conference on Image Analysis (SCIA'97)*, pages 567–572, Lappeenranta, Finland, 1997. Pattern Recognition Society of Finland. 445
4. G. Borgefors, I. Nyström, and G. Sanniti di Baja. Computing skeletons in three dimensions. *Pattern Recognition*, 32(7):1225–1236, July 1999. 445, 446, 447

5. G. Borgefors, I. Nyström, G. Sanniti di Baja, and S. Svensson. Simplification of 3D skeletons using distance information. Accepted for publication in Proceedings of SPIE International Symposium on Optical Science and Technology: Vision Geometry IX (vol. 4117), 2000. 450
6. L. Dorst and P. W. Verbeek. The constrained distance transformation: a pseudo-Euclidean recursive implementation of the Lee algorithm. In I. T. Young, J. Biemond, R. P. W. Duin, and J. J. Gerbrands, editors, *Signal Processing III: Theories and Applications*, pages 917–920. Elsevier Science Publishers B. V. (North-Holland), 1986. 443
7. G. Malandain, G. Bertrand, and N. Ayache. Topological segmentation of discrete surfaces. *International Journal of Computer Vision*, 10(2):183–197, 1993. 445, 446
8. J. Piper and E. Granum. Computing distance transformations in convex and non-convex domains. *Pattern Recognition*, 20(6):599–615, 1987. 447
9. I. Ragnemalm and S. Ablameyko. On the distance transform of line patterns. In K. A. Høgda, B. Braathen, and K. Heia, editors, *Scandinavian Conference on Image Analysis (SCIA'93)*, pages 1357–1363. Norwegian Society for Image Processing and Pattern recognition, 1993. 443, 446
10. P. K. Saha and B. B. Chaudhuri. 3D digital topology under binary transformation with applications. *Computer Vision and Image Understanding*, 63(3):418–429, May 1996. 445, 446
11. G. Sanniti di Baja and S. Svensson. Surface skeletons detected on the D^6 distance transform. In F. J. Ferri, J. M. Iñetsa, A. Amin, and P. Pudil, editors, *Proceedings of SSSPR 2000 - Alicante: Advances in Pattern Recognition*, pages 387–396, Alicante, Spain, 2000. Springer-Verlag, Berlin Heidelberg. Lecture Notes in Computer Science 1121. 446, 450
12. J.-I. Toriwaki, N. Kato, and T. Fukumura. Parallel local operations for a new distance transformation of a line pattern and their applications. *IEEE Transactions on Systems, Man, and Cybernetics*, 9(10):628–643, 1979. 443, 446

The Envelope of a Digital Curve Based on Dominant Points*

David E. Singh[1], María J. Martín[2], and Francisco F. Rivera[1]

[1] Univ. Santiago de Compostela, Dept. of Electronics and Computer Science, Spain
david@dec.usc.es, fran@dec.usc.es
[2] Univ. A. Coruña, Dept. of Electronics and Systems, Spain
mariam@udc.es

Abstract. In this work, we present an optimal solution to the following problem: given a Freeman chain–code curve with n elements, and m points of it, find the minimum envelope of the curve by a set of line segments. This segments are obtained modifying the coordinates of these m points up to a distance h. The complexity of this algorithm is $O(nh + mh^2)$, and it needs a storage of $O(mh)$ data. In addition, we propose a greedy approximation algorithm that provides good results with lower complexity $O(nh)$ in the worst case, and memory requirements $O(h)$. A pre–processing with $O(mn)$ is also needed for both algorithms. Some experimental results are shown.

1 Introduction

Computing minimal enclosures for geometric curves is a fundamental geometric optimization problem. This problem arises in applications in which one wishes to find an approximation to a curve that is simpler in the sense of having a smaller number of elements. The main difference respect of the polygonization of digital curves is that we impose the enclosure restriction. The problem of computing minimum convex polygonal enclosures of a given number of sides for a convex polygon was efficiently solved by Aggarwal and Park [1]. They studied monotone matrix–searching techniques for this optimization problem.

Related work has been done on finding a minimal enclosing triangle of a convex polygon [2], a minimal enclosing k–gon [3], a minimal enclosing equiangular k–gon [4] and minimum polygon enclosures with specified angles [5].

Several heuristics are available for "approximating" a contour by a set of line segments [6],[7] according to various criteria. All these methods are computationally expensive (usually between O(n^2) and O(n^3)). Although most of these works try to produce good results, few attempts have been made so far to actually yield strictly optimal solutions under well–defined optimization criteria.

In this work we present an optimal solution to the following problem:

Given a sequence of n integer–coordinate points and assuming that the curve is represented using the Freeman chain–code, we describe a digital curve C as:

* This work was supported by the Xunta de Galicia grant PGIDT99PXI20602B

G. Borgefors, I. Nyström, and G. Sanniti di Baja (Eds.): DGCI 2000, LNCS 1953, pp. 453–463, 2000.
© Springer-Verlag Berlin Heidelberg 2000

$C = \{p_j = (x_j, y_j), j = 1, ...n\}$
Where p_{j+1} is a neighbour of p_j. The purpose of this work is finding another sequence of m points that we will call $E = \{d_j, j = 1, ...m\}$ where $m < n$. The digital curve related to E will be the one conformed by joining these points by means of segments. This curve is defined as a set of linear functions: $F = \{f_k, k = 1, ...m-1\}$ so that f_k is the segment that joins d_k and d_{k+1}. The sequence of points E has to fulfill the following two conditions:

1. Its associated closed digital curve F is envelope of C.
2. Defining $S(C)$ and $S(F)$ as the number of inner points to C and to F respectively. The goal is to get is the minimum value for $\mid S(C) - S(F) \mid$ that is referred to as error area.

As a starting point, we define E as a subset of m representative points of C. This subset should offer the maximum possible information on the shape of C, so that its digital associated curve be as similar as possible to C. It can be proved [8] that the most characteristic points in C will be those that show the biggest curvature, which are called dominant points. However, the digital curve associated to E may intersect C, not fulfilling the first demanded condition to be an envelope. Therefore, we propose an algorithm that carries out slight modifications on the position of the points in E so that their associate digital curve can verify both conditions.

In this work two different techniques are suggested to obtain the envelope. One of them estimates the best solution in terms of minimizing the error area; since that starting from a initial E set, the coordinates of their elements are modified until an associated curve is obtained with the smallest error area. The second is a greedy heuristic solution that although doesn't secure the best result, finds a close estimate with less computational and memory requirements.

The digital curves that we have used show the following restriction: the values of x_j are monotonous regarding j, that is to say, $0 \leq x_{j+1} - x_j \leq 1$. In case that the digital curve doesn't fulfill this restriction, it can be broken down in tracts that do fulfill it and each can be analyzed independently. In order to obtain the envelope, and to maintain this property of the digital curves, we have imposed a limitation, that is the only coordinate that could be modified in the position of the dominant points is its y coordinate.

For the calculation of the dominant points of C we have used the Kankanhalli algorithm [9]. This means an improvement on the Teh–Chin algorithm [10] in which computational work is strongly reduced. Also, it maintains two important characteristics of the original algorithm: the fact that the entry parameters are not required, and the determination of the dominant points based on the local curvature. We have modified the algorithm allowing the selection of the dominant points based on their importance. This will allow us to select a smaller number of points knowing that these they will be the most significant in the curve. Anyway, the algorithms could be applied to another set of points. An alternative proposed by [11] use naive lines to obtain this points.

Once obtained the first initial set E with an arbitrary number of dominant points of the curve, the following step, common to both proposals, is to obtain information on the characteristics of the C curve among serial elements of E. The following section describes how to obtain that information.

2 Obtaining Information of the Curve

This preprocessing operates on the *tract* of C between two consecutive dominant points $\{d_i, d_{i+1}\}\epsilon E$ that we will call d_l and d_r respectively. The analysis of each tract is independent of the rest of them. If we define as d_l^k the point $(x_l, y_l + k)$, the purpose of this step is to obtain, for each d_l^k, the points of tangency $t_l^k \epsilon C$. These are defined as the points belonging to the intersection of C with the straight line that goes through element d_l^k and that is tangent to C in the segment $\{d_l, d_r\}$. A representation of these points can be seen in Figure 1. For a given d_l^k, and its corresponding t_l^k, by means of an extrapolation of the tangent straight line, the position of the associated element $d_r^{k'}$ can be obtained. Notice that the minimum value of k is zero, and the maximum is the one in which the corresponding point $d_r^{k'}$ belongs to C, or equivalently $k' = 0$.

It is useful some structure like the one shown on Table 1 to store all the relevant information on the process. The values that are shown are those corresponding to the situation given in Figure 1. This structure has a column for each possible value of k.

In order to obtain this table, the entry data are the digital curve C evaluated in the segment $\{d_l, d_r\}$ and, inside that interval, the element with the maximum value of y coordinate: $p_{max} = (x_{max}, y_{max})$. This point can be obtained without additional complexity during the calculation of the dominant points.

Table 1. Data structure associated to a tract of C

Level0	Level1	Level2	Level3	Level4
d_l^0	d_l^1	d_l^2	d_l^3	d_l^4
t_l^0	t_l^1	t_l^2	t_l^3	t_l^4
d_r^4	d_r^3	d_r^2	d_r^1	d_r^0

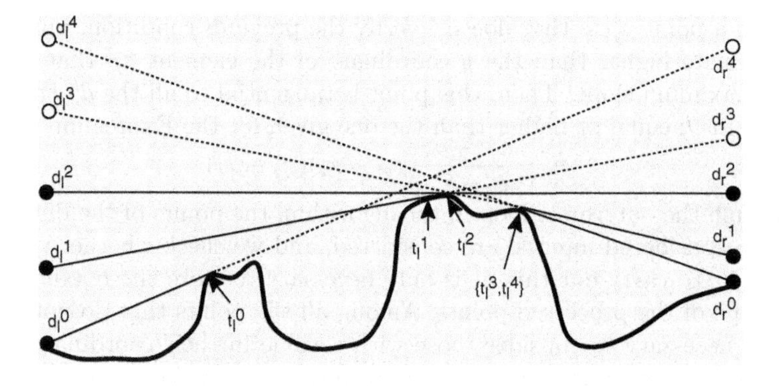

Fig. 1. Representative points of a tract of C

The process is divided in two equivalent problems: obtain the t_l^k in the interval $[x_l, x_{max}]$ for different d_l^k values, and $t_r^{k'}$ in the interval $[x_{max}, x_r]$ for different $d_r^{k'}$ values. The values of k and k' allowed will be those that verify respectively, the following conditions: $k\epsilon[y_l, y_{max}]$ and $k'\epsilon[y_r, y_{max}]$. We should point out that, although the problem has been broken into two halves, it is only necessary one access to the points of C, since these two halves are disjointed. If the curve has more than a maximum value, then the first and the last of them have to be considered, being irrelevant in the calculations the intermediate points of C.

An outline of the used algorithm is shown below. Since the resolution of both problems is equivalent, the shown algorithm corresponds to the resolution of the tract of C among d_l and p_{max} that we call C'. The other part of the tract is solved in the same way going over curve C from d_r to p_{max}.

```
Algorithm TANGENCY_POINTS
input
          C': Interval of C between d_l and p_max
output
          List of entries {d_l^k, t_l^k}
begin
          for each p_j ∈ C'
                a_j <== slope of the straight line that links d_l with p_j
                if a_j is maximal
                        t_l^k <== p_j ∀k
                else
                        h <== evaluate_increment(d_L, p_j)
                        t_l^k <== p_j ∀k ≥ h
                end if
          end for
end algorithm
```

This algorithm is based on two properties[12]:

1. Given a point $p_j\epsilon C$ and the slope a_j of the straight line that joins it with d_l. If the value of this slope is higher than the one obtained for the previous elements with a lower abscissa, then none of these previous elements is tangential. Therefore, the considered point is a tangential point for all d_l^k.
2. Given a point $p_j\epsilon C$ that doesn't verify the previous condition, but with a y coordinate higher than the y coordinate of the element $p_{j'}$ that has given the maximum slope. Then, this point is tangential to all the d_l^k that have a value for k equal or higher than the one given for the Expression 1.

$$h = \frac{a_{j'} - a_j}{x_j - x_{j'}}(x_{j'} - x_l)(x_j - x_l) \tag{1}$$

Through the outermost loop in the algorithm, the points of the digital curve inside the considered interval are considered, and we check whether they verify property 1. To carry out this, it is only necessary to store the maximum value of the slope of the processed points. Among all the points that do not verify it, it is only necessary to consider those with a value in their coordinate y higher than the one of the point which has given the maximum slope. For these points, equation 1 is used to obtain the value for k for which they become tangential.

There is a question that is not mentioned in the algorithm due to lack of space, that is checking that the rest of the entries t_l^k with $k > h$ are not associated to another tangential point. If it is noticed that any t_l^k is assigned to another element $p_{j'}$ different from the one of the maximum slope (another tangential point exists) and it is necessary to evaluate again expression 1 between p_j and the new element.

Carrying out a similar process on the interval $[x_{max}, x_r]$ one can obtain the position of the tangency points for the different d_l^k. Combining both results and extrapolating the values, all the d_l^k and $d_r^{k'}$ positions higher than p_{max} can be obtained as shown in Figure 1. In this figure, the dashed lines link the points that are extrapolated.

3 Obtaining the Optimal Envelope

The problem lies in figuring out the sequence of d_i^k points whose associate digital curve F is a minimum envelope. This method makes use of an exhaustive search among all the possible envelopes to the curve. For each d_i^k element belonging to the tract, the neighbour of the previous tract is searched so that when joined, the associate curve has a minimum error area. An example can be seen in Figure 2, in which two serial tracts of C are represented that span between points $\{d_1, d_2, d_3\}$. Point d_3^0 can be joined with d_2^2, so that the segment is tangent to C. However, it can also be joined to d_2^3 or the d_2^4 satisfying the condition of being a envelope, and although locally it is not the optimum, it can be part of the global optimum.

The number of combinations for joining serial dominant points would grow exponentially with the number of tracts unless the condition of minimizing the error area was demanded. Due to this, the search space in our algorithm decreases considerably and it is possible to enunciate the following property: for a certain dominant point d_i^k there is at least one neighbour $d_{i-1}^{k'}$ with which it can form a minimum envelope [12].

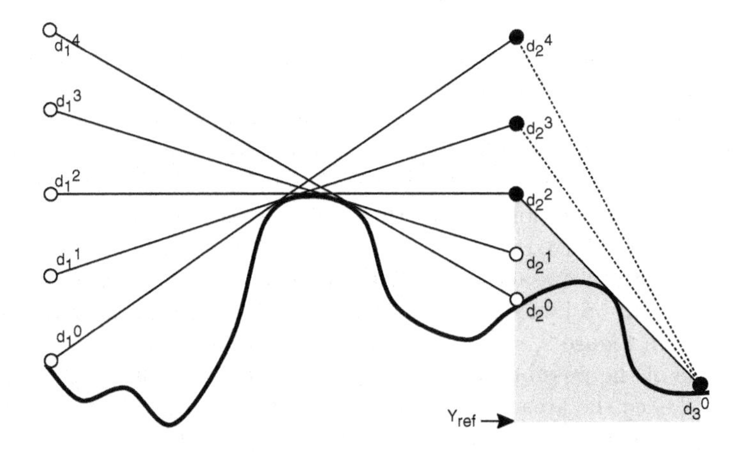

Fig. 2. Possible links between d_3^0 and d_2

Each tract is limited by two elements $\{d_i, d_{i+1}\}\epsilon E$ and has associated a table with all the possible tangency points. This table is obtained with the method described in the previous section. Each level contains the following fields: $\{d_i^k, t_i^k, d_{i+1}^{k'}, area, link\}$. The field *area* stores the error area accumulated in that tract, and *link* indicates the element of the structure immediately previous to the one the current level is connected to.

```
Algorithm OPTIMAL_ENVELOPE
input
        {C ,E}: Digital curve and dominant point set
output
        E: Enclosed polygon points
begin
        for i=1,m-1
             tangency[i] ⇐ tangency_points(d_i, d_{i+1}, p_max, C)
             for each level ϵ tangency[i]
                 local_area ⇐ evaluate_local_area(p_l^k, p_r^k')
                 S ⇐ add_candidates(tangency[i − 1])
                 if S ≠ ∅
                    level.link ⇐ select_minimun_area(S)
                 else
                    level.link ⇐ add_new_entry(tangency[i − 1])
                 end if
                    level.area ⇐ local_area+add_accum_area(level.link)
             end for
        end for
        take_minimun_envelope(tangency[i]) m − 1 ≥ i ≥ 0
end algorithm
```

The OE algorithm is shown above. It successively processes the different tracts of C. For each one, the algorithm obtains the table with the possible tangency points, storing them in the structure called *tangency*. Next, it goes over all the levels of the generated table. For each level of the table the function *evaluate_local_area*() is executed. This function obtains the area of the surface embraced by the tangent segment to C regarding a coordinate of reference y_{ref}. The value of this area is proportional to the error area, so that the entry that obtains the lowest value will represent the tangential tract that minimizes the area of error of the interval. In Figure 2, the tangent segment to C associated to point d_3^0 is the one that joins it to d_2^2. The local area appears shadowed in the figure for that situation. Notice that this calculation is independent of the element to which is linked.

For each d_i^k it is necessary to calculate the set $\mathcal{S}$ of levels of the previous table with which it can be joined. This is carried out by function *add_candidates*(). For example, in Figure 2, the group associated to d_3^0 would be $\mathcal{S} = \{d_2^2, d_2^3, d_2^4\}$. For each one of the possible candidates the accumulated area is evaluated. For the union $d_3^0 - d_2^2$ the area will be the one that this last point has accumulated plus the local one to d_3^0. However, for the union $d_3^0 - d_2^4$ the area of the triangle formed by points $d_3^0 - d_2^2 - d_2^4$ has to be added to the above-mentioned one.

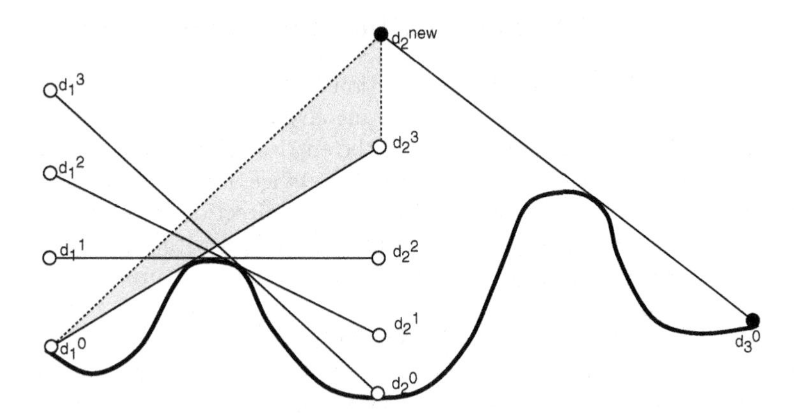

Fig. 3. New level in d_2 due to d_3^0

Of all the possibilities of union, the one that obtains the smallest accumulated area is selected since it minimizes the area of global error of that level. Applying this procedure iteratively, it is sure that, when the possible values for d_i^k are analyzed, the accumulated areas of all the d_j with $j < i$ will be the minimum.

A particular situation that is necessary to consider is when there is no element of the previous table that can be joined to the entry under consideration. This is the case, in Figure 3, of point d_3^0. In this situation, and in order to maintain the coherence of the information, by means of *add_new_entry*() a new entry is created in the previous table that contains the new value of d_{i-1}^k with which it can be joined. The rest of the fields of this new entry are the same that of the entry d_{i-1}^{k-1} immediately lower. In the figure, the entry d_2^{new} associated to the point d_3^0 would be created. The only exception is the accumulated error area, since it is necessary to bear in mind the increment in area in a similar way to how we did in the previous section. Again, in the figure it would be necessary to add to the accumulated area associated to point d_2^3, the area of triangle $d_1^0 - d_2^3 - d_2^{new}$. An important property is that the link of this new entry is the same that the one of the immediately lower entry. This is because the area difference accumulated among both entries (the triangle) is independent of the element to which is linked. As the union of the previous entry (d_2^3 in the figure) is the one with the smallest accumulated area of all the possibilities, that will also be the case for the new entry.

When all the intervals have been processed, only the entry of the last tract with the smallest accumulated area is taken. Next, we go back over the previous tracts and by means of the field *link*, we obtain the sequence of points E whose associated digital curve has the smallest error area.

4 An Heuristic Method of Obtaining the Envelope

The method described in the previous section makes it certain obtaining a minimum error area. However it has several inconveniences: the structures called *tangency* of each tract must be stored, all the entries of each one of them should be analyzed and it is necessary to go over them twice, one to obtain all the possible solutions and the other one, in the opposite direction, to extract the best one. In this section we show a greedy heuristic method with much less computational work, less demanding in terms of memory and that provides a solution close to the best one.

The HE algorithm is shown below. The different intervals of C are processed consecutively calculating for each of them the tangency points. The difference regarding the previous method is that now a local reduction is intended in the error area. For each interval, only the central dominant point d_i, and the two immediate neighbours d_{i-1} y d_{i+1}, are considered. Next, an increase in the coordinate y_i of d_i is conducted, this might imply a decrease of the y coordinate of their two neighbours as it is shown in Figure 4.

```
Algorithm HEURISTIC_ENVELOPE
input
        {C ,E}: Digital curve and dominant point set
output
        E: Enclosed polygon points
begin
        for i=1,m-1
            tangency[i] ⟸ tangency_points(d_i, d_{i+1}, p_max, C)
            Δerror_area ⟸ evaluate_error_area(y_i + 1)
            while (Δerror_area < 0)
                   y_i = y_i + 1
                   Δerror_area ⟸ evaluate_error_area(y_i + 1)
            end while
        end for
end algorithm
```

The increment experienced in the error area derives from the expression $\Delta error\_area = (A2 + A3) - (A0 + A1 + A4)$, where the first term takes into account the gained error area and the second the lost one. For each modification in d_i starting from the tables $tangency[i]$ and $tangency[i-1]$ the value of the new positions of d_{i-1} and d_{i+1} are obtained. Once known, we check whether a net reduction in the error area has taken place and we move on to evaluate the following increase in y_i. It can be easily shown that the previous relationship can be rendered in a function that is dependent only on the coordinates x of the points[12]. Expression (2) is verified if only if $\Delta error\_area < 0$.

$$(x_\beta - x_\alpha)(x_2 - x_\alpha)(x_\beta - x_2) < (x_\alpha - x_1)(x_\alpha - x_0)(x_\beta - x_2) - (x_3 - x_\beta)^2(x_2 - x_\alpha) \quad (2)$$

Where x_α and x_β correspond to the abscissas of the point of intersection of the segment with the initial position and the one that has the final position.

This expression allows us to determine, in a quick way, if there is a reduction in the error area or not. This process verifies the following property: when for some value y_i the previous inequality can no longer be verified, for the remaining higher than y_i values it will not be verified either [12].

Out of this property we can conclude that starting from the last d_i that verifies the inequality there will not be any other one that produces a smaller error area. This way the algorithm of local minimizing of the area should only evaluate the inequality for point y_i+1. If this is verified, then y_i is incremented. This process will be repeated until the inequality no longer is satisfied.

5 Results

The computational complexity of the OE algorithm is $O(nh+mh^2)$. The number of entries that are needed to store the information according to Table 1 is $O(mh)$. By contrast, in the case of the HE algorithm, the complexity is $O(nh)$, and the amount of memory needed to store the information is $O(h)$. Several *rugged* curves were used as benchmark to prove and compare both algorithms. We use a curve of 1666 points to illustrate the results, and similar behaviours were obtained on other curves. A section of this curve with different number of points and its envelope is shown in Figure 5, notice the irregularity of its shape. A pre–processing with $O(mn)$ is also needed for both algorithms.

Figure 6(a) shows the execution time consumed by both methods when they are run on a SUN Enterprise 250 computer. Notice that the reduction in runtime is important; in some cases is even less that half. Figure 6(b) shows the maximum amount of memory needed to store the *tangency* tables. The great improvement in this issue is clear. Finally, figures 6(c) and 6(d) show the comparison in terms of number of error pixels between both methods. The difference is important when m is low (40% with 3 points) but it fastly lows to 1% or less.

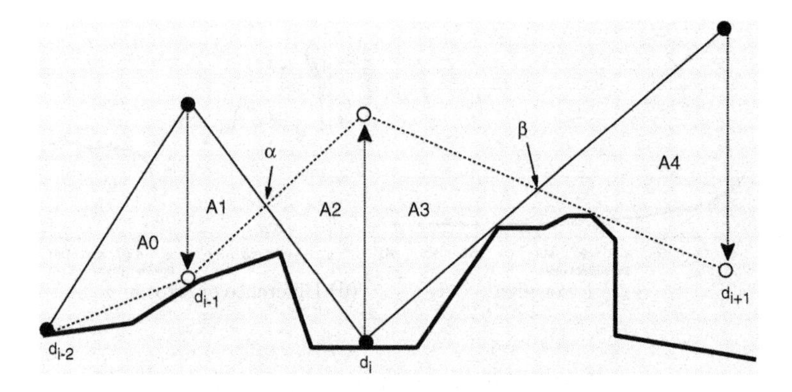

Fig. 4. Effect of increasing y_i

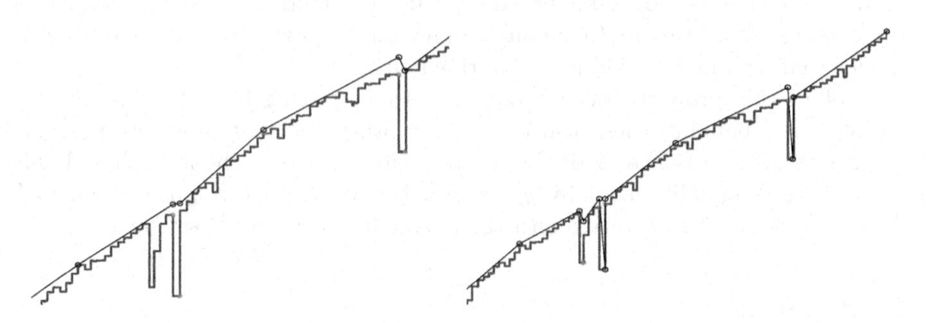

Fig. 5. Curve examples

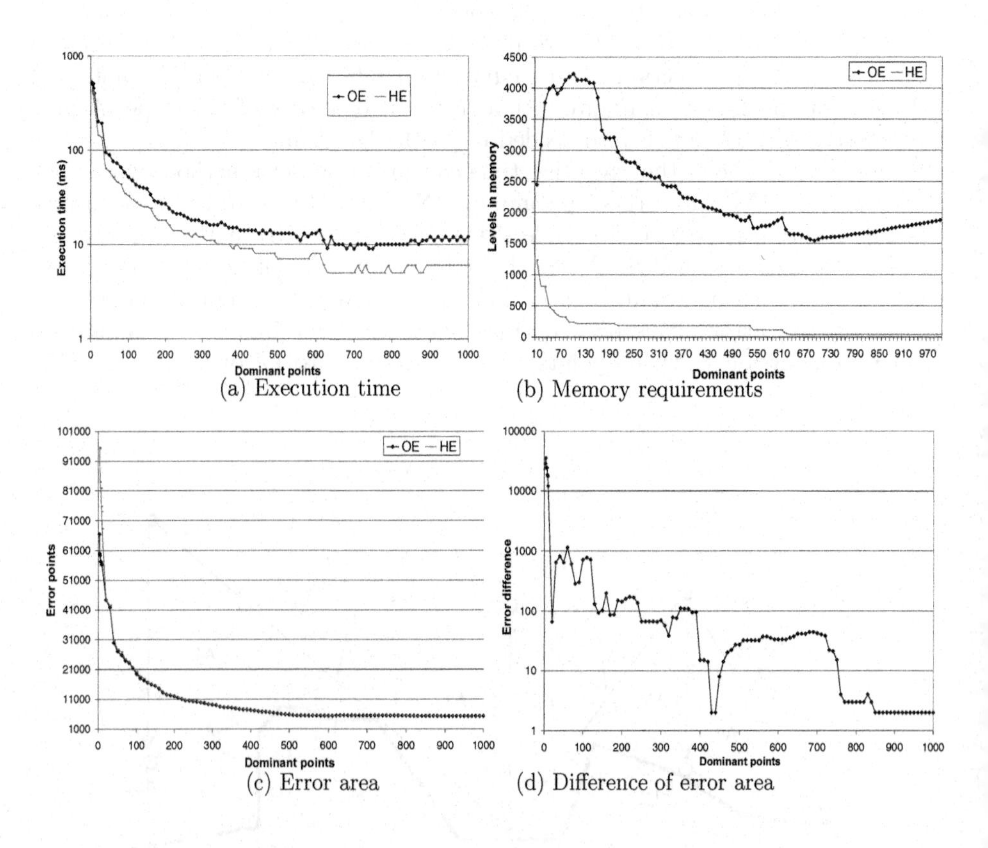

Fig. 6. Efficiency of both approaches

References

1. A. Aggarwal, J. Park: Notes on Searching in Multidimensional Monotone Arrays. Proc. 29th IEEE Symp. on Found. of Comp. Sci. (1988), 497–512 453
2. J. O'Rourke, A. Aggarwal, S. Maddial, M. Baldwin: An optimal algorithm for finding enclosing triangles. J. Algorithms, **7** (1986) 258–269 453
3. A. Aggarwal, J. Chang, C. Yap: Minimum Area Circumscribing Polygons. Visual Computer, **1** (1985), 112–117 453
4. A. DePano, A. Aggarwal: Finding restricted k–envelopes for convex polygons. Proc. 22nd Allerton Conf. on Comm. Control and Computing, (1984) 81–90 453
5. David M. Mount, Ruth Silverman: Minimum enclosures with specified angles. Technical Report, CS–TR–3219 (1994) 453
6. J. G. Dunham: Optimum Uniform Piecewise Linear Approximation of Planar Curves. IEEE Trans. Pattern Anal. and Machine Intelligence, V. 8-1, (1986) 67–75 453
7. Paul L. Rosin: Techniques for Assessing polygonal approximations of curves. IEEE Trans. on Pattern Anal. and Machine Intelligence, Vol 19, No 6 (1997) 659–666 453
8. F. Attneave: Some informational aspects of visual perception. Psychol. Review. vol 61, no 3, (1954) 183–193 454
9. Mohan S. Kankanhalli: An adaptive dominant point detection algorithm for digital curves. Pattern Recognition Letters. 14 (1993) 385–390 454
10. Cho–Huak Teh, Roland T. Chin: On the detection of dominant points. IEEE Trans. on Pattern Anal. and Machine Intelligence. Vol 11. No 8 (1989) 859–872 454
11. Isabelle Debled–Rennesson, Jean–Pierre Reveillès: A linear algorithm for segmentation of digital curves. Int. J. of Pattern Recog. and Artificial Intelligence. 9-6 (1995) 454
12. David E. Singh, María J. Martín, Francisco F. Rivera: Propiedades geométricas en la obtención de la envolvente a una curva digital. Tech. report. (1999) 456, 457, 460, 461

Shape Understanding

Minimum-Length Polygons in Simple Cube-Curves

Reinhard Klette[1] and Thomas Bülow[2]

[1] CITR, University of Auckland, Tamaki Campus
Building 731, Auckland, New Zealand
r.klette@auckland.ac.nz

[2] GRASP Laboratory, University of Pennsylvania
3401 Walnut Street, 330 C, Philadelphia, PA 19104-6228, USA
thomasbl@grasp.cis.upenn.edu

Abstract. Simple cube-curves in a 3D orthogonal grid are polyhedrally bounded sets which model digitized curves or arcs in three-dimensional euclidean space. The length of such a simple digital curve is defined to be the length of the minimum-length polygonal curve fully contained and complete in the tube of this digital curve. A critical edge is a grid edge contained in three consecutive cubes of a simple cube-curve. This paper shows that critical edges are the only possible locations of vertices of the minimum-length polygonal curve fully contained and complete in the tube of this digital curve.

1 Introduction

The estimation of the length of a simple digital curve in three-dimensional euclidean space may be based on the calculation of the shortest polygonal curve in a polyhedrally bounded compact set [11,12]. This paper presents an analysis of possible locations of vertices of such a polygonal curve. This analysis has been used in [2] for the design of an iterative algorithm approximating such curves with measured time complexity in $\mathcal{O}(n)$, where n denotes the number of grid cubes of the given digital curve.

1.1 Digital Curves in 3D Space

Any grid point $(i, j, k) \in \mathcal{R}^3$, i, j, k integers, is assumed to be the center point of a *grid cube* with *faces* parallel to the coordinate planes, with *edges* of length 1, and *vertices*. *Cells* are either cubes, faces, edges or vertices. The intersection of two cells is either empty or a joint *side* of both cells. We consider a non-empty finite set K of cells such that for any cell in K it holds that any side of this cell is also in K. Such a set K is a special finite *euclidean complex* [9]. Let $dim(a)$ denote the dimension of a cell a, which is 0 for vertices, 1 for edges, 2 for faces and 3 for cubes. Then $[K, \subset, dim]$ is also a *cell complex* [5,7,9,13] with properties such as **(i)** $\subset$ is transitive on K, **(ii)** dim is monotone on K with respect to $\subset$, and **(iii)** for any pair of cells $a, b \in K$ with $a \subset b$ and $dim(a) + 1 < dim(b)$ there

G. Borgefors, I. Nyström, and G. Sanniti di Baja (Eds.): DGCI 2000, LNCS 1953, pp. 467–478, 2000.
© Springer-Verlag Berlin Heidelberg 2000

exists a cell $c \in K$ with $a \subset c \subset b$. Cell b *bounds* cell a iff $a \subset b$, and b is a *proper side* of a in this case. Two cells a and b are *incident* iff cell a bounds b, or cell b bounds a.

We define *digital curves* g in 3D space with respect to such a euclidean complex as special sequences $(z_0, z_1, ..., z_m)$ of cells where z_i is incident with z_{i+1}, and $|dim(z_i) - dim(z_{i+1})| = 1$, for $i + 1$ $(mod\ m + 1)$. There are (at least) three different options which may depend upon an application context, or upon a preference of either a grid-point model or a cellular model which are dual approaches [4]. Let $n \geq 1$.

(i) An *edge-curve* is a sequence $g = (v_0, e_0, v_1, e_1, ..., v_n, e_n)$ of vertices v_i and edges e_i, for $0 \leq i \leq n$, such that vertices v_i and v_{i+1} are sides of edge e_i, for $0 \leq i \leq n$ and $v_{n+1} = v_0$. It is *simple* iff each edge of g has exactly two bounding vertices in g. It follows that a vertex or edge is contained at most once in a simple edge curve. [1]

(ii) A *face-curve* is a sequence $g = (e_0, f_0, e_1, f_1, ..., e_n, f_n)$ of edges e_i and faces f_i, for $0 \leq i \leq n$, such that edges e_i and e_{i+1} are sides of face f_i, for $0 \leq i \leq n$ and $e_{n+1} = e_0$. It is *simple* iff $n \geq 4$, and for any two faces f_i, f_k in g with $|i-k| \geq 2$ $(mod\ n+1)$ it holds that if $f_i \cap f_k \neq \emptyset$ then $|i-k| = 2$ $(mod\ n+1)$ and $f_i \cap f_k$ is a vertex.

(iii) A *cube-curve* is a sequence $g = (f_0, c_0, f_1, c_1, ..., f_n, c_n)$ of faces f_i and cubes c_i, for $0 \leq i \leq n$, such that faces f_i and f_{i+1} are sides of cube c_i, for $0 \leq i \leq n$ and $f_{n+1} = f_0$. It is *simple* iff $n \geq 4$, and for any two cubes c_i, c_k in g with $|i - k| \geq 2$ $(mod\ n + 1)$ it holds that if $c_i \cap c_k \neq \emptyset$ then either $|i - k| = 2$ $(mod\ n + 1)$ and $c_i \cap c_k$ is an edge, or $|i - k| = 3$ $(mod\ n + 1)$ and $c_i \cap c_k$ is a vertex. A *tube* **g** is the union of all cubes contained in a cube-curve g. It is a polyhedrally-bounded compact set in $\mathcal{R}^3$, and it is homeomorphic with a torus in case of a simple cube-curve.[2]

1.2 MLP in 3D Space

This paper deals exclusively with simple cube-curves. The cube-curve on the left of Fig. 1 is simple, and the cube-curve on the right is not. The latter example shows that the polyhedrally-bounded compact set **g** of a cube-curve g is not necessarily homeomorphic with a torus if each cube of this cube-curve g has exactly two bounding faces in g. A (Jordan) curve is *complete in* **g** iff it has a non-empty intersection with any cube contained in g.

Definition 1. *A* minimum-length polygon *(MLP) of a simple cube-curve g is a shortest polygonal simple curve $\mathcal{P}$ which is contained and complete in tube* **g**.

Following [11,12], the *length* of a simple cube-curve g is defined to be the length $l(\mathcal{P})$ of an MLP of g.

---

[1] This definition is consistent with, e.g., the definition of a *4-curve* in [10] (see proposition 2.3.3) for 2D grids where our edges are 'hidden' in a neighborhood definition, or of a *closed simple path* in [14] (see page 7) for undirected graphs.

[2] *Closed simple one-dimensional grid continua* [11,12] are defined such that each cube of g has exactly two bounding faces in g.

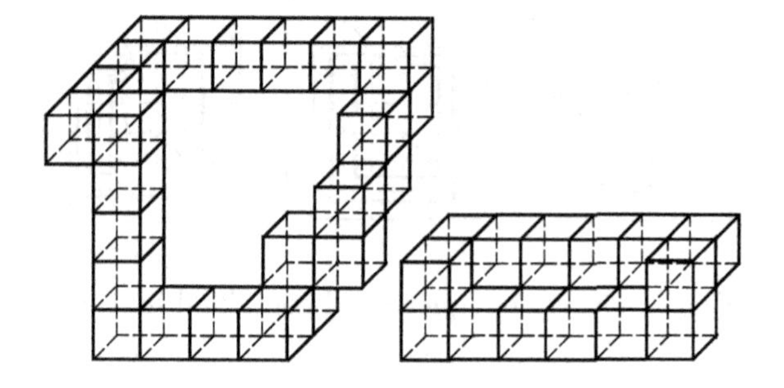

Fig. 1. Two cube-curves in 3D space

A simple cube-curve g is *flat* iff the center points (i, j, k) of all cubes contained in g are in one plane parallel to one of the coordinate planes. A non-flat simple cube-curve in $\mathcal{R}^3$ specifies exactly one minimum-length polygonal simple curve (MLP, minimum-length polygon) which is contained and complete in its tube [11]. The MLP is not uniquely specified in flat simple cube-curves. Flat simple cube-curves may be treated as square-curves in the plane, and square-curves in the plane are extensively studied, see, e.g. [6]. It seems there is no straightforward approach to extend known 2D algorithms to the 3D case. An important reason for that may be that 2D algorithms for (multigrid-convergent) perimeter estimation [6] may be such that all calculated vertices are grid points or vertices, but in the 3D case we are faced with a qualitatively new situation for the calculated vertices. The minimum-length polygon considered in this paper leads to vertices with real coordinates (not just multiples of integers as in the 2D case), i.e. the model of cell complexes is considered as being embedded into the euclidean space. However, independent upon the dimension global information has to be taken into account for length calculation of digital curves to ensure multigrid convergence.

2 Simple Cube-Curves

This section contains fundamental definitions and properties related to simple cube-curves.

2.1 Non-contractible Curves in g

Let g be a simple cube-curve, and $\mathcal{P} = (p_0, p_1, ..., p_m)$ be a polygonal curve complete and contained in $\mathbf{g}$, with $p_0 = p_m$.

Lemma 1. *It holds $m \geq 3$, for any polygon $\mathcal{P} = (p_0, p_1, ..., p_m)$ complete and contained in a simple cube-curve. Two line segments alone cannot be complete in any simple cube-curve.*

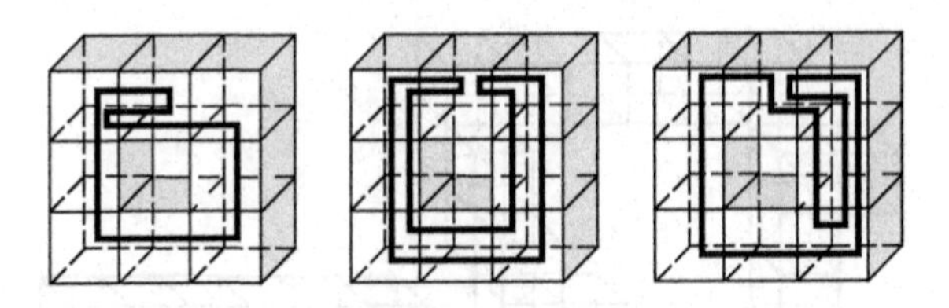

Fig. 2. Curves complete and contained in a tube

Proof. Cases $m = 0$ and $m = 1$ would be a point, and $m = 2$ would be a straight line segment. Both cases are excluded because our simple cube-curves are homeomorphic to the torus. The case $m = 3$ (a triangle) is possible, e.g. for the simple cube-curve shown in Fig. 2. However, in this minimum case of $m = 3$ it holds that no side of the triangle may be completely contained within one of the cubes. □

A curve contained in **g** is *contractible (into a point) in* **g** iff there are continuously repeated topological (i.e. continuous and bijective) transformations of this curve into a family of disjoint curves, all contained in **g**, which converge towards a single point [8]. The curves on the left and on the right in Fig. 2 are *non-contractible in* **g**, and the curve in the middle is contractible in **g** by continuous contractions into a single point.

A Jordan curve γ *passes through* a face f iff there are parameters t_1, t_2, T such that $\{\gamma(t) : t_1 \leq t \leq t_2\} \subseteq f$, and $\gamma(t_1 - \varepsilon) \notin f$, $\gamma(t_2 + \varepsilon) \notin f$, for all ε with $0 < \varepsilon \leq T$. During a traversal along curve γ we *enter* a cube c at point $\gamma(t_1) \in c$ if $\gamma(t_1 - \varepsilon) \notin c$, and we *leave* c at point $\gamma(t_2) \in c$ if $\gamma(t_2 + \varepsilon) \notin c$, for all ε with $0 < \varepsilon \leq T$. A traversal is defined by the starting vertex p_0 of the curve and the given orientation.

We consider polygonal curves $\mathcal{P}$. Let $\mathcal{C}_\mathcal{P} = (c_0, c_1, ..., c_n)$ be the sequence of cubes in the order how they are entered during curve traversal. If $\mathcal{P}$ is complete and contained in a tube **g** then it follows that $\mathcal{C}_\mathcal{P}$ contains all cubes of g, and there are no further cubes in $\mathcal{C}_\mathcal{P}$.

Lemma 2. *For an MLP $\mathcal{P}$ of a simple cube-curve g it holds that $\mathcal{C}_\mathcal{P}$ contains each cube of g just once.*

Proof. Assume that $\mathcal{P}$ enters the same cube c of g twice, say at point q_1 first and at point q_2 again. Both points may be on one face of c, see Fig. 2 on the left and on the right, or on different faces of c, see Fig. 2 middle.

First consider the case that both entry points q_1 and q_2 of c are on one face f of cubes c and c'. Assume the number of passes of $\mathcal{P}$ through f is odd. We insert points q_1 and q_2 into $\mathcal{P}$ as new vertices which split the resulting polygonal curve into two polygonal chains, $\mathcal{P}_1 = (q_2, ..., q_1)$ and $\mathcal{P}_2 = (q_1, ..., q_2)$ such that the union of both is $\mathcal{P}$. The length of $\mathcal{P}_i$ exceeds the length of the straight line segment q_1q_2, for $i = 1, 2$. W.l.o.G. let $\mathcal{P}_1$ be the chain which does not pass through f. It follows that $\mathcal{P}_1$ is complete in **g**. Because the cube c is convex it also contains the straight line segment q_1q_2. We replace the polygonal sequence $\mathcal{P}_2$ by q_1q_2, i.e. we replace $\mathcal{P}$ by $\mathcal{Q} = (q_1, q_2, ..., q_1)$. Curve $\mathcal{Q}$ is still complete and contained in **g**, but shorter than $\mathcal{P}$ which contradicts our assumption that $\mathcal{P}$ is an MLP of g.

Now assume that the number of passes of $\mathcal{P}$ through f is even, it enters c at q_1, then it passes f and enters c' at r_1, then it passes f again and enters c at q_2, then it passes f again and enters c' at r_2. There may be a further even number of passes of $\mathcal{P}$ through f before the curve returns to q_1. We insert points q_1, r_1, q_2, r_2 into $\mathcal{P}$ as new vertices which split the resulting polygonal curve into four polygonal chains, $\mathcal{P}_1 = (q_1, ..., r_1)$, $\mathcal{P}_2 = (r_1, ..., q_2)$, $\mathcal{P}_3 = (q_2, ..., r_2)$ and $\mathcal{P}_2 = (r_2, ..., q_1)$ such that the union of all four is $\mathcal{P}$. It follows that

$$\mathcal{C}_{\mathcal{P}_1} \subseteq \mathcal{C}_{\mathcal{P}_3} \vee \mathcal{C}_{\mathcal{P}_3} \subseteq \mathcal{C}_{\mathcal{P}_1} \ ,$$

and an analog conclusion for $\mathcal{P}_2$ and $\mathcal{P}_4$. W.l.o.g. let $\mathcal{C}_{\mathcal{P}_1} \subseteq \mathcal{C}_{\mathcal{P}_3}$. Then we replace in $\mathcal{P}$ the polygonal chain $\mathcal{P}_1$ by the straight line segment q_1r_1 which is in f. The length of $\mathcal{P}_1$ exceeds the length of the straight line segment q_1r_1. Thus the resulting polygonal curve is still complete and contained in **g**, but shorter than $\mathcal{P}$ which also contradicts our assumption that $\mathcal{P}$ is an MLP of g.

We consider the second case that both points q_1 and q_2 are on different faces of cube c, say q_1 on face f_1 and q_2 on face f_2. Because q_2 is a re-entry point to cube c there must be a point q_{ex} in f_2 where we leave c before entering c again at q_2. If there is another re-entry point on face f_2 then we are back to case one. It follows that $\mathcal{P}$ leaves c once and enters c once. Assume that f_2 is also a face of cube $c' \neq c$ of g. If $\mathcal{P}$ would not intersect the second face of c' contained in g then we may replace the polygonal subsequence $(q_{ex}, ..., q_2)$ (which is contained in c' but not in f_2) by the shorter straight line segment $q_{ex}q_2$ which is contained in f_2 and thus in c', i.e. the resulting polygonal curve would be shorter and still contained and complete in g. It follows that the curve $\mathcal{P}$ has to leave cube c' through its second face contained in g. Tracing g around means that we arrive at the cube $c'' \neq c$ which is also incident with face f_1, and we leave c'' (and enter c) at a point which may be equal to q_1, and we enter c'' again through f_1. Thus $\mathcal{P}$ contains two polygonal subsequences which are both contained and complete in g. This contradicts the shortest-length constraint. □

2.2 Iterative Modifications

Now we consider a special transformation of polygonal curves. Let $\mathcal{P} = (p_0, p_1, ..., p_m)$ be a polygonal curve contained in a tube **g**. A polygonal curve $\mathcal{Q}$ is a **g**-*transform* of $\mathcal{P}$ iff $\mathcal{Q}$ may be obtained from $\mathcal{P}$ by a finite number of steps, where each step is a replacement of a triple a, b, c of vertices by a polygonal sequence $a, b_1, ..., b_k, c$ such that the polygonal sequence $a, b_1, ..., b_k, c$ is contained in the same set of cubes of g as the polygonal sequence a, b, c. The case $k = 0$ characterizes the deletion of vertex b, the case $k = 1$ characterizes a move of vertex b within **g**, and cases $k \geq 2$ specify a replacement of two straight line segments by a sequence of $k + 1$ straight line segments, all contained in **g**.

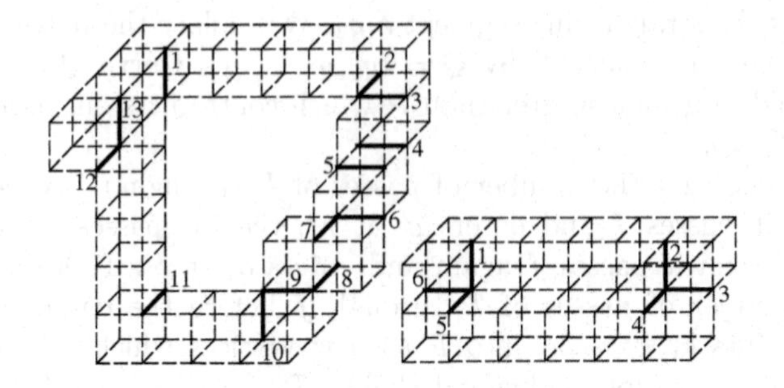

Fig. 3. Critical edges of two cube-curves

Lemma 3. *Let $\mathcal{P}$ be a polygonal curve complete and contained in the tube* $\mathbf{g}$ *of a simple cube-curve g such that $C_{\mathcal{P}}$ is without repetitions of cells. Then it holds that any* $\mathbf{g}$*-transform of $\mathcal{P}$ is also complete and contained in* $\mathbf{g}$.

Proof. By definition of the $\mathbf{g}$-transform it follows that this curve is also contained in $\mathbf{g}$. Because $C_{\mathcal{P}}$ is without repetitions of cells it holds that $\mathcal{P}$ traces $\mathbf{g}$ cell by cell, starting with one vertex in one cell and returning to the same vertex. From Lemma 1 we know that $\mathcal{P}$ has at least three vertices, i.e. at least three line segments, and that for the minimum case of $m = 3$ it holds that two line segments cannot be complete in g, i.e. there is at least one cube not intersected by these two line segments. Thus a replacement of two line segments (within the same set of cells of g) cannot transform $\mathcal{P}$ into a curve contractible in $\mathbf{g}$, i.e. the curve remains complete in $\mathbf{g}$. □

3 Critical Edges

An edge contained in a tube $\mathbf{g}$ is *critical* iff this edge is the intersection of three cubes contained in the cube-curve g. Figure 3 illustrates all critical edges of the cube-curves shown in Fig. 1. Note that simple cube-curves may only have edges contained in three cubes at most. For example, the cube-curve consisting of four cubes only (note: there is one edge contained in four cubes in this case) was excluded by the constraint $n \geq 4$.

Theorem 1. *Let g be a simple cube-curve. Critical edges are the only possible locations of vertices of a shortest polygonal simple curve contained and complete in tube* $\mathbf{g}$.

Proof. We consider arbitrary (flat or non-flat) simple cube-curves g, i.e. the MLP may not be uniquely defined.

Let $\mathcal{P} = (p_0, p_1, ..., p_m)$ be a shortest polygonal simple curve contained and complete in tube $\mathbf{g}$, with $p_0 = p_m$ and $m \geq 3$. We consider w.l.o.g. the polygonal subsequence (p_0, p_1, p_2) of such a shortest polygonal simple curve contained and complete in tube $\mathbf{g}$. We will show that p_1 is on a critical edge. According to

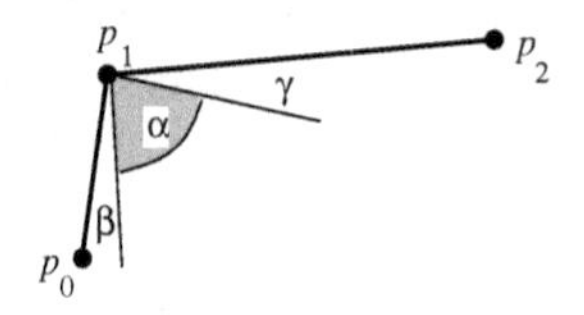

Fig. 4. Sketch of point p_1

Lemma 2 we know that $\mathcal{C}_\mathcal{P}$ is without repetitions, i.e. we may apply Lemma 3 for this curve $\mathcal{P}$ and tube $\mathbf{g}$.

We can exclude the case that p_1 is collinear with p_0 and p_2, because p_1 would be no vertex of a polygon in such a case. Three non-collinear points p_0, p_1, and p_2 define a triangular region $\triangle(p_0, p_1, p_2)$ in a plane $\mathcal{E}$ in $\mathcal{R}^3$. The following considerations are all for geometric configurations within this plane $\mathcal{E}$. In this proof, a *boundary point* is a point on the boundary $\partial\mathbf{g}$.

At first we ask whether p_1 may be moved into a new point p_{new} within the triangle $\triangle(p_0, p_1, p_2)$ towards line segment p_0p_2 such that a resulting polygonal subsequence $(p_0, ..., p_{new}, ..., p_2)$ remains to be contained in $\mathbf{g}$. This describes a $\mathbf{g}$-transform of $\mathcal{P}$, and the resulting curve would be complete and contained in $\mathbf{g}$. It can be of shorter length if the intersection of an ε-neighborhood of p_1 with $\triangle(p_0, p_1, p_2)$ is in $\mathbf{g}$, for $\varepsilon > 0$. It follows that such a move of p_1 is impossible, i.e. it follows that for any $\varepsilon > 0$ there is at least one boundary point q in an ε-neighborhood of p_1 and on one of the line segments p_0p_1 or p_1p_2, avoiding such a move of p_0 into the triangle $\triangle(p_0, p_1, p_2)$. It follows that p_1 itself is a boundary point.

The situation of an ε_0-neighborhood at point p_1 is illustrated in Fig. 4. Angle α represents the region not in $\mathbf{g}$. Angles β and γ are just inserted to mention that they may be zero, and their actual value is not important in the sequel. It holds $\alpha < \pi$ because it is bounded by an inner angle of the triangle $\triangle(p_0, p_1, p_2)$.

A boundary point may be a point within a face, or on an edge. Assume first that boundary point p_1 is within a face f. Plane $\mathcal{E}$ and face f either intersect in a straight line segment, or face f is contained in $\mathcal{E}$. The straight line situation would contradict that $\alpha < \pi$ in the ε_0-neighborhood at point p_1, and $f \subset \mathcal{E}$ would allow to move p_1 into a new point p_{new} within $\triangle(p_0, p_1, p_2)$ towards line segment p_0p_2 which contradicts our MLP assumption.

There are three different possibilities for an edge contained in $\mathbf{g}$: we call it an *uncritical edge* if it is only in one cube contained in g, it is an *ineffective edge* if it is in exactly two cubes contained in g, and it is a critical edge (as defined above) in case of three cubes. Point p_1 cannot be on an ineffective edge such that it is also not on a critical or uncritical edge, because this corresponds to the situation being within a face as discussed before. Point p_1 also cannot be on an uncritical edge such that it is also not on a critical edge. Figure 5 illustrates an intersection point q with an uncritical edge in plane $\mathcal{E}$ assuming that this edge is not coplanar with $\mathcal{E}$. The resulting angle $\alpha > \pi$ (region not in $\mathbf{g}$ in an

ε-neighborhood of q) does not allow that p_1 is such a point. If the uncritical edge is in $\mathcal{E}$ then angle α would be equal to π, what is excluded at p_1 as well. So there is only one option left. Point p_1 has to be on a critical edge (in fact, the angle α is less than π for such an edge). □

Note that this theorem also covers flat simple cube-curves with a straightforward corollary about the only possible locations of MLP vertices within a simple square-curve in the plane (see Fig. 6): such vertices may be convex vertices of the inner frontier or concave vertices of the outer frontier only because these are the only vertices incident with three squares of a simple square-curve.

4 Application of the Theorem

Our algorithm [2] is based on the following physical model: Assume a rubber band is laid through the tube **g**. Letting it move freely it will contract to the MLP which is contained and complete in **g** (assumed the band is slippery enough to slide across the critical edges of the tube). The algorithm consists of two subprocesses: at first an initialization process defining a simple polygonal curve $\mathcal{P}_0$ contained and complete in the given tube **g** and such that $\mathcal{C}_{\mathcal{P}_0}$ contains each cube of g just once (see Lemma 2), and second an iterative process (a **g**-transform, see Lemma 3) where each completed run transforms $\mathcal{P}_t$ into $\mathcal{P}_{t+1}$ with $l(\mathcal{P}_t) \geq l(\mathcal{P}_{t+1})$, for $t \geq 0$. Thus the obtained polygonal curve is also complete and contained in g. This algorithm uses the fact that critical edges are the only possible locations of vertices of the desired polygonal curve. This allowed us to achieve linear running time.

4.1 Initialization on Critical Edges

We sketch the initialization procedure to illustrate the importance of the proved theorem. The initial polygonal curve will only connect vertices which are end points of consecutive critical edges. For curve initialization, we scan the given curve until the first pair (e_0, e_1) of consecutive critical edges is found which are not parallel or, if parallel, not in the same grid layer (see Fig. 1 (right) for a non-simple cube-curve showing that searching for a pair of non-coplanar edges would be insufficient in this case). For such a pair (e_0, e_1) we start with vertices (p_0, p_1), p_0 bounds e_0 and p_1 bounds e_1, specifying a line segment p_0p_1 of

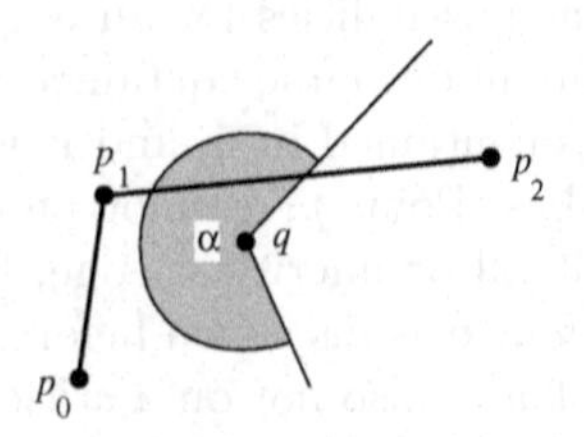

Fig. 5. Intersection with an uncritical edge

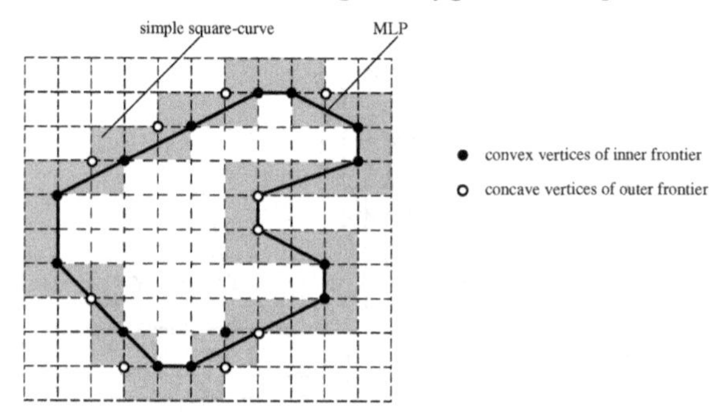

Fig. 6. Convex vertices of inner and concave vertices of outer frontier of the tube of a simple square curve in the plane

minimum length (note that such a pair (p_0, p_1) is not always uniquely defined). This is the first line segment of the desired initial polygonal curve $\mathcal{P}_0$.

Now assume that $p_{i-1}p_i$ is the last line segment on this curve $\mathcal{P}_0$ specified so far, and p_i is a vertex which bounds e_i. Then there is a uniquely specified vertex p_{i+1} on the following critical edge e_{i+1} such that p_ip_{i+1} is of minimum length. Length zero is possible with $p_{i+1} = p_i$; in this case we skip p_{i+1}, i.e. we do not increase the value of i. Note that this line segment p_ip_{i+1} will always be included in the given tube because the centers of all cubes between two consecutive critical edges are collinear. The process stops by connecting p_n on edge e_n with p_0 (note that it is possible that a minimum-distance criterion for this final step may actually prefer a line between p_n and the second vertex bounding e_0, i.e. not p_0). See Table 1 for a list of calculated vertices for the cube-curve on the left in Figs. 1 and 3. The first row lists all the critical edges shown in Fig. 3. The second row contains the vertices of the initial polygon shown in Fig. 7 (initialization = first run of the algorithm). For example, vertex b is on edge 2 and also on edge 3, so there is merely one column for (2/3) for these edges.

This initialization process calculates a polygonal curve $\mathcal{P}_0$ which is always contained and complete in the given tube. Note that traversals following opposite orientations or starting at different critical edges may lead to different initial polygons. For example, a 'counterclockwise' scan of the cube-curve shown in Fig. 1 (left), starting at edge 1, selects edges 11 and 10 to be the first pair of

Table 1. Calculated points on edges ('D' in stands for 'deletion', i.e. there is no polygon vertex on this edge anymore)

critical edge	1	2/3	4	5	6/7	8	9	10	11	12/13
1st run (initialization)	a	b	c	d	e	f	g	h	i	j
2nd run	a	b	D	D	e	D	D	h	i	j

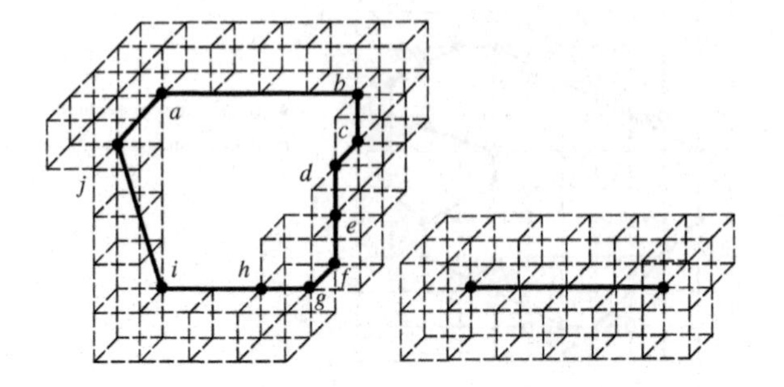

Fig. 7. Curve initializations ('clockwise')

consecutive critical edges, and the generated 'counterclockwise' polygon would differ from the one shown in Fig. 7. Figure 8 shows a curve where the defined initialization does not return to the starting vertex.

Initialization results are shown in Figs. 8 and 9. Note that in case of flat cube-curves the process will fail to determine the specified first pair of critical edges, and in this case a 2D algorithm may be used to calculate the MLP of a corresponding square-curve.

4.2 Non-grid-point Vertices of MLP's

This initialization procedure is followed by an iterative procedure [2] where we move pointers addressing three consecutive vertices of the (so far) calculated polygonal curve around the curve, until a completed run $t+1$ does only lead to an improvement which is below an a-priori threshold τ i.e. $l(\mathcal{P}_t) - \tau < l(\mathcal{P}_{t+1})$.

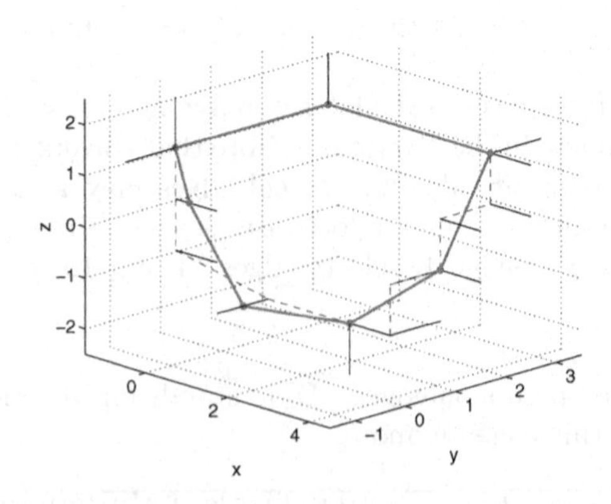

Fig. 8. Initial polygon (dashed) and MLP. Initialization starts below on the left, and the final step of the initialization process would prefer the second vertex of the first edge if a shortest-distance criterion would be used only

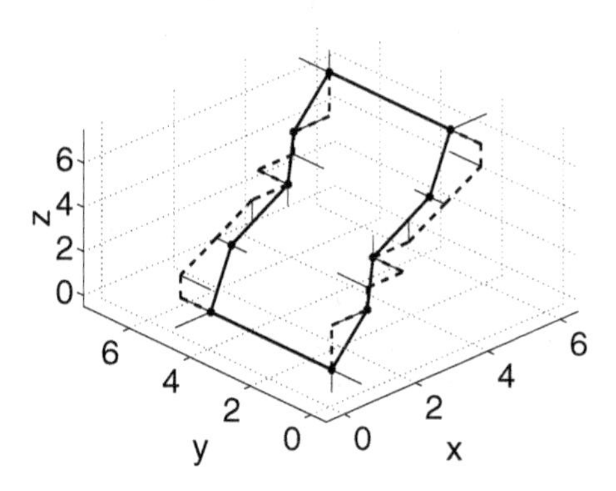

Fig. 9. Initial polygon (dashed) and MLP. Critical edges are shown as short line segments. The rest of the tube is not shown

Figures 8 and 9 show the initial polygon $\mathcal{P}_1$ dashed, and the solid line represents the final polygon. The short line segments are the critical edges of the given tube. For details of the algorithm see [2]. Note that MLP vertices may also be in "non-vertex" positions dividing a critical edge into two segments of non-zero (rational) length.

5 Conclusions

Length-estimators for digital curves in the plane are a well-studied subject. See, for example, [6], for references in this field. There are methods for length estimation in the digital plane which are provable convergent to the true length value for specific convex planar sets, assuming finer and finer grids in the plane. Such a correct convergence behavior is also supported by experimental evidence [6]. Methods showing multigrid convergence in the digital plane are, for example, the 'classical' digital straight line approximation technique (DSS method), the grid-continua based minimum-length polygon method (GC-MLP method), see [11], and a minimum-length polygon approximation method based on so-called *approximation sausages* (AS-MLP method), see [1] for this AS-MLP method.

Length-estimators for digital curves in the 3D space may be designed by following 2D design principles. The discussed method in this paper expands the GC-MLP method, see [12]. Due to this fact it follows that the discussed method satisfies multigrid convergence for the special case of flat curves as the 2D method does for planar curves. This might be sufficient theoretical evidence for the convergence behavior of the discussed curve length estimation method.

The paper has specified an important geometric property of such minimum-length polygons in 3D space which has been used in [2] for designing an efficient algorithm for calculating such a polygon. The given theorem has been used in [2] but without proof and accompanying definitions. In this sense the paper has provided fundamentals for the algorithm discussed in [2]. The given theorem might be also of interest in the context of 3D curves in general.

References

1. T. Asano, Y. Kawamura, R. Klette, K. Obokata: A new approximation scheme for digital objects and curve length estimations. CITR-TR 65, CITR Tamaki, The University of Auckland, August 2000. 477
2. Th. Bülow and R. Klette. Rubber band algorithm for estimating the length of digitized space-curves. Accepted paper for ICPR'2000, Barcelona, September 2000. 467, 474, 476, 477, 478
3. R. Busemann and W. Feller. Krümmungseigenschaften konvexer Flächen. *Acta Mathematica*, 66:27–45, 1935.
4. R. Klette. M-dimensional cellular spaces. Tech. Report TR-1256, Univ. of Maryland, Computer Science Dep., March 1983. 468
5. R. Klette. Cellular complexes through time. *SPIE Conference Proceedings Vision Geometry IX*, 4117:to appear, 2000. 467
6. R. Klette, V. Kovalevsky, and B. Yip. On the length estimation of digital curves. *SPIE Conference Proceedings Vision Geometry VIII*, 3811:117–129, 1999. 469, 477
7. V. Kovalevsky. Finite topology and image analysis. *Advances in Electronics and Electron. Physics*, 84:197–259, 1992. 467
8. J. B. Listing: Der Census räumlicher Complexe oder Verallgemeinerungen des Euler'schen Satzes von den Polyëdern. Abhandlungen der Mathematischen Classe der Königlichen Gesellschaft der Wissenschaften zu Göttingen **10** (1861 and 1862) 97–182. 470
9. W. Rinow. *Topologie.* Deutscher Verlag der Wissenschaften, Berlin, 1975. 467
10. A. Rosenfeld. *Picture Languages.* Academic Press, New York, 1979. 468
11. F. Sloboda and Ľ. Bačík. *On one-dimensional grid continua in R^3*. Report of the Institute of Control Theory and Robotics, Bratislava, 1996. 467, 468, 469, 477
12. F. Sloboda, B. Zaťko, and R. Klette. On the topology of grid continua. *SPIE Conference Proceedings Vision Geometry VII*, 3454:52–63, 1998. 467, 468, 477
13. A. W. Tucker. An abstract approach to manifolds. *Annals of Math.*, 34:191–243, 1933. 467
14. K. Voss. *Discrete Images, Objects, and Functions in Z^n*. Springer, Berlin, 1993. 468

Planar Object Detection under Scaled Orthographic Projection

Julián Ramos Cózar, Nicolás Guil Mata, and Emilio López Zapata

Dept. of Computer Architecture, University of Málaga,
{julian,nico,ezapata}@ac.uma.es

Abstract. In this work a new method to detect objects under scaled orthographic projections is shown. It also calculates the parameters of the transformations the object has suffered. The method is based on the use of the Generalized Hough Transform (GHT) that compares a template with a projected image. The computational requirements of the algorithm are reduced by restricting the transformation to the template edge points and using invariant information during the comparison process. This information is obtained from a precomputed table of the template that is directly transformed and compared with the image table. Moreover, a multiresolution design of the algorithm speeds-up the parameters calculation.

1 Introduction

Planar object recognition and three-dimensional (3D) pose estimation are the most important tasks in computer vision. Some applications related to these methods are object manipulation, autonomous vehicle driving, etc.

Perspective projection is one of the most suitable models for real camera image formation. However, it introduces non-linear relations. Simpler camera models can be used [1] under more relaxed situations which produce a negligible error. When the size of the object is relatively small (in comparison to its depth) and placed near the camera optical axis, the orthoperspective projection (or, in general, the subgroup of affine transformations [2]) is a good approximation to the perspective projection [3]. Here, the scaled orthographic projection will be studied. This projection only incorporates the distance and the foreshortening effects, but it eliminates the non-linear dependencies.

There are several methods to find the transformations between a planar template and an image where this template is included and viewed from an arbitrary position [4]. Selection of the most suitable method will be based on considerations such as the kind of shapes we are dealing with and the ability to cope with practical problems such as occlusion, noise, etc.

A simple approach to object detection is to find, for every possible orientation and position, the template transformation that produces a better matching with the image shape. However, the search space can become overwhelmingly large. An efficient evaluate and subdivide search algorithm is carried out in [5] in order

G. Borgefors, I. Nyström, and G. Sanniti di Baja (Eds.): DGCI 2000, LNCS 1953, pp. 479–490, 2000.
© Springer-Verlag Berlin Heidelberg 2000

to find the affine transformation that brings the larger number of model features close to image features. A variant of the Hausdorff distance is used as a similarity measure between the two point sets.

Search space can be reduced if invariant features are used. Geometric invariants are shape descriptors that remain unchanged under geometric transformations such as changing the viewpoint. Algebraic invariants are obtained using the classical results derived from perspective geometry of algebraic curves. The fundamental invariant in projective geometry is the cross-ratio (CR). Guo Lei [6] uses the CR to recognize 3D views of different polygons. B.S. Song et al. present in [7] a target recognition method based in CR that selects stable points for complex scenes. More general algebraic invariants can be derived from configurations of conics [8], points [9] and lines [10]. Besides the algebraic invariants, Rothwell et al. use a canonical frame, invariant to perspective transformations of the plane, to implement index functions that select models from a model database, as part of a recognition system [11].

Differential invariants can also be applied. In [12], affine invariants requiring one point correspondence and second order derivative, or requiring two points correspondences and first order derivative, are used to determine whether one curve may be a perspective projection of another one.

Other methods based on Fourier Descriptors [13] are aimed to exploit viewpoint invariants, but they are not fully invariant in perspective transformations. Furthermore, they are not robust when occlusion appears.

Several methods have been proposed to detect the 3D pose of a planar object based on the Hough transform (hashing methods). Using the HT we can take advantage of its useful properties, like relative insensivity to noise, and robustness to occlusions. However, it needs high computational and storage requirements.

In this paper we undertake a new approach to planar object detection based on the Generalized Hough Transform (GHT) that compares template and image information in order to calculate the transformations between the template and the corresponding image shape. Computational and storage requirements are greatly reduced by using invariant template and image information derived from the gradients and positions associated with the edge points. Invariant information is stored in template and image tables which are compared during the detection process. Template (or model) table information will be directly modified during algorithm application without needing further computationally expensive shape processing.

The rest of the paper is organized as follows. Next section introduces the mathematical expressions involved in an orthoperspective transformation. Section 3 presents the new method for planar shape detection and the expressions that allows us to speed-up the generation and comparison of the tables. In section 4, several real experiments have been carried out in order to test the algorithm's behaviour. Finally, in section 5 several related works are analyzed more deeply.

2 Scaled Orthographic Projection

The necessary transformations to project a planar object into the image plane are presented here, where f is the focal distance of the camera lens, d is the distance between the focal point and the intersection of the object plane with the $z - axis$ (the optical axis), and $\boldsymbol{n}$ being the normal vector to the object plane. The projection of this vector onto $x - y$ and $x - z$ planes allows us to determine the pan, τ, and the tilt, δ, angles, respectively. Thus, τ is the angle between the projection of $\boldsymbol{n}$ onto the $x - y$ plane and the $x - axis$. On the other hand, δ is the angle between the projection of $\boldsymbol{n}$ onto the $x - z$ plane and the $z - axis$.

The relationship between an object point (x_i, y_i, z_i) and its corresponding image point (u_i, v_i, f) can be expressed using the orthoperspective transformation:

$$u_i = f \cdot \frac{x_i \cdot cos\delta \cdot cos\tau - y_i \cdot sin\tau}{d} \tag{1}$$
$$v_i = f \cdot \frac{x_i \cdot cos\delta \cdot sin\tau + y_i \cdot cos\tau}{d}$$

3 Planar Object Detection

The detection process must find the occurrence of a template in an image. Six different groups of parameters indicate the transformation to be applied to the template to generate the object in the image. 1) *Scaling* of the template in the object plane represented by matrix S. 2) *Displacement* along the object plane, d_x and d_y, represented by matrix D. 3) *Rotation* in the object plane, β, represented by matrix R_β. 4) *Tilt* angle around object plane $x - axis$, δ, represented by matrix R_δ. 5) *Pan* angle around object plane $z - axis$, τ, represented by matrix R_τ. 6) *Scaling* of the scaled orthographic projection, represented by matrix S_o.

After application of the previous transformations to the template, the new coordinates of the template in the image plane are given by:

$$(u_i, v_i)^t = S_o \cdot R_\tau \cdot R_\delta \cdot R_\beta \cdot D \cdot S \cdot (x_i, y_i)^t \tag{2}$$

Note that the only transformation that introduces a distortion in the image object is the application of the tilt angle. The rest of transformations only change either the size (scaling), the orientation or the position.

3.1 Invariant Transformation

Recent work with the GHT [14] shows a new method to detect bidimensional shapes that uses invariant information to displacement and scaling.

The edge points of the image are characterized by the parameters $< x, y, \theta >$ where x and y are the coordinates of the points in a two-dimensional space and

θ is the angle of the gradient vector associated with this edge point. An angle, ξ, called *difference angle* is also defined. Its value indicates the positive difference between the angles of the edge point gradient vectors that will be paired.

From this description we can derive a transformation from the original image that generates new invariant information for the displacement and the scaling, based on paired points. Thus, let p_i and p_j be two edge points, $< x_i, y_i, \theta_i >$, $< x_j, y_j, \theta_j >$ their associated information, and ξ the difference angle to generate the pairing. Then, the transformation $\mathcal{T}$ can be expressed as follows:

$$\mathcal{T}(p_i, p_j) = \begin{cases} (\theta_i, \alpha_{ij}) & : \quad \theta_j - \theta_i = \xi \\ \emptyset & : \quad elsewhere \end{cases} \tag{3}$$

where

$$\alpha_{ij} = \left(\arctan \frac{y_i - y_j}{x_i - x_j} \ \angle \ \theta_i \right) \tag{4}$$

that is, α_{ij} is the positive angle formed by the line that joins p_i and p_j and the gradient vector angle of the point p_i.

The information generated by the application of the previous transformation is stored in a table in order to improve the detection process speed. The multivalued characteristics of the previous transformation will be apparent during table building. Next, we show the contents of this table:

Orientation table ($\mathcal{OT}$). The information generated by the $\mathcal{T}$ transformation is contained by this bidimensional table. The α_{ij} and θ_j values are stored in rows and columns, respectively. When a pairing with α_{ij} and θ_j value is calculated, the $OT[\alpha_{ij}][\theta_j]$ position is incremented. Because different pairings might coincide with the same α_i and θ_{ij} values, the content of $OT[\alpha_{ij}][\theta_j]$ will indicate how many of them have these values. The information stored in this table is invariant to scale and displacement. Note that a rotation τ of an image in a plane causes a rotation of τ columns of its $\mathcal{OT}$.

Then, we can use the $\mathcal{OT}$ table to compare invariant information of the template and the image under scaled orthographic projection. Before starting the generation of the $\mathcal{OT}$ tables, the template must be transformed as follows:

$$(u_i, v_i)^t = R_\delta \cdot R_\beta \cdot (x_i, y_i)^t \tag{5}$$

In this way, the projected template will be similar to the image shape except for a different scaling, displacement and orientation. Then the template and image $\mathcal{OT}$s will be different by a rotation of τ columns. This displacement can be calculated applying a matching process for both tables [14].

In general situation, where the values for the (δ,β) angles are unknown, a $\mathcal{OT}$ table for the template needs to be generated for each (δ, β) value. Each template table is compared with the $\mathcal{OT}$ table of the image and a value for τ is calculated, see Fig. 3.1. The maximum voting will indicate the correct solution.

The table generation and comparison for each δ and β values may require a high computational complexity. However, several improvements have been carried out to obtain a good performance.

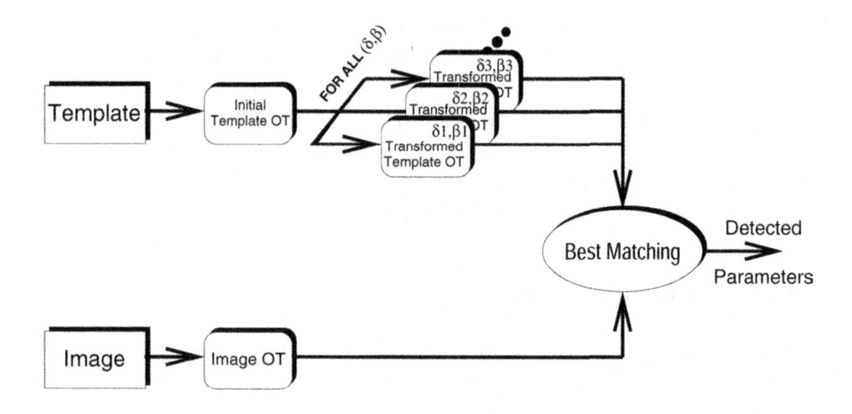

Fig. 1. Block-Diagram of the system

3.2 Table Generation and Comparison

The new $\mathcal{OT}$ table generation is accomplished by only using the edge points of the template shape. Normally, the number of edge points in the template is lower than in the image, where several different shapes can appear. This makes the computational complexity of our method to be lower than other algorithms using the image shape [15].

On the other hand, the $\mathcal{OT}$ table construction is based on both the calculation of the gradient vectors of the edge points (θ) and the value for the difference angles (α). This calculation can take a high computational time. In order to reduce this time, we have studied the modifications in an angle value that arise under a scaled orthographic projection. This allows to generate a $\mathcal{OT}$ table for a concrete δ and β value by directly transforming the original $\mathcal{OT}$ template table.

Modification of the Gradient and Difference Angles

Lemma 1. *Let θ_i be the gradient angle of the template edge point (x_i, y_i). Then, the transformed gradient vector, θ_{Ti}, after application of expression (5) is :*

$$\theta_{Ti} = atan(\tan(\beta + \theta_i)/cos\delta) \tag{6}$$

An important consequence of the previous lemma is that if (x_i, y_i) and (x_j, y_j) are two paired points in the template using a pairing angle of ξ, the transformed points after applying expression (5) are also paired.

Lemma 2. *Let α_i be the difference angle value between two paired points using a pairing angle of ξ. Then, the transformed difference angle, α_{Ti}, after application of expression (5) is:*

$$\alpha_{Ti} = atan(\tan(\beta + \theta_i + \alpha_i)/\cos\delta) - \theta_{Ti} \tag{7}$$

The demonstrations of the previous lemmas are straightforward, so we have omitted them due to space limitations. The two previous lemmas allow the creation of the new $\mathcal{OT}$ tables for each δ and β angles without needing the calculation neither the projected edge points nor the new difference angle for the paired points. The $\mathcal{OT}$ tables are sparse, so only the positions with values different from zero are stored. This reduces the required storage and allows the implementation of a more efficient transformation and comparison process.

3.3 Pose and Position Estimation

A matching algorithm has to compare, by columns, the transformed template $\mathcal{OT}$ table – for each possible tilt, pan value – and the $\mathcal{OT}$ of the image for different shifting values in order to calculate the pose (tilt, pan and rotation angles). The scale and the displacement values are obtained, after eliminating the previous transformations in the contour of the tested image, in the same way as in [14].

4 Experimental Results

As we showed in the previous sections, the detection process is based on the use of the gradient and difference angles associated with the edge points. Then, the accuracy of our method will be limited by the accuracy of the angle detection. Good initial values for the angles and edge points are obtained by applying a Canny operator.

We have used different images in order to test the behaviour of the whole detection process. All the examples have been executed in a SGI Workstation with a R-10000 processor at 225 MHz. The range of the tilt, rotation and pan angles for the experiments are (0°, 60°), (-30°, 30°), and (0° , 360°), respectively. Images are 8 bit greyscale with half PAL size (384x288 pixels). Model images are acquired in a fronto-parallel view. There is no need for any camera calibration procedure. The pairing angles have been chosen taking into account the template's shape [14], in a way that a significant subset of the contour points of the template are paired. For this purpose the gradient's histogram of the shape must be studied.

In addition, the algorithm uses a multipass approach to accelerate the program execution. Starting with a coarse estimation of the pose angles, successive steps focus in a narrower interval around the solution estimated in the previous step. Finally, the scale and displacement values are calculated.

The accuracy of the detection process is checked by using the likelihood function stated in [16]. First, this function measures the similarity, attending to edge points localization and gradient angle values, between transformed template and image contours. A visual checking is also carried out by superimposing the projected template shape on the image one.

First, we have applied the detection to B&W model images that have been projected using a warping process (Fig. 4). In this manner, we can know the

Table 1. Parameters of the warping deformation for the test images

Template	T_{POINTS}	I_{POINTS}	(δ, β, τ)	Scale	Likelihood (%)
Moon	801	602	(45,10,-20)	0.90	87.37
Indalo	1231	1312	(39,13,29)	1.15	83.58
Clover	568	599	(50,19,-29)	1.20	82.39
Key	622	531	(34,-28,13)	0.92	92.43

exact tilt, rotation, pan and scale values which have been used and check the accuracy of the detection. This allows us to study the theoretical behaviour of our programs. The template properties and experiment configurations are showed in Table 1. The first and second columns indicate the number of points detected by the Canny operator in both the template and transformed template. The following two columns show the parameters of the projection of the template by a warping process. Finally, in the last column, the value of the maximum theoretical likelihood is expressed. This value is calculated comparing the projected template with the warped image for the optimal pairing angles using the exact transformation parameters in pose and scale (the displacement has to be estimated). A 100% likelihood is not reached due to the discrete nature of the image transformation.

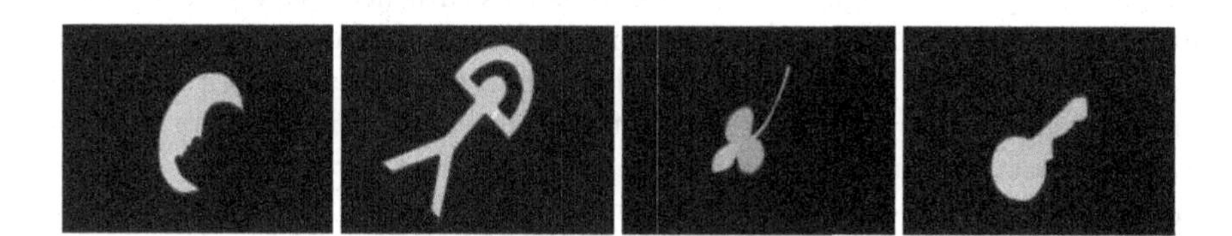

Fig. 2. Warped images

Table 2 summarizes the parameters used in program execution and the results obtained for the detection proposes. (ξ_1, ξ_2) are the used pairing angles. When two pairing angles are used, the achieved accuracy is better in general. The (δ, β, τ), scale and the likelihood for the detected parameters are depicted in the subsequent columns. Finally, the computational time for the multipass and sequential algorithm is shown in the last two columns of Table 2. An important speed-up is achieved for the multipass approach.

Finally, several real images, obtained using a CCD camera, are presented in Fig. 3 in order to show the behaviour of the proposed algorithms in practical situations. The images in the left column are the templates used and the images in the right column shows a visual checking of the solution, by superimposing the projected template on the image.

Table 2. Results obtained for warped images

Image (points)	(ξ_1, ξ_2)	(δ, β, τ)	Scale	Like. (%)	Time$_{Multi}$($s.$)	Time$_{Seq}$ (s.)
$Moon_1$	(180, -)	(45,10,-20)	0.90	87.37	0.20	3.73
$Moon_2$	(180, 90)	(45,10,-20)	0.90	87.82	0.48	9.10
$Indalo_1$	(180, -)	(39,10,-32)	1.15	71.80	0.51	9.72
$Indalo_2$	(175, 125)	(39,13,-29)	1.15	83.58	0.96	18.91
$Clover_1$	(180, -)	(50,18,-30)	1.20	78.68	0.14	2.74
$Clover_2$	(45, 90)	(50,25,-27)	1.12	63.74	0.37	7.13
Key_1	(180, -)	(34,-30,11)	0.92	94.63	0.12	2.36
Key_2	(180, 150)	(33,-25,17)	0.92	88.37	0.30	5.63

Table 3. Practical situation

Image (points)	(ξ_1, ξ_2)	(δ, β, τ)	Like. (%)	Time (s.)
Pliers (1803)	(180, 135)	(20,-19,-40)	73.13	1.41
Jewish Harp (1483)	(180, 45)	(43,-24,8)	81.27	0.91
Cutter (1650)	(180, 135)	(20,-6,9)	73.46	0.47
Scissors (2561)	(45, -)	(12,35,4)	80.47	0.67

Table 3 summarizes the more important configuration values. Now, for real images, two pairing angles, except for *Scissors*, are necessary if a reasonable precision is looked for. The images were acquired from an unknown arbitrary point of view, so the detected pose in the second column is only for informative purposes. The values for the likelihood function reported by the algorithm show that good accuracy have been achieved, in comparison with the values reported for the warped images. Finally, the computational times are presented for the multipass algorithm.

5 Related Works

In this section, a brief review of another works that use the HT to estimate the 3D orientation and position of a planar patch is presented. Others papers, using the HT for detecting 2D transformations are not covered here [17,18].

The method described in [19] claims to be the first one in using the HT for planar object 3D pose estimation. It consists in mapping the classical 2D Hough space into another Hough space of the same dimensions using only geometrical considerations. The new space is related to the former through a 3D transformation that takes into account the orientation and location of an object. In this way, data and model are transformed in a common representation space, where a direct matching can be carried out by determining the pose of the actual object with respect to the known pose of the model. The method is based on the classical HT to detect straight lines, and hence it deals with rigid planar

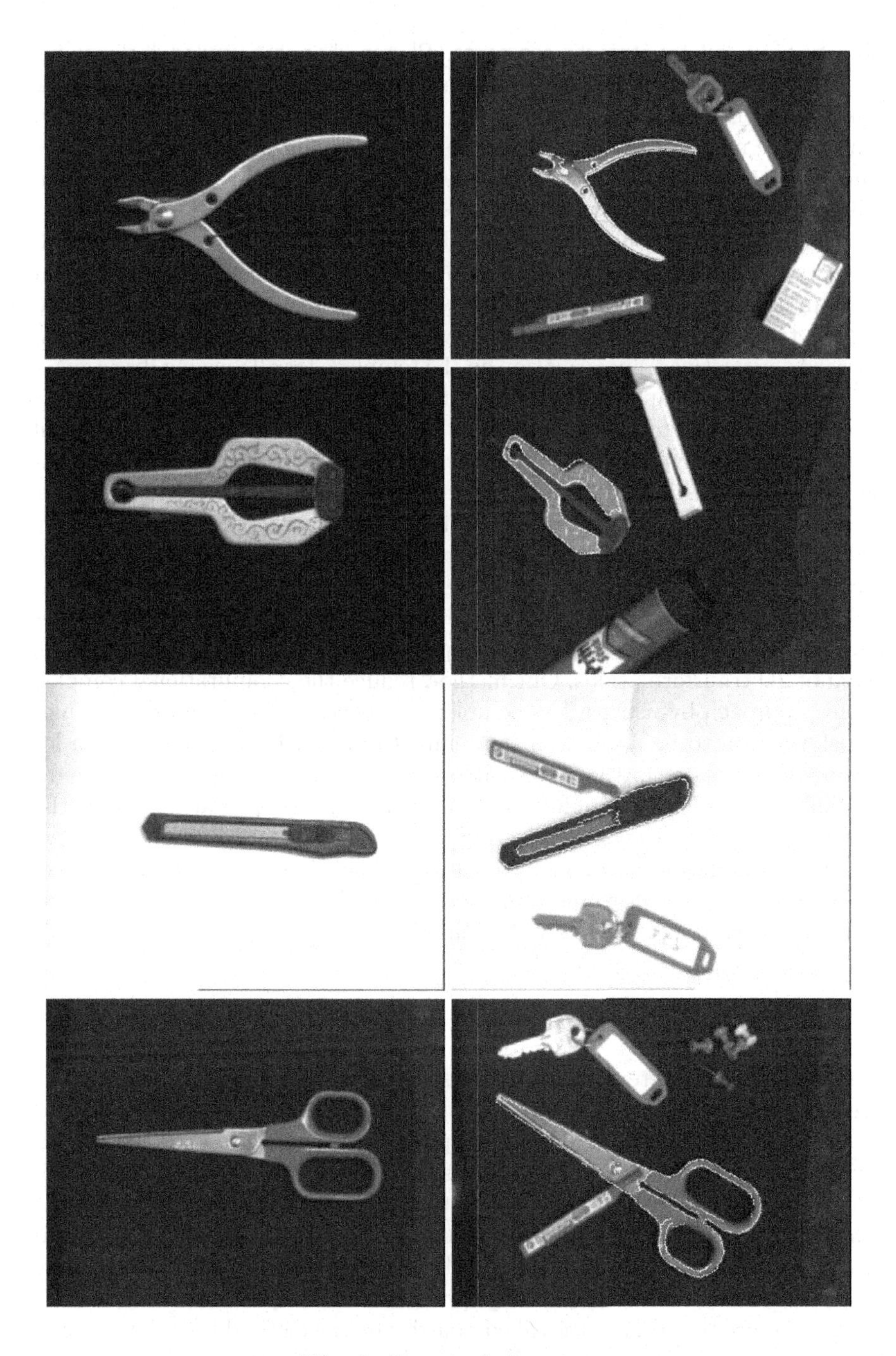

Fig. 3. Practical situation

objects bounded by straight segments. Planar shapes bounded by curvilinear edges, polyhedra and scenes containing several objects can also be tackled, in an indirect and more complex fashion, but they are not addressed in the work presented. The main limitation of this approach, as pointed by the authors, lies in the computational complexity of the matching process, which involves exploring a 6-dimensional space for pose estimation.

A new version of the GHT, called PTIGHT [20], is proposed for detecting a perspectively transformed planar shape in a perspective image that is taken from an unknown viewpoint. In order to build a *perspective reference table* from a given template, all possible perspective shape images are derived by applying forward perspective transformation on the template for all viewing directions and positions. The overlapping of this information imposes two restrictions. First, the positions in the HT with votes greater than a threshold are considered potential candidates. This hypothesis must be verified by back-projecting the image shape over the template in order to find the best solution. The resolutions than can be achieved are poor, $5°$ for the tilt and pan angles in the reported results. Although the images tested there are not complex, the computational time is high. Also, a camera calibration is required in order to know the focal distance.

The algorithm presented in [15] detects planar objects that are rotated a tilt and pan angle, and distorted by a perspective projection before applying a chain of 2D transformations. Our method reduce the computational complexity of this approach by using a less complex perspective model and restricting the transformation to be applied to the template table which, usually, has a lower number of entries than the image table.

Finally, a technique dealing with parallel projected planar objects with rotational symmetry is presented by Yip [21]. The main contribution of the paper is that it provides a method to reduce the dimensionality of the HT space by breaking it down into several lower order ones. The problem is the shape form and projection model restrictions it imposes.

6 Conclusions

A new method to detect planar shapes under orthoperspective projection has been presented. The method copes with a situation in which the shape of the projected image presents a different displacement, orientation and scaling in relation to a template. The method is based on the GHT and generates invariant information, by using gradient information, that allows us to uncouple the parameter calculation and, in this way, reduce computational complexity. A new table generation is accomplished by only using the edge points of the template shape. Generally, the number of edge points in the template is lower than in the image, making the computational complexity of our method to be lower than algorithms using the image shape. Important improvements have been introduced to save gradient angle calculation of the projected points. Several examples, that show the accuracy of the algorithm with real images, have also been presented.

References

1. J. Y. Aloimonos. Perspective approximations. *Image and Vision Computing*, 18(3):179–192, August 1990. 479
2. Yu Cheng. Analysis of affine invariants as approximate perspective invariants. *Computer Vision and Image Understanding*, 63(2):197–207, March 1996. 479
3. S. C. Pei and L. G. Liou. Finding the motion, position and orientation of a planar patch in 3D space from scaled-orthographic projection. *Pattern Recognition*, 27(1):9–25, 1994. 479
4. Isaac Weiss. Review. Geometric invariants and object recognition. *International Journal of Computer Vision*, 10(3):207–231, June 1993. 479
5. W. J. Rucklidge. Efficiently locating objects using the Hausdorff distance. *IJCV*, 24(3):251–270, September 1997. 479
6. G. Lei. Recognition of planar objects in 3-D space from single perspective views using cross ratio. *RA*, 6:432–437, 1990. 480
7. B. S. Song, I. D. Yun, and S. U. Lee. A target recognition technique employing geometric invariants. *PR*, 33(3):413–425, March 2000. 480
8. C. A. Rothwell, A. Zisserman, C. I. Marinos, D. A. Forsyth, and J. L. Mundy. Relative motion and pose from arbitrary plane curves. *IVC*, 10:250–262, 1992. 480
9. D. Oberkampf, D. F. DeMenthon, and L. S. Davis. Iterative pose estimation using coplanar feature points. *CVIU*, 63(3):495–511, May 1996. 480
10. M. I. A. Lourakis, S. T. Halkidis, and S. C. Orphanoudakis. Matching disparate views of planar surfaces using projective invariants. *Image and Vision Computing*, 18(9):673–683, June 2000. 480
11. C. A. Rothwell, A. Zisserman, D. A. Forsyth, and J. L. Mundy. Planar object recognition using projective shape representation. *IJCV*, 16(1):57–99, September 1995. 480
12. R. J. Holt and A. N. Netravali. Using affine invariants on perspective projections of plane curves. *CVIU*, 61(1):112–121, January 1995. 480
13. Hannu Kauppinen, Tapio Seppänen, and Matti Pietikäinen. An experimental comparison of autoregressive and Fourier-based descriptors in 2D shape classification. *IEEE Transactions on Pattern Anal. and Machine Intell.*, 17(2):201–207, February 1995. 480
14. N. Guil, J. M. Gonzalez-Linares, and E. L. Zapata. Bidimensional shape detection using an invariant approach. *Pattern Recognition*, 32(6):1025–1038, 1999. 481, 482, 484
15. N. Guil, J. R. Cózar, and E. L. Zapata. Planar 3D object detection by using the generalized Hough transform. In *10th Intl. Conf. on Image Analisys and Processing*, pages 358–363, September 1999. 483, 488
16. A. K. Jain, Y. Zhong, and S. Lakshmanan. Object matching using deformable templates. *PAMI*, 18(3):267–278, March 1996. 484
17. S. C. Jeng and W. H. Tsai. Scale- and orientation-invariant generalized Hough transform: A new approach. *PR*, 24:1037–1051, 1991. 486
18. D. C. W. Pao, H. F. Li, and R. Jayakumar. Shapes recognition using the straight line Hough transform: Theory and generalization. *PAMI*, 14(11):1076–1089, November 1992. 486
19. Vittorio Murino and Gian Luca Foresti. 2D into 3D Hough-space mapping for planar object pose estimation. *Image and Vision Computing*, 15(6):435–444, 1997. 486

20. R. C. Lo and W. H. Tsai. Perspective-transformation-invariant generalized Hough transform for perspective planar shape detection and matching. *Pattern Recognition*, 30(3):383–396, March 1997. 488
21. Raimond K. K. Yip. A Hough transform technique for the detection of parallel projected rotational symmetry. *Pattern Recognition Letters*, 20(10):991–1004, November 1999. 488

Detection of the Discrete Convexity of Polyominoes

Isabelle Debled-Rennesson[1], Jean-Luc Rémy[2], and Jocelyne Rouyer-Degli[3]

[1] LORIA – Laboratoire LOrrain de Recherche en Informatique et ses Applications
Institut Universitaire de Formation des Maîtres de Lorraine
Campus Scientifique, B.P. 239, F54506 Vandœuvre-lès-Nancy
{debled,remy,rouyer}@loria.fr
[2] LORIA
Centre National de la Recherche Scientifique
Campus Scientifique, B.P. 239, F54506 Vandœuvre-lès-Nancy
[3] LORIA
Université Henri Poincaré, Nancy 1
Campus Scientifique, B.P. 239, F54506 Vandœuvre-lès-Nancy

Abstract. The convexity of a discrete region is a property used in numerous domains of computational imagery. We study its detection in the particular case of polyominoes. We present a first method, directly relying on its definition. A second method, which is based on techniques for segmentation of curves in discrete lines, leads to a very simple algorithm whose correctness is proven. Correlatively, we obtain a characterisation of lower and upper convex hulls of a discrete line segment. Finally, we evoke some applications of these results to the problem of discrete tomography.

Keywords: Discrete Convexity, Segmentation, Discrete line, Polyominoes, Discrete tomography

1 Introduction

Discrete convexity intervenes in numerous domains with regard to geometry and particularly to image processing [4]. It is an important property of plane figures which permits, for instance, methods for geometrical shapes regularisation. The notion of discrete convexity is strongly linked with the paradigm of discrete lines. This is underlined in this article where a simple and efficient algorithm for detection of polyomino convexity is presented. Polyominoes are objects in which any couple of cells may be linked through a path containing only horizontal and vertical moves (4-connectivity). After having checked the hv-convexity (cells of each column and each row are consecutive) of a polyomino P, convexity is checked on the curve points of the border characterising it.

The proposed method uses a variant of the linear algorithm for segmentation of curves in straight lines given in [7,6]. The eventual non-convexity of P is detected by the algorithm, during its scanning of curves of the border of P; if the whole border is scanned, P is convex. The proof of this algorithm uses a result on convex hull of a discrete segment presented in this article.

G. Borgefors, I. Nyström, and G. Sanniti di Baja (Eds.): DGCI 2000, LNCS 1953, pp. 491–504, 2000.
© Springer-Verlag Berlin Heidelberg 2000

This technique and methods directly coming from definitions of discrete convexity have been used for the study of the reconstruction of 2-dimensional discrete sets from their horizontal and vertical projections. In particular, we studied the reconstruction of convex polyominoes. The approaches we use are presented at the end of this article. This work enters in the more general framework of discrete tomography whose applications are numerous, particularly in data compression, image processing, or still in medical imagery for medical diagnosis help in radiography.

In the second section, we introduce miscellaneous definitions of discrete convexity. Then a first simple method, deduced from a definition, is proposed. In the following section, fundamental elements of discrete geometry are given so that we obtain a very efficient algorithm for convexity detection on hv-convex polyominoes. Then, we propose a use of these methods in discrete tomography. At last, a conclusion and further research prospects are given.

2 Discrete Convexity

Convexity is well defined in the continuous case but in the discrete one, several definitions exist. The studied discrete figures are finite 8-connected subsets of discrete points in the plane and are named **discrete regions**.

In 1970, Slansky [18] defines a discrete region as convex if and only if there is a convex (Euclidean) region whose image (after digitizing) is this discrete region. This definition depends on the digitizing process used.

On the other hand, Minsky and Papert [14] gave the following definition of the convexity of a discrete region R: R is convex if and only if there is no triplet of colinear discrete points (c_1, c_2, c_3) such as c_1 and c_3 belong to R and c_2 belongs to the complementary of R.

Then, Kim and Rosenfeld [12,13,11] have given several equivalent definitions. They have shown that a discrete region R is convex

– on the one hand, if and only if its convex (Euclidean) hull does not contain any discrete point of the complementary of R.
– on the other hand, if and only if it fulfils the **area property** i.e. if and only if, for all points p1 and p2 of R, P(R,p1,p2) does not possess any point of the complementary of R, where P(R,a,b) represents the polygon whose edges are made by the segment ab and the edges of R (see Fig. 1).

These last two definitions shall be used in the study of the convexity of polyominoes. We refer to Kim and Rosenfeld [12,13,11] for a comparison of their definitions with Minsky and Papert's one.

A hv-convex polyomino is a polyomino whose cells of each column are consecutive (v-convexity) as well as those of each row (h-convexity).
Some immediate properties may be deduced from this definition:
- a hv-convex polyomino is a discrete region "without hole",
- a convex polyomino is hv-convex.
The convexity study of a polyomino so starts by checking its hv-convexity, this

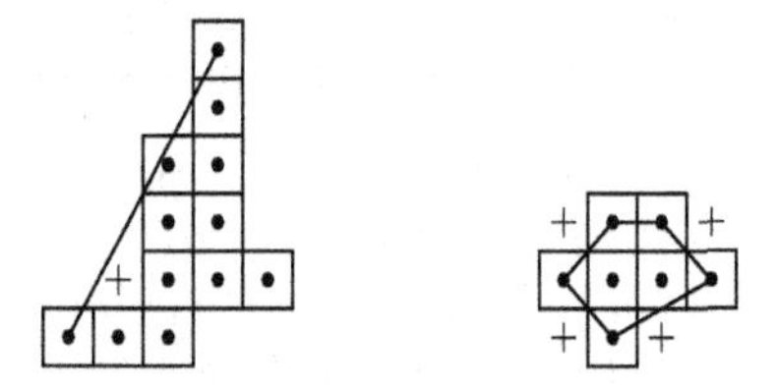

Fig. 1. On the left, a non-convex discrete region. On the right, a convex discrete region

is done through a simple scanning of each row [4]. In the following, we study the convexity of hv-convex polyominoes. It is clear that this study may therefore be reduced to the convexity study of the limit or border of the hv-convex polyominoes (see Sect. 3 and 4).

3 Direct Use of the Area Property to Detect the Convexity of a *hv*-Convex Polyomino

Let T be a hv-convex polyomino, anchored at (k, l), i.e., containing in its first column a cell at row k and in the last column a cell at row l. We call left limit of T, and we note L, the set of cells with minimal column index of each line. We define in the same way the right limit R of T. We distinguish in L the higher part L1, included between the first line and the left anchorage k, and the lower part L2, included between the line k and the last line. In the same way, we distinguish in R the parts R1 and R2 limited by the right anchorage l (see Fig. 2).

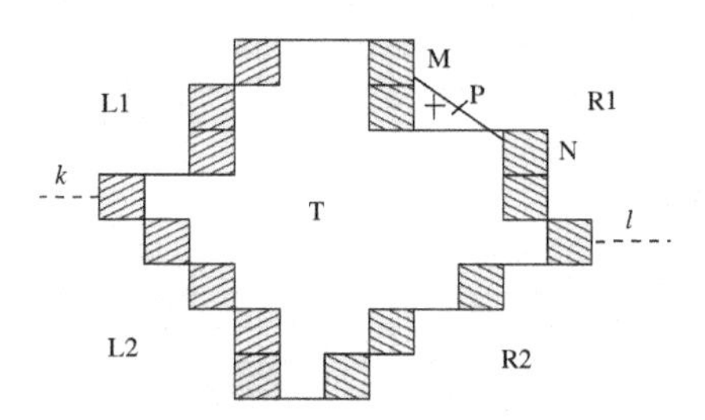

Fig. 2. Limits of the polyomino T

To check that T is convex, we only have to apply the area property on L1, L2, R1, and R2. So T is convex if and only if, for every couple (M, N) of points of L1, L2, R1, or R2, and every y included between the row indexes of M and N, there is no discrete point whose row index y is located between the segment MN (inclusive) and the polyomino T (exclusive). Let (x_1, y_1) be the coordinates of M, (x_2, y_2) the coordinates of N. The coordinates of the point P of the segment

MN with row index y, $y_1 < y < y_2$, are (x, y), with $x = x_1 + \frac{(y-y_1)(x_2-x_1)}{y_2-y_1}$. In order that T be convex, it is necessary and sufficient that:

i. if M and N belong to L1 or to L2, the discrete point with coordinates $(\lceil x \rceil, y)$, located to the right of the point P, belongs to T,
ii. if M and N belong to R1 or to R2, the discrete point with coordinates $(\lfloor x \rfloor, y)$, located to the left of the point P, belongs to T.

If the hv-convex polyomino T is represented by a 0-1 matrix whose dimension is $m*n$, the computation of its limits is in $O(m+n)$, m and n being the numbers of rows and columns of T, respectively, and the tests on couples of points are in $O(\min(m^3, n^3))$. Indeed, when $m > n$, it is permitted to exchange the roles of columns and rows.

4 Use of Discrete Lines to Detect the Convexity

Let us consider a hv-convex polyomino included in a minimal rectangle Rec whose size is $m*n$. Let ([A,A'], [B,B'], [C,C'], [D,D']) be the intersection between the border of the polyomino and Rec (see Fig. 3). By considering the pixels of the border, the points A, A', B, B', C, C', D and D' delimit 8-connected curves of pixels which characterise the hv-convex polyomino. These 4 curves c1, c2, c3, and c4 are made of the points of the polyomino border being respectively located between A' and B for c1, B' and C for c2, C' and D for c3 and between D' and A for c4, respectively.

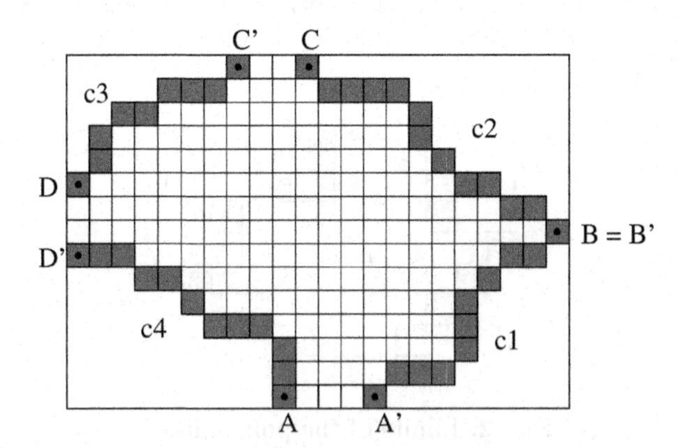

Fig. 3. A hv-convex polyomino, gray pixels represent the curves c1, c2, c3, and c4

As it has been indicated in section 2, to check the convexity of a hv-convex polyomino, we only have to consider the border; curves c1, c2, c3, and c4 must therefore be studied. The principle consists in segmenting curves c1, c2, c3, and c4, the eventual non-convexity of a polyomino shall be detected very simply through this operation.

In the following paragraph, the notion of discrete line [16,15,7] is recalled as well as some properties of these object types with, particularly, a new result on the construction of the convex hull of a discrete line segment. Moreover, an algorithm of discrete line segment recognition, with its use in the detection of convex polyominoes and its adaptation to the problem are presented.

4.1 Discrete Lines

The arithmetic definition of a discrete line was introduced by J.P. Reveillès [16,15,7]:

Definition 1. *A* **discrete line** *with a slope* $\frac{a}{b}$ *with* $b \neq 0$ *and* $pgcd(a,b) = 1$, *with lower bound* μ, *arithmetical thickness* ω, *is the set of points* (x, y) *of* $\mathbb{Z}^2$ *which satisfies the double diophantian inequation*

$$\mu \leq ax - by < \mu + \omega$$

with all integer parameters.

We note the preceding discrete line $\mathcal{D}(a, b, \mu, \omega)$. We are mostly interested in **naïve** lines which verify $\omega = sup(|a|, |b|)$ (see Fig. 4), we shall note them $\mathcal{D}(a, b, \mu)$. To simplify the writing, we shall suppose in the following that the slope coefficients verify $0 \leq a$ and $0 \leq b$, therefore $\omega = max(a, b)$. Real straight lines $ax - by = \mu$ et $ax - by = \mu + \omega - 1$ are named the **leaning lines** of the discrete naïve line $\mathcal{D}(a, b, \mu)$. An integer point of these lines is named **a leaning point**.

The leaning line located above (resp. under) $\mathcal{D}$ in the first quadrant ($0 \leq a$ and $0 \leq b$) respects the following equation $ax - by = \mu$ (resp. $ax - by = \mu + \omega - 1$), it is named **upper leaning line** (resp. **lower leaning line**) of $\mathcal{D}$. All points of $\mathcal{D}$ are located between these two lines at a distance strictly lower than 1. It is clear that every segment of a discrete line is a convex region.

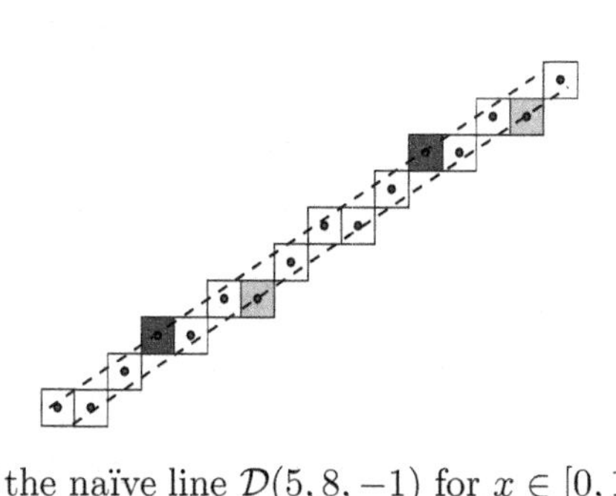

Fig. 4. A segment of the naïve line $\mathcal{D}(5, 8, -1)$ for $x \in [0, 15]$, the upper leaning points are in dark grey and the lower leaning points are in light grey, leaning lines are dotted lines

A naïve line $\mathcal{D}$ may be seen as a set of the integer points taken on the union of real lines $ax - by = c$ with $c = \mu$, $\mu + 1$,..., $\mu + \omega - 1$. We name α**-levelled**

dotted line of the line $\mathcal{D}$, the set of integer points located on the real line $ax - by = \alpha$.

Definition 2. *$r(M)$ is named the* **remainder at point** $M(x_M, y_M)$ **with respect to** $\mathcal{D}$ *and is defined by:*

$$\mathbf{r(M) = ax_M - by_M}.$$

Proposition 1. *Let M_0M_1 be a segment of the line $\mathcal{D}(a, b, \mu)$,* **the lower convex hull** *of the M_0M_1 points is the polygonal curve going through the following points:*

- *The first and the last lower leaning point of the segment, named L_F et L_L.*
- *Between M_0 and L_F : the (N_i) sequence with $x_{M_0} \leq x_{N_i} \leq x_{L_F}$, $N_0 = M_0$ such that $x_{N_i} < x_{N_{i+1}}$ and $r(N_i) < r(N_{i+1}) \leq \mu + \omega - 1$*
- *Between M_1 and L_L : the (P_i) sequence with $x_{M_1} \geq x_{P_i} \geq x_{L_L}$, $P_0 = M_1$ such that $x_{P_i} > x_{P_{i+1}}$ and $r(P_i) < r(P_{i+1}) \leq \mu + \omega - 1$.*

(see Fig. 5 for an example).

Remark 1.

1. Colinear points of the N_i and P_i sequences may simply be omitted in the characterisation of the lower convex hull.
2. To construct the upper convex hull of a discrete line segment, the N_i and P_i sequences are determined in the same way from the upper leaning points.

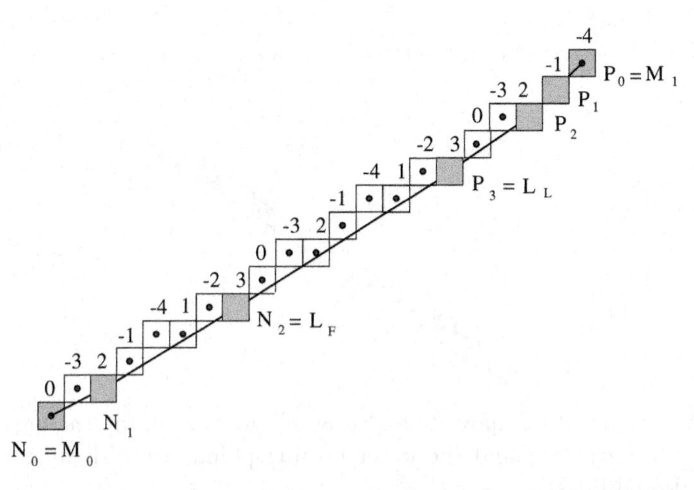

Fig. 5. Segment of the naïve line $\mathcal{D}(5, 8, -4)$, the lower convex hull of the segment is drawn, the value of the remainder is indicated for each point

Proof. All points of $\mathcal{D}$ are located above the lower leaning line, the segment points located between L_F and L_L are therefore above the segment $L_F L_L$ and the points L_F and L_L belongs to the lower convex hull of the segment.

Let us consider now the part of the segment between M_0 and L_F. Between these two points, we find at most $\omega - 2$ points, with all different remainders. Let N_i be a point of the sequence different of L_F, and N_{i+1} the first point encountered since N_i such that N_{i+1} belongs to the interval $[r(N_i), \mu + \omega - 1]$, the segment points between N_i and N_{i+1} are located above the dotted line going through N_i and therefore above the segment $N_i N_{i+1}$.
Let N_i, N_{i+1}, N_{i+2}, be three consecutive points of the sequence. Let us prove that N_{i+2} is located above or on the straight line $N_i N_{i+1}$, thus we shall have demonstrated the convexity of the sequence of points N_i. Two cases must be studied:
Case 1. If $r(N_{i+1}) - r(N_i) \leq r(L_F) - r(N_{i+1})$ and $x_{L_F} \geq 2x_{N_{i+1}} - x_{N_i}$ (see Fig. 6), the point N_i', symmetrical point of N_i with respect to N_{i+1}, is a point on the segment located between N_{i+1} and L_F. Let us suppose that N_{i+2} is a point located between the dotted lines of level $r(N_{i+1})$ and $r(N_i')$ such that $x_{N_{i+1}} < x_{N_{i+2}} < x_{N_i'}$. The symmetrical point N_{i+2}' of N_{i+2} with respect to N_{i+1} is a point of the segment whose remainder is included between $r(N_i)$ and $r(N_{i+1})$ and whose x-coordinate is included between those of N_i and N_{i+1}, which is contradictory with the hypotheses on N_i and N_{i+1}. Necessarily $N_{i+2} = N_i'$. Case $r(N_{i+1}) - r(N_i) \leq r(L_F) - r(N_{i+1})$ and $x_{L_F} < 2x_{N_{i+1}} - x_{N_i}$ may not occur (property related to Klein Theorem and discrete lines [6]).
Case 2. If $r(N_{i+1}) - r(N_i) > r(L_F) - r(N_{i+1})$, N_{i+2} is, by construction, a point located between the dotted lines of level $r(N_{i+1})$ and $r(L_F)$ such that $x_{N_{i+1}} < x_{N_{i+2}} \leq x_{L_F}$. Let us suppose that N_{i+2} is located under the $N_i N_{i+1}$ line, by using the symmetrical point of N_{i+2} with respect to N_{i+1} we also get a contradiction. A similar reasoning may be applied to the sequence (P_i) between M_1 and L_L.

□

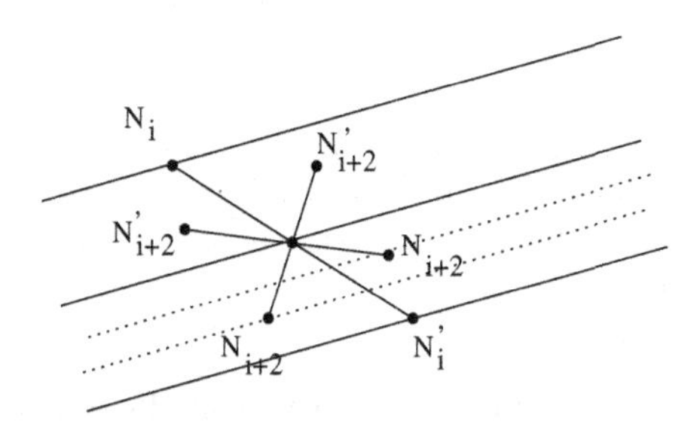

Fig. 6. Figure illustrating proof of Proposition 1 (Case 1)

4.2 Discrete Line Segment Recognition

Let us consider $\Sigma(M_0, M_1)$ a segment of $\mathcal{D}$, discrete naïve line with characteristics a, b, μ, $0 \leq a$ and $0 \leq b$, M_0 and M_1 are respectively the first and the last point of the segment. Let us suppose that the point $M(x_M, y_M)$ (with $M = M_1 + (1,0)$, $M = M_1 + (1,1)$ or $M = M_1 + (0,1)$) is added to Σ. Is $\Sigma' = \Sigma \cup \{M\}$ a straight line segment and, if it is the case, what are its characteristics a', b', μ'?
This problem is solved in [7,6] and relies on some particular leaning points of a discrete line. They are the leaning points located at the extremities of the segment currently being recognised. We note $\mathsf{U_F}$ (resp. $\mathsf{L_F}$) the upper (resp. lower) leaning point whose x-coordinate is minimal. In the same way, we note $\mathsf{U_L}$ (resp. $\mathsf{L_L}$) the upper (resp. lower) leaning point whose x-coordinate is maximal.

Theorem 1 ([7,6]). *Let us consider* $r(M)$, *the remainder at point* $M(x_M, y_M)$ *with respect to* $\mathcal{D}$ ($r(M) = ax_M - by_M$).

(*i*) *If* $\mu \leq r(M) < \mu + \omega$, *then* $M \in \mathcal{D}(a, b, \mu)$, $\Sigma \cup \{M\}$ *is a segment of the straight line* $\mathcal{D}$.
(*ii*) *If* $r(M) = \mu - 1$, *then* $\Sigma \cup \{M\}$ *is a segment of the straight line whose slope is given by the vector* $\mathsf{U_F M}$.
(*iii*) *If* $r(M) = \mu + \omega$, *then* $\Sigma \cup \{M\}$ *is a segment of the straight line whose slope is given by the vector* $\mathsf{L_F M}$.
(*iv*) *If* $r(M) < \mu - 1$ *or* $r(M) > \mu + \omega$, *then* $\Sigma \cup \{M\}$ *is not a segment of a discrete line.*

Remark 2. In the case (*iv*) M is called **strongly exterior** to $\mathcal{D}$ and:
- If $r(M) < \mu - 1$ (M is above Σ), there is an integer point between $\mathsf{U_F M}$ and Σ.
- If $r(M) > \mu + \omega$ (M is under Σ), there is an integer point between $\mathsf{L_F M}$ and Σ.
The existence of these points is demonstrated at page 644 of [7].

Theorem 1 allows us to obtain an incremental algorithm (see Fig. 7) of discrete line segments recognition by scanning a sequence of 8-connected pixels named **discrete path**. A linear segmentation algorithm for 8-connected curves [7,6] is immediately deduced from this result by considering the longest segments, last point of a segment being the first one of the next segment. To detect the convexity of a polyomino, we use a variant of this algorithm which is presented in the following section.

4.3 Use of a Segmentation Algorithm for the Detection of Convex Polyominoes

Detecting the convexity of a hv-convex polyomino P (see Fig. 3) consists in studying its convexity at the neighbourhood of each section of its borders c1, c2, c3, and c4. In this whole paragraph, convexity is described for the discrete curve c1, detection for the other curves is deduced by symmetry.

```
Algorithm AddPoint
        remainder = ax_M − by_M ;
        If μ ≤ remainder < μ + ω
                If remainder = μ then U_L = M Endif
                If remainder = μ + ω − 1 then L_L = M Endif
        else
                If remainder = μ − 1 then
                        L_F = L_L ;
                        U_L = M ;
                        a = |y_M − y_{U_F}| ;
                        b = |x_M − x_{U_F}| ;
                        μ = ax_M − by_M ;
                else
                        If remainder = μ + ω then
                                U_F = U_L ;
                                L_L = M ;
                                a = |y_M − y_{L_F}| ;
                                b = |x_M − x_{L_F}| ;
                                μ = ax_M − by_M − sup(|a|, |b|) + 1 ;
                        else
                                the new point may not be added to the segment
                        Endif
                Endif
        Endif
```

Fig. 7. Algorithm adding a point M to the front extremity of a segment of a naïve line $\mathcal{D}(a, b, \mu)$ in the first quadrant

A curve is said **lower convex** if there is no discrete point between itself and its lower convex hull.
We must therefore prove that c1 is lower convex.

The algorithm for discrete line segments recognition is used on c1, it scans points of c1 one after another and stops when the added point may not belong to a discrete line segment containing all points already scanned plus the one added. The recognition shall not continue for a new segment at this added point but at the last lower leaning point of the segment which has just been recognised. When such a reject forces a segment change, the non convexity may be detected. The rejected point is located strongly above or strongly under the segment. Thanks to Remark 2, there is, in the second case, an integer point located between the discrete segment already scanned and the real segment from M to the first lower leaning point. This segment to which M is added is therefore not lower convex and so neither is c1. In this case, the algorithm stops on a non-convexity statement of the polyomino.

Let us consider the procedure Recognizesegment(M_0) whose input is a point M_0 of the curve c1 and which then adds the next points of c1 from M_0 as long as the scanned set of points is a discrete line segment. This procedure outputs M, the last point tested, which is either the rejected point, or the last point of c1, the characteristics($\mathsf{a}, \mathsf{b}, \mu$) of the scanned segment with a and b positive, and the last lower leaning point $\mathsf{L_L}$ as well as a boolean variable end, equal to false while c1 has still not been completely scanned.

Definition 3. *Let S be a discrete line segment obtained by the procedure Recognizesegment, we call* **reduced** *segment of S and we note S', the segment containing the points of S located before the last lower leaning point existing in S, inclusive of this point.*

```
Algorithm SegConv
        Input : c1, 8-connected sequence of points of the first quadrant
        convex = true
        end = false
        firstpoint= first point of c1
        While convex and not end Do
                Recognizesegment(firstpoint)→ (M, a, b, μ, L_L,end)
                If ax_M − by_M > μ + max(a, b) then
                        convex=false
                Else
                        firstpoint=L_L
                EndIf
        EndWhile
```

Fig. 8. Algorithm testing convexity of the part of a polyomino in the first quadrant

Theorem 2. *A curve c1 of the first quadrant is lower convex if and only if the SegConv algorithm completely scans c1 (convex=true at the end).*

Proof. Let us consider a point M which may not be added to the current segment with characteristics a, b, μ such that $ax_M - by_M > \mu + max(a, b)$; then, according to Remark 2, there is an integer point between the current discrete segment and the straight line segment $L_F M$; therefore, according to the area property, c1 is not convex. In other words, when the algorithm stops because the variable convex has been set to false, c1 is not convex.

Let us suppose that c1 is completely scanned by the algorithm SegConv (see Fig. 8), then, all changes of straight line segments have been done on points M which verify $ax_M - by_M < \mu - 1$. Each point is therefore located above the segment previously scanned.

Let S1 and S2 be two segments successively recognised by SegConv during the scanning of c1. S1 characteristics are a_1, b_1, μ_1 and its slope is α_1 $(=\frac{a_1}{b_1})$. Let us consider L_{1L} the last lower leaning point of S1 and M the rejected point which does not belong to S1.

Let S'1 be the reduced segment of S1. Let us prove that the last edge of the convex hull of S'1 has a slope which is lower than the one of the first edge of the convex hull of S2.

As S'1 ends at the lower leaning point L_{1L} then, according to Proposition 1, if at least two lower leaning points are present in the segment, the slope of the last edge of the convex hull of S'1 is α_1 otherwise, the slope is lower than α_1.

Let $L_{1L}K$ be the first edge of the convex hull of S2, whose slope is β, we must consider two cases:

- If $K \in$ S1, necessarily K is located above the lower leaning line of S1 therefore $\beta > \alpha_1$.

– If $K \notin$ S1, let us suppose that $\beta \leq \alpha_1$, i.e., that K is under, or on, the lower leaning line of S1 (see Fig. 9). The points L_{1L}, M, and K belong to the segment S2. However that is impossible because the distance between M and the leaning line of S1 is greater than 1, and so, *a fortiori*, the one between M and $L_{1L}K$. Therefore these three points may not belong to the same discrete line segment and $\beta > \alpha_1$.

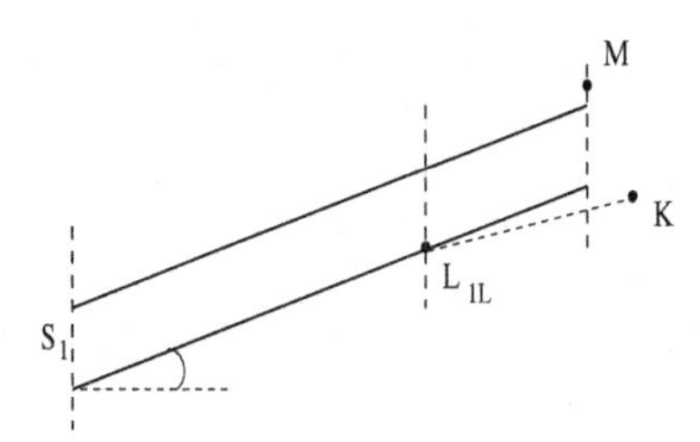

Fig. 9. Figure illustrating proof of Theorem 2

This property implies that the lower convex hull of the union of S'1 and S'2 is the union of the lower convex hulls of these both segments. As a generalisation, the lower convex hull of c1 is the union of convex hulls of the S'i. However each discrete segment is, by definition, lower convex. Therefore c1 is also lower convex, according to the definition of convexity by Kim and Rosenfeld (see Sect. 2) based on the convex hull of the set. □

Complexity of the construction of curves c_i, $i = 1..4$, is in $O(m + n)$, m and n being the numbers of rows and columns of the polyomino. Let us say a few words about the complexity of the SegConv algorithm. We first notice that the algorithm actually terminates. Indeed, at each step, the variable firstpoint is incremented by at least one element (in a non-void segment, L_L is always different of the first element). On the other hand, the segmentation algorithm described in [7,6] is linear in the number of points of the curve to scan. However, the variant we propose obliges to re-start every time at the last lower leaning point encountered, that leads to some overlaps. This algorithm is actually linear despite its overlaps. The proof of its linear complexity will be published in a next article.

5 Applications to Discrete Tomography

The objectives of discrete tomography is the reconstruction of a finite set of points from a number of projections. Several authors have been interested in this subject by using miscellaneous approaches to obtain exact, approached, or random solutions [9].

The particular case of polyominoes reconstruction from their horizontal and vertical projections is a NP-complete problem [19]. Moreover, if we suppose h-convexity (or v-convexity), the problems remains NP-complete [2]. On the other hand, if we consider hv-convex polyominoes, we get today several polynomial algorithms to decide if there is or not a polyomino corresponding to given projections [2,3,5]. The best existing algorithm is the one of Chrobak and Dürr [5]. This algorithm determines, for all couples of rows k and l, if there is a hv-convex polyomino anchored in k and l, with given horizontal and vertical projections. It is done by translating the existence of hv-convex polyominoes anchored in k and l as a 2SAT problem (conjunction of disjunctive clauses with at most two boolean variables), whose size is in $O(mn)$. A 2SAT problem may be solved linearly in its size [1] and the number of possible choices for k and l is $\min(m^2, n^2)$, as roles of rows and columns may be swapped. Therefore, it gives a complexity in $O(mn \min(m^2, n^2))$ for the reconstruction of a hv-convex polyomino with given vertical and horizontal projections. In fact, we may limit choices of k and l, and then obtain an even more efficient implementation [5].

If we add to the algorithm of Chrobak and Dürr a convexity test, by using any method presented above, the complexity of the algorithm remains in $O(mn \min(m^2, n^2))$. On the other hand, if we want to decide if there is a convex polyomino with given projections by first seeking the hv-convex polyominoes having these projections, we shall have at worst an exponential complexity, because we can have an exponential number of hv-convex solutions [8].

We also have started to seek for the convex polyominoes with given orthogonal projections by using the paradigm of constraint programming. Let us give a short description of it:

1. A constraint program is the input
   - of *a priori* **domains** in which variables of the problem are authorised to take their values,
   - of **constraints** to which these variables are submitted,
   - and of a **strategy** of assigning these variables.
2. The running of a constraint program consists in the repetition until a solution emerges, or as long as possibilities exist, of the following iteration:
   - select, as a function of the prescribed strategy, a variable to assign, and an assignment value, among those in the domain of the variable which satisfy the constraints,
   - update all constraints related to the variable which has just been assigned.

   This iteration is associated, in case of a dead end (none of the variables remaining for assignment may be assigned), with a backtracking process, back to the last choice point.

This approach had already been used for the detection of the existence of hv-convex polyominoes with given orthogonal projections [20]. We have [17] modified the program written at this occasion, by adequately enriching the set of constraints.

This new program has successfully run on a first sample of data. Moreover, we have observed the following: if we just take care of how to express the constraints, the adopted process allows us to take them into account globally, rather than successively. So, we avoid to apply first constraints related to the hv-convexity, and then the ones related to the convexity itself. This global treatment allows us to obtain convex polyominoes more quickly, when they exist, without having to enumerate previously the hv-convex polyominoes, which may be more numerous. And in the same way, it is possible to decide more rapidly that none of them exists, when it is the case.

These first results are encouraging and push us to deepen this approach.

6 Conclusion

Our work has permitted a step forward in two directions.

On the one hand, we obtain a characterisation of lower and upper convex hulls of a discrete line segment based on the arithmetic interpretation of these objects.

On the other hand, we obtain a linear algorithm to test the convexity of a polyomino, which is very simple to express. We prove the correctness of this algorithm. After this paper was prepared, the authors' attention was called by a reviewer to Hübler, Klette and Voß' paper [10]. We will compare their approach and ours in our oral presentation.

Moreover, we have explored an alternative approach of the convexity test, in particular by using the methods of constraint programming. We have also applied the convexity test to some problems of discrete tomography. In both cases, the first results obtained urge us to continue on this way.

Acknowledgements

We thank the members of the PolKA research group of LORIA for their review and constructive remarks.

References

1. B. Aspvall, M. F. Plass, and R. E. Tarjan. A linear-time algorithm for testing the truth of certain quantified boolean formulas. *Information Processing Letters*, volume 8, number 3, pages 121–123, 1979. 502
2. E. Barcucci, A. Del Longo, M. Nivat, and R. Pinzani. Reconstructing convex polyominoes from horizontal and vertical projections. *Theoretical Computer Science*, volume 155, number 2, pages 321–347, 1996. 502
3. E. Barcucci, A. Del Longo, M. Nivat, and R. Pinzani. Medians of polyominoes: A property for the reconstruction. *International Journal of Imaging Systems and Technology*, volume 8, pages 69–77, 1998. 502
4. J-M. Chassery, A. Montanvert. Géométrie discrète en analyse d'images. *Traité des nouvelles technologies, série Images*, Hermes, 1991. 491, 493

5. M. Chrobak and C. Dürr. Reconstructing hv-convex polyominoes from orthogonal projections. *Information Processing Letters*, volume 69, pages 283–289, 1999. 502
6. I. Debled-Rennesson. Etude et reconnaissance des droites et plans discrets. Thèse. Université Louis Pasteur, Strasbourg, 1995. 491, 497, 498, 501
7. I. Debled-Rennesson and J. P. Reveillès. A linear algorithm for segmentation of digital curves. In *International Journal of Pattern Recognition and Artificial Intelligence*, volume 9, pages 635–662, Décembre 1995. 491, 495, 498, 501
8. A. Del Lungo and M. Nivat and R. Pinzani. The number of convex polyominoes reconstructible from their orthogonal projections. *Discrete Mathematics*, volume 157, pages 65–78, 1996. 502
9. G. T. Herman and A. Kuba. *Discrete Tomography.* Birkhauser, 1999. 501
10. A. Hübler, R. Klette and K. Voß. Determination of the Convex Hull of a Finite Set of Planar Points Within Linear Time. Elektronische Informationsverarbeitung und Kybernetik, pages 121–139, 1981. 503
11. C. E. Kim. Digital convexity, straightness and convex polygons. In *IEEE Transactions on Pattern Analysis and Machine Intelligence*, volume PAMI-4, pages 618–626, 1982. 492
12. C. E. Kim and A. Rosenfeld. On the convexity of digital regions. In *Pattern Recognition*, volume 5, pages 1010–1015, 1980. 492
13. C. E. Kim and A. Rosenfeld. Digital straight lines and convexity of digital regions. In *IEEE Transactions on Pattern Analysis and Machine Intelligence*, volume PAMI-4, pages 149–153, 1982. 492
14. M. Minsky and S. Papert. Perceptrons. In *M. I. T. Press*, 1969. 492
15. J. P. Reveillès. Géométrie discrète, calculs en nombre entiers et algorithmique. Thèse d'état. Université Louis Pasteur, Strasbourg, 1991. 495
16. J. P. Reveillès. Structure des droites discrètes. In *Journées mathématique et informatique*, Marseille-Luminy, Octobre 1989. 495
17. I. Sivignon. Reconstruction de polyominos convexes par programmation par contraintes. Rapport du stage de Magistère d'Informatique 1ère année, ENS Lyon, effectué au LORIA, 1999. 502
18. J. Slansky. Recognition of convex blobs. In *Pattern Recognition*, volume 2, pages 3–10, 1970. 492
19. G. J. Woeginger. The reconstruction of polyominoes from their orthogonal projections. Technical report, Technische Universität Graz, 1996. 502
20. T. Zajac. Reconstructing Convex Polyominoes using Constraint Programming. Master Thesis, Wroclaw University of Technology, Pologne. 502

An Efficient Shape-Based Approach to Image Retrieval*

Ioannis Fudos and Leonidas Palios

Department of Computer Science, University of Ioannina
GR45110 Ioannina, Greece
{fudos,palios}@cs.uoi.gr

Abstract. We consider the problem of finding the best match for a given query shape among candidate shapes stored in a shape base. This is central to a wide range of applications, such as, digital libraries, digital film databases, environmental sciences, and satellite image repositories. We present an efficient matching algorithm built around a novel similarity criterion and based on shape normalization about the shape's diameter, which reduces the effects of noise or limited accuracy during the shape extraction procedure. Our matching algorithm works by gradually "fattening" the query shape until the best match is discovered. The algorithm exhibits poly-logarithmic time behavior assuming uniform distribution of the shape vertices in the locus of their normalized positions.

Keywords: image retrieval, shape-based matching, image bases

1 Introduction

The last few years, there is an emerging need to organize and efficiently use large pools of images that have been collected over the last decades and contain information potentially useful to areas such as medicine, journalism, weather prediction, environmental sciences, art, fashion and industry. It is estimated that there are more than 20 million pages containing hundreds of millions of images on world wide web pages alone [7]. Traditionally, images were retrieved by their filename, other technical characteristics such as date and size or through text keywords, in the case of manually annotated images. Manual annotation, except for being a time consuming and not real-time process, can describe only a very small percentage of the information that an image contains.

Recently, there is an increasing effort to organize and retrieve images by content based on characteristics such as color, texture, and shape. A number of methods in the literature perform indexing and retrieval based on global image characteristics such as color, texture, layout, or their combinations. QBIC [10,15], a system developed at IBM Almaden supports retrieval by color histograms, texture samples (based on coarseness, contrast and directionality), and shape. QBIC

* This work was supported in part by a GSRT (General Secretariat of Research and Technology) Bilateral Research Cooperation Grant between Greece and the Czech Republic

G. Borgefors, I. Nyström, and G. Sanniti di Baja (Eds.): DGCI 2000, LNCS 1953, pp. 505–517, 2000.
© Springer-Verlag Berlin Heidelberg 2000

uses R^* trees to process queries based on low-dimensionality features, such as, average color and texture. Shape matching is supported using either dimensionality reduction, which is sensitive to rotation, translation and scaling [17], or by clustering using nonlinear elastic matching [9,4], which requires a significant amount of work per shape and some derived starting points as a matching guide. QBIC also supports video queries.

Ankerst et al [1] present a pixel-based shape similarity retrieval method that allows only minor rotation and translation. Their similarity criterion assumes a very high dimension (linear to the number of pixels in the image), therefore dimensionality reduction is performed.

Gary and Mehrotra [16,12,11] present a shape-based method, which stores each shape multiple times. More specifically, the shape is positioned by normalizing each of its edges. The space requirements of this method impose a significant overhead. The method is quite susceptible to noise, thus the authors present a sophisticated preprocessing phase to eliminate the noise effects. Finally, the method favors those shapes of the shape base, which have almost the same number of vertices as the query shape.

Hierarchical chamfer matching (see [6] for hierarchical chamfer matching and [3,5] for chamfer matching) creates a distance image using information from the edges, and then tries to minimize the root mean square average of the values in the distance map that a contour hit. Hierarchical chamfer matching gives quite accurate results rather insensitive to random noise, but involves lengthy computations on every extracted contour per query. In the hierarchical version a resolution pyramid is used to improve the performance of the matching algorithm.

In this work, we present a shape-based method, where information regarding the boundary of objects is automatically extracted and organized to support a poly-logarithmic (in the number of shape vertices) algorithm based on a novel similarity criterion. Specifically, this paper makes the following technical contributions:

- introduces a new similarity criterion for shapes, which works better in the context of image retrieval than traditional similarity criteria;
- describes a novel way of storing shapes which is tolerant to distortion;
- presents an efficient algorithm for finding the closest match to a given query shape; its time complexity is poly-logarithmic in the number of vertices of the shape base assuming uniform distribution of the vertices in the locus of their possible locations;
- the algorithm can be easily extended to retrieve the k best matches instead of the single best match.

The rest of this paper is organized as follows. Section 2 presents our similarity criterion for shapes and compares it with existing criteria. Section 3 describes the organization of the data describing the shapes, and the algorithm to retrieve the best match of a query shape. Section 4 presents some experimental results, while Section 5 concludes the paper and discusses future work.

2 Similarity Criteria

The Hausdorff distance is a well studied similarity measure between two point sets A and B. The directed Hausdorff distance h and the Hausdorff distance H are defined as follows:
$h(A, B) = \max_{a \in A} \min_{b \in B} d(a, b)$, $H(A, B) = \max(h(A, B), h(B, A))$,
where d is a point-wise measure, such as, the Euclidean distance. An inherent problem with the Hausdorff distance is that a point in A that is farthest from any point in B dominates the distance. To overcome this problem, Huttenlocher and Rucklidge have defined a *generalized discrete Hausdorff distance* (see e.g. [14]), given by the k-th largest distance rather than the maximum:
$h_k(A, B) = \text{kth}_{a \in A} \min_{b \in B} d(a, b)$, $H_k(A, B) = \max(h(A, B), h(B, A))$.

This metric eliminates somehow the farthest-point domination disadvantage of the Hausdorff metric, but works only for a finite set of points (it is mainly used for $k = m/2$ where m is the size of the point set). The generalized Hausdorff distance does not obey the metric properties.

An interesting alternative measure, called *nonlinear elastic matching*, is presented in [9]. This measure does not obey the traditional metric properties but a relaxed set of metric properties instead. In practice, this provides the same advantages as any metric, and therefore can be used for clustering. However, the arbitrary number of points distributed on the edges, the need of determining certain starting matching points and the complexity of computing such a match ($O(mn)$ using dynamic programming [2]) makes this measure inappropriate for very large data sets.

In our algorithm we use a new similarity criterion based on the average of minimum point distances:

$$h_{avg}(A, B) = \text{average}_{a \in A} \min_{b \in B} d(a, b)$$

This measure behaves nicely (gives intuitive results) and it can be computed quite efficiently, as is shown in the next Section. The metric properties do not hold for this measure either, but in some sense they hold for a representative average set of points probably different from the original point set. Figure 1 (left) illustrates an example where, using Hausdorff distance, the shape Q is matched with A instead of B (B is intuitively the closest match). According to the similarity measure used in our work, B is indeed closer to Q than A.

3 Efficient Retrieval of Similar Shapes

The algorithm is based on two key ideas: normalizing a shape about its diameter and the notion of the ϵ-envelope.

Normalizing about the diameter. In order to match a query shape to the shapes in the database, some kind of "normalization" is applied so that the matching is translation-, rotation-, and scaling-independent. In [12], Mehrotra and Gary normalize each shape about each of its edges: they translate, rotate, and scale the shape so that the edge is positioned at $((0, 0), (1, 0))$. Although this

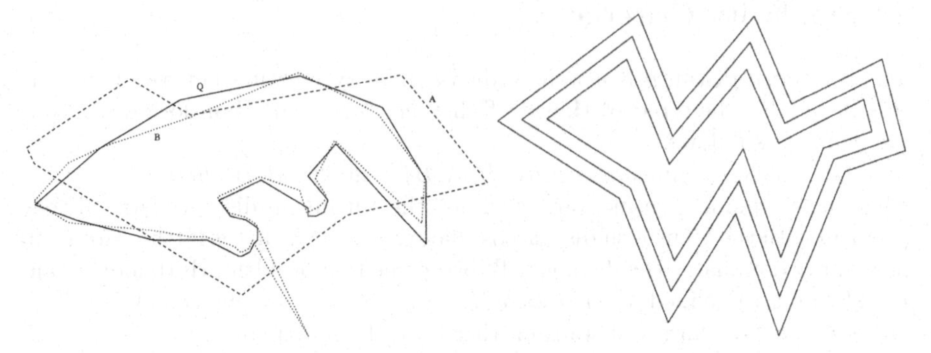

Fig. 1. (left) Depending on the similarity criterion, the query shape Q may be matched with A or B; (right) the ϵ-envelope

approach gives good results in many cases, it would fail to retrieve the distorted shape on the right of Figure 2, if the shape on the left of the figure was used as the query shape. In our retrieval system, instead of normalizing about the edges, we normalize about the diameter of the shape, i.e., by translating, rotating, and scaling so that the pair of shape vertices that are farthest apart are positioned at $(0,0)$ and $(1,0)$. This ensures better results, because the diameter is less susceptible to local distortion (like the one shown in Figure 2), which is very common in shapes extracted via automated image processing techniques.

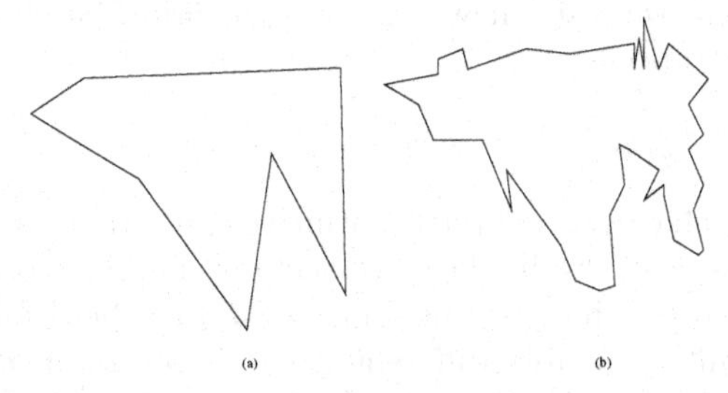

Fig. 2. (a) the query shape; (b) a distorted shape extracted from an image

The ϵ-envelope. The algorithm works by considering a "fattened" version of the query shape which is computed by taking lines parallel to the query shape edges at some distance ϵ on either side (Figure 1 (right)); we call this fattened shape the *ϵ-envelope*. The good matches are expected to fall inside or at least have most of their vertices inside the ϵ-envelope even for small ϵ. Therefore, if we start by using a small initial value of ϵ and keep increasing it, we expect to collect the good matches after a few iterations of this procedure.

The ϵ-envelope can be seen as a collection of trapezoids of height 2ϵ, one for each edge of the query shape. (For simplicity, we assume that ϵ is such that no two trapezoids are overlapping; the method can be extended to handle overlapping trapezoids.)

3.1 Populating the Shape Database

Populating the database of shapes is done by processing each available shape, a polygon or polyline extracted from an image, as follows. First, we compute the diameter of the shape, i.e., the pair of vertices that exhibit the longest Euclidean distance. In order to achieve even better tolerance to distortion, we will not simply normalize the shape about its diameter, as we alluded earlier; instead, we will normalize it about all its *α-diameters*, i.e., all pairs of vertices whose distance is at least $1-\alpha$ times the length of the diameter ($0 \leq \alpha < 1$). For each α-diameter, we scale, rotate, and translate the shape so that the α-diameter is positioned at $((0,0),(1,0))$; each shape is stored twice for each α-diameter by taking both ways to match the two vertices defining the α-diameter to the points $(0,0)$ and $(1,0)$. All these "normalized" copies of the shape constitute the *shape base*, the database of shapes.

Of course, a shape with s vertices may have $\Omega(n)$ α-diameters, which would effectively result in an $O(n^2)$-size database to store shapes of $O(n)$ total size. However, this happens to fairly regular shapes; shapes extracted via automated image processing techniques are unlikely to be regular. In fact, experiments have indicated that for $\alpha = 0.15$ the number of copies of each shape is about 12 on the average, including the doubling due to the double storage of a shape for a given α-diameter (the average number of edges per shape in the test set was 20).

3.2 Outline of the Matching Algorithm

The algorithm works by considering ϵ-envelopes of the query shape for (appropriately) increasing values of ϵ; for each such ϵ, the polygons that have most of their vertices inside the ϵ-envelope are determined and for each of them the value of the similarity measure to the query shape is computed. The algorithm stops whenever the best match has been found, or ϵ has grown "too large" implying that no good matches exist in the shape base. In the latter case, we revert to an alternative but compatible geometric hashing method which is outlined in the technical report version, due to space limitations.

In more detail, the basic steps of the algorithm for the retrieval of the database shape that best matches the query shape are:

1. We compute an initial value $\epsilon_1 = \epsilon_s$ such that the ϵ_1-envelope is likely to contain at least one shape of the shape base (see Section 3.3). We set $\epsilon_0 = 0$ and we signal that we are in the first iteration by assigning $i = 1$.
2. We collect the vertices of the database shapes that fall in the difference (ϵ_i-envelope $-$ ϵ_{i-1}-envelope); this can be achieved by partitioning this difference into triangles and preprocessing the vertices so that inclusion in a query triangle can be answered fast (simplex range searching). (If no vertices are found then the difference $\epsilon_i - \epsilon_{i-1}$ is increased geometrically.) Additionally, each time we find that a vertex of some shape is inside the above envelope difference, we increase a counter associated with that shape that holds the number of its vertices that are inside the ϵ_i-envelope.

3. If no shape of the shape base has at least a fraction $1-\beta$ of its vertices inside the ϵ_i-envelope (for a parameter β such that $0 \leq \beta < 1$), a new larger ϵ is computed (Section 3.5) and we go to step 5.
4. If there are shapes of the shape base that have at least a fraction $1-\beta$ of their vertices inside the ϵ_i-envelope (these are the *candidate shapes*), we process them as described in Section 3.4. During the processing, we may either conclude that the best match has been found, in which case it is reported to the user and the execution is complete, or a new larger value of ϵ is computed.
5. We increment i and set $\epsilon_i = \epsilon$. If ϵ_i does not exceed $\frac{A}{2pl_Q}\log^3 n$, we go to step 2 and repeat the procedure (A is the area of the locus of the normalized shapes (Section 3.3), p is the number of shapes in the shape base, n is the total number of vertices of the p shapes, and l_Q is the length of the perimeter of Q); otherwise, we report the best match so far (if any) and exit. If no match has been found, we employ geometric hashing.

The method converges and if there exist similar shapes it retrieves the best match.

3.3 Computing the Initial Width $2\epsilon_s$ of the ϵ-Envelope

We first compute an initial estimate $\hat{\epsilon}$ of the width of the ϵ-envelope based on an estimate $\tilde{K}_{\hat{\epsilon}}$ of the number of vertices that fall inside the envelope. Then, we calculate the actual number $K_{\hat{\epsilon}}$ of vertices of the database shapes. If $K_{\hat{\epsilon}}$ is at least half and no more than twice $\tilde{K}_{\hat{\epsilon}}$, we set $\epsilon_s = \hat{\epsilon}$; otherwise, we adjust ϵ by performing binary search in the values of K_{ϵ}.

The computation of $\hat{\epsilon}$ is done as follows. Let us compute the area A of the locus of the vertices of the normalized shapes. If the shapes were normalized with respect to their diameter only, then all the vertices would fall in the lune defined by two circles of radius 1 whose centers are at distance 1 apart. Since we store copies of each shape normalized for all pairs of vertices whose distance is at least $1-\alpha$ times the length of the diameter ($0 \leq \alpha < 1$), the area A is equal to the area of the lune defined by two circles of radius $\frac{1}{1-\alpha}$ whose centers are at distance 1 apart. This implies that $A = \frac{2}{(1-\alpha)^2} cos^{-1}(\frac{1-\alpha}{2}) - \sqrt{\frac{1}{(1-\alpha)^2} - \frac{1}{4}}$.

By assuming uniform distribution of the vertices inside this lune, the average number of vertices inside an ϵ-envelope around the query shape Q is estimated to:

$\bar{K}_{\epsilon} = \frac{2\epsilon l_Q}{A} n$, where l_Q and n are the length of the perimeter of the query shape Q and the total number of vertices of all the shapes of the shape base, respectively.

In order that the initial $\hat{\epsilon}$-envelope contains at least enough vertices for a candidate shape (at least a fraction $1-\beta$ of its vertices lie inside the envelope), we derive that:

$\bar{K}_{\hat{\epsilon}} \geq (1-\beta)\frac{n}{p}$ where p is the total number of shapes in the shape base.

The above estimate may yield the necessary number of vertices, but the probability that all of them belong to the same shape is very small. So, we determine experimentally a $\gamma(n)$, and set $\tilde{K}_{\hat{\epsilon}} = (1-\beta)\frac{n}{p}\,\gamma(n)$ which implies that $\bar{K}_{\hat{\epsilon}} = \tilde{K}_{\hat{\epsilon}} = (1-\beta)\frac{n}{p}\,\gamma(n) \;\Rightarrow\; \hat{\epsilon} = \frac{(1-\beta)A}{2\,p\,l_Q}\,\gamma(n)$.

Through experimentation, we have determined that a good choice for $\gamma(n)$ for relatively small shape bases (see Section 4) is: $\gamma(n) = 5\log n$.

3.4 Processing the Shapes

We first process all the new candidate shapes, that is, the shapes that have more than a fraction $1-\beta$ of their vertices inside the current ϵ-envelope but did not do so in the previous envelopes. For each such shape P_j, we compute the value of the similarity criterion of P_j with respect to the query shape Q, which we will call the *cost* c_j of P_j:

$c_j = \text{average}_{a\in P_j} \min_{b\in Q} d(a,b) = \frac{\int_{a\in P_j} \min_{b\in Q} d(a,b)}{length(P_j)}$, where by $length(P)$ we denote the length of the perimeter of P. The computation is done by intersecting each edge of P_j with the Voronoi diagram of Q; the boundary of P_j is thus split into segments that are close to either a vertex or an edge of Q. The contribution of each such segment s in $\int_{a\in s} \min_{b\in Q} d(a,b)$ can then be easily computed:

- if s is in the Voronoi region of an edge e of Q and s does not cross e, then the contribution of s is equal to $c(s) = \frac{d_1+d_2}{2}\,length(s)$, where d_1 and d_2 are the distances of the endpoints of s from e;
- if s is in the Voronoi region of an edge e of Q and s crosses e, then the contribution of s is equal to $c(s) = \frac{d_1^2+d_2^2}{2(d_1+d_2)}\,length(s)$, where d_1 and d_2 are the distances of the endpoints of s from e;
- if s is in the Voronoi region of a vertex v of Q, then the contribution c(s) is given by a more complicated expression, which (of course) only depends on the coordinates of the endpoints of s and of the vertex v.

Then, $c_j = \frac{\sum_s c(s)}{length(P_j)}$. If c_j is less than the cost c_{max} of the best match so far, then P_j becomes the current best match and c_{max} is set equal to c_j.

Next, we process all the shapes that are not yet candidates (and have at least a vertex other than $(0,0)$ and $(1,0)$ in the current ϵ-envelope), in order to determine whether we have found the best matches, and if not to produce a new larger ϵ for the ϵ-envelope. So, for each of these shapes, say, S_j, we compute the contribution $\int_{a\in e} \min_{b\in Q} d(a,b)$ of each of its edges e that has at least one endpoint inside the ϵ-envelope. Let the sum of all these contributions be t_j and let the total length of all these edges be d_j. We check whether $\frac{1}{length(S_j)}(t_j + \frac{\epsilon}{2}(length(S_j) - d_j)) > c_{max}$. If yes, the cost of S_j will not be less than c_{max}; this is so, because the edges of S_j with at least one endpoint in the current ϵ-envelope contribute t_j in $\int_{a\in e} \min_{b\in Q} d(a,b)$, whereas the remaining edges will contribute more than $\epsilon/2$ times their length (each edge has both endpoints at distance larger than ϵ away from Q). So, if the above inequality holds, we ignore S_j from now on. Otherwise, we compute, $\epsilon_j = \frac{2(length(S_j)\,c_{max} - t_j)}{length(S_j) - d_j}$ which turns the previous inequality into equality. Note that ϵ_j is larger than the current width ϵ of the envelope.

After all the S_js have been processed, we consider the set of collected ϵ_js. If the set is empty, then we have found the best match and we stop. Otherwise, we select the smallest element of the set and we use it as the new width ϵ of the envelope.

3.5 Increasing ϵ in the Absense of Candidate Shapes

In this case, all the shapes of our shape base have less than a fraction $1 - \beta$ of their vertices inside the current ϵ_i-envelope. Then for each of the shapes that have a vertex other than $(0, 0)$ and $(1, 0)$ in the envelope, we do the following.

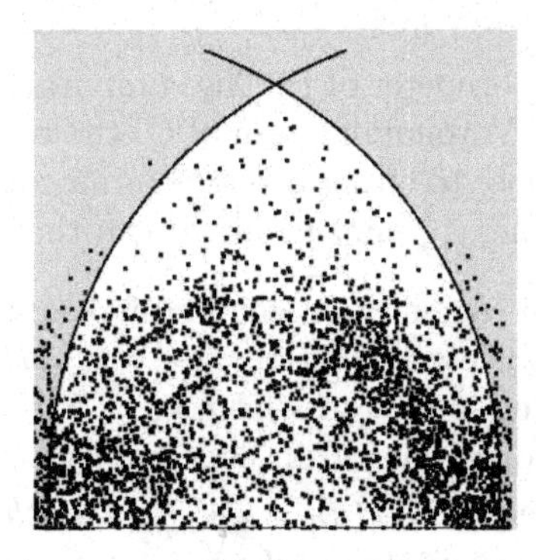

Fig. 3. Distribution of the vertices inside the upper part of the lune; the lower part is symmetric

Let P_j be such a shape with n_j vertices and let V_j be the set of its vertices that are inside the envelope. Consider the set V_j' of vertices of P_j that are adjacent to vertices in V_j; for each vertex v_k in $V_j' - V_j$, we compute the shortest distance $d_Q(v_k)$ of v_k from the query shape Q. If the total number of vertices in $V_j \cup V_j'$ exceeds $(1 - \beta)n_j$ (i.e., there are more than a fraction $1 - \beta$ of P_j's vertices in $V_j \cup V_j'$), then we set ϵ_j' equal to the $\big((1 - \beta)n_j - |V_j|\big)$-th smallest distance $d_Q(v_k)$; this implies that P_j will be a candidate shape in the ϵ_j'-envelope. In the case that the total number of vertices in $V_j \cup V_j'$ does not exceed $(1 - \beta)n_j$, then we set ϵ_j' equal to $\frac{(1-\beta)n_j}{|V_j|}\epsilon$ (i.e., we use linear interpolation in order to estimate the width of the envelope for which P_j will be a candidate shape).

After all these ϵ_j's have been computed, we collect the smallest among them and use it as the new ϵ of the envelope.

3.6 Time Complexity of the Matching Algorithm

Before analyzing the time complexities of each of the steps of the algorithm, we recall that in order to compute the similarity measure, we make use of the

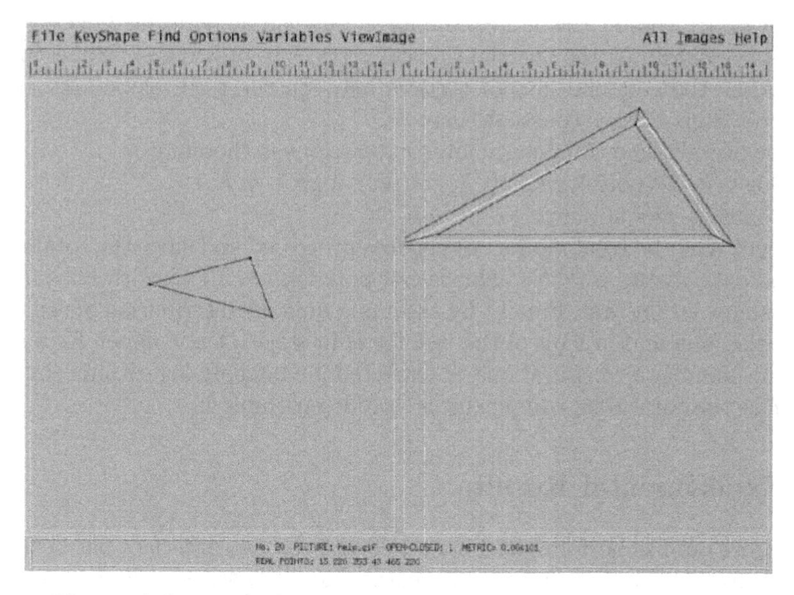

Fig. 4. A best match within the initial guess of the envelope

Voronoi diagram of the query shape Q. This can be computed in $O(m \log m)$ time, where m is the size of Q.

Step 1 of the algorithm begins with the computation of $\hat{\epsilon}$ which takes $O(1)$ time. Then, the number of vertices that fall inside the $\hat{\epsilon}$-envelope is computed; this can be done in $O(\text{poly-log}n)$ time using simplex range counting algorithms and quadratic or near-quadratic space data structures. If the computed number greatly differs from the expected number, $O(\log n)$ repetitions of the previous computation are done, resulting in $O(\text{poly-log}n)$ total time for this step.

In step 2, we need to compute the vertices of the shapes in our database that fall in the difference of the ϵ_i-envelope $-$ ϵ_{i-1}-envelope (this ensures that a vertex will not be processed or counted multiple times). The difference of the m trapezoids (one for each of the m edges of the query shape) can be decomposed into $O(m)$ triangles which can be used with simplex range reporting data structures of near-quadratic space complexity that take $O(\log^3 n + \kappa)$ time per query triangle, where n is the total number of vertices of the shape base and κ is the number of vertices that fall in the triangle [13]. (There are also quadratic-size data structures that allow for $O(\log n + \kappa)$ query time by employing fractional cascading [8].) Thus completing the i-th iteration of step 2 takes $O(m \log^3 n + K_i)$ time in total, where K_i is the number of vertices in all the query triangles for that iteration.

The i-th iteration of step 3 takes $O(mK_i)$ time, where K_i is again the number of vertices between the ϵ_i-envelope and the ϵ_{i-1}-envelope.

Step 4 involves processing the new candidate shapes and the non-candidate shapes. The former takes time $O(m|P_i|)$, where $|P_i|$ denotes the number of vertices of P_i. Processing the non-candidate shapes can be performed in $O(1+mK_i)$ time, by maintaining the contributions of vertices in previous envelopes and simply adding the contributions of vertices between the ϵ_i-envelope and the ϵ_{i-1}-envelope. Step 5 takes constant time.

The overall time complexity after r iterations is therefore,
$O(m \log m) + O(\text{poly-log}n) + \sum_{i=1}^{r} O(\text{poly-log}n + m\, K_i) =$
$O(m \log m) + O(r\, \text{poly-log}n) + O(m\, K)$

where K is the total number of vertices processed and since the total number of candidate shapes is $O(K)$. This is $O(r\, \text{poly-log}n + K)$ since the size m of the query shape is constant. Finally, by assuming uniform distribution of the vertices inside the lune and in light of the test for ϵ_i in step 5, the number K of vertices and the number r of iterations is expected to be poly-logarithmic in n, and therefore the total time complexity is poly-logarithmic in n.

4 Experimental Results

The algorithm has been implemented in C and the user interface has been developed using Tcl/Tk. We currently have a stable version running on a Sun Solaris platform. The software is easily portable to various platforms. The user is first presented with a workspace where she/he can draft a query sketch, which is subsequently presented first to our matching algorithm and in case of failure to the geometric hashing approach.

We have performed experiments with around 100 images and 350 actual shapes. Each shape is stored on the average approximately 12 times resulting in a shape base populated with 3000 normalized shapes. The total number of vertices was $n = 30000$. The distribution of the vertices was not uniform because of the specialized nature of the images (see Figure 3).

In the experiments we used $\alpha = 0.15$, $\beta = 0.15$ and $\gamma(n) = 5 \log_2 n$. The initial estimation of K_ϵ was usually very close to the actual number of vertices that fell in the envelope. In the case of Figure 4, we started with an initial estimation of 244 for K_ϵ which gave an $\epsilon_1 = 0.0030$ with an actual 180 vertices inside the envelope. In this case we obtain immediately a best match with cost $c = 0.0013$. Similarly in the case of Figure 5, we find a best match with the first iteration with slightly larger cost $c = 0.0792$ since two edges of the matched shapes are partially outside the envelope. Finally, in Figure 6 we find a best match after a second refinement iteration; in this case, the returned shape for a triangular query shape is a polygon with 10 vertices that has a cost around $c = 0.0008$. Even though the vertices are not uniformly distributed in the lune, the algorithm behaves as expected in terms of number of iterations and time complexity.

5 Conclusions and Future Work

We have presented an efficient noise tolerant shape-based approach to image retrieval. The algorithm currently yields the best match, but it can be easily

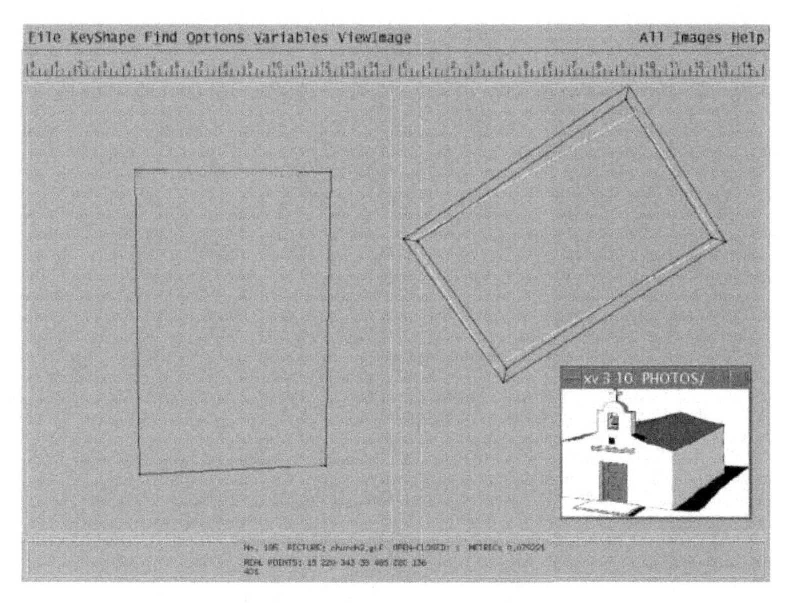

Fig. 5. Querying with an orthogonal shape; the door of the church is the best match

extended the k best matches; in this case, a heap of size k is used to hold the k current best candidates.

We are currently incorporating this algorithm in a video retrieval system, which will allow us to experiment with larger shape bases. Other future research directions include finding alternative ways to do the range searching (whose space requirement is high), ensuring robust calculations, adding 3D awareness support, and using relative position information for allowing more complicated queries such as containment and tangency.

References

1. M. Ankerst, H. P. Kriegel, and T. Seidl. Multistep approach for shape similarity search in image databases. *IEEE Transactions on Knowledge and Data Engineering*, 10(6):996–1004, 1998. 506
2. E. M. Arkin, L. P. Chew, D. P. Huttenlocher, K. Kedem, and J. S. B. Mitchell. An efficiently computable metric for comparing polygonal shapes. *IEEE Transactions on Knowledge and Data Engineering*, 13(3):209–216, 1997. 507
3. H. G. Barrow, J. M. Tenenbaum, R. C. Bolles, and H. C. Wolf. Parametric correspondence and chamfer matching: Two new techniques for image matching. In *Proc. of the 5th IJCAI*, pages 659–663, Cambridge, MA, 1977. 506
4. A. Del Bimbo and P. Pala. Visual image retrieval by elastic matching of user sketches. *IEEE Transactions on Knowledge and Data Engineering*, 19(2):121–132, 1997. 506

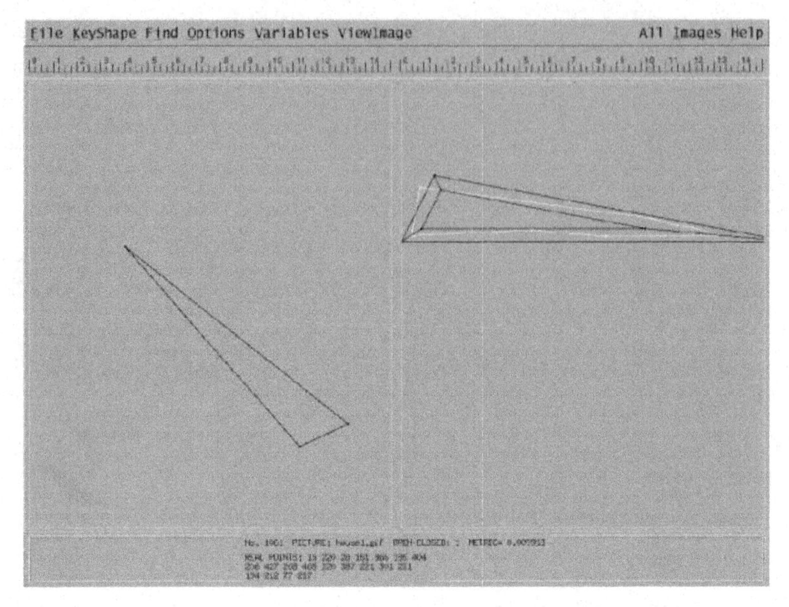

Fig. 6. The best match to this triangular query is a polygon with 10 vertices

5. G. Borgefors. An improved version of the chamfer matching algorithm. In *ICPR1984*, pages 1175–1177, 1984. 506
6. G. Borgefors. Hierarchical chamfer matching: A parametric edge matching algorithm. *IEEE Transactions on Pattern Analysis and Machine Intelligence*, 10(6):849–865, 1988. 506
7. C. Carson and V. E. Ogle. Storage and retrieval of feature data for a very large online image collection. *IEEE Bulletin of the Tech. Comm. on Data Engineering*, 19(4):19–27, 1996. 505
8. B. Chazelle and L. J. Guibas. Fractional cascading: I. a data structuring technique; II. applications. *Algorithmica*, 1:133–191, 1986. 513
9. Ronald Fagin and Larry Stockmeyer. Relaxing the triangle inequality in pattern matching. *International Journal of Computer Vision*, to appear, 1999. 506, 507
10. M. Flickner, H. Sawhney, W. Niblack, J. Ashley, Q. Huang, B. Dom, M. Gorkani, J. Hafner, D. Lee, D. Petkovic, D. Steele, and P. Yanker. QBIC: Query by image and video content. *IEEE Computer*, 28(9):23–32, 1995. 505
11. J. E. Gary and R. Mehrotra. Similar shape retrieval using a structural feature index. *Information Systems*, 18(7):527–537, 1993. 506
12. J. E. Gary and R. Mehrotra. Feature-index-based similar shape retrieval. In S. Spaccapietra and R. Jain, editors, *Visual Database Systems*, volume 3, pages 46–65, 1995. 506, 507
13. J. E. Goodman and J. O'Rourke. *Handbook of Discrete and Computational Geometry*. CRC Press LLC, 1997. 513
14. D. P. Huttenlocher and W. J. Rucklidge. A multi-resolution technique for comparing images using the hausdorff distance. Technical Report TR92-1321, CS Department, Cornell University, 1992. 507
15. IBM. Ibm's query by image content (QBIC) homepage. http: //wwwqbic.almaden.ibm.com. 505

16. R. Mehrotra and J. E. Gary. Similar-shape retrieval in shape data management. *IEEE Computer*, 28(9):57–62, 1995. 506
17. W. Niblack, R. Barber, W. Equitz, M. Flickner, E. Glassman, D. Petkovic, and P. Yanker. The QBIC project: querying images by content using color, texture and shape. In *Proc. SPIE Conference on Storage Retrieval for Image and Video Databases*, volume 1908, pages 173–181. SPIE, 1993. 506

Towards Feature Fusion – The Synthesis of Contour Sections Distinguishing Contours from Different Classes

Dag Pechtel and Klaus-Dieter Kuhnert

University of Siegen
57068 Siegen, Germany
{pechtel,kuhnert}@pd.et-inf.uni-siegen.de
http://www.pd.et-inf.uni-siegen.de

Abstract. In real world problems, where the objects are in general complex and deformed, the automatic generation of local characteristics is necessary in order to distinguish different object classes.
This paper presents an approach towards the automatic synthesis of significant local contour sections of closed, discrete, complex, and deformed 2D-object contours for the distinction of different classes. Neighboring contour points are determined and synthesized (feature fusion) into feature groups. Exclusively with the help of these feature groups the method distinguishes between different 2D-object contour classes of a certain domain. The basic idea is to get only the necessary information of a contour or a contour class for recognition.

1 Introduction

A basic problem in the field of pattern recognition is the automatic synthesis of elementary features to higher level features (feature fusion) , that can distinguish between certain pattern classes of a task domain.

This paper presents an approach for closed discrete 2D-object contours, whose base features, the contour points, are nearly automatically fused to local and significant feature groups, called the significant contour sections. These significant contour sections distinguish a contour from the contours of other contour classes in a task domain. E.g. a significant contour section of a bottle could be its bottle neck and significant contour sections of a fish could be its fins. The method even works with contours that are complex and/or badly deformed.

There are many approaches in literature concerned with the analysis of contour similarities [L97]. There are alignment approaches [U96], the use of invariant properties (basics in [PM47]) that is often a global approach, and the use of parts ([HS97], [LL98b], [KTZ94], [SIK95], [STK95]) that is sometimes based on shape simplification [LL98a], or structural descriptions ([KW96], [P77]). Some works deal with deformation and minimize a cost function: [BCGJ98] define 'elastic energy' needed to transform one contour to the other. [SGWM93] is concerned with shape interpolation, also known as morphing, or the sequence of contour points

G. Borgefors, I. Nyström, and G. Sanniti di Baja (Eds.): DGCI 2000, LNCS 1953, pp. 518–529, 2000.
© Springer-Verlag Berlin Heidelberg 2000

is interpreted as a string in which points are substituted (shifted), inserted, and deleted ([PK00], [SK83]). This work is based on [PK00]. It uses shares of the alignment approach and isn't psychophysically motivated. This work is concentrated on the nearly automatic synthesis of **local** contour sections for recognition and is a step towards the autonomous creation of abstract local characteristics of a contour class consisting of a variety of similar, but not equal objects (eg. the birds in fig. 8). Besides the variety, we also interpret outer influences (eg. occlusion, bending) to an object, autonomous motions (eg. stretching out an arm) of an object, and different projections of an object all as deformations.

Basic notations used in this article are presented in section 2. In section 3 the determination of important contour sections, which distinguish between two different 2D-object contours, is briefly sketched. Section 4 shows how the important contour sections of one contour are combined to significant contour sections of that contour. In section 5 experimental results are presented based on a data set of deformed contours of reusable material collections (plastic bottles, beverage cartons (fig. 7) and a data set of birds, fishes, and plastic bottles (fig. 8). With a summary in section 6 this paper concludes.

2 Notation

The task domain D consists of N classes K_i:

$$D = \{K_1, \ldots, K_i, \ldots, K_N\}. \tag{1}$$

Each class K_i consists of m_i closed, discrete 2D-object contours $C_{i,j}$:

$$K_i = \{C_{i,1}, \ldots, C_{i,j}, \ldots, C_{i,m_i}\}. \tag{2}$$

Each contour $C_{i,j}$ consists of $n_{i,j}$ neighboring successive points $P_{i,j,k}$ with equal Euclidean distances d. Each contour consists also of $n_{i,j}$ sections $\tilde{C}_{i,j,k}^{l_{i,j,k}}$:

$$C_{i,j} = \{P_{i,j,1}, \ldots, P_{i,j,k}, \ldots, P_{i,j,n_{i,j}}\} \tag{3}$$

$$C_{i,j} = \{\tilde{C}_{i,j,1}^{l_{i,j,1}}, \ldots, \tilde{C}_{i,j,k}^{l_{i,j,k}}, \ldots, \tilde{C}_{i,j,n_{i,j}}^{l_{i,j,n_{i,j}}}\}. \tag{4}$$

Each section $\tilde{C}_{i,j,k}^{l_{i,j,k}}$ consists of $(2l_{i,j,k} + 1), l_{i,j,k} \geq 1$ points $P_{i,j,k}$:

$$\tilde{C}_{i,j,k}^{l_{i,j,k}} = \{P_{i,j,k-l_{i,j,k}}, \ldots, P_{i,j,k}, \ldots, P_{i,j,k+l_{i,j,k}}\}. \tag{5}$$

Therefore, neighboring sections have at least 2 common points $P_{i,j,k}$. Finally each point $P_{i,j,k}$ consists of a x- and a y-coordinate:

$$P_{i,j,k} = \{x_{i,j,k}, y_{i,j,k}\}. \tag{6}$$

Each section $\tilde{C}_{i,j,k}^{l_{i,j,k}}$ has an orientation vector $o_{i,j,k}$:

$$o_{i,j,k} = P_{i,j,k+l_{i,j,k}} - P_{i,j,k-l_{i,j,k}}. \tag{7}$$

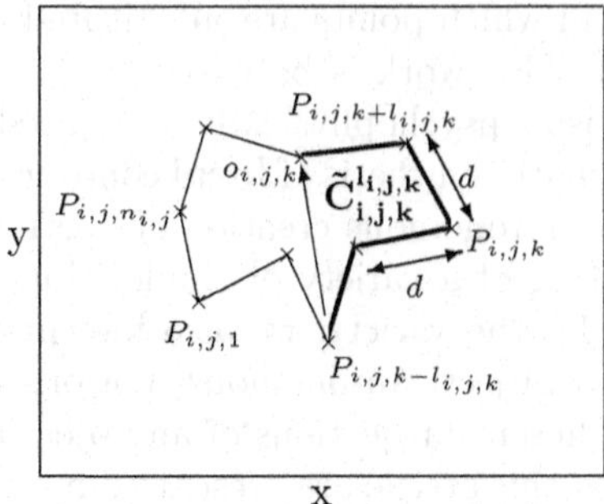

Fig. 1. Example of a closed discrete contour $C_{i,j}$ with one of $n_{i,j}$ sections $\tilde{C}^{l_{i,j,k}}_{i,j,k}$ (marked bold; $l_{i,j,k} = 2$) and the corresponding orientation vector $o_{i,j,k}$. Successive points have all equal Euclidean distances d

An example of such a discrete contour is shown in fig.1.

Two contours are distinguished with the following indices: $C_{i,j}$ and $C_{r,s}$. Two points or sections are distinguished as follows: $P_{i,j,k}, P_{r,s,t}$ and $\tilde{C}^{l_{i,j,k}}_{i,j,k}, \tilde{C}^{l_{r,s,t}}_{r,s,t}$, respectively.

A local comparison of two sections results in a local similarity $\lambda_{(i,j,k),(r,s,t)}$:

$$f_1: \quad \lambda_{(i,j,k),(r,s,t)} = f_1(\tilde{C}^{l_{i,j,k}}_{i,j,k}, \tilde{C}^{l_{r,s,t}}_{r,s,t}) \tag{8}$$

and is entered in the kth row and in the tth column of a local similarity matrix $\Lambda^{(i,j),(r,s)}$:

$$\Lambda^{(i,j),(r,s)}_{k,t} = \lambda_{(i,j,k),(r,s,t)} \tag{9}$$

which results from a local similarity analysis of all possible section pairs of two contours $C_{i,j}$ and $C_{r,s}$:

$$f_2: \quad \Lambda^{(i,j),(r,s)} = f_2(C_{i,j}, C_{r,s}). \tag{10}$$

The point indices have to be calculated modulo, because the contours are closed.

3 Determination of Important Contour Sections

In [PK00] a method is presented for the determination of important contour sections that distinguish between 2 different contours. That method is based on a local similarity analysis of all possible section pairs $\tilde{C}^{l_{i,j,k}}_{i,j,k}, \tilde{C}^{l_{r,s,t}}_{r,s,t}$ with a constant number of points $(2l_{i,j,k} + 1) = (2l_{r,s,t} + 1) = const$ and is shortly recapitulated here in a little changed form.

In fig. 2 is shown an example for the local comparison of two sections $\tilde{C}^2_{i,j,k}$, $\tilde{C}^2_{r,s,t}$ with $l_{i,j,k} = l_{r,s,t} = 2$. First, the sections were translated, that their starting points $P_{i,j,k-2}, P_{r,s,t-2}$ coincided with the origin of the Carthesian coordinate system. Second, the sections were rotated around the origin that their orientation vectors $o_{i,j,k}, o_{r,s,t}$ coincide with the positive x-axis. And third, section $\tilde{C}^2_{r,s,t}$

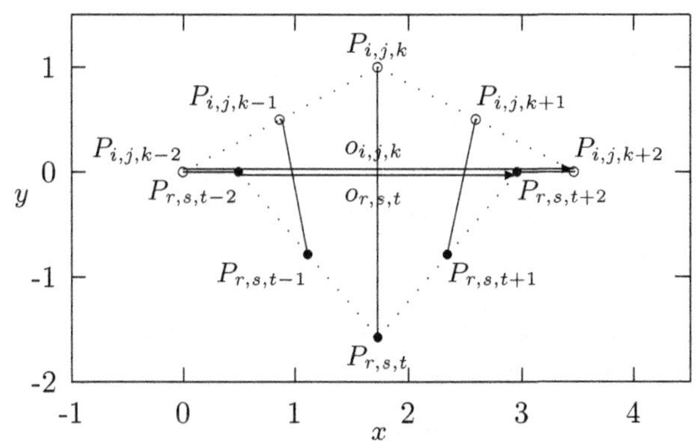

Fig. 2. Example for the comparison of two local contour sections $\tilde{C}^2_{i,j,k}$, $\tilde{C}^2_{r,s,t}$ with $l_{i,j,k} = l_{r,s,t} = 2$ after translation, rotation (orientation vectors $o_{i,j,k}$, $o_{r,s,t}$ coincide with the positive x), and minimizing the sum of Euclidean distances of relating contour points. The 5 relating contour points are connected with 5 straight lines

was translated along the x-axis, that the sum of squared Euclidean distances of relating contour points, eg. starting points or ending points of the sections, is minimized. The 5 relating contour points are connected with 5 straight lines in fig. 2.

Here, we search for the maximum number of points $l_{i,j,k} = l_{r,s,t}$ of the section pair $\tilde{C}^{l_{i,j,k}}_{i,j,k}, \tilde{C}^{l_{r,s,t}}_{r,s,t}$ around the points $P_{i,j,k}, P_{r,s,t}$, that the local distance (dissimilarity) $\epsilon \leq E$. E is the given maximum permitted distance (dissimilarity) between 2 sections. ϵ is calculated with an approximate one-to-one point matching ([PK00]; a survey can be found in [AG96]).

With this definitions and the notation in section 2, the steps for the calculation of the $\Lambda^{(i,j),(r,s)}_{k,t}$ are combined in the algorithm shown in fig. 3. This results in a $n_{i,j} \times n_{r,s}$ - matrix $\Lambda^{(i,j),(r,s)} = [\Lambda^{(i,j),(r,s)}_{k,t}]$ of local similarities. The elements $\Lambda^{(i,j),(r,s)}_{k,t}$ give the maximum length of the sections $\tilde{C}^{l_{i,j,k}}_{i,j,k}$ and $\tilde{C}^{l_{r,s,t}}_{r,s,t}$ around the points $P_{i,j,k}$ and $P_{r,s,t}$, with respect to a given maximum dissimilarity E. In other words, $\Lambda^{(i,j),(r,s)}_{k,t}$ gives the osculating length of two contour sections that can be treated equal in respect to this fixed maximum error E.

In order to get the transformation that possesses minimum cost for the deformation from $C_{i,j}$ to $C_{r,s}$ exactly one **monotone** discrete path (list of neighbored elements) $\hat{P}^{(i,j),(r,s)}_{\psi^{\min}}$ is searched in $\Lambda^{(i,j),(r,s)}$. For that, the cost function

$$G_{(i,j),(r,s)} = \sum_{z=1}^{Z} \hat{P}^{(i,j),(r,s)}_{\psi_z} \quad , \forall\ \psi \in \{1, 2, \ldots, \psi^{\min}, \ldots, \Psi\} \quad \rightarrow \max \tag{11}$$

with $max(n_{i,j}, n_{r,s}) \leq Z < (n_{i,j} + n_{r,s})$ is maximized with the help of *dynamic programming technique* (e.g. [B57], [SK83]). Because of the existence of

for $k = 0 \ldots n_{i,j}$
  for $t = 0 \ldots n_{r,s}$
    $\Lambda_{k,t}^{(i,j),(r,s)} = 0$
    while $\epsilon \leq E$
      $\Lambda_{k,t}^{(i,j),(r,s)} = \Lambda_{k,t}^{(i,j),(r,s)} + 1$
      $\tilde{C}_{i,j,k}^{l_{i,j,k}} = \{P_{i,j,k-\Lambda_{k,t}^{(i,j),(r,s)}}, \ldots, P_{i,j,k}, \ldots, P_{i,j,k+\Lambda_{k,t}^{(i,j),(r,s)}}\}$
      $\tilde{C}_{r,s,t}^{l_{r,s,t}} = \{P_{r,s,t-\Lambda_{k,t}^{(i,j),(r,s)}}, \ldots, P_{r,s,t}, \ldots, P_{r,s,t+\Lambda_{k,t}^{(i,j),(r,s)}}\}$
      // Calculate the translation of $\tilde{C}_{i,j,k}^{l_{i,j,k}}$ and $\tilde{C}_{r,s,t}^{l_{r,s,t}}$ into the origin
      // of the Cartesian coordinate system
      $\check{C}_{i,j,k}^{l_{i,j,k}} = \{0, \ldots, P_{i,j,k} - P_{i,j,k-\Lambda_{k,t}^{(i,j),(r,s)}}, \ldots, P_{i,j,k+\Lambda_{k,t}^{(i,j),(r,s)}} - P_{i,j,k-\Lambda_{k,t}^{(i,j),(r,s)}}\}$
      $\check{C}_{r,s,t}^{l_{r,s,t}} = \{0, \ldots, P_{r,s,t} - P_{r,s,t-\Lambda_{k,t}^{(i,j),(r,s)}}, \ldots, P_{r,s,t+\Lambda_{k,t}^{(i,j),(r,s)}} - P_{r,s,t-\Lambda_{k,t}^{(i,j),(r,s)}}\}$
      // Calculate the orientation vectors of $\check{C}_{i,j,k}^{l_{i,j,k}}$ and $\check{C}_{r,s,t}^{l_{r,s,t}}$
      $o_{i,j,k} = P_{i,j,k+\Lambda_{k,t}^{(i,j),(r,s)}} - P_{i,j,k-\Lambda_{k,t}^{(i,j),(r,s)}}$
      $o_{r,s,t} = P_{r,s,t+\Lambda_{k,t}^{(i,j),(r,s)}} - P_{r,s,t-\Lambda_{k,t}^{(i,j),(r,s)}}$
      rotate $\check{C}_{i,j,k}^{l_{i,j,k}}$ **and** $\check{C}_{r,s,t}^{l_{r,s,t}}$ // that the orientation vectors $o_{i,j,k}$ and $o_{r,s,t}$
        // coincide with the positive x-axis
        // resulting in $\hat{C}_{r,s,t}^{l_{r,s,t}}$ and $\hat{C}_{r,s,t}^{l_{r,s,t}}$
      // Calculate the minimum sum of squared Euclidean distances ϵ
      // of contour points belonging together (1-to-1 match)

$$\epsilon = \sum_{a=-\Lambda_{k,t}^{(i,j),(r,s)}}^{+\Lambda_{k,t}^{(i,j),(r,s)}} \left(\left(\hat{x}_{i,j,k+a} - \overline{\hat{x}_{i,j,k}} - (\hat{x}_{r,s,t+a} - \overline{\hat{x}_{r,s,t}}) \right)^2 + \right.$$
$$\left. + \left(\hat{y}_{i,j,k+a} - \hat{y}_{r,s,t+a} \right)^2 \right)$$

      // with $\overline{\hat{x}_{i,j,k}}, \overline{\hat{x}_{r,s,t}}$ are the means of the
      // $(2\Lambda_{k,t}^{(i,j),(r,s)} + 1)$ x-coordinates $\hat{x}_{i,j,k+a}, \hat{x}_{r,s,t+a}$
    end
  end
end

Fig. 3. Algorithm for calculating the elements $\Lambda_{k,t}^{(i,j),(r,s)}$ of the local similarity matrix $\Lambda^{(i,j),(r,s)}$ of 2 contours $C_{i,j}, C_{r,s}$. Comments are marked with //

$\Psi = min(n_{i,j}, n_{r,s})$ possible start elements for $\hat{P}_{\psi}^{(i,j),(r,s)}$ in $\Lambda^{(i,j),(r,s)}$ (rotation), that path is selected over all ψ causing the lowest cost (invariance against rotation).

Sections of $\hat{P}_{\psi_{\min}}^{(i,j),(r,s)}$ that contribute a little to $G_{(i,j),(r,s)}$, i.e. the osculating length is short, map those contour sections of $C_{i,j}$ and $C_{r,s}$ onto each other, that are different. Thus, those contour sections are important contour sections I for the distinction of the contours $C_{i,j}$ and $C_{r,s}$.

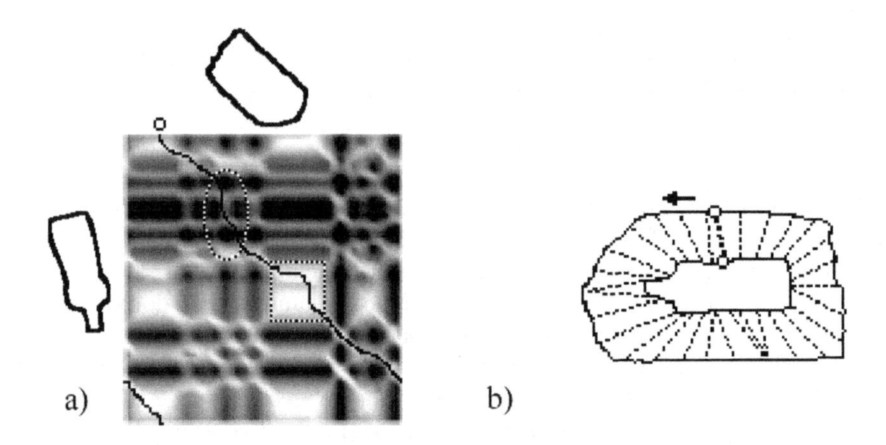

Fig. 4. Mapping of contours $C_{1,5}$ and $C_{2,3}$ (fig. 7) onto each other with lowest cost: a) Visualized similarity matrix $\Lambda^{(1,5),(2,3)}$ of contours $C_{1,5}$ (plastic bottle) and $C_{2,3}$ (beverage carton) with path $\hat{P}^{(1,5),(2,3)}_{\psi_{\min}}$ (marked in black), b) Visualized contour points mapped onto each other with respect to $\hat{P}^{(1,5),(2,3)}_{\psi_{\min}}$ in fig. 4a

An example of such a path $\hat{P}^{(1,5),(2,3)}_{\psi_{\min}}$ for the contours $C_{1,5}$ and $C_{2,3}$ (fig. 7) is shown in fig. 4. The local similarity matrix $\Lambda^{(1,5),(2,3)}$ with $\hat{P}^{(1,5),(2,3)}_{\psi_{\min}}$ is shown in fig. 4a. Elements of $\Lambda^{(1,5),(2,3)}$ are scaled between 0 and 255. Thus, $\Lambda^{(1,5),(2,3)}$ can be visualized as an image. Light (marked with a dotted square) and dark (marked with a dotted ellipse and belongs to the bottle neck) sections respectively in $\Lambda^{(1,5),(2,3)}$ or on $\hat{P}^{(1,5),(2,3)}_{\psi_{\min}}$ (marked in black) map contour sections with high and low local similarity respectively onto each other. $\hat{P}^{(1,5),(2,3)}_{\psi_{\min}}$ starts at an element in the first row marked with a little circle and ends in the last row ($\hat{P}^{(1,5),(2,3)}_{\psi_{\min}}$ have to be always "closed" - imagine that the visualized $\Lambda^{(1,5),(2,3)}$ is bent to a torus). Corresponding circles are found in fig. 4b, too. Here, the mapping of every 20th contour point is visualized with respect to contour $C_{1,5}$.

For further details see [PK00].

4 Synthesis of Significant Contour Sections

With the method described in section 3 important contour sections I can be determined, that distinguish between two 2D-object contours.

This section describes an approach how to generate significant contour sections S for a fixed contour $C_{A,B}$, belonging to a fixed contour class K_A, from all important sections I, that came from the comparison with all contours $C_{r,s}$ of the other classes $K_r, r \neq A$.

First, all local similarity matrices $\Lambda^{(A,B),(r,s)}, A \neq r$ and all paths $\hat{P}^{(A,B),(r,s)}_{\psi^{\min}}$, $A \neq r$ have to be calculated (see section 3). Those sections of $C_{A,B}$, which belong to elements z of $\hat{P}^{(A,B),(r,s)}_{\psi^{\min}}$ satisfying the condition

$$\hat{P}^{(A,B),(r,s)}_{\psi^{\min}_z} < \Theta \tag{12}$$

are the important contour sections of $C_{A,B}$. Θ is constant for all I and is called **importance value**. I.e., now there is a list of all I of the contour $C_{A,B}$ with respect to all contours $C_{r,s}$. In general, this list includes both I, that overlap each other, i.e. they have also equal contour points, and I, that do not overlap each other, i.e. they do not have a single contour point in common.

An example in principle is shown in fig. 5. The task domain D consists of 3 classes. Class K_1 consists of 3 rectangles, class K_2 consists of 2 ellipses, and class K_3 consists of 1 triangle. In this example the fixed contour $C_{A,B} = C_{1,2}$. The important contour sections of $C_{1,2}$ and the corresponding important contour sections of all contours of the other classes K_2, K_3 are marked in bold. This results in the marked 4 groups of overlapping important contour sections.

A reasonable assumption is that contours of the same class are more similar to themselves than contours of different classes. Another reasonable assumption is that contours of the same class are more similar to themselves because they have significant contour sections S. The last assumption is reasonable because human beings are able to recognize objects even if the objects are occluded. And there is another assumption: many overlapping important contour sections I with only a little variation in their position on the contour give a hint that this sections I are significant contour sections S of this contour with respect to the task domain D. But which I is the representative one? There is no general answer to this question. If there are Q_p overlapping important sections $I_q, 1 \leq q \leq Q_p$ a better way is to answer the follwing question:

How are Q_p overlapping I_q combined to 1 S_p?

There are many possible approaches to do that. One possibility is to take the minimum starting point index $\check{s}_q^{\min} = \min\{\check{s}_q\}$ and the maximum ending point index $\check{e}_q^{\max} = \max\{\check{e}_q\}$ as starting and ending point indexes $\check{S}_p, \check{E}_p$ of S_p. But this results in very long S_p. Good results were obtained with the calculation of the arithmetic means $\overline{\check{s}_q}$, $\overline{\check{e}_q}$ and the variances $\overline{\overline{\check{s}_q}}$, $\overline{\overline{\check{e}_q}}$ both of the Q_p starting point indexes $\check{s}_q$, and the Q_p ending point indices $\check{e}_q$ of all overlapping I_q. Then, this results in the following starting and ending point indices $\check{S}_p$, $\check{E}_p$ of the S_p:

$$\check{S}_p = round(\overline{\check{s}_q} - \sqrt{\overline{\overline{\check{s}_q}}}) \quad , \quad \check{E}_p = round(\overline{\check{e}_q} + \sqrt{\overline{\overline{\check{e}_q}}}) \quad , 1 \leq p \leq \Pi. \tag{13}$$

Π is the total number of S_p of a contour and *round*(...) means: round to the nearest integer. A simple example for the combination (fusion) of two I_q is shown in fig. 6.

Since contour $C_{A,B}$ could have several S_p, another question arises:

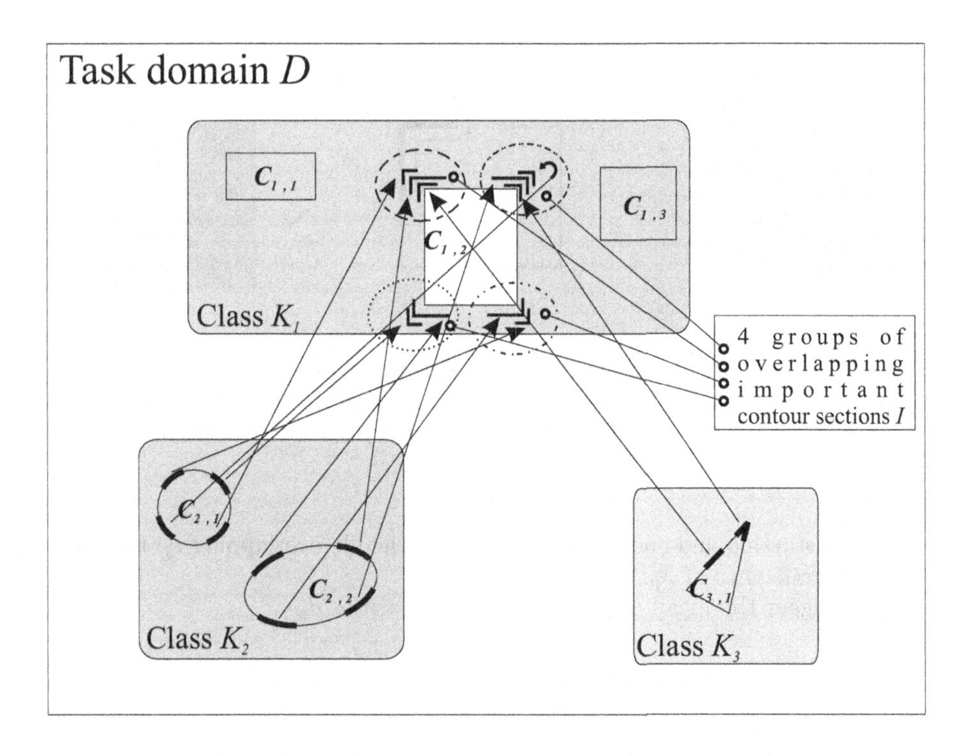

Fig. 5. Example for overlapping and not overlapping important contour sections: 3 classes consisting of 3 rectangles, 2 ellipses, and 1 triangle (K_1, K_2, K_3) are the task domain. Here, contour $C_{1,2}$ is the fixed contour. All important contour sections of $C_{1,2}$ and all corresponding contour sections of the contours of the other classes K_1, K_2 are marked in bold. This results in 4 different groups each consists of overlapping important contour sections (marked with different dotted ellipses), but no important contour sections of two different groups are overlapping each other

On what depends the significance of the generated S_p and what is a reasonable significance measure σ_p?

In general the contour $C_{A,B}$ have several significant sections S_p. It's reasonable to limit the number of S_p. Because of equation 12 there are S_p that comes from not overlapping important sections I_q with path elements $\hat{P}^{(A,B),(r,s)}_{\psi_z^{\min}}$ near Θ. These S_p have only a few points. Therefore a reasonable significance measure σ_p of S_p should have at least 3 dependencies:

- The bigger the number Q_p of overlapping I_q, the bigger the significance of S_p.
- The smaller the variation

$$v_p = \overline{\overline{\check{s}_q}} + \overline{\overline{\check{e}_q}} \tag{14}$$

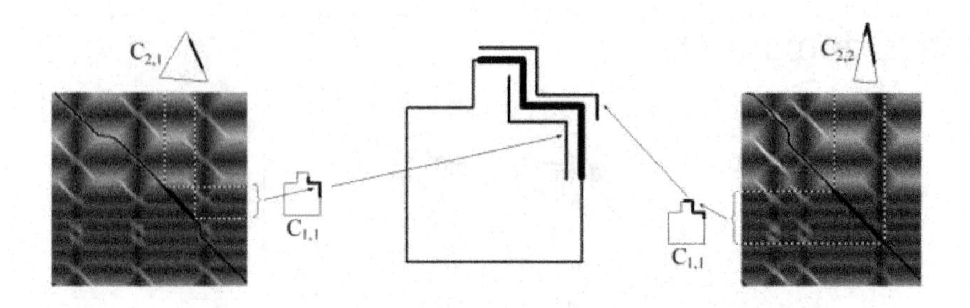

Fig. 6. Example for the fusion of 2 I_q of a sketched bottle $C_{1,1}$ with respect to a class consisting of 2 triangles $C_{2,1}, C_{2,2}$. Important path sections are marked in black. Corresponding important contour sections I are marked bold on the contours

of the starting and ending point indices of the Q_p overlapping I_q, the bigger the significance of S_p.

– The bigger the mean of the sum of dissimilarities d_p

$$d_p = \frac{1}{Q_p} \sum_{q=1}^{Q_p} \frac{1}{w_q} \quad with \quad w_q = \sum_{z=\check{s}_q}^{\check{e}_q} \hat{P}_{\psi_z^{\min}}^{(A,B),(r,s)} \tag{15}$$

between the Q_p individual overlapping I_q and the Q_p corresponding contour sections of the contours $C_{r,s}$, the bigger the significance. w_q is the similarity of two individual corresponding important contour sections of $C_{A,B}$ and $C_{r,s}$.

One possibility to combine this 3 named dependencies in a significance measure is as follows:

$$\sigma_p = \frac{Q_p^\alpha \; d_p^\beta}{1 + {\upsilon_p}^\gamma} \quad , \quad \alpha, \beta, \gamma \in R^+ \tag{16}$$

that is abbreviated called **significance** σ_p of S_p. α, β, γ are weightings for the 3 dependencies. Now, with the equations (12)-(16) one S_p is completely described.

With equation 16 it's possible to limit the number of S_p on those S_p, that satisfy the condition

$$\sigma_p \geq \Phi \; \sigma_p^{\max} \quad , 0.0 \leq \Phi \leq 1.0. \tag{17}$$

$\sigma_p^{\max} = max\{\sigma_p\}$ means the maximum significance of all S_p of $C_{A,B}$. Those S_p, that are more significant than other S_p with respect to equation (17) are called **more significant contour sections** $M_v, 1 \leq v \leq V$ and the S_p that is the one with the highest significance is called **most significant contour section** $\hat{M}$.

5 Experiments

The presented method was tested with a task domain D_1 consisting of 2D-object contours of packagings of reusable plastic material (fig. 7). The collecting process

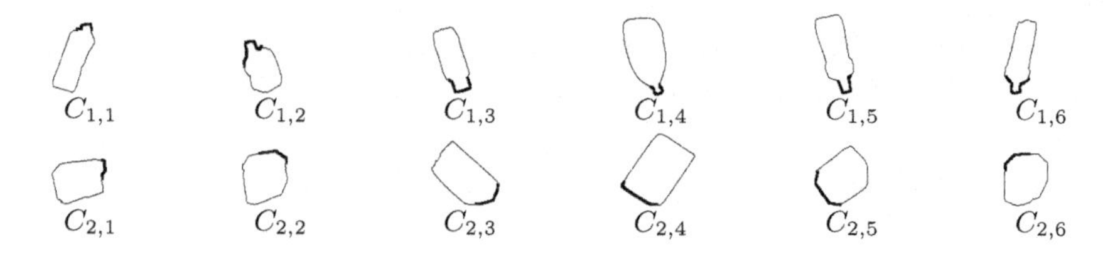

Fig. 7. Contours of 6 plastic bottles ($C_{1,1} - C_{1,6}$) and 6 beverage cartons ($C_{2,1} - C_{2,6}$) with their most significant contour sections $\hat{M}$ marked in bold

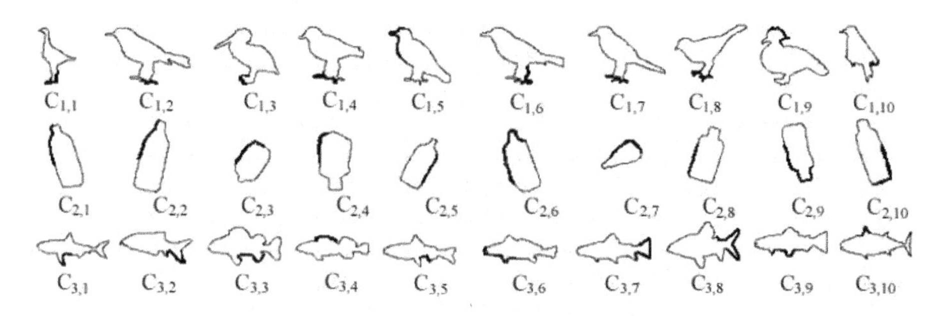

Fig. 8. Contours of 10 birds ($C_{1,1} - C_{1,10}$), 10 bottles ($C_{2,1} - C_{2,10}$), and 10 fishes ($C_{3,1} - C_{3,10}$) with their most significant contour sections $\hat{M}$ marked in bold

and the mechanical preparation in a sorting plant make the packagings strongly dirty and deformed. The method was also tested with a more complex task domain D_2 consisting of 3 classes: fishes, birds and plastic bottles (fig. 8).

All contours are scaled so that their biggest generalized diameters are approximately equal. The contours have from 400 up to 600 contour points and successive contour points have all equal Euclidean distances of 1.2 pixels, because it's a compromise between 1.0 and $\sqrt{2}$ (horizontal and diagonal pixel distance on a square pixel grid). The experiments are carried out with heuristically found parameters. In both cases it seems to be good if the weighting of Q_p is high in respect to d_p, v_p.

The results of the experiments are shown in fig. 7 and fig. 8. The $\hat{M}$ of the contours are marked in bold.

For the bottle class in the task domain D_1 shown in fig. 7 the most significant contour sections $\hat{M}$ are the bottle necks. Here, it is possible to distinguish this two classes only with the help of the bottle necks. In the more complex task domain D_2 the $\hat{M}$ of the bottles are not only the bottle necks. Also the sections around one corner are now significant to distinguish the bottles from the birds and the fishes. The most significant sections $\hat{M}$ of the birds are in most cases the claws but there is also a beak. Interesting is bird $C_{1,10}$ which has no claws.

Here, a tail feather is the $\hat{M}$. The fishes have their most significant sections at their fins. Because of the great variety of the fins the presented method find most significant fins at different positions on the fish contours.

6 Conclusion

A method was presented in this paper, that nearly automatically, supervised learns those contour sections M_v of a certain contour, with which this contour can be distinguished from contours out of other contour classes. We call this generation of feature groups **feature fusion**: Base features or feature primitives (contour points) are synthesized to feature groups (are fused to more significant contour sections M_v). Now, the fused more significant contour sections M_v can serve as knowledge base of a classifier for assigning an unknown contour to a certain contour class.

Experimental results were made with two different task domains. The method extracted that the class of the plastic bottles can be distinguished from the class of the beverage cartons only with the help of the contour sections we call bottle necks. In the other task domain D_2 the method extracted that the most significant sections $\hat{M}$ of the birds are the claws and the beaks, the $\hat{M}$ of the fishes are the fins, and the $\hat{M}$ of the bottles are the corners and the bottle necks. This seems a little like a human being would do.

A further combination of the more or most significant contour sections $M_v, \hat{M}$ of a certain contour class is desirable in order to create **abstract contour sections** $\hat{A}$. E.g., combining all $\hat{M}$ at the position of the bottle necks into an $\hat{A}$ "BOTTLE NECK". Thus, a problem still to be solved is the problem of abstraction for the single classes. First experiments with a hierarchical classifier based on the most significant sections $\hat{M}$ were made and give encouraging results.

References

AG96. Alt, H., Guibas, L. J. : Discrete Geometric Shapes: Matching, Interpolation, and Approximation - A Survey. Technical Reports of FU Berlin (1996), Institut für Informatik u. Mathematik. 521

BCGJ98. Basri, R., Costa, L., Geiger, D., Jacobs, D. : Determining the similarity of deformable shapes. Vision Research **38** (1998), 2365-2385. 518

B57. Bellman, R. : Dynamic Programming. Princeton University Press, Princeton, New Jersey (1957). 521

HS97. Hoffman, D. D., Singh, M. : Salience of visual parts. Cognition **63** (1997), 29-78. 518

KTZ94. Kimia, B. B., Tannenbaum, A. R., Zucker, S. : Shapes, shocks, and deformations 1: The components of shape and the reaction-diffusion space. Int. Journal of Computer vision **15(3)** (1995), 189-224. 518

KW96. Kupeev, K., Wolfson, H. : A new method of estimating shape similarity. Pattern Recognition Letters **17** (1996), 873-887. 518

LL98a. Latecki, L. J., Lakämper, R. : Discrete Approach to Curve Evolution. In Levi, P., Ahlers, R.-J., May, F. and Schanz, M. (Edts.): Mustererkennung 1998, 20. DAGM-Symposium, 85-92. 518

LL98b. Latecki, L. J., Lakämper, R. : Shape decomposition and shape similarity measure. In Levi, P., Ahlers, R.-J., May, F. and Schanz, M. (Edts.): Mustererkennung 1998, 20. DAGM-Symposium, 367-376. 518

L97. Loncaric, S. : A survey of shape analysis techniques. Pattern Recognition **31** (1997), 983-1001. 518

P77. Pavlidis, T. :Structural Pattern Recognition. Springer-Verlag, Berlin (1977). 518

PK00. Pechtel, D., Kuhnert, K.-D. : Generating Automatically Local Feature Groups of Complex and Deformed Objects. Appears in Gaul, W., Decker, R. (eds.): Classification and Information Processing at the Turn of the Millenium. Proc. 23th Ann. Conf. of GfKl, Bielefeld, March 10-12, 1999. Springer, Heidelberg, 2000, 237-244. 519, 520, 521, 523

PM47. Pitts, W., McCulloch W. : How we know universals: the perception of auditory and visual forms. Bulleting of Mathematical Biophysics **9** (1947), 127-147. 518

SGWM93. Sederberg, T., Gao, P., Wang, G., Mu, H. : 2D shape blending: An Intrinsic Solution to the Vertex Path Problem. Computer Graphics **38** (1993) 15-18. 518

SIK95. Siddiqi, K., Kimia, B. : Parts of Visual Form: Computational Aspects. IEEE Transactions On Pattern Anal. And Mach. Intell. **17** (1995) 239-251. 518

STK95. Siddiqi, K., Tresnes, K., and Kimia, B. : Parts of visual form: ecological and psychophysical aspects. In Proc. IAPR's Int. Workshop on Visual Form (1994) . 518

SK83. Sankoff, D., Kruskal, B. K. (Eds.): Time warps, String Edits, and Macromolecules: The Theory and Practice of Sequence Comparison. Addison-Wesley, Reading (1983). 519, 521

U96. Ullman, S. : High-level Vision. A Bradford Book, The MIT Press , Cambridge, Massachusetts (1996). 518

Parallel Line Grouping Based on Interval Graphs

Peter Veelaert

Hogent
Schoonmeersstraat 52, 9000 Ghent, Belgium
Peter.Veelaert@hogent.be

Abstract. We use an interval graph to model the uncertainty of line slopes in a digital image. We propose two different algorithms that group lines into classes of lines that are parallel, or almost parallel. This grouping is strongly based on the Helly-type property of parallelism in the digital plane: a group of lines is digitally parallel if and only if each pair of lines is digitally parallel. As a result, the extraction of parallel groups is reduced to the extraction of cliques in the interval graph generated by the slope intervals. Likewise, the extraction of lines that are almost parallel becomes equivalent to the detection of subgraphs that resemble cliques.

1 Introduction

In this paper we consider the graph-theoretical properties of parallel line grouping in a digital image, mostly from the computational viewpoint. This work is part of an ongoing effort to define a sound mathematical basis for the extraction of geometric structure from an image while taking into account the uncertainty that is present. In previous work we already considered some basic geometric properties such as collinearity, parallelism and concurrency [8], and we studied some of the algorithmic aspects of the theory [9]. In this paper we propose algorithms that classify lines into groups of parallel lines, taking into account the uncertainty of edge positions and the fact that parallelism is not an equivalence relation in the digital plane.

Any algorithm that extracts geometric structure must take into account the uncertainty on the position of the pixels. This uncertainty has several causes: the digitization process, the localization errors introduced by the edge detector, and the possible errors introduced by the line extraction algorithms, e.g., RANSAC or the Hough Transform. In fact, uncertainty can even be an artificial parameter, e.g., to detect automatically in an image vanishing points of the perspective projection, a first step is to extract groups of lines that "almost" parallel.

In the approach proposed here, the uncertainty of point positions is modeled by defining for each point a small uncertainty region in which the point can lie, without knowing its exact position. This may be compared to the more general approach as advocated by Durrant-Whyte and others where one defines for each position a probability distribution [1]. These distributions are then used to derive new geometric distributions, e.g., for the parameters of a line passing

G. Borgefors, I. Nyström, and G. Sanniti di Baja (Eds.): DGCI 2000, LNCS 1953, pp. 530–541, 2000.
© Springer-Verlag Berlin Heidelberg 2000

through two points. By contrast, instead of computing precise distributions, we only calculate in which region a distribution is non-zero. The advantage of this simplification is that we can go much further in the extraction of geometric information, without computing complicated probability measures. By comparison, also Lowe makes simplifying assumptions regarding the distribution of lines in an image and introduces an empirical significance measure for parallelism [5]. Lowe's interesting view is inspired by concepts from perceptual organization, but it remains empirical, without a strict formal introduction of uncertainty.

In euclidian geometry grouping into parallel classes would be trivial, as euclidian parallelism is an equivalence relation. In a digital image uncertainty on positions causes uncertainty on the slopes of lines, and parallelism is therefore no longer an equivalence relation. We must find partitions that "resemble" equivalence classes. In this work we explain how this problem can be reformulated as finding groups of lines that resemble cliques in the so-called graph of parallel pairs. This reformulation is possible because parallelism in the digital plane turns out to be a Helly-type property: a collection of lines is digitally parallel if each its pairs of lines is digitally parallel [8]. That is, the Helly-type property holds for the entire collection as soon as it holds for each of the collection's n-membered subcollections, where $n = 2$.

In this paper we show how a classification into parallel groups depends on two factors: first, a measure for the uncertainty on pixel positions, which we call the acceptable thickness, and second, an optimization criterion, which determines how much parallelism is required in a group of parallel lines. Thus, we may require that each group consists of lines that are all parallel to each other, or we may impose the weaker requirement that it suffices that each group contains at least one line to which all other lines in the group are parallel.

In Section 2 we give a brief overview of the concepts of digital straightness, collinearity, parallelism, and concurrency as they are used in this work. Section 3 introduces the so-called graph of parallel pairs, which represents all parallel relations between line segments. In Section 4 two different algorithms are proposed that use the graph of parallel pairs to extract large groups of parallel line segments. We give some concluding remarks in Section 5.

2 Brief Overview of Geometric Properties

In this section we briefly review results obtained in previous work [8,9]. Since image pixels lie on a rectangular grid, we assume that all pixels are part of the digital plane, although this is not strictly necessary for what follows. The digital plane $\mathbb{Z}^2$ consists of points that have integral coordinates. A digital set S is a subset of the digital plane.

Several issues have influenced the development of mathematical discretization schemes. For computer generated images discretization schemes are used to obtain a digital representation of a euclidian object. For natural images they provide a simplified model of the complicated discretization process of digital image formation, i.e., a model for a CCD camera. Finally, for natural as well as computer images, the discretization scheme can be used to examine the inverse process, i.e., given a digitized image How can we reconstruct the original image and the geometric structure contained within it?

In this paper we are mostly interested in the last issue. During discretization the precise knowledge about position of geometric objects is lost. Or in other words, given the digital image we are uncertain about the position of objects in the original image. We shall model this uncertainty by an *uncertainty region* that we associate with each grid point. The discretization process that coincides naturally with this notion of uncertainty is the *discretization by dilation* scheme developed by Heijmans and Toet [2]. Let U denote a set in $\mathbb{R}^2$, let A be a second subset of $\mathbb{R}^2$, called the structuring element, and let A_p be the translate of A by p. Then the discretization by dilation of U consists of all points $p \in \mathbb{Z}^2$ for which $A_p \cap U$ is non-empty [2,7].

Furthermore, to simplify the exposition, in this paper we shall restrict ourselves to one particular form of discretization by dilation, that is, we use a simple variant of grid-intersect discretization. To model the uncertainty of its position, for each digital point $p = (x, y)$, we introduce as translate of the structuring element the vertical line segment $C_p(\tau)$, which comprises all points $(x, b) \in \mathbf{R}^2$ that satisfy $y - \tau/2 \leq b < y + \tau/2$. Here τ is a positive real number, called the *acceptable thickness*. Up to a certain multiplication factor, this notion of acceptable thickness coincides with Réveillès's notion of arithmetical thickness of a discrete straight line [6]. Fig. 1 shows several digital sets, their vertical segments, and straight lines that transverse these segments. To simplify what follows even further, we shall only discuss the extraction of straight lines of the form $y = \alpha x + \beta$, where the slope α satisfies $-1 < \alpha < 1$. We assume that each set S contains at least two points with distinct x-coordinates.

Although the foregoing restrictions do not change our results in a fundamental way, they allow us to discard several special cases discussed more thoroughly in previous work [8], where we show, for example, that for lines whose slope lies between -1 and 1, vertical segments are sufficient to model position uncertainty. Also in previous work, other cells than vertical line segments have been considered in a more general framework where we examine the modeling of uncertainty in the multitude of parameter spaces that arise during the hierarchical grouping of line segments [10].

Definition 1. A digital set S is called *digitally straight* if all segments $C_p(\tau)$, $p \in S$ can be transversed by a common straight line.

Note that the above definition of straightness is more general than the standard definition of a digitally straight line, where one chooses $\tau = 1$, and where additional connectivity constraints must be satisfied.

Definition 2. Let $S_1, \ldots, S_n$ a finite collection of finite digitally straight sets, and let A_i denote a euclidian straight line. We define the following digital geometric properties, illustrated in Fig. 1:

- The sets S_i are called *digitally collinear* if there exist n collinear line segments $A_1, \ldots, A_n$ where each A_i transverses all the segments of S_i;

- The sets in S_i are called *digitally parallel* if there exist n parallel straight lines $A_1, \ldots, A_n$ where each A_i transverses all the segments of S_i;
- The S_i are called *digitally concurrent* if there exist n straight lines $A_1, \ldots, A_n$ that meet in a common point and where each A_i transverses all the segments of S_i.

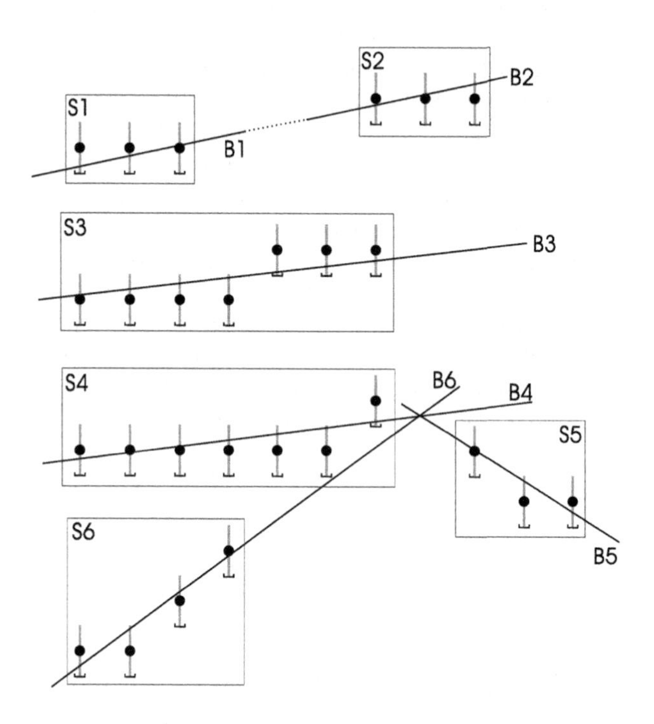

Fig. 1. S_1 and S_2 are digitally collinear; S_3 and S_4 are digitally parallel; S_4, S_5 and S_6 are digitally concurrent

The following result is crucial [8]:

Theorem 1. *Let $\mathcal{S} = \{S_1, \ldots, S_n\}$ a finite collection of finite digitally straight sets. Then the collection $\mathcal{S}$ is digitally parallel if each subcollection containing two sets of $\mathcal{S}$ is digitally parallel.*

According to Theorem 1 all knowledge about parallelism can be derived from the parallelism of pairs. Although we focus in this paper mainly on the graph-theoretical algorithms, we give a brief overview of the geometric algorithms that are needed to determine whether a pair of lines is digitally parallel [9].

Definition 3. Let S be a finite digital set that contains at least two points with distinct x-coordinates, and let τ be a chosen acceptable thickness with $0 < \tau$.

Then the *domain* of S, denoted as $\text{dom}_x(S;\tau)$, is the set of all parameter points $(\alpha,\beta) \in \mathbb{R}^2$ that satisfy the following system of inequalities:

$$-\tau/2 < \alpha x_i + \beta_i - y_i \leq \tau/2, \quad (x_i, y_i) \in S. \tag{1}$$

In other words, the domain contains the parameters of all the euclidian lines that transverse the vertical segments of S. In general, the domain of a large and sufficiently elongated set is small, since the parameters of the lines transversing all these segments can vary only within a small range.

Consider for example the collection of digital sets representing lines in Fig. 2, whose domains are shown in Fig. 3, for an acceptable thickness $\tau = 2$. The long line segment O has a small domain, while the short segment K has a large domain. In fact, since each point of a set corresponds to two halfplanes in the parameter space, the domain of a large set is the intersection of a large number of halfplanes, and, in general, the domain will therefore be small. Furthermore, a domain of a digital set with N points is a convex bounded set, and can be computed in $O(N \log N)$ time as an intersection of $2N$ halfplanes [9]. When we let τ vary, a domain gets larger for increasing values of τ. If the acceptable thickness is too small, however, the domain of a set will be empty. For given set S, let $D \subset \mathbb{R}$ be the set of acceptable thicknesses for which the domain of S is non-empty. The *thickness* of S is defined as the infimum of D [8].

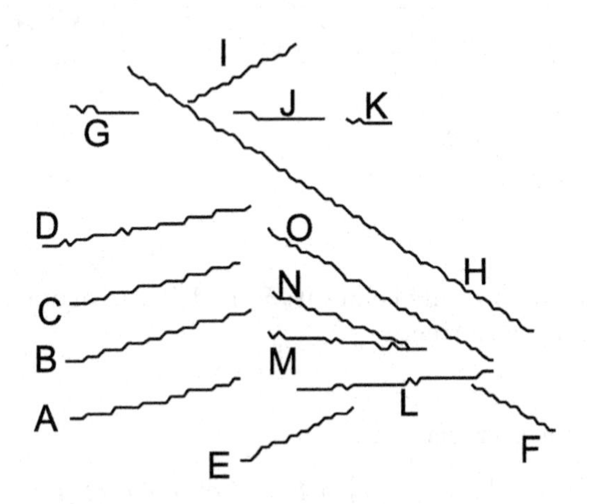

Fig. 2. A collection of lines

To reformulate the digital geometric concepts of Definition 2 in terms of domains, we need one more notion, i.e., the projection of a domain upon the α-axis in the (α,β) parameter plane. For a subset S of $\mathbb{R}^2$, let $\pi_\alpha(\text{dom}_x(S;\tau))$ denote the interval that results by projecting the domain of S upon the α-axis. The interval $\pi_\alpha(\text{dom}_x(S;\tau))$ has the following geometrical meaning: $\pi_\alpha(\text{dom}_x(S;\tau))$ is equal to the open interval $]\alpha^1, \alpha^2[$, where α^1 is the infimum of the slopes of the

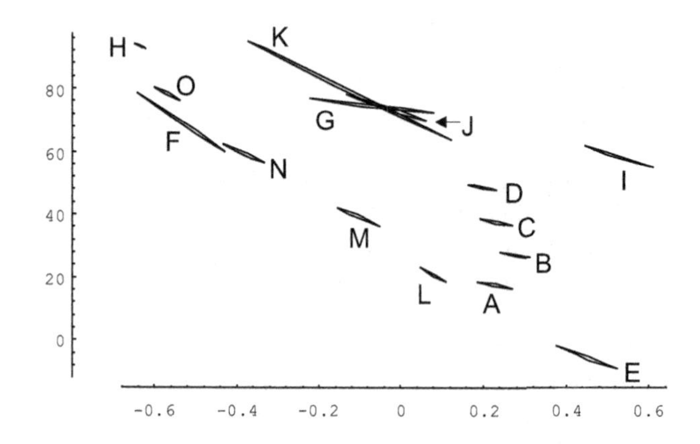

Fig. 3. Domains of lines for the acceptable thickness $\tau = 2$

euclidian lines that transverse the vertical segment associated with the points in S, and where α^2 is the supremum. Fig. 4(a) shows the slope intervals that result by projecting the domains in Fig. 3 upon the α-axis. It follows almost immediately that two lines are digitally parallel if and only if their slope intervals have a non-empty intersection [8,9].

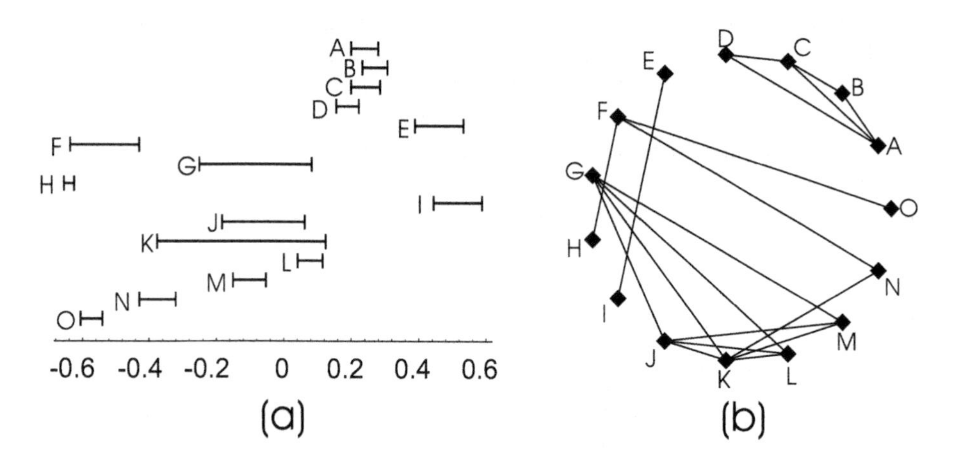

Fig. 4. (a) The intervals that result from the projection of the domains, and (b) the corresponding graph of parallel pairs

3 Graph of Parallel Pairs

We shall represent parallelism by the graph shown in Fig. 4(b), called the *graph of parallel pairs.* In this graph two vertices are connected by an edge if the corresponding lines are digitally parallel. Theorem 1 remains essential here, since it allows us to conclude that any complete subgraph (or clique) is digitally parallel, i.e. the cliques $GJKL$ and $GJKM$ in Fig. 4(b), . In other words, this conclusion is valid because parallelism in the digital plane is a Helly type property, that is, we can draw a conclusion for the entire set by examining all pairs, i.e, the edges of the graph.

Before proceeding we must briefly review the definitions and properties of graphs that we need (see, e.g., [11]). Let G be a simple graph, i.e, a graph that contains no loops or double edges. *An independent set* is a set of vertices that does not contain any pair of adjacent vertices. *A dominating set* is a set of vertices such that each vertex in G is adjacent to at least one vertex in the dominating set. *A vertex cover* is a set of vertices such that each edge in G is incident with at least one vertex in the vertex cover. *A coloring* assigns an index to each vertex so that two adjacent vertices always receive distinct indices. An *optimal coloring* uses a minimum number of indices, i.e., the *chromatic number* of the graph. A graph is *perfect* if every induced subgraph has an independent set meeting all its maximum cliques, that is, cliques of maximal size.

In the graph of parallel pairs, two vertices are joined by an edge when the slope intervals of the two lines intersect. Therefore, this graph falls into the category of *interval graphs.* Similarly, the *complement of an interval graph* belongs to the category of *comparability* graphs. Interval graphs as well as comparability graphs are perfect [11], which has important consequences from the computational viewpoint. In general, finding an optimal coloring, a minimum dominating set, a minimum vertex covering, minimum clique covering, or maximum dependent set are NP-hard problems for graphs in general [3,11]. As soon as we have a perfect graph, however, each of these sets and coverings can found in polynomial time [3]. Interval graphs possess even stronger properties, and the above sturctures can often be found in linear time.

4 Classification of Lines into Parallel Groups

In euclidian geometry a graph of parallel pairs would consist of disjoint cliques. Since parallelism in the digital plane is not an equivalence relation we must look at techniques that extract subgraphs that are similar to cliques, and in this way approximate the equivalence relation of euclidian geometry. Fig. 5 shows an application, where long lines with slope between -1 and 1 have been extracted from the image of a building. Fig. 5(b) shows the graph of parallel pairs for $\tau = 2$. In this particular application we wanted to detect automatically some of the vanishing points in the image. The idea is to look for three or more lines that meet at a common point and whose slopes are almost equal. To reduce the number of possible combinations, one of the essential steps is to extract good candidate groups of lines that are almost parallel.

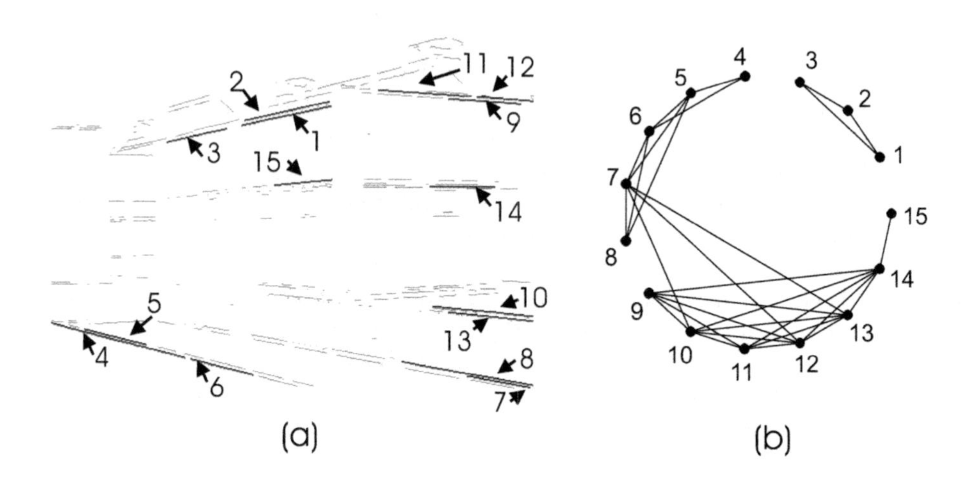

Fig. 5. (a) Lines detected in image of building with horizontal Sobel operator. (b)Graph of parallel pairs of longest lines for acceptable thickness $\tau = 2$

We shall propose two algorithms that partition a graph of parallel pairs into subgraphs that resemble cliques. Clearly, the partitioning will depend on the amount of parallelism that is required within each group. The first algorithm looks for a minimal number of groups in which parallelism is maximal, that is, each line in the group must be parallel to each other group member. By contrast, we could look for groups with a minimal amount of parallelism, by extracting the connected components of the graph. The second algorithm, imposes a more natural requirement, it looks for groups in which there is at least one member to which all other lines are parallel.

4.1 Classification Based on Minimum Clique Coverings

Covering the graph of parallel pairs by a minimum number of cliques is equivalent to finding an optimal coloring of its complement which is a comparability graph. The algorithm that we propose here is a modification of an existing coloring algorithm that uses the partial ordering induced by a comparability graph (see [11], page 201). We can use the fact, however, that the comparability graph considered here is the complement of an interval graph, and that the partial ordering can be derived directly from the intervals. Each line S yields an interval $\pi_\alpha(\mathrm{dom}_x(S;\tau)) =]a_i, b_i[$ of slopes. Let $]a_1, b_1[, \ldots,]a_n, b_n[$ the intervals be sorted according to their lowest boundary a_i and assume that we have reassigned indices such that $a_i \leq a_j$ whenever $i < j$. Note that the ordering of this sequence is not unique. Furthermore, the intervals induce an ordering on the complementary graph. To be precise, we let $S_i < S_j$ whenever the slope intervals do not intersect and we have $b_i < a_j$.

Algorithm 1. *Partitioning into parallel subsets (Equivalent to optimal coloring of complement of interval graph).* We start by selecting from the sorted sequence the interval $B_1 =]a_j, b_j[$ for which b_j takes a minimum value. To this interval and to all the intervals that intersect it we assign the first color, and we remove these colored intervals from the sequence. We then proceed iteratively by assigning the ith color to the interval $B_i =]a_k, b_k[$ for which b_k takes the minimum value in the remaining sequence and to all the intervals that intersect it. The algorithm yields a partitioning into a minimum number of cliques. □

In addition, after coloring the vertices, if we select one vertex of each color, then it is clear that we obtain a maximal independent set, that is, an independent set that cannot be extended further. That this independent set is also the largest possible follows from the Perfect Graph Theorem, according to which the maximal size of an independent set of a perfect graph is equal to the minimum number of colors needed to color its complement.

Correctness of the Algorithm. The algorithm produces a valid coloring for the complement of the interval graph. In fact, we claim that the intervals are only assigned the same color when their common intersection is non-empty. After assigning m colors, let $]a_k, b_k[$ be an interval for which b_k takes a minimum value in the remaining sequence. Then the real number b_k must lie in all those intervals in the remaining sequence that intersect $]a_k, b_k[$, since otherwise b_k would not be a minimum. Thus, all intervals that are in the same coloring class have a common non-empty intersection. As a result they are all mutually incomparable, and therefore they form an independent set in the comparability graph. Furthermore, the coloring must be optimal, since if the algorithm uses m colors then there are at least m mutually disjoint intervals, or in other words, there is a clique of size m in the comparability graph.

Example. Fig. 4(a) shows the slope intervals that correspond to the lines in Fig. 2. If we restrict ourselves to the lines of the largest component of Fig. 4(b), we find the following partitioning: $\{H, F\}$, $\{O\}$, $\{N, K\}$, $\{M, G, J\}$, $\{L\}$. This corresponds to a coloring of the complement of the graph of parallel pairs. Since we have a minimum number of color classes, the algorithm yields a partitioning into a minimum number of classes of parallel lines, or in other words a minimum clique partitioning of the interval graph. The lines $\{H, O, N, M, L\}$ form a maximum independent set.

4.2 Classification Based on a Minimum Dominating Set

A partitioning into a minimal number of parallel sets may not always yield the most natural result. Consider the example in Fig. 6. If we apply Algorithm 1, we find the following partitioning: $\{A, B\}$, $\{C, D, E\}$, $\{F\}$. The symmetry apparent in the graph of parallel pairs and the intervals does not occur in the grouping. In fact, in this example any minimum clique partitioning is asymmetric. We now propose a second technique that requests less parallelism in the groups, and often leads to more natural results. It is a covering based on a minimum dominating set of the graph. Recall that a set of vertices is dominating if each vertex is adjacent to at least vertex in the dominating set. We shall group the lines such that each group contains one vertex of a dominating set and all its neighbors. Thus, in each group there will be a line to which all other lines in the group are parallel. Note that distinct groups may not be disjoint.

Let S denote the set of intervals, and assume that the intervals $]a_1, b_1[, \ldots,]a_n, b_n[$ have been assigned indices according to an ordering on their left boundaries, i.e., $a_i \leq a_j$ whenever $i < j$.

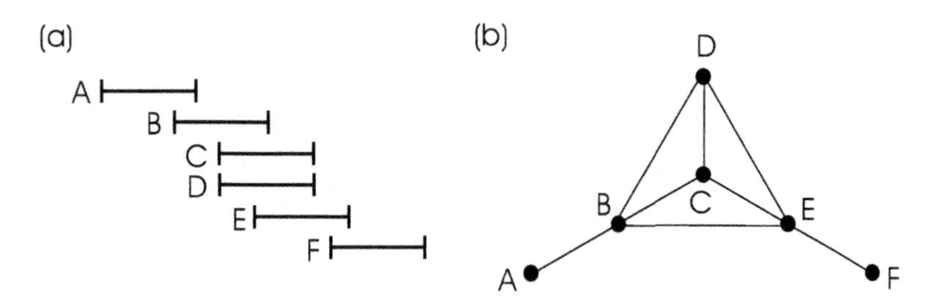

Fig. 6. (a) Slope intervals of a set of lines; (b) The corresponding graph of parallel pairs

Algorithm 2. *Covering with parallel subsets (Equivalent to minimum dominating set for interval graphs).* We first set the dominating set D equal to the empty set. Let $B_j =]a_j, b_j[$ be the interval in the sorted sequence for which b_j takes a minimum value. We look in the sorted sequence for an interval $]a_i, b_i[$ that intersects $]a_j, b_j[$ and for which b_i takes a maximum value. We add $]a_i, b_i[$ to the dominating set D, and we remove $]a_i, b_i[$ and all the intervals that intersect it from the sorted sequence. We proceed iteratively by taking the next interval $]a_k, b_k[$ in the remaining sequence for which b_k takes a minimum value. □

Correctness of the Algorithm. Clearly, the resulting set D is a dominating set, since it includes for each interval at least one interval that intersects it. Furthermore, the dominating set has minimum size, which can be seen as follows. Let $D = \{]p_1, q_1[,]p_2, q_2[, \ldots)\}$ be the dominating set found by the algorithm, and let $]a_j, b_j[$ be the interval for which b_j is a minimum in the starting sequence.

First, for the interval $]a_j, b_j[$ there must be at least one interval in any dominating set that intersects it. Let E^0 be any minimum dominating set, and let $]a_k, b_k[$ be an interval in E^0 that intersects $]a_j, b_j[$. Then we claim that the set $E^1 = (E^0 \backslash]a_k, b_k[) \cup]p_1, q_1[$ is also a minimum dominating set. It suffices to prove that any interval that intersects $]a_k, b_k[$, also intersects $]p_1, q_1[$. Let $]a_m, b_m[$

be an interval that intersects $]a_k, b_k[$. Suppose that $]a_m, b_m[$ also intersects $]a_j, b_j[$. Then, $]p_1, q_1[$ also intersects $]a_m, b_m[$, since b_j is common to all the intervals that intersect $]a_j, b_j[$, otherwise b_j would not be a minimum value. Suppose $]a_m, b_m[$ does not intersect $]a_j, b_j[$, then $b_j < a_m$. It follows that $]a_m, b_m[$ must intersect $]p_1, q_1[$, because among the set of all intervals $]a_i, b_i[$ that intersect $]a_j, b_j[$, which includes $]a_k, b_k[$, we have $b_i \leq q_1$.

Next, it is clear that the set $E^1 \backslash]p_1, q_1[$ must be a minimum dominating set for the set formed by all intervals in S that do not intersect $]a_j, b_j[$, because otherwise E^1 itself would not have minimum size. Thus, by the same reasoning as above, we can prove that there exists a minimum dominating set E^2 that contains $]p_2, q_2[$, and therefore there must a minimum dominating set containing both $]p_1, q_1[$ and $]p_2, q_2[$. By proceeding iteratively, it follows that D is indeed a dominating set of minimal size.

If we apply this algorithm to the intervals of Fig. 4(a), then we find $\{F, K\}$ as a minimum dominating set. This dominating set yields the following grouping: $\{F, O, N, H\}$ and $\{K, L, M, J, G, N\}$. Note that the line N belongs to both groups, as it is parallel to F as well as K. Although this grouping is not a partitioning, it leads to a more natural result than the minimum clique partitioning algorithm.

One may expect that the dominating set method generally produces fewer groups than a minimum clique partition. This is indeed the case, as the following properties hold for any graph: (i) An independent set is a dominating set if and only if it is maximal; Thus, the size of minimum dominating set cannot be larger than the size of a maximum independent set; (ii) Any vertex cover is a dominating set; Thus, the size of minimum dominating set cannot be larger than the size of a minimum vertex cover; (iii) The complement of an independent set is a vertex cover. Hence the following relation follows:

$$\#(\text{minimum dominating set}) \leq \\ \min\left(\#(\text{maximum independent set}),\ \ |S| - \#(\text{maximum independent set})\right).$$

For example, for the complete graph K_n, the size of a maximum independent set is 1, and a minimum vertex cover has size $n-1$. For the star shaped graph, defined as the graph with n vertices where there are only edges between one central vertex and each other vertex, the size of the maximum independent set is $n-1$, and the size of a minimum vertex cover is 1. In both cases, however, the size of a minimum dominating set is 1.

5 Concluding Remarks

In future work, we will examine the application of the proposed algorithms to discretization by dilation schemes that use structuring elements that are more general than the vertical line segments used here. This extension seems to be rather straightforward, the major consequence being that the computation of the domains of the line segments becomes more intricate. The extension to other more recent digitization schemes such as the topology preserving scheme proposed by Latecki et al. [4], and the Hausdorff discretization of Ronse and Tajine seems less obvious, however [7]. In the first place both schemes have been developed to have certain desirable properties such as preservation of topology and reconstructability of the original object (when the grid resolution is increased). To capture the notion of position or line slope uncertainty was not the primary goal of these two schemes.

Furthermore, we are also looking how parallel line grouping can be integrated into a larger framework that also involves collinearity, concurrency and proximity of lines. In this regard, we expect that graph theory will remain inportant. In fact, whatever the way uncertainty in a digital image is represented, it is rather natural to represent the uncertainty of line slopes by slope intervals, which has led us to interval graphs. We concluded that algorithms that perform parallel line grouping can do this by extracting groups of vertices that resemble cliques, and we discussed two possibilities: (i) extract the cliques themselves, or (ii) extract neighborhoods of dominating sets. Both techniques have been well examined in graph theory, and we showed how they provide a formal way to extract parallel groups of lines.

References

1. H. F. Durrant-Whyte: Uncertain Geometry. In Geometric Reasoning, Eds. Kapur and Mundy, 447–481 (1989) 530
2. H. J. A. M. Heijmans and A. Toet: Morphological sampling. Computer Vision, Graphics & Image Processing: Image Understanding **54**, 384–400 (1991) 532
3. M. Grötschel, L. Lovász, and A. Schrijver: Geometric Algorithms and Combinatorial Optimization. New York: Springer-Verlag, 2nd Ed. (1993) 536
4. L. J. Latecki, C. Conrad, and A. Gross: Preserving topology by a digitization process. J. Math. Imaging and Vision. **8**, 131–159 (1998) 541
5. D. Lowe: 3-D object recognition from single 2-D images. Artificial Intelligence **31**, 355–395 (1987) 531
6. J.-P. Réveillès: Géométrie discrète, calcul en nombres entiers et algorithmique. Thèse de Doctorat d'Etat. Université Louis Pasteur, Strasbourg (1991) 532
7. C. Ronse and M. Tajine: Discretization in Hausdorff space. J. Math. Imaging and Vision. **12**, 219–242 (2000) 532, 541
8. P. Veelaert: Geometric constructions in the digital plane. J. Math. Imaging and Vision. **11**, 99–118 (1999) 530, 531, 532, 533, 534, 535
9. P. Veelaert: Algorithms that measure parallelism and concurrency of lines in digital images. Proceedings of SPIE's Conference on Vision Geometry VIII, (Denver), SPIE, 69–79 (1999) 530, 531, 533, 534, 535
10. P. Veelaert: Line grouping based on uncertainty modeling of parallelism and collinearity. Proceedings of SPIE's Conference on Vision Geometry IX, (San Diego), SPIE (2000) 532
11. D. B. West: Introduction to Graph Theory. Upper Saddle River: Prentice Hall (1996) 536, 537

Author Index

Lecture Notes in Computer Science

For information about Vols. 1–1893
please contact your bookseller or Springer-Verlag

Vol. 1894: R. Dechter (Ed.), Principles and Practice of Constraint Programming – CP 2000. Proceedings, 2000. XII, 556 pages. 2000.

Vol. 1895: F. Cuppens, Y. Deswarte, D. Gollmann, M. Waidner (Eds.), Computer Security – ESORICS 2000. Proceedings, 2000. X, 325 pages. 2000.

Vol. 1896: R. W. Hartenstein, H. Grünbacher (Eds.), Field-Programmable Logic and Applications. Proceedings, 2000. XVII, 856 pages. 2000.

Vol. 1897: J. Gutknecht, W. Weck (Eds.), Modular Programming Languages. Proceedings, 2000. XII, 299 pages. 2000.

Vol. 1898: E. Blanzieri, L. Portinale (Eds.), Advances in Case-Based Reasoning. Proceedings, 2000. XII, 530 pages. 2000. (Subseries LNAI).

Vol. 1899: H.-H. Nagel, F.J. Perales López (Eds.), Articulated Motion and Deformable Objects. Proceedings, 2000. X, 183 pages. 2000.

Vol. 1900: A. Bode, T. Ludwig, W. Karl, R. Wismüller (Eds.), Euro-Par 2000 Parallel Processing. Proceedings, 2000. XXXV, 1368 pages. 2000.

Vol. 1901: O. Etzion, P. Scheuermann (Eds.), Cooperative Information Systems. Proceedings, 2000. XI, 336 pages. 2000.

Vol. 1902: P. Sojka, I. Kopeček, K. Pala (Eds.), Text, Speech and Dialogue. Proceedings, 2000. XIII, 463 pages. 2000. (Subseries LNAI).

Vol. 1903: S. Reich, K.M. Anderson (Eds.), Open Hypermedia Systems and Structural Computing. Proceedings, 2000. VIII, 187 pages. 2000.

Vol. 1904: S.A. Cerri, D. Dochev (Eds.), Artificial Intelligence: Methodology, Systems, and Applications. Proceedings, 2000. XII, 366 pages. 2000. (Subseries LNAI).

Vol. 1905: H. Scholten, M.J. van Sinderen (Eds.), Interactive Distributed Multimedia Systems and Telecommunication Services. Proceedings, 2000. XI, 306 pages. 2000.

Vol. 1906: A. Porto, G.-C. Roman (Eds.), Coordination Languages and Models. Proceedings, 2000. IX, 353 pages. 2000.

Vol. 1907: H. Debar, L. Mé, S.F. Wu (Eds.), Recent Advances in Intrusion Detection. Proceedings, 2000. X, 227 pages. 2000.

Vol. 1908: J. Dongarra, P. Kacsuk, N. Podhorszki (Eds.), Recent Advances in Parallel Virtual Machine and Message Passing Interface. Proceedings, 2000. XV, 364 pages. 2000.

Vol. 1909: T. Yakhno (Ed.), Advances in Information Systems. Proceedings, 2000. XVI, 460 pages. 2000.

Vol. 1910: D.A. Zighed, J. Komorowski, J. Żytkow (Eds.), Principles of Data Mining and Knowledge Discovery. Proceedings, 2000. XV, 701 pages. 2000. (Subseries LNAI).

Vol. 1911: D.G. Feitelson, L. Rudolph (Eds.), Job Scheduling Strategies for Parallel Processing. VII, 209 pages. 2000.

Vol. 1912: Y. Gurevich, P.W. Kutter, M. Odersky, L. Thiele (Eds.), Abstract State Machines. Proceedings, 2000. X, 381 pages. 2000.

Vol. 1913: K. Jansen, S. Khuller (Eds.), Approximation Algorithms for Combinatorial Optimization. Proceedings, 2000. IX, 275 pages. 2000.

Vol. 1914: M. Herlihy (Ed.), Distributed Computing. Proceedings, 2000. VIII, 389 pages. 2000.

Vol. 1915: S. Dwarkadas (Ed.), Languages, Compilers, and Run-Time Systems for Scalable Computers. Proceedings, 2000. VIII, 301 pages. 2000.

Vol. 1916: F. Dignum, M. Greaves (Eds.), Issues in Agent Communication. X, 351 pages. 2000. (Subseries LNAI).

Vol. 1917: M. Schoenauer, K. Deb, G. Rudolph, X. Yao, E. Lutton, J.J. Merelo, H.-P. Schwefel (Eds.), Parallel Problem Solving from Nature – PPSN VI. Proceedings, 2000. XXI, 914 pages. 2000.

Vol. 1918: D. Soudris, P. Pirsch, E. Barke (Eds.), Integrated Circuit Design. Proceedings, 2000. XII, 338 pages. 2000.

Vol. 1919: M. Ojeda-Aciego, I.P. de Guzman, G. Brewka, L. Moniz Pereira (Eds.), Logics in Artificial Intelligence. Proceedings, 2000. XI, 407 pages. 2000. (Subseries LNAI).

Vol. 1920: A.H.F. Laender, S.W. Liddle, V.C. Storey (Eds.), Conceptual Modeling – ER 2000. Proceedings, 2000. XV, 588 pages. 2000.

Vol. 1921: S.W. Liddle, H.C. Mayr, B. Thalheim (Eds.), Conceptual Modeling for E-Business and the Web. Proceedings, 2000. X, 179 pages. 2000.

Vol. 1922: J. Crowcroft, J. Roberts, M.I. Smirnov (Eds.), Quality of Future Internet Services. Proceedings, 2000. XI, 368 pages. 2000.

Vol. 1923: J. Borbinha, T. Baker (Eds.), Research and Advanced Technology for Digital Libraries. Proceedings, 2000. XVII, 513 pages. 2000.

Vol. 1924: W. Taha (Ed.), Semantics, Applications, and Implementation of Program Generation. Proceedings, 2000. VIII, 231 pages. 2000.

Vol. 1925: J. Cussens, S. Džeroski (Eds.), Learning Language in Logic. X, 301 pages 2000. (Subseries LNAI).

Vol. 1926: M. Joseph (Ed.), Formal Techniques in Real-Time and Fault-Tolerant Systems. Proceedings, 2000. X, 305 pages. 2000.

Vol. 1927: P. Thomas, H.W. Gellersen, (Eds.), Handheld and Ubiquitous Computing. Proceedings, 2000. X, 249 pages. 2000.

Vol. 1928: U. Brandes, D. Wagner (Eds.), Graph-Theoretic Concepts in Computer Science. Proceedings, 2000. X, 315 pages. 2000.

Vol. 1929: R. Laurini (Ed.), Advances in Visual Information Systems. Proceedings, 2000. XII, 542 pages. 2000.

Vol. 1931: E. Horlait (Ed.), Mobile Agents for Telecommunication Applications. Proceedings, 2000. IX, 271 pages. 2000.

Vol. 1658: J. Baumann, Mobile Agents: Control Algorithms. XIX, 161 pages. 2000.

Vol. 1756: G. Ruhe, F. Bomarius (Eds.), Learning Software Organization. Proceedings, 1999. VIII, 226 pages. 2000.

Vol. 1766: M. Jazayeri, R.G.K. Loos, D.R. Musser (Eds.), Generic Programming. Proceedings, 1998. X, 269 pages. 2000.

Vol. 1791: D. Fensel, Problem-Solving Methods. XII, 153 pages. 2000. (Subseries LNAI).

Vol. 1799: K. Czarnecki, U.W. Eisenecker, Generative and Component-Based Software Engineering. Proceedings, 1999. VIII, 225 pages. 2000.

Vol. 1812: J. Wyatt, J. Demiris (Eds.), Advances in Robot Learning. Proceedings, 1999. VII, 165 pages. 2000. (Subseries LNAI).

Vol. 1932: Z.W. Raś, S. Ohsuga (Eds.), Foundations of Intelligent Systems. Proceedings, 2000. XII, 646 pages. (Subseries LNAI).

Vol. 1933: R.W. Brause, E. Hanisch (Eds.), Medical Data Analysis. Proceedings, 2000. XI, 316 pages. 2000.

Vol. 1934: J.S. White (Ed.), Envisioning Machine Translation in the Information Future. Proceedings, 2000. XV, 254 pages. 2000. (Subseries LNAI).

Vol. 1935: S.L. Delp, A.M. DiGioia, B. Jaramaz (Eds.), Medical Image Computing and Computer-Assisted Intervention – MICCAI 2000. Proceedings, 2000. XXV, 1250 pages. 2000.

Vol. 1937: R. Dieng, O. Corby (Eds.), Knowledge Engineering and Knowledge Management. Proceedings, 2000. XIII, 457 pages. 2000. (Subseries LNAI).

Vol. 1938: S. Rao, K.I. Sletta (Eds.), Next Generation Networks. Proceedings, 2000. XI, 392 pages. 2000.

Vol. 1939: A. Evans, S. Kent, B. Selic (Eds.), «UML» – The Unified Modeling Language. Proceedings, 2000. XIV, 572 pages. 2000.

Vol. 1940: M. Valero, K. Joe, M. Kitsuregawa, H. Tanaka (Eds.), High Performance Computing. Proceedings, 2000. XV, 595 pages. 2000.

Vol. 1941: A.K. Chhabra, D. Dori (Eds.), Graphics Recognition. Proceedings, 1999. XI, 346 pages. 2000.

Vol. 1942: H. Yasuda (Ed.), Active Networks. Proceedings, 2000. XI, 424 pages. 2000.

Vol. 1943: F. Koornneef, M. van der Meulen (Eds.), Computer Safety, Reliability and Security. Proceedings, 2000. X, 432 pages. 2000.

Vol. 1945: W. Grieskamp, T. Santen, B. Stoddart (Eds.), Integrated Formal Methods. Proceedings, 2000. X, 441 pages. 2000.

Vol. 1948: T. Tan, Y. Shi, W. Gao (Eds.), Advances in Multimodal Interfaces – ICMI 2000. Proceedings, 2000. XVI, 678 pages. 2000.

Vol. 1952: M.C. Monard, J. Simão Sichman (Eds.), Advances in Artificial Intelligence. Proceedings, 2000. XV, 498 pages. 2000. (Subseries LNAI).

Vol. 1953: G. Borgefors, I. Nyström, G. Sanniti di Baja (Eds.), Discrete Geometry for Computer Imagery. Proceedings, 2000. XI, 544 pages. 2000.

Vol. 1954: W.A. Hunt, Jr., S.D. Johnson (Eds.), Formal Methods in Computer-Aided Design. Proceedings, 2000. XI, 539 pages. 2000.

Vol. 1955: M. Parigot, A. Voronkov (Eds.), Logic for Programming and Automated Reasoning. Proceedings, 2000. XIII, 487 pages. 2000. (Subseries LNAI).

Vol. 1960: A. Ambler, S.B. Calo, G. Kar (Eds.), Services Management in Intelligent Networks. Proceedings, 2000. X, 259 pages. 2000.

Vol. 1961: J. He, M. Sato (Eds.), Advances in Computing Science – ASIAN 2000. Proceedings, 2000. X, 299 pages. 2000.

Vol. 1963: V. Hlaváč, K.G. Jeffery, J. Wiedermann (Eds.), SOFSEM 2000: Theory and Practice of Informatics. Proceedings, 2000. XI, 460 pages. 2000.

Vol. 1966: S. Bhalla (Ed.), Databases in Networked Information Systems. Proceedings, 2000. VIII, 247 pages. 2000.

Vol. 1967: S. Arikawa, S. Morishita (Eds.), Discovery Science. Proceedings, 2000. XII, 332 pages. 2000. (Subseries LNAI).

Vol. 1968: H. Arimura, S. Jain, A. Sharma (Eds.), Algorithmic Learning Theory. Proceedings, 2000. XI, 335 pages. 2000. (Subseries LNAI).

Vol. 1969: D.T. Lee, S.-H. Teng (Eds.), Algorithms and Computation. Proceedings, 2000. XIV, 578 pages. 2000.

Vol. 1970: M. Valero, V.K. Prasanna, S. Vajapeyam (Eds.), High Performance Computing – HiPC 2000. Proceedings, 2000. XVIII, 568 pages. 2000.

Vol. 1971: R. Buyya, M. Baker (Eds.), Grid Computing – GRID 2000. Proceedings, 2000. XIV, 229 pages. 2000.

Vol. 1974: S. Kapoor, S. Prasad (Eds.), FST TCS 2000: Foundations of Software Technology and Theoretical Computer Science. Proceedings, 2000. XIII, 532 pages. 2000.

Vol. 1975: J. Pieprzyk, E. Okamoto, J. Seberry (Eds.), Information Security. Proceedings, 2000. X, 323 pages. 2000.

Vol. 1976: T. Okamoto (Ed.), Advances in Cryptology – ASIACRYPT 2000. Proceedings, 2000. XII, 630 pages. 2000.

Vol. 1977: B. Roy, E. Okamoto (Eds.), Progress in Cryptology – INDOCRYPT 2000. Proceedings, 2000. X, 295 pages. 2000.

Vol. 1983: K.S. Leung, L.-w. Chan, H. Meng (Eds.), Intelligent Data Engineering and Automated Learning – IDEAL 2000. Proceedings, 2000. XVI, 573 pages. 2000.

Vol. 1987: K.-L. Tan, M.J. Franklin, J. C.-S. Lui (Eds.), Mobile Data Management. Proceedings, 2001. XIII, 290 pages. 2001.